1994

ZEBRA MUSSELS

BIOLOGY, IMPACTS, AND CONTROL

Edited by

Thomas F. Nalepa
Donald W. Schloesser

Front Cover. Zebra mussels attached to a reed taken from Lake Maarsseveen in the Netherlands. (Photograph by Simon van Mechelen and courtesy of Michiel Kraak, University of Amsterdam, The Netherlands.)

Library of Congress Cataloging-in-Publication Data

Zebra mussels : biology, impacts, and control / edited by Thomas F.
Nalepa and Don Schloesser.
p. cm.
Includes bibliographical references and index.
ISBN 0-87371-696-5
1. Zebra mussel. 2. Zebra mussel--Control. 3. Zebra mussel--Control--Environmental aspects. I. Nalepa, T. F. II. Schloesser, Donald W.
QL430.7.D8Z45 1992
594'.11--dc20

92-10333
CIP

PRINTED IN THE UNITED STATES OF AMERICA
2 3 4 5 6 7 8 9 0

Printed on acid-free paper

Preface

When the zebra mussel, *Dreissena polymorpha,* was first discovered in North America in Lake St. Clair of the Laurentian Great Lakes in June 1988, few people realized the importance of this discovery in terms of its consequences and, ultimately, its challenges. Within just a few months after the first individuals were found, mussels were being found in great numbers along the southern shoreline of Lake St. Clair and along the northern shoreline of western Lake Erie, covering all hard surfaces such as water intakes, rocks, piers, and breakwalls. Water plants drawing water from the lakes experienced a 50% decline in efficiency because of mussels attaching to the insides of intake pipes. Such a large and rapid increase in populations of a species in North America may not be unprecedented, but certainly none have been more visible or have had such an immediate impact. As populations spread and numbers continued to increase in 1989, so did the concerns of ecologists, water users, and the public at large. Disruptions, or certainly changes, in the web of aquatic life, ranging from bacteria and algae to fish, seemed imminent, and disruptions in water supplies became a reality.

Considering these startling events in the Great Lakes region and the impending spread of the mussel to other regions of North America, the need and urgency of compiling information about this species became apparent. The primary purpose of this book then is to document the initial impacts of the zebra mussel in North America and to summarize the initial response of various water users in seeking control measures. The book also includes contributions by non-North American authors who provide their own insights and perspectives based on, in some cases, decades of research. The range of topics presented is diverse and, in actuality, reflect concerted efforts to include contributions dealing with a variety of mussel-related subjects. Some contributions are rather focused, while others are broad in scope and purpose. Contributions feature descriptions of life history, studies of distribution, case histories of impacts, ecosystem responses to the mussel, mussel responses to the ecosystem, comparisons between North American and European populations, and control measures and strategies. Other contributions provide the basis for understanding the form, function, and behavior of the mussel at levels of both the individual and the population.

As research on the biology, impacts, and control of the zebra mussel in North America progresses, it is hoped that the contents of this book will provide a basis for these future research efforts. At the very least, this book documents the efforts, thoughts, and interpretations of scientists and engineers during the early years of the zebra mussel invasion in North America.

Thomas F. Nalepa
Donald W. Schloesser

About the Editors

Thomas F. Nalepa is a research biologist with the Great Lakes Environmental Research Laboratory, National Oceanic and Atmospheric Administration, in Ann Arbor, Michigan. He graduated from Michigan State University with a B.S. and M.S. in aquatic biology/limnology. He was a biologist with the Environmental Protection Agency for three years before coming to the Great Lakes Environmental Research Laboratory in 1975. His research interests include long-term trends in benthic communities, role of benthic invertebrates in the cycling of contaminants and nutrients, and trophic interactions between benthic communities and the upper food web. He has published over 40 papers on topics related to benthic invertebrates. He is leader of the Non-Indigenous Species Program at the laboratory and is involved with several projects dealing with zebra mussels, including a major effort assessing impacts of zebra mussels on the Saginaw Bay, Lake Huron ecosystem. He belongs to numerous professional societies and was recently president of the International Association for Great Lakes Research. He is currently an Associate Editor for the *Journal of Great Lakes Research* and Vice-Chair of the Great Lakes Panel on Non-Indigenous Species.

Donald W. Schloesser is a fisheries scientist with the U.S. Fish and Wildlife Service at the National Fisheries Research Center - Great Lakes located in Ann Arbor, Michigan. He received a B.S. and M.S. from Michigan State University in fisheries and limnology. In his position he has been responsible for performing habitat and fishery resources research aimed at protecting and utilizing the fish of the Laurentian Great Lakes. His research interests have been intense and varied, having published 30 peer-reviewed research papers, presented 40 requested papers, and published 30 reports on subjects such as macrozoobenthos, aquatic botany, fish predation, habitat and fish interactions, habitat and waterfowl interactions, habitat restoration, water and substrate quality, taxonomic notes, aerial photography, and exotic species.

He has an interest in exotic species that spans most of his professional career. He has published nine papers addressing unusual and exotic species, including four papers on exotic mussels, such as the zebra mussel, *Dreissena polymorpha*. His interest in the zebra mussel started immediately after he attended the first organized meeting concerning this taxon in North America. He and Thomas Nalepa were two of three United States representatives attending this meeting where a total of twelve people (i.e., the dirty dozen) gathered to assess the impacts of zebra mussels on water users and native species in North America. The twelve scientists at this meeting correctly assessed the dangers of the zebra mussel. Don subsequently helped organize the First North American Conference on Zebra Mussels held in Ann Arbor, Michigan in June 1989 when few people knew about this exotic species. His research specialty, concerning the zebra mussel, is the assessment of impacts of zebra mussel infestation on native unionid bivalves and the long-term effects on the ecology of waters in North America.

Acknowledgments

The editors thank their many colleagues who contributed to this volume with either their expertise, encouragement, or time: Al Beeton, Cathy Darnell, Dave Fanslow, John Fenton, Mark Ford, Wendy Gordon, Gerald Gostenik, Jeff Lefevre, Norma Lojewski, Dave Reid, Karen Sparks, Marijo Wimmer, and Jim Wojcik from the Great Lakes Environmental Research Laboratory, National Oceanic and Atmospheric Administration, and Marilyn Murphy, John Gannon, Jon Stanley, Glen Black, Joy Love, and Ann Zimmerman from the National Fisheries Research Center — Great Lakes, U.S. Fish and Wildlife Service. The editors also thank the reviewers who took the time and effort to provide constructive comments and suggestions to improve the various manuscripts. The prompt response and cooperation of individual authors is appreciated. The editors would like to recognize Robert (Skip) DeWall, of Lewis Publishers, for planting the seed for this book, and Andrea Demby, of CRC Press, Inc., for her patience, understanding, and perseverance.

Publication of the color plates was funded by the Great Lakes Sea Grant Network Zebra Mussel Outreach Program. The editors thank John Schwartz, Michigan Sea Grant, for his efforts. The Great Lakes Sea Grant Network consists of university-based programs in Illinois, Indiana, Michigan, Minnesota, New York, Ohio, and Wisconsin. Funds were transmitted through the Michigan Sea Grant College Program from grant number NA164G0272-01 from the National Sea Grant College Program, National Oceanic and Atmospheric Administration, U.S. Department of Commerce, and from the State of Michigan. The Michigan Sea Grant College Program is a cooperative program of the University of Michigan and Michigan State University.

"The *Dreissena* is perhaps better fitted for dissemination by man and subsequent establishment than any other fresh-water shell; tenacity of life, unusually rapid propagation, the faculty of becoming attached by string byssus to extraneous substances, and the power of adapting itself to strange and altogether artificial surroundings have combined to make it one of the most successful molluscan colonists in the world."

-Kew (1893)

"The possibility of...the zebra mussel being introduced [to the United States] is very great. There is entirely too much reckless dumping of aquaria into our ponds and streams. A number of foreign fresh-water shells, etc., have been introduced in this way. Why not the mussel?"

-Johnson (1921)

"There is the real possibility that *Dreissena polymorpha* will eventually become established in the North American continent despite all efforts to prevent introduction of exotic species."

-Sinclair (1964)

"...the occurrence of its veliger larvae in the plankton... greatly enhances its potential for introduction to the Great Lakes in ship ballast water. If introduced, *Dreissena* could establish itself in North America."

-Bio-Environmental Services, Ltd. (1981)

Quotes compiled by Jim Carlton.

Contents

SECTION I
Ecology and Life History

SECTION III
Effects

SECTION IV
Mitigation

SECTION V
General

SECTION I

Ecology and Life History

CHAPTER 1

Thirty Years of Studies of *Dreissena polymorpha* Ecology in Mazurian Lakes of Northeastern Poland

Anna Stańczykowska and Krzysztof Lewandowski

Zebra mussel ecology was studied over a 30-year period in the lakes of northeastern Poland. The zebra mussel (*Dreissena polymorpha*) was very common, being found in 60–100% of four groups of lakes studied. Examination of environmental parameters in different lakes indicates that zebra mussel densities were not strongly dependent on any one factor, but could be related with the trophic status of the water. Zebra mussels either were not found or were found in low densities in lakes that were usually shallow, strongly eutrophic, and polymictic. In general, densities of mussels were highest in mesotrophic lakes. Densities ranged from a few individuals per square meter to a maximum of a few thousand per square meter. Mussel densities differ between and within individual lakes over time. Within most lakes, densities of mussels were determined primarily by the mortality of planktonic veligers during settlement and in the post-veliger stage. Mortality of adult mussels by predation of fishes and waterfowl was believed to be relatively small. Typically, in lakes where unusually high densities build up over several years, there was a population crash to low densities that may continue for more than 10 years. The reason for these unexpected population crashes is not known. Over the 30 years of the present study, zebra mussel densities and distributions decreased in Polish lakes; this phenomenon is attributed to increased eutrophication in the areas studied.

0-87371-696-5/93/$0.00 + $.50

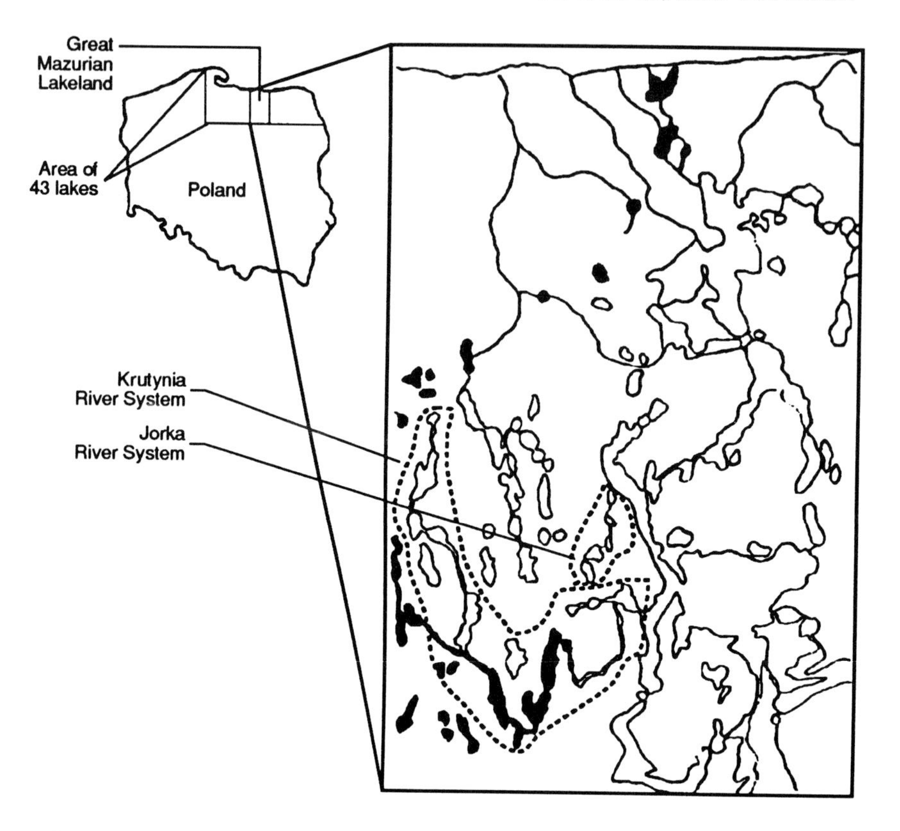

Figure 1. Locations of the areas of 43 lakes, the Great Mazurian Lakes, and Jorka and Krutynia River Systems within the Mazurian lakeland area of Poland.

INTRODUCTION

Zebra mussels, *Dreissena polymorpha* (Pallas), have occurred in lakes and reservoirs of the study region for about the past 150 years (Wahl, 1855; Hensche, 1861, 1862, 1866; Polinski, 1917; Berger, 1960). This region, in the broad sense, has been known as Mazurian Lakeland or Mazurian Lakes. By the 1950s, the zebra mussel was the most common and abundant mussel found in both rivers and lakes of this region (Berger, 1960).

Our studies of the Mazurian Lakes began in 1959 with the description of occurrence, distribution, and abundance of zebra mussels in the region known as the complex of Great Mazurian Lakes (Figure 1). In addition, we studied the aggregation and "condition" (i.e., size and weight of individuals of a population) of many mussel populations of different densities in lakes of varying trophic status (Stańczykowska, 1961; 1964; 1966). The studies that were initiated in the late 1950s and early 1960s were repeated, especially in Mikolajskie Lake (Stańczykowska, 1975; Stańczykowska et al., 1975b; several times throughout the 1970s and 1980s. During this latter period we also began to study the ecology of *Dreissena polymorpha* in additional lakes in other regions such as those in the Krutynia and Jorka river basins and in a group of 43 lakes of different mictic and trophy status (Lewandowski, 1982a and 1982b; Stańczykowska et al., 1983a and 1983b; unpublished data).

Studies were further expanded to include examining the role zebra mussels played in seston removal from the water column. In situ experiments were carried out in Mikolajskie Lake in which daily measurements of filtration, consumption, food assimilation, and feces production during the ice-free season were determined (Mattice et al., 1972; Stańczykowska et al., 1975a and 1976). Based on these studies we obtained the proportions of primary production that were utilized by different populations of zebra mussels and also obtained the proportion of tripton production that was attributable to mussels in individual lakes (Stańczykowska et al., 1975a; Stańczykowska, 1987).

We also supplemented our information with studies of how the zebra mussel fits into the food chains in Polish lakes by examining mussel predation by fish (Prejs et al., 1990) and waterfowl (Stańczykowska et al., 1990).

In recent years, the growing importance of water quality problems in Poland prompted us to study the role zebra mussels play in acting as: (1) a biofilter, in which the filtering activities of mussels remove suspensions from the water (Stańczykowska, 1968; Stańczykowska and Jawacz, 1971; Stańczykowska et al., 1975a); (2) an organism that affects nutrient cycling (Stańczykowska, 1984a; Stańczykowska and Planter, 1985); and (3) an indicator of the trophic status of lakes (Stańczykowska et al., 1983a; Stańczykowska, 1984b).

The current study presents some selected results of the 30 years of studies of *Dreissena polymorpha* performed in approximately 100 lakes of Mazurian Lakeland, with emphasis placed on the abundance, distribution, and occurrence of zebra mussels over time. This information will be useful in determining the similarities and differences between populations of zebra mussels found in Poland and North America.

AREA OF MAZURIAN LAKELAND

The area of Mazurian Lakeland is the largest lake district in Poland (Figure 1). Descriptive information about the lakes in Mazurian Lakeland are discussed in numerous other studies (Hillbricht-Ilkowska, 1989; Gliwicz and

Kowalczewski, 1981; Gliwicz et al., 1980; Kajak and Zdanowski, 1983; Bajkiewicz-Grabowska, 1983; Kajak, 1978; Pieczyńska et al., 1988); therefore, only a brief description of the lakes is given in the current study.

All the lakes studied are of the same geological origin (i.e., postglacial) and climatic exposure (i.e., temperate). Lakes differ in their morphology, surface area, depth, and flushing rate. Natural basin differences and varying impacts of agriculture, deforestation, urban sprawl, and recreational uses have led to marked differences in the trophic status of these lakes. Various littoral and pelagic data indicate that most of the lakes have undergone some eutrophication and are at present classified as mesotrophic (very few), eutrophic (most), and hypereutrophic (rare) lakes.

Studies of the zebra mussel were conducted in four groups of lakes, all contained within the geographical area of Mazurian Lakeland (Figure 1). The four groups are designated as (see Appendices 1–4): Great Mazurian Lakes (36 lakes), lakes of the Jorka River watershed (5 lakes), lakes of the Krutynia River watershed (19 lakes), and lakes of different mictic and trophic characteristics (43 lakes).

The Great Mazurian Lakes are a complex of interconnected lakes with a total area of 310 km^2 and water flow of low intensity. The lakes differ in their flushing rates; the smallest lakes (i.e., 26–42 ha) have flushing rates up to 10 times greater than the largest lakes (i.e., 2,400–10,000 ha). The whole area is placed within the extent of the last glaciation (i.e., the Baltic glaciation); thus it bears the marked features of a young postglacial landscape with many ramparts and moraines. Many of the lakes are trough lakes, or lakes made by depressions with no surface outlet. The entire area is heavily utilized for recreational purposes. Increased rates of eutrophication began in the 1950s with an intensification of agriculture, recreation, and general expansion of the towns in the basin. Eutrophication is characterized by high phosphorus concentrations, high summer phytoplankton biomass, low Secchi disk visibility (Figure 2), high oxygen deficits in lake hypolimnions (Figure 3), and high caloric values and organic matter contents of bottom sediments (Gliwicz et al., 1980).

Lakes of the Jorka River watershed are relatively small (i.e., less than 170 ha), have a mean depth of between 3.7 and 11.8 m, and are located in a topographically distinct watershed (Hillbricht-Ilkowska and Jawacz, 1983). The watershed is a typical elementary component of a mosaic-lake hilly lakeland landscape of Mazurian Lakeland. Lake Majcz Wielki, the lake occurring in the highest altitude of the watershed, is of mesotrophic character. Several lakes within the area receive intermittent and continual inputs of water from rivers and streams that are highly polluted (i.e., high phosphorus concentrations).

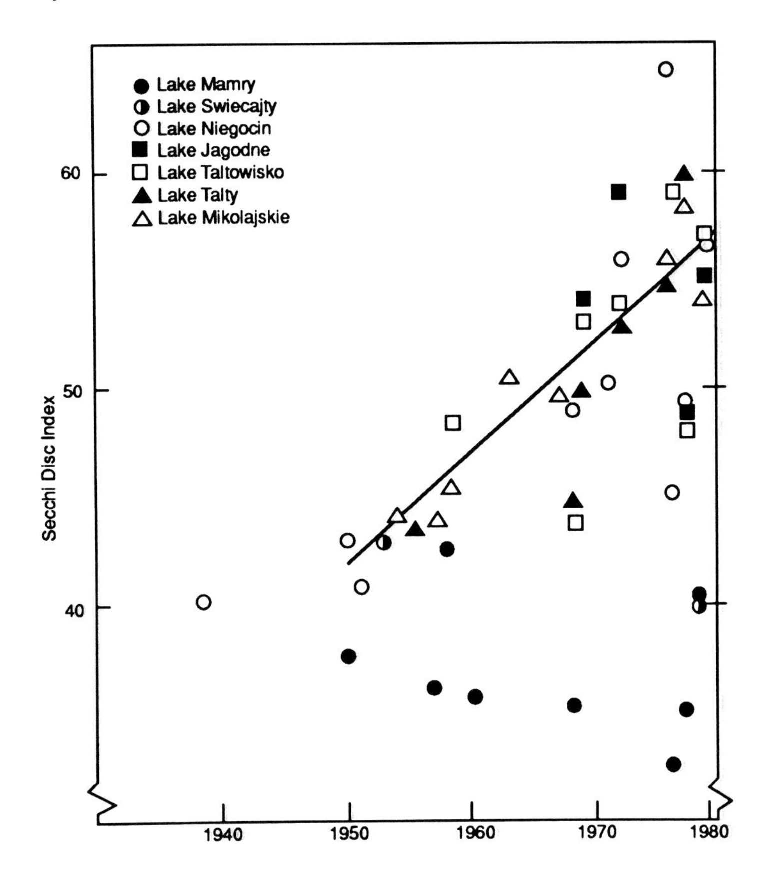

Figure 2. Average summer Secchi disk index (i.e., transparency) of Great Mazurian Lakes between 1940 and 1980. In addition, Secchi disk indices of 40 were obtained in 1902 for Lake Talty and Lake Jagodne (From Gliwicz, Z. M. et al. *Wyd. Inst. Ksztatr. Srodow., Warszawa* [1980] p. 104. With permission.)

The Krutynia River system consists of 19 lakes located on the Krutynia River, which is the longest waterway in this area. Most of these lakes have a strong flow of water and are shallow and moderately eutrophic (i.e., 51–100 μg/L total phosphorus). Although Krutynia River lakes are susceptible to eutrophication because of loading by the Krutynia River and because of their shallow depth, most limnological features are typical of low productive lakes. It is, however, to be emphasized that almost every lake contains hydrogen sulfide in stratified waters near the bottom in the summer.

The last group of 43 lakes do not belong to a distinct watershed or river system, but are representative of lakes of varying mictic and trophic levels in northeastern Poland. These lakes range in size from 24 to 382 ha, but most

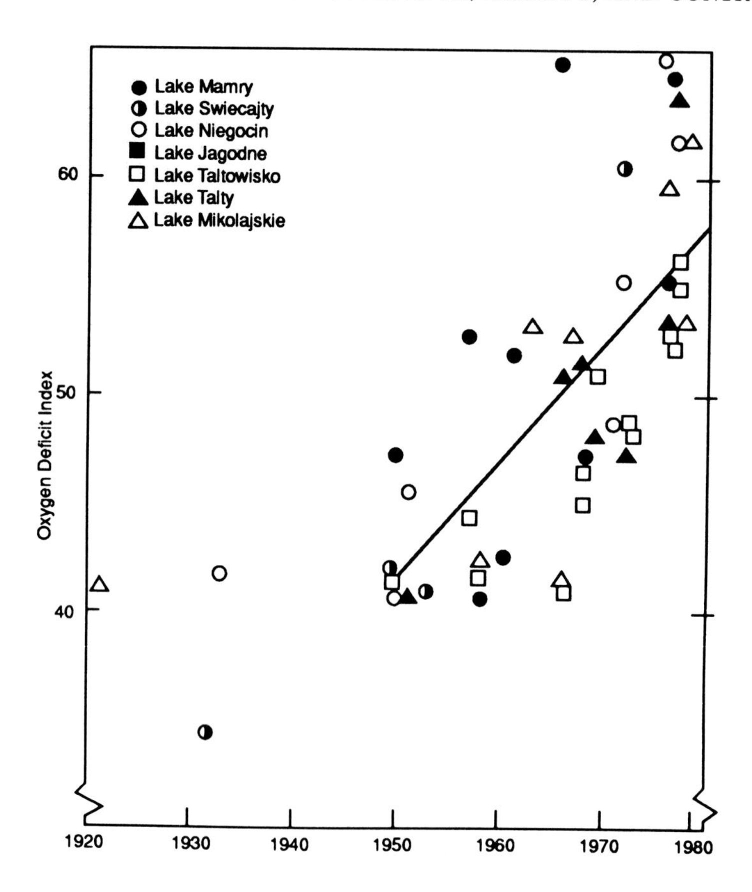

Figure 3. Yearly changes in values of the summer absolute oxygen deficit in hypolimnion (transformed into trophic state values) for seven Great Mazurian Lakes between 1930 and 1980 (From Gliwicz, Z. M. et al. *Wyd. Inst. Ksztatr. Srodow., Warszawa* [1980] p. 104. With permission.)

are between 50 and 150 ha. Dimictic lakes overturn twice per year (in the spring and fall), are stratified in summer, and are inversely stratified in winter, whereas polymictic lakes overturn frequently or continuously during the entire year. Dimictic and polymictic lakes were divided into groups based on total phosphorus content of the water during summer. Ranges of total phosphorus content in each mictic type of lake were grouped as: 20–50 μg/L, 51–100 μg/L, 101–300 μg/L, and >300 μg/L for dimictic lakes; and 51–100 μg/L, 101–300 μg/L, and 301–950 μg/L for polymictic lakes.

OCCURRENCE OF ZEBRA MUSSELS IN LAKES

Zebra mussels were found in 64–81% of the Great Mazurian Lakes (23 to 29 of 36), 100% of the Jorka River lakes (5 of 5), 90% of the Krutynia

Table 1. Number and Percent (in Parentheses) of Lakes where *Dreissena polymorpha* Were Found in the Mazurian Lakeland of Northeastern Poland Between 1959 and 1989

Group of Lakes	Years Studied	Number of Lakes
Great Mazurian (n = 36)	1959–1962	29 (81)
Great Mazurian (n = 36)	1972	26 (73)
Great Mazurian (n = 36)	1988	23 (64)
Jorka River (n = 5)	1977–1978	5 (100)
Krutynia River (n = 19)	1989	17 (90)
43 Lakes (n = 43)	1976–1977	37 (88)

River lakes (17 of 19), and 88% of the 43 representative lakes (38 of 43) (Table 1). The two lakes in the Krutynia River system that did not have zebra mussels were shallow, reservoir-type lakes with high rates of siltation. The other 17 lakes were relatively deep (7–34 m), had high flushing rates, and had low phosphorus concentrations. In the Jorka River system, all of the lakes had zebra mussels. These lakes were relatively small in size but quite deep (maximum 7–34 m), and were characterized by a high flushing rate and low phosphorus concentrations.

The lakes of the Great Mazurian region were sampled three times, in 1959–1962, in 1972, and in 1988. Of the 36 lakes sampled each time, the number with zebra mussels declined from 29 in 1959–1962, to 26 in 1972, and to 23 in 1988 (Table 1). Lakes with no zebra mussels in 1959–1962 were strongly eutrophic and, according to Stangenberg (1936), had a character of eutrophic lake-ponds. Most were shallow, having an average depth of only 1–3 m. A few of the lakes were deeper (i.e., Guzianka Duza and Guzianka Mala; Appendix 1), but rather small in surface area and strongly eutrophic in the early 1960s. The largest of the lakes with no zebra mussels, Warnolty Lake (i.e., 470 ha), might be classified as hypereutrophic. In 1972, no mussels were found in three more lakes of the Great Mazurian Lakes Region (i.e., Kaczerajno, Seksty, and Kotek). These three lakes also exhibited a eutrophic to hypereutrophic state in the 1970s. By 1988, zebra muscles disappeared from another three lakes (i.e., Rynskie, Talty, and Beldany). These lakes all exhibit a high degree of eutrophy. The environmental studies that have recently been conducted in these lakes show well-marked oxygen deficits. The visibility of the Secchi disk is greatly reduced, and the Vollenweider criterion of total phosphorus content is exceeded by a factor of two or more (Hillbricht-Ilkowska, 1989; Kajak, 1978; Pieczynska et al., 1988). A sharp decrease in the zebra mussel density has also been recently observed in Niegocin Lake

Table 2. Percent Occurrence, Mean Density (Number/m²) and Percent *Dreissena polymorpha* Mussels of the Total Mollusc Community in 43 Dimictic and Polymictic Lakes of Different Total Water Phosphorus Concentrations

		Total Phosphorus Concentration (μg/L)			
		20–50	51–100	101–300	301–950
Percent Occurrence	Dimictic	100	100	86	0
	Polymictic	0	84	86	60
Mean Density	Dimictic	247	436	498	0
	Polymictic	0	438	4	<1
Percent of Mollusks	Dimictic	68	49	60	0
	Polymictic	0	37	2	<1

(another large Mazurian lake) which is becoming more eutrophic due to inputs from municipal and industrial sewage (Cydzik and Soszka, 1988).

Among the 43 lakes of various mictic and trophic characteristics examined in the years 1976–1977, zebra mussels were found in 88% of them (Table 1; Stańczykowska et al., 1983a). More specifically, mussels were found in 100% of the dimictic lakes with a phosphorus content of 20–50 and 51–100 μg/L, in 86% of the dimictic lakes of 101–300 μg/L, in 84–86% of the polymictic lakes of 51–100 and 101–300 μg/L, and in only 60% of the polymictic lakes of highest phosphorus content (Table 2).

DISTRIBUTION OF ZEBRA MUSSELS WITHIN LAKES

In all Mazurian lakes where zebra mussels were found, the mussels formed a continuous belt around the shores, usually covering the littoral and upper sublittoral zones. The extent of the area colonized varied depending upon the width and slope of the littoral.

In most lakes, densities of zebra mussels in regions closest to shore (i.e., 0.3–0.6 m depth) were very low (Figures 4 and 5; Stańczykowska, 1964). The density of mussels increases with depth, reaching a maximum at a depth of 2–4 m, and then declines. Generally, the depth of occurrence of zebra mussels in most lakes does not exceed between 6 and 8 m (Stańczykowska et al., 1975; 1983a; 1983b), but in a few lakes mussels occur at depths as great as 12 m. Similar distributions and ranges of occurrence have been noted by most European authors (Wesenburg-Lund, 1939; Dunn, 1954; Mikulski and Gizinski, 1961). A deeper range was noted by Lundbeck (1962), Grim (1971), and Walz (1974); the latter author found zebra mussels at depths of up to 45 m in some German lakes.

Investigations in Lake Majcz Wielki have shown that postveligers and young individuals live mostly near the shore, whereas adult individuals are

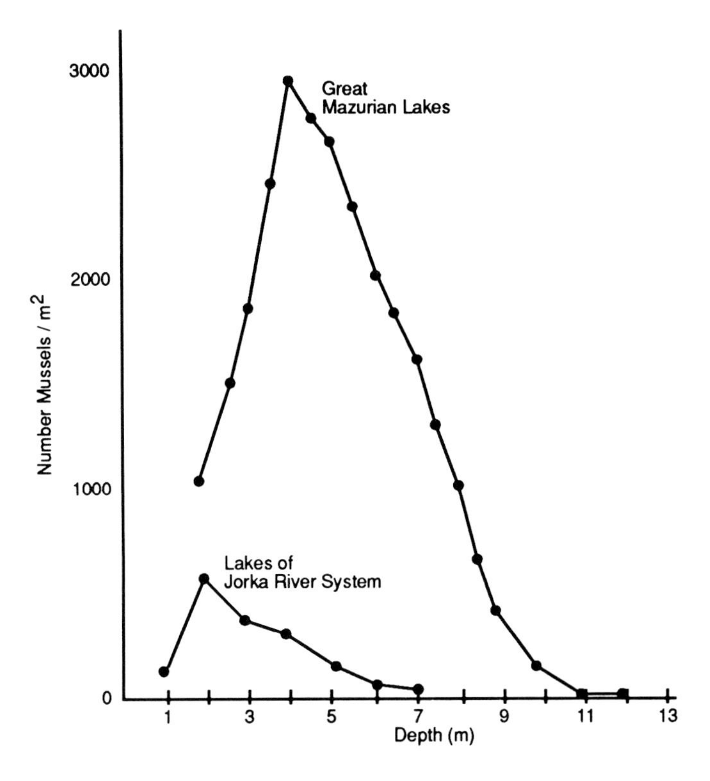

Figure 4. Depth distribution of *Dreissena polymorpha* in 36 Great Mazurian Lakes and 5 lakes in the Jorka River system of northeastern Poland (1959–1962).

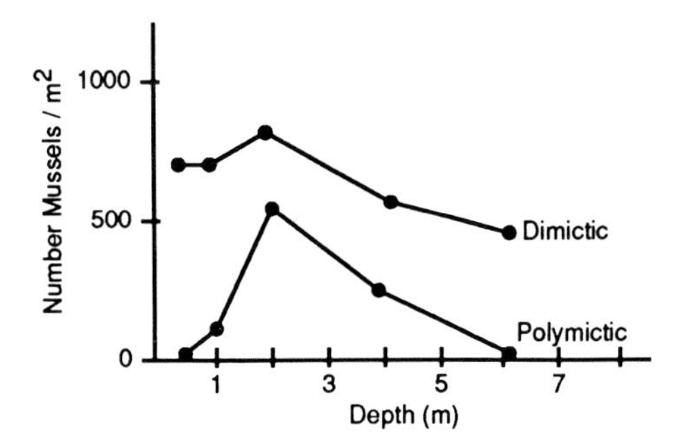

Figure 5. Depth distribution of *Dreissena polymorpha* in a total of 43 dimictic and polymictic lakes of northeastern Poland.

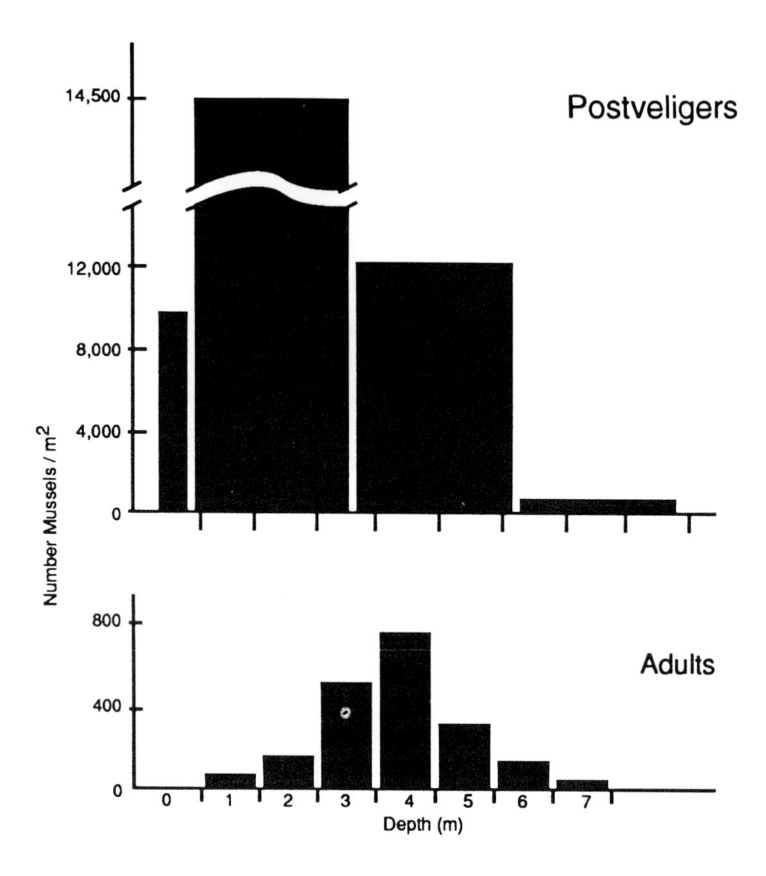

Figure 6. Occurrence of postveligers (September 1978) and adult *Dreissena polymorpha* at different depths in Lake Majcz Wielki. (Modified from Lewandowski, K. *Ekol. Pol.* 30:223–286 [1982b]. With permission.)

more numerous at greater depths (Figures 6 and 7; Lewandowski, 1982b; Stańczykowska, 1964). For example, postveligers are most abundant at depths of 1.0–2.5 m in areas with growths of Characeae, while adult mussels are most abundant at a depth of 4 m (Figure 6). The 4-m depth in Lake Majcz Wielki marks the border between stands of Characeae in shallower water and deeper areas with declining vegetation or with no vegetation at all, and where colonies of mussels dominate the benthos. The relationship between the population age structure of adult zebra mussels and water depth structure was also observed in Lake Taltowisko (Figure 7). In areas near shore at less than 1-m depth, 1- and 2-year-old mussels make up the majority of the population, whereas at deeper sites the percentage of mussels older than 2 years make up the majority of the population.

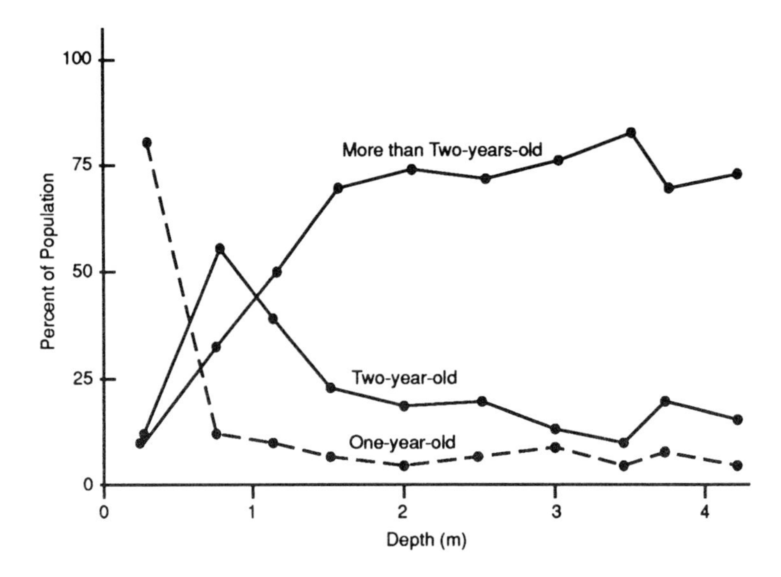

Figure 7. Population age structure of *Dreissena polymorpha* at different depths in Lake Taltowisko. (Modified from Stańczykowska, A. *Ekol. Pol., Ser. A* 12:653–690 [1964]. With permission.)

Within an individual lake, densities of mussels not only vary with depth, but also vary between sites of the same depth; however, in most lakes these differences are small (Figures 8 and 9) (Stańczykowska et al., 1983b). In recent years, zebra mussels have disappeared from certain locations in lakes with increasing eutrophy, such as Lakes Mikolajskie and Niegocin, so that there is now a discontinuous band of mussels along the shores. Moreover, in Mikolajskie Lake from 1959 to 1976, zebra mussels were found at depths ranging from 0.2 to 7.5 m; but in the period from 1977 to 1990, they were not found deeper than 4 m. The change is associated with the increased eutrophy and low oxygen conditions (Stańczykowska and Lewandowski, unpublished data).

Veligers passing from the plankton to the sedentary stage settle most readily on submerged vegetation (Table 3; Figure 10; Lewandowski, 1982b). Veligers settle on annuals and perennials at different rates (Table 4; Lewandowski, 1982b). In addition, the rate of settling varies by plant species depending on its particular vegetative structure. In the majority of the Mazurian lakes, plant-dwelling zebra mussels represent over 85% of the population. For a given square meter of area in the littoral, densities of mussels settled on vegetation may be many times higher than the density of those settled directly on other substrate (Table 5; Lewandowski, 1982b). As many as 100,000/m^2 of lake bottom may be attached to submersed vegetation.

Mussels that settle beyond plants, stones, and colonies of existing adult zebra mussels have little chance for survival to become adults (Tables 6 and

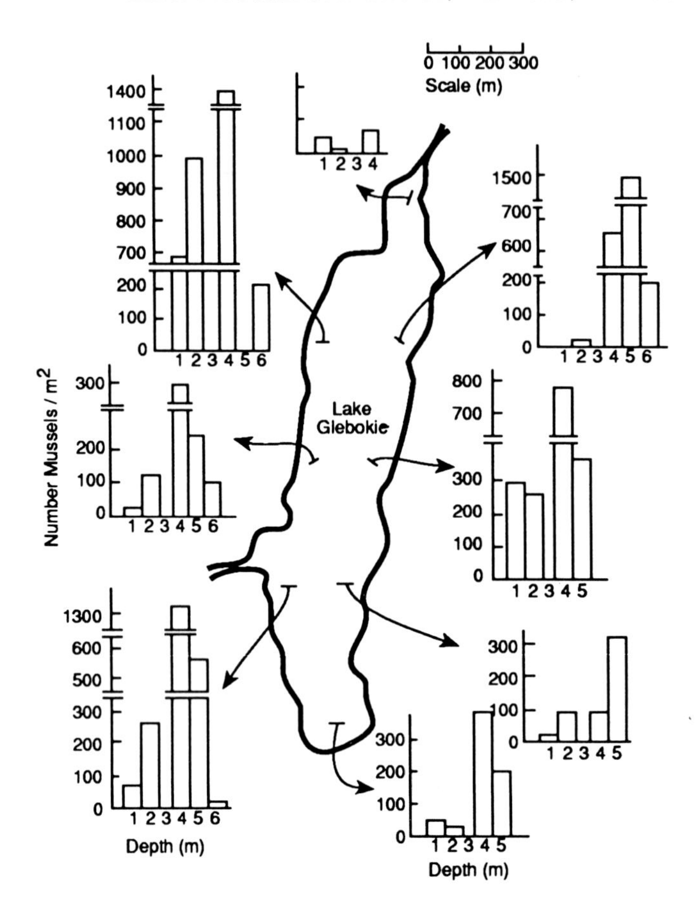

Figure 8. Numbers of *Dreissena polymorpha* at various depths and locations in Lake Glebokie. (Modified from Stańczykowska, A. et al. *Ekol. Pol.* 31:761–780 [1983b]. With permission.)

7; Lewandowski, 1982b). Colonies of adult mussels (found in zones perennially occupied by mussels) consist of several age classes with both young and adults being present (Stańczykowska, 1964 and 1977). This is attributable to juvenile individuals that settle in separate aggregations and experience high mortality when they settle in unsuitable environments. For example, mussels that settle on sand and mud and in shallow water that freezes in the winter die due to exposure.

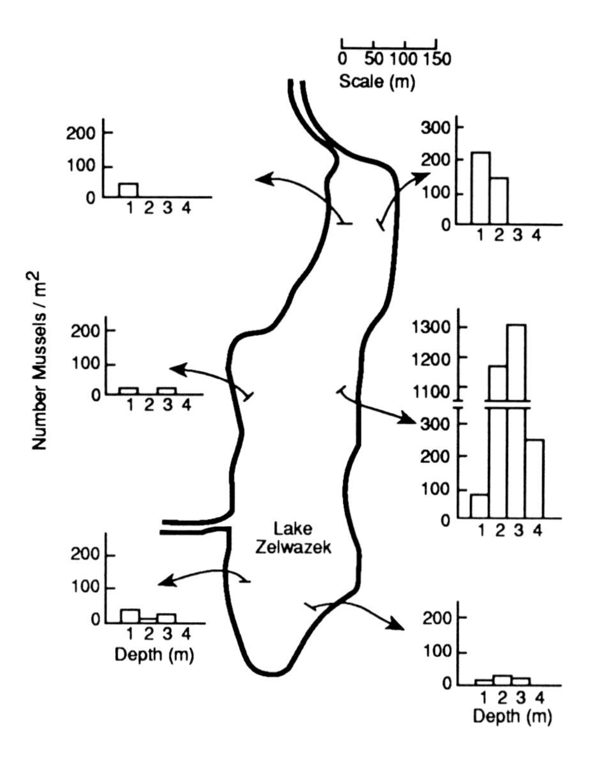

Figure 9. Numbers of *Dreissena polymorpha* at various depths in Lake Zelwazek. (From Stańczykowska, A. et al. *Ekol. Pol.* 31:761–780 [1983b]. With permission.)

DENSITY OF ZEBRA MUSSELS — BETWEEN LAKE COMPARISONS

Densities of zebra mussels in lakes of the various regions ranged from 1 to 2,000–3,000/m^2 (Figure 11). We attempted to correlate observed densities in the lakes to various environmental parameters, such as trophic status, size, depth, visibility, pH, and calcium content of the water. This comparative analysis did not give any positive associations. The only generalization is that mesotrophic and relatively large and deep lakes have higher densities of zebra mussels than strongly eutrophic and relatively small and shallow lakes. No direct correlation was found between densities of mussels and visibility, pH, and calcium concentrations of the water (Stańczykowska, 1964).

Examinations of the 43 lakes of various mictic and trophic status have shown that zebra mussels dominated the mollusc community in dimictic lakes at all three ranges of total phosphorus concentrations between 20 and 300 μg/L (Table 2). Zebra mussels were absent in dimictic waters where total phosphorus concentrations exceeded 300 μg/L. In polymictic lakes, zebra

Table 3. Percent Frequency of Samples by Density Classes (Number/m²) of *Dreissena polymorpha* Settled on Submerged Vegetation and Other Substrates in 24 Mazurian Lakes

	Settled Individuals	
Density Class	**On Other Substrates**	**On Vegetation**
Less than 100	22.1	8.8
101–1,000	36.4	31.0
1,001–5,000	32.4	26.6
5,001–20,000	9.1	21.2
20,001–100,000	0	8.0
Over 100,000	0	4.4

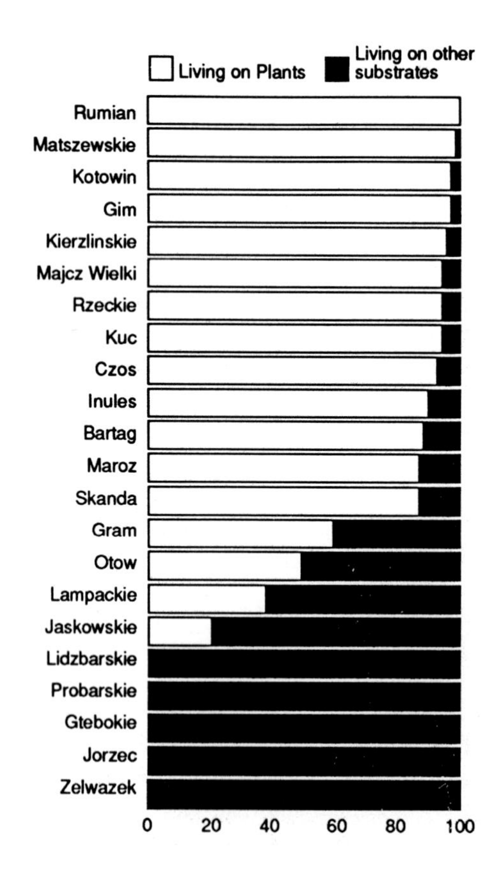

Figure 10. Percent *Dreissena polymorpha* individuals living on plants and other substrates in the Mazurian Lakes. (Modified from Lewandowski, K. *Ekol. Pol* 30:81–110 [1982b]. With permission.)

Table 4. Maximum Numbers of Settled *Dreissena polymorpha* per 100 g of Plants in the Lakes under Study

Plants	Number of *D. polymorpha*
Perennials	
Fontinalis antipyretica	118,000
Ceratophyllum demersum	110,000
Elodea canadensis	105,000
Characeae	34,000
Annuals	
Potamogeton pectinatus	104,000
P. lucens	19,000
P. perfoliatus	3,500
Myriophyllum spicatum	1,200

Table 5. Maximum Numbers of Different Age Classes of *Dreissena polymorpha* Settled on Plants and Other Substrates of Lake Rumian (August 1977)

Age Class of *D. polymorpha* (yr)	Numbers of *D. polymorpha*/m² of Littoral Bottom	
	On Plants	On Bottom
0	1,700,000	1,800
1	3,300	2,100
2	250	1,600
3	80	900
4	0	280
5	0	40

Table 6. Densities of (Number/m²) *Dreissena polymorpha* Larvae on Different Natural Substrates in Lake Majcz Wielki from May to September 1978

Kind of Substrate	Mean	Range
Characeae	1727	604–2720
Dreissena polymorpha	455	167–697
Stones	61	25–109
Sand	13	7–23
Mud	13	3–38

Table 7. Percent Dead *Dreissena polymorpha* on Different Natural Substrates in Lake Majcz Wielki from May to September 1978

Kind of Substrate	Mean	Range
Characeae	2.6	1.3–3.5
Dreissena polymorpha	3.8	2.8–6.5
Stones	6.1	1.8–16.6
Sand	31.6	4.7–66.6
Mud	46.3	13.3–70.7

mussels were a large proportion of the mollusk community only where the phosphorus content was relatively low (i.e., between 51 and 100 μg/L). In polymictic lakes of higher phosphorus concentrations, zebra mussels were absent or did not occur in substantial numbers.

DENSITY OF ZEBRA MUSSELS — WITHIN LAKE COMPARISONS

More thorough studies on the variation of zebra mussel densities were carried out in Lake Mikolajskie and in numerous other Great Mazurian Lakes (Figures 12–14). Studies in these lakes were repeated several times during the 30-year period between 1959 and 1989 (Stańczykowska, 1961; 1964; 1966; 1975; 1976; 1977; 1984). Lake Mikolajskie is one of the lakes in Poland whose hydrology has been extensively studied. It is a dimictic, eutrophic, tunnel-valley lake with a surface area of 460 ha, situated in the central part of the Great Mazurian Lakes region. It has natural connections with Lakes Talty, Sniardwy, and Beldany. Land use around the lake includes forests, agricultural land, and municipal buildings in about equal proportions. The lake is subjected to various eutrophication influences, of which the strongest is the disposal of municipal sewage. Many years of studies in Lake Mikolajskie have shown various environmental changes. Eutrophication is indicated in various environmental parameters: reduced Secchi disk values of 3 m in 1954 to 1 m in 1977 (Szczepanski, 1968; Gliwicz et al., 1980); a reduction of submersed macrophyte biomass from 2.1 ton/ha in 1963 to 0.2 ton/ha in 1980 (Ozimek and Kowalczewski, 1984); an increase of pelagic phytoplankton biomass from 18 mg/L in 1963 to 55 mg/L in 1972 (Spodniewska, 1976); an increase of littoral algae biomass from 1 g/L in 1966 to 894 g/L in 1982 (Pieczynska et al., 1988); and a decrease in the distribution of unionid mollusks from 1972 to 1987 (Lewandowski and Stańczykowska, 1975).

Samples taken during 14 sampling periods between 1959 and 1989 show a high year-to-year variation in zebra mussel densities in Mikolajskie Lake (Figure 12). Densities measured during this period ranged from 1 to 2200/m^2.

Densities were highest in 1972 and 1976. A marked decline in densities occurred in 1960 and again in 1977; both these years were preceded by years of extremely high densities of mussels (Stańczykowska, 1961; 1977). This decline was observed in high density populations having relatively large and heavy individuals.

In general, zebra mussels were in better condition (i.e., size and weight) in lakes characterized by a low average total abundance than in lakes where the total number of mussels was higher (Figure 13). This trend was not apparent, however, in lakes with relatively low densities. As mentioned before, in 1960–1962 the total number in these lakes was lowest. Prior to 1959,

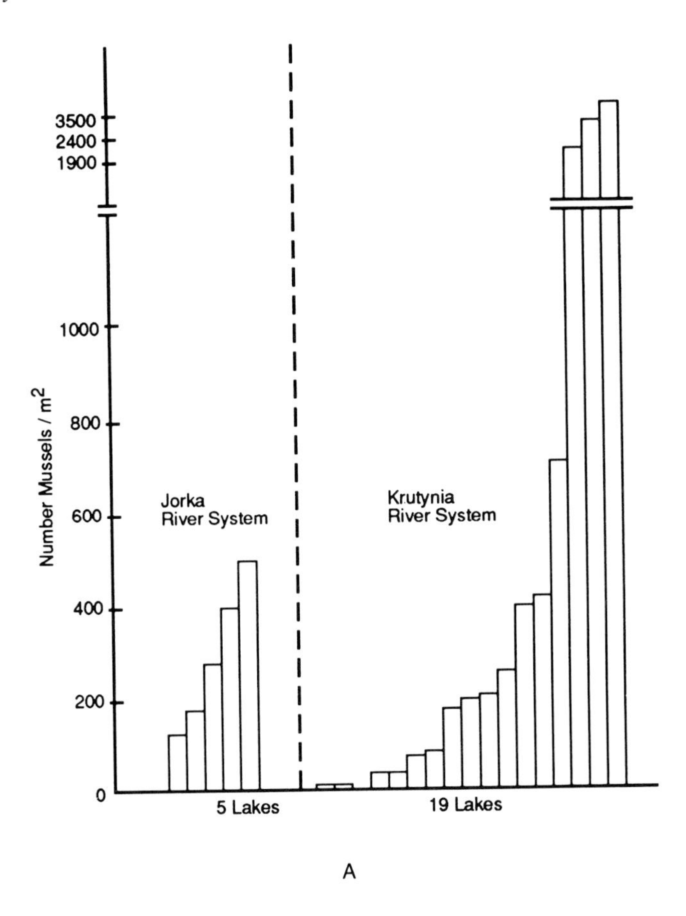

Figure 11. Ranked density (from lowest to highest) of *Dreissena polymorpha* in lakes (A) from the Jorka and Krutynia River systems and (B) in 43 lakes of different mictic and trophic status.

as well as in that year, the abundance of mussels was very high in Mikolajskie, Beldany, Talty, and Rynskie; and in the winter of 1959–1960, the abundance of mussels rapidly declined. However, the condition of mussels in these four lakes was relatively good before and after the sudden reduction in densities (Figure 12). Rapid reductions of population numbers were observed in Mazurian Lakes in different lakes and periods. These reductions occurred for all age classes of mussels. This phenomenon is generally seen in extremely dense populations consisting of unusually large and heavy individuals. The reasons for extreme density reductions of zebra mussels populations are not known. A gradual degradation of water quality does not usually cause sudden decreases of population densities in other organisms. Possible explanations for sharp population declines following a period of high densities include: de-

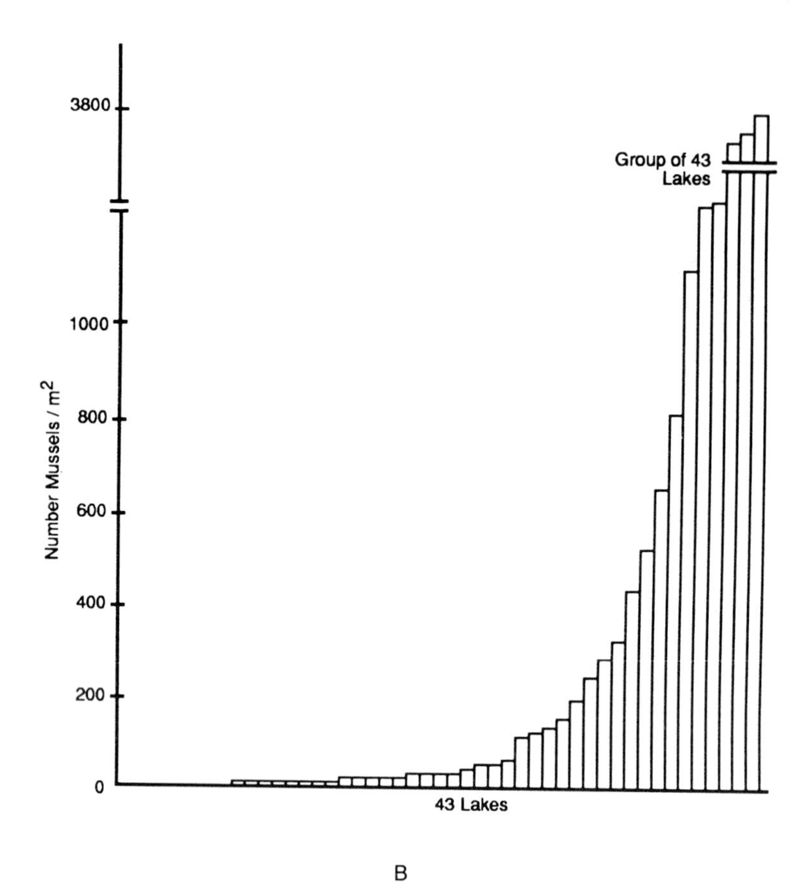

B

teriorating feeding conditions (i.e., lack of sufficient food to support the population), invasion of parasites, viral or bacterial diseases, or multiple factors acting collectively in direct and indirect ways. After the population crash in Lake Mikolajskie in the late 1970s, mussel densities remained low throughout the 1980s and disappeared from some parts of the lake (Figure 12).

Additional long-term density changes of zebra mussels in 12 other Great Mazurian Lakes were conducted in 1959, 1960–1962, 1972, 1982, 1985, and 1988 (Figure 14). Although lakes were not studied as intensively as Lake Mikolajskie, they exhibited a wide range of density changes over time. In one group of lakes, the density of mussels was relatively stable. Most of these lakes are large, deep lakes of mesotrophic status or in a moderate state of eutrophy. In the second group of lakes, densities of mussels tended to fluctuate from year to year indicating relatively unstable populations. Although densities decreased and increased markedly between the 1960s and early 1980s, they tended to stabilize at relatively high densities in the late 1980s. It is

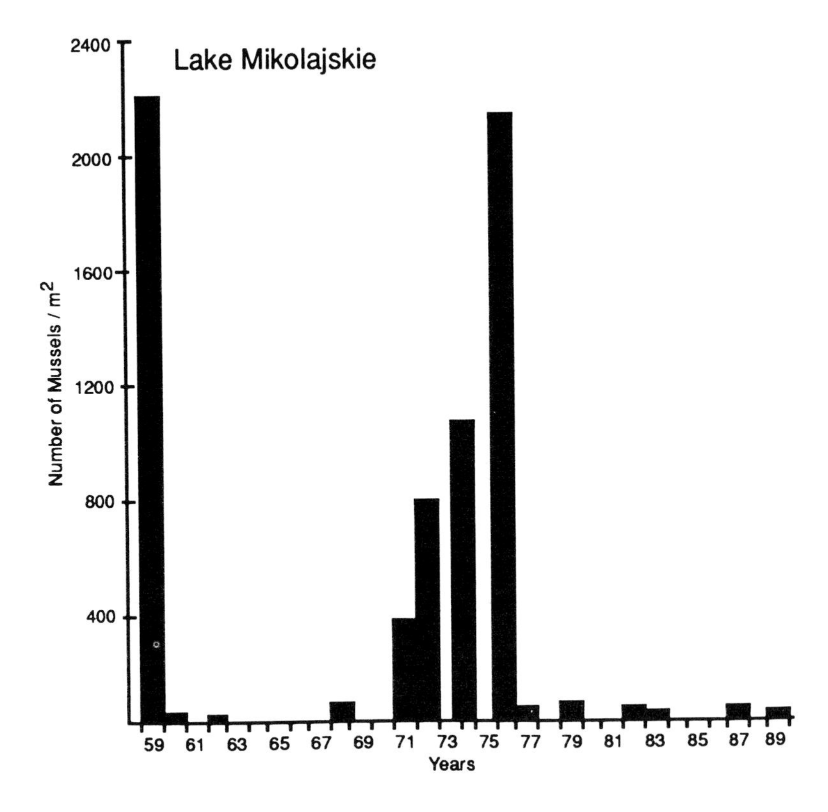

Figure 12. Numbers of *Dreissena polymorpha* in Lake Mikolajskie between 1959 and 1989. Samples were collected during 14 of the 31-year time span.

important to note that the density of mussels in the late 1980s was lower than in the early 1960s. Increased eutrophy of these lakes may be the cause of such a decrease in densities. In a third group of lakes, decreased densities of mussels were associated with marked changes of environmental conditions. In this group of lakes, densities of mussels in the 1980s were greatly reduced compared to earlier years. An analysis of environmental factors (Gliwicz et al., 1980) and data of zebra mussel densities over the past 30 years showed that eutrophication increased dramatically in periods when zebra mussel densities substantially declined. These lakes were eutrophic to hypereutrophic throughout most of the 1980s (Kajak, 1978; Gliwicz et al., 1980; Gliwicz and Kowalczewski, 1981; Ozimek and Kowalczewski, 1984; Pieczynska et al., 1988; Hillbricht-Ilkowska, 1989; Cydzik and Soszka, 1988).

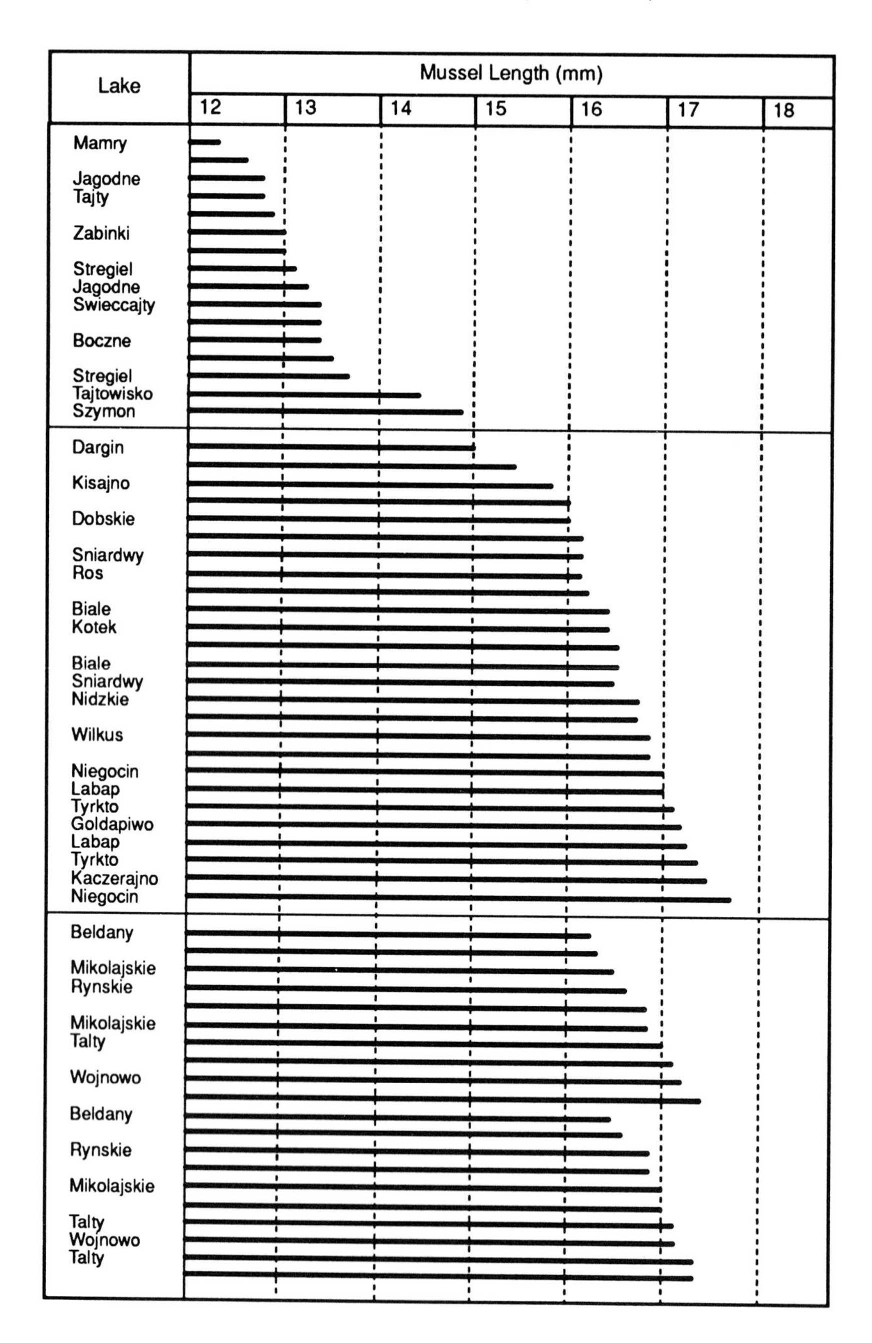

Figure 13. Average length of *Dreissena polymorpha* shells in lakes of northeastern Poland. (Modified from Stańzykowska, A. *Ekol. Pol., Ser. A* 12:653–690 [1964]. With permission.)

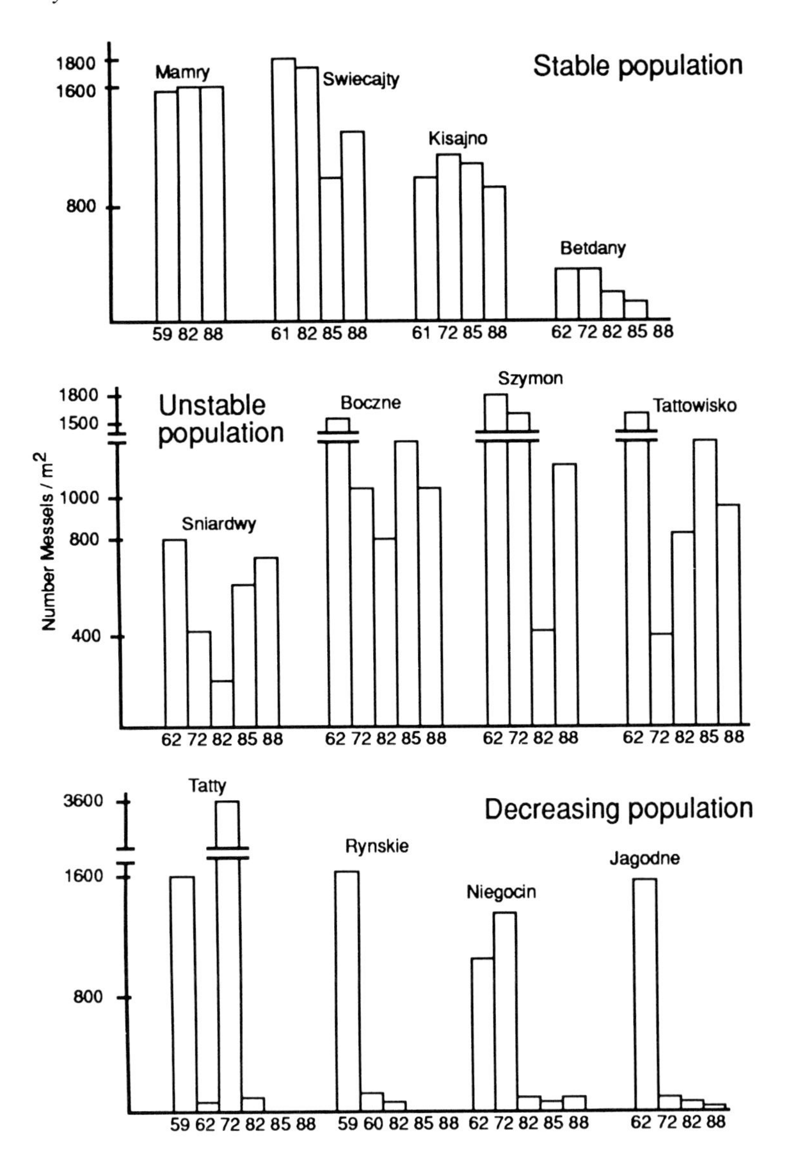

Figure 14. Numbers of *Dreissena polymorpha* per square meter in lakes of stable, unstable, and decreasing population patterns as determined for several sampling periods between 1959 and 1988.

CHARACTERISTICS OF POPULATION CHANGES

Changes in densities of zebra mussel populations can be divided in two categories: changes associated with the invasion period and those associated

Table 8. Characteristics of Population Changes of Zebra Mussels Discussed in the Text

I. Associated with invasion period
  1. Fast spread and high densities sometimes leads to low population densities
  2. Slow spread and low densities

II. Not associated with invasion period
  1. Fluctuations regulated by relatively low mortality of veligers and high mortality of immediate postveliger stage
  2. Decreases of selected populations or age classes due to predation
  3. Sudden reduction of densities in high-density populations
  4. Decreases of two- to threefold due to slow environmental changes

with an already established population (Table 8). Assessment of changes in zebra mussel populations directly after the invasion of the mussel into Polish lakes is not possible because there is little documentation during the period of time that invasions occurred. Zebra mussels invaded and have been part of the aquatic fauna of this region for approximately 150 years. Fluctuations not associated with the invasion period have been observed in Polish lakes and elsewhere. Although the exact mechanisms responsible for population changes are still unclear, we believe that most of the fluctuations can be attributable to environmental factors including suitable substrate for postveligers and, to a lesser extent, to the activity of predators.

Mortality in zebra mussels varies depending on the different life cycle stages (Table 8, II.1). The mortality of free-swimming veligers for about the first 10 days does not exceed 20%, as shown by the previously discussed data and by Wiktor (1969) for the Szczecin Lagoon (Kirpichenko, 1964; Shevtsova, 1968; Hillbricht-Ilkowska, 1989; Stańczykowska, 1977). Although the veligers are partly consumed by fish fry (Wiktor, 1958) and carried by currents into other water bodies (Wiktor, 1969), these factors are believed to make a relatively insignificant contribution to veliger mortality.

The postveliger stage, however, is critical to the developing zebra mussel population. Veligers begin to settle out of the water and must find suitable substrate for survival (Table 8, II.1). A detailed analysis of the body size of planktonic larvae and the newly settled postveligers has shown that the transition from planktonic to sedentary habits may occur over a fairly wide range of body size. Most zebra mussel larvae settle at a body size of over 200 μm but some settle at 140 μm (Figure 15). Studies on Mazurian Lakes have shown that mortality in the postveliger stage can be as high as 99% (Stańczykowska, 1977). Only a small part of the veligers settle in the mid- and sublittoral area and thus have a high chance for survival. Hence, survival is limited by the area with suitable substrate on which veligers can settle. Survival of postveligers is dependent on hardness of the substrate, position of the substrate in relation to the bottom, and chemical and physical properties of the water (Liakhov, 1964; Stern, 1965; Kirpichenko, 1971; Walz, 1975).

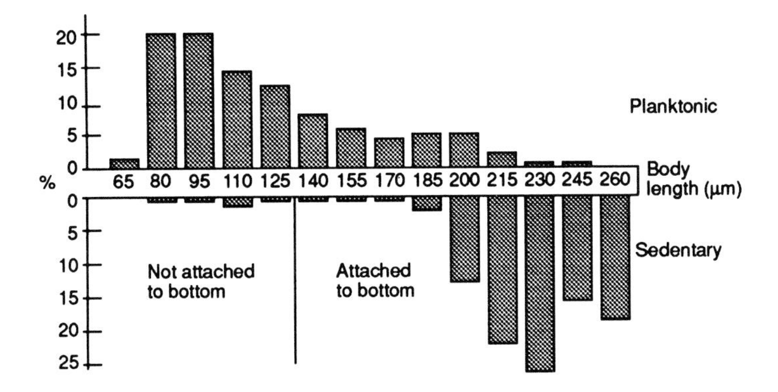

Figure 15. Body size of planktonic and sedentary *Dreissena polymorpha* larvae in Lake Majcz Wielki from June to August 1978. A total of 1118 planktonic and 966 sedentary larvae were measured. (Modified from Lewandowski, K. *Ekol. Pol.* 30:223–286 [1982b]. With permission.)

Several factors determine the percent of veligers occurring above suitable substrates during the period of their transformation into postveligers. These include weather conditions (i.e., direction and strength of winds: Figure 16), lake morphometry (i.e., area and character of the shoreline), circulating currents, and possibly water flow in and out of the lake. All these factors result in substantial differences noted in the density of the mussels at different locations in the same lake (Figures 8 and 9) (Siessegger, 1969; Schalekamp, 1971; Wisniewski, 1974; Stańczykowska et al., 1983b). The beginning and length of larvae and postveliger stages of zebra mussels are determined by water temperature that may vary substantially between years and water bodies (Figure 17; Lewandowski, 1982a). Consequently, particular cohorts of veligers have different chances of survival. As a whole, environmental variables as they affect veliger settling explain extreme variations of density and age structure in different water bodies with similar characteristics (Stańczykowska, 1975 and 1976; Stańczykowska et al., 1975b and 1983b).

Analysis of empty mussel shells has shown that, as a rule, the population is composed of a relatively large number of older mussels, indicating that once mussels live for 1 year they have good chances for continued survival. However, older 1- and 2-year-old mussels can be fed upon by predators such as fish and birds (Poddubnyj, 1966; Pliszka, 1953 and 1956; Prejs, 1976; Prejs et al., 1990; Leuzinger and Schuster, 1970; Stempniewicz, 1974; Borowiec, 1975). On the basis of ornithological studies, we believe that these sources of mortality are relatively small and do not reduce the density and distribution of zebra mussels in Mazurian Lakes (Table 9, II.2) (Borowiec, 1975; Halba, 1975). Moderate impacts of waterfowl on mussel populations can be totally different in other ecological situations than those of Mazurian

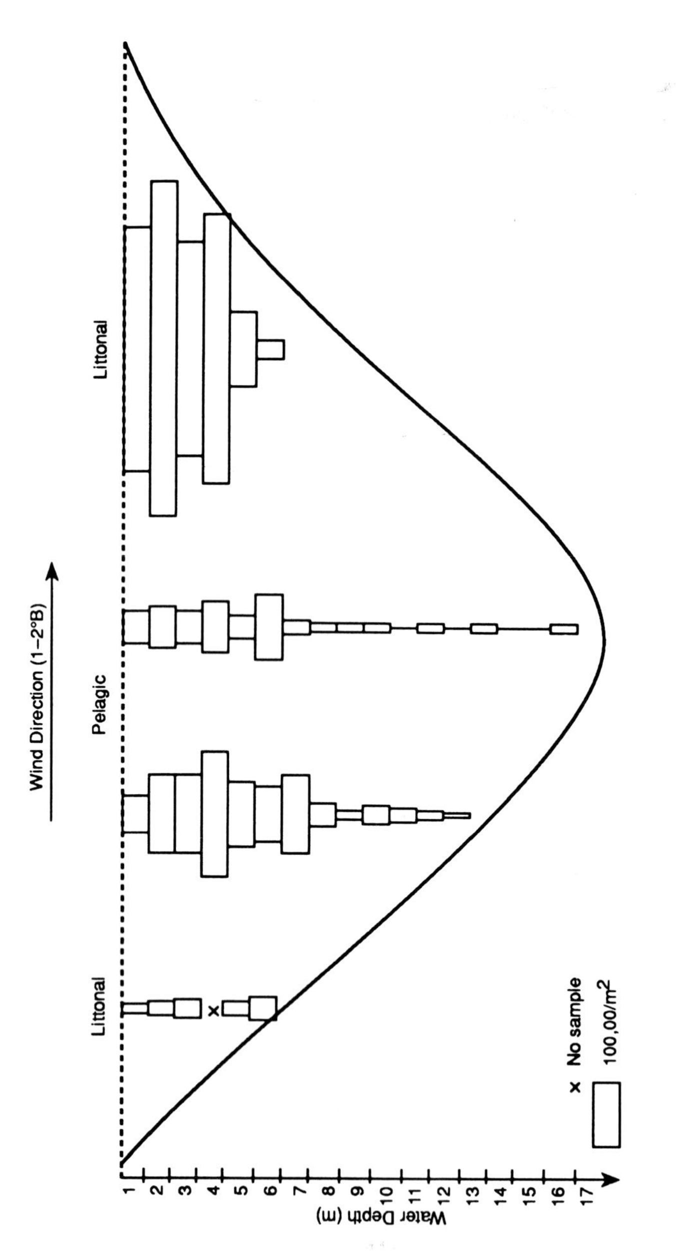

Figure 16. Vertical distributions of *Dreissena polymorpha* veligers at two littoral and two pelagic stations at Lake Zarnowieckie in August 1980.

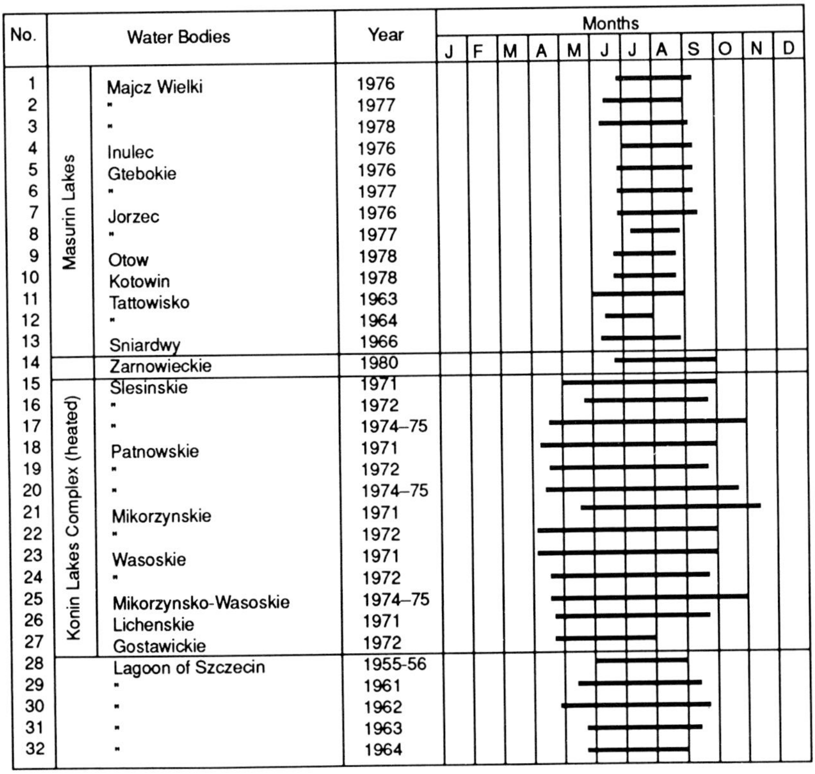

No.		Water Bodies	Year
1	Masurin Lakes	Majcz Wielki	1976
2		"	1977
3		"	1978
4		Inulec	1976
5		Gtebokie	1976
6		"	1977
7		Jorzec	1976
8		"	1977
9		Otow	1978
10		Kotowin	1978
11		Tattowisko	1963
12		"	1964
13		Sniardwy	1966
14		Zarnowieckie	1980
15	Konin Lakes Complex (heated)	Slesinskie	1971
16		"	1972
17		"	1974–75
18		Patnowskie	1971
19		"	1972
20		"	1974–75
21		Mikorzynskie	1971
22		"	1972
23		Wasoskie	1971
24		"	1972
25		Mikorzynsko-Wasoskie	1974–75
26		Lichenskie	1971
27		Gostawickie	1972
28		Lagoon of Szczecin	1955-56
29		"	1961
30		"	1962
31		"	1963
32		"	1964

Figure 17. Different periods of occurrence of *Dreissena polymorpha* larvae in plankton in Polish lakes. Data sources include: 1–10 and 14 in the present study; 11–13 from Hillbricht-Ilkowska and Stańczykowska (1969); 15, 18, and 23 from Stańczykowska (1976); 16, 19, 21, 22, 24, 26, and 27 from Lewandowski and J. Ejsmont-Karabin (1983); 17, 20, and 25 from Kornobis (1977); 28 from Wiktor (1958); and 29–32 from Wiktor (1969). (Modified from Lewandowski, K. *Ekol. Pol.* 30:81–110 [1982a]. With permission.)

Lakes. In Goplo Lake of central Poland, waterfowl consumed about 32% of the mussel biomass in the summer and about 93% of it in the winter (Stempniewicz, 1974). In Swiss lakes only recently invaded by zebra mussels, significant increases in the number of waterfowl occurred, especially in the autumn migrations in connection with increasing densities of mussels (Leuzinger and Schuster, 1970). In addition, no substantial effect on the population of mussels by parasites has been observed (Raabe, 1956; Dobrzanska, 1958 and 1961; Breitig, 1965).

Sudden reductions of the densities of adult zebra mussel populations have been observed in Polish lakes (Table 8, II.3). These sudden reductions include all age classes of adult mussels (Stańczykowska, 1961 and 1964; Stańczykowska et al., 1975b). As noted earlier, this phenomenon is generally observed in extremely dense populations, consisting of bigger and heavier individuals than usually are found in similar populations of high densities. It is probable that sudden density reductions occur when the population density surpasses an optimum level for environmental conditions. In water bodies, such as dam reservoirs of gulfs, densities of zebra mussels may be much higher than observed in Mazurian Lakes. For example, densities of mussels in dam reservoirs and lagoons may exceed 100,000/m^2 (Markovskij, 1954; Wiktor, 1969). It is difficult to determine whether population crashes after a period of high densities are caused by deteriorating feeding conditions (i.e., lack of sufficient food for numbers of bigger and heavier individuals), by invasion of parasites, by viral or bacterial diseases, or by a combination of these or other factors. Although the trend of sudden reductions in almost the entire adult mussel population density is similar in many lakes, the rate of density increases after sudden population shifts is different between lakes (Stańczykowska, 1975). Differentiated dynamics of recovery of zebra mussel populations may result in different age structure between lakes, even if the lakes are interconnected (Stańczykowska, 1964 and 1976; Stańczykowska et al., 1975a).

Characteristics of population decreases (i.e., on the order of two- to threefold) of mussels due to slow environmental changes are well documented (Table 8, II.4). Density fluctuations occur in lakes where the environmental conditions change as a result of trophic status change often caused by man's activities. Comparative studies conducted in many Great Mazurian Lakes, including Lake Mikolajskie, have shown that a great increase in their eutrophication, which has been observed in the whole lakeland, not only leads to disappearance of the zebra mussel from many water bodies, but also may lead to reduced densities. In some lakes, mussels decreased considerably and finally disappeared. In most lakes in the 1980s, the average and maximum density observed was much lower than in the 1960s.

In many Mazurian Lakes, the factor responsible for the long-term reduction of zebra mussel densities is attributed to the displacement of zebra mussels close to polluted shorelines. This effect was thoroughly described for example in Lake Mikolajskie (Pieczynska et al., 1975). The total or partial disappearance of zebra mussels due to degradation of the environment, contamination, and eutrophication is increasingly reported in the literature. For examples, Swierczynski et al. (1987) observed these effects in Szczecin Lagoon and Schiller (1990) observed them in the Rhine River.

In summary, we believe that in lakes where zebra mussels are a dominant component of the fauna, population changes determine the functioning of many ecosystems and, therefore, the monitoring of their condition is essential to understanding lake ecosystems. Zebra mussels are important water filterers that result in a large biomass and volume of pelagic energy being deposited as food for benthic organisms and empty shells for substrate integration. This acts as a purifier of water, but the rapid increase of eutrophication of lakes caused by man during the last 30 years has caused decreases in the zebra mussel population.

ACKNOWLEDGMENTS

We thank our colleagues of the Department of Hydrobiology, Institute of Ecology PAS and of the Department of Hydrobiology in Warsaw University for help and advice during these studies. We also thank Dr. Joanna Królikowska who took field samples on many occasions.

REFERENCES

Bajkiewicz-Grabowska, E. "Ecological Characteristics of Lakes in North-Eastern Poland Versus Their Trophic Gradient. II. Lake Catchment Areas — Physico-Geographical Environment. Description of the Region and 43 Lakes," *Ekol. Pol.* 31:257–286 (1983)

Berger, L. "Investigations on Mollusc in the Mazurian Lake District," *Badania Fizjogr. Pol. Zachod.* 6:7–49 (1960).

Borowiec, E. "Food of the Coot (*Fulica atra* L.) in Different Phenological Periods," *Pol. Arch. Hydrobiol.* 22:157–166 (1975).

Breitig, G. "Beitrage zur Biologie, Verbreitung und Bekampfung von *Dreissena polymorpha* (Pall. 1771 Lam.)," *Dissertation Math*, Matisch-Naturwissenschaftlichen Universität, Grieswald (1965).

Cydzik, D., and H. Soszka. "Atlas stanu czystosci jezior Polski badanych w latach 1979–1983," *Wyd. Geol. Warszawa* (1988) p. 610.

Dobrzańska, J. "Sphenophrya Dreissenae sp. n. (Ciliata, Holotricha, Thigmotrichida) Living on the Gill Epithelium of *Dreissena polymorpha* Pall. 1754," *Bull. Acad. Pol. Sci., Ser. Sci. Biol.,* 6:173–178 (1958).

Dobrzańska, J. "Further Study on Sphenophyra dreissenae Dobrzanska 1958 (Ciliata, Thigmotricha)," *Acta Parasitol. Polonica* 9:117–140 (1961).

Dunn, D. R. "Notes on the Bottom Fauna of Twelve Danish Lakes," *Vidensk. Meddr. Dansk. Naturh. Foren.* 116:251–268 (1954).

Gliwicz, Z. M., and A. Kowalczewski. "Epilimnetic and Hypolimnetic Symptoms of Eutrophication in Great Mazurian Lakes, Poland," *Freshwater Biol.* 11:425–433 (1981).

Gliwicz, Z. M., A. Kowalczewski, T. Ozimek, E. Pieczynska, A. Prejs, K. Prejs, and J. I. Rybak. "An Assessment of the State of Eutrophication of the Great Mazurian Lakes," *Wyd. Inst. Ksztatr. Srodow.,* (1980) p. 104.

Grim, J. "Tiefenverteilung der Dreikantmussel — *Dreissena polymorpha* (Pallas) im Bodensee," *Gas-Wasserfach. Wasser Abwasser* 112:437–441 (1971).

Halba, R. "Role of *Fulica atra* L. and *Anas platyrhynchos* L. in Mazurian Lakes Biocenosis," Thesis, University of Warsaw (1975) (In Polish).

Hensche, A. "Preussen's Molluskenfauna," *Schrift. Phys.-Ökonom. Ges. Königsberg* 2:73–96 (1861).

Hensche, A. "Zweiter Nachtrag zur Molluskenfauna Preussen's," *Schrift. Physik.-Ökonom. Ges. Königsberg* 3:9 (1862).

Hensche, A. "Dritter Nachtrag zur Mollusken-Fauna Preussen's," *Schrift. Physik. Ökonom. Ges. Königsberg* 7:99–106 (1866).

Hillbricht-Ilkowska, A., Ed. "Lakes of the Mazurian Landscape Protected Area. State of Eutrophication and Measures of Protection," *Ossolinium, Wroclaw* (1989) p. 168.

Hillbricht-Ilkowska, A., and W. Lawacz. "Biotic Structure and Processes in the Lake System of R. Jorka Watershed (Mazurian Lakeland, Poland). I. Land Impact, Loading and Dynamics of Nutrients," *Ekol. Pol.* 31:539–585 (1983).

Hillbricht-Ilkowska, A., and A. Stańczykowska. "The Production and Standing Crop of Planktonic Larvae of *Dreissena polymorpha* (Pall.) in Two Mazurian Lakes," *Pol. Arch. Hydrobiol.* 16:193–203 (1969).

Kajak, Z. "The Characteristics of a Temperate Eutrophic, Dimictic Lake (Lake Mikolajskie, Northern Poland)," *Int. Rev. Ges. Hydrobiol.* 63:451–480 (1978).

Kajak, Z., A. Hillbricht-Ilkowska, and E. Pieczynska. "The Production Processes in Several Polish Lakes, in *Productivity Problems of Freshwater,* Z. Kajak and A. Hillbricht-Ilkowska, Eds. (*Warszawa-Krakow: PWN,* 1972) pp. 129–147.

Kajak, Z., and B. Zdanowski. "Ecological Characteristics of Lakes in Northeastern Poland Versus Their Trophic Gradient. I. General Characteristics of 42 Lakes and Their Phosphorus Load," *Ekol. Pol.* 31:239–256 (1983).

Kirpichenko, M. J. "Phenology, Abundance, and Growth of *Dreissena* Larvae in the Kujbyshev Reservoir, in Biologia *Dreissena* i borba s nej," B. S. Kuzin, Ed. *Izdat. Nauka, Moskva* (1964) 19–30.

Kirpichenko, M. J. "On Ecology of *Dreissena polymorpha* Pallas in the Zimljansk Reservoir," *Tr. Inst. Biol. Vnutr. Vod. Akad. Nauk SSSR* 21:142–154 (1971).

Kornobis, S. "Ecology of *Dreissena polymorpha* (Pall.) (Dreisseniclae, Bivalvia) in Lakes Receiving Heated Water Discharges, *Pol. Arch. Hydrobiol.* 24:531–546 (1977).

Leuzinger, H., and S. Schuster. "Auswirkungen der Massenvermehrung der Wandermuschel *Dreissena polymorpha* auf die Wasservögel des Bodensees," *Ornit. Beob.* 67:269–274 (1970).

Lewandowski, K. "The Role of Early Development Stages in the Dynamics of *Dreissena polymorpha* (Pall.) population in lakes. I. Occurrence of Larvae in the Plankton," *Ekol. Pol.* 30:81–110 (1982a).

Lewandowski, K. "The Role of Early Development Stages in the Dynamics of *Dreissena polymorpha* (Pall.) Population in Lakes. II. Settling of Larvae and the Dynamics of Numbers of Settled Individuals," *Ekol. Pol.* 30:223–286 (1982b).

Lewandowski, K., and J. Ejsmont-Karabin. "Ecology of Planktonic Larvae of *Dreissena polymorpha* (Pall.) in Lakes with Different Degree of Heating," *Pol. Arch. Hydrobiol.* 30:89–101 (1983).

Lewandowski, K., and A. Stańczykowska. "The Occurrence and Role of Bivalves of the Family Unionidae in Mikolajskie Lake," *Ekol. Pol.* 23:317–334 (1975).

Liakhov, S. "Research of Biological Institute for Inland Waters on Protection of Hydrotechnical Systems Against *Dreissena* fouling, in Biologia *Dreissena* i borba s nej," B. S. Kuzin, Ed. *Izdat. Nauk. Moskva* (1964) pp. 3–18.

Lundbeck, J. "Die Bodentierwelt norddeutscher Seen," *Arch. Hydrobiol., Suppl.* 7:1–473 (1926).

Markovskij, Yu, M. "Invertebrate Fauna in Lower Parts of Ukraina Rivers," *Kiev. Jzdat. Akad. Nauk USSR* (1954) (In Russian).

Mattice, J., A. Stańczykowska, and W. Lawacz. "Feeding and Assimilation of *Dreissena polymorpha* in Mikolajskie Lake, Poland," *Am. Zool.* 12:209 (1972).

Mikulski, J., and A. Gizinski. "Bottom Fauna of Wdzydze Lake," *Roczn. Nauk. Roln. Ser. D. Monogr.* 93:141–162 (1961).

Ozimek, T., and A. Kowalczewski. "Long-term Changes of the Submerged Macrophytes in Eutrophic Lake Mikolajskie (North Poland)," *Aquat. Bot.* 19:1–11 (1984).

Pieczynska, E. Ed. "Selected Problems of Lake Littoral Ecology," Wyd. Uniw. Warsz., Warszawa (1976) 238.

Pieczynska, E., T. Ozimek, and J. I. Rybak. "Long-Term Changes in Littoral Habitats and Communities in Lake Mikolajskie (Poland)," *Int. Rev. Ges. Hydrobiol.* 73:361–378 (1988).

Pieczynska, E., U. Sikorska, and T. Ozimek. "The Influence of Domestic Sewage on the Littoral Zones of Lakes," *Pol. Arch. Hydrobiol.* 22:141–156 (1975).

Pliszka, F. "Dynamics of Feeding Relations in Fish of the Lake Harsz," *Pol. Arch. Hydrobiol.* 1:271–300 (1953).

Pliszka, F. "Importance of the Aquatic Animals as a Food Base of Fish after Polish Investigations," *Pol. Arch. Hydrobiol.* 3:429–458 (1956).

Poddubnyj, A. G. "Adaptation of *Rutilus rutilus* to Variable Environmental Conditions," *Biol. Volzhsk. Vodokhran.* 10:131–138 (1966).

Polinski, W. "Materialy do fauny malakozoologicznej Krolestwa Polskiego, Litwy i Polesia," *Pr. Tow. Nauk. Warszaw.* 27:1–130 (1917).

Prejs, A. "Fishes and Their Feeding Habits, in *Selected Problems of Lake Littoral Ecology,*" E. Pieczynska, Ed. *Wyd. Uniw. Warsz., Warszawa* (1976) 155–171.

Prejs, A., K. Lewandowski, and A. Stańczykowska-Piotrowska. "Size-Selective Predation by Roach *(Rutilus rutilus)* on Zebra Mussel (*Dreissena polymorpha*): Field Studies," *Oecologia (Berlin)* 83:378–384 (1990).

Rabbe, Z. "Investigations on the Parasitofauna of Freshwater Molluscs in the Brackish Waters," *Acta Parasitol. Pol.* 4:375–406 (1956).

Schalekamp, M. "Neuste Erkenntnisse über die Wandermuschel *Dreissena polymorpha* Pallas und ihre Bekampfung," *Gas-Wasserfach. Wasser Abwasser* 51:329–336 (1971).

Schiller, W. "Die Entwicklung der Makrozoobenthonbesiedlung des Rheins in Nordrhein-Westfalen im Zeitraum 1969–1987," *Limnol. Aktuell* 1:259–275 (1990).

Shevtsova, L. "Peculiarities of the Reproduction and Development of *Dreissena* in the Canal the Dnepr-Krivoj-Rog,"*Gidrobiol. Zh.* 4:70–72 (1968).

Siessegger, B. "Vorkommen und Verbreitung von *Dreissena polymorpha* Pallas im Bodensee," *Gas-Wasserfach. Wasser Abwasser* 110:814–815 (1969).

Spodniewska, I. "Changes in the Structure and Production of Phytoplankton in Mikolajskie Lake 1963–1972," *Limnologica* 10:299–306 (1976).

Stańczykowska, A. "Die gewaltige Zahlenverminderung *Dreissena polymorpha* Pall. in einigen Mazurischen Seen neben Mikolajki," *Ekol. Pol., Ser. B*, 7:151–153 (1961).

Stańczykowska, A. "On the Relationship Between Abundance Aggregations and "condition" of *Dreissena polymorpha* Pall. in 36 Mazurian Lakes," *Ekol. Pol., Ser. A* 12:653–690 (1964).

Stańczykowska, A. "Einige Gesetzmassigkeiten des Vorkommens von *Dreissena polymorpha* Pall." *Verh. Int. Ver. Theor. Angew. Limnol.* 16:1761–1776 (1966).

Stańczykowska, A. "The Filtration Capacity of Populations of *Dreissena polymorpha* Pall. in Different Lakes, as a Factor Affecting Circulation of Matter in the Lake," *Ekol. Pol., Ser. B,* 14:265–270 (1968).

Stańczykowska, A. "Ecosystem of Mikolajskie Lake. Regularities of the *Dreissena polymorpha* Pall. (Bivalvia) Occurrence and its Function in the Lake," *Pol. Arch. Hydrobiol.* 22:73–78 (1975).

Stańczykowska, A. "Biomass and Production of *Dreissena polymorpha* (Pall.) in some Mazurian Lakes," *Ekol. Pol.* 24:103–112 (1976).

Stańczykowska, A. "Ecology of *Dreissena polymorpha* (Pall.) (Bivalvia) in lakes," *Pol. Arch. Hydrobiol.* 24:461–530 (1977).

Stańczykowska, A. "Role of Bivalves in the Phosphorus and Nitrogen Budget in Lakes," *Verh. Int. Ver. Theor. Angew. Limnol.*22:982–985 (1984a).

Stańczykowska, A. "The Effect of Various Phosphorus Loading on the Occurrence of *Dreissena polymorpha* (Pall.)," *Limnologica* 15:535–539 (1984b).

Stańczykowska, A. "The place mussel *Dreissena polymorpha* (Pall.) in the food web of lakes ecosystems," *Haliotis* 16:129–135 (1987).

Stańczykowska, A., E. Jurkiewicz-Karnkowska, and K. Lewandowski. "Ecological Characteristics of Lakes in Northeastern Poland Versus Their Trophic Gradient. X. Occurrence of Molluscs in 42 Lakes," *Ekol. Pol.* 31:459–475 (1983a).

Stańczykowska, A., and W. Lawacz. "Field Experiment on the Role of *Dreissena polymorpha* in Removing Seston from Water," *Konf. Inst. Gospodarki Wodnej* 1:27 (1971).

Stańczykowska, A., and W. Lawacz. "Caloric Value of the *Dreissena polymorpha* (Pall.) Dry Body Weight in Some Mazurian Lakes," *Pol. Arch. Hydrobiol.* 23:271–275 (1976).

Stańczykowska, A., P. Zyska, A. Dombrowski, H. Kot, and E. Zyska. "The Distribution of Waterfowl in Relation to Mollusc Population in the Man-made Lake Zegrzynskie," *Hydrobiologia* 191:233-290 (1990).

Stańczykowska, A., W. Lawacz, and J. Mattice. "Use of Field Measurements of Consumption and Assimilation in Evaluating of the Role of *Dreissena polymorpha* (Pall.) in a Lakes Ecosystem," *Pol. Arch. Hydrobiol.* 22:509–520 (1975a).

Stańczykowska, A., W. Lawacz, J. Mattice, and K. Lewandowski. "Bivalves as Affecting Circulation of Matter in Lake Mikolajskie (Poland)," *Limnologica* 10:347–352 (1976).

Stańczykowska, A., and K. Lewandowski. In preparation.

Stańczykowska, A., K. Lewandowski, and J. Ejsmont-Karabin. "Biotic Structure and Processes in the Lake System of R. Jorka Watershed (Mazurian Lakeland, Poland). IX. Occurrence and Distribution of Molluscs with Special Consideration to *Dreissena polymorpha* (Pall.)," *Ekol. Pol.* 31:761–780 (1983b).

Stańczykowska, A., and M. Planter. "Factors Affecting Nutrient Budget in Lakes of the R. Jorka Watershed (Mazurian Lakeland, Poland). X. Role of Mussel *Dreissena polymorpha* (Pall.) in N and P cycles in a Lake Ecosystem," *Ekol. Pol.* 33:345–356 (1985).

Stańczykowska, A., H. J. Schenker, and Z. Fafara. "Comparative Characteristics of Populations of *Dreissena polymorpha* (Pall.) in 1962 and 1972 in 13 Mazurian Lakes," *Bull. Acad. Pol. Sci. Ser. Sci. Biol.* 23:383–390 (1975b).

Stangenberg, M. "Szkic limnologiczny na tle stosunkow hydrochemicznych pojezierza suwalskiego," *Rozpr. Spraw. Inst. Bad. Las. Panstw.* 19:7–85 (1936).

Stempniewicz, L. "The Effect of Feeding of Coot (*Fulica atra* L.) on the Character of the Shoals of *Dreissena polymorpha* Pall. in Lake Goplo," *Acta Univ. Nicolai Copernici* 34:84–103 (1974).

Stern, E. "Damages of *Dreissena* Fouling and Its Control, in N. A. Dzjuban, Ed., Soveshchanie po biologii drejsseny i zashchite gidrotekhnicheskikh sooruzhenij ot ee obrastanii," *Tr. Inst. Biol. Vnutr. Vod. Akad. Nauk SSSR Togljatti* (1965) pp. 7–10.

Swierczynski, M., P. Kadela, and K. Kolasa. "Biologia i ekologiczna rola racicznicy (*Dreissena polymorpha* Pall.) w doczyszczaniu wod Roztoki Odrzanskiej i poludniowej czesci Zalewu Wielkiego (Zalew Szczecinski)," *Spektrum* 2:91–109 (1987).

Szczepanski, A. "Scattering of Light Visibility in Water of Different Types of Lakes," *Pol. Arch. Hydrobiol.* 15:51–76 (1968).

Wahl, E. "Die Süsswasser-Bivalven Livlands," *Arch. Naturk. Liv-Ebst- Kurlans, 2 Ser.* 1:75–148 (1855).

Walz, N. "Ruckgang der *Dreissena polymorpha* — Population im Bodensee," *Gas-Wasserfach. Wasser Abwasser* 115:20–24 (1974).

Walz, N. "Die Besiedlung von Kuntlichen Substraten durch Larven von *Dreissena polymorpha*," *Arch. Hydrobiol., Suppl.* 47:423–431 (1975).

Wesenberg-Lund, C. *Biologie der Süsswassertiere.* (Wien: Springer-Verlag, 1939) p. 817.

Wetzel, R. G. *Limnology* (Philadelphia, PA: W. B. Saunders Company, 1975).

Wiktor, J. "The Biology of *Dreissena polymorpha* (Pall.) and Its Ecological Importance in the Firth of Szczecin," *Stud. Mat. Morsk. Inst. Ryb. Gdynia, Ser. A,* 5:1–88 (1969).

Wiktor, K. "Larvae of *Dreissena polymorpha* Pall. as a Food for Fish Spawn," *Przegl. Zool.* 2:182–184 (1958).

Wisniewski, R. "Distribution and Character of Shoals of *Dreissena polymorpha* Pall. in the Bay of Goplo Lake," *Acta Univ. Nicolai Copernici* 34:73–81 (1974).

Appendix 1.

Names of Lakes, Area, Maximum Depth, Trophic Type, and Densities of *Dreissena polymorpha* in Great Mazurian Lakes of Northeastern Poland

Lake Name	Area (ha)	Maximum Depth (m)	Trophic Type	Densities (Number/m²)
Guzianka Duza	72	29.4	Strong eutrophy	0
Guzianka Mala	42	14.0	Strong eutrophy	0
Kirsajty	212	7.0	Eutrophy	0
Pozezdze	124	4.0	Eutrophy	0
Przylesne	26	1.5	Eutrophy	0
Szymonieckie	154	2.9	Mesotrophy	0
Warnolty	470	5.5	Eutrophy	0
Beldany	780	31.0	Eutrophy	0–50
Rynskie	620	47.0	Eutrophy	0–100
Kotek	42	2.5	Eutrophy	100–500
Nidzkie	1724	25.0	Strong eutrophy	300–700
Dargin	1988	37.0	Eutrophy	400–800
Dobskie	1760	21.0	Eutrophy	400–800
Wilkus	94	5.5	Eutrophy	400–800
Seksty	792	7.0	Eutrophy	400–800
Sniardwy	10598	25.0	Eutrophy	400–800
Tyrklo	215	24.5	Early eutrophy	400–900
Kisajno	2536	24.0	Eutrophy	500–900
Labap	364	—	Eutrophy	500–900
Biale	7	2.0	Eutrophy	500–900
Kaczerajno	364	—	Eutrophy	600–1000
Ros	2212	28.0	Strong eutrophy	600–1000
Wojnowo	190	16.5	Eutrophy	0–1300
Niegocin	2499	40.0	Eutrophy	500–1300
Goldapiwo	1070	24.5	Mesotrophy	300–1400
Jagodne	936	34.0	Early eutrophy	100–1500
Mamry Pn.	2478	40.0	Mesotrophy	700–1600
Taltowisko	323	35.0	Mesotrophy	700–1600
Stregiel	412	10.0	Eutrophy	200–1800
Zabinki	41	42.5	Mesotrophy	300–1800
Boczne	190	15.0	Eutrophy	1100–1800
Tajty	251	34.0	Eutrophy	1100–1800
Szymon	204	34.0	Eutrophy	1100–1800
Swiecajty	814	28.0	Eutrophy	1100–1800
Mikolajskie	460	27.8	Eutrophy	0–2200
Talty	1162	37.5	Eutrophy	0–3600

Source: Stańczykowska, 1977.

Appendix 2.

Names of Lakes, Area, Maximum and Mean Depth, Trophic Type, and Densities of *Dreissena polymorpha* in Lakes of the Jorka River System of Northeastern Poland

Lake Name	Area (ha)	Maximum Depth (m)	Mean Depth (m)	Trophic Type	Densities (Number/m²)
Jorzec	42	16.4	6.0	Eutrophy	130
Zelwazek	12	7.4	3.7	Eutrophy	175
Glebokie	47	34.3	11.8	Eutrophy	280
Inulec	178	10.1	4.6	Eutrophy	400
Majcz Wielki	163	16.4	6.0	Mesotrophy	510

Source: Stańczykowska et al., 1983b.

Appendix 3.

Names of Lakes, Area, Maximum and Mean Depth, and Densities of *Dreissena polymorpha* in Lakes of the Krutynia River System of Northeastern Poland

Lake Name	Area (ha)	Maximum Depth (m)	Mean Depth (m)	Densities (Number/m²)
Zyzdroj Maly	51	12.8	3.8	0
Spychowskie	49	7.7	2.3	0
Zyndaki	39	10.3	4.0	6
Zdruzno	250	25.9	5.3	8
Uplik	61	9.2	2.8	8
Warpunskie	49	6.9	2.6	28
Gardynskie	83	11.5	2.4	30
Kujno	24	6.0	2.8	66
Krutynskie	55	3.2	1.7	72
Mokre	841	51.0	12.7	162
Biale	341	31.0	7.3	183
Jerzewko	23	2.7	—	196
Zyzdroj Wielki	210	14.5	4.8	242
Dluzec	123	19.8	6.3	384
Malinowka	16	2.7	—	404
Gieladzkie	475	27.0	6.8	686
Gant	75	28.3	9.4	1920
Lampasz	88	21.7	4.8	2392
Lampackie	198	38.5	11.1	3453

Appendix 4

Names of Lakes, Area, Maximum Depth, Trophic Status, and Densities of *Dreissena polymorpha* in 43 Lakes in Northeastern Poland

Lake Name	Area (ha)	Maximum Depth (m)	Trophic Type	Densities (Number/m²)
Kraksy Duze	44	4.0	Polymictic	0
Siercze	55	2.0	Polymictic	0
Sedanskie	168	6.1	Polymictic	0
Czos	279	42.6	Dimictic	0
Jaskowskie	152	16.5	Dimictic	1
Wobel	24	15.0	Dimictic	3
Hartowiec	69	5.2	Polymictic	6
Elckie	382	55.8	Dimictic	6
Lampackie	198	38.5	Dimictic	7
Pilakno	259	56.6	Dimictic	7
Burgale	79	7.4	Polymictic	7
Brajnickie	186	5.2	Polymictic	8
Tuchel	43	5.1	Polymictic	11
Kolowin	78	7.2	Polymictic	16
Ranskie	291	7.8	Polymictic	16
Barlewickie	64	8.5	Polymictic	20
Sambrod	128	4.3	Polymictic	22
Maroz	332	41.0	Dimictic	23
Lidzbarskie	122	25.5	Dimictic	25
Moj	116	4.1	Polymictic	25
Dlugie	62	5.4	Polymictic	35
Rumian	306	14.4	Dimictic	38
Malaszewskie	202	16.9	Dimictic	41
Warpunskie	49	6.9	Polymictic	44
Ilawskie	154	2.6	Polymictic	67
Liwieniec	81	2.4	Polymictic	109
Kokowo	37	11.8	Dimictic	131
Szoby Male	319	3.7	Polymictic	133
Bartag	72	15.2	Dimictic	162
Probarskie	201	31.0	Dimictic	212
Stryjewskie	67	6.2	Polymictic	246
Grom	240	15.8	Dimictic	287
Rzeckie	56	29.0	Dimictic	332
Sztumskie	50	24.6	Dimictic	439
Olow	61	40.1	Dimictic	514
Kierzlinskie	93	44.5	Dimictic	676
Kuc	98	28.0	Dimictic	840
Badze	150	6.7	Polymictic	1197
Skanda	51	12.0	Dimictic	1312
Sarz	77	15.0	Dimictic	1324
Gim	176	25.8	Dimictic	2441
Juno	381	33.0	Dimictic	2454
Szelag Maly	84	15.2	Dimictic	3792

Source: Bajkewicz-Grabowska, 1983.

CHAPTER 2

The Other Life: An Account of Present Knowledge of the Larval Phase of *Dreissena polymorpha*

Martin Sprung

This chapter gives an overview of current knowledge of the biology of *Dreissena polymorpha* Pallas larvae and points out that there is a disproportionate amount of information on this larval stage. On one hand, the morphology of its developmental stages is well documented as well as its occurrence at many locations. On the other hand, many basic data about the importance of environmental factors (e.g., food requirements, causes of mortality, tolerance limits) are imprecise or lacking. This is also true of nearly all its physiological rates (e.g., filtration and respiration).

INTRODUCTION

When the species *Dreissena polymorpha* Pallas is considered, one generally depicts a semisessile, filter-feeding bivalve of several centimeters in length. However, each one of these individuals has spent a certain period of

0-87371-696-5/93/$0.00 + $.50

time in the plankton as a microscopically small larva. During this period, it differed not only in size and shape from the adult, but also in its physiology and ecology.

The discovery of this planktonic larval stage has been ascribed to Korschelt (1891) and Blockmann (1891), and the first detailed description was published by Korschelt and Weltner (1892). Most marine bivalves have a similar planktonic larva in their life cycle. This mode of reproduction is exceptional for freshwater bivalves. All other freshwater bivalves either have specialized types of larvae such as parasites (Unionidae) or develop directly in the mantle cavity of the adults (Pisididae). Only *Corbicula fluminea* still passes through a planktonic larval stage as does *Mytilopsis leucophaeta* (= *Congeria cochleata*); however, the latter prefers slightly brackish waters.

Dreissena is of separate sex; the animals release gametes in enormous quantities into the water column. In contrast to earlier findings by Katchanova (1961) who reported fecundity figures of 30–40,000 eggs per female, Walz (1978) and Sprung (1991) obtained much higher numbers. Depending on its size, one female produces up to more than 1 million eggs during one spawning event. One must see this high fecundity in the context of not protecting the gametes further.

DEVELOPMENT AND GROWTH

The Gametes

According to my own observations, an egg has a diameter of about 70 μm: this size agrees with most other authors (e.g., Meisenheimer, 1901; Breitig, 1965); however, Feigina (1959) reported egg diameters of up to 112 μm and Walz (1973) found a size range of 110–190 μm. Walz (1978) estimated that the carbon content of an egg was 21.2 ng. I found the ash-free dry weight of an egg was 34.4 ± 9.9 ng (x ± S.D., from 30 females) and that of a sperm was 5.8 ± 1.3 ng (x ± S.D., from 8 males) (Sprung, 1989 and 1991). The head of the sperm is 4.5–5.0 μm long (Karpevich, 1961). Once shed, the eggs have the tendency to sink to the bottom. For successful fertilization the temperature must be higher than 10°C. Within a temperature range of 12–24°C, the eggs can be fertilized 2.5–4.75 h after release, while the sperm normally remain motile much longer (2–22 h). High sperm concentrations (up to 10^6 sperm per milliliter) do not have an adverse effect on development. This may indicate an effective mechanism to prevent polyspermy (Sprung, 1987).

Lecithotrophic Period

The development of a *Dreissena* egg has been described by Meisenheimer (1901). This description remains one of the most comprehensive ever written

Table 1. *Dreissena polymorpha*, Embryos: Time Sequence in Hours of Characteristic Stages of Development Described by Meisenheimer (1901); Also Indicated are the Q_{10} Values Calculated Between the Temperature Levels

Temperature (°C)	12		15		18		21		24
Stage of Development (Figure in Meisenheimer, 1901)									
Polar bodies (1)	0.28		0.20		0.20		0.16		0.15
Two-cell stage (3,4)	3.0		2.3		1.8		1.5		1.4
Four-cell stage (5)	5.0		3.8		2.8		2.5		2.4
d_1-Cell (6–8)	5.5		4.1		3.5		3.0		2.9
More than 4 micromeres (10–12)	6.6		4.1		4.1		4.0		3.9
Micromeres cover macromeres completely (23)	8.2		5.5		4.1		4.0		4.4
Larva begins to swim	20.0		12.6		9.5		7.5		6.3
Apical tuft of two cilia (46)	53		34		30		25		?
D-shaped shell (49)	90		55		45		35		31
Q_{10} values		5.2		2.0		2.3		1.5	

Source: Revised from Sprung, 1987.

for a bivalve embryo. Some characteristic events which can be easily observed are formation of two polar bodies; an unequal cleavage to a macromere and a micromere; an equal cleavage of both to a four-cell stage; the formation of a fifth d_1-cell by the D-cell; a cleavage of the other cells (A, B, C); and a further cleavage of all cells, so that more than four micromeres can be observed. During further development, the micromeres begin to overgrow the macromeres which is a signal of gastrulation. The rate of development depends upon temperature (Table 1). From 6 to 20 h after fertilization, the larva begins to swim by means of cilia. An ectoderm is formed which produces an apical plate with a tuft of only two cilia (in contrast to Meisenheimer's drawing) and a velum. This velum has two functions: it enables the larva to swim more efficiently and it is the organ that filters food out of the water. This becomes essential when a shell gland produces a D-shaped shell. By that time the larva has formed an intestine.

To this point in its development, the metabolism of the embryo has been exclusively fueled by the reserve substances which the egg contains. Now the larva has to take up food to develop further, i.e., it enters into its planktotrophic period. According to Werner (1940), this special type of shelled veliger is called the veliconcha. The transition from lecithotrophy to planktotrophy may be gradual. This has been demonstrated for *Mytilus edulis* larvae (Lucas et al., 1986). In the absence of food, *Dreissena* larvae remain active for 1-2 weeks and then die shortly thereafter. This time period declines with increasing temperature (Sprung, 1989).

Planktotrophic Period

According to various authors (e.g., Meisenheimer, 1901; Feigina, 1959; Katchanova, 1961; Breitig, 1965; Hillbricht-Ilkowska and Stanczykowska, 1969; Waltz, 1973; Sprung, 1989) the larva has a shell size of 70–100 μm when it enters the planktotrophic period. Subsequently, shell material is secreted by the mantle edges, which not only increases the size of the larva but also changes its shape. Walz (1973) and Lewandowski (1982a) point out that some call the length of a larva the dimension which normally is called the height in a more advanced stage. I consider the longest dimension as the length, as most others do. At the end of the planktotrophic period, the shell length is generally about 200 μm (Kirpichenko, 1965; Wiktor, 1969; Szlauer, 1974; Lewandowski, 1982a and 1983b; Sprung, 1989). However, there is one report by Lewandowski (1983b) of settled individuals exceptionally smaller than 200 μm. There are also reports of larger larvae which stay in the plankton: shell length of 225–255 μm (Kirpichenko, 1971), up to 270 μm (Lewandowski, 1983b) or even 300 μm (Walz, 1973). The phenomenon of smaller larvae that already metamorphose is difficult to explain. Larger larvae may be the result of a lack of suitable substrates which stimulates the larvae to delay their metamorphosis. This has been reported for marine bivalve larvae (e.g., *Mytilus edulis*; Bayne, 1965) and is probably true for *Dreissena polymorpha*.

Hillbricht-Ilkowska and Stanczykowska (1969) tried to approximate the weight of a larva by the formula:

$$W = 58.207 - 2.636\, l + 0.037\, l^2$$

W: weight (probably fresh weight, ng)
l: shell length (μm)

Consequently, a larva of 90-μm shell length would have a weight of 0.12 μg and a larva shell length of 220 μm would have a weight of 1.27 μg. This means that the larva increases roughly ten times in weight during its planktotrophic period. This is the only weight estimate of *Dreissena* larvae that I am aware of, and it compares fairly well with those I obtained for *Mytilus edulis* larvae; in this case, *Mytilus* larvae of 90 and 220 μm, respectively had a wet weight of 0.13 and 1.74 μg and a dry weight of 0.08 and 0.77 μg (Sprung, 1984a).

Without presenting any data, Korschelt (1891) indicated that *Dreissena* larvae stay in the plankton for 8 days. This number has been widely referenced in the literature. Subsequently, Katchanova (1961) confirmed this estimate in her own studies, and Shevtsova (1968) observed a range between 5 and 16 days. These estimates are clearly shorter than the 5 weeks indicated by Walz (1975) and 18, 28, and 33 days that I observed when following larval cohorts on three occasions (Sprung, 1989).

Shell growth rates and residence time in the plankton are closely linked. If the residence time is only 8 days, this would mean that the larva increases about 20 μm in shell length every day — a figure clearly above estimates obtained from laboratory cultures of marine bivalve larvae which are in the range of 10–15 μm/day (Sprung, 1989). When examining data from cohort growth, I calculated an increase of 1–4 μm/day when temperatures ranged from 14 to 21°C. This growth rate, in turn, would imply mean residence times between 30 and 100 days, which are considerably longer than actually observed (Sprung, 1989). Part of this inconsistency can be explained by assuming for the extrapolation a linear model for the relationship between shell length vs time. Besides a linear model, there has also been evidence for sigmoid and exponential models to describe the shell growth of marine bivalve larvae (Bayne, 1983). The latter would imply a faster increase in shell length for a more advanced veliconcha stage. Also, a high variation in growth rates has been documented for *Mercenaria mercenaria* larvae under identical conditions (Loosanoff et al., 1951). This implies that a correct interpretation or model of growth is difficult to obtain. Since *Dreissena* larvae generally have high mortality rates during their larval life (see below), it is essential to know whether an average growth rate or a maximum growth rate (thus shortest residence time) is most representative for larvae reaching the settling stage.

Tolerance Limits

Conditions needed for successful development of *Dreissena polymorpha* embryos have been examined by Sprung (1987). Basically, the temperature must be between 12 and 24°C (optimum at about 18°C), the pH must be between 7.4 and 9.4 (optimum at about pH 8.5), and calcium ions must be present in the water. For calcium, a concentration below 40–60 mg Ca^{2+} per liter will increase the number of crippled larvae, and the absence of these ions will cause the embryo to crumble into separate cells. All other ions are not essential. Reduced levels of oxygen affect developmental success only when levels drop below 20% at 18°C. The embryo is quite sensitive to turbulence.

The earliest stages are the most sensitive in the life cycle of a bivalve (Bayne, 1976). As the larva develops, its tolerance limits for various environmental conditions increases. Shevtsova (1968) indicated that the larvae of *Dreissena* can tolerate a temperature range between 0 and 30°C. Tolerance of temperatures in the lower range should enable larvae from autumn to overwinter in the plankton at low metabolic rates. Lewandowski and Ejsmont-Karabin (1983) reported that they found high numbers of dead *Dreissena* larvae when temperatures exceeded 29°C, which implies an upper tolerance limit of this temperature.

Some of the tolerance studies have been performed when testing methods of effectively killing the larvae before they infest a location, e.g., waterworks: an electric current of 380–400 V/cm^2 should work even when the shells of the veligers are closed (Kirpichenko et al., 1963). Breitig (1965) found that ultrasound of 22–189 kHz for 3 min and chlorine at an initial concentration of 2.9–3.3 mg/L applied over 10 min. successfully killed the larvae.

Mortality

Data collected in the field have given different estimates of when larval mortality occurs. In Szczecin lagoon (Wiktor, 1969) and in Masurian lakes (Stanczykowska, 1978; Lewandowski, 1982b) mortality during the planktotrophic period was rather low, rarely exceeding 20%. Most of the larvae (99%) died during the process of settlement because a suitable substrate was not encountered. In contrast, I calculated from regular samples in two lakes near Cologne that only 11–40% of the unshelled larvae found in the plankton developed into the D-shell stage, and that only 0.0–0.7% of the D-shaped larvae attained a settling size of 220 μm (Sprung, 1989).

There are probably several causes of mortality during the planktotrophic period. Wiktor (1958) stressed the importance of predation by fish larvae (*Osmerus esperlanus, Lucioperca lucioperca, Acerina arnua, Rutilus rutilus*) while Kornobis (1977) indicated that fish from 10 to 16 mm preyed heavily on *Dreissena* larvae. Karabin (1978) reported that some *Cyclops* species feed on *Dreissena* larvae, and Mikheev (1967) discussed mortality being caused by the filtering activity of adult *Dreissena polymorpha*. I observed high mortality in larvae within 20L cages placed in a lake; the cages had a mesh size of 49 μm which should have excluded any predators (Sprung, 1989). Consequently, other causes of mortality may be important, such as bacterial infection, shortage of food (see later), or, in this special case, confining the larvae and impeding free-swimming activity.

Abundance

Numerous studies have documented abundances of *Dreissena polymorpha* larvae. Larvae are first found in the plankton when the surface temperature rises above 12°C in the spring. The temperature when larvae were first observed and the length of time they were found in the plankton in different regions of Europe are given in Table 2. At times, larvae show distinct abundance peaks during the course of the year. These peaks are related to the spawning behavior of the adults, which, in turn, is influenced to a great extent by the temperature at the depth they inhabit (Borcherding, 1991). Consequently, every lake has a unique and different seasonal pattern of abundance. For example, Shevtsova (1968) found in the Dnepr-Krivoj-Rog a great peak in abundance in May, another in July, and a minor peak in August/September.

Table 2. Literature Data on the Occurrence of *Dreissena polymorpha* Recorded by Various Authors

Author	Time Span When Larvae Were Found in the Plankton	Temperature at the First Recording (°C)	Locality and Year
Katchanova (1961)	June/July-October	15–16	Uchinsk Reservoir (SU) 1956–1959
Kirpichenko (1964)	June/July-October	15	Kujbyshev Reservoir (SU) 1960/1961/1963
Shevtsova (1968)	April/May-October	12–15	Dnepr-Krivoj-Rog (SU) 1965/1966
Wiktor (1969)	April-September	12–14	Firth of Szczecin (PL)
Hillbricht-Ilkowska and Stanczykowska (1969)	June-September	16–21	Mazurian Lakes (PL) 1963/1964/1966
Kornobis (1977)	April/May-August/ October	12–13	Mikorzynsko-Wasoskie-Lake (heated lake near Konin, PL) 1973–1975
Lewandowski (1982a)	June-September	17–19	7 Masurian Lakes (PL) 1973–1975
Lewandowski and Ejsmont-Karabin (1983)	April-September	?	Heated lakes near Konin (PL) 1970–1978
Breitig (1965)	April/June	10–15	Pohlitzer See and Oder-Spree-Kanal (D) 1956–1964
Walz (1973)	June-September July-October	17 14	Gnadensee (D) 1971 Uberlinger See (D) 1971
Einsle (1973)	June-October	14–15	Lake Constance (D) 1971
Siller (1983)	April-October	10	Fühlinger See (D) 1981
Sprung (1989)	May-September/ October	12	Fühlinger See and Heider Bergsee (D) 1985–1987

Source: Revised from Sprung, 1989.

Roughly similar seasonal variations in abundances are reported by Kirpichenko (1964, 1965, 1971) and Lewandowski and Ejsmont-Karabin (1983), and in some places by Breitig (1965) and Hillbricht-Ilkowska and Stanczykowska (1969). In other water bodies, only one great peak has been observed or no detectable peak at all (Einsle, 1973; Lewandowski, 1982a; Sprung, 1989; Breitig, 1965; and Hillbricht-Ilkowska and Stanczykowska, 1969). Maximum reported abundances of larvae are in the range of 7–700 larvae per liter (e.g., Kirpichenko, 1964 and 1965; Breitig, 1965 and 1969; Shevtsova, 1968; Hillbricht-Ilkowska and Stanczykowska, 1969; Wiktor, 1969; Szlauer, 1974; Stanczykowska, et al., 1983). Kornobis (quoted by Stanczykowska, 1977) reported 9,000 larvae per liter directly over a *Dreissena* population. Despite the high sporadic advances, Hillbricht-Ilkowska and Stanczykowska (1969) calculated that these larvae contribute only a few percent to the total zooplankton production during summer months in two Masurian lakes.

Larval abundance may vary considerably from year to year at a particular site. The two longest data records are those of Breitig (1965) with 9 years in GroBen Pohlitzer See (N. Germany) and Einsle (1989) with 13 years in Lake Constance. The former study implied a 4-year cycle in the annual abundance of larvae, but this could not be confirmed in the latter study.

The distribution of larvae within a given lake is often uneven. This can be attributed to the presence of the adult populations (as mentioned before), to wind conditions (Einsle, 1973), and also to the depth at which samples are taken. Breitig (1965) reported a maximum larval abundance in the top 1–4 m; Einsle (1973) reported a maximum abundance in the upper 10 m; while Hillbricht-Ilkowska and Stanczykowska (1969) and Lewandowski and Ejsmont-Karabin (1983) refer to the epilimnion as their preferred site. Thus, for many waterbodies, population figures based on water surface rather than on water volume are more meaningful. For instance Walz (1973) gave figures of 195,000 or 1,037,000 larvae per square meter for Lake Constance, and Suter-Weider (1976) gave figures of 210,000 larvae per square meter for Lake Zürich.

Various authors have reported vertical migrations by the larvae (Serafimova-Hadishche, 1954; Siller, 1983). In Lake Constance, maximum larval numbers were found at about 5 m (lower limit of epilimnion) at noon. At dusk, maximum numbers were found at the water surface; at dawn, maximum numbers were again found at 5 m (Einsle and Walz, 1972).

PHYSIOLOGY

Food and Feeding

The veliconcha takes up food by means of its velum. Most likely the description given by Strathmann et al. (1972) for an ''opposed band system'' in marine forms is also applicable to *Dreissena* larvae. The food is transported by means of a band of long cilia to an adjacent food groove. The other margin of the food groove is marked by a band of shorter cilia. Once in the groove, the food particles are encased in mucus and transported to the gut. Food selection according to size is effected by means of the larger cilia. Other mechanisms for food selection occur when a bolus of several cells is formed at the entry of the esophagus, or when cells are ejected from the esophagus by muscular constractions before entering the stomach (Strathmann and Leise, 1979; Gallager, 1988).

The size spectrum *Dreissena* larvae can feed on is rather narrow when compared with data of other bivalve larvae or planktonic filter feeders; the size spectrum of particles ingested by *Dreissena* larvae is only 1–4 μm in diameter. Consequently, potential food must consist of bacteria, blue-green algae, small green algae, and very fine detritus particles. In tests to isolate food for laboratory cultures of larvae, I examined 50 different food items consisting mostly of algae and blue-green algae, but also of bacteria and yeast. Although most items of the appropriate size were ingested (as to be concluded by the color of the digestive gland), I had only very sporadic success with a *Synechococcus elongatus* culture (Sprung, 1989).

No data exist on filtration rates of *Dreissena* larvae. When using data of *Mytilus* larvae as the basis for an extrapolation, the rate should lie in the range of 10–40 μL/h, depending on the size (Sprung, 1984b). There is, however, some uncertainty in this extrapolation because *Mytilus* larvae retain particles of a different size spectrum than *Dreissena* larvae. Marine bivalve larvae can take up dissolved organic substances directly from the seawater; the rate depends on concentrations actually encountered. However, under natural conditions they can probably meet only a few percent of their metabolic requirements by dissolved organic material (Manahan, 1983 and 1990) — a figure that is probably also true for *Dreissena* larvae.

Metabolism

To my knowledge, respiration rates of *Dreissena* larvae have never been estimated. Extrapolating from data on the oxygen consumption rates of *Mytilus edulis* larvae (Sprung, 1984c), rates between 0.08 and 1.00 nL of O_2 per hour may be expected for *Dreissena* larvae. Usually lipids play an important role in the metabolism of planktonic larvae because of their high energy content and their low specific weight (Holland, 1978). The average lipid content of a *Dreissena* egg was estimated to be 29.6% of the ash-free dry weight (Sprung, 1989); however, there was a high degree of variation between eggs from different females. Lipids are, of course, important as fuel for the first lecithotrophic period. Yet, lipid reserves may also play an important role during periods of food deprivation and settlement. Consequently, levels of lipid have been used as a condition index for marine bivalve larvae (Gallagher et al., 1986). Despite this, I could not show any correlation between the rearing success and the lipid content of *Dreissena* eggs (Sprung, 1989) as demonstrated for *Ostrea* and *Mytilus* (Bayne, 1972; Helm et al., 1973).

Little is known about the energetic relation and metabolic demands of *Dreissena* larvae. No correlation could be found between the number of particles between 1 and 4 μm and the growth rate of larval cohorts in the surface water of two lakes (Sprung, 1989). This may imply that the ingestion capacity of *Dreissena* larvae (in the sense of Sprung and Rose, 1988) is very low when compared to its metabolic demands. If this were true, the high particle concentration in the lake would have meant only an excess of food to the larvae. Another nonexclusive explanation could be that *Dreissena* larvae have very specific qualitative food demands. This fits in with the results of the laboratory study of food preferences mentioned earlier, but certainly would be a strange property of a filter-feeding organism.

Settlement

The process of settlement is a very critical event in the *Dreissena* life cycle (Stanczykowska, 1978). During this period, the larva gradually undergoes many morphological changes. It develops into what is called the pediveliger stage. In this stage the larva can already crawl by means of a foot and also swim by means of its velum. Bayne (1971) described the most important changes during this period for *Mytilus edulis*: the secretion of byssus threads, the reorientation of organs in the mantle cavity, the collapse and disappearance of the velum, and the formation of labial palps. During this last event the larva cannot take up food because the labial palps transport food to the gut which is filtered out by the newly formed gills.

Conditions which have a negative impact on settling success are summarized by Stanczykowska (1978); they include unfavorable oxygen conditions in the hypolimnion, unsuitable bottom structures, and (especially in the shallow littoral) strong water movements which do not allow the larvae to settle to the bottom and impede feeding due to a significant amount of large mineral particles in suspension.

In general, the substrate where the larvae can settle is not necessarily very specific (Stanczykowska, 1964). Settled *Dreissena* larvae have been found on the underside of the leaves of *Nuphar luteum* (Weltner, 1891; Meisenheimer, 1899); Characeae: *Ceratophyllum, Fontinalis,* and *Elodea* (Lewandowski, 1982b and 1983a); and *Myriophyllum spicatum* (Mikheev, 1966). Lewandowski and Stanczykowska (1987) calculated that in Lake Zarnowieckie (Poland) about 94% of the young *Dreissena* settled on plants. Settling *Dreissena* also show a certain preference for the shells of adult *Dreissena* and Unionids, or stones (Lewandowski, 1976; Hebert et al., 1989). Mud and sand are, however, unsuitable (Lewandowski, 1982b). Also, artificial substrates have been used to test substrate preferences tested for settlement (Leentvaar, 1971; Walz, 1973). Both the color of the substrate as well as the structure (smooth or rough) seemed to be unimportant. Larvae prefer to settle on the underside of objects in a location with a certain amount of water current. Settling was most intense on iron and PVC; reasonable on tin, concrete, and plexiglas, but very poor on brass and copper, and absent on zinc, which is toxic. Shevtsova (1968) and Lewandowski (1982b) monitored the settlement on glass plates. Szlauer (1979) reported on settlement with preference to the hard knots of stylon nets.

The depth at which larvae generally settle is indicated by Shevtsova (1968) as between 0.5 and 4.5 m. This varies with the season; in the spring larvae prefer shallower water of 2.5 m, but later in the year deeper water of 4.5 m is preferred. Leentvaar (1971) observed the maximum at 1.4 to 1.6 m depth and Morton (1969) recorded settlement on asbestos plates between 2 and 7 m; he found only a few young mussels at 1 and 10–12 m.

The season for settlement should correlate with the larval abundance in the plankton. Piesik (1983) recorded three peaks: in May, in the last days of July, and in August/September. The last peak, although the weakest, showed the highest survival rates of the young spat. I observed a settlement in a lake near Cologne only during June following a huge peak in larval abundance; all the other cohorts later in the year did not reach settling size (Sprung, 1989 and unpublished observations).

CONCLUSIONS

Since the first studies of the larval stages of *Dreissena* and continuing to the present time, there has been a disproportionate amount of information concerned with detailed descriptions of developmental aspects of this stage in contrast to nearly complete ignorance of the physiology. Also, numerous studies that report variation in larval abundance in time and space provide little insight into the significance of environmental factors limiting or promoting successful development of the larvae. This will persist as long as no techniques are available to facilitate testing the effect of environmental factors on larval development (e.g., food requirements, significance of bacterial infection, capacity to withstand oxygen deficiencies) and to estimate larval physiological parameters (e.g., filtration and ingestion rates, and metabolic demands). This is a compelling task not only for mere scientific reasons. Because of high larval abundances and mortality rates, the population dynamics of *Dreissena* is at no other phase of its life cycle influenced so effectively as during its larval stage. Bearing in mind the possible ecological and economic consequences of the presence of *Dreissena* in the environment (as outlined in other chapters of this book), a deeper insight into its larval life is desirable.

ACKNOWLEDGMENT

I am grateful to W. Hollander for revising my English.

REFERENCES

Bayne, B. L. "Growth and the Delay of Metamorphosis of the Larvae of *Mytilus edulis L. Ophelia* 2:1–47 (1965).

Bayne, B. L. "Some Morphological Changes that Occur at the Metamorphosis of the Larvae of *Mytilus edulis*," in *Fourth European Marine Biology Symposium*, D. J. Crisp, Ed., (London: Cambridge University Press, 1971) 203–229.

Bayne, B. L. ''Some Effects of Stress in the Adult on Larval Development of *Mytilus edulis*,'' *Nature (London)* 237:459 (1972).

Bayne, B. L. ''The Biology of Mussel Larvae, in *Marine Mussels: Their Ecology and Physiology*,'' B. L. Bayne, Ed., (London: Cambridge University Press, 1976) pp. 81–120.

Bayne, B. L. ''The Physiological Ecology of Marine Molluscan Larvae,'' in *The Mollusca III: Development,* N. H. Verdonk, J. A. M. van den Biggelaar, and A. Tompa, Eds., (New York: Academic Press, 1983) pp. 199–243.

Blochmann, F. ''Eine frei schwimmende Muschellarvae im Süsswasser,'' *Biol. Zentralbl.* 11:476–478 (1891).

Borcherding, J. ''The Annual Reproductive Cycle of the Freshwater Mussel *Dreissena polymorpha* Pallas in Lakes,'' *Oecologia (Berlin)* (Submitted).

Breitig, G. ''Beiträge zur Biologie, Verbreitung und Bekämfung von *Dreissena polymorpha* Pall.,'' Dissertation, Math.-Nat. Fak., University Greifswald (1965).

Breitig, G. ''Das Molluskenplankton und seine Rolle in der Besiedlung der Binnengewässer,'' *Wasserwirtsch.-Wassertech.* 19:116–118 (1969).

Einsle, U. and N. Walz. ''Die tägliche Vertikalwanderung der Larven von *Dreissena polymorpha* Pallas im Bodensee-Obsersee (1971),'' *Gas-Wasserfach, Wasser Abwasser* 113:428–430 (1972).

Einsle, U. ''Zur Horizontal- und Vertikalverteilung der Larven von *Dreissena polymorpha* Pallas im Pelagial des Bodensee-Obersees (1971),'' *Gas-Wasserfach, Wasser Abwasser* 114:27–30 (1973).

Einsle, U. ''Das Vorkommen der Larven von *Dreissena polymorpha* (Pallas), der Dreikantmuschel, im Pelagial des Bodensee-Obersees 1972–1985,'' *Veröff. Naturschutz Landschaftspflege Bad.-Württ.* 64/65:435–437 (1989).

Feigina, S. S. ''Thermal Action Against *Dreissena* Fouling in Conditions Thermal Power Stations,'' *Elektr. Stanktl. Moskva* 10:38–40 (1959) (In Russian).

Gallager, S. M., R. Mann, and G. C. Sasaki. ''Lipid as an Index of Growth and Viability in Three Species of Bivalve Larvae,'' *Aquaculture* 56:81–103 (1986).

Gallager, S. M. ''Visual Observations of Particle Manipulation During Feeding in Larvae of Bivalve Mollusc,'' *Bull. Mar. Sci.* 43:344–365 (1988).

Hebert, P. D. N., B. W. Muncaster, and G. L. Mackie. ''Ecological and Genetic Studies on *Dreissena polymorpha* (Pallas): A New Mollusc in the Great Lakes,'' *Can. J. Fish. Aquat. Sci.* 46:1587–1591 (1989).

Helm, M. M., D. L. Holland, and R. R. Stephenson. ''The Effect of Supplementary Algal Feeding of a Hatchery Breeding Stock of *Ostrea edulis* L. on larval vigour,'' *J. Mar. Biol. Assoc. U.K.* 53:673–684 (1973).

Hillbricht-Ilkowska, A., and A. Stanczykowska. *''The Production and Standing Crop of Planktonic Larvae of Dreissena polymorpha* Pall. in Two Mazurian Lakes,'' *Pol. Arch. Hydrobiol.* 16:193–203 (1969).

Holland, D. L. ''Lipid Reserves and Energy Metabolism in the Larvae of Benthic Marine Invertebrates,'' in *Biochemical and Biophysical Perspectives in Marine Biology 4,* D. C. Malins and J. R. Sargent, Eds., (New York: Academic Press, 1978) pp. 85–123.

Karabin, A. ''The Pressure of Pelagic Predators of the Genus *Mesocyclops* (Copepoda, Crustacea) on Small Zooplankton,'' *Ekol. Pol.* 26:241–257 (1978).

Karpevich, A. F. ''Adaptive Nature of the Morphology of Spermatozoa and Eggs of Bivalves,'' *Zool. Zh.* 40:340–350 (1961) (In Russian).

Katchanova, A. A. "Some Data on the Reproduction of *Dreissena polymorpha* Pallas in the Uchinsk Reservoir, *Tr. Vses. Gidrobiol. Ova* 11:117–121 (1961) (In Russian).

Kirpichenko, M. Ya., V. P. Mikheev, and E. P. Shtern. "Action of Electric Current on *Dreissena polymorpha* Larvae and Planktonic Crustaceans with Short Exposures," *Akad. Nauk SSSR, Moskov, Leningr.* (1963) pp. 76–80 (In Russian).

Kirpichenko, M. Ya. "Phenology, Abundance, and Growth of *Dreissena* Larvae in the Kujbyshev Reservoir," in *S: Biol. Drejsseny borba nej, Moskva.* (1964) pp. 19–30 (In Russian).

Kirpichenko, M. Ya. "The Ecology of Early Stages of *Dreissena polymorpha* Pallas Ontogenesis," Ph.D. Thesis, Dneproppetrovsk, Dnepropetrovskij Gosud University (1965) (In Russian).

Kirpichenko, M. A. "On the Ecology of *Dreissena polymorpha* in the Zimljansk Reservoir," *Tr. Inst. Biol. Vnut. Vod.* 21:142–154 (1971) (In Russian).

Kornobis, S. "Ecology of *Dreissena polymorpha* (Pall.) (Dreissenidae, Bivalvia) in Lakes Receiving Heated Water Discharges," *Pol. Arch. Hydrobiol.* 24:531–546 (1977).

Korschelt, E. "Entwicklung von *Dreissena polymorpha* Pallas," *Gesellsch. Naturforsch. Freun.* Berlin, Sitzung vom 21:131–146 (1891).

Korschelt, E., and W. Weltner. "Die Lebensverhältnisse der *Dreissena polymorpha*," *Naturwiss. Wochenzeitschr.* 7:391–393 (1892).

Leentvaar, P. "Das Vorkommen von *Dreissena polymorpha* in den Niederlanden," in: Internationale Arbeitsgemeinschaft der Wasserwerke im Rheineinzugsgebiet, Bericht über die 2, Arbeitstagung, Rotterdam 1971, *RIN Bericht* 37:133–148 (1971).

Lewandowski, K. "Unionidae as a substratum for *Dreissena polymorpha* Pall," *Pol. Arch. Hydrobiol.* 23:409–420 (1976).

Lewandowski, K. "The Role of Early Developmental Stages in the Dynamics of *Dreissena polymorpha* (Pall.) (Bivalvia) Populations in Lakes. I. Occurrence of Larvae in the Plankton," *Ekol. Pol.* 30:81–110 (1982a).

Lewandowski, K. "The Role of Early Development Stages in the Dynamics of *Dreissena polymorpha* (Bivalvia) Populations in Lakes. II. Settling of Larvae and the Dynamics of Numbers of Settled Individuals," *Ekol. Pol.* 30:223–286 (1982b).

Lewandowski, K. "Occurrence and Filtration Capacity of Young Plant-Dwelling *Dreissena polymorpha* (Pall.) in Majcz Wielki Lake," *Pol. Arch. Hydrobiol.* 30:255–262 (1983a).

Lewandowski, K. "Formation of Annuli on Shells of Young *Dreissena polymorpha* (Pall.)," *Pol. Arch. Hydrobiol.* 30:343–351 (1983b).

Lewandowski, K., and J. Ejsmont-Karabin. "Ecology of Planktonic Larvae of *Dreissena polymorpha* (Pall.) in Lakes with Different Degree of Heating," *Pol. Arch. Hydrobiol.* 30:89–101 (1983).

Lewandowski, K., and A. Stanczykowska. "Molluscs in Lake Zarnowieckie," *Pol. Ecol. Stud.* 12:315–330 (1987).

Loosanoff, V. L., W. S. Miller, and P. B. Smith. "Growth and Setting of Larvae of *Venus mercenaria* in Relation to Temperature," *J. Mar. Res.* (1951) pp. 59–81.

Lucas, A., L. Chebab-Chalabi, and D. A. Aranda. "Passage de l'endotrophie á l'extrophie chez les larves de *Mytilus edulis*," *Oceanologica Acta* 9:97–103 (1986).

Manahan, D. T. "The Uptake and Metabolism of Dissolved Amino Acids by Bivalve Larvae," *Biol. Bull. Mar. Biol. Lab., Woods Hole, Mass.* 164:236–250 (1983).

Manahan, D. T. "Adaptations by Invertebrate Larvae for Nutrient Acquisition from Seawater," *Am. Zool.* 30:147–160 (1990).

Meisenheimer, J. "Zur Eiablage der *Dreissena polymorpha*," *Forschungsber. Biol. Stat. Plön.* 7:25–28 (1899).

Meisenheimer, J. "Entwicklungsgeschichte von *Dreissena polymorpha*," *Z. Wissensch. Zool.* 69:1–137 (1901).

Mikheev, V. P., "*Dreissena* as Food for Fish Reared in Floating Nets," *Tr. Vses. Nauchno-Issled. Inst. Prudov. Ryb. Khozy.* 14:157–167 (1966) (In Russian).

Mikheev, V. P. "Filtration Nutrition of the *Dreissena*," *Tr. Vses. Naucho-Issled. Inst. Prudov. Ryb. Khozy.* 15:117–129 (1967) (In Russian).

Morton, B. "Studies on the Biology of *Dreissena polymorpha* Pall. III. Population Dynamics," *Proc. Malacol. Soc. London* 38:471–482 (1969).

Piesik, Z. "Biology of *Dreissena polymorpha* (Pall.) Settling on Stylon Nets and the Role of This Mollusc Eliminating the Seston and the Nutrients from the Water-Course," *Pol. Arch. Hydrobiol.* 30:353–361 (1983).

Serafimova-Hadishche, J. "Vertical Migrations of the Zooplankton in Lake Prespa (in Kroatic)," *Recl. Trav. Stn. Hydrobiol. Ohrid. II* 1(8) (1954).

Shevtsova, L. V. "Peculiarities of the Reproduction and Development of *Dreissena* in the Canal the Dnepr-Krivoj-Rog," *Gidrobiol. Zh.* 4:70–72 (1968) (In Russian).

Siller, M. "Abundanzdynamik, Tiefenverteilung und tägliche Vertikalwanderung der Larve der Wandermuschel *Dreissena polymorpha* Pallas. Staatsexamensarbeit, University of Cologne (1983).

Sprung, M. "Physiological Energetics of Mussel Larvae (*Mytilus edulis*). I. Shell Growth and Biomass," *Mar. Ecol. Prog. Ser.* 17:283–293 (1984a).

Sprung, M. "Physiological Energetics of Mussel Larvae (*Mytilus edulis*). II. Food Uptake," *Mar. Ecol. Prog. Ser.* 17:295–305 (1984b).

Sprung, M. "Physiological Energetics of Mussel Larvae (*Mytilus edulis*). III. Respiration," *Mar. Ecol. Prog. Ser.* 18:171–178 (1984c).

Sprung, M. "Ecological Requirements of Developing *Dreissena polymorpha* Eggs," *Arch. Hydrobiol., Suppl.* 79:69–86 (1987).

Sprung, M., and U. Rose. "Influence of Food Size and Food Quantity on the Feeding of the Mussel *Dreissena polymorpha*," *Oecologia (Berlin)* 77:526–532 (1988).

Sprung, M. "Field and Laboratory Observations of *Dreissena polymorpha* Larvae: Abundance, Growth, Mortality and Food Demands," *Arch. Hydrobiol.* 115:537–561 (1989).

Sprung, M. "Costs of Reproduction: A Study on Metabolic Requirements of the Gonads and Fecundity of the Bivalve *Dreissena polymorpha*," *Malacologia* 33:63–70 (1991).

Stanczykowska, A. "On the Relationship Between Abundance, Aggregation and "Condition" of *Dreissena polymorpha* Pall. in 36 Mazurian Lakes," *Ekol. Pol., Ser. A.* 12:653–690 (1964).

Stanczykowska, A. "Ecology of *Dreissena polymorpha* (Pall.) (Bivalvia) in Lakes," *Pol. Arch. Hydrobiol.* 24:461–530 (1977).

Stanczykowska, A. "Occurrence and Dynamics of *Dreissena polymorpha* (Pall.) (Bivalvia)," *Verh.-Int. Ver. Theor. Angew. Limnol.* 20:2431–2434 (1978).

Stanczykowska, A., K. Lewandowski, and J. Ejsmont-Karabin. "Biotic Structure and Processes in the Lake System of R. Jorka Watershed (Masurian Lakeland, Poland). IX. Occurrence and Distribution of Molluscs with Special Consideration to *Dreissena polymorpha* (Pall.)," *Ekol. Pol.* 31:761–780 (1983).

Strathmann, R. R., T. L. Jahn, and J. R. C. Fonseca. "Suspension Feeding by Marine Invertebrate Larvae: Clearance of Particles by Ciliated Bands of a Rotifer, Pluteus, and Trochophore," *Biol. Bull. Mar. Biol. Lab., Woods Hole, Mass.* 142:505–519 (1972).

Strathmann, R. R., and E. Leise. "On Feeding Mechanisms and Clearance Rates of Molluscan Veligers," *Biol. Bull. Mar. Biol. Lab., Woods Hole, Mass.* 157:524–535 (1979).

Suter-Weider, P. "Über das Vorkommen der Larven von *Dreissena polymorpha* Pallas im unteren Zürichsee in den Jahren 1971 bis 1975," *Gas. Wasser. Abwasser* 56:371–374 (1976).

Szlauer, L. "Use of Steelon-Net Veils for Protection of the Hydroengineering Works against *Dreissena polymorpha* Pall.," *Pol. Arch. Hydrobiol.* 21:391–400 (1974).

Szlauer, L. "On the Application of Barriers to Protect Hydrotechnic Installations from *Dreissena* Settling and to Nutrient Removal from Water," *Zesz. Nauk. Akad. Roln. Szczecinie Ser. Ryb. Morsk. Tech. Zywn.* 75:29–38 (1979) (In Polish).

Walz, N. "Untersuchungen zur Biologie von *Dreissena polymorpha* Pallas im Bodensee," *Arch. Hydrobiol., Suppl.* 42:452–482 (1973).

Walz, N. "Die Besiedlung von künstlichen Substraten durch die Larven von *Dreissena polymorpha*," *Arch. Hydrobiol., Suppl.* 47:423–431 (1975).

Walz, N. "The Energy Balance of the Freshwater Mussel *Dreissena polymorpha* Pallas in Laboratory Experiments and in Lake Constance. II. Reproduction," *Arch. Hydrobiol., Suppl.* 55:106–119 (1978).

Weltner, W. "Zur Entwicklung von *Dreissena*," *Zool. Anz.* 14:447–451 (1891).

Werner, B. "Über die Entwicklung und Artunterscheidung von Muschellarven des Nordseeplanktons, unter gesonderter Berücksichtigung der Schalenentwicklung," *Zool. Jahr. (Anat.)* 66:1–54 (1940).

Wiktor, K. "Larvae of *Dreissena polymorpha* Pall. as a Food for Fish Spawn," *Przeglad Zool.* 2:182–184 (1958) (In Polish).

Wiktor, J. "The Biology of *Dreissena polymorpha* Pall. and Its Ecological Importance in the Firth of Szczecin," *Stud. Mater. Morski Inst. Ryb., Gdynia Poland Ser. A* 5:1–88 (1969) (In Polish).

CHAPTER 3

Colonization, Ecology, and Positive Aspects of Zebra Mussels *(Dreissena polymorpha)* in The Netherlands

Henk Smit, Abraham bij de Vaate, Harro H. Reeders, Egbert H. van Nes, and Ruurd Noordhuis

The colonization, ecology, and potential positive benefits of the zebra mussel, *Dreissena polymorpha*, in some lakes and rivers of The Netherlands is described. Population levels in Dutch Lakes are limited most by the scarcity of solid substrates and consumption by ducks. Mussel densities in the recovering Rhine River ranged between 0/m^2 in 1973 to nearly 2500/m^2 in 1989. Sudden temperature changes are identified as a possible limiting factor for the occurrence of mussels in small shallow (depth <2 m) water bodies. In eutrophic lakes, water movements rather than algal concentrations seem to determine shell growth. In Lake IJsselmeer, zebra mussel production was 41–102 $\times$ 10^{12} J, of which 15–26 $\times$ 10^{12} J was consumed by ducks and 1.3–2.1 $\times$ 10^{12} J was consumed by fish. In Lake Volkerakmeer, management authorities plan to use zebra mussels to increase water transparency and to prevent contaminated suspended matter from entering the lake. To this end, the density of zebra mussels was artificially increased by adding marine shells as a substrate for attachment. Colonization of different substrate types in hanging cultures was tested as a first step toward a biological filter.

0-87371-696-5/93/$0.00 + $.50

INTRODUCTION

The recent introduction of the zebra mussel, *Dreissena polymorpha*, in North America has focused attention on its biofouling and ecological properties (Griffiths et al., 1989; Hebert et al., 1989). In The Netherlands, during the 1970s, attention was mainly focused on the nuisance qualities of this species. However, in the last decade, applied research activities have been increasingly concentrated on its positive applications for water management and its role in the ecosystem (Van Eerden and bij de Vaate, 1984; Heuck van der Plas, 1984; Reeders et al., 1989). This chapter reviews the colonization and ecology of the zebra mussel in The Netherlands and gives an example of its use in water quality management.

RESULTS AND DISCUSSION

Colonization Patterns

The zebra mussel invaded The Netherlands in the 1830s and now commonly occurs in most larger rivers, lakes, and canals (Leentvaar, 1975). Suitable habitat increased considerably during the 20th century because several estuaries were converted into freshwater lakes by damming them from the sea. For example, Lake IJsselmeer (3500 km^2) was created in 1932 by closing off the former ''Zuiderzee'' from the sea (Figure 1). Chloride concentrations gradually dropped from 6 g/L in 1932 to 0.6 g/L in 1935, and to 0.4 g/L in 1936. Prior to 1932, zebra mussels, which commonly occurred in the River Rhine, were restricted in the River IJssel delta where it emptied into the Zuiderzee. Zebra mussels were recorded in Lake IJsselmeer in 1936; by 1937, zebra mussels were found in the northern part of the lake and by 1938, were distributed throughout the lake (Figure 2).

A recent example of zebra mussel colonization was observed in Lakes Volkerakmeer and Zoommeer (60 km^2) (Figure 3). These lakes were dammed from the Eastern Scheldt estuary in April 1987 and, consequently, chloride concentrations dropped from 10 g/L in April 1987 to 0.8 g/L in September 1987. The main source of zebra mussel veligers entering Lake Volkerakmeer was water from the Rhine and Meuse Rivers that entered the lake through the Volkerak sluices. The first adult mussels were reported in Lake Volkerakmeer in October 1987. Colonization probably occurred in September, because veligers only occur in plankton of both rivers between May and September. In January 1988, zebra mussels had spread throughout Lake Volkerakmeer and Eendracht Canal. Lake Zoommeer, which has somewhat higher chloride concentrations (0.7 g/L) compared to Lake Volkerakmeer (0.3–0.4 g/L), was colonized 2 years later; these higher chloride concentrations probably slowed the colonization process. Monthly monitoring of nine

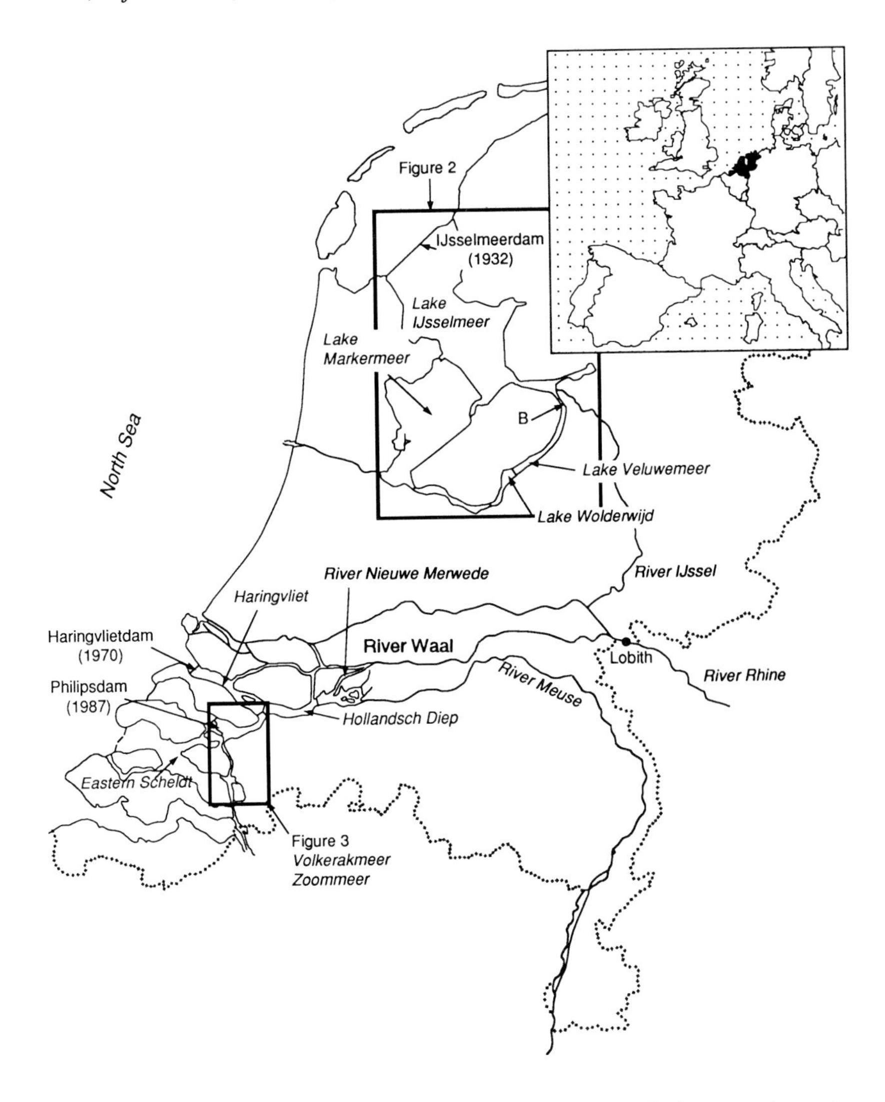

Figure 1. Location of water bodies in The Netherlands where *Dreissena polymorpha* has been studied.

sampling sites indicated that both density and biomass increased to peak levels 2 years after settlement (Figure 4).

Severe pollution probably caused the disappearance of zebra mussels from the Lower Rhine in the late 1960s (Wolff, 1969; Peeters and Wolff, 1973; Van Urk, 1976). Annual monitoring of macrozoobenthos on stone blocks of the IJssel River (one of the branches of the Rhine River) (Van Urk and bij de Vaate, 1990) has given insight into the recolonization by zebra mussels,

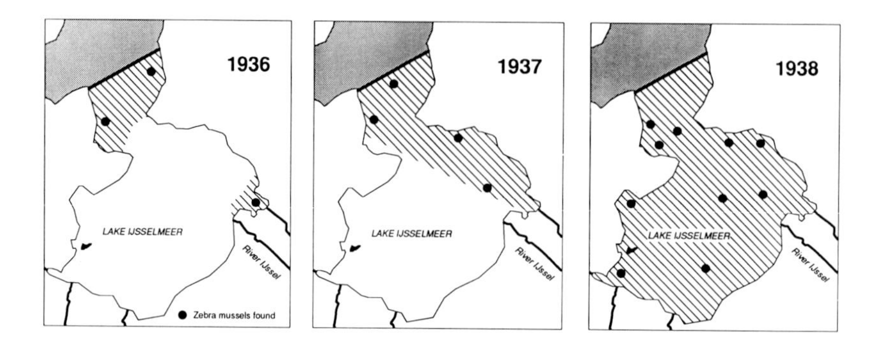

Figure 2. Colonization of Lake IJsselmeer by *Dreissena polymorpha*. (From Van Benthem Jutting, W. S. S., "Changes in the Flora and Fauna of the Zuiderzee (at present Lake IJsselmeer) after the Embankment in 1932," Report of the Zuiderzee Commission, The Nederlandse Dierkundige Vereniging, The Netherlands [1954], pp. 233–253 [In Dutch]. With permission.)

which started in the second half of the 1970s (Figure 5). While the zebra mussel was absent from the Lower Rhine, a considerable reproducing population was present in Lake Constance at the Swiss-German border (Walz, 1978). Since the average residence time of the water between this lake and The Netherlands part of the Lower Rhine is only about 9 days, environmental conditions rather than the supply of veligers probably limited colonization. Recolonization of the IJssel River took place at the same time (late 1970s) that improvements in water quality of the Rhine River occurred. Concentrations of several heavy metals (e.g., cadmium and mercury), organic pollutants, ammonium, and nitrite decreased considerably, and oxygen concentrations showed a substantial increase (Van Broekhoven, 1987).

Veliger Larvae

Seasonal patterns of veliger density in the Rhine River at the German-Dutch border (Lobith) showed that larvae were present from May through September 1987–1989 (De Ruyter van Steveninck et al., 1990; Figure 6). A first peak was observed in May and a second peak in June or July. One or two other peaks were observed in August or September. Peaks in May and June or July are probably the result of first and second spawning periods (Borcherding, 1991) of mussels 1 year old or older. Peaks in August and September may originate from populations which spawn later, or may be the result of a third or even a fourth spawning period. Borcherding (1991) found up to four spawning periods per season in Lake Fühlinger See (Germany). Borcherding and De Ruyter van Steveninck (1992) indicate that different veliger peaks originate from different parts of the Rhine River system. However, it cannot be excluded that young of the year settled in May, and had

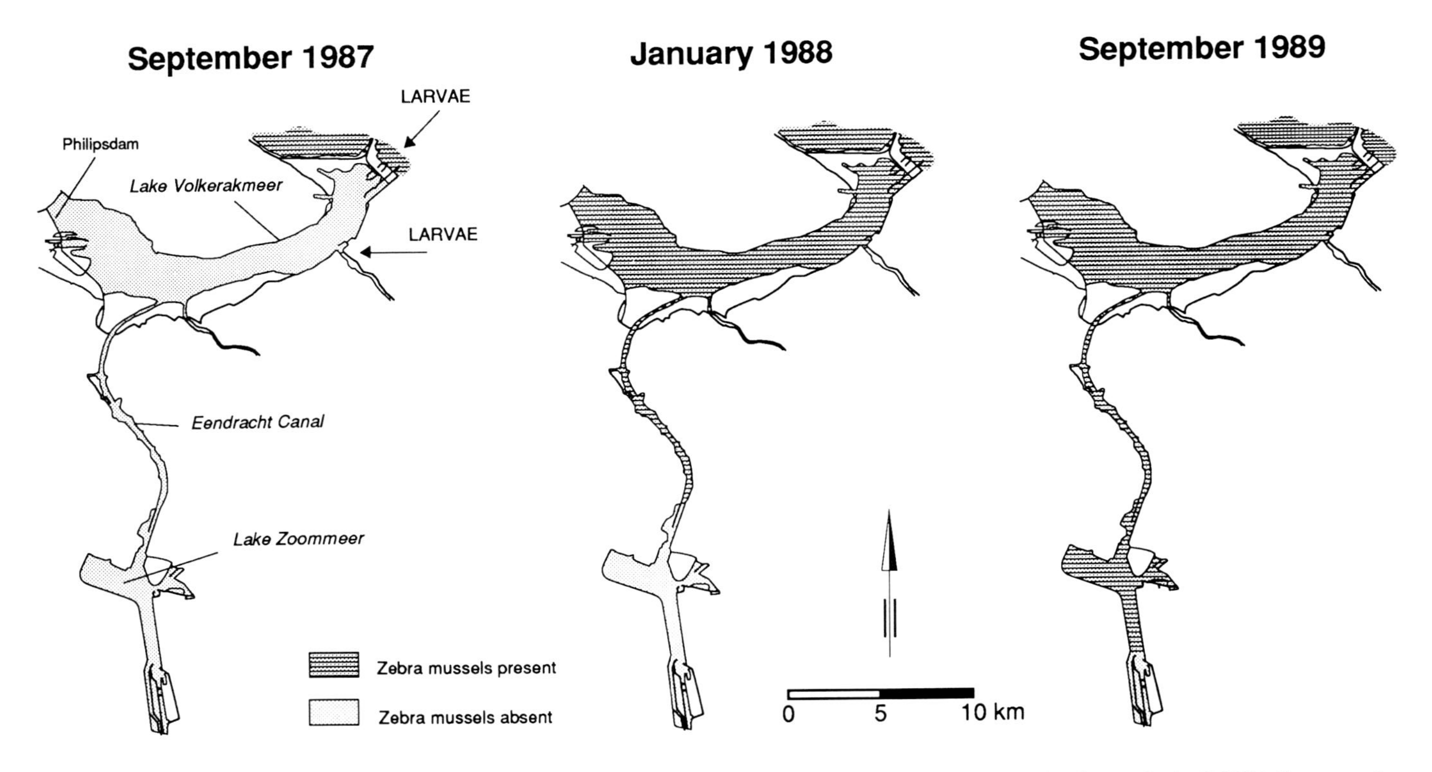

Figure 3. Colonization of Lakes Volkerakmeer and Zoommeer by *Dreissena polymorpha* after the embankment in April 1987. Sources of larvae are indicated with an arrow.

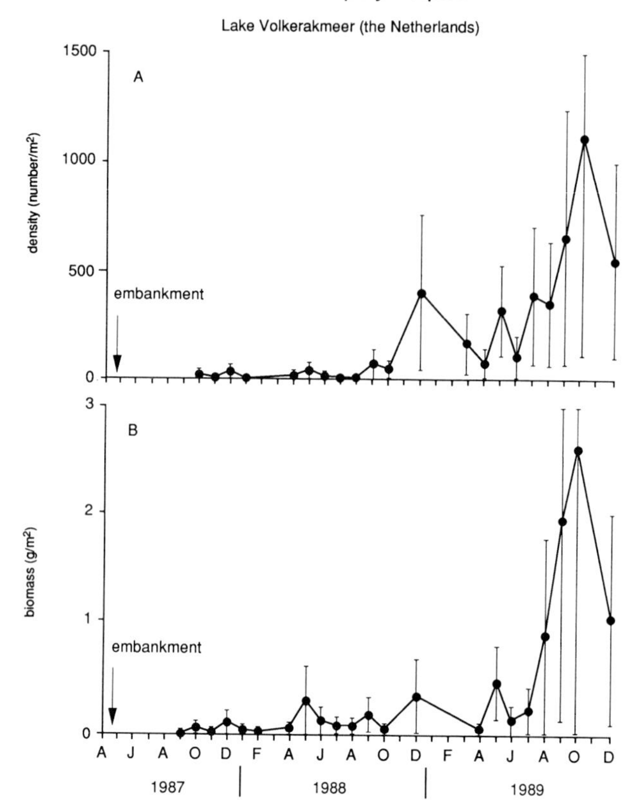

Figure 4. Mean (± S.E.) density and biomass (AFDW) of *Dreissena polymorpha* in Lake Volkerakmeer during the first 2 years after embankment. (Biomass calculated according to Smit, H., and E. Dudok van Heel. *The Zebra Mussel, Dreissena polymorpha: Ecology, Biological Monitoring and First Applications in the Water Quality Management. Limnologie Aktuell.* [Stuttgart, Germany: Fischer-Verlag, 1992.])

already initiated reproduction in September. Veliger growth (shell length) in the Rhine River was 7.6 μm/day (range 6.0–11.0 μm/d) at water temperatures between 18 and 21°C. Because of the short residence time (i.e., 10 days) in the river, most veligers are unable to reach their size of settlement in the river itself (Borcherding and De Ruyter van Steveninck, 1992).

Settlement and Growth

An important part of veliger settlement probably takes place in Lake IJsselmeer and the Hollandsch Diep/Haringvliet; highest densities were ob-

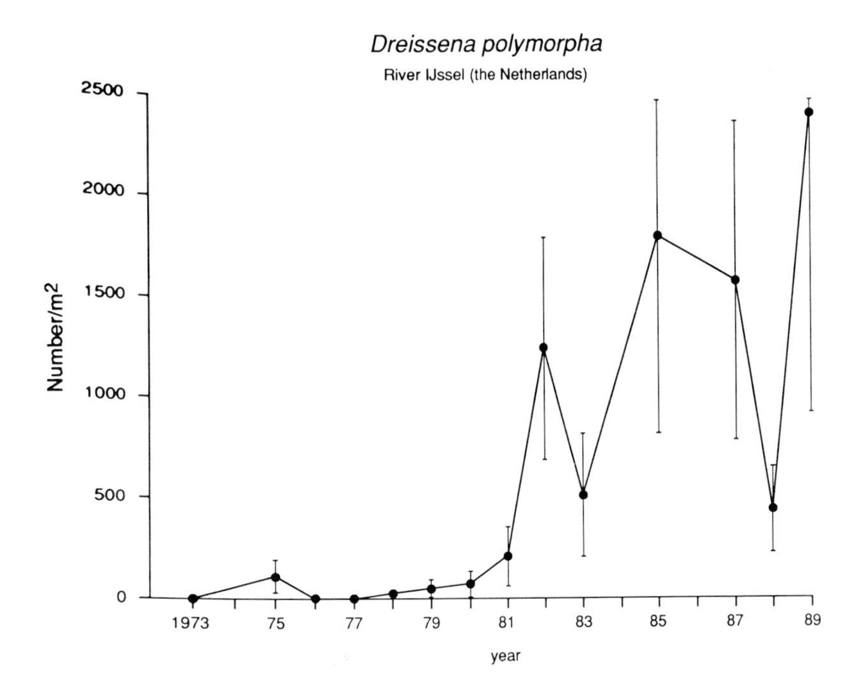

Figure 5. Mean (± S.E.) density of *Dreissena polymorpha* on solid substrates (bricks and stones) in the littoral zone of the River IJssel, September 1973-1989.

served at 10–20 km from the outflow of the IJssel and Nieuwe Merwede Rivers (bij de Vaate, 1991; H. Smit, unpublished data). In Lakes IJsselmeer and Markermeer, settlement of young mussels occurs between July and October, with a maximum in July. In Lake Maarseveen, a small oligotrophic lake in the center of The Netherlands, settlement started in May and showed a peak in September (bij de Vaate, 1991). In the Haringvliet, a settlement peak was found in September–October 1988, which resulted from a pronounced veliger peak found in September of the same year (H. Smit, unpublished data).

Young-of-the-year (0+) mussels in Lake IJsselmeer reach a maximum shell length of about 6 mm, whereas in the Rhine River shell lengths up to 16 mm are not unusual (Figure 7). This large difference in the first year of growth can be attributed to two factors. First, young-of-the-year mussels have a shorter period of growth in Lake IJsselmeer than in the Rhine River; veligers settle later and shell growth ceases earlier. Second, shell growth in Lake IJsselmeer is lower than in the Rhine River (Smit et al., 1992).

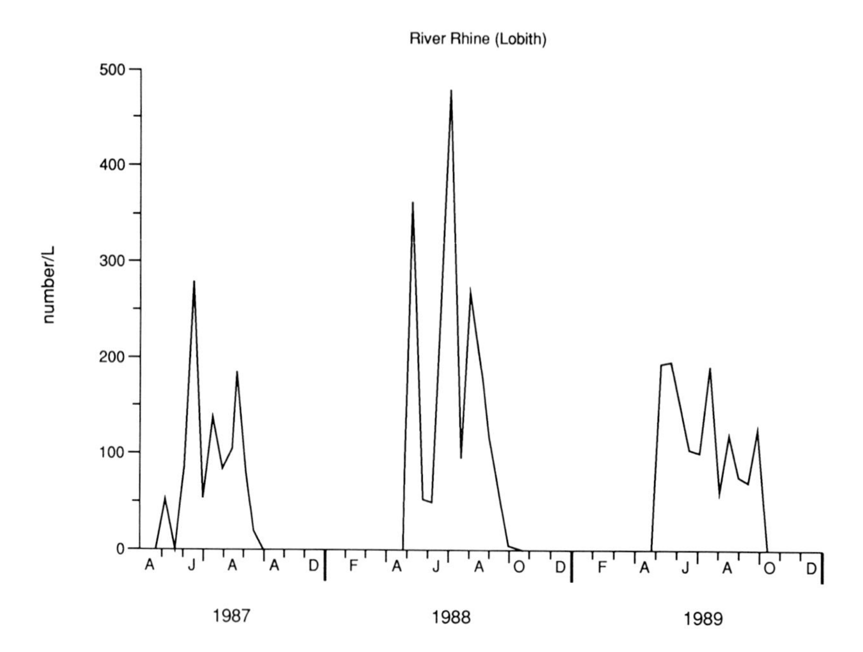

Figure 6. Density of *Dreissena polymorpha* veligers (number/L) in zooplankton samples from the Rhine River at the German-Dutch border (Lobith), 1987-1989. (From De Ruyter van Steveninck, E. D., W. Admiraal, and B. van Zanten. *Regul. Rivers* 5:67–71 [1990]; Admiraal, W. Unpublished data [1989]. With permission.)

Mortality

Cage experiments were performed in various lakes to determine factors affecting mortality (Smit et al., 1992). Annual mortality of 1- and 2-year-old mussels in experimental cages was 12–35% in Lake Volkerakmeer, 34–39% in the Hollandsch Diep, and 75–85% in the Rhine River. Seasonal mortality (i.e., March-November) of caged mussels in Lake IJsselmeer was less than 15% (Table 1). These mortality rates are lower than those found for natural populations (Table 2). Annual mortality rates of natural populations in Lakes IJsselmeer and Markermeer (bij de Vaate, 1991), and Hollandsch Diep/Haringvliet (Figure 7) were based on length-frequency distributions. In lakes, a strong intraspecific competition for space, predation by waterfowl, and sediment movement may be responsible for higher mortalities. These processes did not occur in the cages, where mussels had enough space and were free from predation and sediment. Only in the Rhine River did the annual mortality of caged mussels approach the high mortality of natural populations (bij de Vaate et al., 1992). High mortality rates observed in the Rhine River may be due to large amounts of sediment found in cages, although other processes like toxic stress may not be excluded.

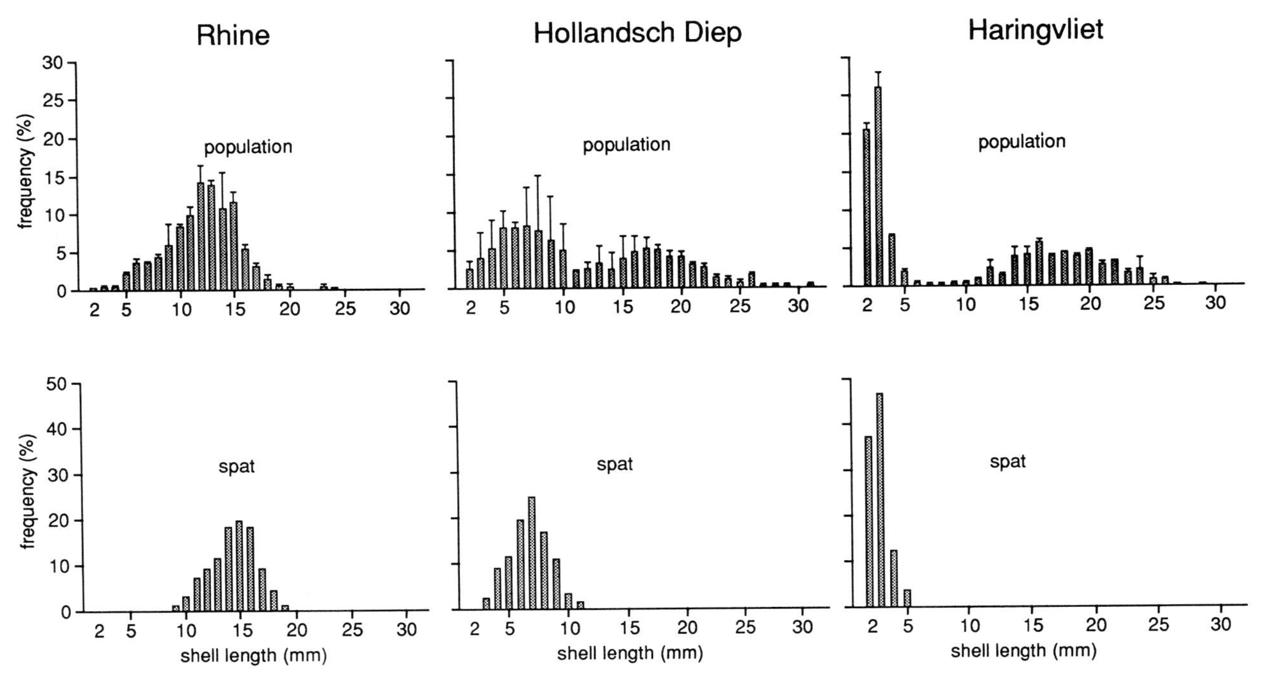

Figure 7. Length-frequency distributions of *Dreissena polymorpha* populations (mean ± S.D.; n = 2) and young-of-the-year only (spat) in autumn in the Rhine River, Hollandsch Diep, and Haringvliet. Population structure for the Rhine River was obtained from mussels collected from the Waal River branch, September 1989 and spat were collected from recreational buoys in the River Nieuwe Merwede, October 1989. Population structure in Hollandsch Diep and Haringvliet was determined from mussels collected at Willemstad and Stellendam, respectively, in early November 1987, and spat were collected from cages at the same localities in late October 1987. (Adapted from Greijdanus-Klaas, M. Report 90.046X, Institute for Inland Water Management and Waste Water Treatment, The Netherlands [1990] [In Dutch]. With permission.)

Table 1. Percent Mortality Rates (Mean) of Zebra Mussels, *Dreissena polymorpha* in Cages in Lake IJsselmeer, Lake Volkerakmeer I and II, Hollandsch Diep, and Rhine River over the Periods March–November 1988 (i.e., Growth Period) and November 1988–March 1989 (i.e., Nongrowth Period)

Age (years) Season	1		2	
	Growth Period	Nongrowth Period	Growth Period	Nongrowth Period
Lake IJsselmeer[a]	3–12	—	5–15	—
Lake Volkerakmeer I	21	14	15	14
Lake Volkerakmeer II	12	0	14	0
Haringvliet	23	—	20	—
Hollandsch Diep	27	7	32	7
Rhine River	61	16	69	16

Note: Age of mussels deduced from their shell length with site-specific seasonal growth equations. Age determined in March at the beginning of the growing season.

[a] Data from 1981–1985 as reported by bij de Vaate, 1991.
— Not determined

Table 2. Age, Length, and Frequency Distributions and Percent Mortality Rates (Mean ± S.D.) of Two Populations of *Dreissena polymorpha* (a and b) Estimated from Shell Length Frequency Distributions and Seasonal Growth Rates in Haringvliet and Hollandsch Diep in 1987

Water Body	Age (yr)	Length (mm)	Frequency (%) a	Frequency (%) b	Mortality (%)
Haringvliet	0	<9	53.7	55.6	55.6 ± 7.4
	1	7–18	26.7	21.8	31.0 ± 10.2
	2	18–23	16.5	16.6	71.7 ± 9.3
	3	23–27	2.9	5.8	97.4 ± 1.2
	>3	>26	0.1	0.1	
Hollandsch Diep	0	<12	59.6	45.2	28.4 ± 33.2
	1	10–20	28.7	43.0	70.3 ± 8.2
	2	20–25	10.2	10.3	88.0 ± 4.7
	3	25–28	1.4	0.9	71.5 ± 30
	>3	>27	0.1	0.45	

Population Structure

Population structure in the Rhine River in 1989 was strongly dominated by young-of-the-year mussels (bij de Vaate et al., In 1992). Spat, observed in autumn on recreational buoys which are removed from the water every winter, were of the same length as the major part of the population found on stones (Figure 7). In Haringvliet, spat were only a few mm long and were easily recognized as a separate, young-of-the-year cohort. In Hollandsch Diep, spat grew to an average shell length of 7 mm by fall 1989, and the young-of-the-year cohort already showed some overlap with older cohorts. Two-year-old and older mussels were frequently found in Hollandsch Diep, but were rare in the Rhine River (bij de Vaate et al., 1992). In Lake IJsselmeer and Haringvliet, up to 3-year-old mussels contribute significantly to population structure (bij de Vaate, 1992; Table 2). In all water bodies, 4-year-old and older mussels were rare. In Lake IJsselmeer, spat were not observed at all sites in successive years, which caused a considerable variation in population structure among sites.

PHYSICAL-CHEMICAL PARAMETERS

Chloride

In the former tidal reaches of the Rhine and Meuse Rivers, zebra mussels were confined to freshwater reaches (Wolff, 1969). Colonization patterns in Lakes IJsselmeer, Volkerakmeer, and Zoommeer indicate that the maximum chloride concentration in which the zebra mussel was able to complete its entire life cycle is between 0.4 and 0.7 g/L. Although mussels are frequently recorded from water bodies with higher chloride concentrations (e.g., Van Benthem Jutting, 1943; Wolff, 1969), these individuals probably developed from veligers transported from outside the water body itself.

Temperature

In the Rhine River and some associated lakes, shell growth rates were positively correlated with average water temperature in the growing season. Shell growth rates of mussels with an initial length of 5 mm at the beginning of the growing season were correlated with water temperature between March and May. The equation $G = 0.0063 \times (T - 3)^2$ ($n = 39$, $r^2 = 0.82$, $3.0 < T < 18.0$), where G = shell growth in millimeters per day and T = temperature in °C, best described this correlation. A minimum temperature of 3°C was needed for shell growth to occur (Smit et al., 1992). This minimum temperature for growth was much lower than the 11 and 12°C reported by Katchanova (1962) and Morton (1969), respectively. However, this 3°C lower limit is supported by observations of filtration rates; Reeders and bij de Vaate (1990) found a strong increase in filtration rate at this temperature, while Smirnova (1990) reported a lower limit of 3°C for filtration.

The rate of change in water temperature may be a key factor in the reproduction of zebra mussels. Field investigations in Lake Veluwemeer in 1990 revealed that a small population of 2- and 3-year-old mussels were present near the mouth of a canal that supplies water to the lake; in 1989 and 1990, these individuals produced few spat. An increase in water temperature in spring, which can be relatively fast in this shallow lake, may be responsible for this disruption in reproduction (Borcherding, 1991). In May 1989, mussels from Lake IJsselmeer were stocked in Lake Veluwemeer at least 1 week after the sharp temperature rise. These mussels produced large quantities of spat, whereas the natural population of Lake Veluwemeer did not. In the much larger Lake IJsselmeer (1190 km^2, mean depth 4.5 m) water temperatures rose much slower and did not reach high values measured in Lake Veluwemeer. Borcherding (1991) also suggested that disturbances in the reproduction of zebra mussels in Lake Heider Bergsee were due to a sharp rise in temperature. This might also explain why the zebra mussel is confined to larger and deeper water bodies and canal systems where water temperature changes occur more slowly.

Water Current and Chlorophyll *a*

In eutrophic lakes in The Netherlands, water movement seems to have a larger influence on growth than the amount of algal food in the water column. Small mussels placed in cages on a shallow wind-exposed site in Lake Wolderwijd showed an enhanced shell growth compared to mussels placed in cages in all other lakes (Figure 8). On the other hand, shell growth of mussels in Lakes IJsselmeer, Markermeer, and Volkerakmeer did not show any correlation with average chlorophyll *a* concentrations. Only in the Rhine River did shell growth of young mussels (initial length 6 mm) correlate positively with average chlorophyll *a* content. Shell growth in the Rhine River exceeded that in several associated lakes, in spite of lower chlorophyll *a* concentrations; the range of mean chlorophyll *a* concentrations, measured at the water surface, was 10–42 μg/L in the Rhine River in summer of 1987–1988 compared to 34–106 μg/L in Lakes IJsselmeer and Markermeer in summer 1981–1987. Differences in seasonal biomass growth were even more pronounced: a comparison of length-AFDW regression equations from the Rhine River and several associated lakes showed that mussels with similar shell lengths were heavier in the Rhine River (Smit et al., 1992).

Substrate and Dispersion

Zebra mussels need to attach to solid substrates with byssus threads, and therefore their occurrence is limited to localities with solid structures. In The Netherlands lake bottoms consist of sand, silt, clay, and a mixture of these substrates, which are unsuitable for mussels. Zebra mussels are only found on submerged vegetation, stones, and shells of dead and live mussels (i.e., Unionidae, and other zebra mussels). In former estuaries, marine shells form an important solid substrate on which zebra mussels settle. In the course of time, shells of zebra mussels become the most important solid substrate. Three negative factors that have been identified as affecting zebra mussel settlement follow.

Silt and detritus — Covering of hard substrates by silt and soft detritus occurs in most water bodies where hydrological conditions allow settlement

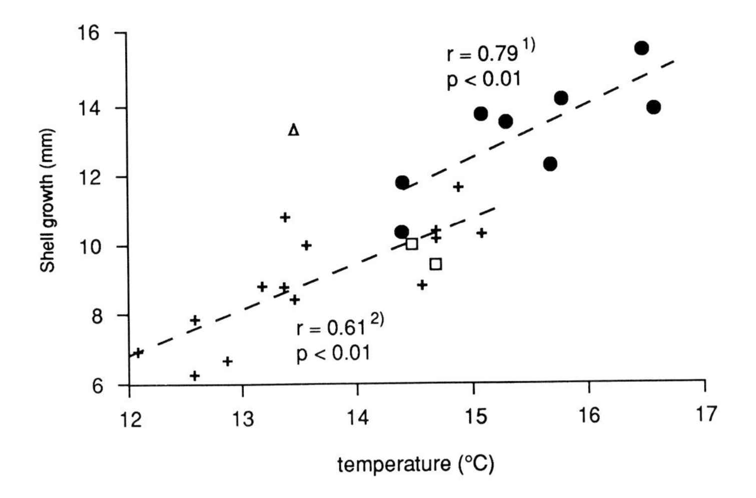

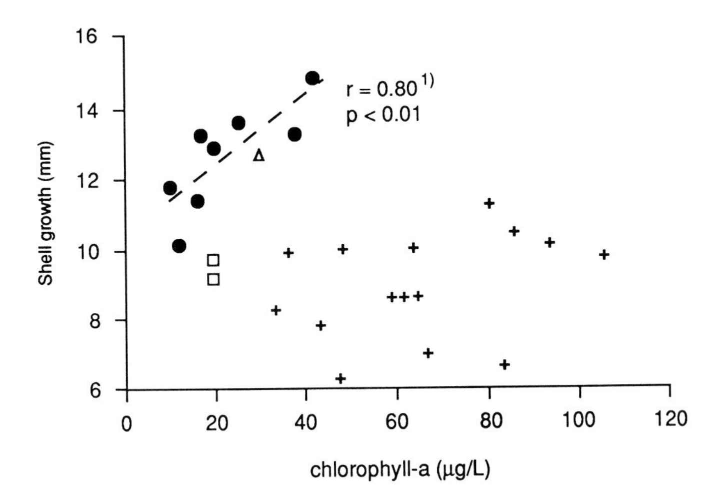

Figure 8. Seasonal shell growth (mm/year) of *Dreissena polymorpha* with initial length of 6 mm in the Lakes IJsselmeer and Markermeer (+), Lake Wolderwijd (△), Lake Volkerakmeer (□), and the Rhine River (●) in relation to average chlorophyll *a* content and average water temperature measured at the water surface. Period of averaging: March 1-September 30; 1) = Rhine River localities, 2) = all lake localities. (From Smit, H., A. bij de Vaate, and A. Fioole. *Arch. Hydrobiol.* 124:257-280. With permission.)

of silt. For example, in Lake IJsselmeer, erosion and sedimentation processes strongly influence the amount of solid substrates on the bottom. Wind-induced resuspension and erosion processes have covered some settled populations of zebra mussels, but also have uncovered previously buried substrates for further colonization. Consequently, mussel density patterns vary over time. This also occurs in the closed-off Rhine-Meuse estuary (Haringvliet and Hollandsch Diep), where the river discharge induces similar sediment dynamics.

Competition with sessile macroalgae and other macrozoobenthos species — Observations show that zebra mussels were absent from parts of stones in the Rhine and Meuse Rivers which were covered with macroalgae. Interspecific competition with other zoobenthos species has become manifest after the recent invasion of *Corophium curvispinum* (Amphipoda:Corophiidae) in the Rhine River (Van den Brink et al., 1991). This Amphipod prevents settlement of young zebra mussels by forming dense colonies on stones of breakwaters and river banks, and its silt tubes make the stones less suitable for zebra mussel settlement. A decline in the zebra mussel population was recently observed (bij de Vaate, 1991) and was also reported by fishermen (Van den Brink et al., 1991). *Cordylophora caspia* (Hydrozoa:Clavidae) was also observed to compete for space with zebra mussels on stone substrates in the River Rhine and Lake Markermeer (bij de Vaate, 1991).

Water level fluctuations — Densities of zebra mussels on stones in the Rhine River generally increase with water depth (bij de Vaate, 1991). Water levels in this river may fluctuate several meters, and the normal period of low discharge (i.e., August-October) partly overlaps with the period of zebra mussel reproduction. Consequently, the probability of any stone being colonized will depend on its position relative to the lowest water level of that year. Further, mussels settled above the lowest water level during higher discharges will probably die, once water levels fall to minimum levels.

Ecosystem Interactions

In terms of biomass, the zebra mussel is the most abundant macro-invertebrate species for predator species in the Rhine River including Haringvliet and Hollandsch Diep, and Lakes IJsselmeer and Markermeer. In these waters, large numbers of mussel-consuming ducks are mainly present from October to April. Annual zebra mussel production in Lake IJsselmeer was estimated to be $41–102 \times 10^{12}$ J, of which $15–26 \times 10^{12}$ J was consumed by diving ducks. Important zebra mussel-consuming ducks are the tufted duck *(Aythya fuligula)*, pochard *(Aythya ferina)*, scaup *(Aythya marila)*, coot *(Fulica atra)*, and goldeneye *(Bucephala clangula)*. In Lake IJsselmeer, $1.3–2.1 \times 10^{12}$ J of zebra mussels were consumed by fish (bij de Vaate, 1982; Van Eerden and Zijlstra, 1986). Among the fish species that prey on zebra mussels, roach *(Rutilus rutilus)* and eel *(Anguilla anguilla)* are important (De Nie, 1982). In Lake IJsselmeer, roach accounted for nearly all the mussels consumed by fish (Van Eerden and Zijlstra, 1986). In the Rhine River, however, predation by waterfowl is probably less important as a density-regulating factor because mussels attach to stones in single layers which seem to make them unattractive prey for diving ducks (bij de Vaate et al, 1992). Leeches *(Glossiphonia complanata)*, which mainly prey on young mussels, and crayfish *(Orconectes limosus)* are important invertebrate predators of the zebra mussel (personal observations). Piesik (1974) also reported heavy predation of small zebra mussels (shell length <12 mm) by crayfish on hanging nets in Poland. Parasites may also have a considerable influence on the survivorship of mussels. Histopathological analysis showed that more than 75% of the young-of-the-year mussels in Hollandsch Diep were infested by a protozoan-like unidentified parasite; in Lake Volkerakmeer, 26% of young-of-the-year mussels showed extensive and putatively lethal infestations of an ascetosporidian protozoan parasite (Bowmer and van der Meer, 1991).

Higher densities and biomass of the following oligochaete species were found in areas where zebra mussels occurred in Hollandsch Diep: *Limnodrilus claparedeanus, Limnodrilus hoffmeisteri, Quistadrilus multisetosus, Potamothrix moldaviensis, Psammoryctides barbatus,* and *Tubifex tubifex* (Dudok van Heel, 1992). Spaces between shells seem excellent sites where feces and pseudofeces accumulate, and high numbers of tubificids in these areas indicate that this benthic group may play an important role in the processing of feces and pseudofeces.

Application in Water Quality Management — The Lake Volkerakmeer Case

After being dammed from the Eastern Scheldt estuary in 1987, a freshwater ecosystem developed within Lake Volkerakmeer which is characterized by relatively high transparency (i.e., Secchi depth 2–4 m) despite high nutrient loads (±6 g P/m^2/year) (Van Nes, 1991a). This characteristic has resulted in large numbers of macrophytes and waterfowl. In order to fulfill agricultural needs for freshwater, the lake is flushed with water from Hollandsch Diep. However, this water contains toxic pollutants and high levels of nutrients. Since nature conservation and development is a first-priority management objective for this lake, several field experiments with zebra mussels were initiated to help achieve management objectives. The first experiment involved adding suitable substrate throughout the lake to increase zebra mussel densities and, thereby, lower the number of phytoplankton and increase water transparency. The second experiment involved placing nets at the inlet of the lake to serve as a substrate for zebra mussels and thus achieve a "biological filter." Mussels concentrate polluted suspended matter by transforming the very small suspended particles into fecal pellets with much higher sinking rates. With this in mind, studies were conducted to determine ways to keep large numbers of mussels on the nets and to determine what reduction of micropollutant loads could be realized.

Water Transparency

As noted earlier, the amount of solid substrate probably limits the population level of zebra mussels in most lakes in The Netherlands. Field

Table 3. Design of a Marine Shell Stocking Experiment in Lake Volkerakmeer to Increase the Amount of Solid Substrate

Number of Experimental Sites	Shell Type	Density (m^3/ha)
4	*Mytilus edulis*	30
4	*Cerastoderma edule/Macoma balthica*	30
3	*Mytilus edulis*	10
3	*Cerastoderma edule/Macoma balthica*	10
6	None	0

experiments therefore focused on enlarging the amount of solid substrate using natural material. Since the area is of recent estuarine origin, empty marine bivalve shells were used. Two types of shell substrate (*Mytilus edulis* and a mixture of *Cerastoderma edule* and *Macoma balthica*) were added at two different densities (10 and 30 m^3/ha) (Table 3). Twenty experimental sites of 10 ha were randomly selected at depths between 2 and 10 m. During the first spawning period (June 18, through June 27, 1990), shells were spread over each area. Prior to spreading the shells and again after 5 months 40–50 samples were randomly taken in each area with a Van Veen grab (0.023 m^2). Zebra mussels were measured and counted to the nearest millimeter (Van Nes, 1991b). Differences in density before and after shell shocking were examined. A bootstrap estimate (Efron and Tibshirany, 1986) of the standard error of differences was used since this method is more efficient than a normal approximation when distributions are highly aggregated, as was the case here.

Average densities of zebra mussels increased in almost all areas between May and November (Figure 9). Increases were generally higher at sites where shells were sown than at reference sites, but differences were not significant (*t*-test). Further examination revealed that density increases were significantly correlated with the relative number of samples per area of sediment type classified as ''sand'' (Pearsons' correlation coefficient = 0.57 one-tailed $p = 0.007$). Stepwise multiple linear regression yielded the equation:

$$I = 6.7\ FSa + 13.5\ D - 80$$

in which I = increase in individuals (number/m^2); FSa = percentage of samples of which the sediment was classified as sand ($t < 0.001$, $n = 20$); D = amount of shells added (m^3/ha) ($p < 0.001$, $n = 20$).

From this equation, graphically presented in Figure 10, it can be concluded that the mean (+S.E.) increase in density of *Dreissena* was $13.5 \pm 2.9/m^2$ ($n = 20$)/m^3 of material sown to 1 ha of sediment. No difference was found between the two types of shells.

In Lake Volkerakmeer, zebra mussels were small (mean shell length 8 mm in 1990) and consequently a density increase of a few hundred mussels per square meter did not result in a significant increase in water transparency.

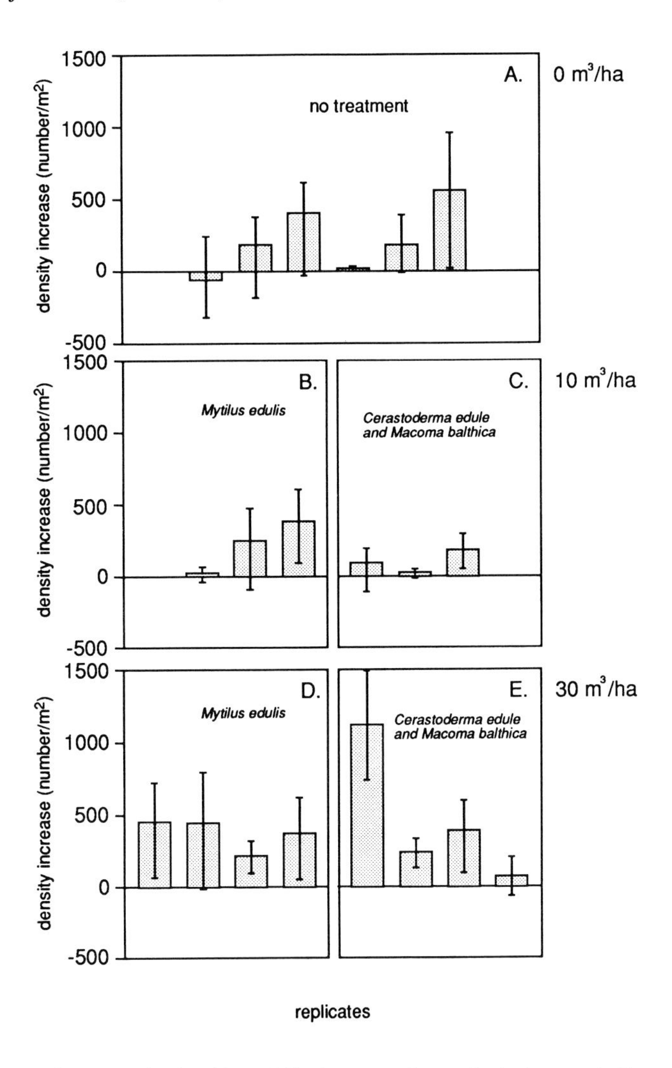

Figure 9. Differences in densities of *Dreissena polymorpha* before and after a marine shell stocking experiment in Lake Volkerakmeer (95% confidence limits are calculated with a bootstrap method).

The average density of 600 mussels per square meter in 1990 was capable of filtering the whole lake within 12 days (calculations with the formulas given by Reeders and bij de Vaate, 1990). An average density of 900 mussels per square meter would be needed to filter the lake every 4 days, assuming an eventual length frequency distribution similar to that in nearby Haringvliet (mean length 11 mm). The eventual effect of shell stocking on mussel population levels and subsequent filtering capacities will become evident within the next few years.

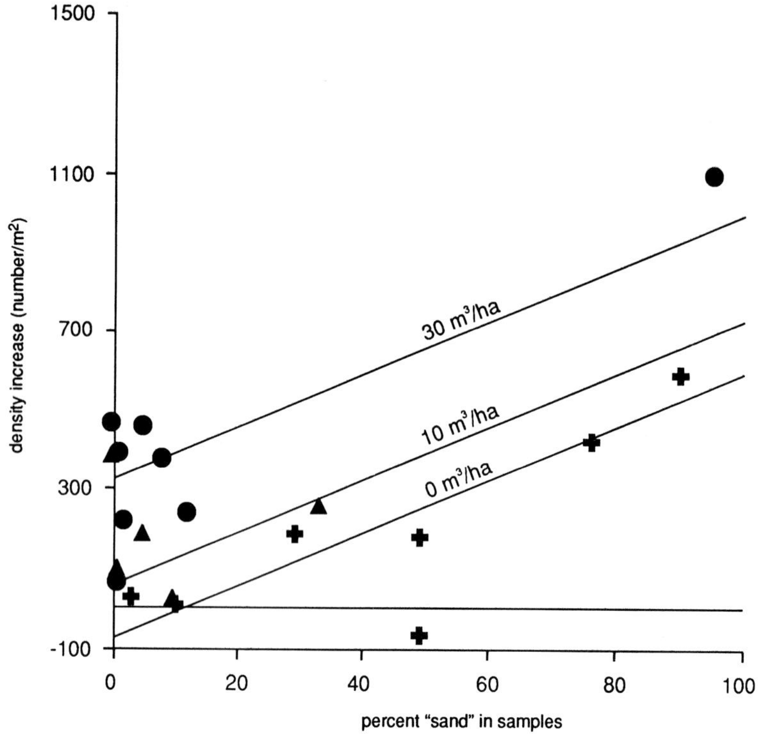

Figure 10. Relation between increase in density of *Dreissena polymorpha*, the fraction of sediment samples that were classified as "sand", and the amount of shells that were sown. See text for linear regression equation.

Biological Filter Principle

Plans are being made to hang vertical nets from the water surface and cover the entire cross-sectional area of the inlet to Lake Volkerakmeer. After being colonized by zebra mussels, polluted suspended matter will be concentrated on the bottom underneath the nets instead of entering the lake. This "biological filter" consists of 24 ha of netting and should contain about 1.2 billion mussels. Calculations based on the relationship between suspended matter content and filtration rate or pseudofeces production (Reeders and bij de Vaate, 1990) predict that this suspension of mussels may remove 49% of the suspended matter and, if densities of 10,000/m^2 can be reached, up to 69% of the suspended matter may be removed. Micropollutants will be reduced to different degrees, depending on their distribution coefficient (K_d); total loads of PCB-153, Cd, and HCB may be reduced by 45, 30, and 10%, respectively. Szlauer (1974) described the use of vertical nets as a substratum for settlement to remove larvae from water. Piesik (1983) discussed the role of colonized nets in eutrophication control. However, the use of a "biological filter" as described above will be the first large-scale application of this idea.

Colonization of Net Substrates

To test the principle of a "biological filter," several field experiments were conducted in 1989 (Reeders, 1990). Three different net substrates were hung over a mussel bank in Haringvliet, just before the spawning period. Colonization with spat was considerable; on one of the substrates (polyamide/PVC/polyester, mesh size 8 × 8 cm) densities amounted to 65,000/m^2 (mussels larger than 2 mm). However, in winter of 1989-1990, several months after colonization, most mussels disappeared from the nets within a short period. Possible causes of mussel loss were (1) wind action (a heavy storm preceded the discovery of the mussels loss), (2) predation by waterfowl and/or fish, and (3) mass migration of mussels off the nets. With regard to migration, both active and passive migration of small zebra muscles is described by Lewandowski (1982), and resettling is known to occur in marine bivalves like *Mytilus edulis* (Bayne, 1964).

In 1990, nets with a mesh size of 15 × 15 cm were again placed over the mussel bank for colonization. Eight nets 0.5 × 0.5 m were placed at depths of 1, 3, 5, and 8 m. On eight dates between October 1990 and March 1991, one sample net was taken out of each row for examination. Densities on the nets, sampled on October 1990, ranged from 29,000–190,000/m^2 (6,500–47,000/m^2 were less than 2 mm). Mean densities ($n = 5$) at 1, 3, and 5 m were 105,000, 135,000, and 71,000/m^2, respectively, while mean density at 8 m ($n = 1$) was 29,000/m^2.

After colonization, one set of nets was suspended from interconnected buoys at the surface; one was moved to calmer water; one was placed within cages of wire netting to reduce predation; one was placed horizontally on the lake bottom (depth of 5 m) to recapture migrating mussels that resettle; and one was suspended just off the bottom by submerged buoys, again to reduce the impact of wave action. Between October 1990 and March 1991, densities gradually decreased on all nets, but a complete disappearance similar to 1989 did not occur. The number of mussels remaining was a function of initial densities, with a higher proportion of initial densities remaining when initial densities were lower. The portion remaining (expressed as a percentage of initial densities) was lowest on nets that were simply left suspended from the surface over the mussel bank where they were subjected to wind and wave action (19–30%). The portion remaining was only slightly higher on nets surrounded by wire netting (30%), which may have been partly due to a lower initial density. The limited impact of predation in this experiment was probably related to the relatively small size of the mussels when compared to 1989 (average length in 1990 was 2–3 compared to 6 mm in 1989). This would have made mussels less attractive to waterfowl and larger fish (De Leeuw and Noordhuis, 1991). The highest portion of mussels remaining occurred on the nets that were moved to calmer water (28–50%), nets suspended just off the bottom (38–61%), and nets placed horizontally on the lake bottom (76%).

These preliminary results indicate that wind and wave action have a considerable impact on the rate of density decrease, either by causing direct mechanical damage or by stimulating mussels to resettle. On most nets, densities in May were still higher than the expected 5000/m^2. The results presented above therefore have been considered promising enough to proceed toward a full-scale implementation of a ''biological filter'' in the near future.

ACKNOWLEDGMENTS

The authors thank Dr. W. Admiraal of the National Institute of Public Health and Environmental Protection for kindly providing the veliger larvae data; Mr. L. Timmer, Mr. G. van Noord, and Mr. P. Block for accurately sampling and measuring numerous mussels; and Prof. G. van der Velde, Drs. H. A. Jenner, and Dr. J. A. van der Velden for their valuable comments on the manuscript.

REFERENCES

Bayne, B. L. ''Primary and Secondary Settlement in *Mytilus edulis* L. (Mollusca),'' *J. Animal Ecol.* 33:513–523 (1964).

bij de Vaate, A. Personal communication (1991).

bij de Vaate, A. ''Quantification of Predation of Zebra Mussels (*Dreissena polymorpha*) by Roach (*Rutilus rutilus*) in the Lake IJsselmeer Area,'' IJsselmeerpolders Development Authority, Lelystad, Report 1982-18 Abw (In Dutch).

bij de Vaate, A. ''Distribution and Aspects of Population Dynamics of the Zebra Mussel, *Dreissena polymorpha* (Pallas, 1771), in the Lake IJsselmeer Area (The Netherlands)'', *Oecologia (Berlin)* 86:40–50 (1991).

bij de Vaate, A. Unpublished data (1991).

bij de Vaate, A., M. Greijdanus, and H. Smit. ''Densities and Biomass of Zebra Mussels in the Dutch Part of the Lower Rhine,'' in *The Zebra Mussel, Dreissena polymorpha: Ecology, Biological Monitoring and First Applications in the Water Quality Management. Limnologie Aktuell,* 4:67-77 (1992), D. Neumann and H. A. Jenner, Eds., (Stuttgart, Germany: Fischer-Verlag).

Borcherding, J. ''The Annual Reproductive Cycle of the Freshwater Mussel *Dreissena polymorpha* Pallas in Lakes,'' *Oecologia (Berlin)* 87:208–218 (1991).

Borcherding, J., and E. D. De Ruyter van Steveninck. Abundance and growth of *Dreissena polymorpha* larvae in the River Rhine during downstream transportation,'' in D. Neumann and H. A. Jenner, Eds. *The Zebra Mussel, Dreissena polymorpha: Ecology, Biological Monitoring and First Applications in the Water Quality Management. Limnologie Aktuell,* 4:29 (1992) (Stuttgart, Germany: Fischer-Verlag).

Bowmer, C. T., and M. van der Meer. ''Reproduction and Histopathological Condition in First Year Zebra Mussels (*Dreissena polymorpha*) from the Haringvliet, Volkerakmeer and Hollandsch Diep Basins, Report R91/132. TNO Laboratory for Applied Marine Research, The Netherlands (1991).

De Leeuw, J., and R. Noorduis. ''Predation of Zebra Mussels by Waterfowl. A Review on Literature and Prognosis of the Impact on the Efficiency of the Biological Filter,'' Report 91.050. Institute for Inland Water Management and Waste Water, The Netherlands (1991) (In Dutch).

De Nie, H. W. ''A Note on the Significance of Larger Bivalve Mollusca (*Anodonta* spp. and *Dreissena* sp.) in the Food of the Eel (*Anguilla anguilla*) in Tjeukemeer,'' *Hydrobiologia* 95:307–310 (1982).

De Ruyter van Steveninck, E. D., W. Admiraal, and B. van Zanten. ''Changes in Plankton Communities in Regulated Reaches of the Lower Rhine,'' *Regul. Rivers* 5:67–71 (1990).

Dudok van Heel, H. C., H. Smit and S. M. Wiersma ''Densities, Biomass and Species Composition of Macrozoobenthos in the Lower Rhine-Meuse (The Netherlands),'' Report 91.051. Institute for Inland Water Management and Waste Water Treatment, The Netherlands (1992) (In Dutch with English summary).

Efron, B., and R. Tibshirany. ''Bootstrap Methods for Standard Errors, Confidence Intervals, and Other Measures of Statistical Accuracy,'' *Stat. Sci.* 1:54–77 (1986).

Greijdanus-Klaas, M. ''Inventory of Zebra Mussels on Solid Substrates in the River Rhine,'' Report 90.046X, Institute for Inland Water Management and Waste Water Treatment, The Netherlands (1990) (In Dutch).

Griffiths, R. W., W. P. Kovalak, and D. W. Schloesser. ''The Zebra Mussel, *Dreissena polymorpha* (Pallas, 1771), in North America: Impact on Raw Water Users,'' in Symposium: Service Water Systems Problems Affecting Safety-Related Equipment (Palo Alto, CA Nuclear Power Division, Electric Power Research Institute, 1989) 11–26.

Hebert, P. D. N., B. W. Muncaster, and G. L. Mackie. ''Ecological and Genetic Studies on *Dreissena polymorpha* (Pallas): A New Mollusc in the Great Lakes,'' *Can. J. Fish. Aquat. Sci.* 46:1587–1591 (1989).

Hueck van der Plas, E. H. ''Heavy Metals in Aquatic Systems, Overview of the Research and Conclusions,'' Report, Institute for Inland Water Management and Waste Water Treatment and Deltadienst, The Netherlands (1984).

Katchanova, A. A. ''The Ecology of *Dreissena polymorpha* in the Uchinsk reservoir,'' *Vopr. Ekol. Vyssh. Shbala, Moscow* 5:94–95 (1962) (In Russian).

Leentvaar, P. ''Geographical Distribution and Biology of *Dreissena polymorpha* Pallas,'' *Hydrobiol. Bull.* 9:120–122 (1975).

Lewandowski, K. ''The Role of Early Developmental Stages in the Dynamics of *Dreissena polymorpha* (Pall.) (Bivalvia) Populations in Lakes. II. Settling of Larvae and the Dynamics of Numbers of Settled Individuals,'' *Ekol. Pol.* 30:223–286 (1982).

Morton, B. S. "Studies on the Biology of *Dreissena polymorpha* Pall. III. Population Dynamics," *Proc. Malacol. Soc. London* 38:417–482 (1969).

Peeters, J. C. H., and W. J. Wolff. "Macrobenthos and Fishes of the Rivers Meuse and Rhine, The Netherlands," *Hydrobiol. Bull.* 7:121–126 (1973).

Piesik, Z. "The Role of the Crayfish *Orconectes limosus* (Raf.) in Extinction of *Dreissena polymorpha* (Pall.) subsisting on Steelon-Net," *Pol. Arch. Hydrobiol.* 21:401–410 (1974).

Piesik, Z. "Biology of *Dreissena polymorpha* (Pall.) Settling on Stylon Nets and the Role of this Mollusc in Eliminating the Seston and the Nutrients from the Water-Course," *Pol. Arch. Hydrobiol.* 30:353–361 (1983).

Smirnova, N. A. Personal communication (1990).

Reeders, H. H. "Mussel Power in Fresh Water, a Natural Filter to Counter Water Pollution," *Land & Water Int.* 67:16–17 (1990).

Reeders, H. H., A. bij de Vaate, and E. Slim. "The Filtration Rate of *Dreissena polymorpha* (Bivalvia) in Three Dutch Lakes with Reference to Biological Water Quality Management," *Freshwat. Biol.* 22:133–141 (1989).

Reeders, H. H., and A. bij de Vaate. "Zebra Mussels (*Dreissena polymorpha*): A New Perspective for Water Quality Management," *Hydrobiologia* 200/201:437–450 (1990).

Smit, H. Unpublished data (1991).

Smit, H., A. bij de Vaate, and A. Fioole. "Shell Growth of the Zebra Mussel (*Dreissena polymorpha* [Pallas]) in Relation to Selected Physico-chemical Parameters in the Lower Rhine and Some Associated Lakes," *Arch. Hydrobiol.* 124:257–280 (1992).

Smit, H., and H. C. Dudok van Heel. "Methodical Aspects of a Simple Allometric Biomass Determination of *Dreissena polymorpha* Aggregations," in *The Zebra Mussel, Dreissena polymorpha: Ecology, Biological Monitoring and First Applications in the Water Quality Management. Limnologie Aktuell.* 4:79–86) (1992) D. Neumann and H. A. Jenner, Eds. (Stuttgart, Germany: Fischer-Verlag).

Szlauer, L. "Use of Steelon-Net Veils for Protection of the Hydro-Engineering Works Against *Dreissena polymorpha* Pall.," *Pol. Arch. Hydrobiol.* 21:391–400 (1974).

Van Benthem Jutting, W. S. S. "*Dreissena polymorpha*," in *Fauna van Nederland, Part 12, Mollusca* (Sijthoff, Leiden, The Netherlands: 1943) (In Dutch).

Van Benthem Jutting, W. S. S. "Mollusca," in *Changes in the Flora and Fauna of the Zuiderzee (at present Lake IJsselmeer)* after the Embankment in 1932," L. F. De Beaufort, Ed. Report of the Zuiderzee Commission, The Nederlandse Dierkundige Vereniging, deBocr, Deuflelder The Netherlands (1954), pp. 233–253 (In Dutch).

Van Broekhoven, A. L. M. "The River Rhine in the Netherlands: State and Developments anno 1987," Report 87.061. Institute for Inland Water Management and Waste Water Treatment, The Netherlands (1987) (In Dutch).

Van den Brink, F. W. B., G. van der Velde, and A. bij de Vaate. "Amphipod Invasion on the Rhine," *Nature (London)* 352:576 (1991).

Van Eerden, M. R., and A. bij de Vaate. "Values of Nature in the Lake IJsselmeer Area. An Inventory of Values of Nature in the Open Water in the Lake IJsselmeer Area," Report of the IJsselmeerpolders Development Authority, Flevobericht 242, Lelystad, The Netherlands (1984) (In Dutch).

Van Eerden, M. R., and M. Zijlstra. ''Values of Nature in the Lake IJsselmeer Area: Forecast for Some Values of Nature in the IJsselmeer Area in Case of the Construction of the Markerwaard polder,'' Report of the IJsselmeerpolders Development Authority, Flevobericht 273, Lelystad, The Netherlands (1986) (In Dutch).

Van Nes, E. H. ''Lake Volkerak-Zoommeer Fresh and Clear,'' Report 91.027. Institute for Inland Water Management and Waste Water Treatment, The Netherlands (1991a) (In Dutch).

Van Nes, E. H. ''Stimulating Zebra Mussels by Stocking Shells in Lake Volkerak-Zoommeer in 1990,'' Report 91.095X. Institute for Inland Water Management and Waste Water Treatment, The Netherlands (1991b) (In Dutch).

Van Urk, G. ''The Zebra Mussel, *Dreissena polymorpha* in the River Rhine,'' *H_2O* 9:327–329 (1976) (In Dutch).

Van Urk, G., and A. bij de Vaate. ''Ecological Studies in the Lower Rhine in The Netherlands,'' in *Biologie des Rheins, Limnologie Aktuell,* R. Kinzelbach and G. Friedrich, Eds. (Stuttgart, Germany: Fischer-Verlag, 1990), pp. 131–145.

Walz, N. ''The Energy Balance of the Freshwater Mussel *Dreissena polymorpha* Pallas in Laboratory Experiments and in Lake Constance. IV. Growth in Lake Constance,'' *Arch. Hydrobiol., Suppl.* 55:142–156 (1978).

Wolff, W. J. ''The Mollusca of the Estuarine Region of the Rivers Rhine, Meuse and Scheldt in Relation to the Hydrography of the Area. II. The Dreissenidae,'' *Basteria* 33:93–103 (1969).

CHAPTER 4

Growth and Population Structure of the Zebra Mussel *(Dreissena polymorpha)* in Dutch Lakes Differing in Trophic State

Jaap Dorgelo

Young zebra mussels, *Dreissena polymorpha*, were placed in baskets in two eutrophic lakes and in a mesooligotrophic lake. Growth rates in 1985 were higher under eutrophic conditions (mean increase in shell length of 0.54–0.59 mm/week) than under mesooligotrophic conditions (0.35 mm/week). Mean increase in shell length of young *Dreissena* in the laboratory, when fed cultured *Chlamydomonas* in a lake water suspension and conducted with a flow rate of 200 mL/hr through a growth chamber, was only 0.13 mm/week at 17°C. Growth rates in situ in 1986 were approximately similar to the values obtained in the laboratory *Chlamydomonas* suspensions, regardless of the trophic state of the lakes. Yearly variation in the composition of the dynamic algal communities is probably the key factor in determining growth rate. Literature data suggest that (1) growth rates of zebra mussels are higher in fast flowing lake water pumped into the laboratory, (2) the nutritional value of algal monocultures is inferior to that of multispecies diets, and (3) turbulently flowing water in situ guarantees a better food supply resulting in higher growth rates. The population structure of the populations of the zebra mussel in the two eutrophic lakes and the mesooligotrophic lake in different seasons and years was documented. The biggest individuals were found under eutrophic conditions. The highest densities occurred in the mesooligotrophic lake.

0-87371-696-5/93/$0.00 + $.50

INTRODUCTION

The ecophysiology (i.e., the study of how an organism is adapted to its environment) of freshwater bivalves is little understood when compared with economically important marine species. This is particularly true with regard to growth and reproduction. The analysis of growth rates under natural conditions is a necessary first step in understanding the population dynamics and energy balance of a given bivalve population.

The zebra mussel, *Dreissena polymorpha*, is highly successful in many types of aquatic ecosystems and can have a major influence on them (e.g., Dorgelo and Smeenk, 1988). In this paper, growth rates of young *Dreissena* are examined in cages placed in selected lakes of different trophic state and also in the laboratory under varying food conditions. Further, the natural size composition of the population of *Dreissena* in these lakes was analyzed. Problems associated with obtaining growth in the laboratory and differences in growth and size composition in situ are discussed, with emphasis on food parameters.

METHODS

Laboratory experiments to determine growth rates of *Dreissena* consisted of three different trials. All mussels used in these laboratory trials were 5–6 mm in length and were collected from the nearshore zone (0.5 m depth) of Lake Maarsseveen I. In the first trial, water from the mesooligotrophic Lake Maarsseveen I, the eutrophic Lake Vechten, and the very eutrophic Lake Maarsseveen II was placed in separate 30-mL vials. One mussel was placed in each vial. The vials were kept in a water bath at 18 ± 0.5°C, and the water containing zebra mussels was changed three times a week. The experiments lasted 6 weeks (May–June). Comparative physical and chemical features of these three lakes are given in Table 1.

In the second trial, water from the lakes mentioned above was siphoned from aerated storage tanks by means of catheter tubes (1-mm diameter) into each of three 10-L aquariums at a flow rate of 200 mL/hr. This flow rate was chosen based on reported filtration rates of 5–200 mL/hr for mussels over 20 mm in length (Hinz and Scheil, 1972; Walz, 1978a; Benedens and Hinz, 1980; Morton, 1971). Each aquarium contained 20 mussels. Experiments were conducted at room temperatures (20°C) and lasted 5 months (May–October).

In the third trial, suspensions of *Chlamydomonas* were mixed with unfiltered water from Lake Maarsseveen I in 25-L holding tanks and then pumped into special growth chambers (Dorgelo and Smeenk, 1988) at a rate of 200 mL/hr. Water temperature was a constant 17°C. Algal densities were 23,100, 46,250, 69,400, and 92,500 cells per milliliter. Preliminary experiments

Table 1. Physical and Trophic Properties of the Lakes Involved in This Study

Lake	Trophic Status	Mean/Max. Depth (m)	Surface Area (ha)[b]	P-PO_4 (mg/L)	N-NO_3 (mg/L)	Maximum Primary Production (g C/m^2/d)
Maarsseveen I	Oligomesotrophic	12.1/30[a]	70	<0.01[a]	0.5[a]	0.5[b]
Vechten	Eutrophic	6/12[b]	5	0–0.03[b]	0–0.26[b]	1.3[b]
Maarsseveen II	Very eutrophic	16.2/25[a]	20	1.0[a]	9.0[a]	8.1[b]
IJssel	Eutrophic	4.5/8–9[c]	123,000	0.01–0.10[d]	0.1–0.2[d]	10–15[d]

[a] Swain et al. (1987).
[b] Dorgelo and Gorter (1984).
[c] Berger (1987).
[d] Anonymous (1990).

indicated that zebra mussels readily fed on this algal species; cell material was found in the stomachs of the mussels after 2 hr of filtering. It was assumed that *Dreissena* is capable of crushing *Chlamydomonas* cells since this species belongs to the order Lamellibranchiata; species in this order have a small chitin plate (gastric shield) in their stomachs for this purpose (Barnes, 1980). All experiments with *Chlamydomonas* were conducted in the dark to prevent the growth of other algal species on the walls of the containers. Each growth chamber contained 20 mussels and experiments lasted 5 months (December–April).

In situ growth rates were studied by placing 30 mussels into stainless steel baskets (85-mm diameter, 65-mm height, 2-mm mesh width) that were attached to the underside of PVC plates (20 × 100 cm). The plates were suspended at a depth of 3 m in the epilimnion of each of the three lakes. Three sets of experiments were conducted. In the first set, mussels from Lake Maarsseveen I were placed in baskets in each of the three lakes in June 1985. Individuals in the baskets were measured in November 1985 and again in June 1986 when the baskets were removed. In the second set, individuals from Lake Maarsseveen I and from eutrophic Lake IJssel (sampling location: northern foot of the dike at Lelystad) were placed in baskets in Lake Maarsseveen I and II in June 1986, measured in December 1986, and removed in June 1987. In the third set, mussels from Lake Maarsseveen I were placed in baskets in this lake in December 1988 and lengths were measured about every 2 months until October 1989. In all experiments, dead mussels were removed from the baskets on the dates when lengths were measured and these individuals were not included in the data analysis.

The size composition of the zebra mussel population in Lakes Maarsseveen I, Maarsseveen II, and Vechten was examined by random sampling individuals attached to styrofoam substrates floating for years in the center of each of the three lakes. The size composition of the populations was estimated by measuring the shell length of at least 150 individuals collected from each lake on each sampling date. In 1985, individuals were collected from nearshore habitats in each of these lakes to compare the size composition of the populations on the styrofoam to those on a more natural substrate.

RESULTS

Growth Under Laboratory Conditions

In the first two trials to determine growth in the laboratory (i.e., in the 10-mL vials and in 10-L aquariums), no growth was observed. After the 6-week and 5-month experimental periods, respectively, there was no apparent increase in the shell length of individuals. However, no mortality occurred over this period, indicating that the food supply was only adequate for maintenance requirements.

In the third trial, shell growth did occur. The lowest mortality was obtained at a *Chlamydomonas* density of 23,100 cells per milliliter (Dorgelo and Smeenk, 1988). The mean increase was 0.13 mm/week over the 5-month experimental period, but variation between individuals was high. Unfortunately, mortality increased as food concentrations increased. At algal densities of 92,500 cells per milliliter, 100% mortality occurred after just 2 weeks; and at densities of 69,400 and 46,250 cells per milliliter, mortality increased after 6 weeks. For these latter two densities, during the first 6 weeks of the experiment when no mortality was apparent, growth rates were similar to rates observed at 23,100 cells per milliliter.

Growth In Situ

In situ growth rates in 1985–1986 over periods of 20 and 52 weeks and expressed as shell length, were significantly higher ($p < 0.01$) in the eutrophic lakes (Maarsseveen II and Vechten) than in the mesooligotrophic lake (Maarsseveen I) (Table 2). For the latter lake, shell growth was not apparent in the fall of 1985, possibly as a result of a decline in food supply at this time.

In 1986–1987, again after 20 and 52 weeks, the mean increase in length of young mussels from Lake Maarsseveen I placed in Lake Maarsseveen II did not significantly differ ($p > 0.05$) from the mean increase in length of mussels in Lake Maarsseveen I (Table 3). This similar overall growth occurred despite the fact that growth did not occur in the fall of 1986 in Lake Maarsseveen I. Overall growth was similar after 26 weeks for mussels from Lake IJssel placed in the two lakes. However, after 52 weeks, the mean length of mussels in Lake Maarsseveen II was significantly ($p < 0.05$) greater than those in Lake Maarsseveen I.

Growth rates of zebra mussels from Lake Maarsseveen I placed in both Lakes Maarsseveen I and II during summer 1986 were significantly ($p < 0.01$) lower than growth rates in summer 1985. The growth rates in 1986 were similar to those obtained in the laboratory when mussels were fed 23,100 cells per milliliter of *Chlamydomonas*.

Seasonal variations in the growth rate of zebra mussels in Lake Maarsseveen I were examined between December 1988 and October 1989 (Table 4). No growth was observed from December through February. Mean increases in shell length were significant ($p < 0.01$) between February and April, April and June, June and September, and September and October. Highest growth rates of 0.38 mm/week occurred from February through June. During summer (from June through October) growth rates of 0.14 mm/week were similar to rates found in the summer of 1986.

Size Composition of *Dreissena*

Lakes Maarsseveen I, Maarsseveen II, and Vechten were sampled in May-June 1983, and in February-March 1985. Highest densities were encountered

Table 2. Mean Shell Lengths (± S.E.) of *Dreissena polymorpha* Collected from Lake Maarsseveen I and Suspended in Baskets at 3-m Depth in Three Different Lakes

	Lake Maarsseveen I			Lake Vechten			Lake Maarsseveen II		
Date	Length (mm)	Growth (mm/wk)	n	Length (mm)	Growth (mm/wk)	n	Length (mm)	Growth (mm/wk)	n
June 1985	5.2 ± 0.02		90	6.3 ± 0.03		60	5.8 ± 0.01		60
November 1985	12.1 ± 0.2	0.35	77	18.1 ± 0.2	0.59	53	16.5 ± 0.3	0.54	57
June 1986	15.8 ± 0.2	0.20	74	21.9 ± 0.4	0.30	25	—		—

Note: June–November 1985 = 20 wk; June 1985–June 1986 = 52 wk. Mean temperatures in Lakes Maarsseveen I and II in June 1985 were 16.2 ± 2.6(S.D.) and 16.0 ± 2.5°C, respectively, and the mean temperature for the June 1985–June 1986 period was 11.4 ± 5.6 and 11.2 ± 5.5°C, respectively. No temperatures were obtained in Lake Vechten.

Table 3. Mean Shell Length (± S.E.) of *Dreissena polymorpha* Collected from Lake Maarsseveen I and Suspended in Baskets at 3-m depth in Lake Maarsseveen I (MI/MI) and in Lake Maarsseveen II (MI/MII), and also of Mussels Collected from Lake IJssel and Suspended in Baskets in Lake Maarsseveen I (IJ/MI) and in Lake Maarsseveen II (IJ/MII)

	MI/MI			MI/MII			IJ/MI			IJ/MII		
Date	Length (mm)	Growth (mm/wk)	n	Length (mm)	Growth (mm/wk)	n	Length (mm)	Growth (mm/wk)	n	Length (mm)	Growth (mm/wk)	n
June 1986	7.6 ± 0.1		30	7.6 ± 0.1		30	7.4 ± 0.1		30	7.4 ± 0.1		30
December 1986	11.3 ± 0.5	0.14	29	10.7 ± 0.3	0.12	27	13.1 ± 0.2	0.22	28	12.4 ± 0.3	0.19	25
June 1987	14.9 ± 0.2	0.14	27	15.3 ± 0.3	0.15	24	15.0 ± 0.3	0.15	21	15.9 ± 0.4	0.16	17

Note: June–December 1986 = 20 wk; June 1986–June 1987 = 52 wk. Mean temperatures in Lakes Maarsseveen I and II for June–December 1986 were 15.0 ± 5.2(S.D.) and 14.4 ± 5.0°C, respectively; for June 1986–June 1987 they were 10.9 ± 6.5 and 10.4 ± 6.1°C, respectively.

Table 4. Shell Lengths (± S.E.) of *Dreissena polymorpha* in Lake Maarsseveen I During December 1988–October 1989; Mean Temperature (± S. D.) at 3-m Depth Between December and February, February and June, June and October

Sampling Date	Mean Shell Length (mm)	Numbers Sampled	Temperature (°C)
December 1988	5.6 ± 0.04	30	
			5.6 ± 1.1
February 1989	5.8 ± 0.06	28	
April 1989	7.8 ± 0.14	27	11.1 ± 5.5
June 1989	11.5 ± 0.17	27	
September 1989	13.5 ± 0.28	26	18.1 ± 2.0
October 1989	14.2 ± 0.28	26	

in Lake Maarsseveen I. In 1983, the population in Lake Maarsseveen I was dominated by small mussels (77% of the individuals were smaller than 9 mm), although a few shells of 30-mm length were present (Figure 1). The populations in Lake Vechten and Lake Maarsseveen II were dominated by larger animals: in Lake Vechten, 98% of the animals were larger than 10 mm (13% were 20 mm); and in Lake Maarsseveen II, 21% of the animals had a length of 20 mm. Repeated sampling in Lake Maarsseveen II (n = 668) and in Lake Vechten (n = 265) after May-June 1983 gave approximately the same relative abundances as in May-June. However, there was a shift to larger individuals, demonstrating the reliability of earlier estimates of the size composition.

In 1985, the mussel population on the styrofoam in Lake Maarsseveen I was characterized by the dominance of large mussels; 81% were longer than 13 mm (11% had a length of 18 mm) (Figure 2). The population in the littoral consisted mainly of small animals (93% smaller than 6 mm). In Lake Vechten, the population was dominated by large animals; 92% were longer than 22 mm. In Lake Maarsseveen II, the populations from the styrofoam and the littoral zone had approximately the same size composition, representing a wide range of size classes (98 and 97% between 5 and 32 mm, respectively).

DISCUSSION

Growth Under Laboratory Conditions

No growth was observed in the first two laboratory trials, in which young mussels were kept in stagnant or slowly flowing lake water. In the third trial, in which growth occurred, the flow rate was similar to that used in the second trial but the algal density was much higher. Thus, insufficient food supply most probably prevented the animals from growing in the first two trials.

The increase in mortality among young mussels in the growth chambers with increasing food concentrations might be caused by particle obstruction

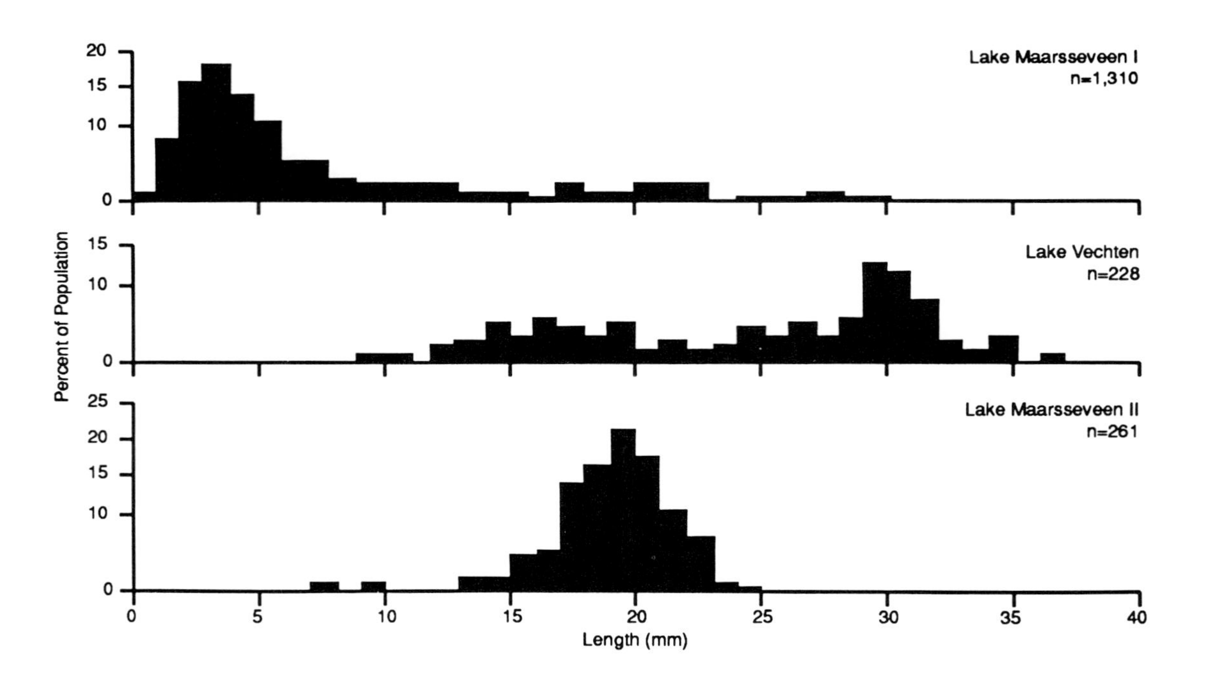

Figure 1. Relative abundance of length classes of *Dreissena polymorpha* in Lakes Maarsseveen I, Vechten, and Maarsseveen II sampled on June 28, 1983, July 5, 1983, and July 1, 1983, respectively. (From Dorgelo, J., and M. Gortner. *Hydrobiol. Bull.* 18:159–163 [1984]. With permission.)

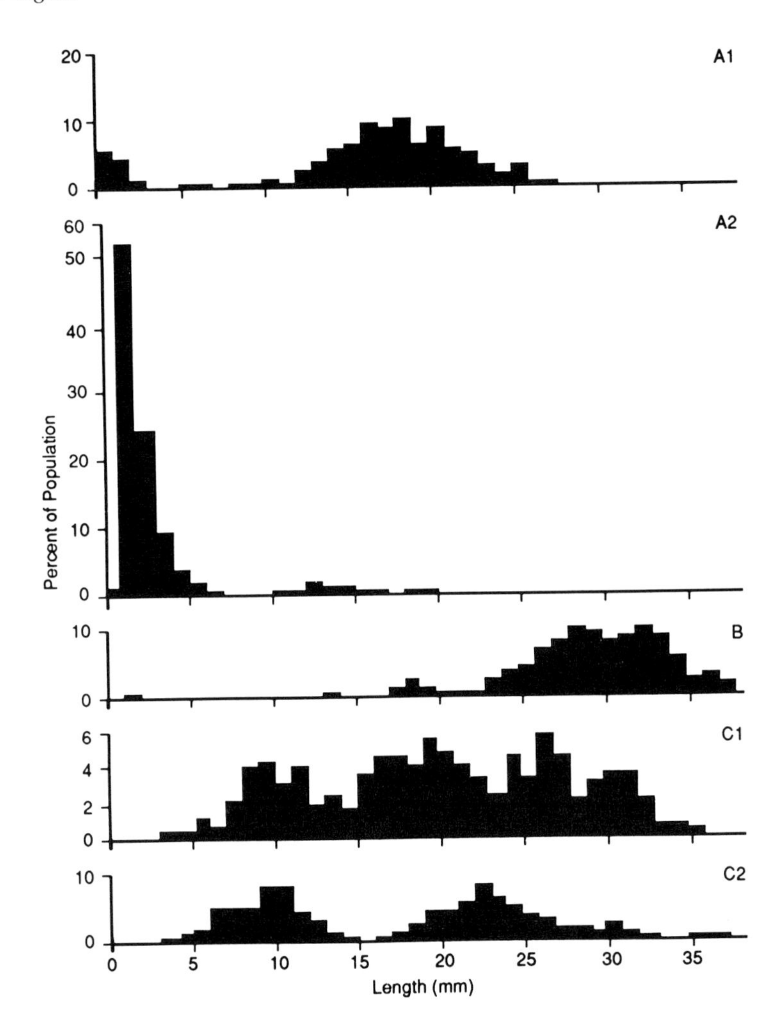

Figure 2. Relative abundance in March of length classes of *Dreissena polymorpha* in 1985 in Lake Maarsseveen I on styrofoam in the center of the lake (A1, n = 217) and in the littoral (A2, n = 954); in Lake Vechten on styrofoam in the center of the lake (B, n = 153); and in Lake Maarsseveen II on styrofoam in the center of the lake (C1, n = 317) and in the littoral (C2, n = 183).

of the filtering system of the mussel. Such obstruction would disrupt food uptake after several weeks. In short-term experiments, a constant filtration rate was measured when *Chlamydomonas* densities ranged from 9,250 to 80,950 cells per milliliter. Filtration rate dropped in a suspension with 92,500 cells per milliliter (Dorgelo and Smeenk, 1988). Maximum growth rate was observed at cell densities of 23,100 cells per milliliter. Similarly, oxygen consumption increased from 9 μg O_2 per gram dry soft body weight per hour at 23,100 cells per milliliter to 15 μg O_2 per gram dry weight per hour at 92,000 cells per milliliter (Unpublished data). Even though high algal densities seem to mechanically interfere with the filtering system of *Dreissena*, other bivalves appear to do well at much higher densities of algal cells. Larvae of *Venus mercenaria*, when fed *Monochrysis lutheri*, grew optimally in concentrations of 250,000 cells per milliliter (Davis and Guillard, 1958).

Tomson (1983) failed to elicit growth in young zebra mussels from Lake IJssel when they were fed blue-green *Oscillatoria agardhii*, bacteria, or a mixture of both in a flow-through system. The results were similar when zebra mussels were kept in the laboratory in Lake IJssel water. Thus, to get growth in young *Dreissena* under strictly controlled laboratory conditions is a difficult task. In the literature to date, only seminatural conditions (fast flowing systems connected with the lake) guaranteed reasonable growth rates (Walz, 1978b). In contrast, successful cultures of marine bivalves have been reported by several authors (e.g., De Pauw and De Leenheer, 1985; Bricelj et al., 1984).

Growth In Situ

Tomson (1983) reported a growth rate of approximately 0.50 mm/week in 4-mm zebra mussels, placed in cages in hypertrophic Lake Wolderwijd at a depth of 1 m. In Lake IJssel, 0.9-mm postveligers (settled on polystyrene disks at 5-m depth) displayed a maximum growth rate of 0.15 mm/week during August-December 1982, with highest growth rates in September-October (Van Diepen, 1983). These data are consistent with the values observed under laboratory conditions as well as in situ (in Lakes Maarsseveen I and II) in 1986 in the present study.

Mollusk growth increases with increasing eutrophication (Eisenberg, 1970; Dorgelo, 1988; Dorgelo, 1991). For instance, Mothes (1985) reported that growth of young zebra mussels was slower in oligotrophic Lake Stechlin than in mesotrophic Lake Nehmitz. bij de Vaate (1991) put different size classes of zebra mussels in cages in lakes of different trophic level, including Lake Maarsseveen I. Lengths ranged from 3 to over 20 mm. He presented length values as a function of age (though being critical about age determination) instead of growth rates of a size class, and concluded that there was no significant difference between the lakes in 1981. As shown, yearly variation in *Dreissena* growth, independent of the trophic state of a lake, can occur.

Yearly variation in growth in a lake is related to yearly variations in food quality and quantity. Many studies have dealt with food parameters controlling the growth of filter-feeding bivalves. As reported by Walz (1978c, 1979), particle concentration must be considered along with intake, assimilation, and biochemical composition of the food. Walne (1970) and Epifanio (1979), however, reported findings which cast doubt upon the role of chemical composition. Morton (1971) discussed adverse effects of exudated algal metabolites on the rate of filtration. Seasonal dynamics and yearly variation of phytoplankton, bacterial flora, and detrital particles in the water column, as well as the many factors that influence filtration rate (Morton, 1983), contribute to the problem of defining the optimally assimilated food type in *Dreissena*. The presence or absence of a cell wall in a particular food item may also influence assimilation by bivalves. For instance, Davis and Guillard (1958) found that genera with naked cells were good food for oyster larvae while Walne (1970) found a low index of food value in *Chlamydomonas coccoides* fed to young *Mercenaria mercenaria*. This algal species has a distinct cell wall. Davids and Gorter (unpublished data) fed young zebra mussels a related alga, *Cryptomonas ovata*, in a flow-through system. No growth and no mortality were observed over a 5-month period. This species also has a distinctive cell wall. However, zebra mussels prefer *Cryptomonas ovata* when offered in combination with *Asterionella formosa* after being prefed with *Chlorella* (to prevent starvation effects), as was evident from stomach analysis (Ten Winkel and Davids, 1982).

The deficiency of a diet based on a single food item as compared to a variety of food items has been illustrated by Davis and Guillard (1958) and discussed by Walz (1978c). The importance of a natural diet with a mixture of food items was emphasized by several authors (Epifanio, 1979; Williams, 1981; Strömgren and Cary, 1984) and is supported by the in situ (mixed diet) vs laboratory (single algal species) findings presented in this study.

Besides food quality, food quantity is also important for *Dreissena* growth. According to Walz (1978b) and Thompson and Nichols (1988), growth in *Dreissena* was related to the quantitative development of phytoplankton in Lake Constance. Improved food supply due to water movements leads to higher densities of estuarine filter feeding macrobenthos (Wildish and Kristmanson, 1979), and food depletion above *Mytilus* beds is readily realized (Frechette and Bourget, 1985). *Dreissena* was found to occur in higher numbers at places with stronger currents (Mothes, 1964) and on sides of sand banks (Wiktor, 1963; Van Soest, 1970). bij de Vaate (1991), referring to the positive effect of currents, observed that *Dreissena* grew better on a buoy than on a cage standing on the bottom. In our laboratory experiments, *Dreissena* clustered at both ends of the growth chambers where the water flow was turbulent, and filtration rates in *Chlamydomonas* suspensions increased at higher flow rates (Dorgelo and Smeenk, 1988). Parada et al. (1989) reported a lower growth rate in a lake population of the freshwater mussel *Diplodon chilensis* than in a stream population. Walne (1972) demonstrated that in marine bivalve species, the filtration rate and growth of juveniles were positively correlated with flow rate.

Size selection of food makes relationships between growth and feeding in *Dreissena* even more complex. Morton (1971) concluded that *Dreissena* preferred to feed on smaller particles; however, Ten Winkel and Davids (1982) mentioned positive selection of particles with lengths and diameters between

15 and 45 μm, and negative selection of particles that were smaller or larger. Mikheev and Sorokin (1966 — in Morton, 1971) showed that *Dreissena* assimilated 50% of the bacteria and 20% of the algae filtered. Tomson (1983) observed filtration of 1-μ bacteria to a small extent. Another factor to be considered is the presence or absence of suspended fine-grained sediment (silt) in the water column. Winter (1976) demonstrated that the growth of *Mytilus edulis* increased when silt was offered while Bricelj et al. (1984) observed growth reduction in *Mercenaria mercenaria* at 44 mg silt per liter. However, Foe and Knight (1985) found growth in *Corbicula fluminea* to be independent of silt concentrations; and Yonge and Campbell (1968) stated that *Dreissena* appears oblivious to the effects of silt.

Mean temperatures in Lakes Maarsseveen I and II (based on weekly measurements) were not significantly different during the in situ growth experiments at the 3-m depth (Tables 1 and 2); thus differences in growth could not be attributed to differences in ambient water temperature.

Size Composition in the Lakes

As found for growth rates, yearly variation in the size composition of mussel populations was considerable. In The Netherlands, spawning of *Dreissena* occurs between late spring and midautumn (Antheunisse, 1963; Van Gool, 1982). When sampling the work platform in Lake Maarsseveen I in 1983 during the spawning period, only young postveligers were found. In 1985 before spawning, very few young postveligers were found underneath this platform, but the sample from stones in the littoral of Lake Maarsseveen I was dominated by this size class. The two sampling stations in Lake Maarsseveen II showed approximately the same size composition. Monthly sampling of the platform in Lake Maarsseveen I during 1983 (Davids and Gortner, Unpublished data) showed that young *Dreissena* predominated until July/August with a very gradual increase of larger individuals, ending in a size composition resembling that in March 1985. Van Diepen (1983) sampled the platform in Lake Maarsseveen I in December 1982 and found a peak in individuals 2–3 mm in length and a smaller peak in individuals around 20 mm in length. Thus, small individuals dominated in Lake Maarsseveen I as compared to the other two lakes.

A positive relationship between trophic state and benthic biomass has been reported by a number of authors (James, 1987; Stanczykowska, 1977). We found the largest mussels in eutrophic lakes, but densities highest in the mesooligotrophic lake, in particular on stones in the littoral. Reduced numbers in the two eutrophic lakes may be caused by lower reproduction and/or higher mortality of postveligers, with the latter being caused either by predators such as roach and diving ducks (Doef, 1982) or by solid substrate limitation. In Lake Maarsseveen II, *Dreissena* was only sporadically found in 1986 and 1987, but common in 1988. The size composition of samples taken in November of 1988 and 1989 (Figure 3) from solid substrates shows a shift to a

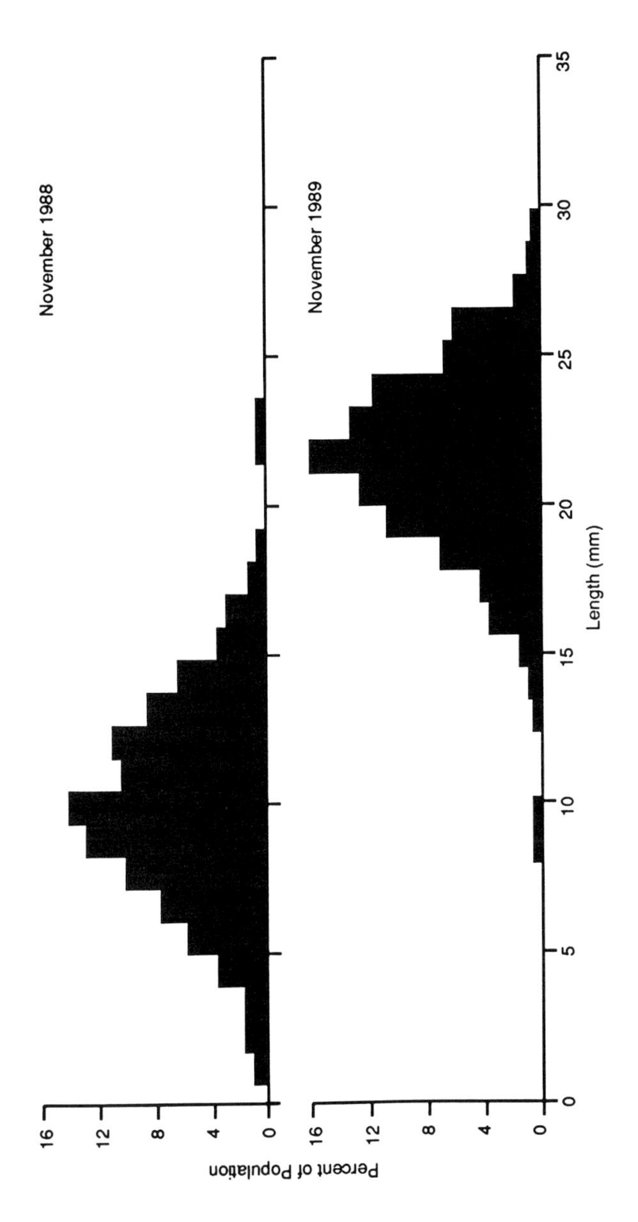

Figure 3. Relative abundance of length classes of *Dreissena polymorpha* in Lake Maarsseveen II in November 1988 and 1989.

larger size class (i.e., growth and no successful reproduction). In Lake Vechten, *Dreissena* was not found in 1986 and was not reencountered afterward.

In conclusion, the population dynamics of *Dreissena* can be very dramatic. More knowledge on food selection and assimilation as part of the energy budget of different life stages under different trophic conditions is needed in order to accurately assess reasons for population fluctuations. These fluctuations cause yearly variation in the impact of *Dreissena* on trophic interactions within an aquatic community. Insight into the underlying mechanisms may be useful for density control in lake management.

ACKNOWLEDGMENTS

The assistance of J.-W. Smeenk, E. Foekema, Ch. de Groot, M. Koper, D. Lindenaar, and M. Rozemeijer is gratefully acknowledged. Dr. C. Davids and M. Gorter kindly provided unpublished data; and A. bij de Vaate, the first cages.

REFERENCES

Anonymous "Algenbloei in het IJsselmeer," Rapport Vakgroep Microbiologie, University of Amsterdam, aan DBW/RIZA (1990).

Antheunisse, L. J. "Neurosecretory Phenomena in the Zebra Mussel *Dreissena polymorpha* Pallas," *Arch. Néerl. Zool.* 15:237–314 (1963).

Barnes, R. D. *Invertebrate Zoology,* 4th ed. (Philadelphia, PA: Saunders College, 1980).

Benedens, H.-G., and W. Hinz. "Zur Tagesperiodizitat der Filtrationsleistung von *Dreissena polymorpha* und *Sphaerium corneum* (Bivalvia)," *Hydrobiologia* 69:45–48 (1980).

Berger, C. "Habitat en ekologie van *Oscillatoria agardhii* Gomont," Ph.D. Thesis, University of Groningen (1987).

bij de Vaate, A. "Distribution and Aspects of Population Dynamics of the Zebra Mussel, *Dreissena polymorpha* (Pallas, 1771), in the Lake IJsselmeer area (The Netherlands)," *Oecologia, (Berlin)* 86:40–50 (1991).

Bricelj, V. M., R. E. Malouf, and C. De Quillfeldt. "Growth of Juvenile *Mercenaria mercenaria* and the Effect of Resuspended Bottom Sediments," *Mar. Biol.* 84:167–173 (1984).

Davis, H. C. and R. R. Guillard. "Relative Value of Ten Genera of Microorganisms as Food for Oyster and Clam Larvae," *Fish. Bull. U.S. Fish Wildl. Serv.* 58:293–304 (1958).

De Pauw, N., and L. De Leenheer. "Outdoor Mass Production of Marine Microalgae for Nursery Culturing of Bivalve Molluscs," *Arch. Hydrobiol. Beih. Ergebr. Limnol.* 20:139–145 (1985).

Doef, R. "Onderzoek aan driehoeksmossels (*Dreissena polymorpha* Pallas) in het kader van de draagkracht van het IJsselmeergebied voor duikeenden," Werkdocument 1982-259 Abw RIJP, Lelystad, The Netherlands (1982).

Dorgelo, J. "Growth in a Freshwater Snail Under Laboratory Conditions in Relations to Eutrophication," *Hydrobiologia* 157:125–127 (1988).

Dorgelo, J. "Growth, Food and Respiration in the Prosobranch Snail *Potamopyrgus jenkinsi* (E. A. Smith) (Hydrobiidae, Mollusca)," *Verh. Int. Verein. Theor. Angew. Limnol.,* 24:2947-2953 (1991).

Dorgelo, J., and M. Gorter. "Preliminary Data on Size Composition and Settlement of *Dreissena polymorpha* (Pallas) (Mollusca:Bivalvia) in Lakes Differing in Trophic State," *Hydrobiol. Bull.* 18:159–163 (1984).

Dorgelo, J., and J.-W. Smeenk. "Contribution to the Ecophysiology of *Dreissena polymorpha* (Pallas) (Mollusca:Bivalvia): Growth, Filtration Rate and Respiration," *Verh. Int. Verein. Theor. Angew. Limnol.* 23:2202–2208 (1988).

Eisenberg, R. M. "The Role of Food in the Regulation of the Pond Snail, *Lymnaea elodes*," *Ecology* 51:680–684 (1970).

Epifanio, C. E. "Comparison of Yeast and Algal Diets for Bivalve Molluscs," *Aquaculture* 16:187–192 (1979).

Foe, C., and A. Knight. "The Effect of Phytoplankton and Suspended Sediment on the Growth of *Corbicula fluminea* (Bivalvia)," *Hydrobiologia* 127:105–115 (1985).

Fréchette, M., and E. Bourget. "Energy Flow Between the Pelagic and Benthic Zones: Factors Controlling Particulate Organic Matter Available to an Intertidal Mussel Bed," *Can. J. Fish. Aquat. Sci.* 42:1158–1165 (1985).

Hinz, W., and H.-G. Scheil. "Zur Filtrationsleistung von *Dreissena, Sphaerium* und *Pisidium* (Eulamellibranchiata)," *Oecologia (Berlin)* 11:45–54 (1972).

James, M. R. "Ecology of the Freshwater Mussel *Hyridella menziesi* (Gray) in a Small Oligotrophic Lake," *Arch. Hydrobiol.* 108:337–348 (1987).

Morton, B. S. "Studies on the Biology of *Dreissena polymorpha* Pallas. V. Some Aspects of Filter-Feeding and the Effect of Micro-Organisms upon the Rate of Filtration," *Proc. Malacol. Soc. London* 39:298–301 (1971).

Morton, S. "Feeding and Digestion in Bivalves," in *The Molluca. Vol. 5, Part 2, Physiology,* A. S. M. Saleuddin, and K. M. Wilbur, Eds. (New York: Academic Press 1983), pp. 65–147.

Mothes, G. "Die Mollusken des Stechlinsees," *Limnol. (Berlin)* 2:411–421 (1964).

Mothes, G. "The Macrozoobenthos," in *Lake Stechlin. A Temperate Oligotrophic Lake,* S. J. Casper, Ed. (The Hague, Netherlands: W. Junk Publishers, 1985) pp. 230–243.

Parada, E., S. Peredo, G. Lara, and T. Valdebenito. "Growth, Age and Life Span of the Freshwater Mussel *Diplodon chilensis chilensis* (Gray, 1928)," *Arch. Hydrobiol.* 115:563–573 (1989).

Stanczykowska, A. "Ecology of *Dreissena polymorpha* (Pall.) (Bivalvia) in lakes," *Pol. Arch. Hydrobiol.* 24:461–530 (1977).

Strömgren, T., and C. Cary. "Growth in Length of *Mytilus edulis* L. Fed on Different Algal Diets," *J. Exp. Mar. Biol. Ecol.* 76:23–34 (1984).

Swain, W. R., R. Lingeman, and F. Heinis. "A Characterization and Description of the Maarsseveen Lake System," *Hydrobiol. Bull.* 21:5–16 (1987).

Ten Winkel, E. H., and C. Davids. "Food Selection by *Dreissena polymorpha* Pallas (Mollusca: Bivalvia)," *Freshwater Biol.* 12:553–558 (1982).

Thompson, J. K., and F. H. Nichols. "Food Availability Controls Seasonal Cycle of Growth in *Macoma balthica* (L) in San Francisco Bay, California," *J. Exp. Mar. Biol. Ecol.* 116:43–61 (1988).

Tomson, A. "De groei en filtratie van *Dreissena polymorpha* in diverse voedings-suspensies," Internal Report, Department of Aquatic Ecology, University of Amsterdam (1983).

Van Diepen, J. "Onderzoek naar de groei van *Dreissena polymorpha* (Pallas) na de broedval en de toestand van het gonadenweefsel in de verschillende wateren," Internal Report, Department of Aquatic Ecology, University of Amsterdam (1983).

Van Gool, P. B. "Voortplanting en groei bij *Dreissena polymorpha* Pallas," Werk-document 1982-35 Abw RIJP, Lelystad, The Netherlands (1982).

Van Soest, R. W. M. "Onderzoek naar aspecten van de oecologie van de driehoeks-mossel *Dreissena polymorpha* (Pallas, 1771) (Lamellibranchiata) in het IJssel-meer," Internal Report, Department of Applied Hydrobiology and Systematic Zoology, University of Amsterdam (1970).

Walne, P. R. "The Seasonal Variation of Meat and Glycogen Content of Seven Populations of Oysters *Ostrea edulis* L. and a Review of the Literature," *Fish. Invest. Ser. II* 26:1–35 (1970).

Walne, P. R. "The Influence of Current Speeds, Body Size, and Water Temperature on the Filtration Rate of Five Species of Bivalves," *J. Mar. Biol. Assoc.* 52:345–374 (1972).

Walz, N. "The Energy Balance of the Freshwater Mussel *Dreissena polymorpha* Pallas in Laboratory Experiments and in Lake Constance. I. Pattern of Activity, Feeding and Assimilation Efficiency," *Arch. Hydrobiol., Suppl.* 55:83–105 (1978a).

Walz, N. "The Energy Balance of the Freshwater Mussel *Dreissena polymorpha* Pallas in Laboratory Experiments and in Lake Constance," IV. Growth in Lake Constance," *Arch. Hydrobiol., Suppl.* 55:142–156 (1978b).

Walz, N. "Growth Rates of *Dreissena polymorpha* Pallas Under Laboratory and Field Conditions," *Verh. Int. Verein. Theor. Angew. Limnol.* 20:2427–2430 (1978c).

Walz, N. "The Energy Balance of the Freshwater Mussel *Dreissena polymorpha* Pallas in Laboratory Experiments and in Lake Constance. V. Seasonal and Nutritional Changes in the Biochemical Composition," *Arch. Hydrobiol., Suppl.* 55:235–254 (1979).

Wiktor, J. "Research on the Ecology of *Dreissena polymorpha* Pall. in the Szczecin Lagoon (Zalew Szczecinski)," *Ekol. Pol., Ser. A* 11:275–280 (1963).

Wildish, D. J., and D. D. Kristmanson. "Tidal Energy and Sublittoral Macrobenthic Animals in Estuaries," *J. Fish. Res. Board Can.* 36:1197–1206 (1979).

Williams, P. "Detritus Utilization by *Mytilus edulis*," *Estuarine Coastal Shelf Sci.*, 12:739–746 (1981).

Winter, J. E. "Feeding Experiments with *Mytilus edulis* L. at Small Laboratory Scale. II. The Influence of Suspended Silt in Addition to Algal Suspensions on Growth," in *Proceedings of the 10th European Symposium on Marine Biology* G. Persoone and E. Jaspers, Eds. (Ostend, Belgium: Universa Press, 1976), pp. 583–600.

Yonge, C. M., and J. I. Campbell. "On the Heteromyarian Condition in the Bivalvia with Special Reference to *Dreissena polymorpha* and Certain Mytilacea," *Trans. R. Soc. Edinburgh* 68:21–43 (1968).

CHAPTER 5

Growth and Seasonal Reproduction of *Dreissena polymorpha* in the Rhine River and Adjacent Waters

Dietrich Neumann, Jost Borcherding, and Brigitte Jantz

Studies of both the growth of individually marked mussels and the reproductive cycle of *Dreissena polymorpha* are presented as a basis for the interpretation of age distribution and population dynamics in both rivers and lakes. During a developmental time of about 18–28 days, the veliger increased its shell length by 6.0–7.4 μm/day while being carried downstream in the Rhine River. Seasonal differences in the settlement between the Upper and Lower Rhine showed that annual recruitment of local populations was derived from different areas. The growth rate of adults in the Rhine River is described by a negative exponential equation in relation to their initial shell length. Fecundity correlated with mussel size; it ranged between 310,000 and 1,610,000 oocytes for a female of about 24-mm shell length. The annual reproductive cycle in *Dreissena* showed that the onset of the spawning season correlated with a temperature threshold of about 12°C. The consequences of such temperature threshold are discussed with regard to both the synchronization of reproduction and geographical spreading of the species.

0-87371-696-5/93/$0.00 + $.50

INTRODUCTION

The invasion of the zebra mussel (*Dreissena polymorpha* Pallas) into lakes and streams of middle Europe began during the first half of the last century. Zebra mussels first settled in the lower and middle reaches of the Rhine River between 1830 and 1840 (Thienemann, 1950). However, Lake Constance, which is situated far upstream above the falls of the Rhine and at the foot of the Alps, was not colonized until the late 1960s when mussels were most probably introduced into this area by motor boat transports (Walz, 1973). In addition, over the last two decades zebra mussels colonized most of the newly formed dredged lakes of the Rhineland. During the early 1970s, the zebra mussel totally disappeared from the Middle and Lower Rhine as a result of increasing sewage polution. With the construction of sewage treatment plants, oxygen conditions in the Rhine generally improved and zebra mussels began to recolonize the Middle Rhine in 1976 and the Lower Rhine in 1981 (Schiller, 1990).

In order to understand both the spreading capability of the zebra mussel and year-to-year variation in recruitment, adaptations of its life cycle to lake and stream environments have to be analyzed by quantitative and experimental methods. In this context, significant attributes of the species are the duration of the planktonic larval stage, growth rate, annual gonad development and fecundity, and the seasonality and synchrony of the reproductive period. In the following, we describe results of our studies on the biology of zebra mussel populations in the Rhineland of Germany. These studies were conducted on man-made dredged lakes near Köln (Fühlinger See, Heider Bergsee) and on the Rhine River.

LARVAL DEVELOPMENT TIME

Plankton samples have indicated that there is a wide temporal range in the duration of the larval stage (8 days — Korschelt, 1891; 8–12 days — Hillbricht-Ilkowska and Stanczykowska, 1969; about 5 weeks — Walz, 1973). Environmental factors which influence larval development are still unclear, since it has been difficult to stimulate appropriate conditions in the laboratory to isolate individual influences (Sprung, 1987 and 1989). The true developmental time of the early lecitotroph *Dreissena* larvae from about 70–80 μm in diameter to a stage of about 220 μm ready for settlement is about 18–28 days according to Sprung (1989) (Figure 1). Thus a maximum daily increase of 7.7 μm might be expected. This growth rate corresponds well with values that were determined in a larval cohort being carried downstream in the Rhine River (Figure 2, Borcherding and de Ruyter van Steveninck, 1991). The means of the larval size distributions increased by 6.0–7.5 μm/

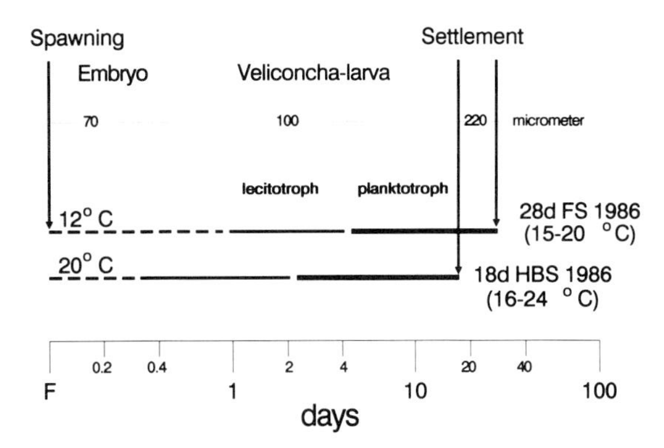

Figure 1. The duration of the planktonic larval stage of *Dreissena polymorpha* between spawning and settlement in days during different temperature conditions (according to data from Sprung, 1987 and 1989). F: date of fertilization. Broken and thin line: data from laboratory experiments. Solid line: data from plankton samples at the Fühlinger See (FS) and the Heider Bergsee (HBS), both lakes near Köln, Germany. (Redrawn from Neumann, D. in *Biologie des Rheins, Limnologie Aktuell, Vol. 1* [Stuttgart, Germany: Fischer-Verlag, 1990], pp. 87–105. With permission.)

day at temperatures between 19–20°C and optimal chlorophyll *a* concentrations (>50 μg/L) in the river. These estimates compare well with values for marine bivalve larvae (about 10 μm/day; Sprung, 1984).

Such an extended plankton stage of a benthic and sessile species has far-reaching consequences on the recruitment and spreading of the population in running waters. Without doubt, the biology of the zebra mussel must have been evolved within lake biotopes. Only in such waters, can most of the progeny of a local population settle in between the parental generation so that selection processes can alter adaptive properties of the population in relation to local conditions. When the gametes are directly released into running waters, the settlement of young mussels will occur far from parental population in a temporal distance of at least 2 weeks. Nevertheless, mass colonization and steady recruitment of local populations on an annual basis also occur in streams and rivers. In the Rhine River, population densities of 30–40,000 young mussels per square meter are observed (Neumann, 1990). Depending on the level of the Rhine, the downstream transportation of a water column from the Upper to the Lower Rhine needs generally less than 12 days. Thus, there exists no chance for any self-recruitment of local *Dreissena* populations in a stream like the Rhine because of the strong discrepancy between the rapid drift of larvae and the relatively slow rate of larval development. Consequently, settling in running waters must be the overproduction of populations from slow-flowing tributaries, adjacent former meanderers, artificial lakes connected to the stream, or lakes upstream.

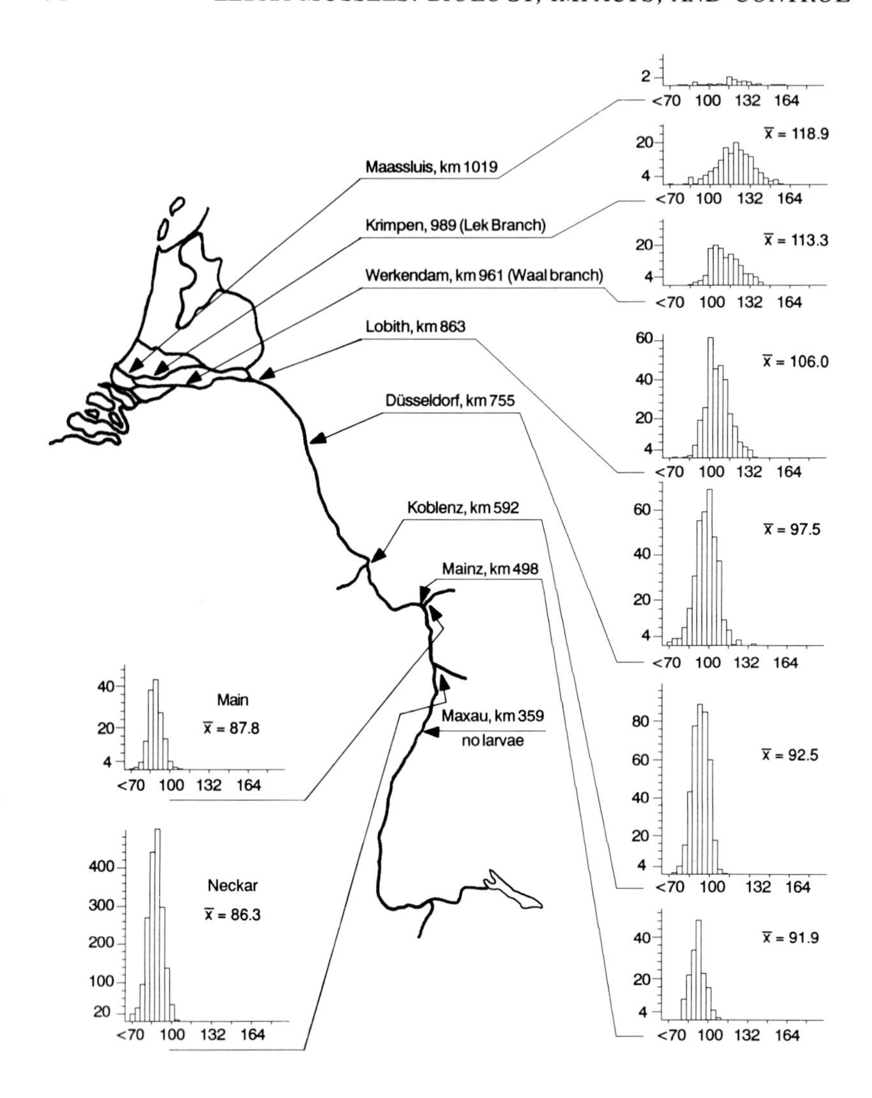

Figure 2. Size distribution (shell length [μm]) of the *Dreissena* larvae in the Rhine River. Samples were taken of a distinct water column moving downstream between May 2 at Maxau and May 11 at Maassluis in 1990. (Redrawn from Borcherding, J., and E. D. de Ruyter van Steveninck in *The Zebra Mussel Dreissena polymorpha, Limnologie Aktuell, Vol. 4* [Stuttgart, Germany: Fischer-Verlag, 1992]. With permission.)

AGE-DEPENDENT GROWTH RATE OF THE SHELL LENGTH

The growth of both individually marked specimens at Bad Honnef (Rhine—642 km) and distinct age groups at Grietherort (Rhine—845 km)

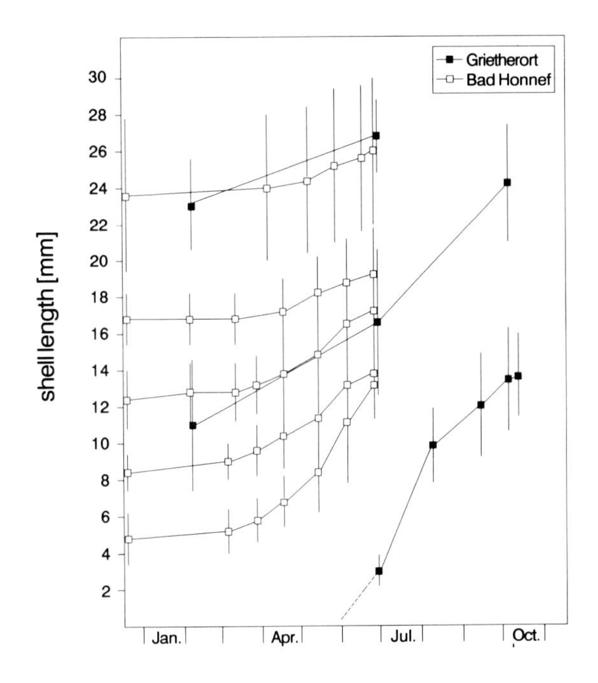

Figure 3. Shell length increase of *Dreissena polymorpha* in groups of different shell size during 1 year. Growth was measured on individuals held in a laboratory basin with a continuous flow of unfiltered Rhine River water (open symbol, Bad Honnef), and at one natural location of the Rhine population (dark symbol, Grieterort). (Redrawn from Jantz, B., and D. Neumann. in *The Zebra Mussel Dreissena polymorpha, Limnologie Aktuell, Vol. 4* [Stuttgart, Germany: Fischer-Verlag, 1992]. With permission.)

were examined (Figure 3) (Jantz and Neumann, 1991). The results indicate an obvious relationship between growth rate and shell size. This relationship can be best described by a negative exponential equation; i.e., the growth rate is considerably larger in the younger mussels than in the older ones. For instance, mussels of 4–7 mm size of the Rhine population in winter can reach a shell length increase of about 14 mm/year, in comparison to only 4-mm increase in mussels of about 24-mm size. Growth of the individually marked specimens at Bad Honnef can be described as $I = 20.6 \times \exp(^{-0.071 \times L})$ ($r = -0.96$) (I: shell length increase, L: initial shell length in winter). The relationship for Rhine populations at Lobith (863 km) was $I = 18.9 \times \exp(^{-0.064 \times L})$ (Smit et al., 1992).

If one correlates these maximum annual increases in younger and older specimens of the Rhine population with the time of year, two further influences on *Dreissena* growth rates are indicated: (1) there was nearly no shell length increase during winter time (i.e., mid-October until mid-March); it is not known if this lack of growth is a result of low temperatures (range: 1–10°C), an inadequate food supply (chlorophyll *a* concentrations below 3 μg/L), or both of these environmental factors; (2) the effective increase per year depends on the time of settlement of the veliconcha larva. When settlement occurs at the end of May or beginning of June, this youngest age group reaches a length of about 12 or even 14 mm by October (Figure 3). However, when reproduction occurs in late summer with settlement in August, this youngest age group reaches a length of less than 4 mm before winter.

GONAD DEVELOPMENT AND FECUNDITY

The annual reproductive cycle of the female zebra mussel was quantitatively described in relation to the local conditions of two lake populations (Borcherding, 1990 and 1991). In this histological study, areas of the gonads as well as size and numbers of the oocytes were measured in about 15 sections per specimen using an image analyzing system. Also, gonad volume and size distribution of oocytes were first calculated for each female, and then extrapolated for females of standardized shell length (Borcherding, 1991).

An example of an annual gonad cycle is shown in Figure 4. Variability can be evaluated by the cycles of three consecutive years at two locations (upper and lower littoral of the Fühlinger See) and one and a half year cycle from a third place (Heider Bergsee) (Figure 5). During the winter months of 1984 and 1985, both gametogenesis and growth of the gonad began in December and January when water temperatures fell from 6 to 3°C (Figure 5). Numbers of oocytes increased from February to April during rising temperatures. Continued increase of the gonad volume was correlated with the final growth of oocytes (about 1×10^6 in 24-mm females). Ripe oocytes ready for release can be recognized when they are detached from the follicle epithelium. Because they were tightly packed in the follicles, their diameters varied between 40 and 65 μm on the cross sections. As already suggested by Sprung (1989), the temperature threshold for the start of spawning events was about 12°C. This occurred at the level of 2-m depth during the first half of May, and at 9-m depth at the end of July (Figure 5, compare gonad cycles of 1985). The temporary and nonsignificant decrease of the extrapolated gonad volume at 9-m depth during the end of June probably was not correlated with an earlier first spawning event.

The progress in the release of gametes can roughly be followed by examining both numbers and stage of the remaining oocytes of the gonad every 4 weeks (Figure 4). These data indicate that mature oocytes of 24-mm females were released three times in cohorts of about 300,000–400,000 per female between early May and early September at this location in 1985. One out of six females studied in September 1985 still contained its last cohort of mature oocytes. This observation demonstrates that the spawning of closely settled

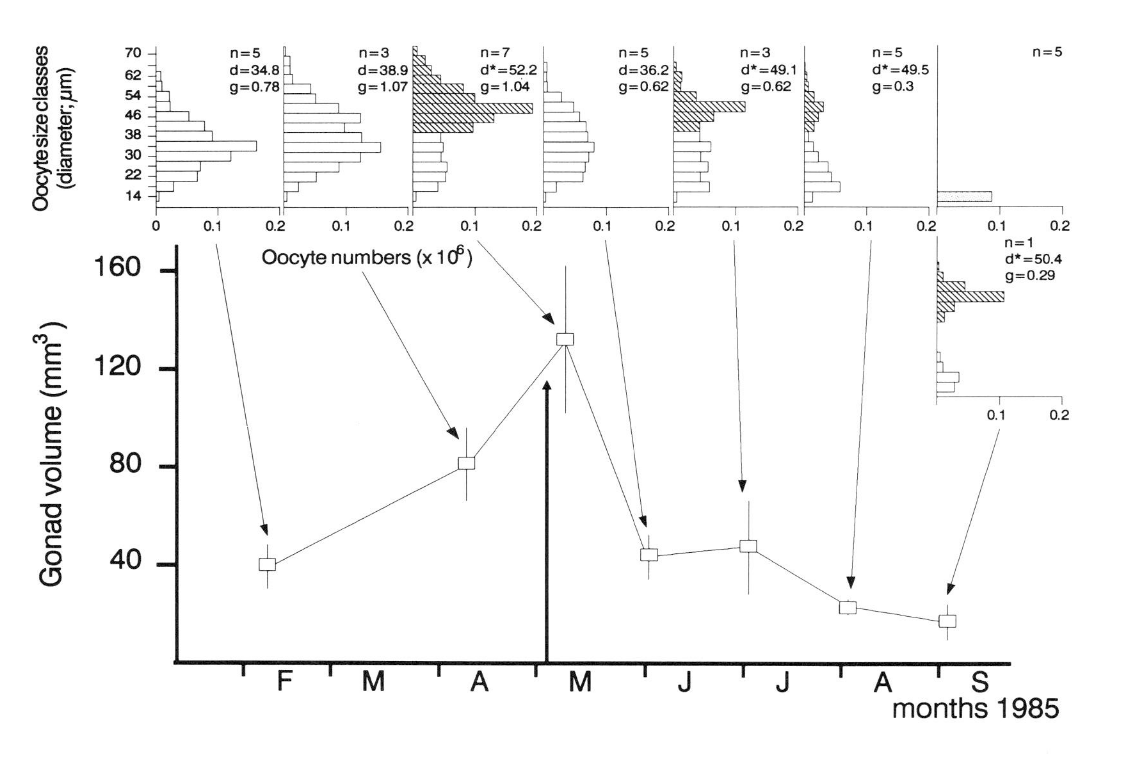

Figure 4. Gonad volume (mean ±95% confidence interval) of *Dreissena polymorpha* (below) in combination with data on both oocyte sizes and numbers (above) in specimens of 24-mm shell length from the Fühlinger See during 1 year. n: Number of 24-mm females quantitatively studied; d: mean oocyte diameter of all oocytes; d*: mean diameter of the ripe groups of the oocytes only; g: calculated number of oocytes per mussel in millions. Hatched bars: oocytes apparently ripe and ready for release. The solid vertical arrow at the beginning of May indicates the temperature rise above 12°C. (From Borcherding, J. *Oecologia (Berlin)* 87:208–218 [1991a]. With permission.)

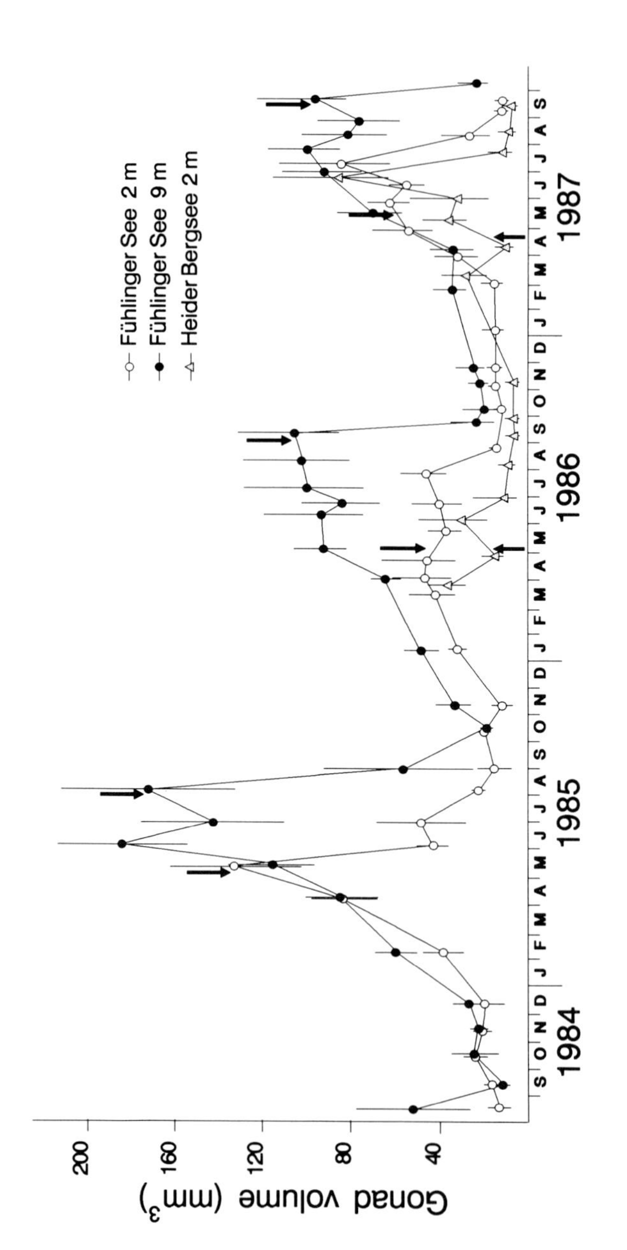

Figure 5. Seasonal changes of the gonad volume in a standardized specimen of 24 mm of *Dreissena polymorpha* at three different locations near Köln, Germany in 1984–1987. Solid arrows: date of the temperature rise above 12°C. (From Borcherding, J. in *The Zebra Mussel Dreissena polymorpha. Limnologie Aktuell, Vol. 4* [Stuttgart, Germany: Fischer-Verlag, 1992]. With permission.)

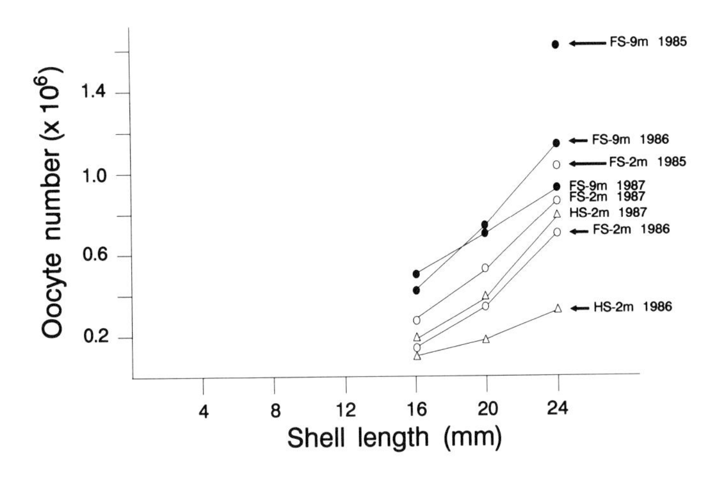

Figure 6. The mean oocyte numbers of *Dreissena polymorpha* females in correlation to the shell length of specimens from different locations from 1985 to 1987. Values of 1985 (long arrows), 1986 (short arrows), and 1987 (without arrows).

specimens must not be strongly synchronized, at least during the end of the spawning season. The empty gonads of other females showed a resting phase with diffuse follicle structures. This developmental inactivity of the gonad during decreasing water temperatures can be relatively short (e.g., FS 9 m, October 1985); or it can last for about 3 months (e.g., FS 2 m, September until December 1984), or even half a year (e.g., FS 2 m, August 1986 until March 1987).

A comparison of annual gonadal cycles from different years and locations shows that the maximum gonad volumes may differ in a remarkably wide range (Figure 5). In 24-mm females, numbers of oocytes varied between 300,000 and 1,700,000 (Figure 6). During all 3 years numbers of oocytes were always greatest at the 9-m depth of the Fühlinger See, where temperatures were lower than at the other sites. This suggests that temperatures below 12°C extend the annual growth phase of the gonad and favor energy input into the gonad. Food supply was the second environmental factor which controlled volume and fecundity of the gonad. Both locations at the Fühlinger See were characterized by a relatively rich food supply; average seston concentrations (ash-free dry weight) between March and June were 1.93 and 0.96 mg/L at depths of 2 and 9 m, respectively. On the other hand, the site in the Heider Bergsee had a poor food supply, especially during early spring; the average seston concentration (ash-free dry weight) between March and June 1987 was 0.43 mg/L. During early spring at the Heider Bergsee site, the gonad volume even decreased when immature oocytes were reabsorbed, probably as a direct result of a poor food supply. Under these environmental conditions, the subsequent growth and release of mature oocytes was delayed until June when water temperatures were between 15 and 20°C, far higher than the normal 12°C threshold. Both effects of temperature and food availability were confirmed in laboratory experiments (Borcherding, In preparation). Any additional influence of photoperiods (LD 8:16, LD 16:8), however, could not be established during the gonad development.

These observations show that an annual temperature cycle with a threshold temperature of about 12°C controls the annual synchronization of the reproduction cycle of zebra mussel populations. During winter at temperatures below 12°C, multiplication of the gametocytes and cytoplasmic growth of the oocytes begin after a refractory phase of the gonads. During spring, when temperatures slowly rise and food supply is sufficient, gonads carry the maximum numbers of gametes. Oocytes and probably also sperms in male gonads do not mature simultaneously. Thus, only one portion can be released during early summer when water temperatures pass the 12°C threshold. The remainder of subsequently maturing gametes is released later in the season. If the mean water temperature at the local site reaches the 12°C threshold not before late summer, most oocytes of the females are already mature and can be spawned in larger portions. This distinct temperature threshold corresponds well with laboratory experiments which show that temperature above 10°C is required for the successful fertilization of the *Dreissena* eggs (Sprung, 1987). Thus, in the range of 10–12°C, distinct temperature-dependent processes became activated which are essential for synchronized and successful reproduction of the zebra mussel.

POPULATION DYNAMICS IN THE RHINE RIVER

In 1989, changes in the shell length of zebra mussels from different local populations of the Rhine River were examined. An example from the Lower Rhine is shown in Figure 7. The sample taken in November 1988 was composed of mainly two age groups. In June 1989, a high number of small young mussels appeared. Younger stages of this cohort were not collected from the riverbed before this date because of high water levels. However, this youngest age group of 1989 probably represents mussels that settled at about mid-May or somewhat later. These mussels reached shell lengths of 3–4 mm by the end of June, 10 mm by mid-August, and 13–14 mm by October. These separate results allowed the following evaluation of growth in this cohort: (1) growth rates of young mussels were in the same range as growth rates of mussels under controlled conditions (Figure 3); (2) negative exponential correlation between initial shell length and shell increment allowed the dating of settlement of 3- to 4-mm young mussels, which was at the end of May; (3) shell structures of very fine growth lines in the periostracum and in the upper calcareous layers continuously followed one another without any strong interruption, as might be found during a first winter period. In December

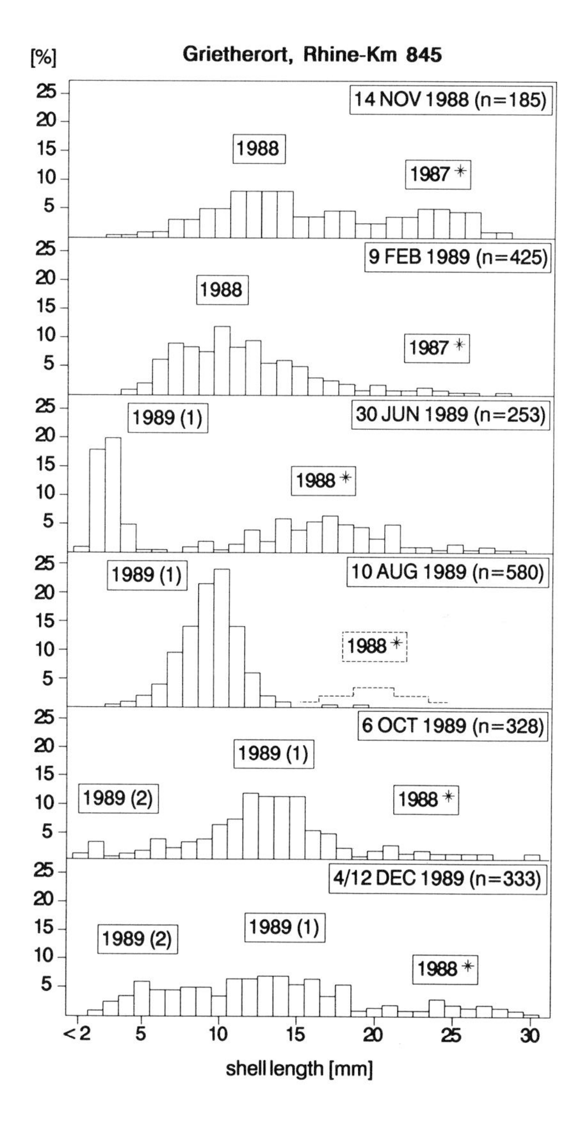

Figure 7. Shell length distributions of *Dreissena polymorpha* in the riverbed of the Lower Rhine during 1 year. The successive cohorts are marked by the year of settlement. * Few older specimens are included, n: number of shells measured. (From Jantz, B., and D. Neumann. in *The Zebra Mussel Dreissena polymorpha. Limnologie Aktuell, Vol. 4* [Stuttgart, Germany: Fischer-Verlag, 1992]. With permission.)

1989, the rapidly grown specimens of the early 1989 cohort were already similar in size to the slowly growing specimens of the 1988 age group with shells characterized by an annual growth interruption of the last winter. The 1989 cohort was obviously complemented by some mussels which did not

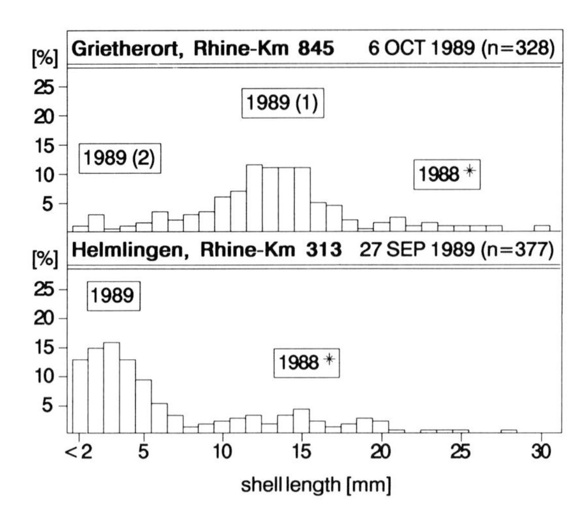

Figure 8. Comparison of two *Dreissena* populations in the Rhine River with phase-shifted recruitment in the course of the year. (From Jantz, B., and D. Neumann. in *The Zebra Mussel Dreissena polymorpha. Limnologie Aktuell, Vol. 4* [Stuttgart, Germany: Fischer-Verlag, 1992]. With permission.)

settle until late summer. In Figure 7, this cohort is indicated as 1989(2) in the October sample.

At other reaches of the Rhine, the shell length distribution can obviously follow a phase-shifted annual pattern when the first recruitment does not occur during May and June, but mainly during late summer. Such a pattern occurred in 1989 at locations of the Upper Rhine (Figure 8). In this case, the youngest cohort achieved shell sizes of only 4–6 mm by the first winter. It can be assumed that for such a population the main increase to sizes of about 14 mm occurs during summer of the following year, and that the first reproductive phase is retarded for 1 year.

With such striking differences between populations of the Upper and Lower Rhine, the annual recruitment of local populations probably is derived from different areas. Considering the 12°C threshold for the start of the reproduction season, it may be suggested that early settling cohorts from May and June came from *Dreissena* in shallow locations where water temperatures rise above this threshold in April and May, and that late settling cohorts in August came from *Dreissena* in colder littoral regions of deep lakes (such as Lake Constance or lakes connected with the River Aare [Switzerland]).

In summary, growth dynamics of a local population of zebra mussels can be determined in areas with sufficient food supply if one considers the size distribution of the shells in relation to the size-specific growth rate, and the occurrence of early or late recruitment during summer. In the Lower Rhine, the *Dreissena* population seems to be composed at the end of a year by the cohorts of the 0+ and 1+ generations. Older specimens above 25 mm were a rarity. A high mortality among the older zebra mussels of the Grietherort population could be obviously established in February 1989 (Figure 7, generation of 1987) when both the ratio of the bigger mussels had decreased since November, and the temporarily remaining clusters of their byssus threads were detected on the stony substrates among the younger generation. A question still remains whether such an exceptional age group structure (i.e., without 2+ and older generations) correlates with ecotoxic influences from the chemical load in the water of the Rhine.

SUMMARY AND CONCLUSIONS

These studies in the Rhine River have revealed some quantitative details on zebra mussel reproduction and growth that are important in understanding the spread and population dynamics of these mussels in lakes and rivers. Larval development in the plankton lasted 4 or even 5 weeks while the larval shell grew from 70 to 220 μm. In this time period, larvae are swept down current; thus, rivers such as the Rhine can be settled only by populations occurring in slowly streaming tributaries or in lakes connected with the river. Fecundity depended on the size of the mussel, and on the food availability during winter and spring. In females of 16-mm shell length, up to 500,000 oocytes were found; and in females of 24-mm shell length, up to 1,700,000 were found.

The shell growth capacity of the mussels is a function of size. When larvae settle about mid-May in the Rhine River, the young mussels can reach sizes of up to 14-mm shell length by their first winter, so that they can start with their first annual reproduction after about 12 months. Larvae settling later in the year (about mid-August), only reach 4 mm in shell length by winter. It is unclear whether these late settling mussels can produce ripe gametes by the following spring.

If sufficient food is available, the annual reproductive cycle of the populations seems to be synchronized by the annual temperature cycle. In the upper and lower littoral of the Fühlinger See a temperature of about 12°C initiates the onset of reproduction but both gametogenesis and gonad growth begin during low winter temperatures. The existence of a temperature threshold already has been reported for many marine bivalve species (e.g., 10–12°C for *Mytilus edulis* and *Mya arenaria*) (Mackie, 1984).

We suppose that an annual temperature cycle around the threshold temperature of 12°C is an important limiting factor for the successful spreading of zebra mussels in lakes and their connected rivers. The threshold temperature enables the synchronization of seasonal maturity processes so that mass concentrations of eggs and sperms will be released. This has significant implications for the geographic range of the species. In subtropical waters with minimum winter temperatures above 12°C, any seasonal synchronization of both oocyte maturing and spawning will be disrupted and low reproductive fitness would limit spreading of the species into warm polymictic lakes of southern latitudes. On the other hand, both spawning and fertilization will be disrupted in cold monomictic lakes with maximum littoral temperatures about 10–12°C; any spreading into northern latitudes would therefore be hindered at temperatures below this threshold. A detailed study of zebra mussel distributions in Europe between 40 and 60° latitude, particularly of lakes in lowland and mountainous countries, is needed to examine the relationship between temperature and geographic range.

ACKNOWLEDGMENTS

Our cordial thanks are due to both editors for improving the English text.

REFERENCES

bij de Vaate, A. "Distribution and Aspects of Population Dynamics of the Zebra Mussel, *Dreissena polymorpha* (Pallas, 1771), in the Lake IJsselmeer Area (The Netherlands)," *Oecologia (Berlin)* 86:40–50 (1991).

Borcherding, J. "Die Reproduktionsleistungen der Wandermuschel *Dreissena polymorpha*," Ph.D. Thesis, Universitat Köln, Germany (1990).

Borcherding, J. "The Annual Reproductive Cycle of the Freshwater Mussel *Dreissena polymorpha* Pallas in Lakes," *Oecologia (Berlin)* 87:208–218 (1991a).

Borcherding, J. "Morphometric Changes in Relation to the Annual Reproductive Cycle in *Dreissena polymorpha* — a Prerequisite for Biomonitoring Studies with Zebra Mussels," in *The Zebra Mussel Dreissena polymorpha, Limnologie Aktuell, Vol. 4,* D. Neumann and H. A. Jenner, Eds. (Stuttgart, Germany: Fischer-Verlag, 1992).

Borcherding, J. "Laboratory Experiments on the Influence of Food Availability, Temperature and Photoperiod on the Gonad Development of the Freshwater Mussel *Dreissena polymorpha*," (In preparation).

Borcherding, J., and E. D. de Ruyter van Steveninck. "Abundance and Growth of *Dreissena polymorpha* Larvae in the Water Column of the River Rhine During Downstream Transportation," in *The Zebra Mussel Dreissena: Ecology, Biological Monitoring and First Applications in Water Quality Management, Limnologie Aktuell, Vol. 4,* D. Neumann and H. A. Jenner, Eds. (Stuttgart, Germany: Fischer-Verlag, 1992).

Hillbricht-Ilkowska, A., and A. Stanczykowska. "The Production and Standing Crop of Planktonic Larvae of *Dreissena polymorpha* Pall. in Two Mazurian Lakes," *Pol. Arch. Hydrobiol.* 16:193–203 (1969).

Jantz, B., and D. Neumann. "Shell Growth and Aspects of the Population Dynamics of *Dreissena polymorpha* in the River Rhine," in *The Zebra Mussel Dreissena polymorpha: Ecology, Biological Monitoring and First Application in Water Quality Management, Limnologie Aktuell, Vol. 4,* D. Neumann and H. A. Jenner, Eds. (Stuttgart, Germany: Fischer-Verlag, 1992).

Korschelt, E. "Über die Entwicklung von *Dreissena polymorpha* Pallas," Sik.-Ber. Ges. Natürf. Freunde, Berlin N71:131-146 (1891).

Mackie, G. L. "Bivalves," in *The Mollusca, Vol. 7, Reproduction,* K. M. Wilbur, Ed. (New York: Academic Press, 1984), pp. 351–418.

Morton, B. "Studies on the Biology of *Dreissena polymorpha* Pall. III. Population Dynamics," *Proc. Malacol. Soc. London* 38:471–482 (1969).

Neumann, D. "Makrozoobenthos-Arten als Bioindikatoren im Rhein und seinen angrenzenden Baggerseen," in *Biologie des Rheins, Limnologie Aktuell, Vol. 1,* R. Kinzelbach and G. Friedrich, Eds. (Stuttgart, Germany: Fischer-Verlag, 1990), pp. 87–105.

Schiller, W. "Die Entwicklung der Makrozoobenthonbesiedlung des Rheins in Nordrhein-Westfalen im Zeitraum 1969–1987," in *Biologie des Rheins, Limnologie Aktuell, Vol. 1,* R. Kinzelbach and G. Friedrich, Eds. (Stuttgart, Germany: Fischer-Verlag, 1990), pp. 259–275.

Smit, H., A. Bij de Vaate, and A. Fioole. "Shell Growth of the Zebra Mussel (*Dreissena polymorpha* Pallas) in Relation to Selected Physico-chemical Parameters in the Lower Rhine, and some Associated Lakes," *Arch. Hydrobiol.* 124:257-280 (1992).

Sprung, M. "Physiological Energetics of Mussel Larvae (*Mytilus edulis*). I. Shell Growth and Biomass," *Mar. Ecol. Prog. Ser.* 17:283–293 (1984).

Sprung, M. "Ecological Requirements of Developing *Dreissena polymorpha* Eggs," *Arch. Hydrobiol., Suppl.* 79:69–86 (1987).

Sprung, M. "Field and Laboratory Observations of *Dreissena polymorpha* Larvae: Abundance, Growth, Mortality and Food Demands," *Arch. Hydrobiol.* 115:537–561 (1989).

Stanczykowska, A. "Ecology of *Dreissena polymorpha* (Pall) (Bivalvia) in Lakes," *Pol. Arch. Hydrobiol.* 24:461–530 (1977).

Thienemann, A. "Die Verbreitungsgeschichte der Süsswassertierwelt Europas," in *Die Binnengewässer, Vol. 18* A. Thienemann, Ed. (Stuttgart, Germany: E. Schweizerbart'sche Verlagsbuchhandlung, 1950).

Walz, N. "Untersuchungen zur Biologie von *Dreissena polymorpha* Pallas im Bodensee," *Arch. Hydrobiol., Suppl.* 42:452–482 (1973).

CHAPTER 6

Seasonal Reproductive Cycles and Settlement Patterns of *Dreissena polymorpha* in Western Lake Erie

David W. Garton and Wendell R. Haag

Seasonal patterns of gametogenesis, planktonic veliger abundance, and postplanktonic larval settlement were monitored at Stone Laboratory in western Lake Erie during 1989 and 1990. In both years, veliger larvae were present in plankton samples from mid-June to mid-October. In 1989, abundance of veligers and postplanktonic larvae settling onto glass slides exhibited distinct bimodal patterns, peaking in late July and late August. However, histological analyses of gonadal maturation revealed highly synchronous spawning activity in the local population, with spawning occurring only once, in late August. In contrast to 1989, there was only a single peak in planktonic veliger abundance and larval settling during 1990. Peak larval abundance occurred in late July, with peak settling occurring several weeks later in August. Spawning patterns were also different, being less synchronized in 1990 as compared to 1989. In 1990, spawning occurred over a 5-week period, from late July to late August. Spawning and veliger abundance were not correlated with temperature, as reported in previous studies.

Annual differences in spawning and recruitment of *Dreissena* in western Lake Erie may be explained by density dependent effects coupled with varying environmental conditions. Peak summer temperatures were 5°C cooler in 1990 than in 1989 (25 vs 30°C, respectively); and many cold fronts moved across Lake Erie during the summer of 1990, resulting in higher rainfall, storm surges, and sediment inputs than in 1989. Increased macrophyte abundance in 1990 may have altered water flow patterns in the study site, reducing inshore-offshore water exchange. Increased density of resident *Dreissena* may have inhibited recruitment by direct predation or physical disturbance (400 mussels/m^2 in 1989 compared to 30,000/m^2 in 1990).

INTRODUCTION

Unregulated ballast water discharge associated with European commercial ship traffic in the Great Lakes has been implicated in the successful introduction of several exotic species during the 1980s. These species include a predatory cladoceran, *Bythotrephes cederstroemi*; fish including the river ruffe, *Gymnocephalus cernuua*, and the tube-nosed goby, *Proterorhinus marmaratus*; and a filter-feeding bivalve mollusk, the zebra mussel, *Dreissena polymorpha* (Hebert et al., 1989; Sprules et al., 1990). Invading species may have negative effects on communities through predation, physical displacement, or competition for limiting resources. *Dreissena*, first discovered in Lake St. Clair in 1988 (Hebert et al., 1989), is well-known as an important biofouling organism in Europe; and in many aquatic systems, the species is now one of the dominant benthic species. The appearance of *Dreissena* in North America has generated alarm about long-term economic and ecological impacts, not only within its current range in the Great Lakes, but also in other inland waterways and reservoirs as it eventually spreads.

The appearance of exotic species offers an opportunity to study the adaptation of an organism to new environments. In order to be successful in its new habitat, an organism must be able to adapt to appropriate seasonal cues regulating important life history events, such as reproduction. Because the ecology of the Great Lakes has been studied in some detail, the impact of these exotic species on other components of the ecosystem can be examined. Assessing potential impacts of *Dreissena* on native communities and developing successful control strategies require a thorough understanding of its basic life history patterns in its new environment.

Dreissena is unusual among freshwater mollusks in having a free-swimming veliger larva. In Europe, larvae typically remain in the plankton for up to 3 weeks, then settle onto substrates where they attach firmly with byssal threads (Lewandowski, 1982a, and 1982b; Sprung, 1989). In contrast, all other freshwater mollusks have either direct development or a brief, parasitic larval stage. The great capacity for dispersal as larvae along with the high fecundity of adults (Stanczykowska, 1977; Sprung, 1987) allows *Dreissena* to numerically dominate the benthic community, and seasonally, the zooplankton community.

Based on European studies, the seasonal occurrence of *Dreissena* veligers in the plankton is highly variable, with larvae typically first appearing in late spring (April-May), peaking in midsummer (July-August), and disappearing in autumn (September-October) (reviewed in Sprung, 1989). Annual variation within lakes or lake systems is also considerable, with veliger abundance more dependent on local environmental conditions than densities of adult mussels. Water temperature has been implicated as the primary factor regulating gonadal maturation and spawning. Temperatures above 12°C have been reported as necessary for spawning to begin (Sprung, 1989). In heated lakes, spawning begins earlier and persists later into the fall than in nearby nonheated lakes (Lewandowski and Ejsmont-Karabin, 1983; Stanczykowska et al., 1988; Sprung, 1989). However, increased water temperature alone may be insufficient to trigger spawning. Walz (1978) observed that increased water temperature was not effective in inducing spawning in the laboratory. Near the southern geographic limit for *Dreissena* (and perhaps also in its original range in the Caspian and Black Sea basins), veliger larvae may be present in the plankton year-round (Zhdanova and Gusynskaya, 1985). More recently, increasing phytoplankton abundance has been identified as an important factor triggering spawning in marine invertebrates (Starr et al., 1990). Geographic and annual variation in environmental factors (water temperature, phytoplankton abundance, and other limnological variables) have been suggested as causing observed variability in *Dreissena* spawning and veliger abundance. The combined effects of multiple environmental factors on reproduction in *Dreissena* have yet to be examined in detail.

In this chapter we present data describing the baseline life history of a *Dreissena* population recently established in western Lake Erie. These data were collected as part of ongoing research on environmental factors regulating *Dreissena* life history patterns (Garton and Haag, 1991; Haag and Garton, 1992), and will be of use in further studies on the biology of *Dreissena* in North America. This chapter describes annual cycles of gametogenesis, seasonal abundance of planktonic veliger larvae, settlement patterns of post-planktonic juveniles onto artificial substrates, and their relationships with seasonal fluctuation of water temperature in western Lake Erie.

METHODS

Study Site

All field work was conducted at The Ohio State University F. T. Stone Laboratory at Put-in-Bay, OH in the western basin of Lake Erie (Figure 1). The site consists of a shallow rocky reef in water approximately 2–3 m deep.

Gametogenesis

In 1989, 20–30 adult mussels were collected monthly from November 1988 to May 1989 and approximately every 10 days from June to October

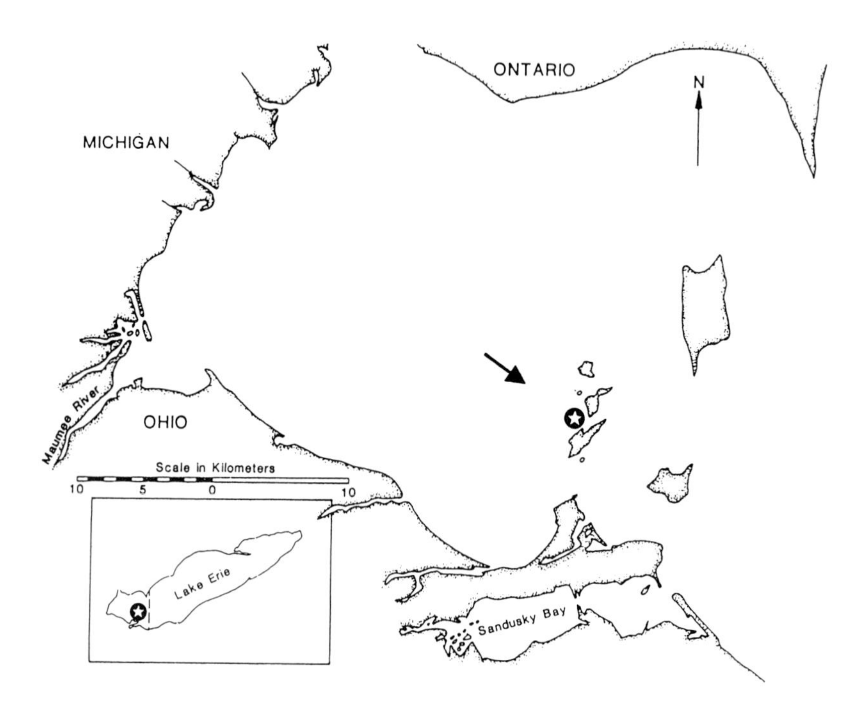

Figure 1. Western Lake Erie. The study site, located at The Ohio State University F. T. Stone Laboratory, is indicated by the star.

1989 and June-September 1990 for determination of gamete development. Specimens were fixed whole in Bouin's preservative, decalcified in 4 *N* formic acid, dehydrated in ethanol, and then embedded in paraffin. Thin sections (12 μm) were cut through the gonad of each specimen (n = 5–18 for each sampling date) and stained using standard methods (Humason, 1962). Permanent slides of each specimen were examined. Then the stage of gamete development was described using a qualitative scale of 0–4 with: 0 = gonad spent with few or no gametes remaining, 1 = immature (no mature gametes present), 2 = intermediate immature (initial stages of gametogenesis evident), 3 = late immature (approximately 50% mature gametes in gonad). This method for staging gametogenesis is similar to that applied by Tourari et al. (1988).

In addition to histological analyses, shell length-dry weight regressions were generated from a sample of 20 adult mussels collected concurrently with histological samples. Mussels were frozen and later their soft tissues were removed, dried, and weighed to the nearest 0.001 g. Shell length was determined with vernier calipers to the nearest 0.1 mm. Dry weights were adjusted to the average shell length (18 mm) using linear regression equations from log-transformed data.

Seasonal Abundance of Veligers

Qualitative plankton samples were taken with a 45 μm mesh net at irregular intervals from March, until the first appearance of veliger larvae in early June. From June 14, 1989 to October 30, 1989 and June 10, to October 6, 1990, two replicate quantitative vertical hauls were taken at 5- to 10-day intervals with a 45-μm mesh plankton net equipped with a flow meter. Vertical hauls were made in the channel separating South Bass and Middle Bass Islands, where average water depth was approximately 6 m and there was little or no thermal stratification (surface and bottom water temperatures usually within 1°C). Surface water temperature was recorded on each date when plankton samples were collected. Plankton samples were preserved in a sugar-formalin solution (40 g sucrose per liter of formalin). Larvae were enumerated from subsamples of each plankton sample and veliger abundance expressed as number of larvae per liter.

Settlement of Postplanktonic Juveniles

Settling plates consisted of 6–12 glass microscope slides (75 × 25 mm) supported in wood and styrofoam frames. Microscope slides were conditioned (i.e., a thin film of periphyton allowed to develop) prior to placement in the frames by soaking in running lake water in the laboratory for 1 week. In 1989, settling plates were suspended from the laboratory docks in water 2–3 m deep, approximately 0.5 m off the bottom. In 1990, settling plates were located at both the dock site and a site offshore from Stone Laboratory (2–3 m water depth). Settling plates at the offshore site were anchored to the bottom, and a float attached to the rack held the settling plates parallel to the bottom. Slides were removed from the frames at 5- to 7-day intervals and inspected under a stereomicroscope for the presence of newly settled juvenile mussels. The number of mussels on each slide was recorded and extrapolated to number of mussels settling per square meter per day for the sampling period. Older juveniles and adults were occasionally encountered on the slides but were presumed to have moved onto the slides from the wooden frames. To exclude these individuals from settling counts, we did not include mussels in the upper 5th percentile for shell length. All mussels were removed from the slides before being returned to the water. During period of heavy settlement, all mussels could not be removed from previously exposed slides, so these were replaced by clean, preconditioned slides that had been allowed to develop a coating of periphyton in the laboratory.

Statistical Analyses

The relationships among water temperature, gametogenic index, veliger abundance, and settling intensity were tested using linear correlation coefficients. The statistical relationships among these variables were calculated for each year independently and for data pooled from 1989 and 1990. Statistical significance is given at the $p < .05$ level.

RESULTS

Gametogenesis and Spawning

1989

A total of 152 adults selected randomly from all mussels collected (n = 8–18 per sampling date, approximately 8–9 mm and greater shell length, 77 males and 75 females) from 16 sample dates were examined. Male and female mussels showed similar patterns of maturation between December 1988 and late August 1989; therefore, the gametogenic indices were combined for both sexes (Table 1). In males, the testis can be identified in immature individuals (December 1988 samples) as a dark strip or layer of undifferentiated tissue lying at the margin of the viscera. Spermatogenesis was apparent in males collected in late spring (primary and secondary spermatocytes, spermatids) and the proportion of mature spermatozoa increased throughout the summer (June-July). The numerous lobes of the testis were completely filled with mature spermatozoa by late July and through mid-August, but were devoid of spermatozoa in samples collected August 29 and later.

Gametogenesis in female mussels followed a pattern similar to males. In the winter and early spring of 1989 immature ova were present and ovarian follicles were poorly developed. Size of ova and follicles increased through the spring and early summer; in mid-July ovarian follicles were filled with ova that appeared mature and were lying free, having detached from the follicular basal membrane. Ova were absent from the follicles in samples collected on August 29 and later.

Spawning, inferred from the presence of spent individuals in the histological samples, occurred between the August 18 and August 29 sampling dates (Table 1). Lake temperature was cooling from a summertime peak of 30°C (from late July to early August), and was approximately 22–23°C at the time of spawning. Mature gametes were not observed in the gonads of mussels collected after August 29.

Mussel body mass, standardized to a common shell length, declined during two periods, the first in June and the second in late August (Figure 2). The second period of weight loss was associated with spawning, during which mussels lost 50% of dry body mass. Adjusted body mass decreased from an average of 20 mg prespawning to 10 mg dry weight postspawning (Figure 2). The period of rapid loss of relative body mass early in the season (June) occurred during a period of rapid increase in the gametogenic index (gametogenesis) and before spawning.

Table 1. Gametogenic Index for *Dreissena polymorpha* Collected from Put-in-Bay, OH Near F. T. Stone Laboratory

		Stage						
	Date	1	2	3	4	Spent	N[a]	Mean Index
1988	Dec. 12	15	2	0	0	0	16	1.19
1989	Apr. 01	2	11	0	0	0	18	2.17
	May 02	0	3	5	0	0	8	2.63
	May 19	0	4	4	0	0	8	2.50
	Jun. 09	0	1	7	0	0	8	2.88
	Jun. 19	0	1	5	2	0	8	3.12
	Jul. 03	0	0	4	4	0	8	3.50
	Jul. 15	0	0	1	7	0	8	3.88
	Jul. 27	0	0	0	8	0	8	4.00
	Aug. 04	0	0	0	12	0	12	4.00
	Aug. 11	0	0	0	10	0	10	4.00
	Aug. 18	0	0	0	8	0	8	4.00
	Aug. 29	0	0	0	0	8	8	0.00
	Sept. 05	0	0	0	0	8	8	0.00
	Oct. 04	1	1	0	0	6	8	0.38
	Nov. 05	4	2	0	0	2	8	1.00
1990	Jun. 19	4	7	0	0	0	11	1.64
	Jul. 16	1	3	5	0	0	10	2.44
	Jul. 25	0	2	3	3	0	8	3.13
	Aug. 01	0	1	1	2	3	8	1.86
	Aug. 17	0	0	1	1	6	8	0.88
	Aug. 22	0	0	2	1	5	8	1.25
	Aug. 29	1	0	0	0	7	8	0.13
	Sept. 01	0	0	0	1	5	8	0.67
	Sept. 06	2	1	0	0	2	6	0.80
	Sept. 14	2	0	2	1	3	8	1.50
	Sept. 18	1	0	0	0	3	8	0.25
	Sept. 21	0	0	0	0	7	8	0.00
	Sept. 29	0	0	0	0	8	8	0.00

Note: Gametogenic index scored as: 1, immature, with no mature gametes present; 2, intermediate immature, with initial stages of gametogenesis visible; 3, late immature, with approximately 50% mature gametes present; 4, mature, with >50% mature gametes present; and 0, spent, with few or no gametes remaining in the gonad.

[a] During 1990, some rows do not sum to sample size because gonadal condition for 11 individuals could not be determined.

1990

A total of 112 adults randomly selected from all mussels collected (n = 6–11 per sample date — 49 males, 52 females, and 11 individuals where sex could not be determined due to the absence of identifiable gametes) from 13 sample dates were examined. As in 1989, seasonal change in the gametogenic index was identical for both sexes. The gametogenic index for pooled males and females increased in June and July, but declined gradually after July 25, reaching a minimum value in late August (Table 1). Average gametogenic index values were lower in 1990 than in 1989 because spawning

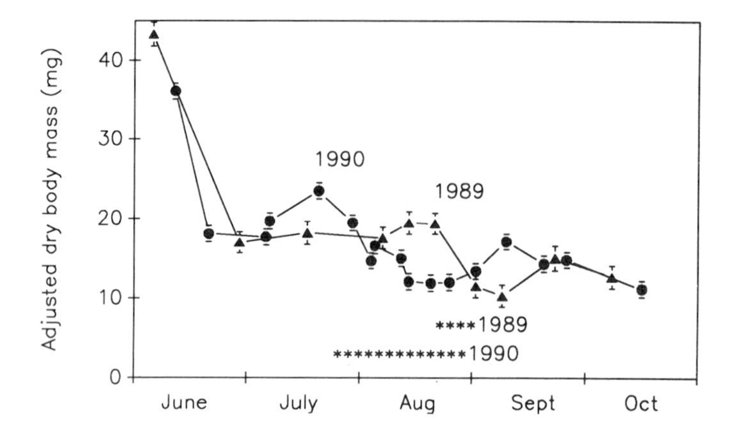

Figure 2. *Dreissena polymorpha*. Dry weight (mg) adjusted to common shell length for early summer-early fall sample dates. Triangles, 1989; circles, 1990. Error bar represents ± one standard error (S.E.) of the mean. Asterisks indicate active spawning period for each year.

began earlier in 1990 than in 1989 and was not highly synchronized. Not all mussels shed gametes at the beginning of the spawning period (late July). Instead, the percentage of spawned-out mussels increased gradually during August until all mussels were spent by the August 29 sample date. Extension of the active spawning phase resulted in lower values for the gametogenic index in 1990 than in 1989 (Table 1).

As in 1989, adjusted mussel weight showed two declines: the first, early in the season prior to spawning and a second decline associated with the spawning period (Figure 2). Approximately 50% of the dry body mass was lost gradually between July 25 (onset of spawning) and August 29 (end of spawning period).

Seasonal Abundance of Larvae

1989

Abundance of *Dreissena* larvae in plankton samples showed a distinctly bimodal distribution (Figure 3). Larvae first appeared in qualitative plankton samples on June 9 (water temperature approximately 18°C). A small peak in abundance was observed on June 14, at which time larvae were present at densities of 42 (±4.6 S.E.) larvae per liter. Densities declined and remained low (<20 larvae per liter) until July 18 when a sudden, large peak of 375 larvae per liter (no error estimate for this date) occurred. Densities subsequently declined to 147 larvae per liter (±46.0) but reached another larger peak of 451 larvae per liter (±51) on August 17. Immediately after this largest peak, densities declined dramatically and remained at less than five larvae per liter, with no larvae being collected after October 17 (water temperature approximately 10°C).

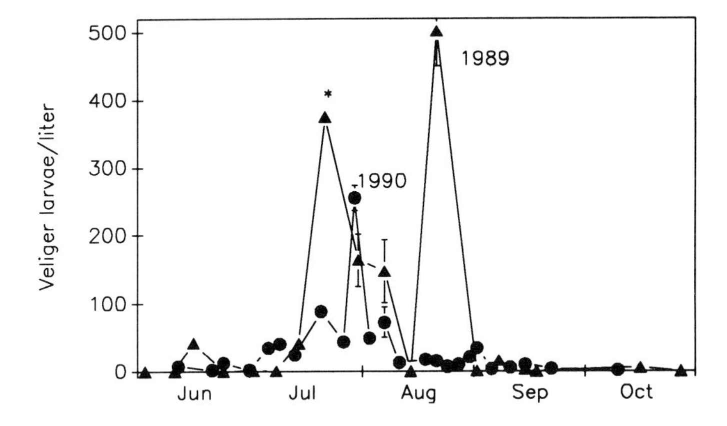

Figure 3. *Dreissena polymorpha.* Veliger abundance (larvae per liter) in plankton samples June to October. Triangles, 1989; circles, 1990. Error bars represent ± one S.E. of the mean. Asterisk indicates no replicate sample. No error bars indicate S.E. less than symbol diameter.

1990

Veliger larvae were first collected on June 10 (7 larvae per liter [±1.8]); abundance gradually increased during June and July, peaking on July 25 at 255 larvae per liter (±18), the period of maximum veliger abundance in 1989 (Figure 3). Unlike 1989, there was only a single peak in veliger abundance during 1990, and veliger abundances were less than 35 larvae per liter after August 8, the period of maximum veliger abundance in 1989 (Figure 3). Veligers were not collected after October 6 (water temperature approximately 15°C).

Settlement of Postplanktonic Juveniles

1989

Dreissena larval settling densities showed a bimodal distribution corresponding with abundance of planktonic larvae (Figure 4). Newly settled larvae were first observed on June 29, settling at a density of 10 (±5) larvae per square meter per day. Settling ceased during the week ending on July 11 and then climbed steadily until a peak was reached on July 25, during which larvae settled at a density of 1231.8 (±341) larvae per square meter per day. Settling density dropped to 314 (±45) larvae per square meter per day and then reached another peak during the week ending on August 18 with a density of 1885 (±171) larvae per square meter per day. Settling density declined steadily throughout the rest of August and September, and ceased during the

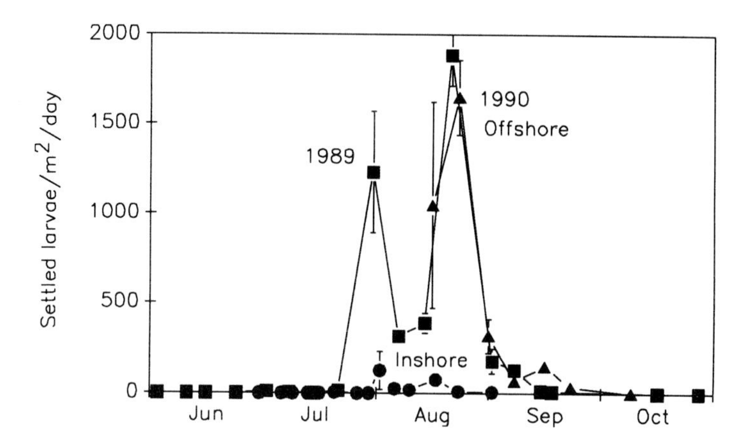

Figure 4. *Dreissena polymorpha.* Recruitment of postplanktonic larvae onto settling plates (larvae per square meter per day). Squares, 1989; circles, 1990 inshore site; triangles, 1990 offshore site. Error bars represent ± one S.E. of the mean. No error bars indicate S.E. less than symbol diameter.

week ending on October 4. A total of approximately 30,000 mussels per square meter were collected on plates during the reproductive season.

1990

During 1990 little or no settling occurred on the inshore plates (dock site) until late July (Figure 4). Water clarity improved between 1989 and 1990, resulting in increased macrophyte growth in the shallow waters around the docks at Stone Laboratory (personal observation). We suspected that increased macrophyte abundance hindered water circulation and transport of larvae into the dock sampling station; therefore, additional sampling plates were placed offshore in midsummer. The offshore sampling plates collected a much greater number of mussel larvae than the inshore plates (Figure 4). Settling at the offshore site peaked at 1651 (±209) larvae per square meter per day on August 21, compared to peak settling at the inshore site of 125 (±107) larvae per square meter per day on July 31. No newly settled larvae were collected after September 20 (offshore site). Approximately 17,000 mussels per square meter were collected on the inshore and offshore settling plates during the reproductive season; however, the figure underestimates larvae which may have settled prior to early August.

Correlation of Spawning with Body Mass

For 1989 and 1990 and both years pooled, there were significant correlations between gametogenic index and adjusted body mass of adult mussels

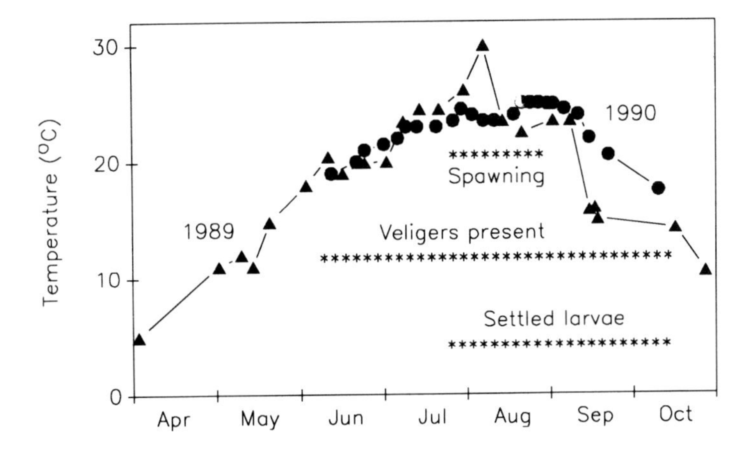

Figure 5. Seasonal surface water temperature records for Stone Laboratory, western Lake Erie. Triangles, 1989; circles, 1990. Periods of spawning in the local population, presence of veligers in plankton samples, and settlement of post-planktonic larvae onto glass slides during both years are indicated by asterisks.

during the spawning season (1989: $r = 0.85$, $p < 0.004$, $n = 9$; 1990: $r = 0.97$, $p < 0.001$, $n = 9$; pooled data: $r = 0.81$, $p < 0.001$, $n = 18$). Midsummer decline of gametogenic index (release of gametes) was highly correlated with loss of body mass, indicating that loss of body mass was caused by spawning.

Correlation of Veliger Abundance and Recruitment with Temperature

In both 1989 and 1990, veligers appeared in plankton samples at the same time (mid-June) and at similar water temperatures (approximately 18°C) (Figure 5). However, the relationship between veliger abundance and water temperature was not significant within years or when years were pooled (pooled data, $r = 0.29$, $p < 0.065$, $n = 42$ sampling dates). Similarly, the number of settled juveniles was not correlated within years or when years were pooled (pooled data, $r = 0.32$, $p < 0.063$, $n = 34$). In general, veligers were present and settling of juveniles occurred whenever open-lake surface water temperature exceeded 18°C (Figure 5).

The intensity of juvenile settlement onto substrates was highly correlated with veliger abundance in the plankton in 1989 ($r = 0.72$, $p < 0.001$, $n = 18$) (Figure 6). In contrast, there was no significant correlation between rate of settlement and veliger abundance during 1990 (Figure 7). Pooling of 1989 and 1990 data results in a significant positive relationship between settling and veliger abundance, although the amount of variation explained was low ($r = 0.45$, $p < 0.008$, $n = 34$). Overall, juvenile settlement intensity was proportional to abundance of veligers in the water column.

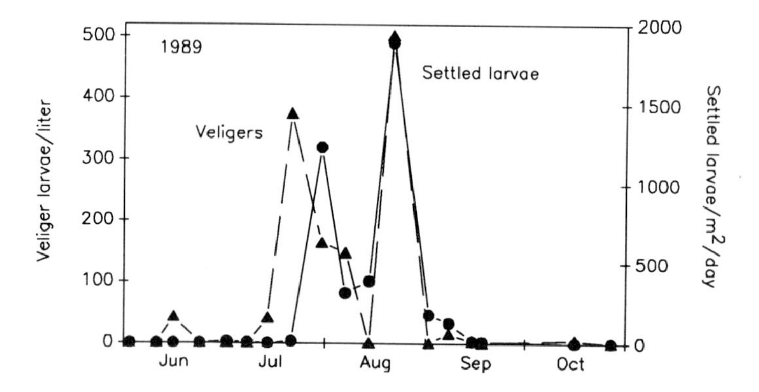

Figure 6. *Dreissena polymorpha.* Abundance of veligers (larvae per liter), triangles and dashed line; and recruitment of postplanktonic larvae onto settling plates (larvae per square meter per day), circles and solid line, at Stone Laboratory in 1989. Error bars not shown for clarity.

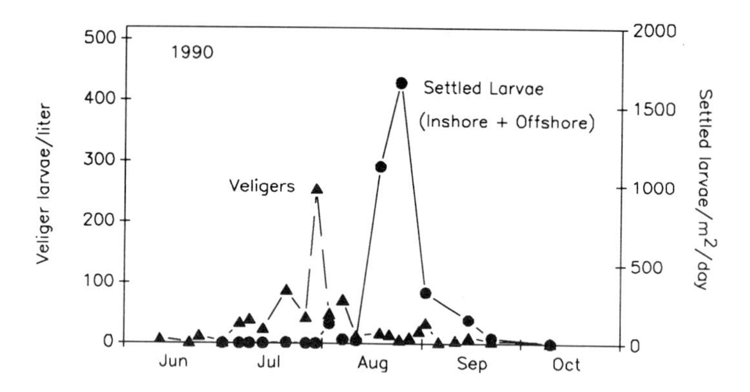

Figure 7. *Dreissena polymorpha.* Abundance of veligers (larvae per liter), triangles and dashed line; and recruitment of postplanktonic larvae onto settling plates (larvae per square meter per day), circles and solid line, at Stone Laboratory in 1990. Error bars not shown for clarity. Data from inshore and offshore settling plates are combined.

DISCUSSION

In western Lake Erie, *Dreissena polymorpha* displayed strong seasonal patterns in its reproductive cycle, abundance of veligers in the plankton, and settling of postplanktonic juveniles onto substrates. These seasonal patterns were sensitive to annual variation, as patterns observed in 1990 were different from patterns observed in 1989. In 1989, spawning was a highly synchronous event within the local population, whereas in 1990 gametes were released over a longer time period (weeks vs days). Veliger abundance patterns did not coincide with spawning in the local study population, and seasonal veliger abundance patterns differed between the 2 years. Settlement of postplanktonic juveniles into the population was dependent upon abundance of veligers in the plankton in 1989, but not in 1990. Veliger abundance and hence recruitment of postplanktonic juveniles were lower in 1990 than in 1989; however, spawning, veliger abundance, and settlement intensity peaked during late July and August for both years. Possible causes of observed annual variation in seasonal reproduction of *Dreissena* include differences in temperature cycles, abundance and timing of phytoplankton blooms, increasing adult mussel density, severe weather, influx of veliger larvae from disjunct populations, and increased macrophyte abundance in the study area.

Annual and geographic variation in temperature has been identified as the primary factor explaining variation in timing of reproduction of *Dreissena* (Galperina, 1978; Sprung, 1989; and others). Commonly, appearance of veligers and hence spawning are reported when water temperature exceeds 12°C. In both years of this study, however, veligers did not appear in plankton samples until water temperature had reached 18°C; and spawning within the local population occurred at 22–23°C. Walz (1978) found that increased temperature alone was ineffective for inducing spawning in *Dreissena*, and *Dreissena* has been observed spawning at temperatures as low as 2.5°C (Apstein, 1896 — cited in Walz, 1978). In this study, abundance of veligers and settling of juveniles were not positively correlated with water temperature. Instead, veligers and settling larvae were present throughout the summer when lake temperature exceeded 18°C. Therefore, it is likely that additional factors other than temperature alone regulate reproductive patterns of *Dreissena* in western Lake Erie.

Intraspecific variation in spawning times among populations is common in marine and freshwater bivalves, and factors regulating synchronous spawning are not well understood. Temperature, photoperiod, and salinity have all been identified as important factors regulating mass spawning; but more recent studies have shown that the presence of abundant phytoplankton alone can induce spawning in the marine mussel, *Mytilus edulis*, and the sea urchin, *Strongylocentrotus droebachiensis* (Starr et al., 1990). Once spawning begins, it can be accelerated by the presence of conspecific sperm. Similarly, sperm suspensions induce spawning in *Dreissena* (Walz, 1978). Therefore, it is more likely that a combination of physical and biological factors in midsummer triggers spawning in *Dreissena*, and presence of sperm from a few individuals encourages synchronous mass spawning within local populations (Haag and Garton, 1992). Geographic and temporal variation in spawning among local populations within a large lake (e.g., Lake Erie) probably results from regional variation in temperature, phytoplankton abundance, current patterns, and water depth.

Overall veliger abundance was much lower in 1990 than in 1989, even though the density of adult mussels at the study site was much greater in the second year of this study than in the first (adult mussel density in June 1989: 400 individuals per square meter; June 1990: 15,000 individuals per square meter). Previous studies have shown poor correlation between number of planktonic veligers and density of adult mussels, and at high densities (above 10,000/m^2) predation by adult mussels becomes an important source of veliger mortality (Stanczykowska, 1977; Lewandowski, 1982a and 1982b; Lewandowski and Ejsmont-Karabin, 1983; Stanczykowska et al., 1988; Sprung, 1989). Competition for limiting resources (space and food) also increases with adult mussel density. Therefore, adult density may have either little or negative influence on veliger abundance and larvae settling intensity.

The summer of 1990 was stormier than 1989, with frequent cold fronts moving across Lake Erie with associated high winds (storm surge and seiches) and rain (increased surface water inputs to lake). These storms may have caused increased mixing of the water column and turbidity in 1990, which may have led to decreased veliger abundance by disrupting synchronization of spawning, increasing mortality of veligers by physical disturbance, altering current patterns in the study area, and reducing larval growth rates as a result of lower phytoplankton production.

Veliger larvae appeared in plankton samples well before the onset of spawning in the local study population, drifting in from other regions in western lake Erie from populations that had spawned earlier in the year. *Dreissena* has attained high densities throughout western Lake Erie, and easterly currents carry larvae into the Bass Island region from Maumee Bay (near Toledo, OH) and the Michigan shore (Detroit River). A water residence time in western Lake Erie of approximately 30 days and larval development period of 10 days for *Dreissena* (Lewandowski, 1982a) imply that larval transport from distant populations can be reflected in local veliger abundance patterns. The larval abundance peak in 1989 that occurred prior to spawning in late August in local populations was probably caused by drift of larvae from Maumee Bay, where adults spawn earlier in the year than in the Bass Island region (Fraleigh, personal communication, 1990). Similarly, water currents disperse larvae derived from spawing adults in the local population, resulting in poor correlation between veliger abundance and spawning in the location population. Therefore, temporal and geographic variation in spawning and veliger abundance patterns results in veliger larvae being present in the plankton over a wider time period than might be predicted from spawning activity in local adult populations.

Settlement of postplanktonic larvae was reduced in 1990 relative to 1989 at the inshore (dock) sampling site. This reduction in settled larvae on the glass plates may have resulted from increased macrophyte abundance during 1990. Previous studies have shown that *Dreissena* may initially settle at high densities in aquatic vegetation (Lewandowski, 1982b). Abundant macrophytes near the inshore site acted as settling substrate for veligers transported from offshore, and in addition may have significantly restricted mixing between inshore and offshore water masses. Higher settling rates on offshore plates established in midsummer provide supporting evidence that abundant macrophytes reduced settlement at the inshore site. Therefore, estimates of total recruitment into mussel populations at the Stone Laboratory study site for 1990 are probably low.

Peak settlement of larvae during 1989 was highly correlated with abundance of planktonic veligers (two peaks of veliger abundance and two peaks of settled larvae), but not during 1990 (single peak of settlers 4 weeks after peak veliger abundance). Although veliger abundance was relatively low during peak settlement (10–20 larvae per liter), the absolute abundance was high enough to account for the observed settling intensity. A veliger density of 20 larvae per liter is equivalent to an area density of 40–60,000 larvae per square meter, which is considerably greater than observed peak settling rates (1,500 larvae per square meter per day). This calculation illustrates several points. First, settling rates are more likely dependent upon maturational stages of larvae than simply on number of planktonic larvae in the water column (a greater percentage of larvae are competent to settle later in the season than earlier). Second, veliger abundance exceeding 10 larvae per liter (which is most of the summer season) can cause very high rates of settlement. Last, the vast majority of planktonic larvae do not survive long enough to be collected on settling plates (peak veliger densities reach approximately 1–1.5 million larvae per square meter; total annual settlement on glass slides is less than 30,000 individuals per square meter).

The gametogenic cycle observed in Lake Erie *Dreissena* is similar to patterns described in European studies, with gametogenesis occurring rapidly after water temperature warms above 10–15°C (Tourari et al., 1988). Rate of gametic development of *Dreissena* is temperature sensitive, with developmental rates increasing with temperature until inhibition begins at water temperatures of 30°C and higher (Tourari et al., 1988). Similarly, gametogenic index (a measure of gametogenesis) increased in early summer with increasing water temperature. Following spawning, gametogenesis commenced again after water temperature began to cool in early autumn. The delay in spawning observed in 1989, relative to 1990, might have been caused by midsummer water temperature near 30°C, occurring at the time mussels were ripe for spawning in mid-July. Spawning was less synchronized within the mussel population in 1990 than in 1989. In 1989, all mussels in the population spawned over an 11-day period, August 18–29, while in 1990 spawning began July 25, but all mussels were not "spent" until August 29. During the 1990 spawning period, several samples contained both fully ripe (gametogenic index = 4) and fully spent (gametogenic index = 0) individuals. This variation in timing of spawning is reflected in the higher values of gametogenic index in 1989 compared to 1990. Walz (1978) observed that not all gametes were released when *Dreissena* was induced to spawn in the laboratory. Instead, individual mussels underwent multiple spawning events (2–6 spawns) until gonads were empty of mature gametes, as confirmed by histological examination (Walz, 1978). Synchronization of spawning in *Dreissena* has not been well studied, although most studies conclude that spawning can occur anytime during a 1- to 2-month period when gametes are fully developed, depending on environmental conditions (Sprung, 1987, Tourari et al., 1988, Borcherding, 1990).

Two periods characterized by rapid declines in adjusted body mass were observed in *Dreissena* from Lake Erie. The first period of mass loss, occurring in late spring, was probably associated with low abundance of phytoplankton. The second decline, occurring in midsummer, was correlated with spawning. Similar periods of body mass loss in *Dreissena* have also been recorded from populations in Lake Constance (Walz, 1978). Walz (1978) attributed the early season body mass loss to phytoplankton concentrations insufficient to meet minimum energetic requirements for maintenance, while the second was attributed to gamete release. Seasonal patterns of body mass gain and loss are similar for *Dreissena* from Lake Constance and Lake Erie.

Although strong selective pressures favor simultaneous spawning in organisms with external fertilization, the pattern of synchronous spawning in *Dreissena* observed in 1989 may also be explained by the preponderance of young individuals in the population (Haag and Garton, 1992). *Dreissena* was absent, or at the very least, extremely rare in the Bass Island region of Lake Erie in 1987; significant colonization apparently occurred in early summer of 1988 as indicated by the size of individuals (up to 10-mm shell length) collected in October 1988. The 1988 cohort was therefore immature at the beginning of this study, and matured and spawned for the first time during the summer of 1989. Uniformity of age and gametogenesis may have favored a single spawning event in this cohort. In 1990, spawning indeed did begin earlier than in 1989, possibly because the breeding population included 2+-year-old mussels. European studies (e.g., Stanczykowska, 1977) have reported that *Dreissena* typically spawns during the second year of life, which is consistent with the results of this study. Assuming that the majority of mussels sampled in this study were recruited in July and August 1988 (the period of peak settlement in 1989), then the 1988 cohort spawned the following year at 12–13 months of age. Rapid growth and maturation of *Dreissena* in the western basin of Lake Erie reflects environmental conditions favoring the establishment of this invading species.

Favorable environmental conditions have resulted in exponential population growth and recruitment in the 1987–1990 period at Stone Laboratory in western Lake Erie. Settlement of postplanktonic larvae in 1989 and 1990 totaled approximately 50,000 mussels per square meter at the study site, resulting in an adult density of approximately 15,000 mussels per square meter. The common pattern of *Dreissena* invasions in Europe is a phase of rapid population growth, followed by sudden decreases to low steady-state population densities (reviewed in Mackie et al., 1989). The colonization of

Lake Erie by *Dreissena* is still in its initial stage, and continued study is necessary in order to determine relationships between environmental factors regulating reproduction and long-term mussel population dynamics.

ACKNOWLEDGMENTS

This study was supported by the Ohio Sea Grant College Program, project number R/ER-15, grant number NA89AA-D-SG132. We owe sincere thanks to John Hageman, laboratory manager at F. T. Stone Laboratory, for logistical support and helpful comments throughout this study; Matthew Misicka for assisting in the collection and processing of field samples; and David Berg and Ann Stoeckmann for assisting in field work. We also would like to acknowledge the following for their various contributions to this study: B. Andres, G. Balogh, T. Cvetnic, L. Hall, D. Jamison, T. Peavy, J. Pushay, and C. Willis.

REFERENCES

Borcherding, J. ''Die Reproduktionleistungen der Wandermuschel, *Dreissena polymorpha*,'' Ph.D. Dissertation, University of Cologne (1990).

Fraleigh, P. Personal communication (1990).

Galperina, G. E. ''The Relationship of Spawning of Some North Caspian Bivalves to the Distribution of Their Larvae in the Plankton,'' *Malacol. Rev.* 11:108–109 (1978).

Garton, D., and W. Haag. ''Heterozygosity, Shell Length and Metabolism in the European Mussel, *Dreissena polymorpha*, from a Recently Established Population in Lake Erie,'' *Comp. Biochem. Physiol.* 99A:45-48 (1991).

Haag, W., and D. Garton. ''Synchronous Spawning in a Recently Established Population of the Zebra Mussel, *Dreissena polymorpha*, in western Lake Erie, USA,'' *Hydrobiologia* 234:103-110 (1992).

Hebert, P. D., B. W. Muncaster, and G. L. Mackie. ''Ecological and Genetic Studies on *Dreissena polymorpha* (Pallas): A New Mollusc in the Great Lakes,'' *Can. J. Fish. Aquat. Sci.* 46:1587–159 (1989).

Humason, G. L. *Animal Tissue Techniques* (San Francisco: W. H. Freeman & Co., 1962).

Lewandowski, K. ''The Role of Early Developmental Stages in the Dynamics of *Dreissena polymorpha* (Pall.) (Bivalvia) in Lakes. I. Occurrence of Larvae in the Plankton,'' *Ekol. Pol.* 30:81–110 (1982a).

Lewandowski, K. ''The Role of Early Developmental Stages, in the Dynamics of *Dreissena polymorpha* (Pall.) (Bivalvia) Populations in Lakes. II. Settling of Larvae and the Dynamics of Numbers of Settled Individuals,'' *Ekol. Pol.* 30:223–286 (1982b).

Lewandowski, K., and J. Ejsmont-Karabin. ''Ecology of Planktonic Larvae of *Dreissena polymorpha* (Pallas) in Lakes with Different Degree of Heating,'' *Pol. Arch. Hydrobiol.* 30:89–101 (1983).

Mackie, G. L., W. N. Gibbons, B. W. Muncaster, and I. M. Gray. ''The Zebra Mussel, *Dreissena polymorpha*: A Synthesis of European Experiences and a Preview for North America,'' Ontario Ministry of the Environment (1989) 76 pp.

Sprules, W. G., H. P. Riessen, and E. H. Jin. ''Dynamics of the *Bythotrephes Invasion of the St. Lawrence Great Lakes,'' J. Great Lakes Res.* 16:346–351 (1990).

Sprung, M. ''Ecological Requirements of Developing *Dreissena polymorpha* Eggs,'' *Arch. Hydrobiol., Suppl.* 79:69–86 (1987).

Sprung, M. ''Field and Laboratory Observations of *Dreissena polymorpha* Larvae: Abundance, Growth, Mortality and Food Demands,'' *Arch. Hydrobiol.* 115:537–561 (1989).

Stanczykowska, A. ''Ecology of *Dreissena polymorpha* in some Masurian Lakes,'' *Pol. Arch. Hydrobiol.* 24:461–530 (1977).

Stanczykowska, A., K. Lewandowski, and J. Ejsmont-Karabin. ''The Abundance and Distribution of the Mussel *Dreissena polymorpha* (Pall.) in Heated Lakes near Konin (Poland),'' *Ekol. Pol.* 36:261–273 (1988).

Starr, M., J. H. Himmelman, and J. Therriault. ''Direct Coupling of Marine Invertebrate Spawning with Phytoplankton Blooms,'' *Science* 247:1071–1074 (1990).

Tourari, A., C. Crochard, and J. Pihan. ''Action de la temperature sur le cycle de reproduction de *Dreissena polymorpha* (Pallas) étude ''in situ'' et au laboratoire,'' *Haliotis* 18:85–98 (1988).

Walz, N. ''The Energy Balance of the Freshwater Mussel, *Dreissena polymorpha* Pallas in Laboratory Experiments and Lake Constance. II. Reproduction,'' *Arch. Hydrobiol., Suppl.* 55:106–119 (1978).

Zhdanova, G. A., and S. L. Gusynskaya. ''Distribution and Seasonal Dynamics of *Dreissena* Larvae in Kiev and Kremenchug Reservoirs,'' *Gidrobiol. Zh.* 3:34–39 (1985) (English translation).

CHAPTER 7

Abundance and Settling of Zebra Mussel *(Dreissena polymorpha)* Veligers in Western and Central Lake Erie

Peter C. Fraleigh, Paul L. Klerks, Gerald Gubanich, Gerald Matisoff, and Robert C. Stevenson

Temporal dynamics, depth distributions, and losses during passage through water intake pipes of zebra mussel (*Dreissena polymorpha*) veligers were studied in 1990 in the western and central basins of Lake Erie near Toledo and Cleveland, OH, respectively. In addition, postveliger settling rates were studied in the western basin. Veligers were first collected in the western basin on May 31, when the water temperature was 18°C. Densities peaked May 31-June 21 and July 24-August 21. Veligers were collected as late as November 7 when the water temperature was 12.6°C. In the central basin, veligers were present June 28-September 10, and were most abundant from late July to early August. Mean veliger densities in July-August were greater in the western than the central basin, with mean densities of 130/L and 26/L at the two sites, respectively. In the western basin, veligers increased with depth in the water column when wind speed was <8 km/hr but were uniformly distributed when wind speeds were higher. In the central basin, veligers were generally most abundant in the epilimnion; within the epilimnion, abundances were higher below the surface layer (0–2 m). Veliger densities decreased during passage through water

0-87371-696-5/93/$0.00 + $.50

intakes, with a 79% decline noted in the Toledo water intake pipe (4.7 km long). Settled postveligers were first found in the western basin on July 16. Monthly mean settling rates were greatest in August (3760/m^2/week) and September (2120/m^2/week). Settling continued through the last sampling period on November 20, 1990. Veliger densities in Lake Erie appear to be high, relative to its overall size, when compared to European lakes.

INTRODUCTION

Following the introduction of the zebra mussel, *Dreissena polymorpha*, into the Great Lakes, one of the first areas to be severely affected was the western basin of Lake Erie (Griffiths et al., 1989; Mackie et al., 1989; McMahon, 1989; Roberts, 1990). To develop a better understanding of the life cycle of the zebra mussel in Lake Erie, a study was undertaken to describe seasonal changes in veliger abundances, settling rates, and vertical depth distributions in the western and central basins. Also, since zebra mussels have a severe impact on raw water users, a study was conducted to determine differences in veliger abundances at the beginning and end of water intake pipes.

METHODS

Sampling sites were located in Lake Erie near the water intake for the City of Toledo, OH and near the Kirtland water intake crib for the City of Cleveland, OH (Figure 1). Samples were also collected at the pump stations of both water intake systems. The Toledo water intake is located in the western basin of Lake Erie near Maumee Bay and the Maumee River; the intake is 4.7 km from shore in about 7 m of water (41°42.0′ N, 83°15.5′ W). Water enters the intake through ports that extend from 2.5 to 5.5 m below the water surface. Water passes through a 2.7-m diameter pipe to the low service pump station located on shore. At the pump station, water enters a surge well about 2 hr after initially entering the intake. The Kirtland water intake crib is located in the central basin of Lake Erie just offshore from the City of Cleveland, OH. The crib is 7.9 km from shore in about 15.2 m of water (41°32.7′ N, 81°45.0′ W). Water enters the crib at a depth of 3.7 m below the surface and passes through a 2.7-m diameter pipe to a pump station on shore.

Veligers were sampled in western Lake Erie near Toledo at approximately weekly intervals from March 1 to November 21, 1990. Sampling was conducted within 100 m of the Toledo water intake, usually in the downwind direction. Duplicate vertical hauls were taken from 5 m to the surface using an 80-μm mesh Wisconsin-style plankton net. One replicate was stored on ice and the other was preserved with 10% buffered ($MgCO_3$) formalin. To measure temperature, water was collected at 4.3 m using a PVC Kemmerer bottle and dispensed into a 300-mL glass bottle, allowed to chill or warm the

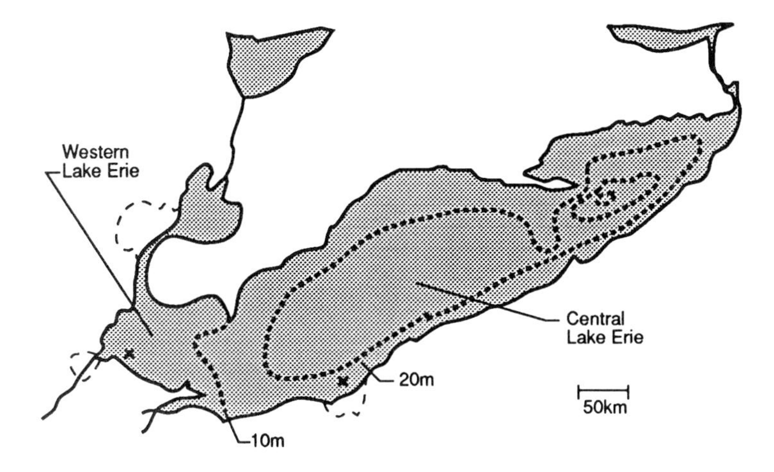

Figure 1. Location of sampling sites (x) in western and central basins of Lake Erie where zebra mussel (*Dreissena polymorpha*) veligers were collected in 1990.

bottle, and then discarded. After the bottle was filled, a second water sample was taken and the temperature was measured with an alcohol thermometer. Depth distributions of veligers in the western basin were determined 10 times during the sampling period. Samples were collected with a Birge closing net (63-μm mesh) by vertical hauls over depth intervals of 0–2 m, 2–4 m, and 4–6 m. Each of the three net samples was preserved with 10% buffered formalin. After sampling near the water intake, samples were collected from the surge well at the Toledo low service pump station. These samples were taken about 2 hr after those taken in the lake near the intake. Thus, the same parcel of water was sampled before and after intake passage. Veliger samples were taken from the surge well by vertical net hauls from a depth of 2–4 m (depending on the water level in the surge well). Water samples were also taken with the Kemmerer bottle for temperature determination. Veligers were counted in a 1-mL Sedgwick-Rafter counting cell, with three replicate subsamples counted for each net sample. This procedure gave a minimum detectable veliger density of 0.1/L. Means and standard errors were calculated using mean densities for the preserved and live net samples. Live net samples were counted upon return to the laboratory.

In the central basin, sampling was conducted at a site about 180 m west of the Kirtland water intake crib at approximately weekly intervals from June 28 to September 10, 1990. Water samples (1 L) were taken with a bottle-type water sampler at 1.5–3.2 m intervals from the surface to just off the lake bottom. In the laboratory, water samples were concentrated to 10 mL using a sand filter (Whipple, 1947). Aliquots of the concentrated samples were counted in a 1-mL Sedgwick-Rafter cell. Depth profiles of temperature and dissolved oxygen were determined on site using a YSI Model 51B DO meter. Samples were also collected at the Kirtland pump station on shore.

Postveliger settling in the western basin was examined using a glass slide sampling device located on the lake bottom within 200 m of the Toledo water intake. The sampling device consisted of a cinder block mounted vertically on a wooden cross base. The upper hole of the cinder block contained a test tube rack containing glass slides (2.5 × 7.6 cm). The glass slides and test tube rack were oriented so that the glass slides were horizontal to the lake bottom. The sampler was first set on May 14, just before the first veligers appeared in the lake. Then, approximately weekly between May 14 and November 21, 3–5 slides were removed and replaced with clean (not conditioned) slides. Each collected slide was individually placed in a jar of lake water and transported to the laboratory on ice. In the laboratory, postveligers attached to both sides of the slide (minus 1.3 cm of the end where the slide had been touched) were counted at (magnification × 40) using a dissecting microscope. The jar and water in the jar were examined for postveligers that may have become detached during transport. Since both sides of a total of six slides were counted (194 cm^2), this method gave a minimum detectable density of settled larvae of $52/m^2$.

RESULTS

Veligers were first observed in the western basin on May 31, when lake water was 18°C (Figure 2). At this time, veliger density increased from undetectable levels on May 14 (when the lake water was 12.8°C) to 123/L on May 31. After this date, veliger densities ranged from 94 to 174/L for the next 3 weeks. After June 21, densities declined until a second peak became apparent on July 24. Veliger density then increased to the observed maximum for the year (360/L) on August 2. After this date and beginning before water temperatures began to decrease, veliger densities declined rapidly at first and then gradually through the end of October. On October 20, when the water temperature was 12.6°C, densities were less than 1/L. No veligers were found on November 7 when the water temperature was 9.5°C.

In the central basin, veligers were already present at a density of 27/L (depth-weighted mean from 0 to 15 m) when sampling began on June 28 (Figure 2). Water temperature at this time was 19.3°C. Veligers were found in the water column at least until the last sampling on September 10, when 7/L were found and the water temperature was 19°C. During the entire sampling period, the mean density ranged from 3 to 96/L. The maximum density occurred on August 8, which was close to the time that the maximum density was found in the western basin.

Vertical distributions of veligers were variable at both sampling sites. In the central basin, few veligers were counted from each depth, and thus vertical

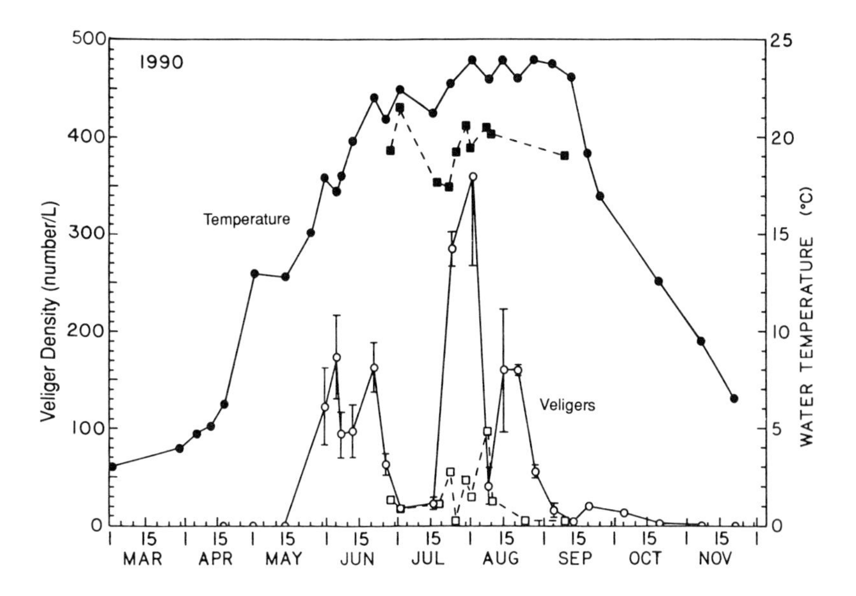

Figure 2. Temporal changes in densities of *Dreissena polymorpha* veligers (open) and in water temperatures (filled) near Toledo (circles) and near Cleveland (squares) in 1990. Densities given as means (±S.E.).

stratification was difficult to detect (Figure 3). However, temperature profiles appeared to have had some impact on vertical distributions. Veligers were absent from warmer surface waters in July. On July 18, densities were greater at depths just below warm surface waters; on July 2, densities were greater at deeper depths. While thermal stratification occurred on several dates, only on July 23 were there well-defined epilimnion, metalimnion, and hypolimnion. On this date, veligers were relatively abundant and appeared to be uniformly distributed in the epilimnion; however, they were absent from the hypolimnion which had little oxygen (Figure 3). When thermal gradients were weak (i.e., July 30 through August 10), veligers tended to be found between 5 m and the bottom.

In the western basin, vertical distributions of veligers appeared to be more related to wind velocity and wind-driven mixing. Data were grouped into three wind speed categories, based on observations made at times of sampling (<8 km/hr, 8–16 km/hr, and >16 km/hr) (Figure 4). Using analyses of variance, a significant vertical distribution was found only for winds <8 km/hr. ($F = 72.3$, $df = 2/9$, $p < 0.001$; Figure 4). At these wind velocities, 5% of the population was found at 0–2 m, 30% at 2–4 m, and 64% at 4–6 m. Thus, when not mixed vertically by wind, veligers were found deeper in the water column. This may be a result of preference for deeper waters or a consequence of passive settling. At any rate, this phenomenon may account

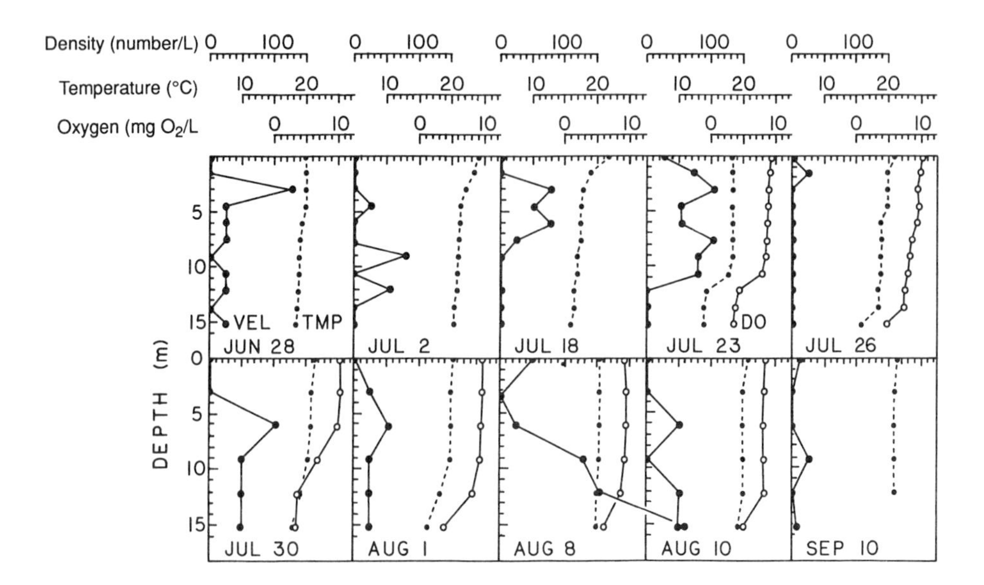

Figure 3. Vertical profiles of densities of *Dreissena polymorpha* veligers (filled circle/solid line), water temperatures (filled circle/dashed line), and dissolved oxygen concentrations (open circle/solid line) on different dates in 1990 in the central basin of Lake Erie near Cleveland.

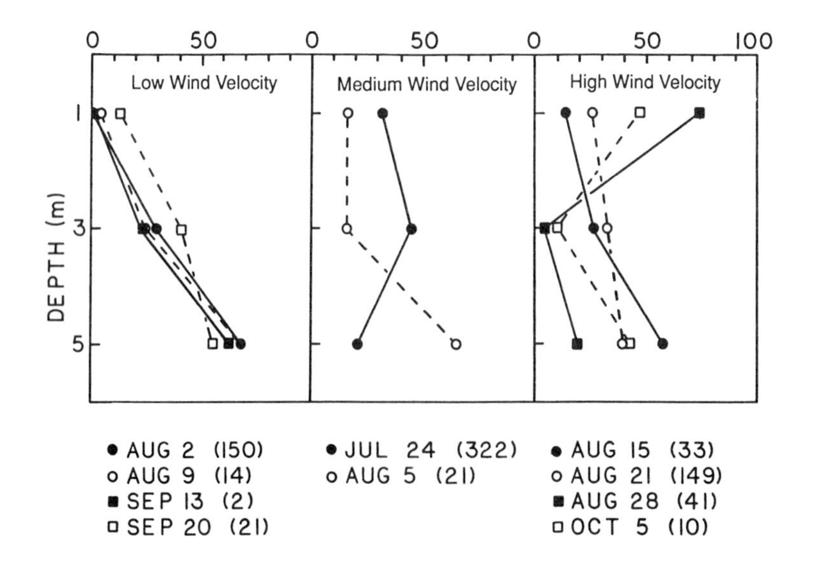

Figure 4. Vertical profiles of densities (expressed as a percentage of total numbers) of *Dreissena polymorpha* veligers in the western basin of Lake Erie near Toledo on different dates in 1990. The data have been grouped into three wind velocity conditions: low wind velocity (<8 km/hr); medium wind velocity (8–16 km/hr); and high wind velocity (>16 km/hr). Numbers in parentheses are mean veliger densities for all depths, for each date, in numbers per liter.

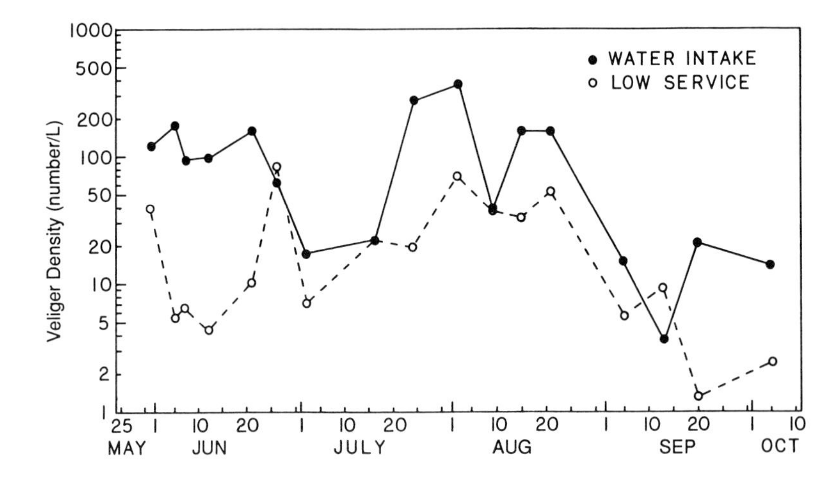

Figure 5. Temporal changes in densities of *Dreissena polymorpha* veligers in the western basin of Lake Erie near the Toledo water intake (filled circle/solid line) and in the surge well of the low service pump station, at the end of the 4.7-km long, 2.7-m diameter pipe from the Toledo water intake (open circle/dashed line) in 1990.

for the low densities in the upper water column of the central basin when thermal stratification indicated the absence of wind-driven mixing.

During transit through the pipe from the water intake to shore, veliger densities decreased significantly. At the surge well of the low service pump station at Toledo, veliger densities from May 31 through August 21 were significantly lower (two-way analysis of variance [ANOVA], $F = 16.5$, df $= 1/12$, $p < 0.001$) than those in the lake near the mouth of the pipe leading to the surge well (Figure 5). Densities at the low service surge well were only 21% of those in the lake. After August 21, chlorine was used to kill mussels in the pipe so comparisons could not be made after this date. Density differences were also apparent at the Kirtland water intake. Few veligers were found at the Kirtland pump station relative to densities in the lake near the Kirtland crib (Table 1). Thus, it appears that there was a substantial settling of veligers in the intake pipes, or adult mussels in the pipes were feeding on veligers, or both. In any case, relative to monitoring veliger densities in the lake, sampling at onshore pump stations would probably underestimate lake population sizes.

In the western basin, larval settlement began 1 month after veligers were first found, and began almost simultaneously with the occurrence of the second veliger peak (Figure 6). Settled postveligers were not found following the first peak in veliger abundance in June. The first spawn either failed to settle or was carried away from the intake site and settled elsewhere.

The first settled postveligers were collected July 16 on slides which had been set July 2, giving a settling rate of $90/m^2/week$. Between July 16 and

Table 1. Time-Weighted Monthly Mean Veliger Densities (Numbers/L), or Ranges of These Means, Calculated from Data Derived from Figures Published in the Literature

Location	Mar.	Apr.	May	June	July	Aug.	Sept.	Oct.	Nov.
Lake Erie									
Western Basin[a]	0	0	36	112	126	142	16	6	<1
Central Basin[a]				—	25	27	—		
Island Region[b]			0	16	154	268	8	1	
Kuibyshev Reservoir[c]			0	<1–8	1–16	2–14	<1–3	<1	<1
Lake Constance[d]				0	0–21	5–75	7–10	<1–1	
Kiev Reservoir[e]			0	66–84	26–54	11–34	11	<1	0
Lake Snairdwy[f]			0	126	75	12	0		
Lake Zarnowieckie[g]			0	2–4	9–11	57–84	10–12	0	
Lake Taltowisko[f]			0	6–98	100–114	1–15	0		
Lake Majczwielki[h]			0	9	18	4	0		
Konin Lakes[i]	0–10	0–16	30–162	47–140	132–320	22–139	8–22	0	
Babinski backwater[j]			0	71–102	66–194	13–82	0	0	

Note: Linear interpolation was used to estimate densities at the beginning and end of each month. Zero values indicate densities less than 0.1–1.0/L. Densities given on an areal basis were divided by the epilimnetic depth to convert to volume.

[a] This study.
[b] Haag and Garton (1992).
[c] Kirpichenko (1964).
[d] Walz (1973).
[e] Zhdanova and Gusynskaya (1985).
[f] Hillbricht-Ilkowska and Stanczykowska (1969).
[g] Lewandowski and Stanczykowska (1986); range of two sites.
[h] Lewandowski (1982b).
[i] Lewandowski and Ejsmont-Karabin (1983).
[j] Biryukov et al. (1964).

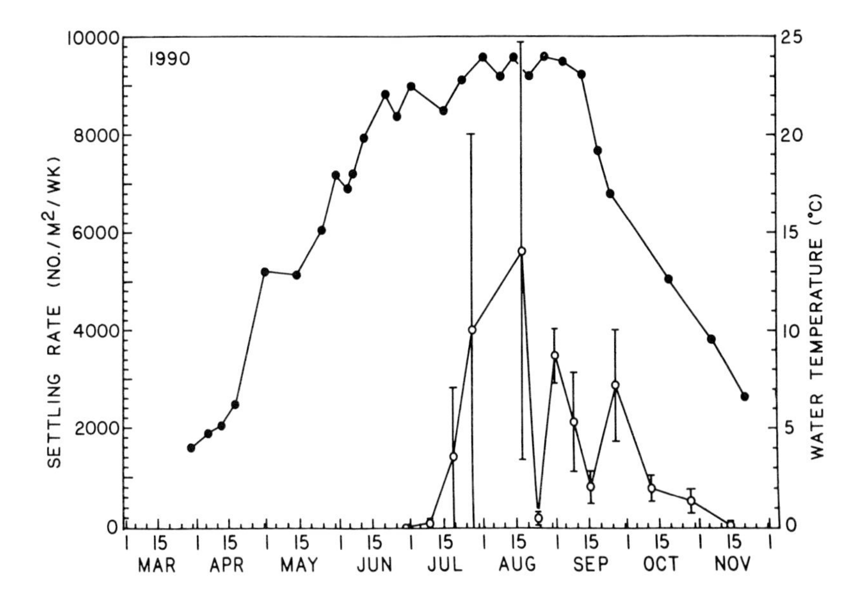

Figure 6. Temporal changes in *Dreissena polymorpha* postveliger settling rates (open circle) and water temperature (closed circle) in the western basin of Lake Erie near the Toledo water intake in 1990. Densities given as means (±S.E.).

24, the settling rate was $1400/m^2/week$; between July 24 and August 2, the settling rate was $4000/m^2/week$. Between mid-July and early October settling rates were relatively constant with a mean of $2550/m^2/week$ (S.E. = 640, n = 8). After the beginning of October, the settling rate declined and only a few were found on the last sampling day, November 21 (water temperature = 6.5°C.

Postveliger settling rates increased with increased veliger density, and the rate of settling per unit density of veligers increased exponentially with time after settling began. From the beginning of the settling period to the end, about 100 days later, the number of postveligers settling per unit density of veligers increased from 10 to 1000 settled postveligers (per square meter per week) per 1000 veliger per liter (Figure 7). This would indicate a maturing veliger population. A regression between time and natural log of the rate of settling per unit density of veligers was highly significant ($F = 47.3$, $df = 1/8$, $p < 0.001$). The August 28 sampling date was omitted from this analysis since an outlier test for this date was significant ($p < 0.01$). It is not known why this sampling date was unusual. The regression gave the following relationship between the settling rate (S, in numbers of settled postveligers per square meter per week) and the density of planktonic veligers (V, in numbers/L) and time (t, in days from the first recorded settling of postveligers, with the first day being designated as day 1):

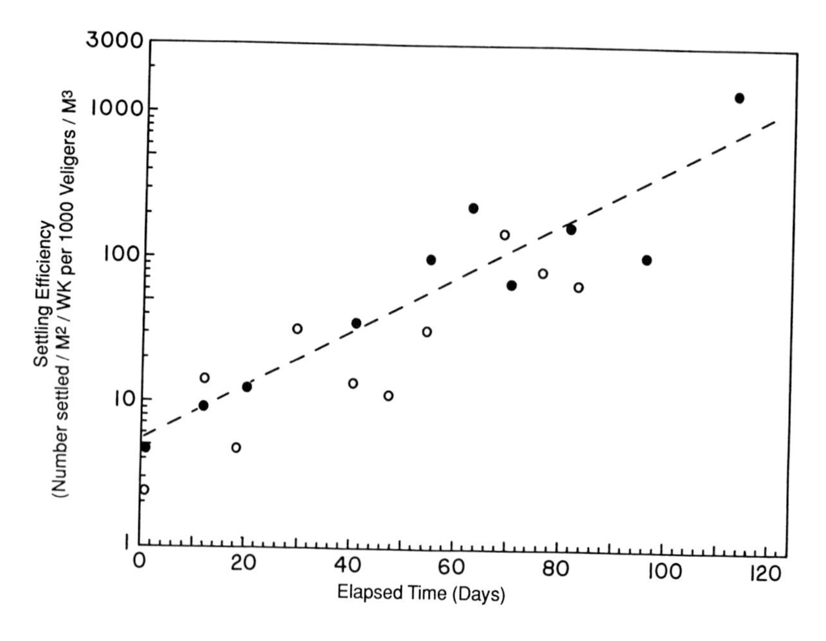

Figure 7. Increase in settling efficiency (the rate of postveliger settling per unit density of planktonic veligers) with time after settling began for the western basin of Lake Erie near the Toledo water intake in 1990 (filled circle), and for the western basin in the Island region in 1989 (open circle; from Haag and Garton, 1991). Regression line fitted to the Toledo data only.

$$S = [5.58*e^{(0.0431*t)}]*V$$

DISCUSSION

Population dynamics of zebra mussel veligers in the western and central basins of Lake Erie were generally similar to the typical pattern observed in European lakes. Reviews of *Dreissena polymorpha* ecology have concluded that, while the dynamics of early developmental stages are variable, several patterns appear to be common (Griffiths et al., 1989; Mackie et al., 1989; McMahon, 1989; Stanczykowska, 1977). In general, European studies indicate that reproduction begins in the spring when water temperature reaches 12°C, and veligers are initially found in the plankton when water temperature is about 14–16°C. Veligers were first collected in the western basin when the water temperature was 18°C, but were not found at the previous sampling when the water temperature was 12.8°C. In the central basin, veligers were present when sampling began and mean water temperature was 19.6°C. European studies also indicate that reproduction continues at least intermittently, if not continuously, through the period that the water temperature is greater than about 12–15°C. Similarly in this study, veligers continued to be found for at least several months following their initial occurrence. In the western basin, veligers were present for 5 months and were last observed when the water temperature was 12.6°C; in the central basin veligers were present for the entire 2-month sampling period, during which time the mean water temperature was always greater than 17°C.

Some characteristics of the vertical distributions of veligers observed in this study have been found in other lakes. The absence of larvae from hypolimnetic waters having low oxygen concentrations is common (Walz, 1973; Lewandowski and Ejsmont-Karabin, 1983; Biryukov et al., 1964). In addition, the effect of wind on creating a uniform vertical distribution of veligers was observed in the Zaporozskoe reservoir, where veligers were most abundant at a depth of 5 m on windless days (Dyga and Galinskij, 1975 — cited in Lewandowski, 1982a). Densities in the western basin, and to a lesser degree in the central basin, tended to be greatest deeper in the water column. Frequently, however, environmental conditions such as hypolimnetic waters having low oxygen concentrations and wind-driven mixing changed this distribution pattern.

The veliger stage typically lasts for about 2 weeks and then the larvae begin to settle (Mackie et al., 1989). In the western basin near Toledo, however, postveligers were not found until mid-July, 7 weeks after veligers were first observed. The reason for the long delay is not known. Veligers were sufficiently dense so that if settling had occurred, postveligers should have been observed. Conceivably, veligers found during late May and early June were produced by a local population and were carried away by currents before they matured and settled. Termination of spawning of this local population, along with transport of veligers away from the sampling area by water currents, could then account for the decline and low veliger densities observed from late June to early July.

Settling began in the western basin after the second period of high veliger densities. As settling continued, the proportion of veligers that settled increased. While this relationship has not been documented specifically, reports of progressive maturing of veligers are not uncommon. For instance, over a 4-week period in the pelagial region of Lake Kolowin, the percentage of veligers less than 0.1 mm decreased gradually from 70 to $<$10%, and the percentage of larvae $>$0.2 mm increased gradually 0 to about 40% (Lewandowski, 1982a). Similarly, in the littoral region of this same lake, the percentage of larvae $<$0.1 mm decreased gradually from 85 to 0% and the percentage of larvae $>$0.2 mm increased gradually from 0 to about 40% (Lewandowski, 1982a). Similarly, in the littoral region of this same lake, the percentage of larvae $<$0.1 mm decreased gradually from 85 to 0% and the percentage of larvae $>$0.2 mm increased gradually from 0 to 30% (Lewandowski and Stanczykowska, 1986). Thus, as the *Dreissena* reproductive season progresses, veligers mature, and reproduction diminishes, the size struc-

Table 2. Morphometric Characteristics of Various Water Bodies Where Veliger Densities Were Obtained for Comparisons to This Study

Water Body	Surface Area (km^2)	Mean Depth (m)	Maximum Depth (m)	North Latitude
Lake Erie[a]	25,750	18.0	64.0	42° 09′
Western Basin[b]	3,110	7.5	10.9	41° 42′
Central Basin[b]	16,300	18.3	25.0	41° 33′
Kuibyshev Reservoir[c]	6,450	8.9	40.0	53° 40′
Lake Constance[a]	540	90.0	252.0	47° 35′
Kiev Reservoir[d]	225	4.1		51° 00′
Lake Sniardwy[e]	106		25.0	53° 46′
Lake Zarnowieckie[f]	14.3	8.4	19.4	55°
Lake Taltowisko[e]	3.2		35.0	54°
Lake Majczwielki[g]	1.6	6.0	16.4	54°
Konin Lakes[h]	1.5–3.8	1.3–11.9	3.0–38.0	50° 30′
Babinski backwater[i]	0.4		11.0	55°

[a] Herdendorf (1982).
[b] FWPCA (1968).
[c] Fortunatov (1979).
[d] Prokhorov (1973).
[e] Hillbricht-Ilkowska and Stanczykowska (1969).
[f] Lewandowski and Stanczykowska (1986).
[g] Lewandowski (1982a).
[h] Lewandowski and Ejsmont-Karabin (1983).
[i] Biryukov et al. (1964).

ture of the veliger population shifts to relatively large individuals. As observed in this study, this maturing probably accounted for the progressive increase in the proportion of veligers settling as time since first settling increased.

To compare veliger densities in Lake Erie to densities in other lakes, time-weighted monthly means were calculated (Table 1). Linear interpolation between data points was used to estimate densities at the beginning and end of each month. Also, morphometric characteristics of most other lakes were obtained from various sources (Table 2). Time-weighted monthly means indicated that veliger densities in Lake Erie were greater in the western basin than in the central basin. During July and August, veliger densities near Cleveland were only 19–20% of those near Toledo (Table 1). This difference may have been because of lower lake productivity in the central basin or because colonization near Cleveland had not proceeded as far as it had near Toledo. While veliger densities in the western basin near Toledo in 1990 were similar to densities in the island region of the western basin in 1989 (Haag and Garton, 1992), veliger densities were more uniform over the reproductive period near Toledo (Table 1). Greater temporal uniformity in 1990 (as compared to 1989) suggests that as the zebra mussel population became more established reproduction became more diverse among subpopulations, leading to a more consistent and extended period of veliger production.

Veliger densities in Lake Erie appear to be high, relative to its overall size, when compared to other lakes (Table 1). These lakes are generally rank-ordered by surface area (Table 2). Overall, with the exception of Lake Erie, the general trend is for large, deep lakes to have lower veliger densities than small, shallow lakes.

The length of time that veligers were found in Lake Erie appears to be greater than in many other lakes (Table 1). This is probably related to the longer time period during which water temperature was high enough for mussel reproduction. Lake Erie is relatively shallow and more southerly than many of the other lakes considered. The only group of lakes that had veligers for as long a seasonal period as Lake Erie were the Konin Lakes (Table 1). These lakes are used for cooling waters for power plants and thus have warmer temperatures than expected from their latitude and morphometry. Hence, Lake Erie emerges as a lake that provides excellent habitat for the zebra mussel.

ACKNOWLEDGMENTS

This research was supported by a contract from the American Water Works Association Research Foundation and a grant from the City of Toledo, OH. Additional support and cooperation were provided by the Cities of Cleveland and Oregon, OH, and by Finkbeiner, Pettis, and Strout, Ltd. (Consulting Engineers), Toledo, OH.

REFERENCES

Biryukov, I. N., M. Ya. Kirpichenko, S. M. Lyakhov, and G. I. Segeeva. "Living Conditions of the Mollusk *Dreissena polymorpha* Pallas in the Babinski Backwater of the Oka River," in Biology and Control of Dreissena. B. K. Shtegman, Ed. *Tr. Inst. Biol. Vnut. Vod,* 7(10):32–38 (1964) (Translated 1968 from Russian, Israel Program for Scientific Translations, Jerusalem, Israel).

Dyga, A. K., and V. L., Galinskij. "Sutoconoe vertikalnoe raspredelenie licinok drejsseny v Zaporozskom vodochranilisce v svjazi s zascitoj sistem vodosnabzenija ot bioobrastanija," *Tech. Gidrobiol.* 16:53–56 (1975).

Federal Water Pollution Control Administration (FWPCA). Lake Erie Report, A Plan for Water Pollution Control, Washington, DC (1968).

Fortunatov, M. A. "Physical Geography of the Volga Basin," in *The River Volga and Its Life,* Mordukhai-Boltovskoi, P. D., Ed. (The Hague, The Netherlands: W. Junk Publishers,1979) pp. 1–29.

Griffiths, R. W., W. P. Kovalak, and D. W. Schloesser. "The Zebra Mussel, *Dreissena polymorpha* (Pallas, 1771), in North America: Impact on Raw Water Users," in Symposium: Service Water Systems Affecting Safety-Related Equipment (Palo Alto, CA: Nuclear Power Division, Electric Power Research Institute, 1989), 11–26.

Haag, W. R., and D. W. Garton. "Synchronous Spawning in a Recently Established Population of the Zebra Mussel, *Dreissena polymorpha*, in Western Lake Erie," *Hydrobiologia,* 234:103–110 (1992).

Herdendorf, C. E. "Large Lakes of the World," *J. Great Lakes Res.* 8:379–412 (1982).

Hillbricht-Ilkowska, A., and A. Stanczykowska. "The Production and Standing Crop of Planktonic Larvae of *Dreissena polymorpha* Pall. in Two Mazurian Lakes," *Pol. Arch. Hydrobiol.* 16:193–203 (1969).

Kirpichenko, M. Ya. "Phenology, Population Dynamics, and Growth of *Dreissena* larvae in the Kuibyshev Reservoir," in Biology and Control of Dreissena, B. K. Shtegman, Ed. *Tr. Inst. Biol. Vnut. Vod,* 7(10):15–24 (1964) (Translated 1968 from Russian, Israel Program for Scientific Translations, Jerusalem, Israel).

Lewandowski, K. "The Role of Early Developmental Stages in the Dynamics of *Dreissena polymorpha* (Pall). (Bivalvia) Populations in Lakes. I. Occurrence of Larvae in the Plankton," *Ekol. Pol.* 30:81–109 (1982a).

Lewandowski, K. "The Role of Early Developmental Stages in the Dynamics of *Dreissena polymorpha* (Pall.) (Bivalvia) Populations in Lakes. II. Settling of Larvae and the Dynamics of Numbers of Settled Individuals," *Ekol. Pol.* 30:223–286 (1982b).

Lewandowski, K., and J. Ejsmont-Karabin. "Ecology of Planktonic Larvae of *Dreissena polymorpha* (Pall.) in Lakes with Different Degree of Heating," *Pol. Arch. Hydrobiol.* 30:89–101 (1983).

Lewandowski, K., and A. Stanczykowska. "VI. Molluscs in Lake Zarnowieckie," *Pol. Ekol. Stud.* 12:315–330 (1986).

Mackie, G. L., W. N. Gibbons, B. W. Muncaster, and I. M. Gray. "The Zebra Mussel, *Dreissena polymorpha*; A Synthesis of European Experiences and a Preview of North America," A report prepared for the Ontario Ministry of the Environment, Water Resources Branch, Great Lakes Section. Queen's Printer, Toronto, Ontario (1989).

McMahon, R. F. "European Freshwater Macrofouling Bivalve, *Dreissena polymorpha* (zebra mussel), Gets a Foothold in the Great Lakes," Liaison Report to Electric Power Research Institute, Service Water Working Group, Richmond, VA (June 1989).

Prokhorov, A. M. *Great Soviet Encyclopedia Vol. 12* 3rd ed. (Moscow: Sovetskaia Entsiklopediia Publication House, 1973) (Translated 1976 from Russian, MacMillan Publishing Company, New York).

Roberts, L. "Zebra Mussel Invasion Threatens U.S. Waters," *Science* 249:1370–1372 (1990).

Stanczykowska, A. "Ecology of *Dreissena polymorpha* (Pall.) (Bivalvia) in Lakes," *Pol. Arch. Hydrobiol.* 24:461–530 (1977).

Walz, N. "Studies on the Biology of *Dreissena polymorpha* Pallas in the Lake of Constance," *Arch. Hydrobiol., Suppl.* 42:452–482 (1973).

Whipple, G. C. *The Microscopy of Drinking Water.* 4th ed. (New York: John Wiley & Sons, Inc., 1947).

Zhdanova, G. A., and S. L. Gusynskaya. "Distribution and Seasonal Dynamics of *Dreissena* larvae in Kiev and Kremenchug Reservoirs," *Hydrobiol. J.* 3:35–40 (1985).

CHAPTER 8

Distribution of Zebra Mussel *(Dreissena polymorpha)* Veligers in Eastern Lake Erie during the First Year of Colonization

Howard P. Riessen, Thomas A. Ferro, and R. Allan Kamman

The zebra mussel, *Dreissena polymorpha*, was first discovered in North America in Lake St. Clair in 1988 and has subsequently spread rapidly eastward through Lake Erie and into Lake Ontario. The seasonal and spatial distribution of zebra mussel veligers was investigated during the initial colonization of this animal in the eastern end of Lake Erie during 1989 and 1990. Veligers were first observed in this part of Lake Erie in late summer 1989 at relatively low densities (≤3/L). During 1990, veligers first appeared in the water column in mid-July, increased to peak densities in August (ranging from 6 to 137/L), and declined sharply during September. Veliger densities were about an order of magnitude higher at inshore stations than in offshore areas of the lake during both July and August. In general, veligers were uniformly distributed in the water column during daytime, both at inshore and offshore stations. While veliger densities increased sharply in eastern Lake Erie during the first year of colonization, densities were less than one-half those observed in the western basin at the same time.

INTRODUCTION

The life cycle of the zebra mussel, *Dreissena polymorpha*, is unusual for a benthic freshwater invertebrate in that it contains a free-swimming planktonic larval stage (veliger). This veliger larva is the primary agent of dispersal for the species (Mackie et al., 1989), and is mainly responsible for the recent rapid spread of zebra mussels throughout the lower Great Lakes. Within 2.5 years after its initial discovery in Lake St. Clair in June 1988 (Hebert et al., 1989), the zebra mussel had spread through passive transport of veliger larvae by water currents downstream throughout all of Lake Erie and into Lake Ontario (Griffiths et al., 1991). In addition to its role in the dispersal of the species within and between bodies of water, the veliger is also the life cycle stage that invades water intake pipes, resulting (after settlement and growth) in obstruction of the pipes and subsequent economic problems for power plants, water treatment facilities, and industries.

In this chapter we document the seasonal and spatial distribution of zebra mussel veligers in eastern Lake Erie during 1989-1990, the first year of colonization in this part of the lake. We specifically examine: (1) the timing and dynamics of the zebra mussel invasion in eastern Lake Erie, (2) seasonal veliger densities immediately following this invasion, (3) differences in veliger densities between inshore and offshore areas of the lake, and (4) vertical distribution of veligers in the water column. This type of information is important to understand the population dynamics of zebra mussels in Lake Erie and the other Great Lakes.

METHODS

Veligers were quantitatively sampled at 10 stations in the eastern end of Lake Erie during the ice-free seasons of 1989 and 1990 (Figure 1, Table 1). Samples were collected at these stations by taking vertical net tows through the entire water column of the lake with a 75-cm diameter, 130-μm mesh plankton net. Mesh size of the net used was larger than the smallest veligers (90–120 μm body length) present, potentially allowing some to escape. This could have caused an underestimate of total veliger density; however, samples collected simultaneously with a 61-μm mesh Schindler-Patalas trap indicated that these small veligers were relatively scarce in the water column.

During 1989, one station (4) was sampled every 2 weeks from May through mid-September and once again in late October. Two other stations were sampled occasionally in 1989: station 5 in mid-August and early November, and station 2 in mid-September. Stations 4 and 5 are at the western edge of the region of Lake Erie sampled in this study (Figure 1), and thus would be the first to have veligers drifting into the area (via water currents) from populations located farther west in the lake. During 1990, stations 1–4

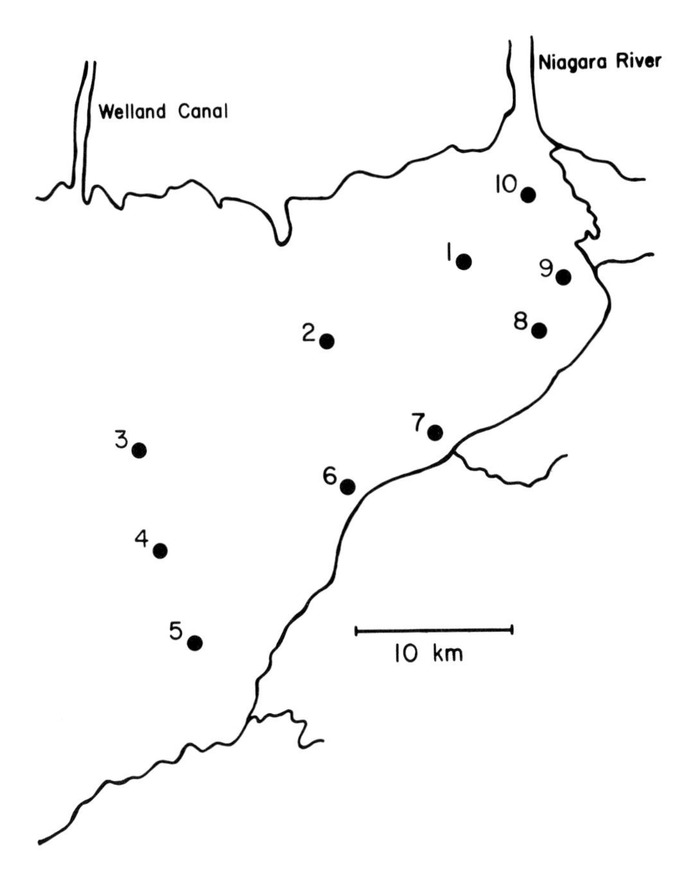

Figure 1. Location of sampling stations in eastern Lake Erie.

Table 1. Stations Sampled in Eastern Lake Erie during 1989–1990

Station	Location (Latitude, Longitude)	Distance from Shore (km)	Depth (m)
1	42° 48.78′, 78° 58.25′	6.7	15
2	42° 46.70′, 79° 04.15′	6.5	19
3	42° 42.97′, 79° 14.30′	14.8	22
4	42° 39.79′, 79° 12.38′	9.3	22
5	42° 36.84′, 79° 10.61′	4.9	18
6	42° 41.82′, 79° 03.99′	1.0	10
7	42° 43.87′, 78° 59.30′	1.5	10
8	42° 46.80′, 78° 54.41′	2.9	10
9	42° 48.50′, 78° 53.90′	1.8	10
10	42° 51.35′, 78° 54.40′	2.4	10

and 6–10 were sampled monthly from May through September. Five of these stations (6–10) were located in inshore areas (10m depth, 1–3 km from shore), while the other four stations (1–4) were in offshore areas (15–22 m depth, 7–15 km from shore) (Figure 1, Table 1). For each sampling date in 1990, differences in veliger densities between inshore ($n = 5$ stations) and offshore ($n = 4$ stations) areas of the lake were analyzed with *t*-tests.

The vertical distribution of veligers in the lake was determined at several stations in daytime (between 0845 and 1630) during summer 1990 by sampling at 2-m intervals in the water column from the surface to near bottom with a 31.7-L Schindler-Patalas plankton trap fitted with a 61-μm mesh net. On July 18, one vertical series was sampled from station 1 (offshore), while on August 16, vertical series were sampled at two offshore stations (1, 3) and two inshore stations (6, 9).

RESULTS

Veligers were not found in the 1989 samples collected at station 4 until mid-September (density = 3/L). They were, however, present in very low densities (<1/L) in samples collected at station 5 in mid-August. The earlier appearance of veligers at station 5 is probably due to its closer proximity to shore, where veliger densities are likely to be higher. Samples collected from station 2 (farther to the east) in mid-September revealed veliger densities of <1/L, an order of magnitude lower than densities at station 4 at the same time. Higher densities would be expected at the more western stations during this initial invasion process because the veligers are passively carried into the area by water currents from their population source located farther to the west in Lake Erie. Densities declined rapidly after the initial appearance of veligers in September. By late October, veliger densities at station 4 had decreased to <1/L; by the first week in November, no veligers could be found in the water column.

Veligers were absent from the water column during May and June 1990, and first appeared in mid-July. They were very common in July, with densities ranging from <1 to 2/L at offshore stations, and 7 to 32/L at inshore stations (Figure 2). Densities peaked during August in 1990 at relatively high densities, ranging from 6 to 20/L at offshore stations, and 27 to 137/L at inshore stations. There was a sharp (2–22×) increase in veliger densities between July and August at 8 of 9 sampling stations and a modest increase at the other station. There was also high variability in densities among sampling stations on a given date, much of it due to differences between inshore and offshore stations. Veliger densities were significantly higher (approximately an order of magnitude difference) at the inshore stations than in offshore areas of the lake during both July and August (*t*-tests, $p < 0.05$). The concentration of veliger larvae inshore and their patchy distribution in the lake also appear to

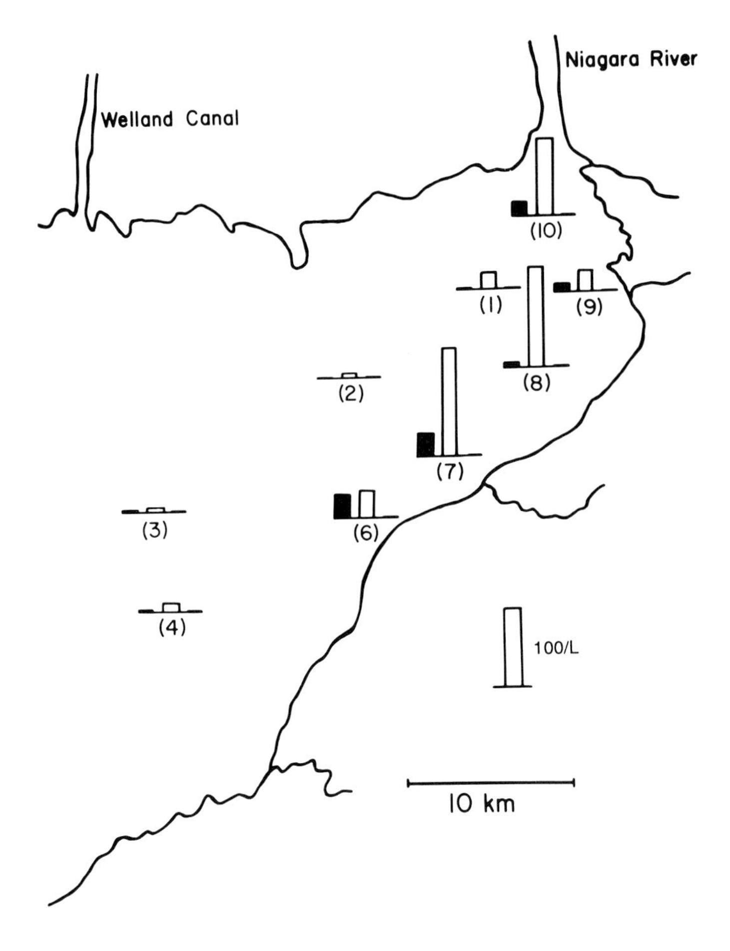

Figure 2. Veliger densities of *Dreissena polymorpha* in eastern Lake Erie in 1990. Vertical bars represent densities from net tows taken at each station on July 18 (left bar), August 16 (middle bar), and September 27 (right bar). Station numbers are shown in parentheses.

be common features in European lakes (Mackie et al., 1989). Densities of veligers declined rapidly at each station during September, reaching very low levels ($<$1/L) in both inshore and offshore areas by the end of this month (Figure 2). There was no significant difference in densities between inshore and offshore areas at this time (t-test, $p > 0.8$).

Veligers were uniformly distributed throughout the water column during daytime, both at inshore and offshore stations (Figures 3–5). Some unevenness in their distribution occurred, but these deviations from uniformity were not usually large or consistent among the various sampling stations. At station 1 on July 18, and August 16, 1990, there were slightly higher densities near

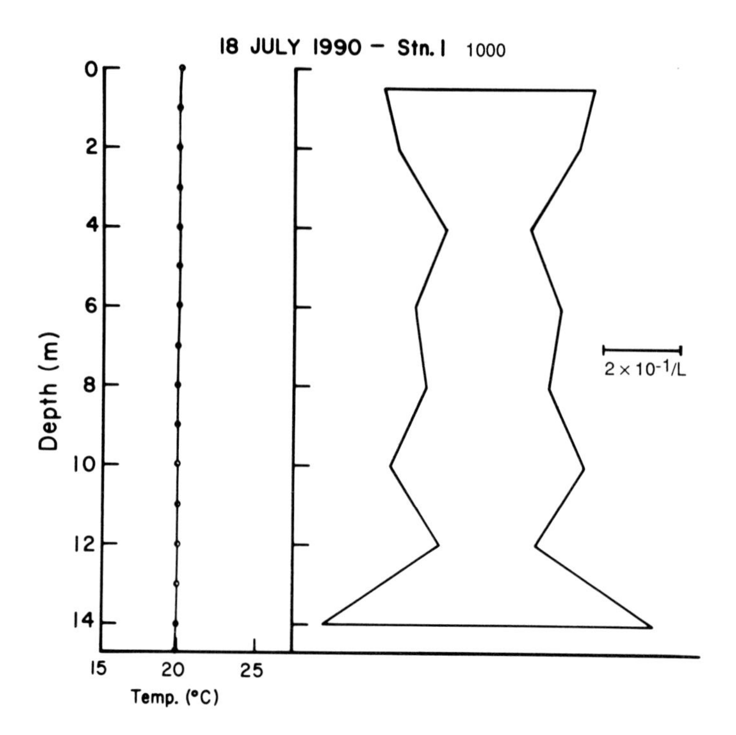

Figure 3. Vertical distribution of *Dreissena polymorpha* veligers and temperature profile in the water column at offshore station 1 on July 18, 1990 (time = 1000).

the bottom (Figures 3–4), while at station 9 densities near the bottom were lower than in the rest of the water column (Figure 5). At stations 1 and 6 on August 16, veliger densities were lower near the surface (Figures 4–5), but this was not observed at the other stations. At the deepest station (3), veliger densities were very uniform from surface to bottom (Figure 4). This fairly uniform vertical distribution pattern was similar to that found in the western basin of Lake Erie when wind velocities exceeded 8 km/hr (Fraleigh et al., 1992). It was different, however, from patterns observed in most European lakes, where veligers are typically found in highest densities in the upper half, but usually not in the top 1–2 m, of the water column (Mackie et al., 1989).

DISCUSSION

Zebra mussel veligers first invaded the eastern end of Lake Erie during late summer 1989 as water currents carried them in from established populations located farther to the west in the lake. This initial appearance occurred

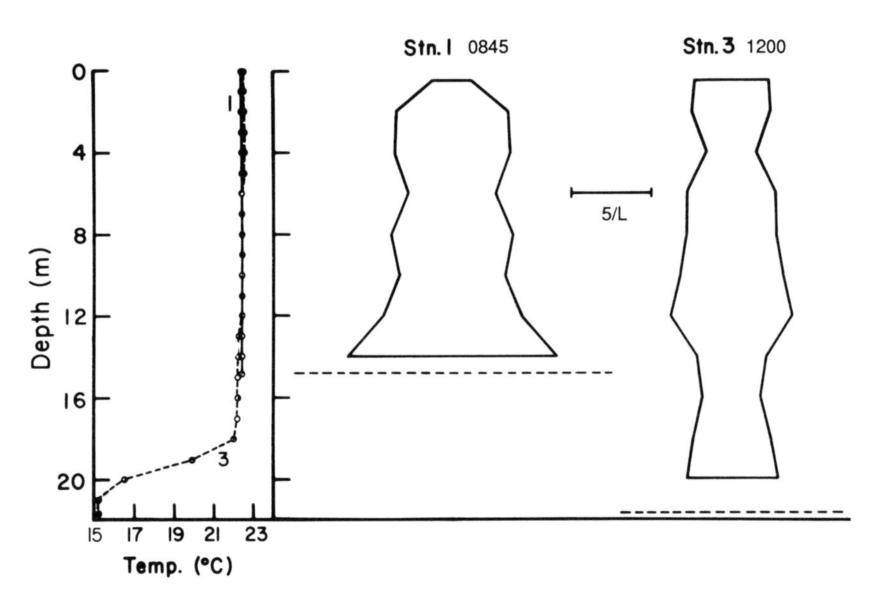

Figure 4. Vertical distributions of *Dreissena polymorpha* veligers and temperature profiles in the water column at offshore station 1 (time = 0845) and offshore station 3 (time = 1200) on August 16, 1990. Horizontal dashed lines indicate location of lake bottom at each station.

about 15 months after the original discovery of zebra mussels in Lake St. Clair in June 1988 (Hebert et al., 1989). Veliger densities in the eastern end of Lake Erie remained low during 1989 ($\leq$3/L) as the initial invasion occurred late in the growing season and there was as yet no established reproducing population in the eastern part of the lake. Veligers introduced from the west at this time resulted in higher densities in the western portion of the study area (station 4) than in areas farther east (station 2). Veliger densities in arriving water currents were probably higher closer to shore than toward the middle of the lake because the source population of zebra mussels that produced these veligers was probably concentrated in shallower inshore areas of the lake. This would explain the earlier arrival of veligers in our study area at station 5, which is located close to shore, than at offshore station 4.

The establishment of a population of settled juvenile zebra mussels in the eastern end of Lake Erie occurred shortly after veligers were first observed in late summer 1989. Settled mussels were first discovered near station 6 (Figure 1) in November 1989 and in the Niagara River area the following month (O'Neill, 1991).

The absence of veliger larvae in the water column until July in 1990 is undoubtedly a result of the age structure of the zebra mussel population

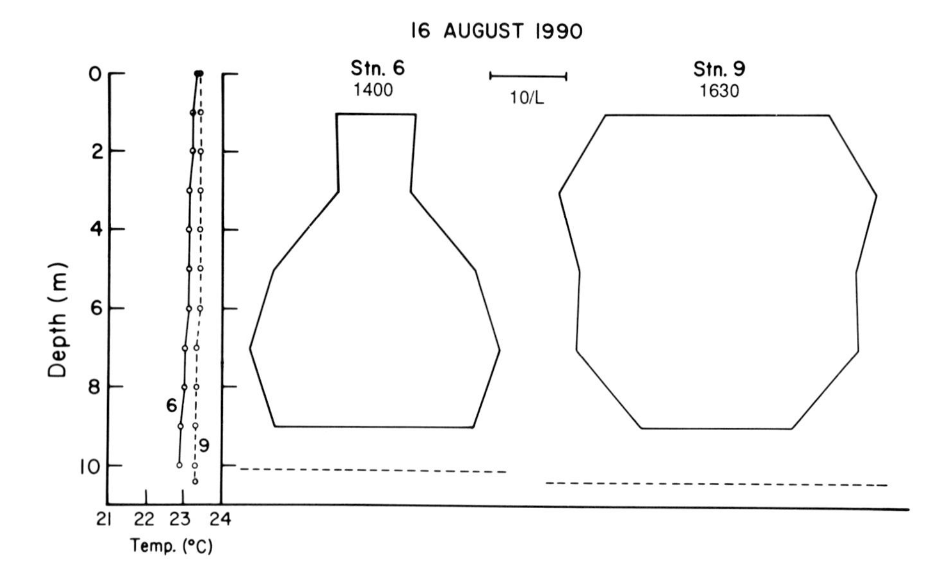

Figure 5. Vertical distributions of *Dreissena polymorpha* veligers and temperature profiles in the water column at inshore station 6 (time = 1400) and inshore station 9 (time = 1630) on August 16, 1990. Horizontal dashed lines indicate location of lake bottom at each station.

(absence of adults until summer 1990) and not of the water temperature in the lake. Veliger production in a lake can begin in late spring once the water temperature reaches 12–16°C (Mackie et al., 1989). In 1990, these temperatures were reached by mid-June in the eastern end of Lake Erie (personal observation), yet the production of veligers did not begin for at least an additional 2–3 weeks. Since it takes about 1 year for most settled zebra mussels to reach maturity (Mackie, 1991), the first cohort in the eastern end of Lake Erie would be expected to begin reproduction in mid- to late summer 1990. Veliger production in 1990 began in July and peaked in August, approximately 10–12 months after the initial introduction of zebra mussels into the area. This is slightly earlier than expected and suggests that the age at first reproduction in this population is less than 1 year (probably about 11 months).

Except for the delay in veliger production until July in eastern Lake Erie during 1990, the seasonal pattern of veliger population dynamics in this part of the lake was similar to that observed in the western basin. In western Lake Erie during 1989 and 1990, veligers were first seen in late May or early June, exhibited maximum densities in early to mid-August, and then declined rapidly in late August and early September (Haag and Garton, In press; Fraleigh et al., 1992). In most European lakes that have been studied, veligers do not become common until June, and exhibit peak densities in either July or August (Fraleigh et al., 1992).

Peak densities of veligers in the eastern end of Lake Erie during 1990 averaged about 85/L at inshore (10m depth) stations, and exceeded 100/L at two of these stations. In the relatively shallow (mostly <10m depth) western basin of Lake Erie during 1989 (the second year of the invasion in this part of the lake), peak veliger densities varied between 200 and 500/L (Leach, 1990; Garton and Haag, 1990; Haag and Garton, In press). Similar peak densities were also found during 1990 in the western basin (Fraleigh et al., 1992). Veliger densities in the eastern end of Lake Erie in 1990 were therefore only about 20–40% of those observed in the western basin. These densities, however, are in the range of those characteristically found in a variety of European lakes (Fraleigh et al., 1992).

Veliger densities increased sharply in eastern Lake Erie during the first year of colonization (1989–1990) in this part of the lake. At station 4 there was a 4-fold increase and at station 2 there was a 21-fold increase in maximum veliger densities between 1989 and 1990. If maximum seasonal densities at all offshore stations are averaged, there has been approximately a 7-fold increase in veliger densities during this time. We expect these densities to continue to increase for at least the next couple of years, but whether they will eventually reach the exceedingly high levels observed in the western basin of Lake Erie is as yet unclear. The western basin is more productive than the eastern basin and contains many shoals and rocky reefs, which provide an ideal substrate for adult zebra mussels. Higher food concentrations and a somewhat more suitable habitat for adults in the western basin may keep veliger densities at higher levels than in the eastern basin of Lake Erie.

ACKNOWLEDGMENTS

We thank Charles Merckel, captain of the R/V Hutchinson, and the following people who helped with the sampling routine: R. Snyder, D. Adrian, E. Kozuchowski, R. Cull, J. Hughes, and B. Johns. We also thank R. Snyder, P. Fraleigh, and two anonymous reviewers for their critiques of the manuscript. Funding for this study was provided by a grant from the New York Sea Grant Institute (R/FO-1-PD) and a National Science Foundation Research Experiences for Undergraduates Site Award (DIR 9000846).

REFERENCES

Fraleigh, P. C. et al. "Abundance and Settling of Zebra Mussel (*Dreissena polymorpha*) Veligers in Western and Central Lake Erie," in T. F. Nalepa and D. Schloesser, Eds. Zebra Mussels: Biology, Impacts, and Control. 143–152. Lewis Publishers, Boca Raton, Florida, (1993).

Garton, D., and W. Haag. "Reproduction and Recruitment of *Dreissena* During the First Invasion Year in Western Lake Erie," Abstract, Zebra Mussels: The Great Lakes Experience, University of Guelph, Guelph, Ontario (February, 1990).

Griffiths, R. W., D. W. Schloesser, J. H. Leach, and W. P. Kovalak. "Distribution and Dispersal of the Zebra Mussel (*Dreissena polymorpha*) in the Great Lakes Region," *Can. J. Fish. Aquat. Sci.* 48:1381–1388 (1991).

Haag, W. R., and D. W. Garton. "Synchronous Spawning in a Recently Established Population of the European Zebra Mussel, *Dreissena polymorpha*, in Western Lake Erie," *Hydrobiologia* (In press).

Hebert, P. D. N., B. W. Muncaster, and G. L. Mackie. "Ecological and Genetic Studies on *Dreissena polymorpha* (Pallas): A New Mollusc in the Great Lakes," *Can. J. Fish. Aquat. Sci.* 46:1587–1591 (1989).

Leach, J. H. "Potential Ecological Impacts of the Zebra Mussel, *Dreissena polymorpha*, in Lake Erie," Abstract, Zebra Mussels: The Great Lakes Experience, University of Guelph, Guelph, Ontario (February, 1990).

Mackie, G. L. "Biology of the Exotic Zebra Mussel, *Dreissena polymorpha*, in Relation to Native Bivalves and its Potential Impact in Lake St. Clair," in M. Munawar and T. Edsall, Eds. Environmental Assessment and Habitat Evaluation of the Upper Great Lakes Connecting Channels. *Hydrobiologia* 219:251–268 (1991).

Mackie, G. L., W. N. Gibbons, B. W. Muncaster, and I. M. Gray. "The Zebra Mussel, *Dreissena polymorpha*: A Synthesis of European Experiences and a Preview for North America," Report prepared for the Ontario Ministry of the Environment, Water Resources Branch, Great Lakes Section, Toronto, Ontario (1989), 76 pp.

O'Neill, C. R. Personal communication (1991).

CHAPTER 9

Biology of the Zebra Mussel *(Dreissena polymorpha)* and Observations of Mussel Colonization on Unionid Bivalves in Lake St. Clair of the Great Lakes

Gerald L. Mackie

The life history of the zebra mussel, *Dreissena polymorpha*, was determined for populations in Lake St. Clair, May to November 1989. Gametological condition, length to dry-weight relationships, growth rate, life span, birth period, age structure, and population size were determined. In addition, observations of the effects of zebra mussels on native species of unionids were made. The population of mussels in Lake St. Clair consists of individuals that are relatively short-lived (i.e., 2 years) and fast-growing (about 20 mm/year), with most individuals generally achieving a total shell length of less than 30 mm. The smallest mussels usually appeared in late June to early July and again in late October to early November, thus providing evidence of two annual recruitment events. On a unit area basis, zebra mussels colonized unionids to a significantly greater extent than rocky substrates ($p < 0.05$). Zebra mussels were commonly found in and around the gape of unionids, preventing normal valve opening and closing. Larve colonies of zebra mussels were observed to completely envelope the siphonal region of some unionids.

0-87371-696-5/93/$0.00 + $.50

INTRODUCTION

The epicenter of the zebra mussel (*Dreissena polymorpha*) invasion in the Great Lakes and North American surface waters is Lake St. Clair (Hebert et al., 1989). If population growth follows patterns reported in Europe (Stanczykowska, 1977), the standing crop of this introduced species will increase initially and then decline to a base level. This chapter reports on standing crops of zebra mussels in their early colonization stages in Lake St. Clair. This information can be used in forming models to help predict the biology, ecology, and impact of zebra mussels in other North American waters. The overall objective of this study was to collect baseline data on the size, age, growth rate, sex ratio, and birth period(s) of the population, and also to document the colonization of zebra mussels on the native species of unionids.

METHODS

At monthly intervals from May 2 to November 15, 1989, three to five rocks were collected from Lake St. Clair at Puce, Ontario. Rocks varied from 5 to 30 cm in diameter. All mussels were scraped off the rocks and 10–20 individuals of all size classes were randomly selected and fixed in Bouin's fluid. Of these preserved specimens 40 were then randomly selected for sex determination and histological examination of gametogenesis. Another 75–100 individuals of all size classes were randomly selected from one rock for determination of length-dry weight (on fresh specimens) relationships. All other specimens were preserved in 70% ethanol for the derivation of length-frequency histograms. Histograms were used to determine: (1) population age and size class structure, (2) mean growth rate of each age class (cohort), (3) life span, and (4) birth period (i.e., period when larvae were present in the water). The growth rate of the 1989 cohort was determined by placing a cement block in 0.6 m of water and measuring the increase in shell length of newly settled mussels every 2 weeks.

Mean population size was determined by counting the mussels on each rock surface. The surface area of the rock was determined after mussel removal by placing a known area of aluminum foil around that part of the rock surface infested with mussels and pressing the foil into crevices and around protuberances. Excess foil was cut off and pressed flat, and its area measured with a planimeter. The standing crop of mussels on each rock (sampling unit) was expressed as number per square meter. A similar method was used to determine the population size (per unit area) of zebra mussels attached to unionids.

Birth periods were determined by sampling for veligers with a student plankton net (mouth 20 cm in diameter, mesh size 70 μm) at the same time rock samples were collected. The plankton sampler was swept back and forth several times in the top 0.3 m of water, and the plankton collected was preserved in 5% formalin.

For the determination of sex ratios, tissues were removed, embedded in paraffin, sectioned at 6 μm using a Spencer model 820 microtome, stained with Harris' modified hematoxylin, and then counterstained with eosin, following methods in Humason (1972). Only the posterior part of the visceral mass containing gonads was sectioned. The period of gametogenesis and length class at which gametogenesis begins were determined from histological examination of specimens in four length classes (0–5 mm, 5–10 mm, 10–15 mm, and >15 mm) of mussels collected between May and September.

The relationship between unionid size and zebra mussel abundance and size was determined by removing and counting all zebra mussels from each unionid shell, measuring the shell length of >50 mussels selected at random, and then plotting their numbers and mean lengths against shell length of the unionid.

RESULTS AND DISCUSSION

Analysis of length-frequency distributions of zebra mussels on rocks in Lake St. Clair indicated the presence of three distinct cohorts throughout most of the year, with the dominant age class being 1 year olds (Figure 1a–g). The smallest individuals appeared on two occasions, in late June to early July and in late October to early November. This suggests two major recruitment events (birth periods) per year. However, the presence of some small individuals throughout the summer suggests that some recruitment occurs continuously during the warmer months. The disappearance of the oldest (largest) age class by midsummer suggests that most individuals live only 3 years in Lake St. Clair.

Growth rate curves of each age class (cohort) are shown in Figure 2, with the dates representing time of birth. The spring 1989 cohort was observed on June 27 (57 days) and grew almost 20 mm by November 15, 1989 (194 days), giving a mean growth rate of 0.126 mm/day. The fall 1988 cohort grew nearly 20 mm in shell length within 194 days, but averaged only 0.100 mm/day. However, maximum growth rates of both fall 1988 and spring 1989 cohorts were similar, averaging nearly 0.5 mm/day. The cohort born in spring 1988 had similar growth rates, but the cohort died at the end of July, indicating a cohort life span of only 1.5 years. The 1987 cohort was probably born in summer or fall 1987, giving this cohort a life span of only 2 years. Hence, these data suggest that the life span of most zebra mussels in Lake St. Clair is only 2 years.

Interpretation of length-frequency data from mussels collected on rocks is supported by similar data from mussels settled on concrete blocks placed in the water during the first week of June 1989 (Mackie, 1991). Based on these latter data, most mussels grew 15–20 mm each year and lived only 1.5–2 years. Growth was most rapid in newly settled and first-year

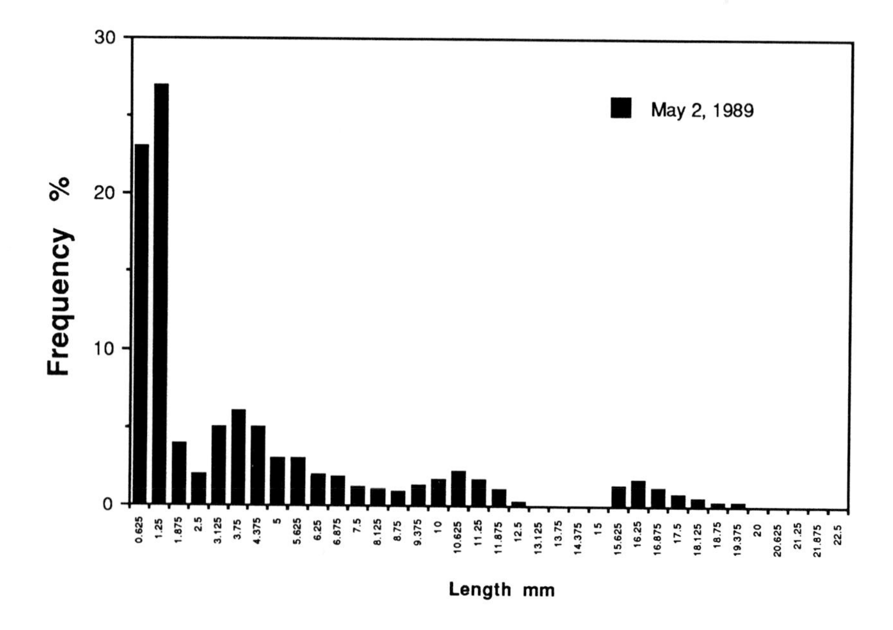

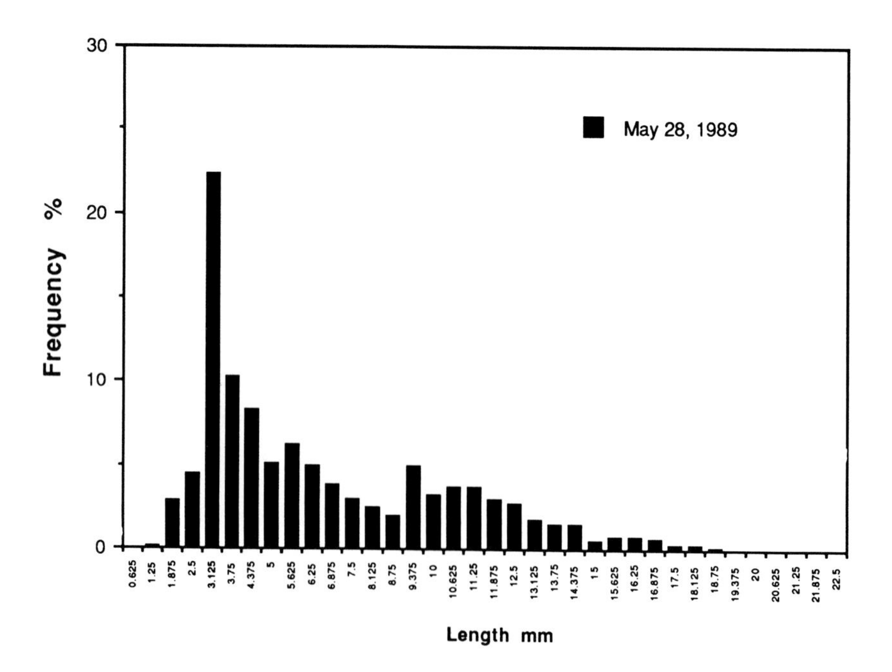

Figure 1. Length-frequency distribution of *Dreissena polymorpha* in Lake St. Clair between May 2 and November 22, 1989.

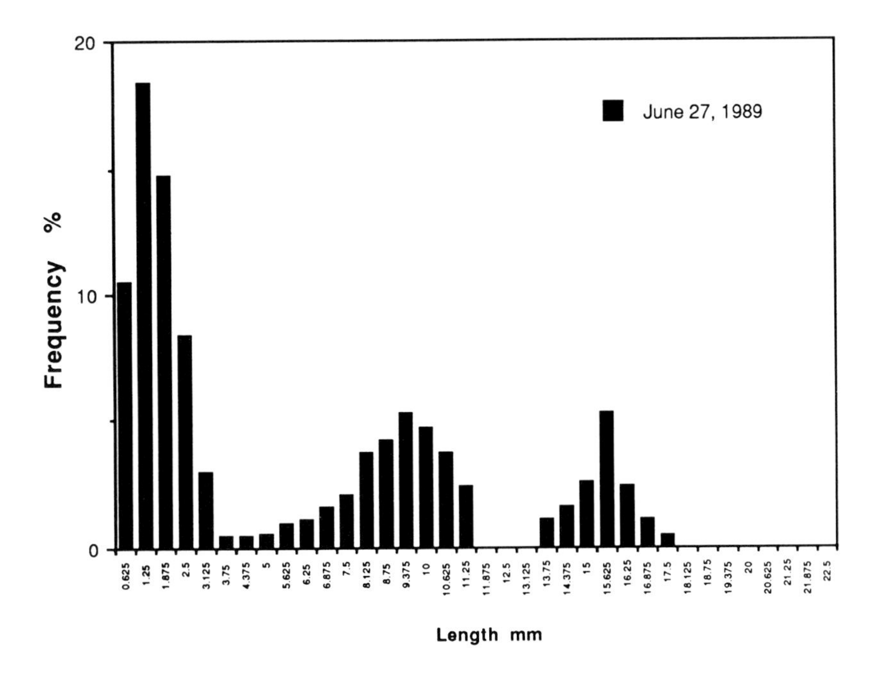

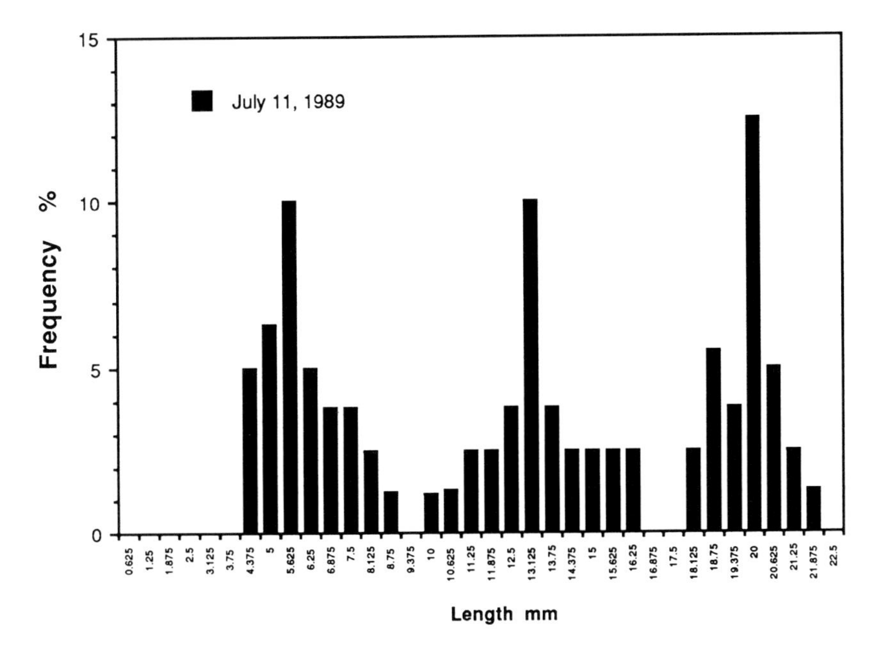

Figure 1 (cont'd)

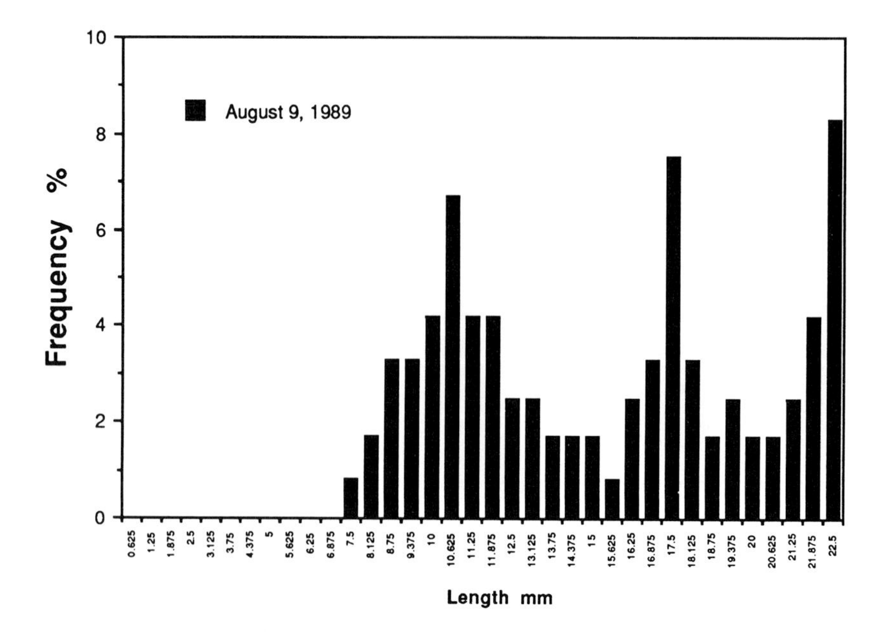

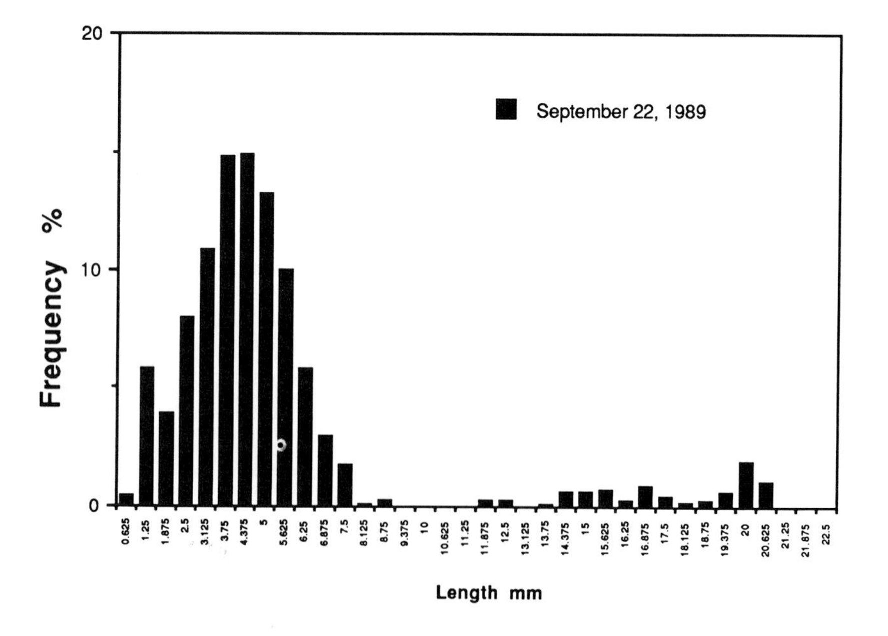

Figure 1 (cont'd)

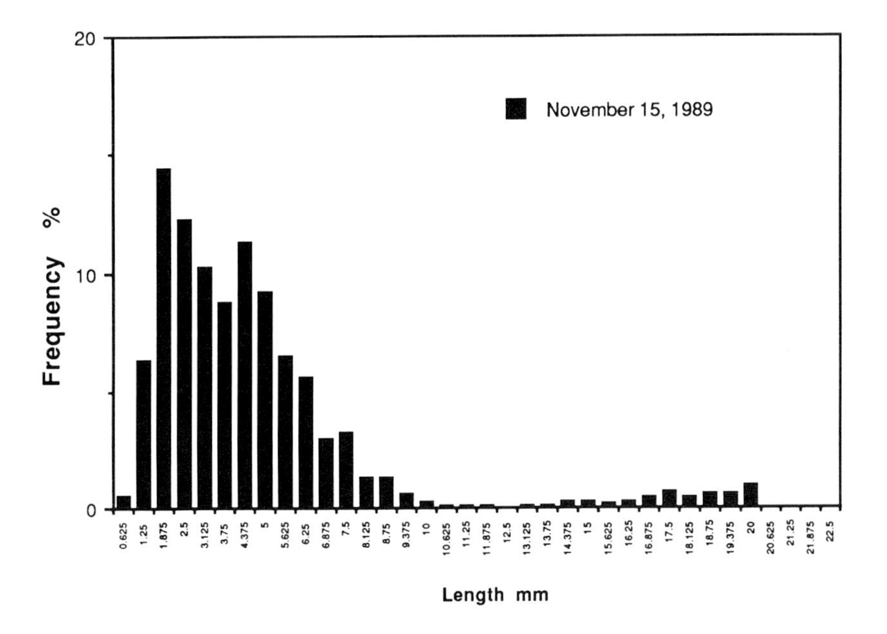

Figure 1 (cont'd)

individuals, and decreased with increasing age, which is typical for the species (Morton, 1969; Stanczykowska, 1977).

The life span and overall growth rate of zebra mussels in Lake St. Clair were different from those reported for European populations. The life span of *D. polymorpha* was found to be 3–5 years in Polish lakes (Wiktor, 1963; Stanczykowska, 1977), 3.5 years in British reservoirs (Morton, 1969), 6–7 years in Swiss lakes (Stanczykowska, 1977), and 6–9 years in some Russian reservoirs (Mikheev, 1964; Lyakhov and Mikheev, 1964). These variable life spans can be categorized into two groups. The first group is slow growing and has a maximum shell length greater than 40 mm. This group is typical of populations in Poland, Austria, Hungary, Switzerland, and Russia (Mackie et al., 1989). The second group is fast growing and has a maximum shell length of less than 35 mm. This group is typical of populations in Britain (Mackie et al., 1989). The population of *D. polymorpha* in Lake St. Clair seems to be most similar to the second group of fast-growing (about 20 mm/year), small adults (generally less than 30-mm maximum size) which live for only 2 years.

The allometry of shell growth is shown in Figure 3. Shell width is described by the relationship: width = 0.048 + 0.488 length (r^2 = 0.99), and shell height is described by the relationship: height = 0.161 + 0.208 length + 0.072 length^2. The length-weight relationship is typical of most animals with a cubic relationship: total weight (mg) = 0.051 length$^{2.996}$; tissue

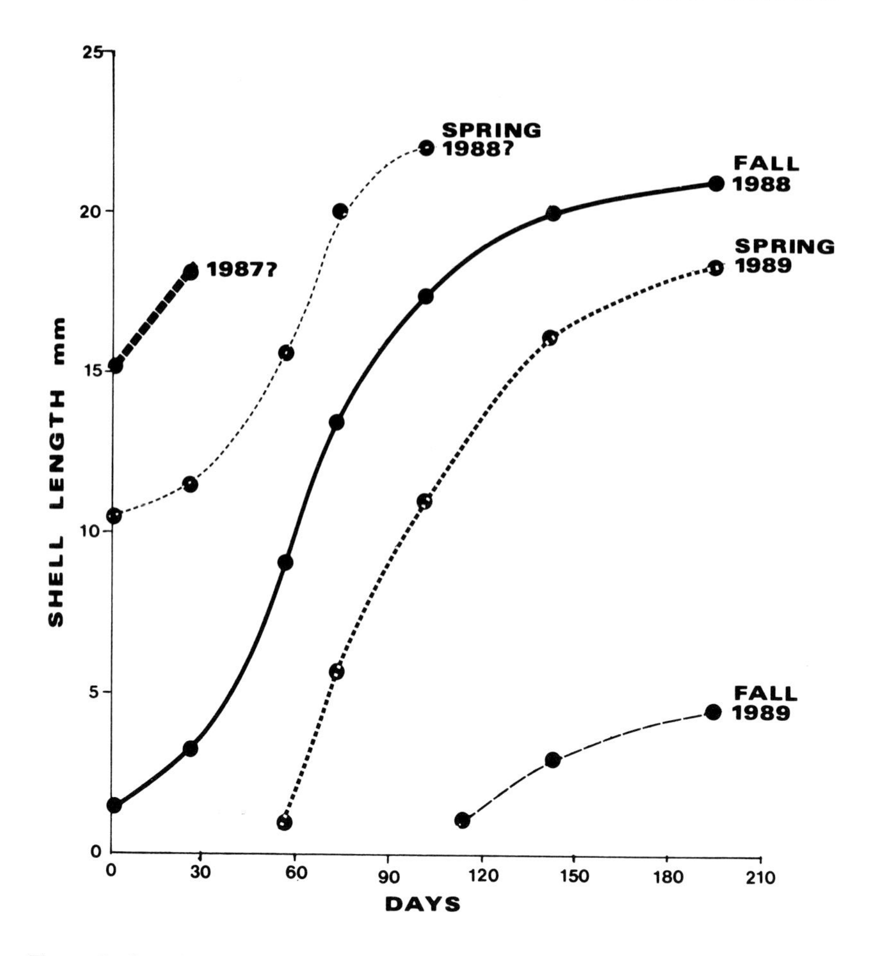

Figure 2. Growth rates of cohorts of *Dreissena polymorpha* on rocks in Lake St. Clair, based on data in Figure 1. Days represent time from birth for the cohort.

weight = 0.007 length$^{2.982}$ (Figure 4). The allometry of shell growth is within variations reported for British (Morton, 1969) and Polish (Stanczykowska, 1977) populations of *D. polymorpha*. In Europe, the length-weight relationship varies considerably among lakes (Lyakhov and Mikheev, 1964; Walz, 1978a and 1978b; Stanczykowska, 1979) and over time (Stanczykowska et al., 1975), and may be an indicator of environmental conditions (Stanczykowska, 1979).

The sex ratio of zebra mussels in Lake St. Clair in 1989 was 3:2 (F:M). This compared to a sex ratio of 1:1 in most populations in Poland (Stanczykowska, 1977). Gametogenesis had already begun and gonads were ripe with both eggs and sperm in the first (May 2) collection of the year (Gillis, 1989) suggesting that gametogenesis began in the winter or early spring. In European

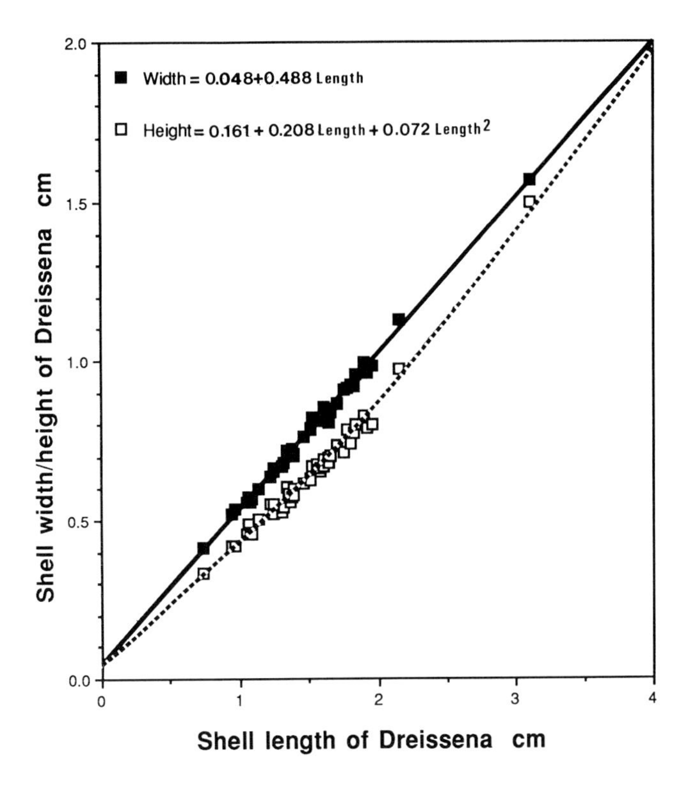

Figure 3. Allometry of shell growth in *Dreissena polymorpha* in Lake St. Clair, 1989. (From Mackie, G., *Hydrobiologia,* 219:251–268, 1991. With permission.)

populations, oogenesis begins in fall, followed by intensive growth of oocytes in early spring (Antheunisse, 1963). In Lake St. Clair, some eggs were released from the ovaries in late May, but the peak period of reproduction was in July and August when most ovaries appeared to be partially spent. The ovaries were completely spent by the end of September. All stages of spermiogenesis were present by early May; that is, the enlarged seminiferous tubules contained primary and secondary spermatocytes, spermatids, and mature spermatozoa (Gillis, 1989). Release of sperm appeared to begin in late May, but most were released by July and August. The seminiferous tubules were completely spent by late September.

Reproduction began when adults were about 5 mm in shell length and was complete before they reached 10 mm. Of 12 specimens examined in the 5–10 mm length class, 9 were sexually mature. An even greater proportion (12 of 13) of individuals in the 10–15 mm length class were ripe. In European populations, gametogenesis begins in adults 8–9 mm in shell length (Morton, 1969b; Stanczykowska, 1977).

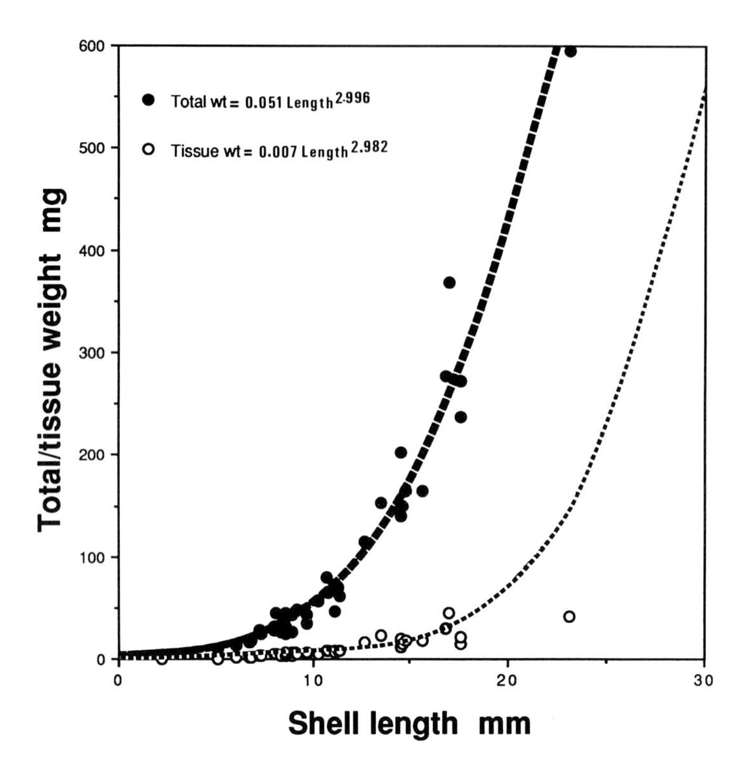

Figure 4. Length-weight relationships of *Dreissena polymorpha* in Lake St. Clair, 1989. (From Mackie, G., *Hydrobiologia,* 219:251–268, 1991. With permission.)

Veligers first appeared in Lake St. Clair (at Puce, Belle River, and Stoney Point) on June 14. No larvae were found in May 28 samples, indicating that veligers appeared sometime between May 28 and June 14 when water temperature was near 18°C. Since veligers require about 1 month to develop and settle (Stanczykowska, 1977), the mussels of <1-mm shell length found in early July had settled from the reproduction event in early June. The large number of mussels of <1-mm shell length found from September to November had probably settled from a reproductive event in late July to mid-August.

Colonization of zebra mussels on unionids increased with exposed surface area and length (Figures 5 and 6) of the unionid in a logarithmic relationship. Living unionids had attached an average of about 210 (±56 S.D.) zebra mussels per individual (estimate based on counts made before the 1989 recruitment event). When the exposed surface of the unionid was covered, new recruits settled on top of older recruits so that the colonization intensity increased greatly in relation to the size of the unionid (Figure 6). Some unionids were prevented from closing their valves because of mussels attached within the individual's gape (Figure 7a), while some unionids were prevented from opening their valves because the byssal threads were attached to both

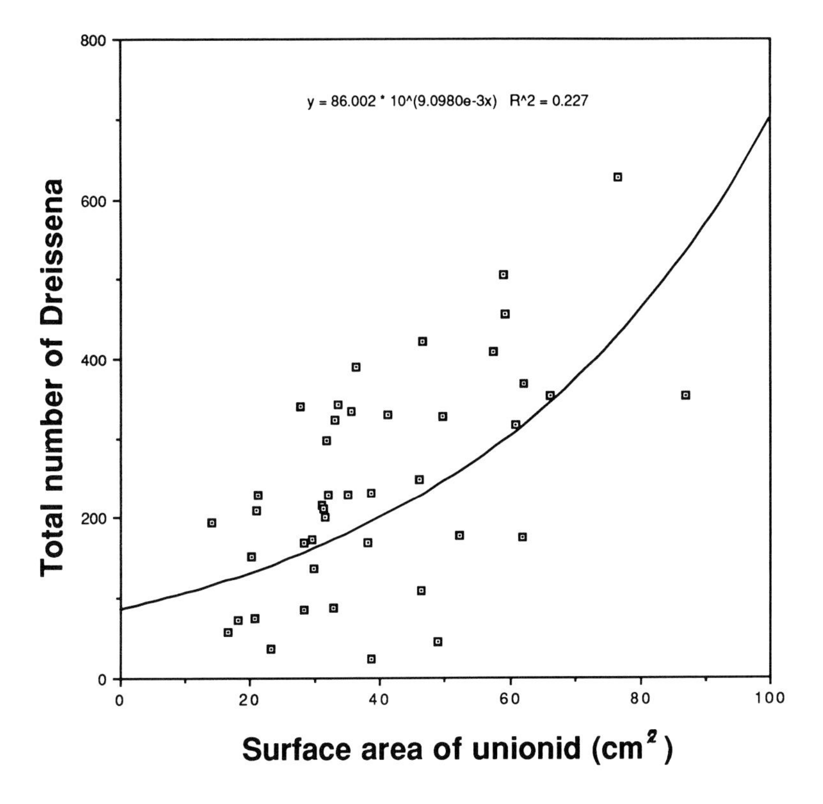

Figure 5. Relationship between exposed surface area of unionid shells (i.e., posterior end) and numbers of attached *Dreissena polymorpha* in Lake St. Clair, 1989. (From Mackie, G., *Hydrobiologia,* 219:251–268, 1991. With permission.)

valves (Figure 7b). These observations show that massive encrustations of zebra mussels on unionids: (1) prevent valve closure, which exposes the unionid to environmental extremes; and (2) prevent or limit valve gaping, which affects normal metabolic functions such as feeding, growth, respiration, excretion and/or reproduction.

ACKNOWLEDGMENTS

This study was funded by the Ontario Ministry of Environment. Ms. Diane Pathy, Ms. Patty Gillis, and Mr. Kevin Wingerden assisted with the field collections and supplied most of the length-frequency data.

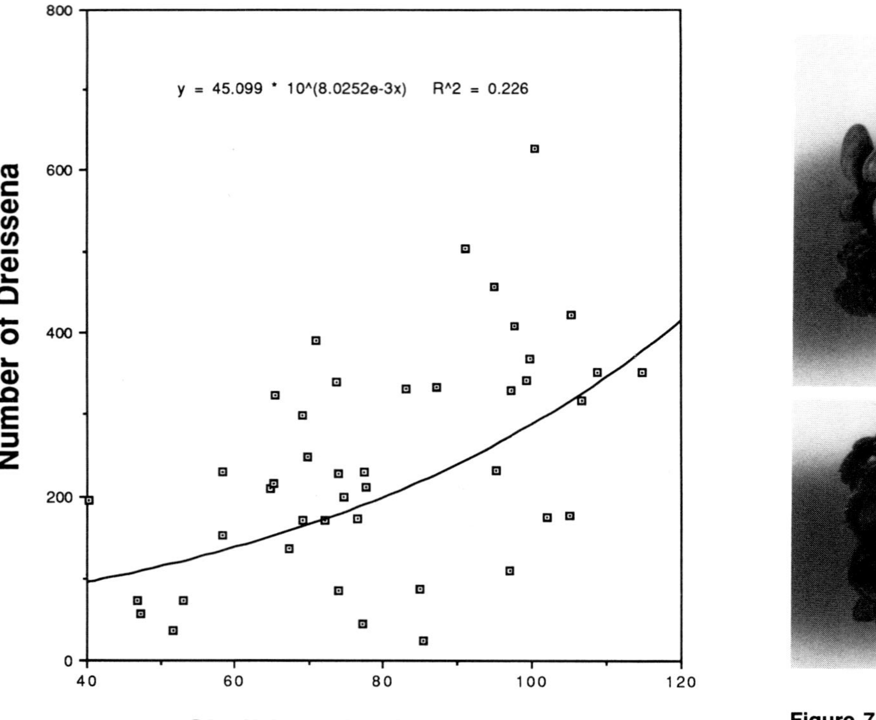

Figure 6. Relationship between shell length of unionid shells and numbers of attached *Dreissena polymorpha* in Lake St. Clair, 1989. (From Mackie, G., *Hydrobiologia,* 219:251–268, 1991. With permission.)

Figure 7. Infestations of *Dreissena polymorpha* that prevented unionid valves from fully opening (a) or closing (b). (From Mackie, G., *Hydrobiologia,* 219:251–268, 1991. With permission.)

REFERENCES

Antheunisse, L. J. ''Neurosecretory Phenomena in the Zebra Mussel *Dreissena polymorpha* Pallas,'' *Arch. Neerl. Zool.* 15:237–314 (1963).

Gillis, P. ''Gametogenesis in *Dreissena polymorpha*,'' Department of Zoology, University of Guelph, Unpublished report (1989).

Hebert, P. D. N., B. W. Muncaster, and G. L. Mackie. ''Ecological and Genetic Studies on *Dreissena polymorpha* (Pallas): A New Mollusc in the Great Lakes,'' *Can. J. Fish. Aquat. Sci.* 46:1587–1591 (1989).

Humason, G. L. *Animal Tissue Techniques* (San Francisco, CA: W. H. Freeman & Company, 1972).

Lyakhov, S. M., and V. P. Mikheev. ''The Population and Distribution of *Dreissena* in the Kuibyshev Reservoir Seven Years after its Construction,'' in Biology and Control of *Dreissena*, B. K. Shtegman, Ed. *Tr. Inst. Biol. Vnut. Vod. Akad. Nauk SSSR* 7:1–145 (1964). (Translated 1968, Israel Program for Scientific Translations, Jerusalem).

Mackie, G. L. ''Biology of the Exotic Zebra Mussel, *Dreissena polymorpha*, in Relation to Native Bivalves and Its Potential Impact in Lake St. Clair,'' in Environmental Assessment and Habitat Evaluation of the Upper Great Lakes Connecting Channels, M. Munawar and T. Edsall, Eds. *Hydrobiologia* 219:251–268 (1991).

Mackie, G. L., W. N. Gibbons, B. W. Muncaster, and I. M. Gray. ''The Zebra Mussel, *Dreissena polymorpha*: A Synthesis of European Experiences and a Preview for North America,'' Ontario Ministry of Natural Resources, Water Resources Branch, Great Lakes Section. Queen's Printer. Toronto, Ontario (1989).

Mikheev, V. P. ''Linear Growth of *Dreissena polymorpha* in Some Reservoirs of the European USSR,'' in Biology and Control of *Dreissena*, B. K. Shtegman, Ed. *Tr. Inst. Biol. Vnut. Vod. Akad. Nauk SSSR* 7:1–145 (1964) (Translated 1968, Israel Program for Scientific Translations, Jerusalem).

Morton, B. S. ''Studies on the Biology of *Dreissena polymorpha* Pall. III. Population Dynamics,'' *Proc. Malacol. Soc. London* 38:471–482 (1969).

Stanczykowska, A. ''Ecology of *Dreissena polymorpha* (Pall.) (Bivalvia) in Lakes,'' *Pol. Arch. Hydrobiol.* 24:461–530 (1977).

Stanczykowska, A. ''Size and Weight of *Dreissena polymorpha* from European Lakes,'' *Hydrobiol. J.* 14:20–23 (1979).

Stanczykowska, A., H. J. Schenker, and Z. Farara. ''Comparative Characteristics of *Dreissena polymorpha* (Pall.) in 1962 and 1972 in 13 Mazurian Lakes,'' *Bull. Acad. Pol. Sci., Ser. Sci. Biol.* 23:383–390 (1975).

Walz, N. ''The Energy Balance of the Freshwater Mussel *Dreissena polymorpha* in Laboratory Experiments and in Lake Constance. III. Growth Under Standard Conditions,'' *Arch. Hydrobiol., Suppl.* 55:121–141 (1978a).

Walz, N. ''The Energy Balance of the Freshwater Mussel *Dreissena polymorpha* in Laboratory Experiments and in Lake Constance. IV. Growth in Lake Constance,'' *Arch. Hydrobiol., Suppl.* 55:142–156 (1978b).

Witkor, L. ''Research on the Ecology of *Dreissena polymorpha* Pall. in the Szczecin Lagoon,'' *Ekol. Pol. Ser. A* 11:275–280 (1963).

CHAPTER 10

Colonization of Different Construction Materials by the Zebra Mussel *(Dreissena polymorpha)*

Bruce W. Kilgour and Gerald L. Mackie

To determine if zebra mussels (*Dreissena polymorpha*) have a preference for different construction materials, we examined the abundance on 16 materials that had been placed in Lake St. Clair for over 3 months. The materials were either 7.7 × 12.7-cm plates or 15 cm long × <7-cm diameter tubes. The abundance of mussels on plates ranged from 0/m^2 on copper to over 21,000/m^2 on asbestos and stainless steel; on tubes the abundance ranged from 22/m^2 on copper to 1467/m^2 on ABS. The following preference was exhibited by zebra mussels on the plates: copper < galvanized iron < aluminum < acrylic < PVS < Teflon™* < vinyl < pressure treated wood < black steel < pine < polypropylene < absestos < stainless steel. For tubes the preference was copper < brass < galvanized iron < aluminum < acrylic < black steel < polyethylene < 5.0-cm i.d. PVC < ABS < 4.0-cm i.d. PVC. Tubes that had a vertical orientation had significantly fewer mussels (mean = 382/m^2) than tubes with horizontal orientation (mean = 871/m^2). The results suggest that the abundance of zebra mussels on structures can be reduced by selection of appropriate construction materials and design.

* Registered trademark of E. I. du Pont de Nemours and Company, Inc., Wilmington, DE.

0-87371-696-5/93/$0.00 + $.50

INTRODUCTION

The zebra mussel, *Dreissena polymorpha*, is a biofouling mollusk that has recently entered the Greak Lakes watershed (Griffiths et al., 1989). There is a great deal of concern that this nuisance invertebrate will have a severe economic impact on both domestic and industrial water users along the Great Lakes system (Mackie et al., 1989). The general impact is a reduction in water flow caused by the attachment of live mussels and accumulation of dead shells in underground pipes, ducts, and channels. Procedures that have been used to control zebra mussels include: (1) chemical treatment with chlorine and other molluscacides; (2) physical treatment such as heating, ultrasonic vibration, electric current, variable water flow, screens, and microcosms; (3) biological treatment such as hormones and predation; (4) mechanical treatment with screens, filters, and scraping devices; and (5) sheet replacement (Mackie et al., 1989).

This study examines the efficacy of sheet replacement for reduction of impacts by *D. polymorpha*. Previous work has suggested that settlement of zebra mussels varies among different materials. For example, Walz (1973) ranked a set of eight materials for their tendency to be settled upon by zebra mussels. In terms of relative settlement, he found copper < brass < Plexiglas™ < concrete < aluminum < iron < PVC. In addition, Van Diepen and Davids (1986) suggested that settlement of zebra mussels could be reduced by replacing PVC with polystyrene.

In this study, the settlement of zebra mussels on flat sheets and tubes made of different materials is examined along with differences in settlement between the inside and the outside surfaces of the tubes. The latter tests the hypothesis that zebra mussels prefer to settle in protected locations.

MATERIALS AND METHODS

Three racks containing 13 plates (approximately 12.7 × 7.7 cm) made of different materials (Figure 1a) were placed in Lake St. Clair, at Puce, Ontario, on July 12, 1989. The plate materials included copper, black steel, aluminum, galvanized iron, PVC, teflon, vinyl, pressure-treated wood, pine, polypropylene, asbestos, acrylic, and stainless steel. These materials were selected for study because they are relatively common construction materials and are inexpensive. An additional three racks containing 20 tubes (approximately 15 cm long × <7.0 cm o.d.) made of 10 different materials (Figure 1b) were placed in Lake St. Clair, also at Puce, on August 22. The materials used for the tubes included copper, brass, aluminum, galvanized iron, acrylic, Teflon™, PVC (4.0 and 5.0 cm i.d.), black steel, and ABS. Racks holding the plates were removed on September 22, 1989, and those holding the tubes were removed on November 5, 1989. When the racks were removed from

a

b

Figure 1. Photographs of racks containing plates (a) and tubes (b) after colonization of zebra mussels in Lake St. Clair at Puce, Ontario (June–September 1989).

the lake, the plates, tubes, and attached mussels were allowed to air dry. This was done to reduce the detachment of mussels that can occur when they are preserved on ice or in ethanol or formaldehyde. All plates and tubes were scraped clean with a knife, and the mussels were sorted from algae and debris and counted under a dissecting microscope.

Data were analyzed separately for the plates and tubes. Prior to any statistical analyses all abundance values were $\log_{10}$ transformed. Analysis of

Table 1. ANOVA Tables to Test for Significant Differences in Abundance of *Dreissena polymorpha* on Different Racks and Plates Made of 13 Different Materials

Source	df	SS	MS	F	p	r^2
Rack	2	0.325	0.163	0.209	0.812	0.007
Error	56	43.648	0.779			
Material	12	39.759	3.313	36.169	0.000	0.904
Error	46	4.214	0.009			

Table 2. Results of Tukey's Multiple Comparisons Test

Material	Mean Number (per m^2)	Number of Complete Plates
Copper	0	1
Galvanized iron	548	2
Aluminum	2,324	2
Acrylic	6,896	3
PVC	7,471	3
Teflon(TM)	8,593	3
Vinyl	12,068	1
Pressure-treated wood	15,255	3
Black steel	15,420	2
Pine	16,117	2
Polypropylene	17,554	2
Asbestos	21,333	3
Stainless steel	21,812	2

Note: Means scored with the same line are not significantly different at the 0.05 level.

variance (ANOVA) was used to determine differences. For the plates, sources of variation included the rack and the material; while for the tubes sources of variation included the rack, orientation (i.e., whether the tube was horizontal or vertical), inside/outside surface, materials, and interaction terms. For both the plates and tubes, Tukey's multiple comparison test was used to determine significant differences between the different materials.

RESULTS

There was a significant difference in the abundance of mussels settling on the 13 different plate materials but there were no differences in abundance among the three racks (Table 1). The multiple comparisons test (Table 2) showed that plates made of copper, galvanized iron, and aluminum had significantly fewer mussels than the majority of the remaining materials.

For mussels on the tubes, significant differences in abundance were found for orientation (horizontal vs vertical), for inside vs outside, and for the different materials (Table 3). Mean density on the vertical tubes was 380/m²

Table 3. ANOVA Tables to Test for Significant Differences in Abundance of *Dreissena polymorpha* on Different Racks and Tubes Made of 10 Different Materials

Source	df	SS	MS	F	p	r^2
Material	9	97.24	10.81	16.25	0.000	0.806
Rack	2	2.69	1.32	1.97	0.144	
Inside/outside	1	14.33	14.33	21.55	0.000	
Orientation	1	13.13	13.13	19.75	0.000	
Error	104	69.16	0.665			
Material	9	40.115	4.457	8.156	0.000	0.756
Inside/outside	1	13.365	13.365	24.156	0.000	
Orientation	1	12.23	12.23	22.102	0.000	
Material × inside/outside	9	2.75	0.305	0.552	0.832	
Material × orientation	9	19.16	2.129	3.848	0.000	
Orientation × inside/outside	1	1.87	1.87	3.385	0.069	
Error	87	48.14	0.553			

Table 4. Results of Tukey's Multiple Comparisons Test

Material	Mean Number of Mussels (per m²)
Copper	22
Brass	59
Aluminum	403
Galvanized iron	420
Acrylic	544
Teflon™	806
PVC (5.0 cm i.d.)	829
Black steel	889
ABS	916
PVC (4.0 cm i.d.)	1467

Note: Means scored with the same line are not significantly different at the 0.05 level.

while mean density on the horizontal tubes was 870/m²; inside surfaces had an average mussel density of 350/m², whereas outside surfaces had an average mussel density of 110/m². Tubes made of copper and brass had significantly fewer mussels attached than did the majority of the other materials (Table 4).

DISCUSSION

Results from this study suggest that materials containing toxic metals are the most suitable for reducing the settlement of zebra mussels. Plates made of copper, galvanized iron, and aluminum; and tubes made of copper, brass, galvanized iron, and aluminum had significantly fewer mussels colonized on their surfaces than the majority of other materials. This finding is similar to the findings of Walz (1973, 1975) who showed that materials such as copper and brass were successful at reducing the number of settled mussels. Reduced settlement on these materials may be related to the toxic properties of these metals. Galvanized iron contains zinc, while brass contains zinc and copper. Zinc, copper, and aluminum are known to have toxic effects on mollusks (Havlick and Marking, 1987). However, these materials may also be causing an avoidance response in the mussels, or bonding properties at their surfaces may make it difficult for mussels to attach.

Although these data suggest that materials containing copper, aluminum, and zinc will have reduced numbers of zebra mussels, there are other considerations. First, although settlement of zebra mussels is reduced on these materials, there is still settlement. These materials will be of little use if the objective is to have no settlement. Second, although settlement is reduced, there is some concern that these materials may lose their ability to reduce settlement in the long term. Data from Van Diepen and Davids (1986) suggest that materials initially good at limiting settlement (such as polystyrene) can lose their effectiveness over time. Information on rate of settlement, or the rate of changes in settlement over a full year or more, would be useful to help answer these questions.

Zebra mussels were more abundant on the inside surfaces of the tubes than on the outside. This could have occurred by two mechanisms: (1) zebra mussels may actually prefer sheltered environments and use criteria for selection of such locations and/or (2) predation by fish and ducks, etc. may have reduced the number of mussels on the outside surfaces of the tubes. The effects of predators on the abundance of zebra mussels is probably minimal. There were no ducks in the immediate vicinity when visits were made to the same locations during incubation (every 2 weeks). Also, these racks were placed in a location where the water depth was less than 1 m and predation by large fish was very unlikely. Predation by invertebrates probably would be the same on both the inside and outside surfaces. As such, these data support the hypothesis that zebra mussels preferentially settle on more sheltered substrates. Understanding the selection criteria that zebra mussels employ before attaching to surfaces would help in the design of structures for minimizing colonization.

In summary, the results from this study indicate that colonization of *Dreissena polymorpha*, over a short term, can be reduced using materials containing aluminum, copper, and zinc. Colonization is also higher on the inside surfaces of tubes than on the outside. The results suggest that the appropriate selection of building materials and design will help reduce settlement of zebra mussels.

ACKNOWLEDGMENTS

We thank Mr. Wade Gibbons for contributing comments toward the design of this study; Ms. Dianne Pathey, Mr. Dean Plummer, Mr. Kevin Wingerden, and Mr. Joe Brador for assistance in the field; and Mr. Desmond Lorente for assistance in sorting the mussels. Two anonymous reviewers provided comments that improved the manuscript.

REFERENCES

Griffiths, R. W., W. P. Kovalak, and D. W. Schloesser. "The Zebra Mussel, *Dreissena polymorpha* (Pallas 1771), in North America: Impacts on Raw Water Users," in Symposium: Service Water System Problems Affecting Safety-Related Equipment (Palo Alto, CA: Nuclear Power Division, Electric Research Power Institute, 1989) 11–27.

Havlick, M. E., and L. L. Marking. "Effects of Contaminants on Naiad Mollusks (Unionidae): A Review," United States Fish and Wildlife Service/Resource Publication No. 164.

Mackie, G. L., W. N. Gibbons, B. W. Muncaster, and I. M. Gray. "The Zebra Mussel, *Dreissena polymorpha*: A Synthesis of European Experiences and a Preview for North America," Ontario Ministry of the Environment, Water Resources Branch, Great Lakes Section. Queen's Printer, Toronto, Ontario (1989).

Van Diepen, J., and C. Davids. "Zebra Mussels and Polystyrene," *Hydrobiol. Bull.* 19:179–181 (1986).

Walz, N. "Studies on the Biology of *Dreissena polymorpha* in Lake Constance," *Arch. Hydrobiol., Suppl.* 42:452–482 (1973).

Walz, N. "The Settlement of Larvae of *Dreissena polymorpha* on Artificial Substrates," *Arch. Hydrobiol., Suppl.* 47:423–431 (1975).

CHAPTER 11

Habitat Selectivity by the Zebra Mussel *(Dreissena polymorpha)* on Artificial Substrates in the Detroit River

Tamara L. Yankovich and G. Douglas Haffner

Habitat selectivity by the zebra mussel, *Dreissena polymorpha*, was assessed by providing artificial substrates (i.e., cement blocks) to mimic a variety of niche "choices" available to mussels in their natural environment. A total of 12 cement blocks were set at a nearshore (0.5 m) and an offshore (1.0 m) site in the Detroit River from May 28 to October 5, 1990 to study colonization patterns of zebra mussels. Mean mussel densities in October were 209/m^2 and 2563/m^2 at nearshore and offshore sites, respectively. Of the total number of mussels found on the blocks at the two sites, 79 and 72% were in block holes, 13 and 24% on sides, and 8 and 4% on tops, respectively. Mean mussel lengths were 6.7 and 6.4 mm, 9.0 and 8.9 mm, and 4.1 and 4.9 mm for the holes, sides, and tops of blocks at the two respective sites. Significant differences among possible colonization sites (ANOVA $p < 0.05$) suggest that zebra mussels exhibit microsite selection behavior.

0-87371-696-5/93/$0.00 + $.50

INTRODUCTION

Previous studies have focused on the macroscale and mesoscale distribution patterns of the zebra mussel, *Dreissena polymorpha*. Macroscale distributions, such as variations in abundances between water bodies, are determined by factors such as invasion routes and lake chemistry (Hebert et al., 1989; Sprung, 1987). Once mussels have invaded a water body, mesoscale distribution patterns of adults and planktonic veligers are influenced by such factors as water currents, boat traffic, drifting debris, predators, and freely floating aquatic macrophyte beds (Mackie et al., 1989). Few studies, however, have examined actual habitat selection. Such behavior may be important at low densities when mussels are first colonizing a new substrate. Habitat selection might be a response to predator avoidance, to preferred feeding positions, or to reduced abrasion caused by suspended matter and water currents. Understanding colonization strategies may enhance our ability to control further invasions. Frequently, it is observed that the summation of many small-scale phenomena can determine large-scale survival and distributions.

This study determines the existence and relative importance of colonization strategies as they relate to habitat selection by zebra mussels. By providing zebra mussels with a range of microhabitats, it is possible to test whether mussels exercise microsite selection. Essentially, two hypotheses were tested. In a system such as the Detroit River, there should be sufficient turbulence to randomly disperse veligers; thus there should be no difference in colonization patterns between nearshore and offshore depth study sites. Second, the distribution of mussels among niche choices should not be different, unless zebra mussels are capable of niche selection.

METHODS

Zebra mussel colonization patterns were investigated by placing cement blocks at each of two sites in the Detroit River perpendicular to the shoreline near Windsor, Ontario (Figure 1). Cement blocks (19.2 × 39.5 × 9.1 cm) were chosen because they are comparable in composition to rock substrates found in the natural system. The blocks provided three possible niche choices: horizontal top (high exposure), vertical sides (moderate exposure), and inside holes (low exposure) (Figure 2). Each niche choice had comparatively different habitat characteristics. Tops were exposed to relatively fast water currents and low turbidity, sides were exposed to lower water currents but higher abrasion by suspended particulates, and holes were exposed to little current or water turbulence.

To test whether physical exposure could regulate successful invasions, 12 blocks were set on May 28, 1990 at nearshore and offshore locations,

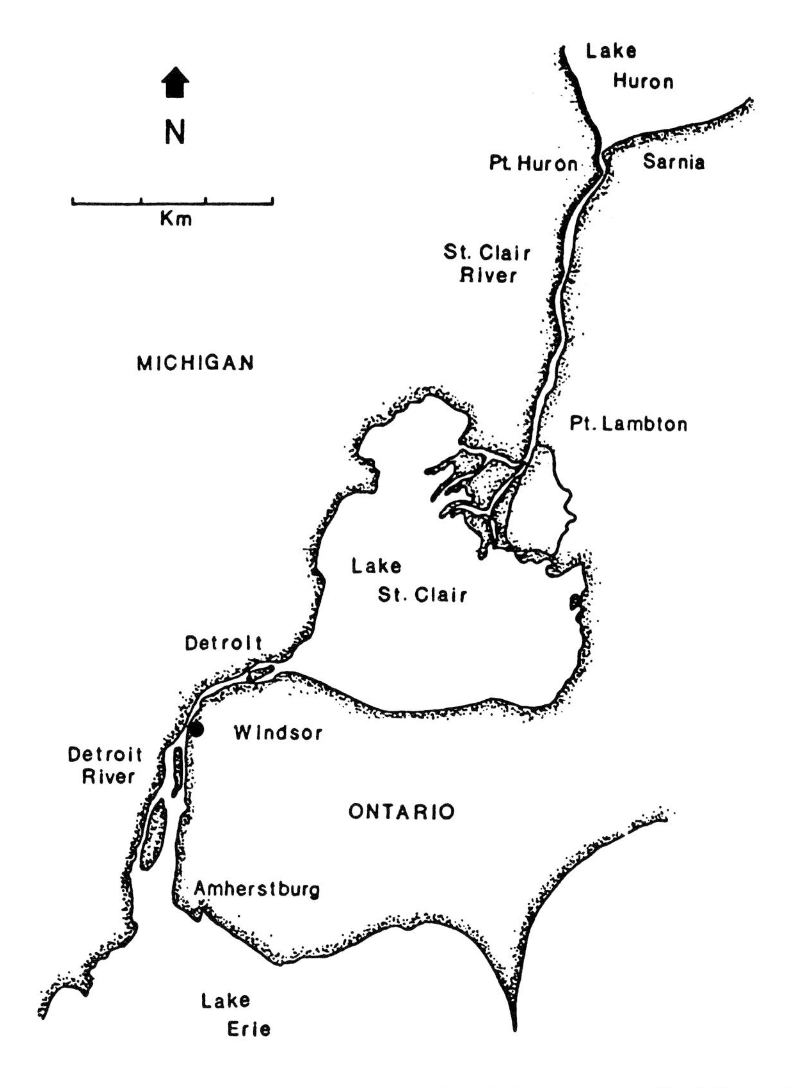

Figure 1. Site at which cement blocks were placed at a nearshore site (0.5-m water depth) and an offshore site (1.0-m water depth) for 130 days to study colonization patterns of *Dreissena polymorpha* in the Detroit River in 1990.

which were designated at depths of 0.5 and 1.0 m, respectively. This difference in depth was not considered important compared to the different physical processes associated with the two locations. The offshore blocks were located at the periphery of a small bay, whereas nearshore blocks were located in the bay itself where waves tended to break and stir up bottom sediments. Flow rates of 0.037 and 0.062 m/sec were measured at the nearshore and offshore sites, respectively. Ponar grabs were taken randomly at both sites to characterize sediments and determine local zebra mussel densities.

Figure 2. Cement block used to study habitat selection by *Dreissena polymorpha* in the Detroit River.

Blocks were left undisturbed until October 5, 1990 (i.e., 130-day period). At this time, mussel numbers and respective lengths on the tops, sides, and holes of each block were recorded. Also recorded was the orientation of mussel siphons relative to position on the block.

A two-level, nested ANOVA (Sokal and Rohlf, 1981) was used to determine significant differences in mussel densities between the two sites (nearshore and offshore) and between block surfaces (top, side, and hole). A similar two-level, nested ANOVA was used to determine significant differences in mean mussel lengths.

RESULTS

There was a highly significant difference in total mussel density between the nearshore and offshore sites ($p < 0.001$, $n = 2$). At the end of the 130-day study period, mean density was 209/m^2 at the nearshore site and 2563/m^2 at the offshore site (Table 1). Despite this density difference, the distribution of mussels on blocks was similar at both sites. There was a significant difference (ANOVA, $p > 0.001$) between the number of mussels found colonizing the top, sides, and holes of blocks; at both sites, densities in holes > sides > tops (Table 1). Considering there was a much greater number of mussels at the offshore site, similar distribution of zebra mussels on the blocks

Table 1. Density and Length of *Dreissena polymorpha* Colonizing Three Microhabitats on Cement Blocks at Two Different Sites in the Detroit River, May 21, to October 5, 1990

Location	Microhabitat	Density[a] (Number/m²)	Length (mm)
Nearshore	Holes	493 ± 63 (79)	6.7 ± 0.5
	Sides	84 ± 12 (13)	9.0 ± 0.6
	Tops	48 ± 22 (8)	4.1 ± 1.0
	Overall	209 ± 32	6.6 ± 0.7
Offshore	Holes	5543 ± 376 (72)	6.4 ± 0.1
	Sides	1825 ± 463 (24)	8.9 ± 0.1
	Tops	319 ± 108 (4)	4.9 ± 0.3
	Overall	2563 ± 315	6.7 ± 0.2

Note: Values given are the mean ±S.E.

[a] Given in parentheses is the percent of the total found in each microhabitat.

suggests that the observed colonization pattern was relatively independent of population size.

Overall, mean lengths of mussels did not vary between the two sites ($p > 0.1$), but lengths did vary depending on block location ($p > 0.001$) (Table 1). Mean size of mussels found in holes (6.6 mm) was significantly larger than those observed on tops (4.5 mm), but smaller than those found on sides (9.0 mm). Sediments surrounding the blocks contained very few adult zebra mussels, and it is assumed that most colonization was a result of veligers settling near and directly on the blocks.

Mussels in holes and tops of blocks indicated no specific orientation, but 62% of those found on sides were oriented such that their external siphons were pointing upward (Table 2).

DISCUSSION

Mussel densities observed during this study were at the lower end of the density range of 40 to 114,000/m² at depths greater than 2 m as reported by Mackie et al. (1989). Although there was a significant difference in population densities between the 0.5 and 1.0 m locations, it is probably not depth which is the most important factor for this difference, but rather turbidity and abrasion caused by waves breaking in shallower water. The somewhat low numbers at both locations in this study, particularly the nearshore site, were probably a result of these physical perturbations.

The external regions (i.e., tops and sides) of the blocks had substantially lower mussel densities than the more sheltered block areas (i.e., holes). Conceivably, this may have been a result of (1) a greater probability of veligers

Table 2. Orientation of Siphons of Mussels Attached to the Sides of Cement Blocks at an Offshore Site (1.0-m Water Depth) in the Detroit River

Orientation[a]	Mean Number[b]
1200	230 ± 10.1 (62.7)
1030	69 ± 2.3 (2.3)
0900	9 ± 0.8 (2.5)
0730	0 ± 0.0 (0.0)
0630	0 ± 0.0 (0.0)
0430	0 ± 0.0 (0.0)
0300	2 ± 0.3 (0.5)
0130	57 ± 3.2 (15.5)

[a] Orientation designations are given according to a clock face (hr) with 1200 symbolizing upward or toward the water surface.

[b] Values given are means (± S.E.) with percent of the total given in parentheses.

settling in the holes because of reduced water turbulence and/or (2) migration of juveniles from tops and sides into the holes after settlement. Clarke (1952) indicated that mussel eggs can become trapped in sheltered microsites, such as within mussel aggregations, and the larvae remain in these shielded areas. Also, following initial colonization, mussel repositioning on substrates is possible (Oldham, 1930). In Lake IJsselmeer, colonization patterns on horizontally oriented artificial plates suggested that microscale distributions could be a function of site selectivity (bij de Vaate, 1991).

Block tops had the lowest density of mussels and also the smallest mean size of individuals. Relatively high rates of exposure to water turbulence and current scour make the top surface a nonpreferred site. Mussels which settle initially on the top surface might search for alternative sites in order to maximize growth and minimize mortality risks. The small size of mussels on the top compared to the other block faces may indicate that energy needed for growth is being used to maintain position in the current and/or that food availability/feeding efficiency is lower in this habitat. Also, sedimentation onto the top surface could obstruct filtering and reduce food intake.

Although densities on block sides were lower than in holes, the mean size of individuals was greater, indicating high growth rates. Suspended matter being swept by the current would begin to settle once in contact with the block side. The orientation of mussel siphons, all facing upward, would indicate an optimization of food availability.

It is not clear whether predation had an impact on mussel densities on the different block microhabitats. Densities of up to 10 crayfish per block were observed in block holes, and predation by crayfish can be an important regulating factor; crayfish feed on mussels 1–5 mm in length and can consume over 100 mussels per day (Piesik, 1974). It is possible that grazing of smaller mussels by crayfish led to a higher mean size of individuals in the holes, but overall densities remained higher than either block tops or sides.

CONCLUSIONS

This study illustrates that colonization success of zebra mussels was site specific; more success was achieved at an offshore site (1.0 m) where conditions were more stable, as compared to a shallower site (0.5 m) where conditions were more turbulent. Within a site, however, there was evidence that microhabitat selection by muscles was occurring, and in some cases resulted in definite orientations on the substrate. Although the initial colonization might be random, it is hypothesized that an adaptive behavior to maximize food intake and growth results in a selection of preferred microsites.

ACKNOWLEDGMENTS

This research was funded by the Environmental Youth Corps Program of the Ontario Ministry of the Environment and the Department of Biological Sciences, University of Windsor. Cement blocks were set by Anita Mudry, and data analysis was aided by Dana Gagnier. Dr. Jan Ciborowski, Diane Laviolette, and Zsolt Kovats provided critical advice throughout the study.

REFERENCES

bij de Vaate, A. B. ''Distribution and Aspects of Population Dynamics of the Zebra Mussel, *Dreissena polymorpha* (Pallas, 1771), in Lake IJsselmeer Area (The Netherlands),'' *Oecologia (Berlin)* 86:40–50 (1991).

Clarke, K. B. ''The Infestation of Waterworks by *Dreissena polymorpha*, a Freshwater Mussel,'' *J. Water Works Eng.* 6:370–378 (1952).

Hebert, P. D. N., B. W. Muncaster, and G. L. Mackie. ''Ecological and Genetic Studies on *Dreissena polymorpha* (Pallas): A New Mollusc in the Great Lakes,'' *Can. J. Fish. Aquat. Sci.* 46:1587–1591 (1989).

Mackie, G. L., W. N. Gibbons, B. W. Muncaster, and I. M. Gray. ''The Zebra Mussel, *Dreissena polymorpha*: A Synthesis of European Experiences and a Preview for North America,'' Report prepared for Ontario Ministry of Environment, Water Resources Branch, Great Lakes Section, Queen's Printer, Toronto, Ontario (1989).

Oldham, C. ''Locomotive Habits of *Dreissena polymorpha*,'' *J. Conchol.* 9:25–26 (1930).

Piesik, Z. ''The Role of the Crayfish *Orconectes limosus* (Raf.) in Extinction of *Dreissena polymorpha* (Pall.) Subsisting on Steelon Net,'' *Pol. Arch. Hydrobiol.* 21:401–410 (1974).

Sokal, R. R., and F. J. Rohlf. *Biometry*, 2nd ed. (New York: W. H. Freeman & Company Publishers, 1981).

Sprung, M. ''Ecological Requirements of Developing *Dreissena polymorpha* Eggs,'' *Arch. Hydrobiol., Suppl.* 79:69–86 (1987).

SECTION II

Morphology and Physiology

CHAPTER 12

The Anatomy of *Dreissena polymorpha* and the Evolution and Success of the Heteromyarian Form in the Dreissenoidea

Brian Morton

The anatomy of *Dreissena polymorpha* is described. Despite the lack of hinge teeth and a superficial similarity to representatives of the Mytiloidea, shell structure and mineralogy suggest that the Dreissenoidea is closely related to the Corbiculoidea (Heterodonta). It is further argued that the heteromyarian Dreissenoidea arose from the isomyarian Corbiculoidea in the late Mesozoic. By the Eocene (Cenozoic), the Dreissenoidea was distinct, recognizable by an apical septum in each valve accommodating the attachments of the anterior adductor and anterior byssal retractor muscles. The various genera of the Dreissenoidea, all of Cenozoic origin, arose rather rapidly with the break up of the Tethys Sea. Infaunal genera, such as *Congeria* and *Dreissenomya*, subsequently became extinct. Species of *Dreissena* and *Mytilopsis* survived; the former was restricted to the Caspian Sea and the latter with a subtropical/tropical north American distribution focused around the Gulf of Mexico. Subsequently, both genera have expanded their distributions with the aid of post-industrial revolution in man. Their success is largely due to the adoption of the heteromyarian form, which is associated with the neotenous retention of the byssus, allowing occupation of hitherto unoccupied (by bivalves) freshwater and estuarine hard substratum habitats. A relatively simple, dioecious sexual strategy, with a free-swimming veliger larva, has facilitated rapid dispersion.

0-87371-696-5/93/$0.00 + $.50

INTRODUCTION

Evolution of the triangular heteromyarian form in the Bivalvia has had a profound influence upon the overall success of the class resulting (importantly) in the adaptive radiation of representatives of various lineages onto hard substrates, with attachment being by means of a byssus (Stanley, 1972; Seilacher, 1984). Members of this class colonize marine, estuarine, and freshwater habitats. Ultimately, adoption of the heteromyarian form facilitated the evolution in some groups (but not the Dreissenoidea) of monomyarian lineages that, in turn, allowed further diversification, including firmer cementation by one valve in the oysters (Ostreoidea). A few monomyarians (e.g., the Pectinoidea) have exploited the swimming mode of life. How the heteromyarian form evolved is critical to our understanding of the success of the Bivalvia in many habitats, particularly with regard to freeing some phylogenies from the constraints imposed by a burrowing mode of life.

Early malacologists lumped all heteromyarian bivalves into the "Anisomyaria." A better understanding of the fossil record has, however, shown that the triangular form has evolved in numerous lineages at different times. Furthermore, far from being a modern evolutionary development, such a form is characteristic of representatives of some of the oldest Paleozoic groups, such as the Pterioida and Mytiloida (Morton, 1992).

The various bivalve lineages in which the heteromyarian form is found have been discussed by Yonge (1962, 1976), Stanley (1972), and Morton (1992) and the necessary morphological sequence of changes from an isomyarian to a heteromyarian form has been described. In all cases but arguably for the most important superfamily, the Mytiloidea, the neotenous retention of the larval byssus into adult life is of major significance. The occurrence of a byssus in diverse phylogenetic groups suggests that retention has occurred independently on a number of occasions. In some byssate bivalve lineages (e.g., the Arcoida) the heteromyarian form has only evolved rarely because of the retention of an anterior inhalant stream such that only a modioliform shell can be obtained (Morton, 1992). The hard surface component of the intertidal environment and probably also of estuarine and fresh waters has been colonized by different epibyssate bivalve lineages at different times. Most recently in the Eocene, exploitation of the lacustrine component of this continuum has been achieved successfully by a modern superfamily of bivalves — the Dreissenoidea.

One of the most characteristic features of the shell of all Dreissenoidea is the possession of an apical shell septum to which the anterior adductor and anterior byssal retractor mussels are attached. An apical septum also occurs in *Septifer* (Mytiloidea), although in representatives of this genus only the anterior adductor muscle has its attachment to it, the anterior byssal retractor attaching to the posterior face of the shell under the ligament (Yonge and Campbell, 1968). It has been suggested by these authors that the possession of a shell septum in these two bivalve phylogenies is the result of convergent evolution by adaptation to similar modes of life. The forward extension of the umbones and the extremes of ventral (morphologically anterior) flattening in *Dreissena* and *Septifer* are possibly associated with the need for firm anchorage (other than the valve itself) of the anterior adductor muscle. The possession of a shell septum in these two genera not only permits retention of the anterior adductor muscle, otherwise lost in some heteromyarian mytiloids (e.g., *Perna viridis*; Morton, 1987), but also enhances its function possibly by reducing shear. Although only occurring in isolated instances in the Mytiloidea (i.e., *Septifer*), a shell septum is a feature of all extinct and living Dreissenoidea. This allows reasonably positive identification of representatives of this superfamily in the fossil record (Nuttall, 1990). The fossil Dreissenoidea can thus be examined in an attempt to explain not only the evolution of its recent, extant representatives but also the success of the heteromyarian form and how it has allowed colonization of estuarine and freshwater systems, even though with the aid of man in modern times.

THE DRESSENOIDEA

Nuttall (1990) has received the classification of the Dreissenoidea and suggests the following scheme:

Family Dreissenidae Gray in Turton, 1840
- Subfamily Dreisseninae Gray in Turton, 1840
  - Genus *Dreissena* van Beneden, 1835
  - Genus *Congeria* Partsch, 1835
  - Genus *Mytilopsis* Conrad, 1858
  - Genus *Prodreissena* Rovereto, 1898
- Subfamily Dreissenomyinae Babak, 1983
  - Genus *Dreissenomya* Fuchs, 1870

The following descriptions of extinct species of *Dreissenomya* and *Congeria* will permit the subsequent interpretation of dreissenoid evolution, importantly with regard to the extant epibyssate species of *Mytilopsis* and *Dreissena*. *Prodreissena* is not discussed because of its close similarity to *Dreissena* (Nuttall, 1990).

Dreissenomya

Dreissenomya aperta possesses a rectangular shell in which the anterior (ventral) margin shows only a small degree of flattening (Figure 1A). Conversely, the posterior margin is extended, with elongation of the ligament and enlargement of the posterior adductor and posterior byssal and/or pedal retractor muscles. There is a small but well-defined pallial sinus, indicating

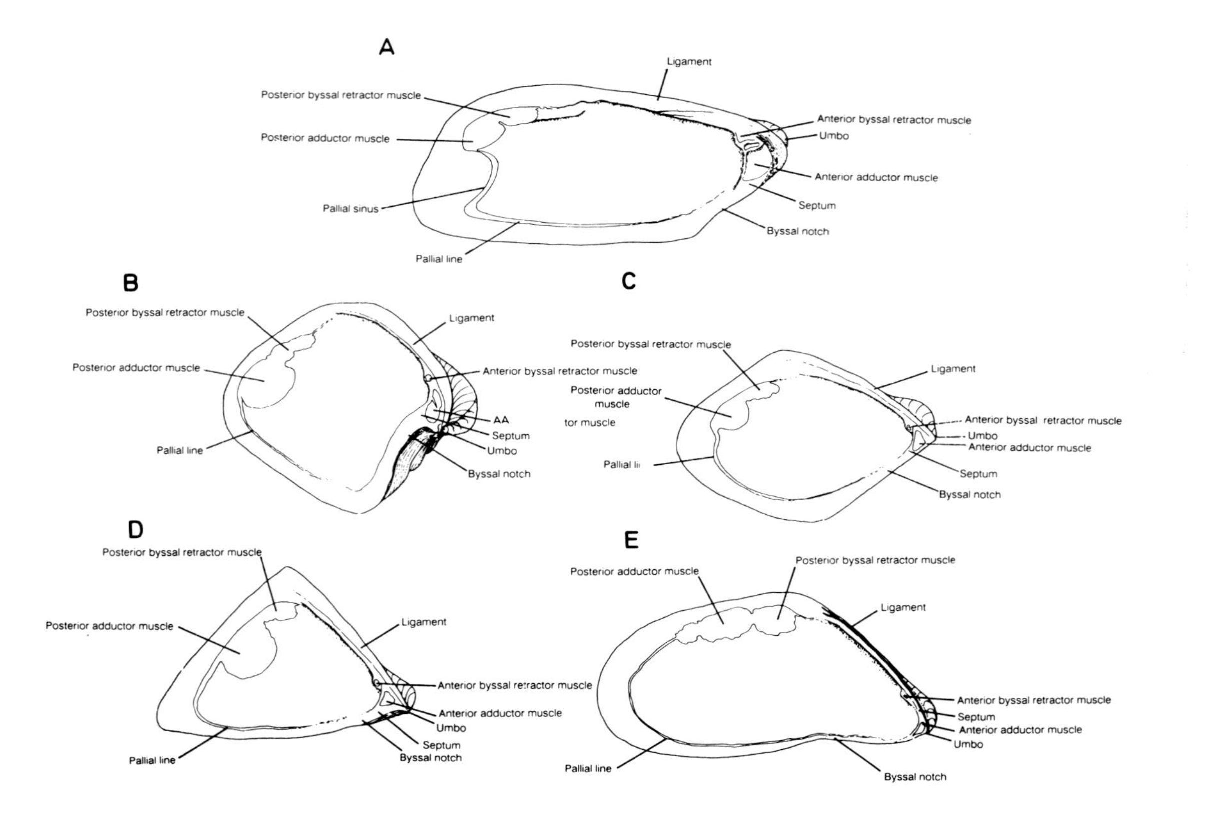

Figure 1. Internal views of the left shell valves of A, *Dreissenomya aperta*; B, *Congeria subglobosa*; C, *C. zsigmondyi*; D, *C. triangularis*; E, *Mytilopsis sallei.* (From Morton, B. *Palaeontology* 13:563–572 [1970]. With permission.)

that *Dreissenomya* had discrete, retractable siphons and probably, therefore, pursued an infaunal mode of life. Anteriorly, the presence of what Morton (1970) has interpreted as a byssal notch suggests that *Dreissenomya* was also attached to the substratum. The coexistence of a pallial sinus and byssal notch can be explained by the animal living byssally attached to stones or sand grains, but either partly or wholly buried, as is typical of many, modern endobyssate representatives of the Mytiloidea (e.g., *Musculista, Modiolus, Arcuatula*; Morton, 1992). Situated on the septum of *Dreissenomya aperta* are the scars of the anterior muscular system, with the anterior adductor muscle lying below the site of insertion of the anterior byssal and/or pedal retractor muscle. Marinescu (1977) suggests that species of *Dreissenomya* were abyssate burrowers, living at a depth of 5–7 cm in the sediment; *D. schroeckingeri* was considered by this author to live at a depth of up to 30 cm.

Congeria

The solid shell of *Congeria subglobosa* is almost square in outline and possesses a distinct byssal notch (Figure 1B). Morton (1970) considered it to be highly unlikely that such a bivalve could lead an epibyssate mode of life since water movement would put a great strain upon the byssal apparatus. *C. subglobosa* was probably attached to stones buried within a soft substratum. Nuttal (1990) envisages a similar lifestyle for this species. The anterior adductor muscle scar is situated on the shell septum while the scar of the anterior byssal and/or pedal retractor muscle is situated slightly farther posteriorly than in *Dreissenomya*. Ventral flattening of a shell like that of *C. subglobosa* would result in the evolution of a shape similar to extant representatives of the Dreissenoidea. Perhaps representing an intermediate stage in the evolution of this form is *C. zsigmondyi*, in which the central margin of the shell is much flatter (Figure 1C). Even so, *C. zsigmondyi* was probably still endobyssate. The anterior adductor muscle scar occupies the familiar position underneath the umbo, while the anterior byssal and/or pedal retractor muscle scar is situated on a lobe projecting down from the septum. Apart from the lack of extreme ventral flattening, *C. zsigmondyi* is similar to *Mytilopsis*. Ventral flattening occurs in *C. triangularis* (Figure 1D), with the mid-dorsal region of the shell accordingly heightened so that a triangular outline has resulted. *C. triangularis* was probably epibyssate since mytiloids with a similar shell form pursue this lifestyle (Morton, 1992). The anterior adductor muscle is small and located on the shell septum with the scar of what would almost certainly be an anterior byssal retractor muscle situated on a separate lobe of the septum, as in *Mytilopsis*.

Mytilopsis

There is an imprecisely understood number of species of *Mytilopsis* (Marelli and Gray, 1983; 1985). This is not only because original and introduced distributions are unknown, but also because constituent species are highly variable in both shell form and color. The shell of *Mytilopsis sallei* (Figure 1E) is fragile, ventrally flattened and anteriorly reduced to create a smoothly triangular outline. The posterior adductor and byssal retractor muscle scars are located middorsally, while the small anterior adductor muscle scar is located on the posterior face of the shell on a lobe of the septum. The anatomy of *Mytilopsis sallei* has been described by Morton (1981). All species of *Mytilopsis* examined so far are epibyssate.

DISTRIBUTION OF THE DREISSENOIDEA

Mytilopsis first appeared in the European Eocene and invaded the Western Hemisphere in the late Oligocene. *Dreissenomya, Congeria,* and *Dreissena* arose in the late Miocene of the Paratethys. Nuttall (1990) provides maps of the known distributions of these genera, and it seems that the Dreissenoidea evolved in the Tethys, most likely from a corbiculoidean ancestor (Morton, 1970), to occupy coastal, estuarine habitats. The extinction of what are interpreted as coastal zone, burrowing, endobyssate relatives (i.e., *Congeria* and *Dreissenomya*) coincides with the break-up of the Paratethys and the Messinian salinity crisis. This left only epibyssate species of *Mytilopsis* occurring in coastal environments where they were probably capable of withstanding wide fluctuations in salinity, as is the case with *Mytilopsis sallei* in Hong Kong (Morton, 1989) and *M. leucophaeta* in low salinity lagoons of the Gulf of Mexico (Britton and Morton, 1989).

The breakup of the Tethys also isolated *Dreissena polymorpha* and other (now extinct) species of *Dreissena* with very restricted distributions, in an area of central Europe and northern Asia Minor as far east as the Aral Sea and the Euphrates River (Babak, 1983). During quaternary glacial epochs, the distribution of *Dreissena* was reduced considerably to small pockets including the slightly brackish areas of the Caspian and Aral Seas, and the freshwater Azov and Black Seas and the Balkan Peninsula. From this area, *D. polymorpha* has recolonized much of its original distribution and has spread throughout western Europe in the rivers and canals of inland waterways interconnected for trade during the industrial revolution (Locard, 1893; Haas, 1929; Archambault-Guezou, 1976) (Figure 2). The occurrence of *Mytilopsis* in North America may represent isolation of the constituent species of this genus from their ancestral stock following breakup of the Tethys or introduction from Europe, prior to historical record taking. It is known, however, that *Mytilopsis sallei* has been spread into the Pacific by the activities of man (Morton, 1989).

Figure 2. The distribution of *Dreissena polymorpha* in Europe in 1800 and its subsequent spread during the Industrial Revolution. (From Haas, F. Bronn, Klassen und Ordungen des Tiers-Reich III. Abteilung: Bivalvia [Muscheln] [B. H. Leipzig: Akademish Verlagsgesellschaftman, 1929]. With permission.)

THE ANATOMY OF *DREISSENA POLYMORPHA*

The anatomy of *Dreissena polymorpha* has been described by Yonge and Campbell (1968) and by Morton (1969a). Toureng (1894a and 1894b) has described the nervous system and the circulatory system, respectively. Wlastov and Kachanova (1959) have described the expression of sexuality in *D. polymorpha* while Meisenheimer (1899, 1900, 1901) and Kirpichenko (1968) have described the developmental stages and considered the species to be unique among freshwater bivalves in the possession of a free-swimming veliger larvae. This is not, however, true; the freshwater mytiloid *Limnoperna fortunei* is similarly dioecious and has a free-swimming veliger (Morton, 1973).

The Shell

The shell of *Dreissena polymorpha* is, as its name implies, highly variable in form and exhibits a dull, similarly highly variable zebra pattern of irregular

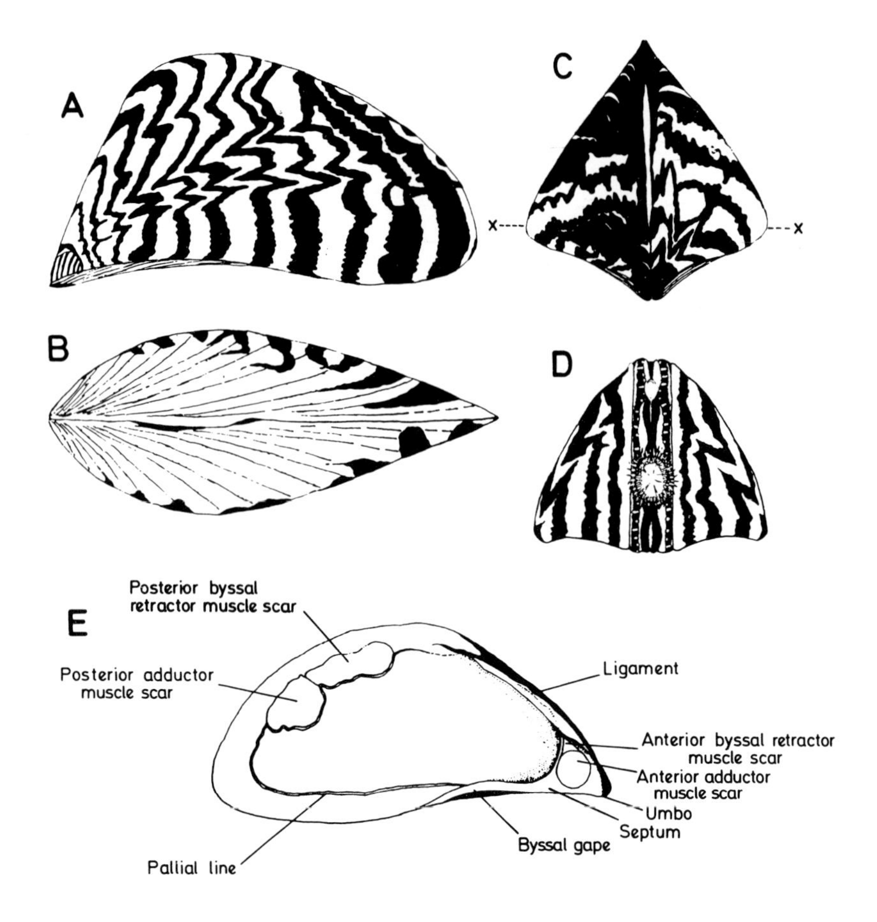

Figure 3. The shell of *Dreissena polymorpha* as seen from A, the left; B, ventral; C, anterior and D, posterior (with siphons extended) views; and E, an internal view of the left shell valve. (From Morton, B. *Proc. Malacol. Soc. London* 38:301–321 [1969a]. With permission.)

brown and cream concentric bands (Figure 3A). Meinhardt and Klingler (1987) have proposed a model for the formation of such a pattern in the shells of numerous mollusks, but do not discuss *Dreissena*. The shell has the typical heteromyarian form in that the anterior end is greatly reduced while the posterior is inflated. Seen from the ventral aspect (Figure 3B), the ventral shell margin is somewhat sinusoidal and emarginated anteroventrally to form a narrow byssal gape. The shell is widest ventrally (Figure 3C,D) below the mid-dorsoventral axis (x----x). Taylor et al. (1973) have examined the shell microstructure of *D. polymorpha* and shown it to comprise an outer crossed-lamellar layer and an inner complex crossed-lamellar layer, thereby most closely resembling the shell of the Corbiculoidea. These authors also note

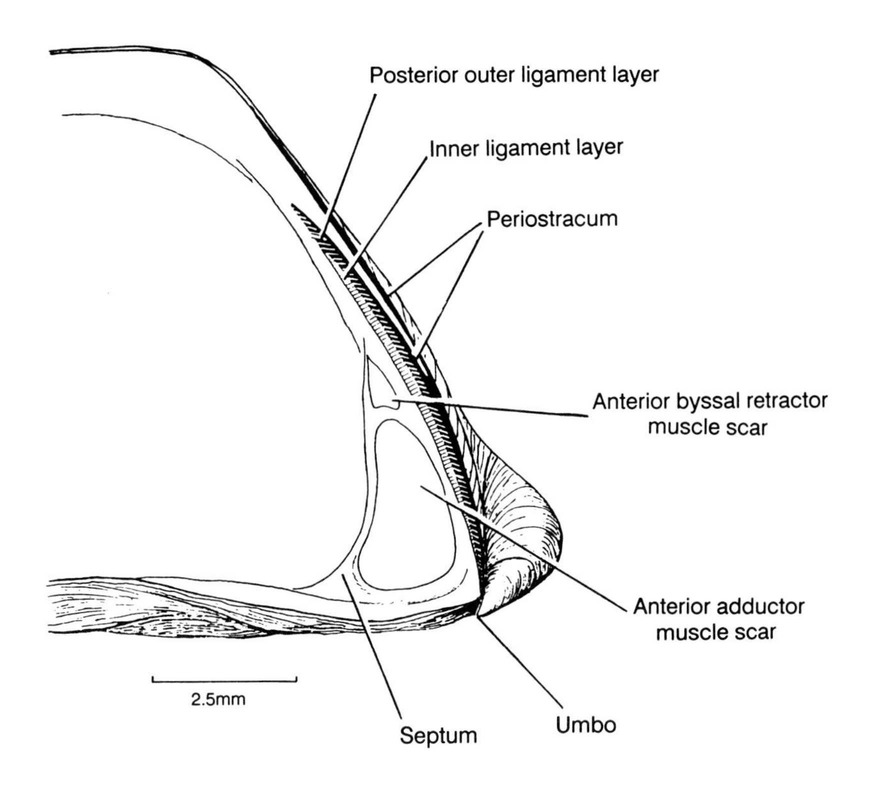

Figure 4. An internal view of the anterior end of the left shell valve and ligament of *Dreissena polymorpha*. (From Yonge, C. M., and J. I. Campbell. *Trans. R. Soc. Edinburgh* 68:21–43 [1968]. With permission.)

that the septum of *D. polymorpha* only comprises the outer crossed-lamellar layer, as does the true hinge of the Corbiculoidea. Thus, there is structural evidence to support the contention of Morton (1970) that the septum is a hinge plate remnant, although this is disputed by Nuttall (1990).

While generally resembling that of the Mytiloidea, the shell of all dreissenoids (as noted before) internally possesses a conspicuous septum occupying the umbonal regions of both valves and upon which the anterior adductor and anterior byssal retractor muscles are inserted (Figure 3E). *D. polymorpha* possesses no hinge teeth. There is a well-defined pallial line (but no pallial sinus), with large insertions posterodorsally for the posterior adductor and posterior byssal and pedal retractor muscles.

Ligament

Yonge and Campbell (1968) describe the opisthodetic, posteriorly elongate ligament of *D. polymorpha* and show it to comprise a "primary" ligament of inner ligament layer and posterior outer ligament layer, with the anterior outer ligament layer being much reduced (Figure 4). This primary ligament is covered by a "secondary" ligament of periostracum which is unusually divided into inner and outer components by a tongue of mantle that secretes it (Figure 5). The dreissenoid ligament differs from that of the Mytiloidea (Yonge and Campbell, 1968).

The Siphons

The siphons of *Dreissena polymorpha* are short and separate (Figures 3D and 6). The inhalant siphon has a relatively large opening surrounded by a crown of 80–100 tentacles arranged in two cycles. The exhalant siphon is conical with a posterodorsally directed opening smaller than that of the inhalant and lacking tentacles. The posterior third of the fused inner mantle folds bear irregularly spaced, colorless sensory papillae. The siphons are formed solely from the fused inner folds of the mantle margins (Yonge, 1982).

Mantle

Mantle fusions occur dorsally above the exhalant siphon, between the exhalant and inhalant siphons, ventrally between the inhalant siphon and pedal gape, and anterior to the pedal gape (Figure 7). Fusion is of the inner folds only (Yonge, 1982), as illustrated for the midventral region posterior to the pedal gape (Figure 8). The extensive pedal gape allows for the extension of the foot and for the large byssus. In transverse section, the mantle margin comprises three folds (inner, middle, and outer) with a divided pallial retractor muscle effecting mantle withdrawal when the valves close (Figure 9). A thin translucent sheet of two-layered periostracum passes from the periostracal groove between the middle and outer folds to cover the shell. The thin outer layer is secreted at the base of the periostracal groove, while the thicker underlying layer is secreted by the inner surface of the outer fold.

Muscular System

The muscular system of *Dreissena polymorpha* is illustrated in Figure 10. The posterior adductor muscle is large and situated posterodorsally. The posterior byssal retractor muscles are attached to the shell in a line parallel to its posterodorsal border and anterior to the posterior adductor muscle. A small posterior pedal retractor muscle is located diffusely among the anterior elements of the posterior byssal retractor muscle. The small anterior adductor muscle and the anterior byssal retractor muscle, again with a small pedal retractor component, are attached to the apical septum of the shell.

Ctenidia, Labial Palps, and Mouth

The lateral ctenidia (Figure 11A) are composed of subequal inner and outer demibranchs giving the characteristic W shape in transverse section

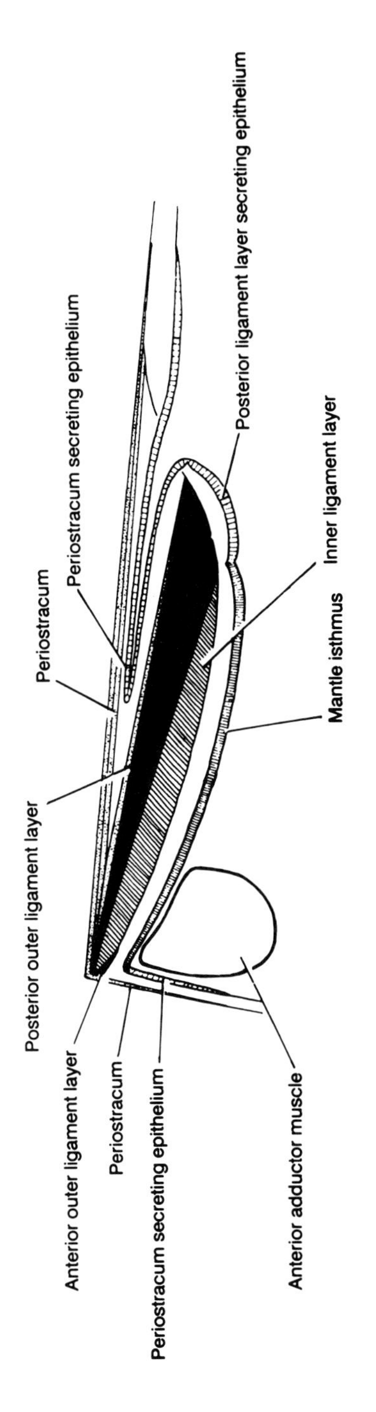

Figure 5. A generalized drawing of the ligament of *Dreissena polymorpha* showing how its various components are secreted. (From Yonge, C. M., and J. I. Campbell. *Trans R. Soc. Edinburgh* 68:21–43 [1968]. With permission.)

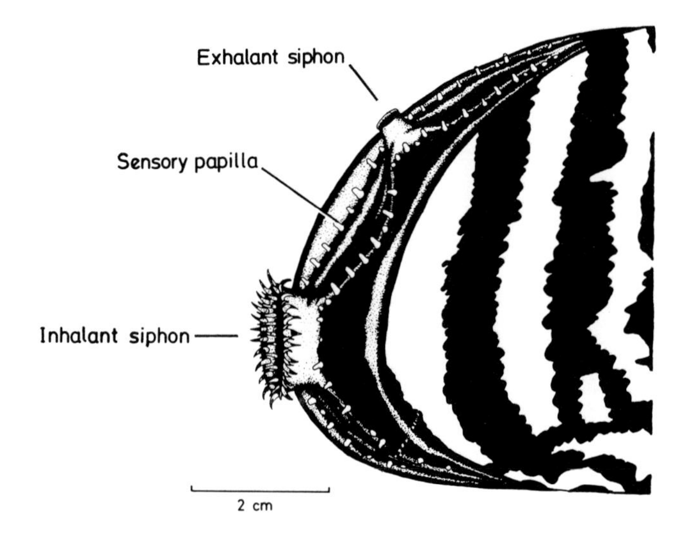

Figure 6. The siphons of *Dreissena polymorpha* as seen from the right posterolateral aspect.

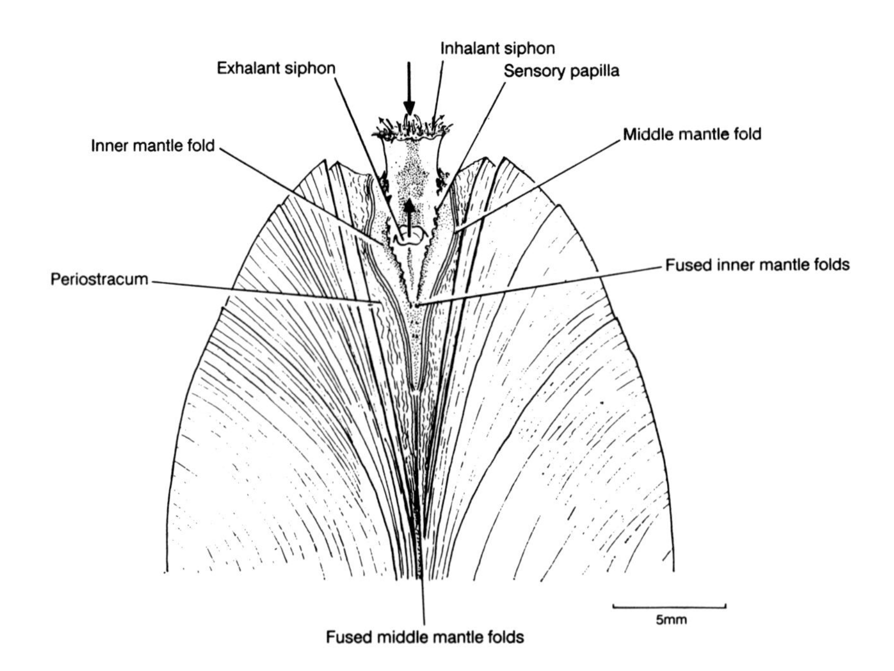

Figure 7. The posterior end of the shell and mantle of *Dreissena polymorpha* as seen from the dorsal aspect. (From Yonge, C. M., and J. I. Campbell. *Trans. R. Soc. Edinburgh* 68:21–43 [1968]. With permission.)

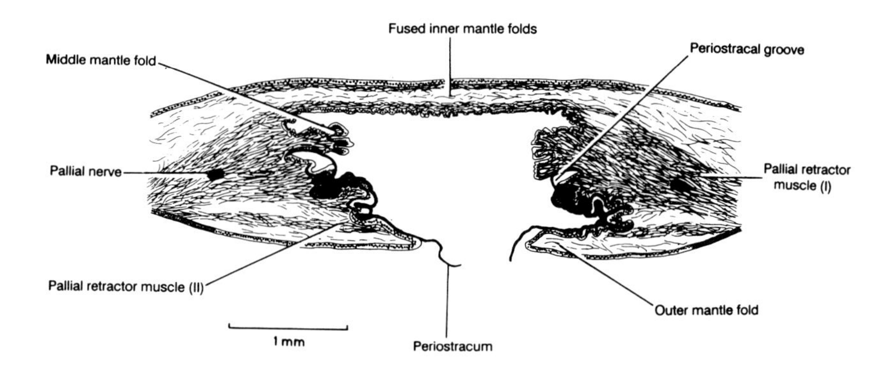

Figure 8. A transverse section through the fused mantle folds of *Dreissena polymorpha*, posterior to the pedal gape.

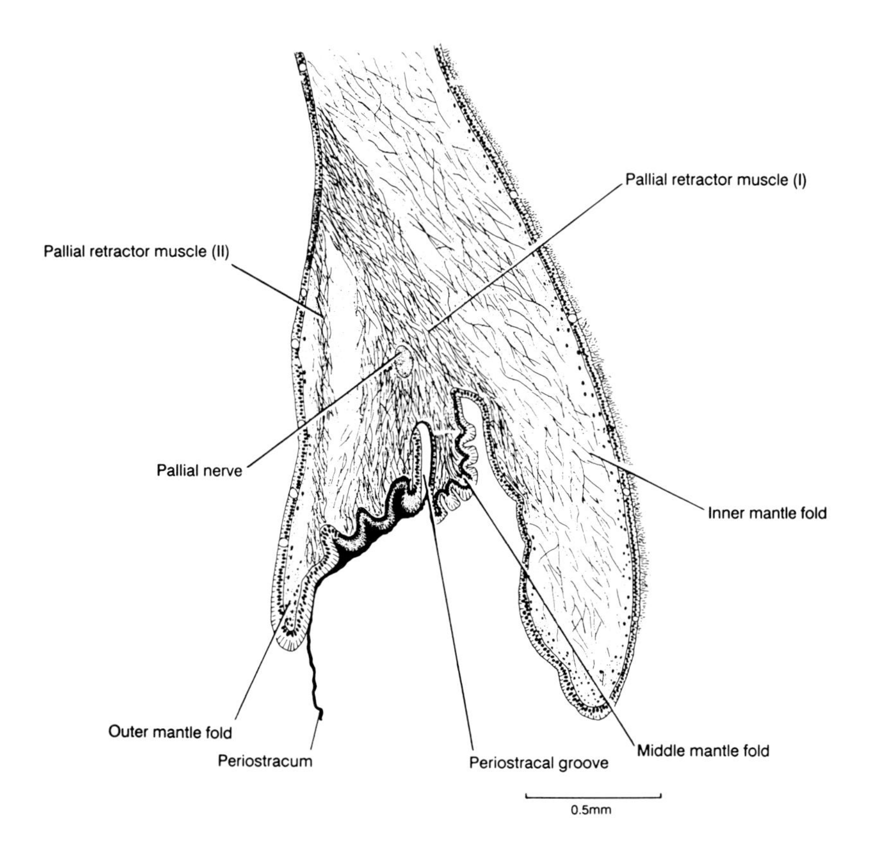

Figure 9. A transverse section through the right mantle lobe of *Dreissena polymorpha* at the pedal gape.

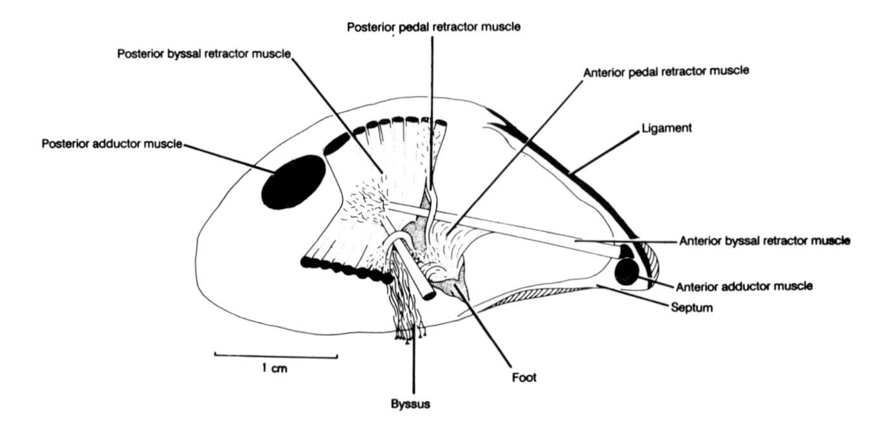

Figure 10. The musculature of *Dreissena polymorpha* as seen from the right side.

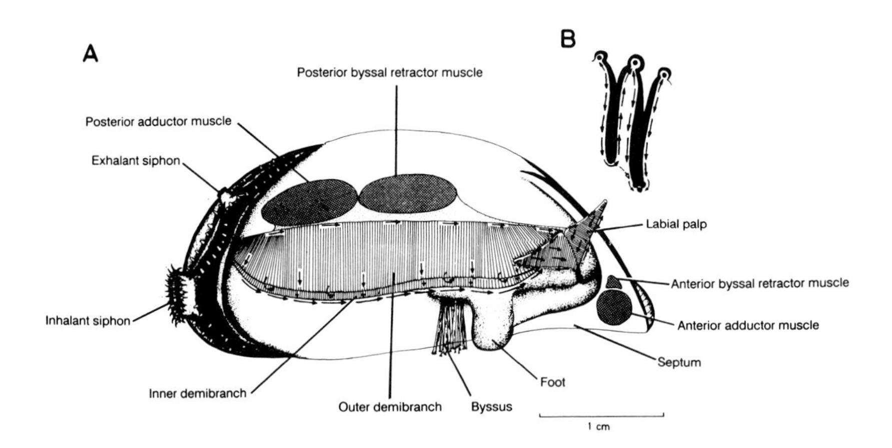

Figure 11. The ciliary currents of the ctenidia of *Dreissena polymorpha* as seen from A, the right side and B, in diagrammatic form in transverse section. (● = Oralward current.)

(Figure 11B). The upper margins of ascending lamella of the outer and inner demibranchs are attached to the mantle and the visceral mass, respectively, by cuticular fusion. The ctenidia are flat, homorhabdic and eulamellibranchiate. The ciliation of the ctenidial surfaces generally directs filtered particles downward except apically and on the inner surface of the outer demibranch where they are directed upward. Orally directed food grooves are thus situated in the junctions of ascending lamellae of the inner and outer demibranch with the visceral mass and mantle, respectively, within the ctenidial axis and in the ventral marginal food groove of the inner demibranch only (Figure 11B).

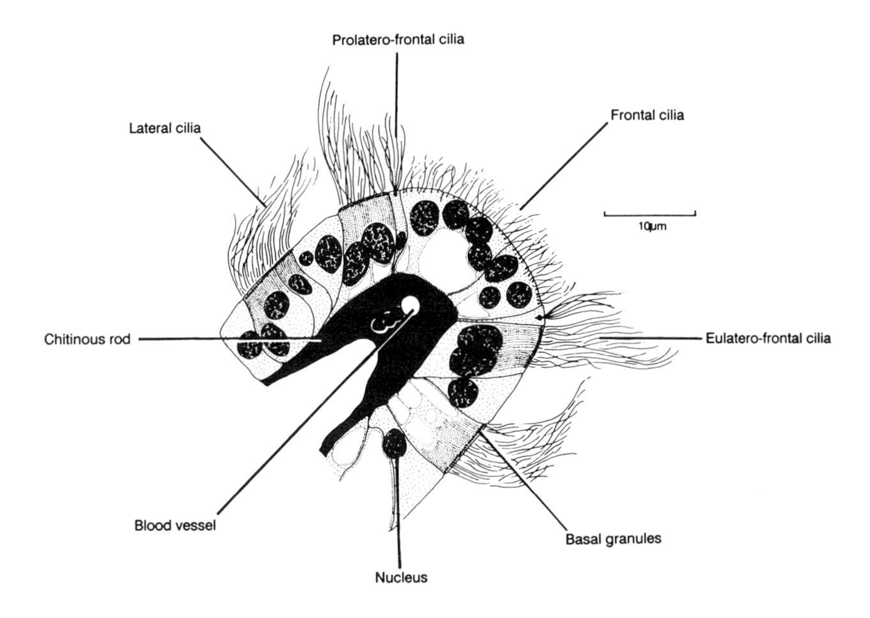

Figure 12. A transverse section through the apex of a single ctenidial filament of *Dreissena polymorpha.* (From Morton, B. *Proc. Malacol. Soc. London* 38:301–321 [1969a]. With permission.)

Ctenidial filaments are about 40 μm broad and are separated by interfilamentar spaces of approximately 25 μm. Each filament head has an array of cilia (Figure 12). Lateral cilia create the flow of water between the filaments. The eulaterofrontal cilia consist of a double row of fused cilia that filter the water and move particles onto the frontal surface of the filament. A single row of prolaterofrontal cilia lies between the eulaterofrontal and the frontal cilia. Short frontal cilia direct particles either up or down the filament toward the oral food grooves, as described above (Figure 11B). On the inner demibranch, near the marginal food groove, the frontal cilia are replaced by larger terminal cilia which transport larger particles to the mouth. Coarse, cirrus-like cilia also occur along the free margin of the inner demibranch and serve to move masses of accumulated large particles which will be ultimately rejected. The ventral marginal food groove is capable of closure, thereby regulating the amount of material entering it, according to need and prevailing particle load in the inflowing water (Morton, 1971; Sprung and Rose, 1988).

The ctenidial-labial palp junction is illustrated in Figure 13. Particles passed anteriorly within the food groove in the junction of the ascending lamella of the inner demibranch and the visceral mass pass onto the surface of the inner labial palp. Small particles that have passed along the ventral marginal food groove of the inner demibranch, however, tend to pass directly into the food groove between the two palps. Larger particles that follow this

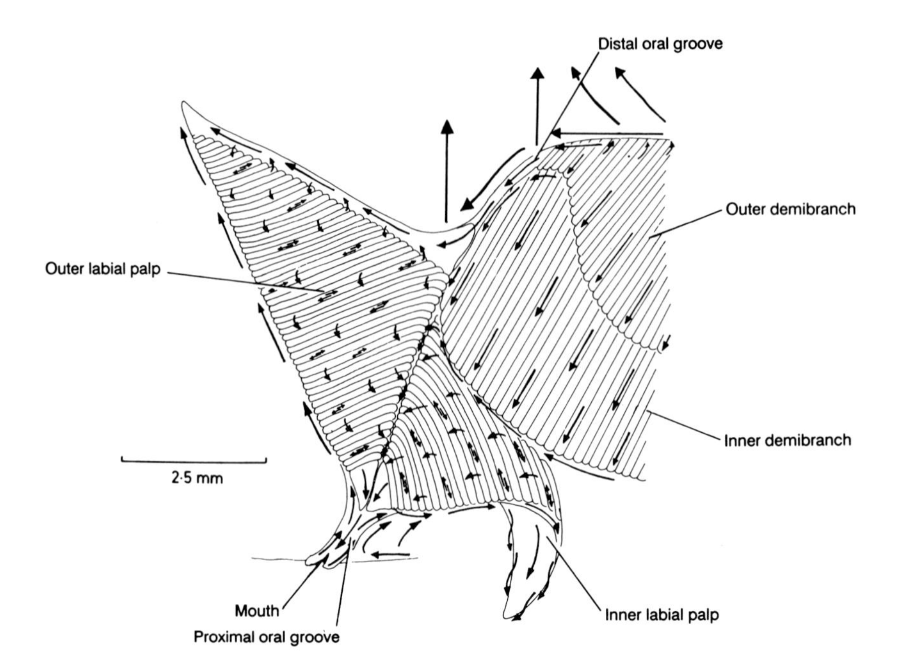

Figure 13. *Dreissena polymorpha*. The anterior end of the ctenidium and the labial palps of the left side, showing the ciliary currents. (From Morton, B. *Proc. Malacol. Soc. London* 38:301–321 [1969a]. With permission.)

latter course usually drop onto the anterior visceral mass or mantle before reaching the palps. Particles arriving at the palps from the ctenidial axis usually pass into the distal oral groove and subsequently into the groove between the palps. Particles passing along the junction between the ascending lamella of the outer demibranch and the mantle, pass onto the opposed ridged surfaces of the palps. This will not occur with large particles, however, since they will be removed by the powerful cleansing currents of the mantle.

The ridged inner surfaces of the labial palps are responsible for sorting those particles arriving from the oralward ciliary tracts of the ctenidia (Figure 14). Small particles pass rapidly over ridges of the palps toward the proximal oral groove. Larger particles fall into troughs between the folds and are then passed to the free, unridged ventral edges of both outer and inner palps and then to rejection currents of the mantle and visceral mass, respectively. In addition to these two distinct acceptance and rejection currents, the palps are strongly selective and possess ''resorting'' currents on the oral and aboral faces of the ridges (Figure 14).

Fine particles may pass along the base of the proximal oral grooves directly into the mouth, but larger particles can be rejected by the lips (Figure 15). Not all the particles that reach the anterior regions of the palps are ingested.

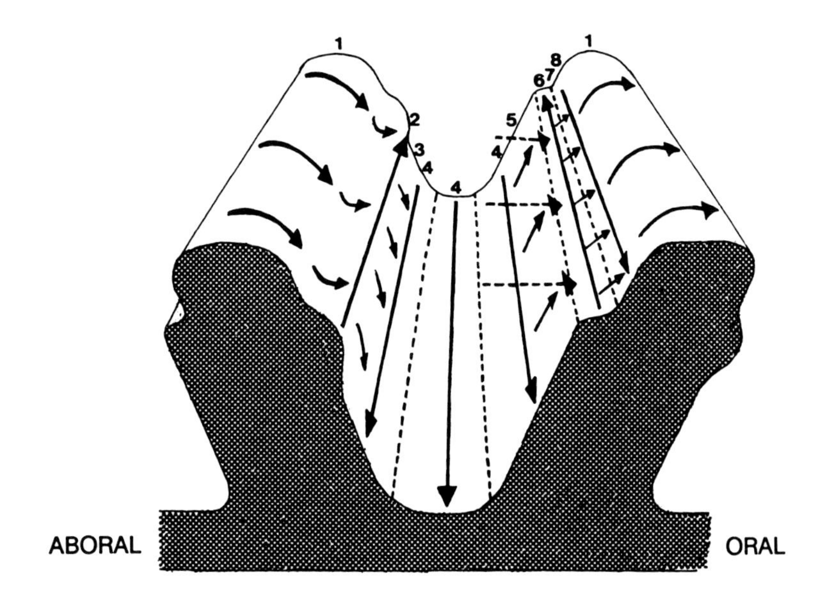

Figure 14. The ciliary currents of two adjoining ridges and their groove of the labial palps of *Dreissena polymorpha*. (From Morton, B. *Proc. Malacol. Soc. London* 38:301–321 [1969a]. With permission.)

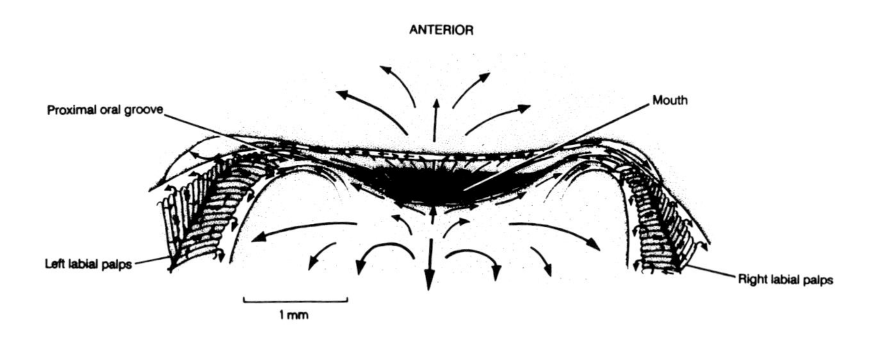

Figure 15. The ciliary currents of the mouth of *Dreissena polymorpha* as seen from the ventral aspect. (From Morton, B. *Proc. Malacol. Soc. London* 38:301–321 [1969a]. With permission.)

Cleansing currents of the visceral mass anterior to the mouth pass rejected material forward and laterally to the outer labial palps and then to the mantle. Posterior to the mouth, currents pass rejected material laterally for disposal by the visceral mass, via the inner labial palps. The anterior lip of the mouth is covered by the muscular posterior lip to enclose food particles.

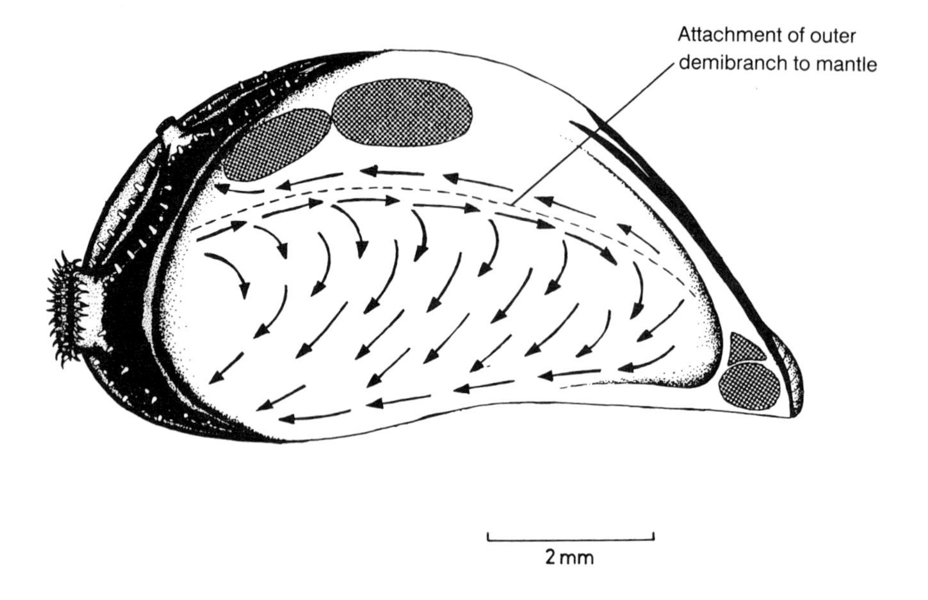

Figure 16. The ciliary currents of the left mantle lobe of *Dreissena polymorpha.*

Ciliary Currents of the Mantle, Visceral Mass, and Foot

The ciliary currents of the mantle are cleansing, except for an oralward current passing just ventral to the line of attachment of ascending lamella of the outer demibranch (Figure 16). Elsewhere, particles are passed posteroventrally to the base of the inhalant siphon, where waste material accumulates as pseudofeces. These are periodically expelled from the inhalant siphon as a bolus by the rapid adduction of the shell valves.

The ciliary currents of the visceral mass are also predominantly used for cleansing, removing material from anterodorsal regions to the posterior edge where they fall onto the mantle and are ejected as pseudofeces (Figure 17). A triangular area on the surface of the visceral mass under the inner labial palps is unciliated, this area being cleansed by outer surfaces of the palps.

The surface of the foot is also ciliated and particles are passed posterodorsally to its base and then to the visceral mass (Figure 17). The foot is used in locomotion, especially in the not yet permanently attached juvenile, and has been described by Oldham (1930). The foot also secretes the attaching byssus. Secretions from the byssus gland are passed into the byssal groove, extending down the ventral surface of the foot where byssal threads are formed one at a time. After detachment, the mass of old byssal threads is rejected and the animal secretes a new series. In addition to these two functions, ciliary currents on the foot can remove large amounts of waste material from the anterior mantle cavity, particularly the mouth.

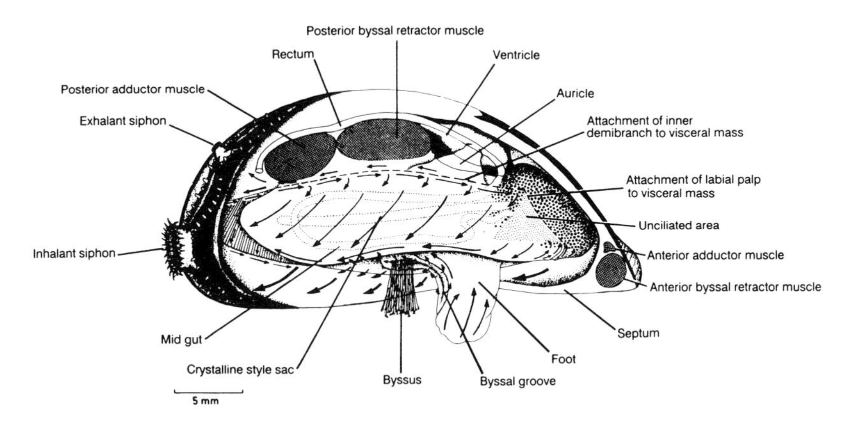

Figure 17. The ciliary currents of the foot and visceral mass of *Dreissena polymorpha* as seen from the right side. Also shown is the course of the alimentary canal.

Alimentary System

In most bivalves, the mouth is located close to the posterior border of the anterior adductor muscle. In *Dreissena polymorpha*, because of the anterior reduction of the shell and thus the anterior space inside the shell, the mouth lies a short distance posterior to it. The esophagus passes upward from the mouth which lies between the bases of the anterior byssal retractor muscles. The esophagus opens into a stomach lying beneath the ligament, close to the surface of the visceral mass. The digestive diverticula lie lateral and posterior to the stomach. The course of the alimentary canal is illustrated in Figures 17 and 19C. The style sac lies on the left side of the midgut and pursues a separate course into the visceral mass, terminating just behind the posterior bend of the midgut.

The midgut leaves the posteroventral wall of the stomach to pass posteroventrally, traversing about three quarters of the length of the visceral mass before turning anterodorsally as the hindgut toward the pericardium. The rectum penetrates the ventricle of the heart and passes between the posterior byssal retractor muscles and then over the dorsal surface of the posterior adductor muscle to terminate in an anal papilla on its posterior face.

The internal structure of the stomach of *Dreissena polymorpha* has been described by Purchon (1960) and Morton (1969a) and is illustrated in Figure 18. The major typhlosole arises on the floor of the stomach and passes into the left cecum through which it pursues a straightforward course (Figure 19D). The left cecum is a simple sac receiving 10 or more ducts from the digestive diverticula into which the major typhlosole sends projections. The major typhlosole emerges from the left cecum, traverses the floor of the stomach from left to right and enters the right cecum through which it pursues a sinuous course (Figure 19D). The right cecum is complicated and, besides the major

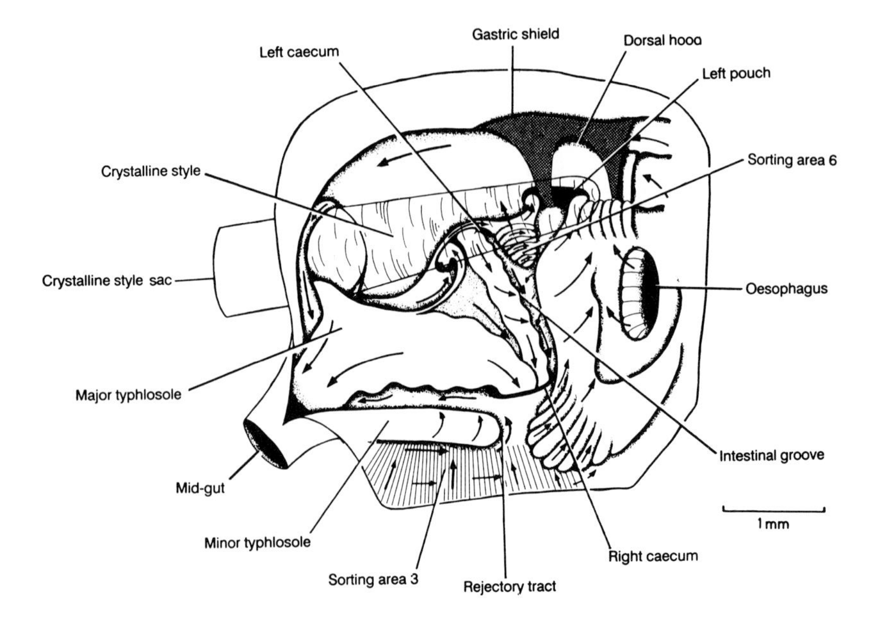

Figure 18. The structure and ciliary currents of the interior of the stomach of *Dreissena polymorpha* as seen from the right side. (From Morton, B. *Proc. Malacol. Soc. London* 38:301–321 [1969a]. With permission.)

typhlosole, receives between 11 and 13 ducts from the digestive diverticula. Four ducts from the digestive diverticula open into the left pouch, which is not invaded by the major typhlosole. No ducts from the digestive diverticula open directly into the stomach. The major typhlosole emerges from the right cecum and passes backward to enter the midgut. Throughout its course, the major typhlosole dictates the course of the intestinal groove. Particles not destined for intracellular digestion in the digestive diverticula pass into the intestinal groove in the stomach and thus enter the midgut for eventual defecation.

In *Dreissena polymorpha*, separation of the midgut and style sac has apparently been achieved by fusion of the major typhlosole with the opposite surface. The portion of the typhlosole remaining within the style sac as a result of this fusion, forms the D cell region, and has a secretory function.

In transverse section, the style sac of *Dreissena polymorpha* comprises four cell regions labeled A, B, C, and D in Figure 20A. The A and B cell regions constitute the main surface area of the style sac. Cells of region A possess more bristlelike cilia than those of region B. Cells of region C occupy the right side of a style sac ''gutter'' (seen in greater detail in Figure 20B) and are short, narrow, and columnar, bearing powerful cilia. These cells are characterized by the distinct conical arrangement of inner fibrils converging

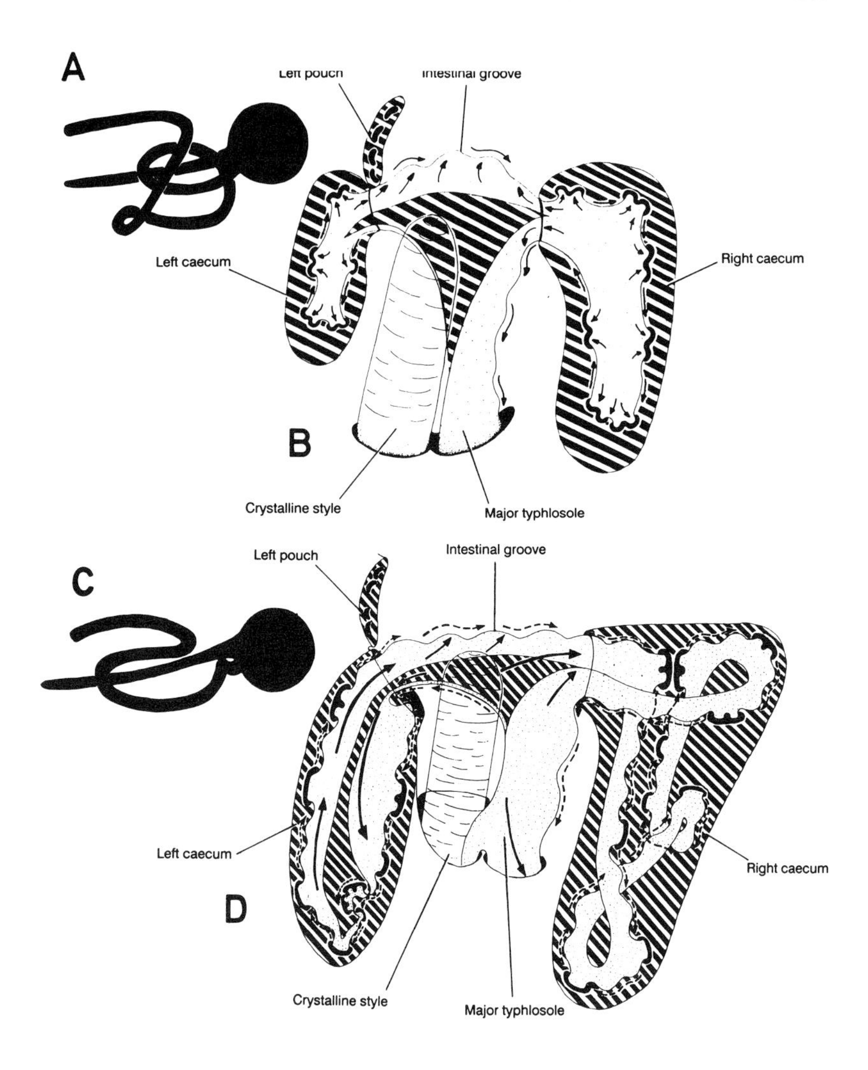

Figure 19. The course of the alimentary canal in the visceral mass and the structure of the left and right caeca of the stomach of A, B, *Mytilopsis leucophaeta* and C, D, *Dreissena polymorpha*. (From Morton, B. *Proc. Malacol. Soc. London* 38:301–321 [1969a]. With permission.)

on the nucleus from the cell margin. Type D cells characterize the opposite side of the style sac gutter and secrete the matrix of the crystalline style. Secretory globules are pinched off from the distal surfaces of these cells and passed into the gutter. The resulting amorphous style material accumulates here and is passed into the style sac proper by the C cell cilia. The cilia of the A and B type cells serve to coat the style with new material and to rotate it against the gastric shield lining the posterodorsal roof of the stomach. The

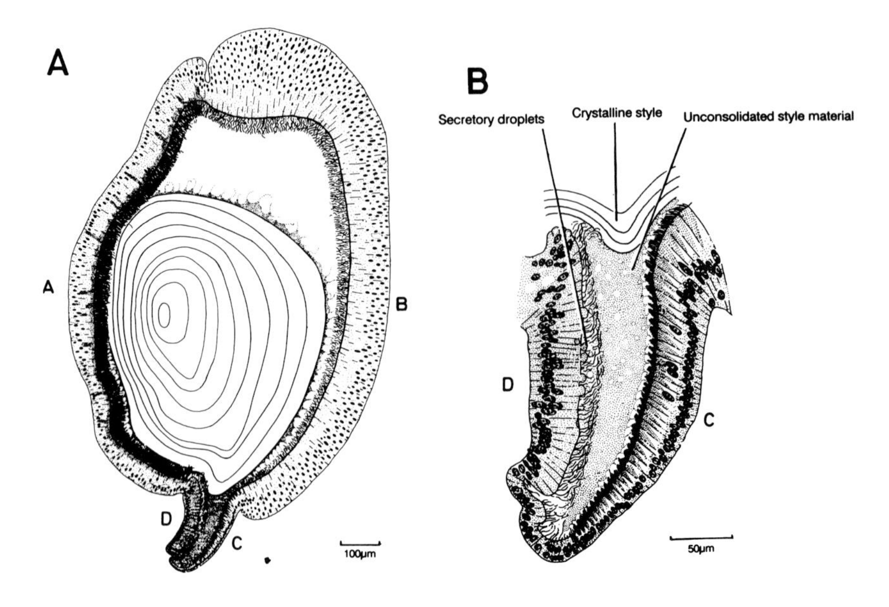

Figure 20. A transverse section through the crystalline style sac (A) and a detail of the secretory epithelia of the style sac "gutter" of *Dreissena polymorpha* (B). (From Morton, B. *Proc. Malacol. Soc. London* 38:301–321 [1969a]. With permission.)

functioning of the digestive system of *Dreissena polymorpha* has been described by Morton (1969b).

Pericardium

The pericardium of *Dreissena polymorpha* is located beneath the ligament and contains the heart which comprises a medial ventricle (penetrated by the rectum), lateral auricles, and anterior as well as posterior aortas (Figure 21) (Toureng, 1894b). Each lateral kidney is a U-shaped tube lying outside the posterior byssal retractor muscles and somewhat beneath the posterior adductor muscle, and extending backward from the floor of the posterior portion of the pericardium. The proximal limb of each kidney is a slender, ciliated, duct which communicates with the pericardium via the renopericardial aperture. The posterior end of the proximal limb opens into the larger distal limb. On the anterior floor of the distal limb, a ciliated funnel leads downward to the excretory aperture which discharges urine into the suprabranchial chamber. The genital aperture lies a little anterior to the excretory aperture. The paired gonads occupy much of the visceral mass. The sexes are separate and fertilization is external, a veliger larva resulting (Wlastov and Kachanova, 1959). The pericardial gland is "pericardial" in position occurring mainly anterior to the heart, but with posterior extensions above the ctenidial axis.

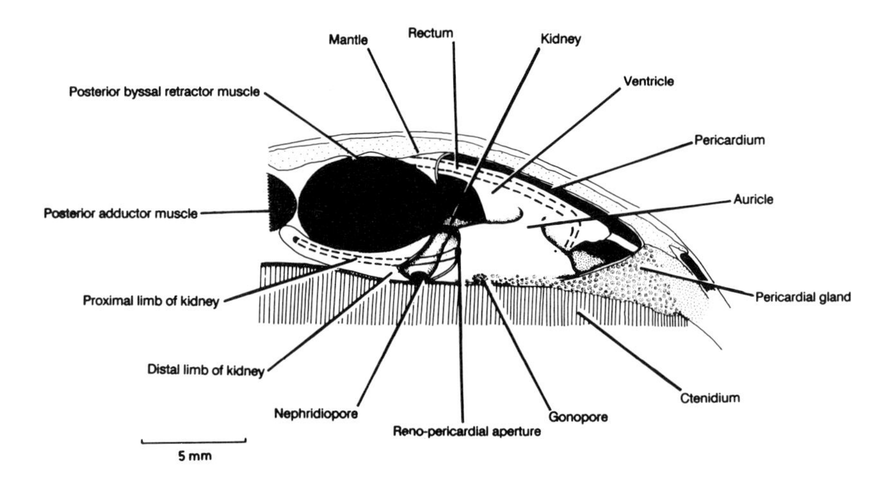

Figure 21. The pericardium and associated organs of *Dreissena polymorpha* as seen from the right side. (From Morton, B. *Proc. Malacol. Soc. London* 38:301–321 [1969a]. With permission.)

It is an organ of ultrafiltration (Morton, 1969b). The pericardial glands of the Lamellibranchia have been reviewed by White (1942), who briefly discussed *Dreissena polymorpha*.

THE EVOLUTION OF THE HETEROMYARIAN FORM IN THE DREISSENOIDEA

Taylor et al. (1973) showed that the shell structure of *Dreissena polymorpha* most closely resembles that of the Corbiculoidea. Inspection of fossil Dreissenoidea similarly suggests the possibility of a common ancestry for these two groups of bivalves. Each valve of *Corbicula fluminea* (Figure 22A) possesses two adductor muscle scars of approximately equal size and two smaller pedal retractor muscles. It is thus a typical isomyarian bivalve in that the shell is equivalve and equilateral, with a small internal opisthodetic ligament and typically with three cardinal teeth lying on the hinge plate immediately below the umbo and anterior and posterior lateral teeth. In the corbiculoid *Villorita cyprinoides* (Figure 22B), the first signs of anterior reduction are seen and are accompanied by extension of the ligament and posterior region of the shell. This has resulted in reduction of the anterior portion of the hinge plate and movement of the anterior adductor and pedal retractor muscles toward the umbo. The extension of the posterodorsal region of the shell has resulted in greater development of the posterior adductor and pedal musculature, relative to the anterior.

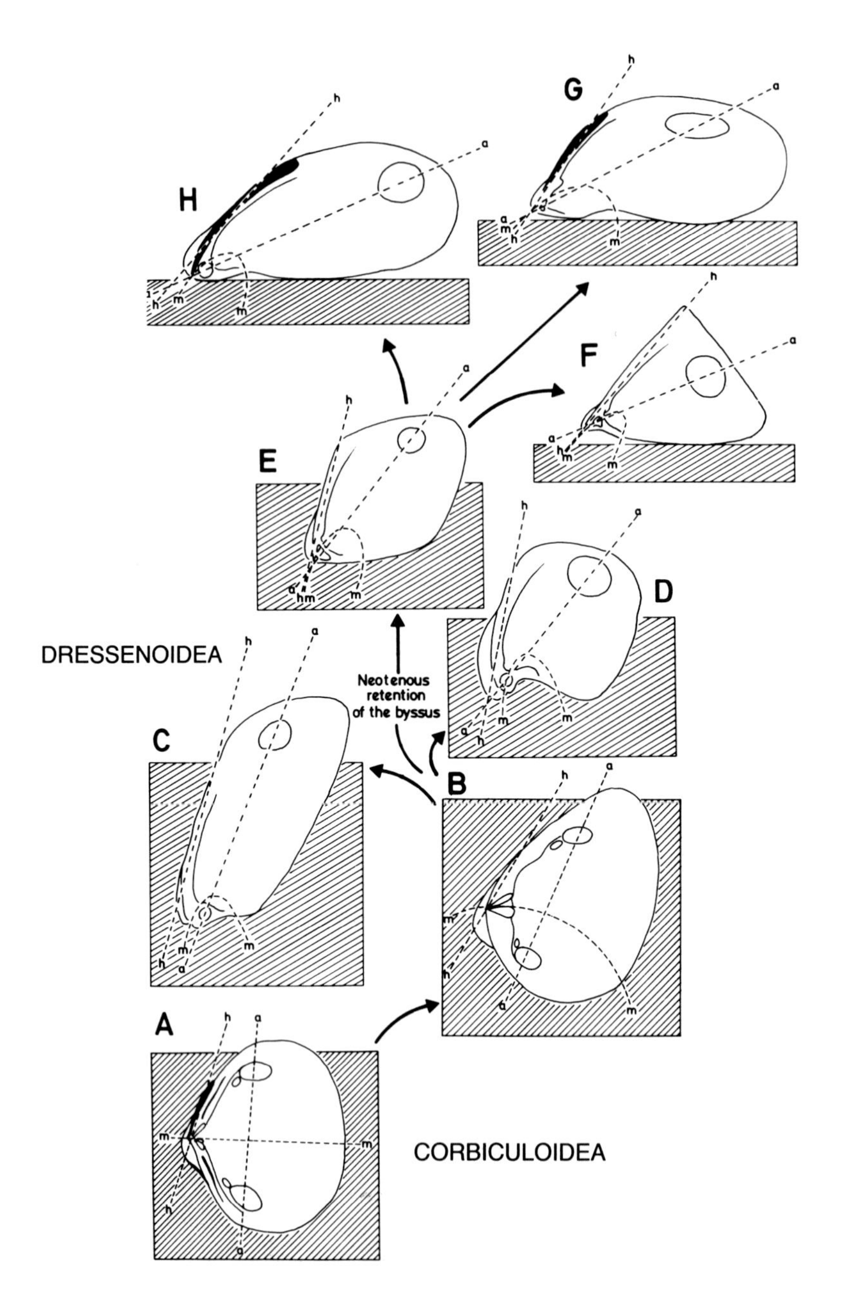

Figure 22. Taxa of the Corbiculoidea and Dreissenoidea used to illustrate changes that must have occurred to obtain the heteromyarian form. A, *Corbicula fluminea*; B, *Villorita cyprinoides*; C, *Dreissenomya aperta*; D, *Congeria subglobosa*; E, *Congeria zsigmondyi*; F, *Congeria triangularis*; G, *Mytilopsis sallei*; and H, *Dreissena polymorpha* (a = axis through the adductor muscles; h = hinge axis; m = the mid-dorsoventral axis).

Yonge (1962) showed that representatives of many bivalve phylogenies have retained into adult life an essentially larval characteristic, the byssus, and that the neotenous retention of this structure has, in many cases, influenced shell form. The effect of byssus retention upon a shell form essentially similar to present-day *Villorita* would have been to produce species in which anterior reduction and ventral flattening became progressively more extreme, associated with a change in mode of life and in functions of the muscular system. The fossil record contains some representatives of the Dreissenoidea in which such changes have occurred. Though extinct, it is suggested that they represent morphological strata through which the ancestors of the present-day Dreissenoidea have passed.

The neotenous retention of the byssus in the ancestors of the Dreissenoidea must have synergistically affected the relative importance of the pedal retractor muscles. In the burrowing Corbiculoidea, pedal retractors are necessary for the movements of the foot associated with burrowing. In the byssate Dreissenoidea, however, the importance of these muscles is diminished as the foot has ceased to be an organ of locomotion (except in the juvenile) and is principally used as a tool for the planting of byssal threads. In order for the byssus to fulfill its function as an anchor, enlargement of the larval byssal retractor muscles has occurred, with concomitant diminution of the pedal retractors such that the latter are now essentially vestigial in *Dreissena polymorpha* (Figure 10). Figure 22 indicates a possible way in which the heteromyarian form could have evolved in the Dreissenoidea to produce a range of animals all basically preadapted for the future adoption of an epifaunal mode of life. The infaunal, endobyssate species of the Dreissenoidea have become extinct, while only the epibyssate *Dreissena* and *Mytilopsis* have survived.

Dreissenomya aperta (Figure 22C) and *Congeria subglobosa* (Figure 22D) are extinct species showing two different ways in which an infaunal life was associated with byssal attachment, the former elongate and the latter more globular, but both anteriorly reduced. Anterior reduction is also seen in *C. zsigmondyi* (Figure 22E); it only requires ventral flattening of such a shell to allow escape from an endobyssate existence and the pursuit of an epibyssate lifestyle, as is clearly the case in *C. triangularis* (Figure 22F). Adoption of the extreme heteromyarian form in *C. triangularis* indicates an intertidal lifestyle attached to rocks, possibly in estuaries, as is now the case with many representatives of the Mytiloidea. *C. triangularis* resembles fossil species of *Lithocardium* (Cardioidea) for which Yonge (1980) also envisaged an epibyssate lifestyle. Other dreissenoids, however, had acquired the powers of osmoregulation necessary for the exploitation of brackish- and freshwaters, notably by species of *Mytilopsis* (Figure 22G) and *Dreissena* (Figure 22H), respectively. In the range of bivalves described, there is a trend toward anterior reduction and posterior inflation of the shell as is seen in the posterior divergence of the adductor axis (a-a) from the hinge axis (h-h) and the curvature of the mid-dorso-ventral axis (m-m).

Evolution of the heteromyarian form in the Dreissenoidea is presumed to have taken place under conditions that differed from those in which *Dreissena polymorpha* now lives. Yonge and Campbell (1968) suggest that evolution probably occurred in the intertidal or shallow sublittoral regions of the sea where some taxa with an ability to osmoregulate invaded river estuaries and ultimately fresh waters. It is envisaged that the marine ancestor of both the Corbiculoidea and the Dreissenoidea underwent adaptive radiation, producing a wide range of species capable of surviving in a variety of coastal, probably intertidal habitats. Modern representatives of the Veneroidea probably constitute such a radiation in the sea. The more recent ancestors of the Corbiculoidea and Dreissenoidea, however, invaded estuaries and underwent further radiation there. Some species were left on tidally influenced shores, in estuaries and in bays (i.e., species of *Mytilopsis*). Species of *Dreissena*, however, invaded fresh waters where they were structurally preadapted for the exploitation of rocky substrata in both lentic and lotic habitats.

The heteromyarian form in the present-day Dreissenoidea is considered to be a direct consequence of the evolutionary trend described earlier and which, it is suggested, proceeded from an established eulamellibranch stock. Available evidence suggests that ancestors of the modern Corbiculoidea gave rise to forms which ultimately produced the various species of *Mytilopsis* and *Dreissena polymorpha*. The neotenous retention in these genera of primitive characters (e.g., the byssus and associated musculature and a free-swimming veliger larva), have made them extremely successful in the exploitation of rocky surfaces of fresh- and estuarine waters. Evolution of the heteromyarian form in the Dreissenoidea mirrors evolution of the same form in the filibranch Mytiloidea, except that the former is a modern, Cenozoic assemblage while the latter are a primitive Paleozoic group.

DISCUSSION

Dreissena polymorpha is ideally suited for the successful exploitation of freshwater systems. The retention of a free-swimming veliger larva and of an active byssal apparatus in adult life enables *Dreissena* to colonize stony locations normally unoccupied by other bivalves. From these areas, *Dreissena* can then colonize the surfaces of adjacent soft deposits by building up colonies on the shells of either dead conspecifics or other species of bivalves (Lewandowski, 1976). In Asia, the mytilid *Limnoperna fortunei* has similar adaptations also enabling it to colonize a wide range of lentic and lotic habitats (Morton, 1973).

Dreissena has been isolated from the Mytiloidea by possession of characteristically eulamellibranch ctenidia (Morton, 1969a); on features of the nervous system (Toureng, 1894a); and possession of a posterior aorta (Toureng, 1894b) and differences in shell (Taylor et al., 1973) and periostracum structure (Beedham, 1959). Yonge and Campbell (1968) have shown that the dreissenoid and mytiloid ligaments are fundamentally different, and the internal architecture of the stomach of *Dreissena polymorpha* is directly comparable with that of numerous representatives of the Heterodonta (Purchon, 1960; Morton, 1969a). These anatomical differences clearly separate the Dreissenoidea from the Mytiloidea (even though Seilacher [1984, p. 224] discusses *Congeria* as a representative of the latter superfamily) and demonstrate that the heteromyarian form has evolved separately, at different times, in these two bivalve phylogenies. These features of the anatomy further suggest that the nearest relatives of *Dreissena* can only be sought among comparatively modern families of bivalves. The fossil record supports this view. Evolution of the Dreissenoidea was rapid, virtually all the constituent genera arising in the early Cenozoic and exploiting a wide range of coastal habitats. Nuttall (1990) has described the lifestyles of many fossil dreissenoidsis and, in most cases, this author agrees with these interpretations. Less heteromyarian species of *Congeria* were, for example, endobyssate. Only with regard to the genus *Dreissenomya* do interpretations differ. Nuttall (1990), quoting Marinescu (1977), suggests that *Dreissenomya* lived deep in sediments and was an active burrower with long separate siphons (Figure 23A) much like a modern tellinoidean (e.g., species of *Tellina*; Yonge, 1949). To accommodate such siphons, however, the pallial sinus would have to be very large, extending far back into the anterior half of the shell. This is not so and the pallial sinus of *Dreissenomya aperta* is small, no bigger than would be seen in a shallow-burrowing cockle (e.g., *Cerastoderma edule*; Yonge, 1980). Thus, I doubt Nuttall's and Marinescu's interpretation of the lifestyle of *Dreissenomya* and suggest that *D. aperta* was a shallow burrower (Figure $23B_1$), probably byssally attached. The arrangement of byssal retractor muscles in *Dreissenomya aperta* is shown in Figure $23B_2$, the likely effect they would have, upon contraction, in pulling the animal into the sediment is illustrated in Figure $23B_3$ and suggests a probable life orientation. This is compared with *Dreissena polymorpha* in Figure $23C_2$ and $23C_3$ to show how byssal retractor contraction would pull the animal down onto a solid substratum, thereby facilitating secure attachment. Such a comparison also explains the differences in the size and orientation notably of the posterior byssal retractor muscles. The endobyssate *Dreissenomya* would have retreated into the sediment while the epibyssate *Dreissena*, with byssal retractor muscles which are dorsally aligned above the byssus, would be pulled onto the rock surface.

Figure 19 highlights differences in the structure of the alimentary canal between *Mytilopsis leucophaeta* (Figure 19A,B) and *Dreissena polymorpha* (Figure 19C,D). Morton (1981) has earlier compared the anatomy of *Mytilopsis sallei* with that of *D. polymorpha*. Species of *Mytilopsis* have large ctenidia and small labial palps, i.e., a higher gill/palp ratio, than seen in *Dreissena*. Conversely, the alimentary canal of *M. sallei* is longer and more

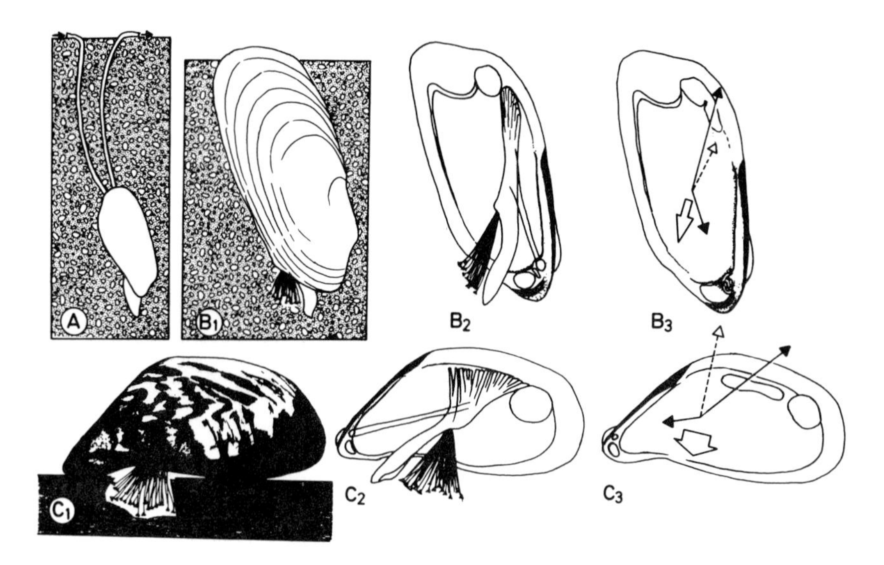

Figure 23. *Dreissenomya* (A) in its postulated position in the sediment (redrawn after Marinescu, 1977); *Dreissenomya aperta* in its postulated position in the sediment (B_1) with the arrangement of the byssal apparatus (B_2) and the direction the animal would be pulled into the sediment when the anterior and posterior byssal retractors contract (B_3). *Dreissena polymorpha* byssally attached to the substratum (C_1), the arrangement of the byssal apparatus (C_2), and the direction the animal would be pulled down onto the substratum when the anterior and posterior byssel retractors contract (C_3).

convoluted, whereas the left and right ceca of the stomach are more complicated in *D. polymorpha*. In terms of their detailed anatomy, therefore, *Mytilopsis* and *Dreissena* are quite different. Species of *Mytilopsis* probably exercise a lesser degree of particle selection in the mantle cavity and process relatively more material in the stomach with relatively smaller amounts of selected material entering the digestive diverticula. *D. polymorpha*, on the other hand, exercises a high degree of selection in both the mantle cavity and the stomach, with the probability of greater feeding efficiency. Such adaptations suit both genera to suspension-feeding modes of life across the broad spectrum of the brackish water-freshwater continuum, a habitat they are progressively coming to dominate worldwide with the help of man.

Evolution of the heteromyarian form by ancestors of the Dreissenoidea was therefore a prelude to a further minor epibyssate adaptive radiation, *Mytilopsis* exploiting coastal situations and *Dreissena* exploiting freshwater ones. Morton (1970) has argued that the position of the anterior byssal retractor muscle upon a lobe of the anterior shell septum in *Mytilopsis*, but not *Dreissena*, is an advanced character making more efficient byssal retractor contraction by reducing shear. In this respect, *Mytilopsis* is more like some species of *Congeria* (e.g., *C. triangularis* and *C. zsigmondyi*) whereas *Dreissena* is more like other species (e.g., *C. subglobosa*). The inescapable conclusion, therefore, is that the relatively recent evolution of modern dreissenoids is from a *Congeria* stock which radiated dramatically in the Cenozoic to exploit a wide range of inshore coastal environments, with the evolution of the heteromyarian form by two generic linkages. However, the question of why the endobyssate ancestors of modern dreissenoids that evolved initially from a heterodont, isomyarian stock have not survived, remains unanswered.

REFERENCES

Archambault-Guezou, J. "Présence de Dreissenidae euxiniques dans les dépôts à Congéries de la vallée du Rhône et sur pourtour du bassin méditerranéen. Implications biogéographiques," *Bull. Soc. Géol France* 18(7):1267–1276 (1976).

Babak, E. V. "Pliocene and Quaternary Dreissenidae of the Euksin Basin," *Tr. Paleontol. Inst. Akad. Nauk S.S.S.R.* 204:104 (1983) (In Russian).

Beedham, G. E. "Observations on the Mantle of the Lamellibranchia," *Q. J. Microsc. Sci.* 99:181–197 (1958).

Britton, J. C., and B. Morton. *Shore Ecology of the Gulf of Mexico.* (Austin, TX: University of Texas Press, 1989).

Haas, F. Bronn, Klassen und Ordnungen des Tier-Reichs III. Abteilung: Bivalvia (Muscheln). (B. H. Leipzig: Akademisch Verlagsgesellschaftman, 1929).

Kirpichenko, M. Y. A. "Phenology, Population Dynamics and Growth of *Dreissena* Larvae in the Kuibyshev Reservoir," in B. K. Shtegman, Ed. Biology and Control of *Dreissena. Tr. Inst. Biol. Vnutr. Vod.* 7(10):15–24 (1968).

Lewandowski, K. "Unionidae as a substratum for *Dreissena polymorpha* Pall," *Pol. Arch. Hydrobiol.* 23:409–420 (1976).

Locard, A. "Les *Dreissensia* du systeme europeen," *Rev. Suisse Zool.* 1:113–185 (1893).

Marelli, D. C., and S. Gray. "Conchological Redescriptions of *Mytilopsis sallei* and *Mytilopsis leucophaeta* (Bivalvia:Dreissenidae) of the Brackish Western Atlantic," *Veliger* 25:185–193 (1983).

Marelli, D. C. and S. Gray. "Comments on the Status of Recent Members of the Genus *Mytilopsis* (Bivalvia:Dreissenidae)," *Malacol. Rev.* 18:177–122 (1985).

Marinescu, F. "Genre *Dreissenomya* Fuchs (Bivalvia, Heterodonta)," *Mém. Inst. Géol. Géophys. Bucharest* 26:75–115 (1977).

Meisenheimer, J. "Entwicklungsgeschichte von *Dreissena polymorpha* Pall. I. Bis zur Ausbildung der jungen Trochophoralarve," *S. B. Ges. Marburg.* 1899:1–43 (1899).

Meisenheimer, J. Vortragüber die Entwicklungsgeschichte von *Dreissena polymorpha* Pall," *S. G. Ges. Marburg.* 1900:93–98 (1900).

Meisenheimer, J. "Entwicklungsgeschichte von *Dreissena polymorpha* Pall," *Z. Wiss. Zool.* 69:1–137 (1901).

Meisenheimer, J., and M. Klinger. "A Model for Pattern Formation on the Shells of Molluscs," *J. Theor. Biol.* 126:63–89 (1987).

Morton, B. "Studies on the Biology of *Dreissena polymorpha* Pall. I. General Anatomy and Morphology," *Proc. Malacol. Soc. London* 38:301–321 (1969a).

Morton, B. "Studies on the Biology of *Dreissena polymorpha* Pall. II. Correlation of the Rhythms of Adductor Activity, Feeding, Digestion and Excretion," *Proc. Malacol. Soc. London* 38:401–414 (1969b).

Morton, B. "The Evolution of the Heteromyarian Condition in the Dreissenacea (Bivalvia)," *Palaeontology* 13:563–572 (1970).

Morton, B. "Studies on the Biology of *Dreissena polymorpha* Pall. V. Some Aspects of Filter-Feeding and the Effect of Micro-Organisms Upon the Rate of Filtration," *Proc. Malacol. Soc. London* 39:280–301 (1971).

Morton, B. "Some Aspects of the Biology and Functional Morphology of the Organs of Feeding and Digestion of *Limnoperna fortunei*, (Dunker) (Bivalvia:Mytilacea)," *Malacologia* 13:265–281 (1973).

Morton, B. "The Biology and Functional Morphology of *Mytilopsis sallei* (Recluz) (Bivalvia:Dreissenacea) Fouling Visakhapatnam Harbour, Andhra Pradesh, India," *J. Moll. Stud.* 47:25–42 (1981).

Morton, B. "The Functional Morphology of the Organs of the Mantle Cavity of *Perna viridis* (Linnaeus, 1758) (Bivalvia:Mytilacea)," *Am. Malacol. Bull.* 5:159–164 (1987).

Morton, B. "Life History Characteristics and Sexual Strategy of *Mytilopsis sallei* (Bivalvia:Dreissenacea), Introduced into Hong King," *J. Zool.* 219:469–485 (1989).

Morton, B. "The Evolution and Success of the Heteromyarian Form in the Mytiloidea," in E. B. Gosling, Ed. *The Mussel Mytilus.* (London: Elsevier, 1992), pp. 21–52.

Nuttall, C. P. "Review of the Caenozoic Heterodont Bivalve Superfamily Dreissenacea," *Palaeontology* 33:707–737 (1990).

Oldham, C. "Locomotive Habit of *Dreissena polymorpha*," *J. Conchol.* 19:25–26 (1930).

Purchon, R. D. "The Stomach in the Eulamellibranchia; Stomach Types IV & V," *Proc. Zool. Soc. London* 135:431–489 (1960).

Seilacher, A. "Constructional Morphology of Bivalves: Evolutionary Pathways in Primary Versus Secondary Soft-Bottom Dwellers," *Palaeontology* 27:207–237 (1984).

Sprung, M., and U. Rose. "Influence of Food Size and Food Quantity on the Feeding of the Mussel *Dreissena polymorpha*," *Oecologia* 77:526–532 (1988).

Stanley, S. M. "Functional Morphology and Evolution of Byssally Attached Bivalve Molluscs," *J. Paleontol.* 46:165–212 (1972).

Taylor, J. D., W. J. Kennedy, and A. Hall. "The Shell Structure and Mineralogy of Bivalvia. Part 2, Chamacea-Poromyacea, Conclusions," *Bull. Br. Mus. (Nat. Hist.), Zool.* 22:255–294 (1973).

Toureng, M. "Sur le systeme nerveux du *Dreissena polymorpha*," *C. R. hebd. Seances Acad. Sci., Paris* 118:544 (1894a).

Toureng, M. "Sur l'appareil circulatoire du *Dreissena polymorpha*," *C. R. hebd. Seances Acad. Sci., Paris* 118:544 (1894b).

White, K. M. "The Pericardial Cavity and Pericardial Gland of the Lamellibranchia," *Proc. Malacol. Soc. London* 25:37–88 (1942).

Wlastov, B. V., and A. A. Kachanova. "Sex Determination in Living *Dreissena polymorpha* and Some Data Concerning the Cycle in this Mollusc," *Zool. Zh.* 38:991–1005 (1959).

Yonge, C. M. ''On the Structure and Adaptations of the Tellinacea, Deposit-Feeding Eulamellibranchia,'' *Philos. Trans. R. Soc. London, Ser. B.* 234:29–76 (1949).
Yonge, C. M. ''On the Primitive Significance of the Byssus in the Bivalvia and its Effects in Evolution,'' *J. Mar. Biol. Assoc. U.K.* 42:113–125 (1962).
Yonge, C. M. ''The 'Mussel' Form and Habit,'' in B. L. Bayne, Ed., *Marine Mussels: Their Ecology and Physiology.* (Cambridge: Cambridge University Press, 1976), pp. 1–12.
Yonge, C. M. ''Functional Morphology and Evolution in the Tridacnidae (Mollusca: Bivalvia:Cardiacea),'' *Rec. Austr. Mus.* 33:735–777 (1980).
Yonge, C. M. ''Mantle Margins with a Revision of Siphonal Types in the Bivalvia,'' *J. Moll. Stud.* 48:102–103 (1982).
Yonge, C. M., and J. I. Campbell. ''On the Heteromyarian Condition in the Bivalvia with Special Reference to *Dreissena polymorpha* and Certain Mytilacea,'' *Trans. R. Soc. Edinburgh* 68:21–43 (1968).

CHAPTER 13

Some Aspects of the Zebra Mussel, *(Dreissena polymorpha)* in the Former European USSR with Morphological Comparisons to Lake Erie

Nataliya F. Smirnova, G. I. Biochino, and Germane A. Vinogradov

INTRODUCTION

The main objectives of this work are to discuss the origin of the zebra mussel, *Dreissena polymorpha*, and to present data on the physiological, morphological, and cytogenetic polymorphism of this species in the former European USSR.

HISTORICAL

Dreissena was first found in the lower course of the Ural River in 1769 and later described as a zoological species in 1771 by the Russian zoologist

0-87371-696-5/93/$0.00 + $.50

Piter Pallas. *Dreissena* became of great interest during the 1820s when it was found at the London docks and then in different places of western Europe. In Germany it acquired the name of the wandering mussel (''Wundermuschel'') because of its ability to spread rapidly to different areas. Later in the same century, this mollusk began to block water supply pipes in Paris, Arlee, Berlin, and many other towns and cities throughout Europe (Zhadin, 1946).

The contemporary family of Dreissenidae is represented by only two genera, *Dreissena* and *Congeria*. The most ancient representatives of *Dreissena* belong to the genus *Congeria*, which appeared in the early Eocene. *Congeria* was most widespread and abundant during the epoch of the first and second Pontic Pier. However, in the Pliocene period *Congeria* almost completely disappeared from Europe and was replaced by the genus *Dreissena*. The largest distribution of *Dreissena* occurred during the Khvalynsk epoch of the Quarternary period. During this period, *Dreissena* was found in the Volga River and its tributaries, in northern areas of Eastern Europe, in Western Europe, and in the Aral Sea (Andrusov, 1897).

During the Quarternary glacial epoch, the geographic range of *Dreissena* declined dramatically. This was probably a result of coarse material suspended in glacial outwash and the negative impact this material had on the sensitive siphons of this species. While the effects of turbid glacial flows were widespread, there were areas in its former range that were not affected and *Dreissena* survived. These areas were the brackish waters of the Caspian and Aral Seas, in the freshwater portions of the Azov and Black Seas, and also in some water bodies of the Balkan peninsula.

PHYSIOLOGICAL, MORPHOLOGICAL, AND CYTOLOGICAL VARIABILITY

Dreissena, as is typical of an organism that is very adaptable and able to occur over a wide range of environmental conditions, can form populations that are locally distinct. Along the Volga River, populations of *Dreissena* have formed distinct ecotypes or races that differ in their tolerances to various environmental parameters, most notably temperature and salinity. Studies have shown that these different tolerance limits are distinctive at both the organism and cell level. For example, populations from six different sites along the Volga (Figure 1) were subjected to thermal tolerance tests. Individuals from the most southern site (Astrahan) and from a site subject to thermal discharges (Kostroma) were more tolerant of elevated temperatures than individuals from two of the more northern sites (Rybinsk and Kuibyshev) (Figure 2). In experiments to determine salinity tolerances, individuals from the site nearest to the Caspian Sea (Astrahan) had a lower mortality in relation to increases in salinity than individuals from the site farthest from the Caspian (Rybinsk) (Figure 3). In addition, survival of ciliary epithelial cells under

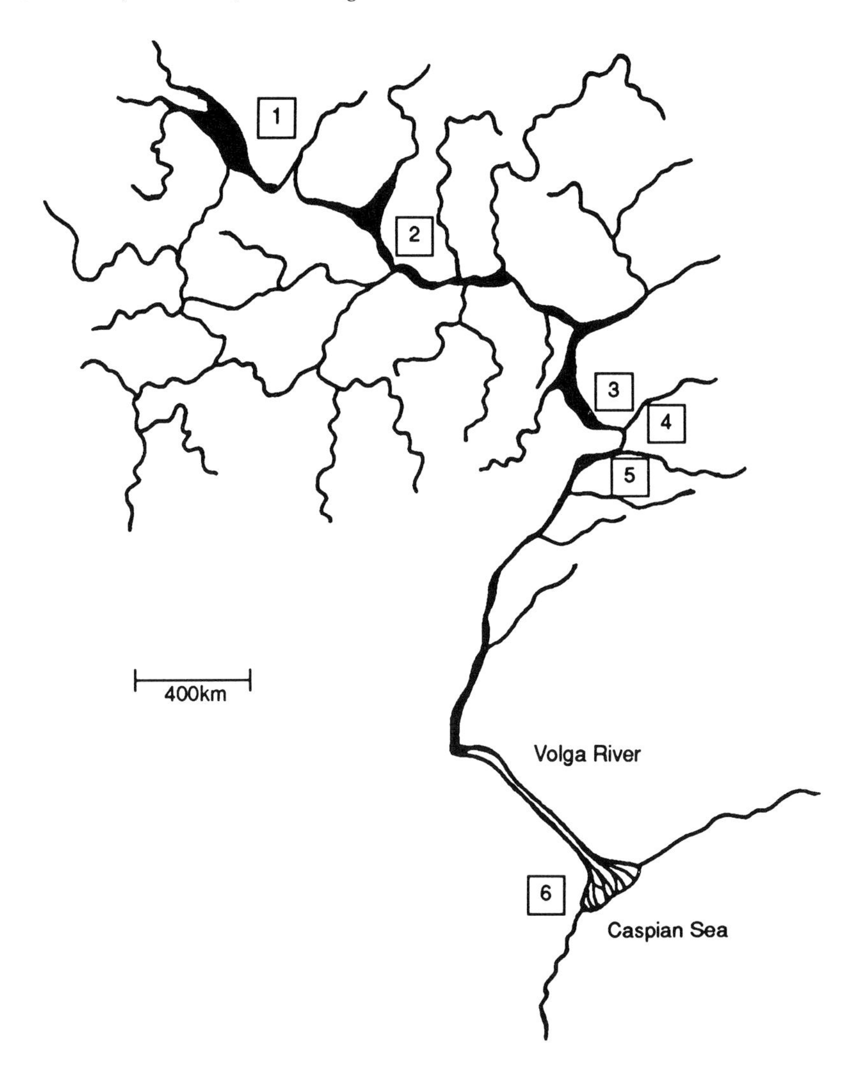

Figure 1. Location of sampling stations along the Volga River; 1 = Rybinsk, 2 = Kostroma, 3 = Kuibyshev, 4 = Samara, 5 = Chapaevsk, 6 = Astrahan. (Redrawn from Shkorbatov, G. L. *System of Integration of Species as a System* [Vilnus, 1986]).

increased salinity conditions (25‰) was much greater in individuals from near the Caspian than from a freshwater site (Kuibyshev) (Figure 4).

An examination of polymorphism in both color and pattern of *Dreissena* shells from different parts of the former European USSR revealed six main varieties or phenotypes (Figure 5). Differences in the frequencies and relative proportions of occurrence of these varieties indicate five main population

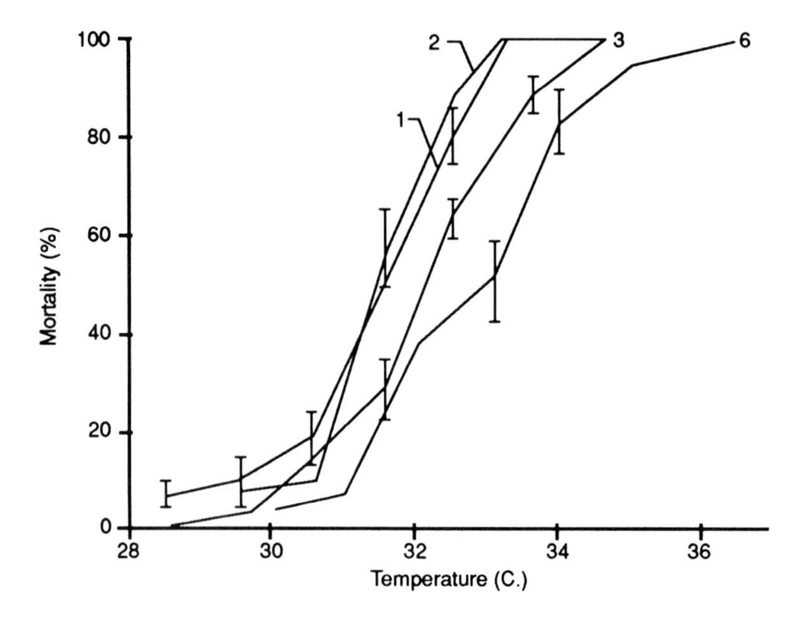

Figure 2. Relationship between mortality (%) and temperature in *Dreissena* from different water bodies in the former European USSR. Numbers correspond to the water bodies as shown in Figure 1. (Redrawn from Shkorbatov, G. L. *System of Integration of Species as a System* [Vilnus, 1986]).

groups: Aral-Caspian, Ponto-Caspian, Middle-Russian, Baltic, and Northeast. The maximum number of varieties and the greatest differences in the ratio of these phenes are found in the Aral-Caspian group, which is indicative of the unique position of this group in the system of intraspecific differentiation of *Dreissena* populations (Biochino and Slynko, 1988 and 1990; Biochino, 1990). These five groups are mostly confined to specific and separate geographical regions that can be differentiated by the time period in which *Dreissena* first colonized that region (Morduhai-Boltovski, 1960). Further, since these regions coincide with the separation of Eurasian mammal fauna (Starobogatov, 1970), it may be assumed that these *Dreissena* groups possess the status of distinct geographical races (Mayr, 1974).

Specimens of *Dreissena* were collected from Lake Erie near Monroe, MI in summer 1990 to compare shell color and pattern of North American individuals to those from the former European USSR. Preliminary analysis indicates that the specimens from Lake Erie are most similar to specimens from the Ponto-Caspian region; that is, only Lake Erie and the Ponto-Caspian group have individuals with the DD phenotype (Table 1). To further assess the degree of similarity between populations from the different regions, the similarity index of Zhivotovsky (1982) was used. This index is calculated as: $r = \Sigma (p_m q_m)^{0.5}$ where p is the occurrence frequency (or proportion) of

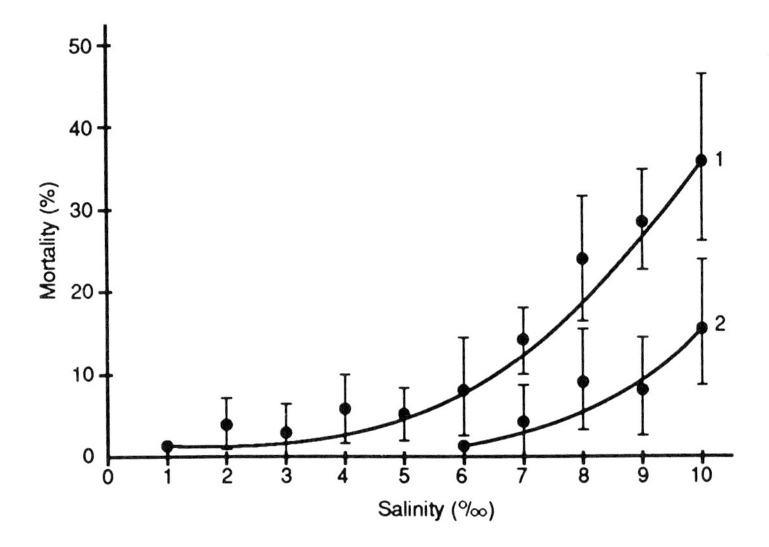

Figure 3. Relationship between mortality (%) and salinity (‰) in *Dreissena* from two different water bodies in the former European USSR. The exposure period was 2 weeks. 1 = Rybinsk (freshwater), 2 = Astrahan (near Caspian Sea, brackish water). (Redrawn from Antonov, P. I., and Shozbatov, G. L. *Species and Its Productivity Within Distribution Area* [Moscow, 1983]).

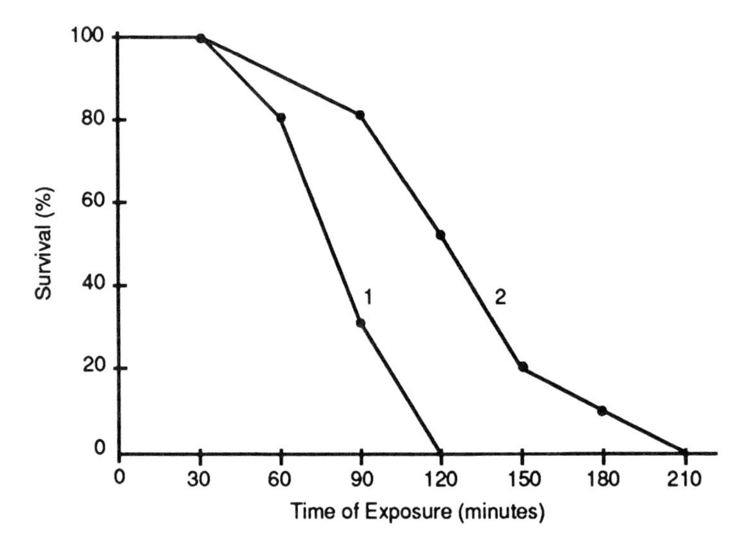

Figure 4. Relationship between survival (%) and time of exposure (min) for ciliated cells from the gill epithelium of *Dreissena* in water of 25‰ salinity. The *Dreissena* cells were taken from individuals from two different water bodies in the former European USSR. 1 = Rybinsk (fresh water), 2 = Astrahan (near Caspian Sea, brackish water). (Redrawn from Antonov, P. I., and Shozbatov, G. L. *Species and Its Productivity Within Distribution Area* [Moscow, 1983]).

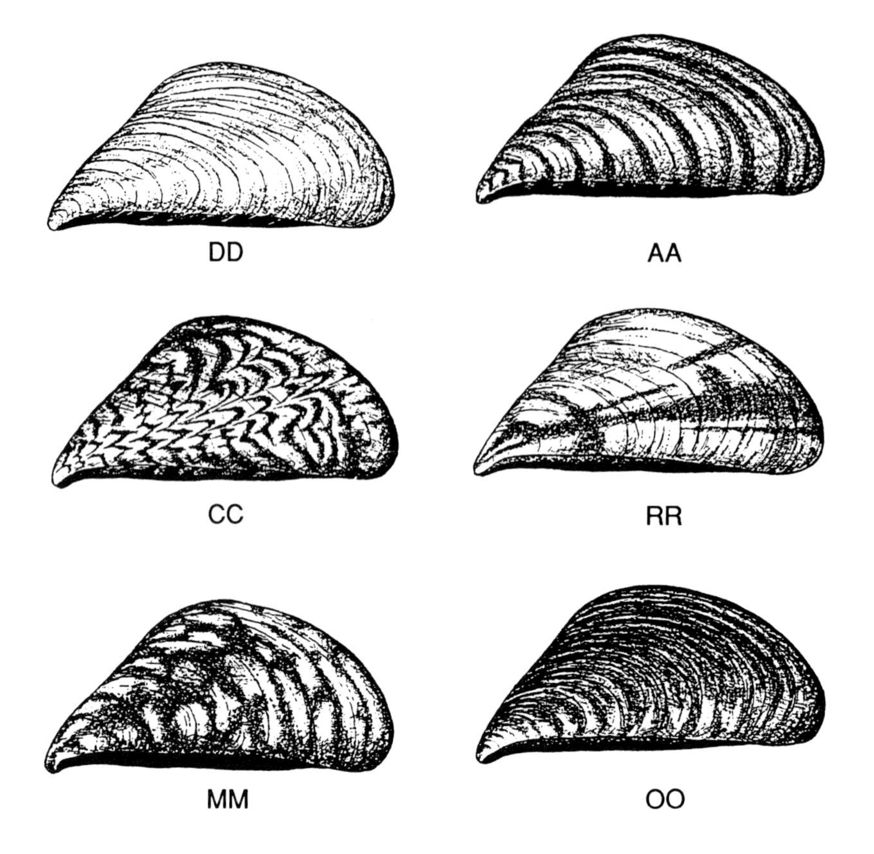

Figure 5. The six varieties (phenotypes) of *Dreissena* found in the former European USSR. Differences based on color and pattern of shells. DD = light, stripeless; AA = arched stripes; CC = mixed, zigzag; RR = radial striped; MM = spotted; OO = dark, stripeless.

Table 1. Proportion of Different Phenotypes in Populations of *Dreissena* from Different Regions in the Former European USSR and from Lake Erie

	Phenotype					
Region	**AA**	**DD**	**OO**	**RR**	**MM**	**CC**
Ponto-Caspian	0.094	0.022	0.056	0.008	0.006	0.814
Middle-Russian	0.170	0.000	0.110	0.002	0.010	0.708
Baltic	0.080	0.000	0.015	0.005	0.001	0.899
Lake Erie	0.120	0.030	0.170	0.110	0.190	0.380

Note: The different phenotypes are shown in Figure 5. Specimens from the various regions were collected from the following water bodies: Ponto-Caspian Region — lower Volga River, Tsymlyansk and Krenenchug reservoirs, and mouth of the Don river; Middle-Russian Region — Kuibyshev, Cheboksarsk, Saratov, and Volgograd reservoirs; Baltic Region — Rybinsk and Ivankovsk reservoirs and Kurshsky Bay.

Table 2. Index of Similarity Between Populations of *Dreissena* from Regions in the Former European USSR and Lake Erie

	Region			
Region	**Ponto-Caspian**	**Middle-Russian**	**Baltic**	**Lake Erie**
Ponto-Caspian	—	0.976	0.980	0.849
Middle-Russian	—	—	0.961	0.830
Baltic	—	—	—	0.760
Lake Erie	—	—	—	—

Note: Samples collected from water bodies as given in Table 1. See text for index derivation.

a phene of the first population and q is the occurrence frequency of the same phene in the second population. Values are calculated for each phene, and then the values are added to get the index r. The index ranges from 0 (no common phene) to 1 (all phenes similar). Based on this index, it is apparent that specimens from the Ponto-Caspian, Middle-Russian, and Baltic regions are more similar to each other than to specimens from Lake Erie (Table 2). Thus, given the extent of these phenotypic differences, it may be proposed that the introduction of *Dreissena* into North America occurred with individuals from outside these particular regions of the former European USSR. It must be noted, however, that Lake Erie specimens were not compared to specimens from the Northeast and Aral-Caspian regions.

In addition to examining the physiological and morphological variability in these populations, it is also useful to examine cytogenetic variability. Peculiarities in the organization of chromosomes are directly linked to formation of reproductive isolation, differences in fecundity and viability of the organisms, formation of geographical races, etc. Investigations of cells of the gill epithelium and gametes of *Dreissena* from the former European USSR have shown that the chromosome number varies from 21 to 32 and that the karyotype of 32 chromosomes is modal (Grishanin, 1990). The modal complex included 24 meta, 4 submetacentric, 2 subtelocentric, and 6 acrocentric chromosomes. The percentage of modal chromosomes, however, is highly variable. For instance, in specimens from cooling waters of the Litovskaya power station, the modal number of chromosomes was found in 87% of the nuclei; but in specimens from the Rybinsk reservoir, the modal number of chromosomes was found in only 23% of the nuclei (Barshene, 1990).

YEAR-TO-YEAR FLUCTUATIONS

Various long-term studies of *Dreissena* larvae in reservoirs of the former European USSR have shown natural year-to-year fluctuations in numbers. In the Dnepropetrovsk reservoir, numbers declined every third year (Dyga, 1966)

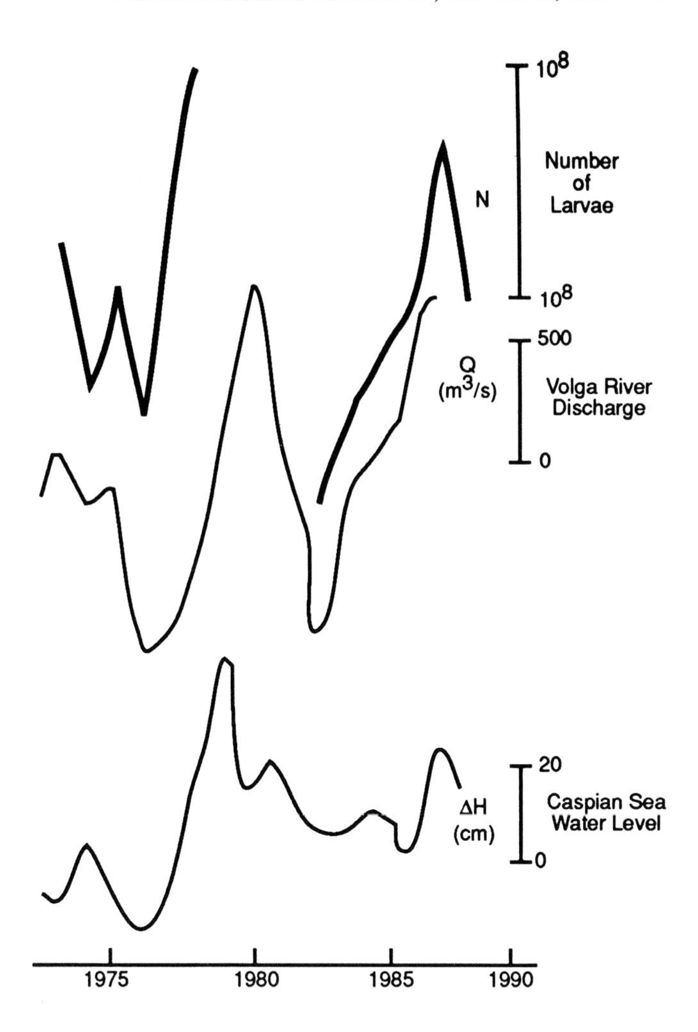

Figure 6. Year-to-year fluctuations in the number of *Dreissena* in the Ivankovsk and Uglich Reservoir relative to fluctuations in the discharge in the Volga River and water levels of the Caspian Sea. N = Mean total number of larvae in the two reservoirs, Q = discharge of the Volga River (m^3/s), ΔH = changes in water levels of the Caspian Sea (cm).

and in the Kyibyshevsk reservoir numbers declined every fourth year (Kirpichenko, 1964). We examined annual fluctuations in larvae number, water temperature, and water transparency in the Ivankovsk and Uglich Reservoirs over the period of 1973–1988. In general, larvae numbers were greater in years when temperatures were higher than normal and water transparency was lower. Since these reservoirs are located on the Volga River, many of their physical and chemical characteristics are influenced by discharge patterns of this river. Therefore, we further examined changes in larval numbers relative to annual changes in water volume of the Volga. Water volume is primarily influenced by changes in the level of the Caspian Sea which, in turn, is related to changes in the earth poles (Smirnov, 1969, Sarukhanyan and Smirnov, 1971). The number of larvae in the reservoirs was clearly related to discharge volume of the river and changes in levels of the Caspian Sea (Figure 6).

ACKNOWLEDGMENT

We would like to acknowledge the contribution of Olga Zhavoronkova who drew the various zebra mussel phenotypes.

REFERENCES*

Andrusov, N. I. ''Extinct and Living Dreissenidae of Eurasia,'' *Tr. SPB. Oba. Estestvoisp., Otd. Geol. Miner.* 27:683 (1897).

Antonov, P. I., and G. L. Shkorbatov. ''Ecological and Morphological Variability of the Volga Populations of *Dreissena polymorpha* (Pallas),'' in *Species and Its Productivity Within Distribution Area* (Moscow, 1983), pp. 116–128.

Barshene, Y. V. ''Cytogenetic Peculiarities of *Dreissenas* and Unionids,'' in *Species in Its Area. Biology, Ecology, and Production of Aquatic Invertebrates* (Minsk., 1990), pp. 126–130.

Biochino, G. I. ''Polymorphism and Geographical Variability of *Dreissena polymorpha* (Pallas),'' in *Microevolution of Freshwater Organisms* (Rybinsk, 1990), pp. 143–158.

Biochino, G. I., and Yu. V. Slynko, ''Interspecific Differentiation of *Dreissena* Within Its Area,'' *Ecology of Populations, Part 1* (Novosibirsk, 1988), pp. 87–89.

Biochino, G. I., and Yu. V. Slynko. ''Population Structure of *Dreissena polymorpha* (Pallas) Within Its Area,'' in *Species in Its Area. Biology, Ecology, and Production of Aquatic Invertebrates* (Minsk, 1990), pp. 130–135.

Dyga, A. K. ''Biological Fouling on Hydrotechnical Structures in the Dnepropetrovsk Reservoir and Their Control,'' Cand. Thesis (Dnepropetrovsk, 1966) 19p.

Grishanin, A. K. ''Kariotype of the Bivalve Mollusc *Dreissena*,'' in *Species in Its Area. Biology, Ecology, and Production of Aquatic Invertebrates* (Minsk, 1990), pp. 121–123.

Kirpichenko, M. Y. ''Phenology, Number, Dynamics, and Growth of *Dreissena* Larva in the Kuibyshevsk Reservoir,'' in *Dreissena Biology and Control* (Moscow and Leningrad, 1964), pp. 19–30.

* Editors' Note: Citations are presented as provided by the authors. The articles are published in Russian with titles translated into English. For further information about a specific citation, contact the senior author: Dr. Nataliya Smirnova, Institute Biology of Inland Waters, Nekouzskiy Raion, Yaroslavskaya, Oblast, Borok, Russia.

Mayr, E. *Populations, Species, and Evolution* (Moscow, 1974). 46 Op.
Sarukhanyan, E. I., and N. P. Smirnov. *Many-Year Fluctuations of the Volga Discharge* (Leningrad, 1971) 164p.
Smirnov, N. P. "Causes of Long-Period Stream Flow Fluctuations," *Sov. Hydrol.* 3:308–314 (1969).
Starobogatov, Y. I. "The Mollusc Fauna and Zoogeographical Division into Districts of the Continental Waters," *Nauka* (1970) 372 p.
Shkorbatov, G. L. "Interspecific Differentiation and Integrity of Species as a System," in *System of Integration of Species* (Vilnus, 1986). pp. 118–137.
Zhadin, A. Y. "The Wandering Mussel *Dreissena*," *Priroda* 5:29–37 (1946).
Zhivotovsky, L. A. "Indices of Population Variability on Polymorphic Characters," in *Population Phenetics* (Moscow, 1982). pp. 38–45.

CHAPTER 14

Genetics of the Zebra Mussel *(Dreissena polymorpha)* in Populations from the Great Lakes Region and Europe

Marc G. Boileau and Paul D. N. Hebert

The spatial patterns of variation in gene frequencies and heterozygosity were examined at 11 polymorphic enzyme loci among zebra mussels collected from 10 sites in the Great Lakes region and Europe, including 3 from both Lakes St. Clair and Erie. Gene frequencies were not uniform among sites in North America and heterogeneity was even detected at half the loci examined among sites in single lakes. North American and European populations generally had similar levels of genetic variation but heterozygosity was lower in a population at the distributional limit in North America. This evidence suggests that population growth results mostly from local recruitment rather than long distance dispersal. Three allelic variants were unique to the Great Lakes populations indicating that broader surveys of the European zebra mussel may reveal the origin(s) of North American populations. In an additional study, conducted to ascertain if a relationship existed between genetic variability and growth, there was no linkage between the number of loci that were heterozygous in young-of-the-year individuals and their growth.

0-87371-696-5/93/$0.00 + $.50

INTRODUCTION

Since the zebra mussel, *Dreissena polymorpha*, was first introduced into the Great Lakes (Hebert et al., 1989), considerable effort has been directed toward examining the ecological and economic implications of their invasion to North America. Genetic studies can play an important role in research on invading species by providing information on the number and origin of colonists, dispersal rates, and developmental processes. However, genetic studies have been limited to descriptions of variation within single localities (Hebert et al., 1989; Garton and Haag, 1991). Geographic examination of gene frequency variation provides a necessary basis for assessing both the origins and gene flow rates. Studies of multiple polymorphisms of individuals within cohorts are required to assess the link between genetic variation and growth.

Genetic studies of other organisms that have expanded their ranges have demonstrated that substantial shifts can occur in the organization of genetic variation among populations. For example, allelic frequencies shift significantly and heterozygosity is often reduced due to the loss of alleles when newer populations are established from a few individuals (Bryant et al., 1981; Boileau and Hebert, 1990). Reduced genetic variation within individuals (i.e., heterozygosity) influences various aspects of life history (Palmer and Strobeck, 1986) and has been demonstrated to specifically effect early survivorship and growth rate among marine mollusks (Zouros and Foltz, 1984).

The existence of many polymorphic loci and allelic variants among North American zebra mussel populations suggests that many colonists were released and became established. If a population bottleneck had been experienced, the European populations must have unprecedented levels of variation because the mussels in Lake St. Clair are among the highest 1% of animals examined for allozyme variation (Hebert et al., 1989) (Figure 1). No genetic studies have been conducted on European zebra mussels, however, so this assertion cannot be evaluated.

The present study examines allozyme frequencies and heterozygosities in *Dreissena polymorpha* from the Great Lakes and Europe in order to assess the effects of colonization and dispersal among these populations. In addition, individuals from a 1991 cohort were studied to assess the relationship between growth and genetic variation in this species.

METHODS

During the summer of 1991, zebra mussels were collected from seven locations in North America (Table 1). Lake St. Clair and Lake Erie were both sampled at three sites in order to determine whether gene frequencies were uniform within each lake. Lake St. Clair was sampled at locations along the southern (Puce River) and eastern shores (Thames River) and at one midlake

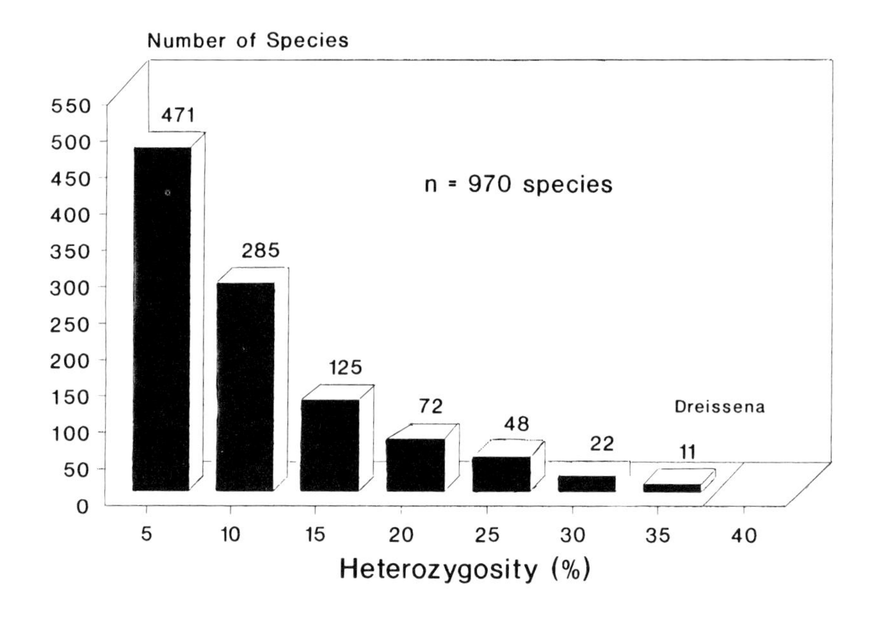

Figure 1. Distribution of heterozygosities for 970 animal species determined by allozyme electrophoresis. (Data from Nevo et al., 1984).

Table 1. List of *Dreissena polymorpha* Populations and the Localities Sampled for Genetic Analysis

Population Name	Localities (Abbreviation)
Great Lakes Region	
Lake St. Clair	Puce River (SC1)
	Site No. 37; Hebert et al., 1991 (SC2)
	Thames River (SC3)
Lake Erie	Colchester Beach (ER1)
	Rondeau Bay (ER2)
	Lowbanks (ER3)
Oneida Lake	Jewell (ON)
Europe	
Stoney Cove	Stoney Stanton, England (UK1)
Triangle Pond	Long Eaton, England (UK2)
Lake Konstanz	Staad Beach, Germany (GDR)

site (site 37). One sample was obtained from each of the three basins of Lake Erie (Colchester Beach — western, Rondeau Bay — central, Lowbanks — eastern). To compare heterozygosities in established Great Lakes populations with those at the range margin, a sample was collected from Oneida Lake, New York soon after their discovery. Two populations of zebra mussels were also obtained from England and a third population from southern Germany (Table 1) to assess whether higher heterozygosities were present in Europe.

Allozyme frequencies were determined using enzyme electrophoresis on cellulose acetate gels and methods outlined in Hebert and Beaton (1989). All indviduals were collected alive and frozen at $-80°C$ until electrophoresis. Individuals were stained for 9 enzymes with variation produced by 11 genetically interpretable loci: phosphoglucose isomerase (PGI), fumerase (FUM), isocitrate dehydrogenase (IDH), malate dehydrogenase (MDH), mannose-phosphate isomerase (MPI), lactate dehydrogenase (LDH), 6-phosphogly-conate dehydrogenase (6 PGDH), phenylalanylproline and leucylglycine peptidases (PP and LG-1,2,3, respectively).

Gene frequency shifts were analyzed using the gene diversity measure of Nei (1977) and tested for significance by X^2 (Workman and Niswander, 1970). Differences in heterozygosities were tested via the paired comparison ANOVA (Sokol and Rohlf, 1981).

Veliger settlement and spat growth at Rondeau Bay, Lake Erie were monitored approximately weekly using settling plates during the 1991 summer reproductive period. Some individuals from this cohort were 5–7 mm by early September. Young of the year (YOY) were collected alive on September 4 and individuals were measured to the nearest 1 mm. Three days later they were returned alive to the collection locality in twenty 22.5 cm^3 enclosures, each with a known number of individuals from a given size class. Seventeen enclosures were recovered on October 13 containing 160 individuals that were measured and electrophoresed. Animals were scored as either heterozygous or homozygous at the same 11 loci as examined in the spatial study.

RESULTS

Allele frequencies varied significantly in the total survey (Table 2) of 10 populations. Of the 36 alleles at the 11 loci examined, two infrequent alleles (IDH^3; LDH^3; Table 2) were unique to European populations while three alleles were unique in the Great Lakes Basin (PGI^5; MPI^5 and PP^3). In addition, marked frequency differences were observed between the two continents. For example, the allele MDH^2 dominated in the Great Lakes, while the MDH^1 allele was most frequent in Europe. Allele frequencies within *Dreissena* populations from Lakes St. Clair and Erie also showed significant heterogeneity at half of the loci studied (Table 3). For example, Lake St. Clair MDH and Lake Erie IDH alleles both varied from near fixation to approximately equal among nearby localities (Table 2).

Heterozygosity varied among populations (Table 4) within the Great Lakes Basin but not between the Lake St. Clair founder population and Europe. Averages within site heterozygosities in Lake St. Clair did not differ significantly from those of the three European populations ($F_{[1,10]} = 3.196$; Table 4). On the other hand, the Oneida Lake population possessed significantly lower mean heterozygosity than those from Lake St. Clair ($F_{[1,10]} = 5.34$; Table 4), largely due to the absence of variation at two loci (MDH and LDH).

Table 2. Allele Frequencies in *Dreissena polymorpha* Populations at 10 Localities in the Great Lakes Watershed and Europe

	Localities										
	Great Lakes Basin							Europe			X^2
Locus	SC1	SC2	SC3	ER1	ER2	ER3	ON	UK1	UK2	GDR	(d.f.)
PGI											
n	24	24	24	24	24	23	48	24	24	24	
1	0.12	0.23	0.12	0.33	0.15	0.15	0.18	0.08	0.08	0.29	122.2
2	0.19	0.35	0.27	0.42	0.42	0.50	0.46	0.35	0.31	0.44	(36)
3	0.35	0.10	0.29	0.25	0.42	0.35	0.33	0.44	0.60	0.23	
4	0.19	0.25	0.23		0.02		0.02	0.12		0.04	
5	0.15	0.06	0.08				0.01				
FUM											
n	24	48	24	24	24	24	28	24	24	24	
1	0.73	0.75	0.56	0.71	0.65	0.67	0.59	0.71	0.81	0.71	34.4
2	0.23	0.25	0.42	0.21	0.33	0.31	0.36	0.10	0.19	0.29	(18)
3	0.04		0.02	0.08	0.02	0.02	0.05	0.19			
IDH											
n	22	23	36	24*	24	24	24	24	22	24	
1	0.66	0.74	0.74	0.62	0.85	0.92	0.79	0.50	0.93	0.06	262.8
2	0.34	0.26	0.26	0.38	0.15	0.08	0.21	0.48	0.07	0.94	(18)
3								.02			
MDH											
n	24	24	24	24*	24	24	24	24	24	24*	
1	0.06	0.48	0.56	0.21	0.12	0.35		0.75	1.0	0.56	183.3
2	0.94	0.52	0.44	0.79	0.88	0.65	1.0	0.25		0.44	(9)
MPI											
n	24	24	24	24	21	23	23	23	24	24	
1	0.15	0.21	0.17	0.15	0.29	0.06	0.06	0.24	0.02		163.3
2	0.31	0.48	0.52	0.67	0.33	0.54	0.54	0.65	0.56	0.31	(36)
3	0.29	0.31	0.31	0.19	0.38	0.39	0.39	0.11	0.42	0.62	
4	0.21									0.06	
5	0.04										

Table 2 (continued). Allele Frequencies in *Dreissena polymorpha* Populations at 10 Localities in the Great Lakes Watershed and Europe

	Localities										
	Great Lakes Basin							Europe			X^2
Locus	SC1	SC2	SC3	ER1	ER2	ER3	ON	UK1	UK2	GDR	(d.f.)
LDH											
n	23	23	21	24	22	23	24	22	24	24	
1	0.26	0.44	0.57	0.46	0.43	0.52		0.70	0.54	0.27	130.5
2	0.74	0.56	0.43	0.54	0.57	0.48	1.0	0.30	0.46	0.71	(18)
3										0.02	
6PGDH											
n	24	24	24	24	36	24	24	24	24	24	
1	0.06	0.15	0.06	0.02	0.04		0.06	0.29	0.17	0.19	81.7
2	0.85	0.67	0.71	0.88	0.89	0.98	0.92	0.54	0.81	0.71	(18)
3	0.08	0.19	0.23	0.10	0.07	0.02	0.02	0.17	0.02	0.10	
PP											
n	24	23	24*	24	24	24	24	24	18	24	
1	0.08			0.17		0.44	0.56	0.56	0.44	0.40	139.7
2	0.52	0.56	0.77	0.56	0.60	0.44	0.42	0.44	0.56	0.60	(18)
3	0.40	0.44	0.23	0.27	0.40	0.12	0.02				
LG-1											
n	24	24	24	24	24	24	24	24	24	22	
1	0.31	0.44	0.29	0.33	0.23	0.44	0.44	0.44		0.41	58.0
2	0.60	0.48	0.56	0.48	0.69	0.44	0.52	0.46	0.75	0.32	(18)
3	0.08	0.08	0.15	0.19	0.08	0.12	0.04	0.10	0.25	0.27	
LG-2											
n	23	24	24	24	24	24	24	24	24	24	
1			0.23	0.60	0.29	0.52	0.21	0.50	0.71	0.23	151.0
2	0.76	0.58	0.56	0.38	0.67	0.46	0.73	0.46	0.27	0.65	(18)
3	0.24	0.42	0.21	0.02	0.04	0.02	0.06	0.04	0.02	0.12	

LG-3											
n	24	24	24	24	24	24	24	24	24	24	
1	0.04	0.27	0.65	0.71	0.73	0.28	0.71	0.58	0.65	0.65	188.2
2	0.65	0.62	0.25	0.25	0.04	0.54	0.21		0.31	0.02	(18)
3	0.31	0.10	0.10	0.04	0.23	0.17	0.08	0.42	0.04	0.33	
H_e	0.462	0.518	0.520	0.491	0.417	0.472	0.374	0.512	0.369	0.472	

Note: Population abbreviations as in Table 1; H_e = mean binomial expected heterozygosity; * = deviations of genotype frequencies from random mating significant $p \leq 0.05$.

Table 3. Chi-Square (Degrees of Freedom in Parentheses) Analysis of Significance of Allele Frequency Variation in *Dreissena polymorpha* Populations at Localities in Lakes St. Clair and Erie

Locus	Lake St. Clair	Lake Erie
PGI	10.98 (8)	9.10 (6)
FUM	10.28[a] (4)	0.88 (4)
IDH	0.13 (4)	13.11[a] (2)
MDH	57.33[a] (2)	6.34 (2)
MPI	12.13 (8)	40.88[a] (6)
LDH	15.69[a] (6)	0.00 (2)
6PGDH	6.35 (4)	5.73 (4)
PP	10.21[a] (4)	21.62[a] (4)
LG-1	2.03 (4)	8.368 (4)
LG-2	14.70[a] (4)	15.86[a] (4)
LG-3	47.25[a] (4)	43.16[a] (4)

[a] $p < .05$.

Table 4. Average Individual Heterozygosity at 11 Loci in Populations *Dreissena polymorpha* from Lake St. Clair, Oneida Lake, and Europe with Paired Difference Tests of the Significance of Differences Between the Lake St. Clair Population and Populations from Oneida Lake and Europe

Locus	Lake St. Clair	Oneida Lake	Europe
PGI	0.759	0.654	0.634
FUM	0.433	0.532	0.398
IDH	0.408	0.337	0.260
MDH	0.369	0.000	0.295
MPI	0.661	0.559	0.519
LDH	0.456	0.000	0.456
6PGDH	0.400	0.155	0.462
PP	0.470	0.520	0.500
LG-1	0.560	0.547	0.552
LG-2	0.479	0.430	0.503
LG-3	0.506	0.457	0.490
Mean	0.500	0.381	0.461
$F_{[1,10]}$		5.34*	3.23

* = $P < .05$

No allelic variants were unique in Oneida Lake among the Great Lakes populations.

Heterozygosities at the 11 loci examined in the enclosed YOY cohort (Table 5) did not differ significantly (paired comparisons) from the wild population examined in the spatial analysis ($t = 0.016$; $p \gg 0.05$). There was no significant correlation between the amount of shell growth by individuals and the number of heterozygous loci they possessed ($r = -0.056$; $t = -0.702$). It should be noted that half of the animals (52.2%) did not

Table 5. Comparison of Individual Heterozygosities (H_e) at 11 Loci in Wild and Experimental *Dreissena polymorpha* from Rondeau Bay, Lake Erie

	Wild		Experimental	
Locus	**n**	**H_e**	**n**	**H_e**
PGI	24	0.631	162	0.660
FUM	24	0.471	161	0.447
IDH	24	0.249	162	0.414
MDH	24	0.219	161	0.279
MPI	24	0.499	162	0.622
LDH	22	0.491	162	0.494
6PGDH	12	0.392	160	0.259
PP	24	0.478	162	0.488
LG-1	24	0.468	160	0.484
LG-2	24	0.469	142	0.358
LG-3	24	0.414	162	0.315
Mean		0.435		0.438

add any new shell length, suggesting an examination of the heterozygosities among animals, which did and did not grow, might yield some insight. However, the distribution of individual heterozygosities (Figure 2) also did not differ significantly (ANOVA, $F_{[1,158]} = 1.46$) between these two classes.

DISCUSSION

Colonization and Dispersal

The patterns of gene frequency and heterozygosity shifts in *Dreissena polymorpha* from the present study both confirm prior explanations for their origins from European populations and suggest unexpected behavior in individuals of this invading species. Similar heterozygosities between European populations and Lake St. Clair are clearly expected by the ballast water discharge explanation for their origins in North America (Hebert et al., 1989). Notably, however, the presence of three unique allelic variants in the Great Lakes populations means that sources for these variants must also exist, and the gene frequency shifts suggest their distribution could be localized.

Although gene frequency shifts are modest in *Dreissena polymorpha*, a considerable degree of genetic differentiation exists among localities in the two Great Lakes first colonized. The Great Lakes distribution of the zebra mussel expanded dramatically (Griffiths et al., 1991) from its probable introduction to Lake St. Clair in 1986 (Hebert et al., 1989), and its rapid spread has been attributed to high fecundity and pelagic veliger larvae (Griffiths et al., 1991). Neither characteristic seems likely to enhance genetic discontinuities, and the present genetic results require closer scrutiny.

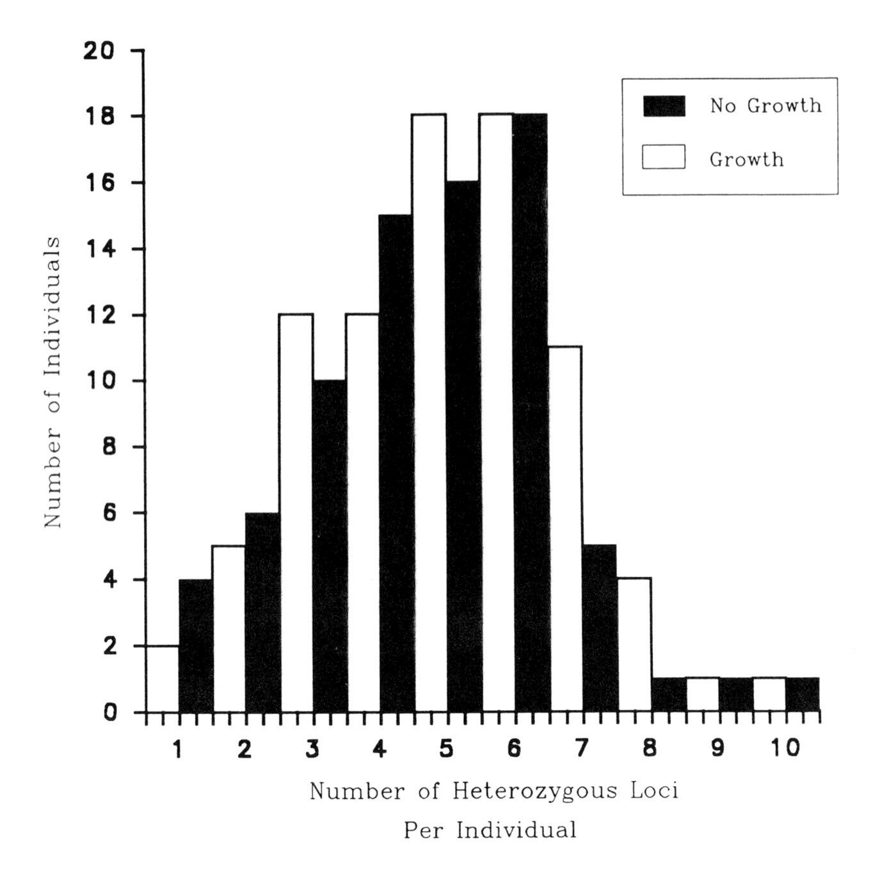

Figure 2. Distribution of the number of loci heterozygous per individual from 160 young of the year in two growth classes.

Gene frequency shifts, such as those observed in the present study, can occur due to the combined effects of sampling founders among localities and selective differences. Alone or in concert, these processes require restricted dispersal. Only substantial selective differences could produce the observed gene frequency shifts unless gene flow is curtailed (Endler, 1973). More simply, colonists numbering 10 or fewer at each locality (Wade and McCauley, 1988) and reduced dispersal distances during the colonization process can generate gene frequency shifts of the magnitude observed among localities. Recently established populations may have started from small numbers of individuals because veliger dispersal appears restricted during the earliest stages of formation (Hebert et al., 1989). If local founder numbers are small, the lower heterozygosities among Oneida Lake individuals, at the leading edge of the inland range expansion, are also expected.

Heterozygosity and Growth Variation

Populations of marine bivalve mollusks often have significant deficiencies of heterozygous genotypes from random mating expectations (Singh and Green, 1984; Zouros and Foltz, 1984), but heterozygous individuals in a specific cohort often have larger shell sizes than homozygous individuals. These patterns have been attributed to the combined effects of heterozygotes having inferior settlement success but superior growth rate, especially during early postsettlement (Zouros and Foltz, 1987). The present study, however, revealed no consistent heterozygote deficits (Table 2) at polymorphic loci in populations of *D. polymorpha*. There was also no evidence of a relation between the degree of heterozygosity and early postsettlement growth. This latter result contrasts with those obtained from a survey of zebra mussel populations in southern Lake Erie (Garton and Haag, 1991); noted was a relationship between size and individual heterozygosity, suggesting that growth, survival, or both may be greater in heterozygous individuals.

In summary, our genetic analyses suggest that dispersal success is restricted despite enormous veliger production and that postsettlement mortality and early growth are not controlled by genetic variability. These results were not expected by previous studies and clearly indicate that expanded studies are required to understand the dynamics of introduced *Dreissena* populations. Interpretations of future work must also consider the implications of these demonstrated gene frequency differences within local lake populations. These differences will likely persist for some years. The decay of genetic differences resulting from colonization will be slow because sessile adults live for 6–8 years (Stanczykowska, 1977), and the ratio of successful migrants to the enormous populations (Hebert et al., 1991; Mackie, 1991) is minute. At the same time, expanded studies of *Dreissena polymorpha* populations in Europe are also likely to provide insight into the origin(s) of North American populations.

ACKNOWLEDGMENTS

This study was funded by grants from the Natural Sciences and Engineering Research Council and the Great Lakes University Research Fund. Neil Billington, Ulrich Einsle, Joe Leach, Ed Mills, and Jason Schwenke kindly assisted with mussel collections. Laboratory space at the Cornell Field Station was provided. Dave Johnston, Tom MacDougall, Rob Van Brunt, and Lori Walker provided field and laboratory assistance.

REFERENCES

Boileau, M. G., and P. D. N. Hebert. "Genetic Consequences of Passive Dispersal in Pond-Dwelling Copepods," *Evolution* 45:721–733 (1990).

Bryant, E. H., H. vanDijk, and W. vanDelden. "Genetic Variability of the Face Fly, *Musca autumnalis* deGeer, in Relation to a Population Bottleneck," *Evolution* 35:872–881 (1981).

Endler, J. A. "Gene Flow and Population Differentiation," *Science* 179:245–250 (1973).

Garton, D. W., and W. R. Haag. "Heterozygosity, Shell Length and Metabolism in the European Mussel, *Dreissena polymorpha*, from a Recently Established Population in Lake Erie," *Comp. Bio-Chem. Physiol.* 99A:45–48 (1991).

Griffiths, R. W., D. W. Schloesser, J. H. Leach, and W. P. Kovalak. "Distribution and Dispersal of the Zebra Mussel (*Dreissena polymorpha*) in the Great Lakes Region," *Can. J. Fish. Aquat. Sci.* 48:1381–1388 (1991).

Hebert, P. D. N., and M. J. Beaton. "Methodologies for Allozyme Analysis Using Cellulose Acetate Electrophoresis," Helena Laboratories, Beaumont, TX (1989).

Hebert, P. D. N., B. W. Muncaster, and G. L. Mackie. "Ecological and Genetic Studies on *Dreissena polymorpha* (Pallas): A New Mollusc in the Great Lakes," *Can. J. Fish. Aquat. Sci.* 46:1587–1591 (1989).

Hebert, P. D. N., C. C. Wilson, M. H. Murdoch, and R. Lazar. "Demography and Ecological Impacts of the Invading Mollusc *Dreissena polymorpha*," *Can. J. Zool.* 69:405–409 (1991).

Mackie, G. L. "Biology of the Exotic Zebra Mussel, *Dreissena polymorpha*, in Relation to Native Bivalves and Its Potential Impact in Lake St. Clair," *Hydrobiologia* 219:251–268 (1991).

Nei, M. "*F*-statistics and Analysis of Gene Diversity in Subdivided Populations," *Ann. Hum. Genet.* 41:225–233 (1977).

Nevo, E., A. Beiles, and R. Ben-Schlomo. "The Evolutionary Significance of Genetic Diversity: Ecological, Demographic and Live History Correlates," in G. S. Mani, Ed. Evolutionary Dynamics of Genetic Diversity. *Lect. Notes Biomath.* 53:13–213 (1984).

Palmer, R. A., and C. Strobeck. "Fluctuating Asymmetry: Measurement, Analysis, Patterns," *Annu. Rev. Ecol. Syst.* 17:391–421 (1986).

Singh, S. M., and R. H. Green. "Excess of Allozyme Homozygosity in Marine Molluscs and Its Possible Biological Significance," *Malacologia* 25:569–581 (1984).

Sokal, R. R., and F. J. Rohlf. *Biometry.* (San Francisco, CA: W. H. Freeman & Company Publishers, 1981).

Stanczykowska, A. "Ecology of *Dreissena polymorpha* (Pall.) (Bivalvia) in Lakes," *Pol. Arch. Hydrobiol.* 24:461–530 (1977).

Wade, M. J., and D. E. McCauley, "Extinction and Recolonization: Their Effects on Genetic Differentiation of Local Populations," *Evolution* 42:995–1005 (1988).

Workman, P. L., and J. D. Niswander. "Population Studies on Southwestern Indian Tribes. II. Local Genetic Differentiation in the Papago," *Am. J. Hum. Genet.* 22:24–49 (1970).

Zouros, E., and D. W. Foltz. "Possible Explanations of Heterozygote Deficiency in Bivalve Molluscs," *Malacologia* 25:583–591 (1984).

Zouros, E., and D. W. Foltz. "The Use of Allelic Isozyme Variation for the Study of Heterosis," in M. C. Ratazzi, J. G. Scanolios, and G. S. Whitt, Eds. *Isozymes: Current Topics in Biological and Biomedical Research, Vol. 13* (New York: Allan R. Liss, Inc., 1987), pp. 1–59.

CHAPTER 15

The Byssus of the Zebra Mussel *(Dreissena polymorpha)*: Morphology, Byssal Thread Formation, and Detachment

Larry R. Eckroat, Edwin C. Masteller, Jennifer C. Shaffer, and Louise M. Steele

Specimens of the zebra mussel, *Dreissena polymorpha*, were collected from Lake Erie during 1990. Byssus stems, threads, and plaques were examined using scanning electron microscopy. The sequence of events and the frequency of byssal thread formation in relation to mussel size were determined. Scanning electron microscope (SEM) observations showed that the byssus was attached to the shell by retractor muscles. In some specimens, cuffs were present on the stems at the bases of the threads. Because of a thread-branching pattern, it is unlikely that the stem lengthens when threads are formed. Threads were composed of a cortex with longitudinal fibers in a matrix and were covered by an outer sheath. The sheath was smooth in the proximal regions and became increasingly rough with longitudinal ridges in the distal regions. Plaques were attached to substrata in rows, which could increase the stability of mussel anchorage. Direct observations showed that in adult mussels there were two types of threads, permanent and temporary. In addition to stabilizing the mussel while it secreted permanent attachment threads, the temporary threads may have had a searching function. The number of threads formed increased

0-87371-696-5/93/$0.00 + $.50

initially and then remained relatively stable. Larger mussels that were stationary formed threads more frequently than the mobile, smaller mussels did; however, if threads of the smaller mussels were temporary, both groups of mussels formed threads at the same rate. Mussels had no preference for vertical or horizontal attachment surfaces. Mussels frequently formed threads, detached, and moved to other locations to form new threads. Detached byssi were sometimes freed from the substratum but remained attached to mussels, while other byssi were detached by the animal. Information concerning byssus morphology and thread formation and detachment may lead to a mechanism of controlling zebra mussels.

INTRODUCTION

The zebra mussel, *Dreissena polymorpha*, was first collected in the Great Lakes in 1988; based upon age analysis, it is believed to have been introduced into the Great Lakes in 1985 or 1986 (Hebert et al., 1989; Griffiths et al., 1991). Since its discovery, the zebra mussel has been a source of considerable concern because it contributes to the biofouling of municipal and industrial water supplies (Griffiths et al., 1989). The primary reason that the zebra mussel contributes to biofouling of water supplies is that it attaches to water intakes profusely by using up to 200 byssal threads per individual (Clarke, 1952). In addition, mussels use byssal threads to attach to other solid substrata such as rocks, boat hulls, anchor chains, and shells of unionid mussels (Mackie, 1991; Schloesser and Kovalak, 1991).

Little information concerning the byssus of zebra mussels is available in the literature. The majority of observations of byssi have been made in marine bivalve molluscan species, specifically in members of the family Mytilidae, including *Mytilus galloprovincialis, M. edulis, M. californianus,* and *Modiolus demissus*. Taxonomically, *D. polymorpha* (subclass Lamellibranchia) is removed from the other byssate mussels (Table 1) (Brusca and Brusca, 1990; Turgeon et al., 1988). *D. polymorpha* is a member of the family Dreissenidae and the superorder Eulamelliobranchia (order Veneroida); most families of the latter are isomyarian (anterior and posterior adductor muscles of equal size), but most families of the former are heteromyarian (anterior and posterior adductor muscles of different size). The heteromyarian condition is found in members of the superorder Filibranchia (order Mytiloida). Although *D. polymorpha* is phylogenetically distinct from the marine *Mytilus* spp. of order Mytiloida, there are many similarities between the two groups. For instance, in addition to showing the heteromyarian condition, both groups are important aquatic fouling animals, both have a larval veliger stage as part of their life cycles, and both have byssi.

Waite (1983a), who studied *Mytilus* spp., described a byssus as "an extraorganismic structure" that consists of threads attached to the animal at the proximal end and to a substratum at the distal end. The byssi of *M. edulis* have the following parts (Brown, 1952):

1. The adhesive plaques (disks), which are located at the distal ends of the threads

Table 1. Phylogenetic Relationships of Byssate Mussels

Phylum Mollusca
- Class Bivalvia (=Pelecypoda)
  - Subclass Lamellibranchia
    - Superorder Filibranchia (characteristics = usually attach by byssal threads, most are hetermyarian, and are primarily found in marine habitats)
      - Order Mytiloida
        1. Mytilidae (mussels)
        2. Ostreidae (oysters)
        3. Arcidae (arc shells)
        4. Anamidae (jingle shells)
    - Superorder Eulamelliobranchia (characteristics = rarely attach by byssal threads, most are isomyarian, and live primarily in freshwater habitats)
      - Order Veneroida
        1. Sphaeriidae (fingernail clams)
        2. Unionidae (freshwater mussels)
        3. Corbiculidae (Asiatic clams)
        4. Dreissenidae (zebra mussels)

Source: Brusca and Brusca, 1990; Turgeon et al., 1988.

2. The threads, which consist of a corrugated proximal part and a smooth distal part
3. The stem, which is continuous with the root, but not embedded in the byssus gland (threads, which have cuffs at their bases, are attached to the stem)
4. The root, which is embedded in the byssus gland

The objectives of this study were to examine the structure of the byssus of *D. polymorpha*, and to determine the frequency of byssal thread formation, the relationship of mussel size to thread formation, and the sequence of events and behavior in thread formation. Because byssal attachment is fundamental to the success of mussels that colonize hard substrata (Lee et al., 1990), information concerning structural characteristics of the byssus, together with knowledge of thread formation and attachment, may lead to a mechanism of zebra mussel control.

METHODS

The morphology of the byssal threads of *D. polymorpha* was examined using scanning electron microscopy. Specimens attached to stones were collected from 1- to 2-m water depth in Lake Erie near Presque Isle, Erie, PA in June 1990. Mussels were left attached to their natural substrate and held in aquariums at 15°C for several days before they were prepared for SEM analysis. In addition, mussels were collected monthly from Presque Isle from August to November 1990. These specimens were removed from the substratum so that they could be observed when forming new byssal attachments in the laboratory. Direct observation with a microscope and video camera was used to study thread formation. Each mussel was maintained in a separate compartment (5.25 × 5.25 × 5.25 cm) of a 24-compartment transparent copolymer plastic box in aquariums maintained at 15°C. Mussels were exposed to fluorescent lights 10 hr/day and fed a combination of dried alga, *(Chlorella* spp.), and water. This diet (3 g/L) was provided in 200 mL frozen aliquots, placed in funnels, and allowed to drip into the tanks as it thawed. The frequency and sequence of byssal thread formation was determined for different size groups of *D. polymorpha*. Specimens maintained in the laboratory were classified into four length categories: 1–3, 8–12, 16–20, and 21–26 mm.

Direct Observations of Thread Formation

Observations were made of byssal thread formation over two 4-week periods. In the first 4-week period, 24 mussels that were 8–12 mm in length and 24 mussels that were 21–26 mm in length were observed. In the second 4-week period, 48 mussels 16–20 mm in length and 48 mussels 21–26 mm in length were observed. Thus, observations were obtained on a total of 144 mussels. During the period of the observations 23% mortality was observed. Mussels that died were replaced each week with mussels of the same size class. The orientation (i.e., horizontal, vertical, or not attached) of each mussel in the copolymer box compartments was observed and recorded. The threads were stained with the vital stain, Janus Green, and counted weekly. Only threads that could be observed on a horizontal surface of the container were counted. Observations of thread uniformity, sequence of events in thread formation, and byssus detachment were made in these laboratory-maintained mussels. Five to eight mussels (1–3 mm) were placed in deep well projection slides (Carolina Biological Supply, catalogue no. 60-3730) with water. A Leitz Laborlux D microscope and a video camera with a color monitor were used to observe the mussels over a 10-hr period as they secreted byssal threads.

SEM Observations of Thread Morphology

Mature specimens observed with a scanning electron microscope (Hitachi model S-570) were prepared using three methods. Specimens prepared by the first method (e.g., Figures 1, 2, and 3, which were byssi that mussels had formed in the laboratory) were immersed in buffered 5% glutaraldehyde (pH 7.2) at 5°C for 24 hr, dehydrated in a graded series of ethanol, and critical-point dried using carbon dioxide as the transitional solvent. Specimens prepared by the second method (e.g., Figures 4–7, which were byssi that mussels had formed in the laboratory) were fixed and dehydrated as described above, immersed in liquid nitrogen, and freeze-fractured (ethanol cryofractured) at various proximal-distal regions (Figure 8) with a cooled razor blade as described by Klomparens et al. (1986). After being returned to 100% ethanol

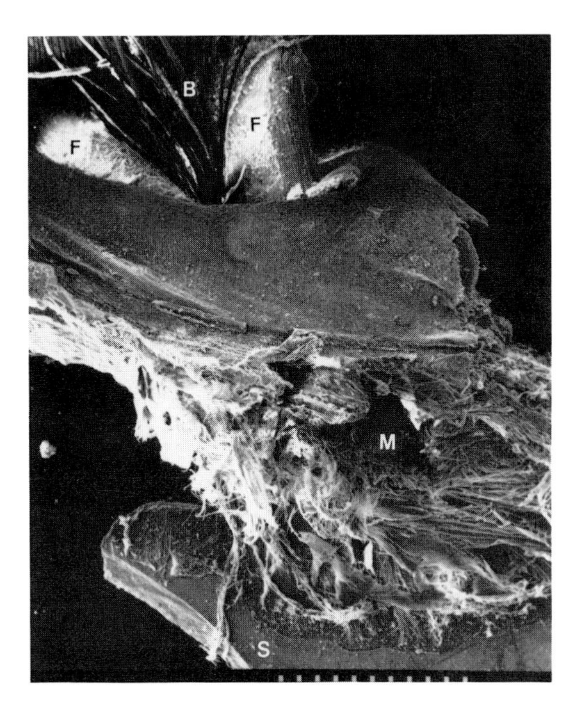

Figure 1. Dissected specimen of *Dreissena polymorpha* showing the byssus (B) emerging from the foot (F) and attaching to the inside surface of the shell (S) via byssal retractor muscles (M). (Line scale = 1.50 mm.)

to remove condensation, specimens were critical-point dried. This technique was advantageous because it permitted observation of internal thread morphology. Specimens prepared by the third method (e.g., in Figures 9 and 10–14A, which were byssi that mussels had formed on stones in their natural environment) were immersed in 5% aqueous glutaraldehyde for 24 hr at 5°C, postfixed with 2% osmium tetroxide vapor at 23°C, and air dried. This procedure was advantageous because the natural byssal attachments could be prepared with minimal manipulation. All specimens were mounted on aluminum stubs, sputter coated with a 10- to 15-nm layer of gold-palladium, and examined and photographed using scanning electron microscopy.

RESULTS

Direct Observations of Thread Formation

Two distinct types of byssal threads, permanent and temporary, were observed in all four size classes of zebra mussels (Figure 15). The two types

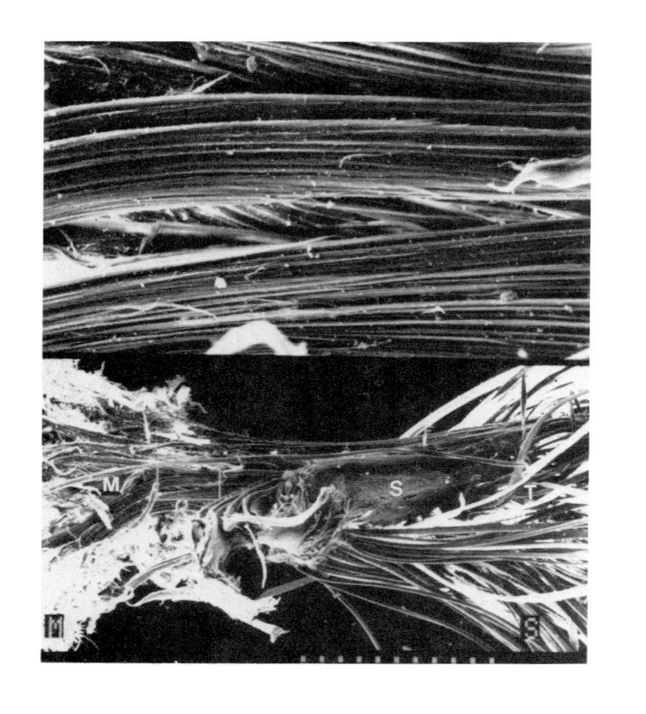

Figure 2. Bottom portion of micrograph: muscles (M) merging with the byssal stem (S) from which threads (T) branch (line scale = 1.2 mm). Top portion of micrograph: detail of muscle fibers (line scale = 0.12 mm).

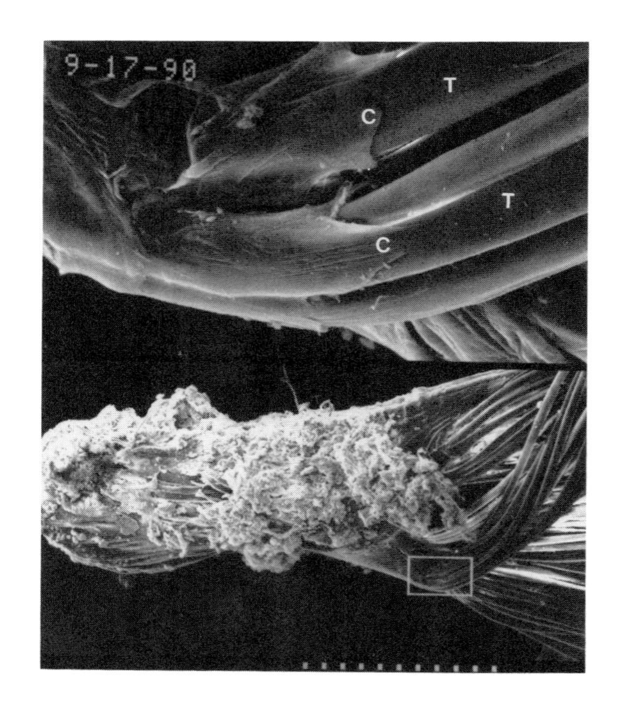

Figure 3. Bottom portion of micrograph: byssus detached by a specimen that formed threads in the laboratory (line scale = 0.75 mm). Top portion of micrograph: detail of threads (T) with cuffs (C) (line scale = 0.075 mm).

Figure 4. Byssal thread of *Dreissena polymorpha* in the smooth, proximal region as prepared by the freeze-fracture technique. (Line scale = 12.1 μm).

Figure 5. Byssal thread of *Dreissena polymorpha* in the very rough, distal region as prepared by the freeze-fracture technique. (Line scale = 20 μm.)

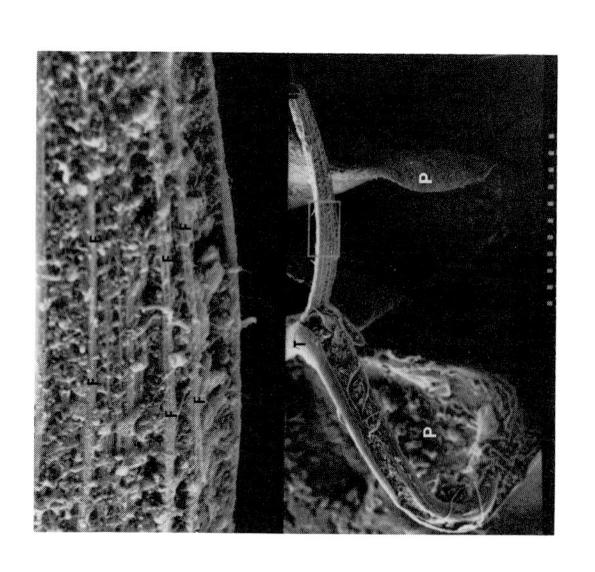

Figure 6. Thread of *Dreissena polymorpha* that fractured longitudinally revealed the outer sheath (S) and inner cortex which was composed of fiber (F) imbedded in a matrix. P = plaque. (Bottom portion of micrograph: line scale = 201 μm; top portion of micrograph: line scale = 20.1 μm.)

Figure 7. Byssal thread (T) with peeled sheath (S) of *Dreissena polymorpha*. (Line scale = 6.0 μm.)

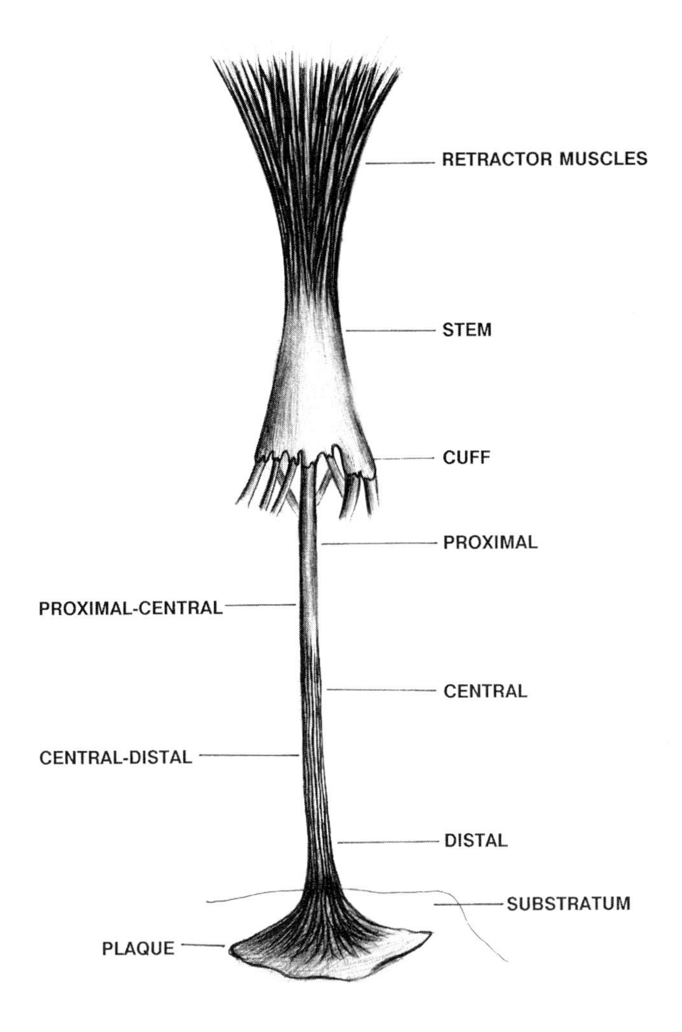

Figure 8. Composite drawing of the byssal structure of *Dreissena polymorpha* showing various proximal-distal freeze-fracture locations where observations using a scanning electron microscope were made. (Not drawn to scale.)

of threads were differentiated based on length, thickness, number, arrangement, and plaque morphology. Permanent attachment threads were formed in clumps or arranged in rows and comprised the majority of the main byssal mass (Figure 16). On the other hand, temporary attachment threads were individual or few in number (1–6); arranged in a tripod pattern; originated individually; and separated spatially from the main byssal mass, which was composed of a group of permanent attachment threads (Figure 17). Temporary threads were not observed in every mussel and were secreted before or after

Figure 9. Stem (S) and threads (T) of *Dreissena polymorpha* from which the shell was pulled after fixation. (Bottom portion of micrograph: line scale = 0.86 mm; top portion of micrograph: line scale = 0.086 mm.)

Figure 10. Threads of *Dreissena polymorpha* formed in a natural habitat attached to a substratum (S). (Line scale = 1.20 mm.)

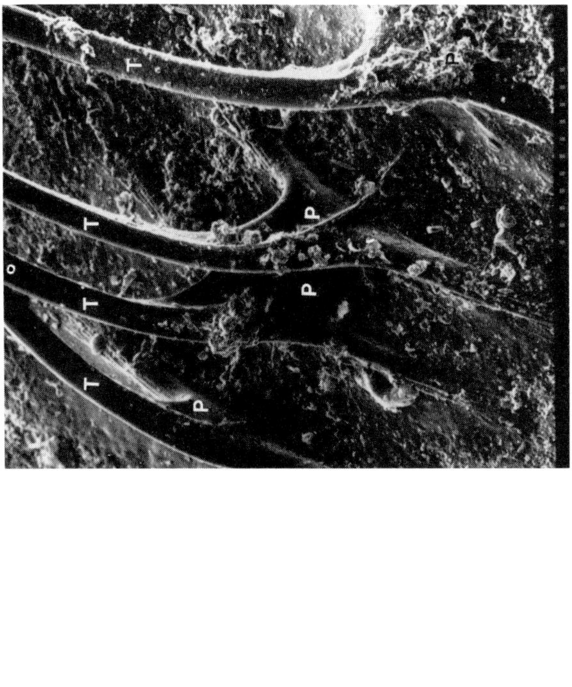

Figure 11. Threads (T) and plaques (P) of *Dreissena polymorpha* formed in a natural habitat. (Line scale = 176 μm.)

Figure 12. Bottom portion of micrograph: plaque of *Dreissena polymorpha* formed in a natural habitat (line scale = 200 μm). Top portion of micrograph: surface of the plaque (line scale = 20 μm).

Figure 13. Plaque (P) of *Dreissena polymorpha* broken free from the substratum. (Bottom portion of micrograph: line scale = 203 μm; top portion of micrograph: line scale = 20.3 μm.)

Figure 14. Permanent (P) and temporary (TT) attachment threads of *Dreissena polymorpha.* (Line scale = 1.36 mm.)

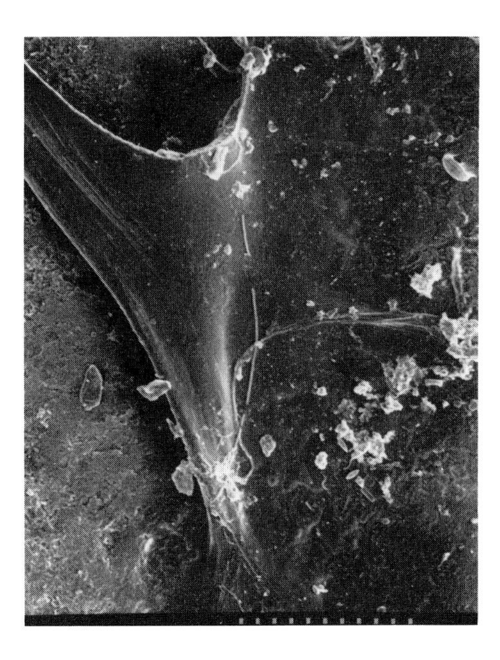

Figure 14A. Detail of Figure 14. Plaque of a temporary thread of *Dreissena polymorpha*. (Line scale = 136 μm.)

initial byssal mass attachment. Differences in thread types became readily apparent when observing and measuring the variability in the lengths of 107 threads from 18 mussels that were 16–20 mm in length. Mussels of this size were the most easily manipulated to observe temporary threads. The number of threads per mussel ranged from 2 to 9 and the overall mean length of the 107 threads was 2.8 ± 1.5 mm. When the mean and standard deviation of the thread length of each individual mussel was examined, it was evident that the majority of threads visually designated as temporary were at least one standard deviation longer than the mean thread length for each individual mussel (Figure 18). Using this criteron, 16 of 18 mussels had temporary threads. Lengths of these threads were 53% greater than the overall mean length. Also, temporary threads were 25–50% thinner than permanent attachment threads. Plaques of temporary threads appeared to be flattened against the substratum (Figure 17), while the distal surface of plaques of permanent threads were convex (Figure 15).

Over the two 4-week examination periods, 11 mussels remained attached horizontally to the containers for the duration of each 4-week period. Observations of the locations of mussels within the box compartments indicated that larger mussels tended to remain attached and thus stationary, while the smaller mussels exhibited greater mobility. Of the 122 mussels that attached,

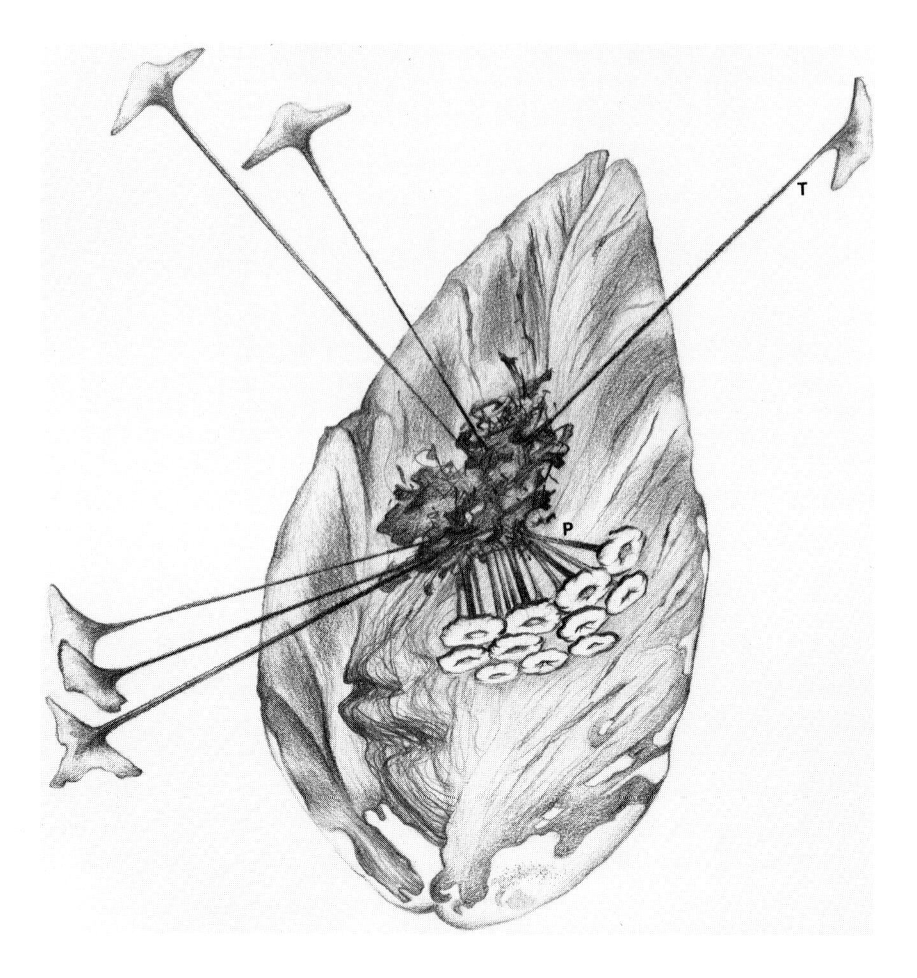

Figure 15. Illustration of *Dreissena polymorpha* attached to a clear surface and oriented in a ventral view. P = permanent byssal threads, T = temporary byssal threads. (Drawn to 1:8.4 scale.)

64 attached horizontally and 58 attached vertically. Therefore, the mussels had no apparent preference for vertical or horizontal attachment surfaces ($\chi^2[122] = 0.29$, $p > 0.5$). The mean number of threads present at each weekly observation ranged from 17 to 27 (Table 2, Figure 19). Mean values indicated that the greatest number of threads were formed in the first week and then remained relatively stable for the remainder of the 4-week period. Mussels 21–26 mm in length showed the same trend (Figure 20). However, because 5 of the 14 attached mussels died during week three, rather than leveling off the number of threads present for mussels decreased. During observations, numerous individual free byssal masses consisting of a detached stem with the plaques and threads intact were found in the boxes.

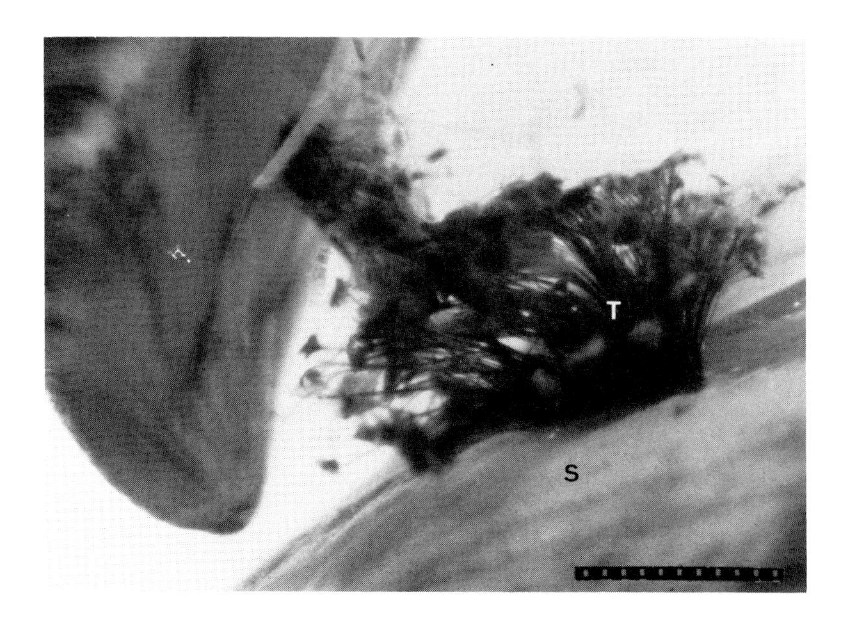

Figure 16. Two *Dreissena polymorpha* with byssal threads (T) emerging from the shell (S). (Line scale = 23.45 cm.)

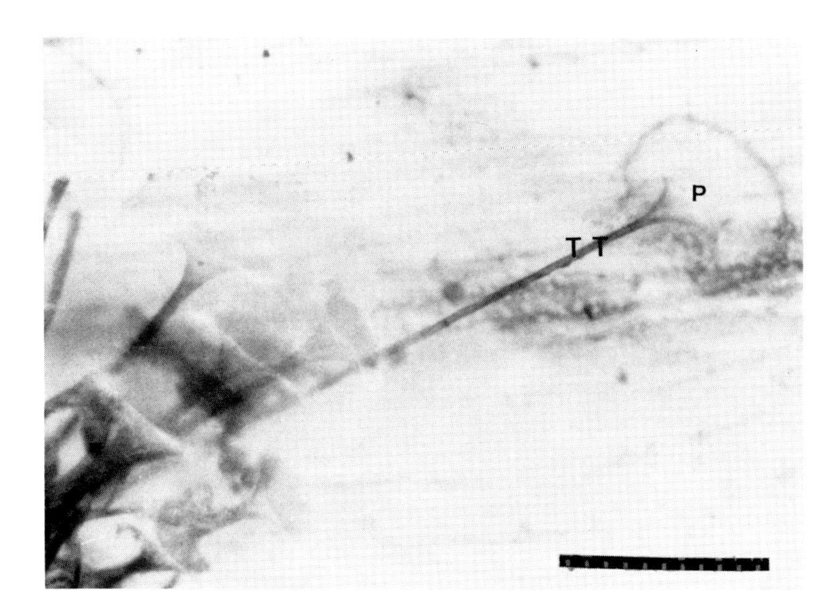

Figure 17. Temporary threads (TT) with plaques (P) of *Dreissena polymorpha*. (Line scale = 5.25 cm.)

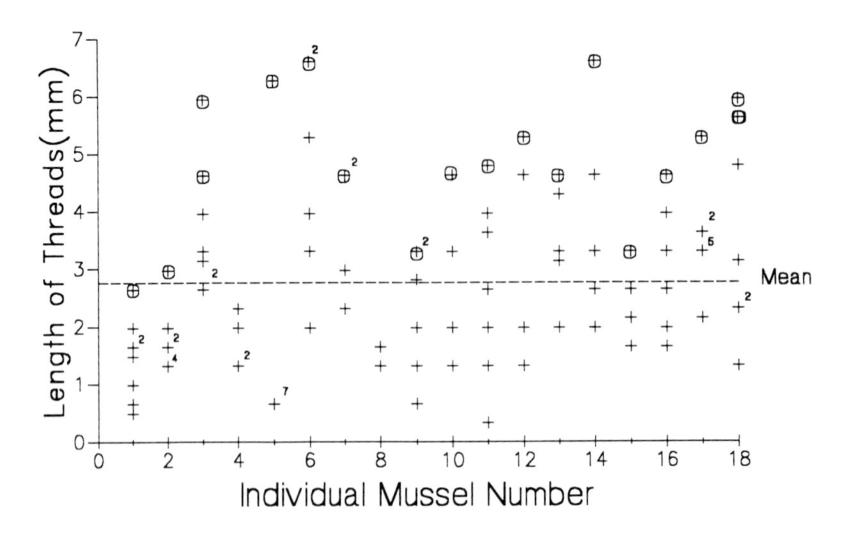

Figure 18. Lengths of newly formed threads of 18 *Dreissena polymorpha* that had a shell length of 16–20 mm. Multiple measurements of the same numerical value are indicated with superscripts; threads with a length greater than one standard deviation of the mean thread length for a particular individual are indicated with a circle; overall mean length of all threads of all individuals is indicated by a dashed line.

Table 2. Number of Threads Formed by *Dreissena polymorpha* during a 4-Week Period

	Number of Threads				
	Week				Size Class
Mussel	1	2	3	4	(Length in mm)
1	16	37	18	29	(21–26)
2	16	32	21	27	(21–26)
3	26	28	25	27	(21–26)
4	18	21	19	5	(21–26)
5	19	17	27	29	(21–26)
6	26	30	36	38	(16–20)
7	25	40	48	59	(21–26)
8	16	16	18	17	(21–26)
9	7	5	20	16	(21–26)
10	10	20	24	25	(21–26)
11	6	5	14	26	(21–26)
Mean	17	23	25	27	
Standard Deviation	7	12	10	14	

The sequence of thread formation in zebra mussels 1–3 mm in length was observed with a compound microscope over a 10-hr period. Thread formation began as the mussel's foot wandered back and forth over the substratum surface. Temporary plaques formed so rapidly that it was not possible

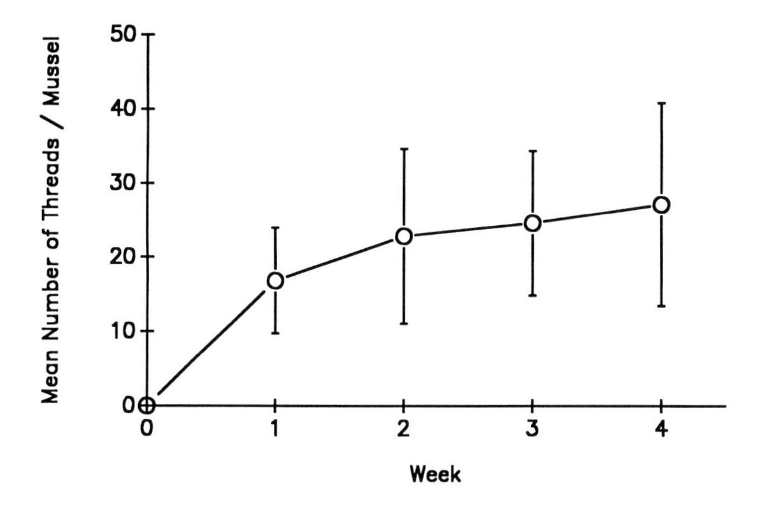

Figure 19. Mean number of byssal threads formed by 11 individuals of *Dreissena polymorpha* during a 4-week period.

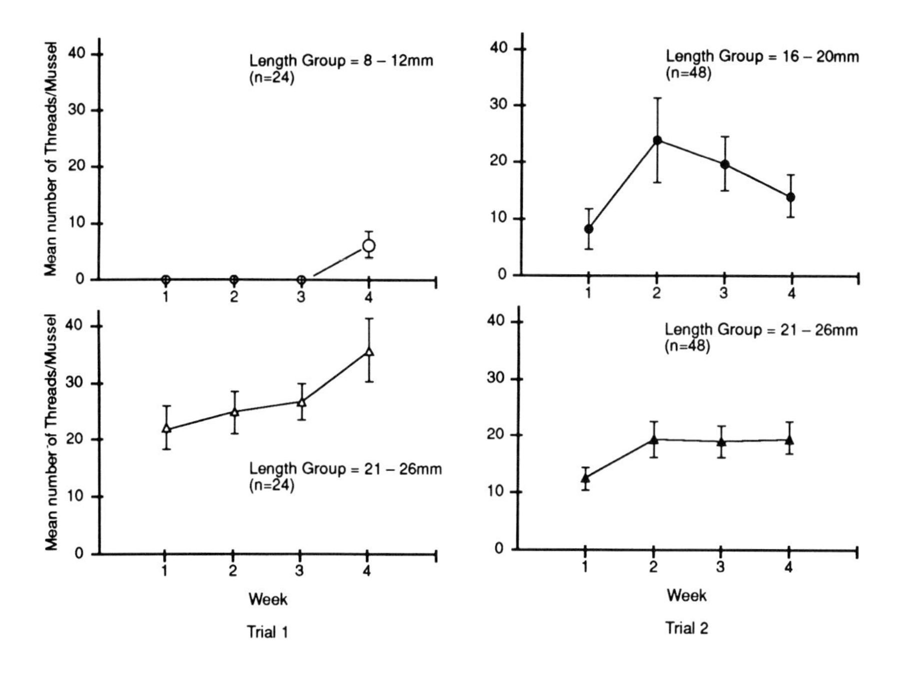

Figure 20. Mean number (±S.E.) of threads formed by three size groups of *Dreissena polymorpha* during two 4-week periods.

to make detailed observations of their formation. As the mussel detached from the substratum to move to another attachment site, temporary threads remained connected to both the substratum and to the mussel. Approximately 7% of the mussels detached within minutes, and then moved to another location where they eventually formed new threads. At least 50% of the early threads formed were temporary. When a permanent attachment site was found, the mussel prepared for the secretion of permanent attachment threads by once again moving its foot over the substratum. The foot was then placed at a 90° angle to the substratum, and a plaque was formed. While continuing to form a liquid thread that hardened on contact with water, the foot partially withdrew. After completing the formation of the thread, which was clear at first, the foot retracted into the shell. A few minutes later, the mussel moved forward slightly and secreted another thread.

SEM Observations of Morphology

Dissection of *D. polymorpha* revealed that the byssus was attached to the inside surface of the shell by byssal retractor muscles (Figure 1). Removal of the foot revealed that muscle fibers merged as they attached to the byssal stem (Figure 2). Therefore, the muscles and byssus formed a continuous structural unit as diagramed in Figure 21. Most of the threads branched from the stem at about the same proximal-distal location (Figures 3 and 9). The outer laminae formed cuffs at the bases of threads of a byssus detached from a laboratory-maintained mussel (Figure 3). No cuffs were observed in a byssus formed in its natural environment and from which the shell was manually removed after fixation (Figure 9).

The freeze-fracture technique revealed internal thread morphology. Although cross-sectional fracture faces were variable and provided little detail (Figures 4 and 5), longitudinal fracture faces indicated that threads had two distinct parts: an interior cortex and an exterior surface sheath. The cortex was composed of internal fibers that ran longitudinally in a coarse matrix (Figure 6). The outer sheath covered the thread and was a separate, flexible layer with longitudinal striations on its inner surface (Figure 7). Observations of the exterior surface characteristics of the different proximal-distal regions of the threads showed that the surfaces of proximal regions were relatively smooth (Figure 4), but the thread surface became increasingly rough distally, with ridges that were parallel to the longitudinal axis of the thread (Figure 5).

A mass of byssal threads is shown attached to a substratum (Figure 10). The threads, which branched from the stem, ended distally in plaques that were arranged in rows (Figures 10 and 11) and had deep ridges (Figure 12).

Observations of a plaque that broke free from the substratum suggested that the plaque may have a hollow cavity between the edges of the plaque (Figure 13, top portion). In another specimen, a single, longer byssal thread

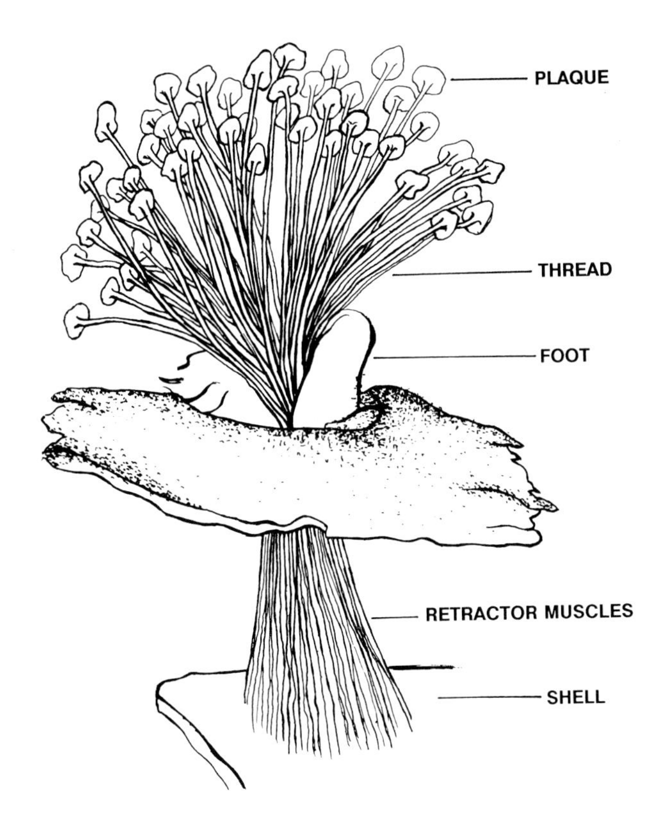

Figure 21. Composite drawing of the structural unit of the byssus of *Dreissena polymorpha.* (Not drawn to scale.)

was attached to the substratum a distance away from the main byssus (Figure 14) and the plaque (Figure 14A) appeared smoother and more elastic than those typically observed on threads of a mature byssus (Figure 12).

DISCUSSION

Thread Types

The present study revealed similarities and differences between the zebra mussel and other, more frequently studied byssate mussels. For instance, direct observation indicated that an additional thread type existed in *D. polymorpha*, but that the frequency of thread formation and the sequence of behavior was similar to that reported in *M. edulis*. Scanning electron microscopy revealed that thread branching patterns and thread topography in *D. polymorpha* differed from that of *Mytilus* spp.

At different stages in the life cycle of *M. edulis*, three types of byssal threads have been previously reported (Lane et al., 1985; Barnes, 1980). Lane et al. (1985) described a filamentous drifting thread in free-swimming larvae and temporary attachment threads in plantigrade larvae; also observed was the fact that the drifting threads had no plaques or internal substructure. Therefore, the drifting threads differed from the tough, adult attachment thread that terminated in an expanded plaque and was made of fibers that were restricted in length (Barnes, 1980). The present study indicates the existence of an additional thread type (i.e., temporary) in adult *D. polymorpha*. These temporary threads differed from the permanently attached byssal threads in length, thickness, number, and arrangement of plaques. These temporary threads became apparent when observing the frequency of thread formation because of their exceptionally long lengths and their arrangement in a tripod manner around the main byssal mass. This arrangement seemed to hold the mussel in place while the permanent attachment threads were secreted. After secretion of permanent attachment threads, the temporary threads either broke or remained attached. A specimen observed with the scanning electron microscope formed its byssal attachment in a natural environment and had a single, long thread that was attached to the substratum at a location distant from the majority of the specimen's other threads (Figure 14). The plaque (Figure 14A) of this distant thread, in contrast to those of the specimen's attachment threads, was smooth. If the sheath of a plaque erodes upon aging, then this distant (perhaps temporary) thread with its smoother plaque may have been formed later than the other attachment threads. Mussels often appeared to be using temporary threads to select a substratum for a different attachment site. Typically, a mussel that appeared to be using temporary threads for this purpose proceeded to move away from the temporary thread plaque with the thread still attached. Plaques of these temporary threads were flattened against the substratum. Flattening may have been caused by tension created on the plaque as the mussel moved. Once at a different site, the mussel secreted an additional thread. From our observations, it seems likely that the temporary threads were functioning as searching threads.

Thread Formation, Attachment, and Frequency of Attachment

Attachment of *M. edulis* and of *D. polymorpha* is similar because in both species, there is a greater frequency of attachment in larger mussels than in smaller mussels. Lee et al. (1990) noted that in *M. edulis* the smaller mussels (<5 mm) exhibited greater mobility, thus exhibiting shorter attachment times for individual byssal threads. Our results and Lee's findings, however, deviated from the results of Van Winkle (1970), who noted that in *M. demissus* the rate of byssal thread secretion decreased with increasing mussel size. From our observations of the 11 mussels that remained attached during each of the two 4-week periods, thread formation was greatest during the first week

(Figure 19). This indicated that mussels formed the majority of their threads during the first week to ensure a strong attachment. After this period, thread formation remained relatively constant for each individual mussel.

The number of threads present (17–27) for mussels 16–26 mm long that remained attached horizontally (Figure 19) was similar to that reported by Lee et al. (1990) for *M. edulis* (average of 16 threads for mussels >3 mm). In *D. polymorpha*, fewer threads were observed in the smaller size class (i.e., 8–12 mm) than in the larger size classes (i.e., 16–20 and 21–26 mm) (Figure 20). Mussels in the smallest size class (i.e., 1–3 mm) that had not established a permanent byssus were observed to detach their threads and then move to a new attachment location. These smaller mussels may have been forming temporary threads, which would suggest that the smaller mussels were forming threads at the same frequency as the larger muscles. Attachments of these smaller mussels, however, would have been temporary rather than permanent. This observation would explain why fewer permanent attachments were observed in the two smallest size classes.

As *D. polymorpha* prepared to form byssal threads, it moved along the surface with its foot, searching for a suitable attachment site. As the mussel moved, a mucus was secreted. This mucus did not resemble a thread, but was secreted from the ventral groove in the foot. Similarly, Tamarin (1977) noted that in *M. edulis*, five different types of gland cells secreted five different compounds and that one of these compounds was a type of mucus. After the plaque was secreted to provide an attachment to the substratum, a clear thread became visible and the foot retracted into the shell. These observations were also noted by Waite (1983a) in *M. edulis*. In addition, Waite (1983b) stated that the plaque was composed of a water-insoluble adhesive polyphenolic protein with an abundance of DOPA residues, which Baier (1990) suggested may also be present in plaques of *D. polymorpha*. In *D. polymorpha*, as in other byssate mussels (Hickman, 1973), the threads, which are of a fluid consistency, are secreted one at a time. As was observed in our study, after one thread has formed, byssate mussels change position slightly and repeat the process to form another thread (Lutz, 1986).

During observations of *D. polymorpha*, numerous individual free byssal masses were found. These byssal masses consisted of a detached stem with plaques and threads intact. In specimens of *D. polymorpha* that voluntarily detached threads, cuffs were observed on the stem at the bases of the threads (Figure 3). In specimens from which the shell was manually removed after fixation, no cuffs were observed (Figure 9). According to Waite (1983a), byssal threads of *M. edulis* are designed to break before the attachment plaque separates from the substratum. This design may be related to the tolerance of this species for water turbulence. In addition, if it becomes advantageous to abandon the anchorage site, *M. edulis* voluntarily detaches the entire byssal mass and, therefore, does not need to rely on chance breakage of threads (Waite, 1983a; Brown, 1952; Van Winkle, 1970; Young, 1983). Waite (1983a) stated that the mechanism by which *Mytilus* spp. can voluntarily detach the entire byssus is unknown. The different appearances of the stems of the voluntarily detached (Figure 3) and manually removed (Figure 9) byssi may suggest that pulling the shell from the organism separates some of the outer laminae from the stem and causes loss of the cuffs. If the cuffs are lost when the shell is pulled from the stem, the mechanism by which *D. polymorpha* voluntarily detaches the byssus would not involve pulling stresses. Although the mechanism used by *D. polymorpha* to voluntarily detach the byssus could involve pulling stresses or some other nonpulling mechanism, the differences noted in Figures 3 and 9 may indicate that a nonpulling detachment mechanism operates. On the other hand, cuff formation may be associated with changes in the byssus, such as degeneration of the outer laminae, that could require or result from the detaching of the byssus. Further investigation of the voluntary detachment mechanism is needed.

SEM Morphological Observations

Scanning electron microscopy showed that the byssus was attached to the inside of the shell via byssal thread retractor muscles, which may help the mussel to push against its anchorage (Brusca and Brusca, 1990). This action would keep byssal threads taut and help to stabilize the byssus attachments to prevent the mussel from being swept away by water currents. The byssal threads of *D. polymorpha* branched from the stem at about the same proximal-distal location, which suggested that the stem did not lengthen in the region where threads emerged. Furthermore, threads branched from all around the circumference of the stem (Figure 9). This pattern is different from the branching pattern observed in *Mytilus* spp. In *M. edulis*, the threads branched from only one side of the stem as it lengthened (Brown, 1952). In *M. galloprovincialis*, the threads emerged in groups from opposite edges along the length of the stem (Bairati and Vitellaro-Zuccarello, 1974).

In *D. polymorpha*, appearance of the surface of the thread sheath differed from that observed in *Mytilus* spp. The threads of *D. polymorpha* appeared cylindrical along their entire lengths, were smooth in their proximal regions, and had longitudinal ridges in their distal regions. In contrast, the proximal portions of the threads of *M. edulis* were flattened and corrugated with ridges around their circumferences and in *M. galloprovincialis* opposite sides of the proximal portions of threads differed; the distal thread regions in *M. edulis* and *M. galloprovincialis* were cylindrical and smooth (Brown, 1952; Bairati and Vitellaro-Zuccarello, 1974).

Bairati and Vitellaro-Zuccarello (1974) stated that the surface folds on threads of *M. galloprovincialis* resulted because fluid thread material was molded to the irregularities of the ventral groove walls of the foot and because pressure was exerted as foot muscles retracted. They explained that the non-corrugated thread portions did not come in contact with groove walls. The different sheath topography in *D. polymorpha* suggests that the surface of ventral groove walls and the action of the foot muscle may differ from that of *Mytilus* spp.

The sheath of *D. polymorpha* was a separate, outer layer covering the cortex that is likely to form as a result of a tanning process. Brown (1952) noted that the sheath enclosing byssus threads of *Mytilus* was quinone tanned. Smyth (1954) explained that en route to the byssus gland, the threads of *M. edulis* were covered by a secretion of the enzyme, polyphenol oxidase, which catalyzed the tanning reaction. A similar tanning process is likely to occur in the threads of *D. polymorpha* because the thin, outer layer of thread material that would come in contact with such an enzyme would undergo tanning, thus forming a separate layer as observed in Figures 6 and 7. This tanned layer (the sheath) would differ in appearance from the deeper, untanned thread material of the cortex. The thread cortex of *D. polymorpha*, like that of *Mytilus* spp. (Vitellaro-Zuccarello et al., 1983; Benedict and Waite, 1986), was composed of longitudinal fibers embedded in a matrix.

The plaques of *D. polymorpha* were attached to the substratum in rows. This arrangement of plaques was observed by Young (1983) in *M. edulis*. Arrangement of the threads in rows may increase the strength of attachment of the byssus to substrata. The threads of *D. polymorpha*, as those of *Mytilus* spp., expanded into plaques. Plaques of mature byssi formed in the natural environment appeared more ridged than those newly formed in the laboratory. Because the threads formed in the natural environment may have been formed a considerable amount of time before specimens were collected and examined, sheaths of plaques may have eroded as threads aged and were exposed to environmental elements. Sheath erosion, if it occurs upon aging or environmental exposure, would reveal the fibrous matrix, which would account for the more ridged appearance of plaques of specimens that formed their threads in a natural environment.

Several plaques of *D. polymorpha* observed by scanning electron microscopy appeared hollow. The cortex, or interior, of plaques of *M. californianus* is a matrix with empty spaces (Tamarin et al., 1976). Benedict and Waite (1986) observed that the inner plaque matrix of *M. edulis* was "spongy" in appearance. Furthermore, Waite (1986) stated that in *M. californianus* and *M. edulis*, plaques are composed of an adhesive foam material. Therefore, it is likely that the interior of plaques of *D. polymorpha* is filled with an adhesive material. In the hollow specimens observed, this material may have been washed away during preparation, or vacant spaces may have formed during secretion of the plaque. Such an absence of adhesive could explain why the plaque in Figure 13 broke free.

CONCLUSION

In this chapter, observations of byssus morphology, byssal thread formation, and byssus detachment in *D. polymorpha* were described and compared with literature reports of *Mytilus* species of Mytiloida. The information presented adds to the current understanding of how mussels attach to firm substrata and could, therefore, be used to develop a mechanism of zebra mussel control.

ACKNOWLEDGMENTS

The authors wish to thank Ann Magenau for preparing the illustrations; Jill Bejarano for maintaining the zebra mussels in the laboratory; Kathy Mauro for typing the manuscript; and Paul E. Barney, Jr., Melissa McClellan, and Andrew Saylor for supplying the computer graphics.

REFERENCES

Baier, R. E. "Control of Bioadhesion by the Zebra Mussel," Abstract, International Zebra Mussel Research Conference, Ohio Sea Grant, Columbus, OH (December 1990).

Bairati, A., and L. Vitellaro-Zuccarello. "The Ultrastructure of the Byssal Apparatus of *Mytilus galloprovincialis*. II. Observations by Microdissection and Scanning Electron Microscopy," *Mar. Biol.* 28:145–158 (1974).

Barnes, Robert D. *Invertebrate Zoology* (Philadelphia, PA: Saunders College, 1980).

Benedict, C., and J. H. Waite. "Composition and Ultrastructure of the Byssus of *Mytilus edulis*," *J. Morphol.* 189:261–270 (1986).

Brown, C. H. "Some Structural Proteins of *Mytilus edulis*," *Q. J. Microsc. Sci., 3rd Ser.* 93:487–502 (1952).

Brusca, R. C., and G. J. Brusca. *Invertebrates* (Sunderland, MA: Sinauer Associates, Inc. 1990), pp. 706–740.

Clarke, K. B. "The Infestation of Waterworks by *Dreissena polymorpha*, a Freshwater Mussel," *J. Inst. Water Eng.* 6:370–379 (1952).

Griffiths, R. W., W. P. Kovalak, and D. W. Schloesser. "The Zebra Mussel, *Dreissena polymorpha* (Pallas, 1771), in North America: Impact on Raw Water Users," in Symposium: Service Water System Problems Affecting Safety-Related Equipment. (Palo Alto, CA: Nuclear Power Division, Electric Power Research Institute, 1989), pp. 11–27.

Griffiths, R. W., D. W. Schloesser, J. H. Leach, and W. P. Kovalak. "Distribution and Dispersal of the Zebra Mussel (*Dreissena polymorpha*) in the Great Lakes Region," *Can. J. Fish. Aquat. Sci.* 48:1381–1385 (1991).

Hebert, P. D. N., B. W. Muncaster, and G. L. Mackie. "Ecological and Genetic Studies on *Dreissena polymorpha* (Pallas): A New Mollusc in the Great Lakes," *Can. J. Fish. Aquat. Sci.* 46:1587–1591 (1989).

Hickman, C. P. *Biology of the Invertebrates* (St. Louis, MO: The C. V. Mosby Company, 1973).

Klomparens, K. L., S. L. Flegler, and G. R. Hooper. *Procedures for Transmission and Scanning Electron Microscopy for Biological and Medical Science: A Laboratory Manual*, 2nd ed. (Burlington, VT: Ladd Research Industries, 1986).

Lane, D. J. W., A. R. Beaumont, and J. R. Hunter. "Byssus Drifting and the Drifting Threads of the Young Post-Larval Mussel *Mytilus edulis*." *Mar. Biol.* 84:301–308 (1985).

Lee, C. Y., S. S. L. Lim, and M. D. Owen. "The Rate and Strength of Byssal Reattachment by Blue Mussels (*Mytilus edulis* L.)," *Can. J. Zool.* 68:2005–2009 (1990).

Lutz, P. E. *Invertebrate Zoology* (Reading, MA: Addison-Wesley Publishing Company, Inc., 1986).

Mackie, G. L. "Biology of the Exotic Zebra Mussel, *Dreissena polymorpha*, in Relation to Native Bivalves and Its Potential Impact on Lake St. Clair," *Hydrobiologia* 219:259–269 (1991).

Schloesser, D. W., and W. P. Kovalak. "Infestation of Native Unionids by *Dreissena polymorpha* in a Power Plant Canal in Lake Erie," *J. Shellfish Res.* 10 (2):355–359 (1991).

Smyth, J. D. "A Technique for the Histochemical Demonstration of Polyphenoloxidase and Its Application to Egg-Shell Formation in Helminths and Byssus Formation in *Mytilus*," *Q. J. Microsc. Sci.* 95:139–152 (1954).

Tamarin, A. "How Mussels Get Attached," *Nat. Hist.* 86:42–47 (1977).

Tamarin, A., P. Lewis, and J. Askey. "The Structure and Formation of the Byssus Attachment Plaque in *Mytilus*," *J. Morphol.* 149:199–222 (1976).

Turgeon, D. D., A. E. Bogan, E. V. Coan, W. K. Emerson, W. G. Lyons, W. L. Pratt, C. F. E. Roper, A. Scheltema, G. F. Thompson, and J. D. Williams. "Common and Scientific Names of Aquatic Invertebrates from the United States and Canada: Mollusks," American Fisheries Society Special Publication 16 (1988).

Van Winkle, W., Jr. "Effect of Environmental Factors on Byssal Thread Formation," *Mar. Biol.* 7:143–148 (1970).

Vitellaro-Zuccarello, L., S. DeBiasi, and A. Bairati. "The Ultrastructure of the Byssal Apparatus of a Mussel: V. Localization of Collagenic and Elastic Components in the Thread," *Tissue Cell* 15:547–554 (1983).

Waite, J. H. "Adhesion in Byssally Attached Bivalves," *Biol. Rev.* 58:209–231 (1983a).

Waite, J. H. "Evidence for a Repeating 3,4-Dihydroxyphenylalanine- and Hydroxyproline-Containing Decapeptide in the Adhesive Protein of *Mytilus edulis* L.," *J. Biol. Chem.* 258:2911–2915 (1983b).

Waite J. H. "Mussel glue from *Mytilus californianus* Conrad: A comparative study," *J. Comp. Physiol., Ser. B* 156:491–496 (1986).

Young, G. A. "Response to, and Selection Between Firm Substrata by *Mytilus edulis*," *J. Mar. Biol. Assoc. U.K.* 63:653–659 (1983).

CHAPTER 16

The Biomechanics of Byssal Adhesion in Zebra Mussels *(Dreissena polymorpha)*: Tests with a Rotating Disk

Josef D. Ackerman, C. Ross Ethier, D. Grant Allen, and Jan K. Spelt

The zebra mussel, *Dreissena polymorpha*, has recently become established in North America, causing serious problems for water users. An important biofouling characteristic of the zebra mussel is its ability to adhere to hard substrates using byssal threads. A rotating disk test system is presented as a method for measuring the adhesion strength of zebra mussels on various surfaces. In addition, the theoretical, biological, and physical processes that influence byssal adhesion are examined along with a theoretical model which uses the Helmholtz free-energy adhesion approach to examine zebra mussel adhesion. Results to date from a rotating disk system indicate that zebra mussel adhesion varies with substrate material and mussel health, but not with mussel residence time.

INTRODUCTION

The ability of zebra mussels (*Dreissena polymorpha* Pallas) to adhere to hard surfaces with a byssus is an important characteristic in the success and

0-87371-696-5/93/$0.00 + $.50

dispersal of this species. Mussel colonization of man-made surfaces has increased operation and maintenance costs of water systems (Roberts, 1990), thereby affecting individual residents, municipalities, and industries.

An understanding of zebra mussel adhesion is important for the development of control and remediation strategies. Unfortunately, scientists and engineers have very little information about adhesion characteristics for designing and implementing non-chemical control strategies. To provide such information, we have examined the theoretical, biological, and physical factors that affect adhesion including: basic biology, form and function of byssal attachment, and techniques used to measure adhesion strength. In addition, we describe a rotating disk test system developed for measuring mussel adhesion. This information will be useful in understanding potential control strategies of zebra mussels that depend on their adhesion.

BIOLOGICAL BACKGROUND

Certain developmental and life history events are considered important to this discussion, as they ultimately affect byssal adhesion. Zebra mussels release gametes which are fertilized externally and then develop into free-swimming larvae; these larvae ultimately settle to become epifaunal on hard substrates. These traits enable zebra mussels to be dispersed long distances and infest pipes used for pumping raw water from aquatic systems.

Developmental biology of zebra mussels is similar to many marine bivalves (Yonge, 1962; Yonge and Campell, 1968; Morton, 1969a, 1969b, 1969c) and therefore parallel observations have been made between marine mussels and zebra mussels. Small postveligers (<0.5 mm) have been shown to recruit preferentially to aquatic plants (Lewandowski, 1982), and new recruits occur at the periphery of mussel beds rather than in the proximity of adults when densities are high (Hebert et al., 1991). This is similar to the marine mussel *Mytilus*, where a primary settlement of "spat" (plantigrades) occurs preferentially on filamentous substrates (e.g., algae and hydroids; Blok et al., 1958) and spat are rarely seen on adult mussel beds (Bayne, 1964). Small zebra mussels change location (Russel-Hunter — cited in Yonge and Campbell, 1968) and later migrate to other substrates, including adult mussel colonies (Lewandowski, 1982). In marine bivalves, secondary settlement occurs via thread drifting (not byssal threads; Lane et al., 1985; Sigurdsson et al., 1976); this is like spider gossamer flight in which long threads increase the viscous drag, thus decreasing terminal velocities and promoting entrainment in faster moving eddies (Lane et al., 1985; Beukema and Vlas, 1989). Biomechanical similarities to this are also seen in filamentous seagrass pollen (Ackerman, 1989). Similarly, "floating" zebra mussels suspended from the water surface by a thread have been observed in laboratory aquariums (personal observations). Given this evidence, it is reasonable to suggest that zebra mussel recruitment into pipes may involve both veliger and postveliger phases, and that this activity may involve byssal detachment and attachment.

BYSSAL ADHESION

Byssus Form and Function

The byssal organ is present in the postlarval stage of most bivalves, where it functions to give stability to the larvae during metamorphosis (Yonge, 1962). Byssally attached bivalves are either epifaunally or infaunally attached, and are thought to have given rise to cemented forms (e.g., oysters), free-swimming forms (e.g., scallops), and borers (Yonge, 1962). For the most part, byssal threads serve as "mooring lines" or "guy wires" securing the adult bivalve to, within, or beneath substrates and/or sediments.

There have been efforts devoted to understanding the general morphology and musculature of the foot and byssus of *Dreissena polymorpha* (Yonge and Campbell, 1968; Morton, 1969a), yet the complete structure of the byssus has not been described. Several factors affect byssal thread formation including: mussel size, mechanical agitation of the water, water chemistry (e.g., salinity, Ca^{2+}, Mg^{2+}, etc.), temperature, and substrate (Van Winkle, 1970; Allen et al., 1976; Meadows and Shand, 1989). In addition, at least in *Mytilus*, attachment plaque size was found to vary directly with sediment size (Meadows and Shand, 1989), and inversely with the surface energy of substrates (Crisp et al., 1985).

Theoretical Considerations

There are various spatial scales by which to consider the adhesion strength of the byssus. At the scale of an individual mussel the adhesive strength per mussel, determined by tensile loadings normal to the substrate, ranges from 28 N in *Arca zebra* to 60 N in *Mytilus californianus*, and 90 N in *Septifer bifurcatus* (Waite, 1983). In *Mytilus edulis* estimates range from 10 to 36 N, depending on the season and to a lesser extent on the study (Waite, 1983). These estimates would include the combined strengths of all the finer-scaled structural components involved in byssal attachment. From the most distal portion of the thread, the attachment plaque fails under a breaking stress of 4×10^5 to 8×10^5 N/m^2 for adhesion on the periostracum and shell, respectively (Allen et al., 1976). This is consistent with the values of 0.12×10^5 to 8.5×10^5 N/m^2 presented in Crisp et al. (1985). The breaking stress for entire byssal threads is 1.6×10^6 N/m^2, but the distal portion (1.9×10^6 N/m^2) is stronger than the corrugated proximal portion (1.1×10^6 N/m^2) (Allen et al., 1976). The stiffness (modulus $= 8.5 \times 10^7$ N/m^2) and tensile strength (ultimate strain $= 0.44$) of byssal threads indicate that they can absorb a great deal of energy before breaking (Smeathers and Vincent, 1979). The number and orientation of byssal threads would have to be known before these separate elements could be integrated into a detailed mechanical model of adhesion (Waite, 1983).

In addition to tensile loading, there are a variety of fluid flow based methods that may prove useful in determining the adhesion properties of zebra mussels. Flow based methods include flow in rotating annular reactors (i.e., Couette flow with τ [shear] constant; Rittman, 1982; Visser, 1970); radial flow chambers (where τ is a function of the radius; Duddridge et al., 1982); rotating disks (Mohandras et al., 1974); parallel plate flow chambers (Mohandras et al., 1974); and flow on inclined plates (Hermanowicz et al., 1989).

Adhesion of one material to another is a result of a combination of mechanisms including: adsorption, mechanical interlocking, and molecular diffusion across the interface (Kinloch, 1987). These same mechanisms are applicable to biological systems such as isolated cells, bacteria, and zebra mussels. In general, more than one of the three adhesion mechanisms is present, and it is not possible to determine the relative importance of each. In all cases, a necessary precondition for adhesion is the establishment of intimate contact between the two materials. In cellular adhesion, this may require that electrostatic repulsion and fluid forces be overcome, whereas at larger scales (e.g., byssal adhesion), microtopography, viscosity, and wetting tendency may be important.

Byssal adhesion in *Dreissena* is likely to be similar in nature to that of *Mytilus* (Crisp et al., 1985). Although the initial contact and adhesion of a mussel to a surface is not well understood, the most probable mechanisms are the mechanical interlocking of secreted adhesive fluids (''glues'') with microscopic pores and crevices, and the action of attractive van der Waals forces. These adhesion mechanisms will only be effective if there is a potential for close molecular contact between the byssal attachment plaque (and/or its secretions) and the surface. One approach to modeling this potential is via the Helmholtz free energy of adhesion (ΔF_{adh}) model:

$$\Delta F_{adh} = \gamma_{SM} - \gamma_{SL} - \gamma_{ML} \qquad (1)$$

where γ is the interfacial tension (energy per unit area) between interfaces of the surface (S), byssal attachment plaque of the mussel (M), and the suspending liquid (L). Equation 1 expresses the change in overall surface energy of the system as a plaque makes contact with a surface when both are initially exposed to the same liquid. Intimate contact (or wetting and spreading) and hence adhesion are favored when ΔF_{adh} is a relatively large negative number, corresponding to a large free-energy decrease due to adhesion. This model has been applied to bioadhesion of bacteria (Absolom et al., 1983); platelets (Neumann et al., 1980); plant and animal cells (Schakenraad et al., 1988; Facchini et al., 1988; Neumann et al., 1979; Stewart et al., 1989); and solid polymer particles (Vargha-Butler, 1985; Omenyi and Neumann, 1976; Spelt et al., 1982 and 1987). The model does not include electrostatic effects, which may influence initial surface contact (Schakenraad et al., 1988; Di Cosmo et al., 1989); and it does not account for possible chemical interactions between the particle and the surface after contact. In these cases, the thermodynamic model is only useful as a predictor of the tendency for contact to occur, which is the first step in the adhesion process. We expect that the adhesion of zebra mussels is probably not affected by either electrostatic forces or specific chemical reactions with the surface.

In some biological systems, application of Equation 1 does not correctly predict the observed behavior, possibly due to the effects previously mentioned. For example, application of Equation 1 to the adhesion strength of endothelial cells shows the expected dependence on the solid surface tension in culture medium but not in buffer (Moussy et al., 1990). The model correctly predicts the spreading of granulocytes on different polymeric surfaces as a function of the surface tension of the suspending liquid, but shows an incorrect dependence on the surface tension of the solid substrate (Stewart et al., 1989). Although surface and interfacial tensions are important in the adhesion process, the free energy of adhesion model does not provide a complete description of adhesion in all cases.

The analysis of zebra mussel adhesion must include the consideration of how organic and microbial biofilms alter the solid surface tension of the underlying substrate. For example, the wettability of protein-coated surfaces was found to be a function of the critical surface tension of the substrate (Baier et al., 1984; Baier, 1986). Moreover, the spreading of cells was shown to be affected by the substrate even when serum proteins were present in the medium (Schakenraad et al., 1988). This suggests that a substrate may influence adhesion directly through a biofilm, or indirectly by modifying the composition and structure of a biofilm. Although further work is required in order to completely answer this question, it is likely that the surface tension of the underlying substrate does play a role in adhesion to biofilm covered surfaces.

With respect to adhesion of larger organisms, such as barnacles and mussels, there has been some application of the free-energy model. For example, the adhesion strength of barnacles has been characterized in terms of polar and nonpolar surface tension components (Becka and Loeb, 1984). Adhesion strength was found to increase as total surface tension of the substrate increased, although there were some complications due to the tearing of the shell plates. Crisp et al. (1985) describe the adhesion strength of mussels (*Mytilus*) and barnacle cyprids (larvae), although their conclusion that adhesion varies with substrate surface tension is not entirely supported by the data. These authors do, however, find that byssal attachment plaques spread more (i.e., have higher water contact angles) on lower energy surfaces, which is consistent with the free energy of adhesion model (Equation 1), assuming that $\gamma_{plaque} < \gamma_{water}$.

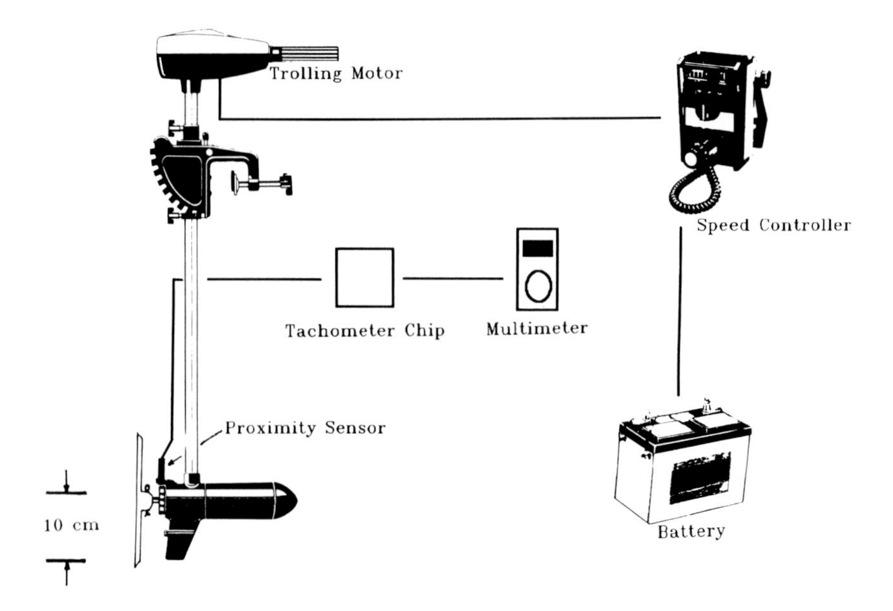

Figure 1. Schematic diagram of the components comprising a rotating disk test system.

In summary, the application of Equation 1 to zebra mussel adhesion will require data on surface tensions of biofilmed substrates and of the byssal attachment plaque. The latter may be inferred by comparing the patterns of adhesion to substrates of known surface tension with the predictions of Equation 1 (e.g., Crisp et al., 1985). While the interpretation of adhesion in terms of surface and interfacial tensions is not appropriate in all cases, it is a useful operating hypothesis for guiding the interpretation of experimental data.

USE OF A ROTATING DISK SYSTEM

A rotating disk test system was developed to measure adhesion strength of zebra mussels (Figure 1). Mussels were "seeded" onto disks of various materials, and disks were spun at increasing speeds until mussels detached. This flow-based method satisfies several criteria needed to measure zebra mussel adhesion strength that could not be met by other mechanical testing methods. First, it allows for the testing of a wide range of mussel sizes. This is important since the relatively small ($\approx$300 μm) postveliger stage is considered a key stage where control strategies may be most successful. Second, the test system is portable, and allows testing both in the laboratory and in locations where zebra mussels occur naturally. Third, the test system components are readily available, and different materials with different surface conditions may be easily tested for screening purposes. Finally, mussel de-

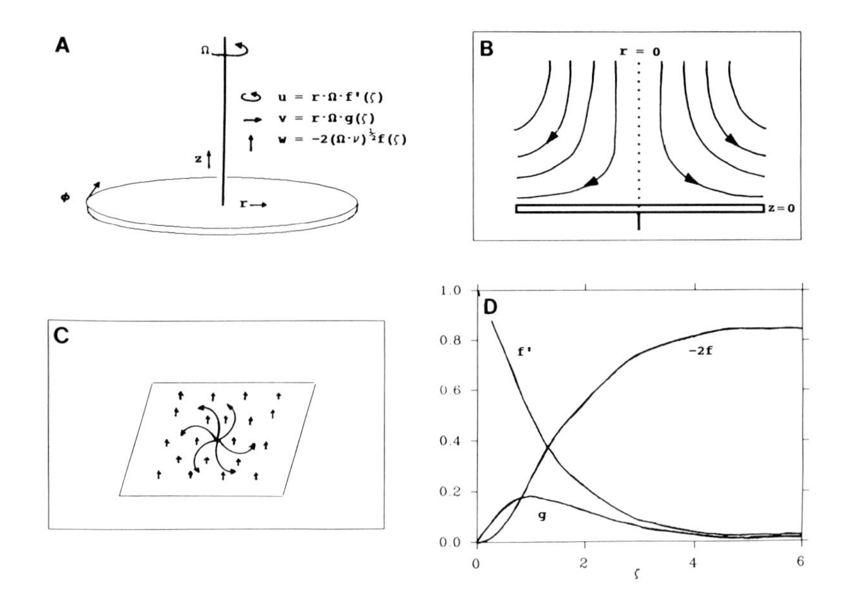

Figure 2. Theoretical flow induced by a rotating disk: (A) rotation direction, notation, and velocity components associated with the rotating disk; (B) streamlines of flow in a plane perpendicular to the face of the disk; (C) streamlines and velocity components in a plane parallel to the disk, indicating that w does not depend on r; (D) plot of nondimensional velocity components vs the nondimensional height (ζ) above the disk for laminar flow. (After Albery, W. J. *Physicochemical Hydrodynamics*, Vol. 1 [London, England: Academic Press, Inc., 1977]. With permission.)

tachment can be related to the flow field over the disk, calculated from a well-known model.

Fluid Dynamic Theory

An infinite disk rotating at angular velocity (Ω) generates a fan-like flow field which can be written as a closed-form exact solution of the Navier-Stokes equations (Figure 2) (Von Karman, 1921; Cochran, 1934; Schlichting, 1979). The angular velocity of the disk generates an inward axial flow perpendicular to the face of the disk (z direction) and outward radial (r direction) and tangential (ϕ direction; or circumferential) flows parallel to the face of the disk. The velocity components in the r, ϕ, and z directions (u, v, and w, respectively) at the disk surface and far from the disk satisfy the boundary conditions:

$$\begin{aligned} &u = 0 \quad v = \Omega \cdot r \quad w = 0 \quad \text{at } z = 0 \\ &u = 0 \quad v = 0 \quad \qquad\qquad \text{at } z = \infty \end{aligned} \tag{2}$$

For laminar flow, u, v, and w are written in terms of functions $f(\zeta)$ and $g(\zeta)$ using the similarity transformation:

$$u = \Omega \cdot r \cdot f'(\zeta) \quad v = \Omega \cdot r \cdot g(\zeta) \quad w = -2(\Omega \cdot \nu)^{1/2} \cdot f(\zeta) \tag{3}$$

where:

$$\zeta = z \cdot (\Omega/\nu)^{1/2} \tag{4}$$

and ν = kinematic viscosity. After substitution into the momentum and continuity equations, a set of ordinary differential equations for f and g are obtained. Cochran's (1934) numerical solution to these equations yields $f(\zeta)$ and $g(\zeta)$ (see Figure 2D), from which u, v, and w can be computed at any point in the domain. The fluid shear stress on the face of the disk under laminar conditions is then given by:

$$\tau_{rz} = 0.51 \cdot \rho \cdot (\nu \cdot \Omega^3)^{1/2} \cdot r \tag{5a}$$

in the radial direction and:

$$\tau_{z\phi} = -0.616 \cdot \rho \cdot (\nu \cdot \Omega^3)^{1/2} \cdot r \tag{5b}$$

in the tangential direction, where ρ is the fluid density. This solution applies to disks of finite radius (R) providing the boundary layer is small compared to the disk radius, which will occur when the disk Reynolds number $Re = R^2 \cdot \Omega/\nu$ is large. Laminar flow occurs for regions of radius r such that $Re_r = r^2 \cdot \Omega/\nu < 3 \cdot 10^5$.

Under turbulent conditions ($Re_r > 3 \cdot 10^5$) an analogous derivation gives the following expression for the fluid shear stresses:

$$\tau_{rz} = 0.0225 \cdot \rho \cdot (\nu/\delta)^{1/4} \cdot c^{7/4} \cdot [1 + \{(r \cdot \Omega)/c\}^2]^{3/8} \tag{6a}$$

in the radial direction, and:

$$\tau_{z\phi} = 0.0225 \cdot \rho \cdot (\nu/\delta)^{1/4} \cdot (r \cdot \Omega)^{7/4} \cdot [1 + \{c/(r \cdot \Omega)\}^2]^{3/8} \tag{6b}$$

in the tangential direction. In the above expressions, the boundary layer thickness (δ) is

$$\delta = 0.526 \cdot r \cdot (\nu/(r^2 \cdot \Omega))^{1/5} \tag{7}$$

and

$$c = 0.162 \cdot r \cdot \Omega \tag{8}$$

The essential feature of the above theory is that, given the rotational rate Ω, the net shear stress acting at any location on the surface in both laminar and turbulent flows can be computed as the vectorial sum of τ_{rz} and $\tau_{z\phi}$.

Apparatus and Protocols

A portable DC-powered trolling motor (Minn Kota Model 35, Johnson Fishing Inc., Mankato, MN) with solid state speed controller (Minn Kota Maximizer, Johnson Fishing) was used to power the rotating disk test system (Figure 1). The rotation of a 15-pin, ferrous gear attached to the propeller shaft was detected by a magnetic proximity sensor (model 3055A, Electro Corporation) and a tachometer chip (model LM2917, National Semiconductor Corporation). The tachometer was calibrated both electronically and stroboscopically. In our design, smooth disks were used that were 20 cm in diameter, approximately 5 mm in thickness, and had 45° beveled edges. Disks were placed over the propeller shaft and held by a cotter pin that extended through a collar on the rear of the disk and the propeller shaft.

Since the fluid dynamic model used in analyzing the experimental data applies for disks rotating in infinite regions, tests should be undertaken in large vessels of water well away from side walls or similar obstructions. In the laboratory, the disk should be placed in the center of a large tank where the ratio of disk diameter to the typical tank dimension is small. Pitot-static tube studies indicated that wall effects were acceptably small in our tank.

Zebra mussels must make byssal attachments to the surface of the disk in one of two ways: (1) pediveligers can be allowed to recruit to the disks or (2) mussels can be detached from other surfaces and allowed to spontaneously form new attachments on disks. In the latter method, mussels need only be collected soon before an experiment, or maintained under favorable culture conditions until an experiment is run. This method is also useful because experiments can be conducted throughout the year and not simply following pediveliger recruitment. In addition, mussels of most sizes can be tested, as opposed to only recruited postveligers. Data presented below were obtained using the second method.

After a desired residence time, the position, length, and orientation of each adhering mussel were marked on an acetate sheet held directly above the disk. The disk was carefully lowered into the center of the tank and slowly accelerated ($\approx$5 rad/sec^2) to 300 rpm where it was held for 30 sec, and then slowly decelerated. The position of each mussel was verified on the disk and any detached mussel was labeled on the acetate sheet. This procedure was repeated at 60 or 120 rpm increments until the maximum speed was achieved ($\approx$1200 rpm in our studies). By the end of the test, the acetate sheet was covered with detachment-speed labels, corresponding to each mussel that had been detached.

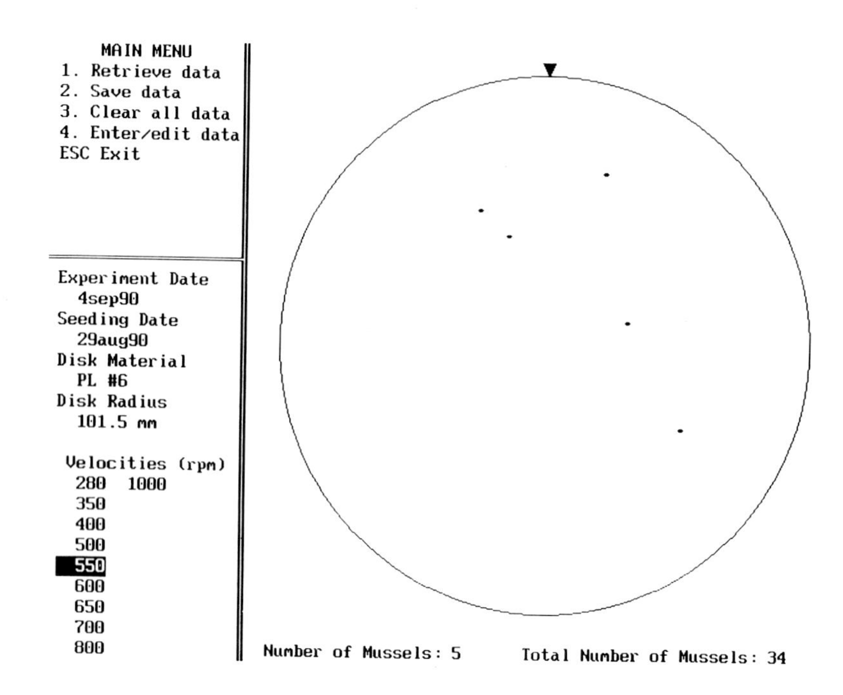

Figure 3. The position and velocity at which *Dreissena polymorpha* detached from a rotating disk as recorded in a digi-pad data entry program. (From Indrigo, R. M. BASc Thesis, University of Toronto, Toronto, Ontario [1991]. With permission.)

The radial position (r) and speed (Ω) at which the mussels detached were recorded by analyzing the acetate sheet. This can be done manually (with a ruler), although it is tedious, or with a digitizing tablet (digi-pad). An example of digi-pad data entry using customized software is given in Figure 3 (Indrigo, 1991). The outline of the disk on the acetate sheet is aligned on the digi-pad using three points on the disk perimeter. Various experimental parameters — including the test date, seeding date (date when mussels where placed on the disk), and the disk material — are entered. The rotational velocity interval is selected (e.g., 550 rpm in Figure 3) and all detachments that occur during that interval (five mussels in Figure 3) are entered using the digi-pad cursor. The size of the mussel can be entered from the keyboard following the cursor entry.

The digitized data were analyzed using the fluid dynamic relationships presented previously. The radial position (r) and speed (Ω) at which detachments occurred were used to calculate the disk Reynolds number (Re), the boundary layer thickness (δ, turbulent flow only), and the laminar or turbulent shear stresses (τ) (Equations 5–8). To avoid inaccuracies due to edge effects, all mussels $< 5 \cdot \delta$ from the edge of the disk were excluded under turbulent

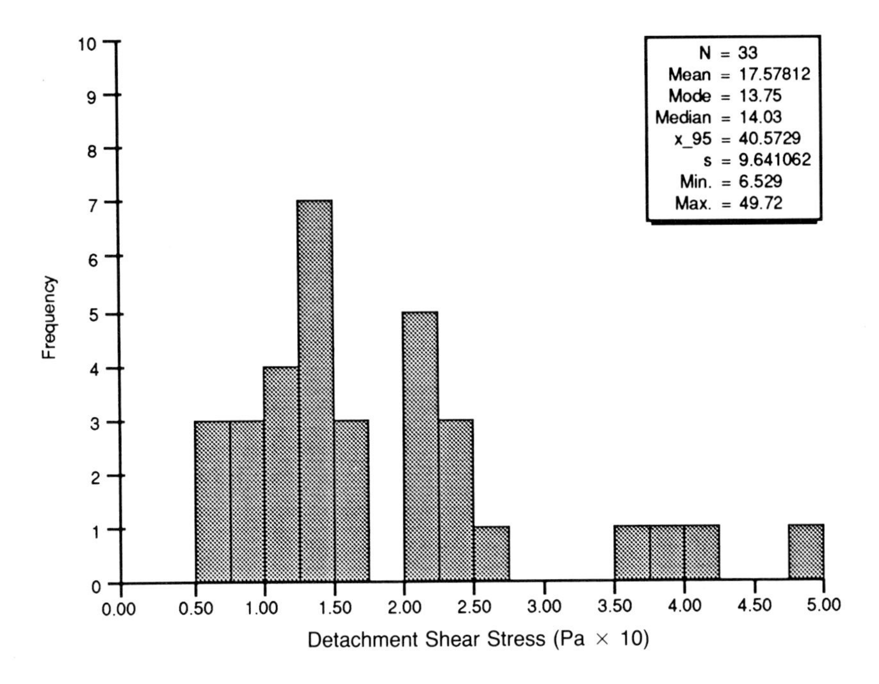

Figure 4. Frequency histogram for *Dreissena polymorpha* detachment shear stresses on a Plexiglass™ (PMMA) disk. (From Indrigo, R. M. BASc Thesis, University of Toronto, Toronto, Ontario [1991]. With permission.)

conditions. Only the radius and net detachment shear stress ($\tau = [\tau_{rz}^2 + \tau_{z\phi}^2]^{1/2}$) were used in the subsequent analysis. The actual fluid forces (e.g., lift, drag, etc.) that cause mussel detachment from a disk are not computed in this procedure. Instead, a characteristic fluid dynamic parameter, in this case the wall shear stress, describes mussel detachment. In reality, τ is the nominal fluid shear stress on the disk at the position where detachment occurred, i.e., the shear stress that would exist in the absence of mussels.

RESULTS AND DISCUSSION

The example presented in Figure 4 is typical of the type of information derived from the rotating disk test system (Ackerman et al., 1992). In this case, 33 of 34 mussels were included in the analysis as 1 mussel was excluded by the $5 \cdot \delta$ criterion. The τ required to detach mussels ranged from 6.5 to 49.7 Pa reflecting the wide range in r and Ω at which mussel detachments occurred. The distribution of τ is skewed to the right and centered between 14 and 18 Pa for all three measures of central tendency (mean, median, and mode). Seven mussels detached at the modal frequency of 14 Pa, which was

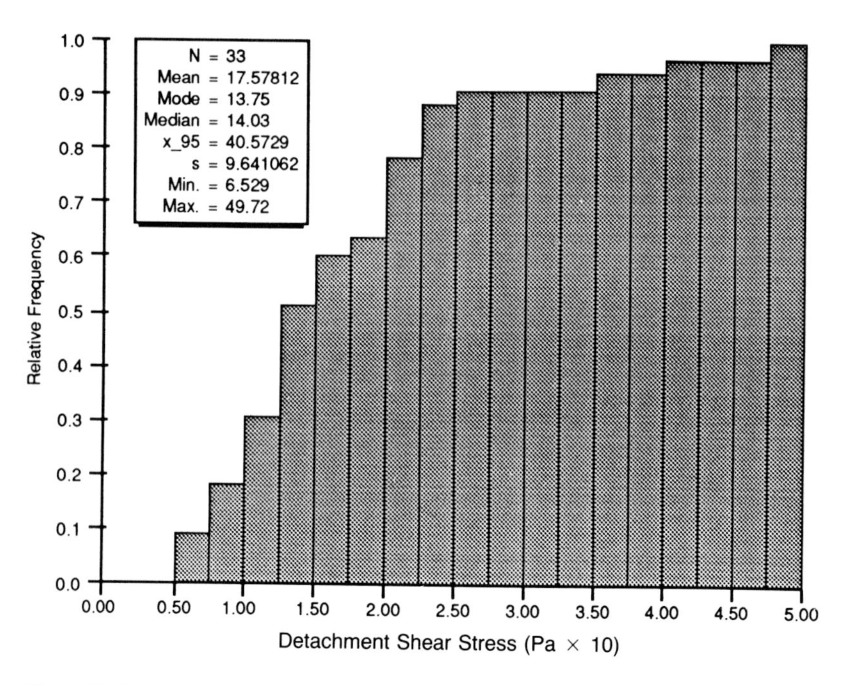

Figure 5. Cumulative relative frequency histogram for *Dreissena polymorpha* detachment shear stresses on a Plexiglass™ disk. (From Indrigo, R. M. BASc Thesis, University of Toronto, Toronto, Ontario [1991]. With permission.)

the same as the median of 14 Pa. The mean was slightly higher at 18 Pa with a standard deviation of 9.6 Pa, giving a standard error of the mean of 1.7 Pa.

For practical purposes, the mean τ tends to underestimate the τ required to remove mussels. A second metric unit, the shear stress that removed 95% of the mussels on a disk ($\tau_{95\%}$), is a more reasonable estimate of the adhesion strength of zebra mussels. The $\tau_{95\%}$ of 40.5 Pa in this example can be determined: (1) numerically by calculating the 95% median rank or (2) graphically by plotting a cumulative relative frequency histogram of τ and finding the frequency bin that intersects a frequency of 0.95 (40.5 Pa in Figure 5). The results of these methods converge when the sample size is large.

Over 70 rotating disk test system experiments involving zebra mussels ($\leq$2 mm) adhering to disks of different materials and having different mussel residence times were conducted (Ackerman et al., 1992). Disk materials were polyvinylchloride (PVC), polymethylmethacrylate (PMMA = Plexiglass™), stainless steel (SS), and aluminum (AL). Initial findings are given in Figures 6 and 7. For the most part, plots of mean τ and $\tau_{95\%}$ are similar although, of course, the $\tau_{95\%}$ values are larger. While the pattern of plotted points is similar, there is no simple mapping of mean τ to $\tau_{95\%}$. There does

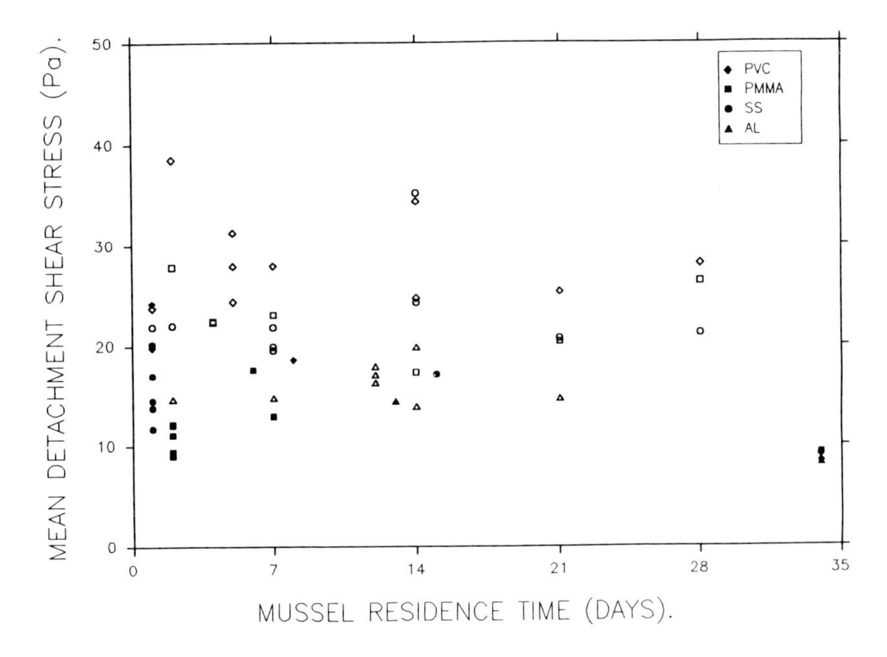

Figure 6. Plot of zebra mussel mean detachment shear stress on different disk materials vs mussel residence time. Open symbols = data from Ackerman et al. (1992), solid symbols = new data. PVC = polyvinylchloride, PMMA = Plexiglass™, SS = stainless steel, AL = aluminum.

not appear to be any correlation in the mean τ and $\tau_{95\%}$ data with residence time, a finding which has been confirmed statistically. There are statistically significant differences in the mean τ and $\tau_{95\%}$ on different disk materials. Generally, the adhesion strength of mussels on the materials ranks as follows: PVC > PMMA ≈ SS > AL.

Experimental results can be affected by the health of the mussels. For example, the five PMMA values on day 2 below 20 Pa (Figures 6 and 7) represent tests conducted on mussels which had been kept in laboratory culture for 5–6 months. These observations suggest that tests be conducted soon after mussels are collected, or certainly within the first several months of laboratory culture. The four tests at 34 days were conducted within the first month of culture, during which time the mussels were not fed. Visual examinations of mussels in the cultures indicated that most were dead prior to testing. Both data sets indicate the importance of testing healthy mussels. It is interesting to note that these results indicate that dead mussels are able to adhere to surfaces, implying that chemical control methods that cause mussel mortality will need to be augmented with a cleaning procedure to remove the attached but dead mussels. Long-term biological (bacterial, fungal, etc.) degradation of the foot will probably leave only the byssal attachment plaque and threads attached to the fouled surface.

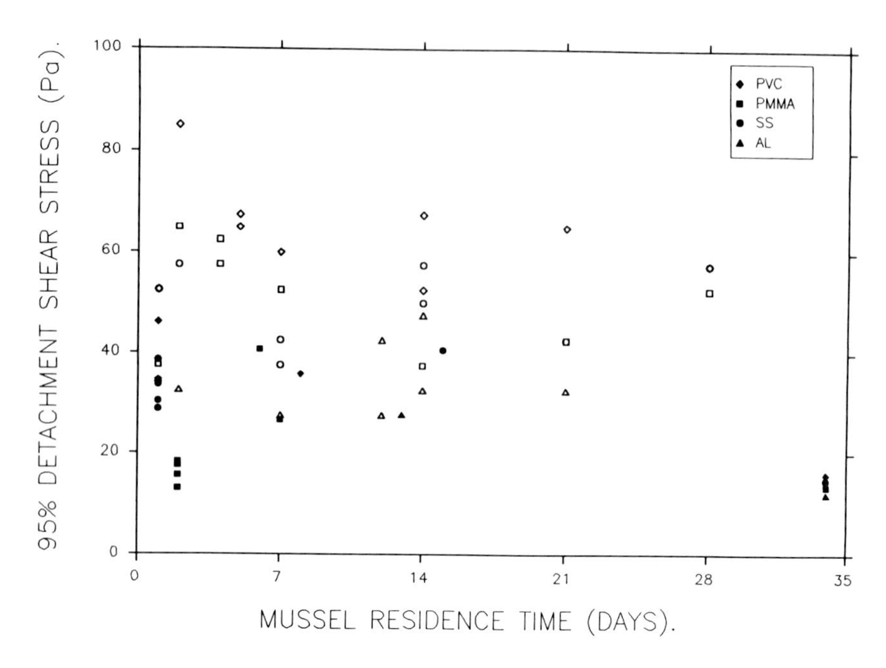

Figure 7. Plot of *Dreissena polymorpha* 95% detachment shear stress on different disk materials vs mussel residence time. Symbols as in Figure 6.

SUMMARY

We recognize the limitations of the rotating disk test system in that it cannot completely replicate the environment inside, for example, a heat exchanger. However, the system provides a reliable and simple method by which to measure the adhesion strength of zebra mussels on various materials under relatively well-controlled conditions. Various tests involving new combinations of mussel sizes, growth conditions, and disk materials are envisioned. In addition, there are two applications where a rotating disk test system will be useful: (1) testing the health of mussels in laboratory culture, and (2) testing the effectiveness of various ''antifouling'' and ''easy-clean'' coatings. The latter experiments could be conducted by comparing the adhesion of mussels on coated disks and uncoated control disks.

As in most fluid flow methods, the adhesion strength of zebra mussels was estimated from a nominal fluid dynamic parameter. More precisely, it was based on the nominal wall shear stress at the position where the mussel detached, ignoring the presence of the mussel. A measure of the net force acting on the mussel will require an understanding of the effects of mussel shape and orientation on fluid flow (i.e., lift and drag coefficients; e.g., Denny, 1988). Approximations of these effects on some radially symmetrical organisms under certain monitored flow conditions are possible (Eckman et al., 1990). Future research may determine these effects on non-radial forms such as zebra mussels using computational fluid mechanics and/or physical modeling. Mechanical tests (e.g., tensile loading) may prove useful in confirming these findings on larger mussels.

Future adhesion strength testing using a flow-based method known as a wall jet may also be useful in investigating zebra mussel adhesion. The wall jet is useful because it can provide the high shear stress required to detach zebra mussels. It produces a well-defined flow field and can be used with a wide range of substrates and surface conditions. Moreover, it will be possible to visualize and videotape the flow effects on mussels and, therefore, determine the mechanics of the detachment process. Ultimately, in addition to understanding the basic biology of adhesion, experiments such as these may present guidelines for material selection and optimization of flow conditions so that the zebra mussel settlement will be minimized and mussels can be easily removed from surfaces.

ACKNOWLEDGMENTS

We acknowledge the support of the following individuals: Renata Claudi, Eileen Leitch, Fred Spencer, Robert Indrigo, Peter Frank, Sandy McLeod, Blair Sim, Fleur Ely, Dave Lowther, Bruno Polewski, and Miro Kalovsky. This research was funded in part by research funds provided by Ontario Hydro, the University Research Incentive Fund of the Ontario Ministry of Colleges and Universities, and the Ontario Petroleum Association.

REFERENCES

Absolom, D. R., F. V. Lamberti, Z. Policova, W. Zingg, C. J. Van Oss, and A. W. Neumann, "Surface Thermodynamics of Bacterial Adhesion," *J. Appl. Environ. Microbiol.* 46:90–97 (1983).

Ackerman, J. D. "Biomechanical Aspects of Submarine Pollination in *Zostera marina* L.," PhD Thesis, Cornell University, Ithaca, NY (1989).

Ackerman, J. D., C. R. Ethier, D. G. Allen, and J. K. Spelt. "Investigation of Zebra Mussel Adhesion Strength Using a Rotating Disk," *J. Environ. Eng.* 118, 708–724 (1992).

Albery, W. J. "Developments and Applications of the Rotating Disk System," in *Physicochemical Hydrodynamics, Vol. 1* D. B. Spalding, Ed. (London, England: Academic Press, Inc., 1977), pp. 413–433.

Allen, J. A., M. Cook, D. J. Jackson, S. Preston, and E. M. North. "Observations on the Rate of Production and Mechanical Properties of the Byssus Threads of *Mytilus edulis* L.," *J. Molluscan Stud.* 42:279–289 (1976).

Baier, R. E. "Modification of Surfaces to Meet Bioadhesive Design Goals: A Review," *J. Adhesion* 20:171–186 (1986).

Baier, R. E., A. E. Meyer, J. R. Natiella, R. R. Natiella, and J. M. Carter. "Surface Properties Determine Bioadhesive Cutcomes: Methods and Results," *J. Biomed. Mat. Res.* 18:337–355 (1984).

Bayne, B. L. "Primary and Secondary Settlement in *Mytilus edulis* L. (Mollusca)," *J. Anim. Ecol.* 33:513–523 (1964).

Becka, A., and G. Loeb. "Ease of Removal of Barnacles from Various Polymeric Materials," *Biotechnol. Bioeng.* 26:1245–1251 (1984).

Beukema, J. J., and J. De Vlas. "Tidal-Current Transport of Thread-Drifting Postlarval Juveniles of the Bivalve *Macoma balthica* from the Wadden Sea to the North Sea," *Mar. Ecol. Prog. Ser.* 52:193–200 (1989).

Blok, J. W. de, and H. J. F. M. Geelen. "The Substratum Required for the Settling of Mussels (*Mytilus edulis* L.)," *Neth. J. Zool.* 13:446–460 (1958).

Cochran, W. G. "The Flow Due to a Rotating Disc," *Proc. Cambridge Philos. Soc.* 30:365–375 (1934).

Crisp, D. J., G. Walker, G. A. Young, and A. B. Yule. "Adhesion and Substrate Choice in Mussels and Barnacles," *J. Colloid Interface Sci.* 104:40–50 (1985).

Denny, M. W. "Biology and the Mechanics of the Wave-Swept Environment," (Princeton, NJ: Princeton University Press, 1988).

Dicosmo, F., P. J. Facchini, and A. W. Neumann. "Plant Cell Adhesion to Polymer Surfaces as Predicted by a Thermodynamic Model and Modified by Electrostatic Interaction," *Colloids Surf.* 42:255–269 (1989).

Duddridge, J. E., C. A. Kent, and J. F. Laws. "Effect of Surface Shear Stress on the Attachment of *Pseudomonas fluorescens* to Stainless Steel Under Defined Flow Conditions," *Biotechnol. Bioeng.* 24:153–164 (1982).

Eckman, J. E., W. B. Savidge, and T. F. Gross. "Relationship Between Duration of Cyprid Attachment and Drag Forces Associated with Detachment of Balanus Amphitrite Cyprids," *Mar. Biol.* 107:111–118 (1990).

Facchini, P. J., A. W. Neumann, and F. Dicosmo. "Thermodynamic Aspects of Plant Cell Adhesion to Polymer Surfaces," *Appl. Microbiol. Biotechnol.* 29:346–35 (1988).

Hebert, P. D. N., C. C. Wilson, M. H. Murdoch, and R. Lazar. "Demography and Ecological Impacts of the Invading Mollusc *Dreissena polymorpha*," *Can. J. Zool.* 69:405–409 (1991).

Hermanowicz, S. W., R. E. Danielson, and R. C. Cooper. "Bacterial Deposition on and Detachment from Surfaces in Turbulent Flow," *Biotechnol. Bioeng.* 33:157–163 (1989).

Indrigo, R. M. "Data Acquisition and Processing for the Rotating Disk Test System: A Technique to Measure Adhesion of Zebra Mussels," BASc Thesis, University of Toronto, Toronto, Ontario (1991).

Kinloch, A. J. *Adhesion and Adhesives — Science and Technology* (London, England: Chapman and Hall, 1987).

Lane, D. J. W., A. R. Beaumont, and J. R. Hunter. "Byssus Drifting and the Drifting Threads of the Young Post-Larval Mussel *Mytilus edulis*," Mar. Biol. 84:301–308 (1985).

Lewandowski, K. "The Role of Early Development Stages, in the Dynamics of *Dreissena polymorpha* (Pall.) (Bivalvia) Populations in Lakes," *Ekol. Pol.* 30:223–286 (1982).

Meadows, P. S., and P. Shand. "Experimental Analysis of Byssal Thread Production by *Mytilus edulis* and *Modiolus modiolus* in Sediments," *Mar. Biol.* 101:219–226 (1989).

Mohandras, N., R. M. Hochmuth, and E. E. Spaeth. "Adhesion of Red Cells to Foreign Surfaces in the Presence of Flow," *J. Biomed. Mat. Res.* 8:119–136 (1974).

Morton, B. "Studies on the Biology of *Dreissena polymorpha*. Pall. I. General Anatomy and Morphology," *Proc. Malacol. Soc. London* 38:301–321 (1969a).

Morton, B. "Studies on the Biology of *Dreissena polymorpha* Pall. II. Correlation of the Rhythms of Adductor Activity, Feeding, Digestion and Excretion," *Proc. Malacol. Soc. London* 38:401–414 (1969b).

Morton, B. "Studies on the Biology of *Dreissena polymorpha* Pall. III. Population Dynamics," *Proc. Malacol. Soc. London* 38:471–482 (1969c).

Moussy, F., A. W. Neumann, and W. Zingg. "The Force of Detachment of Endothelial Cells from Different Solid Surfaces," *Am. Soc. Artif. Intern. Organ Trans.* 36:568–572 (1990).

Neumann, A. W., D. W. Absolom, C. J. Van Oss, and W. Zingg. "Surface Thermodynamics of Leukocyte and Platelet Adhesion to Polymer Surfaces," *Cell Biophys.* 1:79–92 (1979).

Neumann, A. W., O. S. Hum, D. W. Francis, W. Zingg, and C. J. Vanoss. "Kinetic and Thermodynamic Aspects of Platelet Adhesion from Suspension to Various Substrates," *J. Biomed. Mat. Res.* 14:499–509 (1980).

Omenyi, S. N., and A. W. Neumann. "Thermodynamic Aspects of Particle Engulfment by Solidifying Melts," *J. Appl. Physiol.* 47:3956–3962 (1976).

Rittman, B. E. "The Effects of Shear Stress on Biofilm Loss Rate," *Biotechnol. Bioeng.* 24:501–506 (1982).

Roberts, L. "Zebra Mussel Invasion Threatens U.S. Waters," *Science* 249:1370–1372 (1990).

Schakenraad, J. M., H. J. Busscher, CH. R. H. Wildevuur, and J. Arends. "Thermodynamic Aspects of Cell Spreading on Solid Substrata," *Cell Biophys.* 13:75–91 (1988).

Schlichting, H. *Boundary Layer Theory*, 7th ed. (New York: McGraw-Hill Book Company, 1979).

Sigurdsson, J. B., C. W. Titman, and P. A. Davies. "The Dispersal of Young Post-Larval Bivalve Molluscs by Byssus Threads," Nature (London) 262:386–387 (1976).

Smeathers, J. E., and J. F. V. Vincent. "Mechanical Properties of Mussel Byssal Threads," *J. Molluscan Stud.* 45:219–230 (1979).

Spelt, J. K., D. R. Absolom, W. Zingg, C. J. Van Oss, and A. W. Neumann. "Determination of the Surface Tension of Biological Cells Using the Freezing Front Technique," *Cell. Biophys.* 4:117–131 (1982).

Spelt, J. K., R. P. Smith, and A. W. Neumann. "Attraction and Repulsion of Solid Particles by Solidification Fronts: Evaluation of Predictions of the Fowkes Equation," *Colloids Surf.* 28:85–92 (1987).

Stewart, M. G., E. Moy, G. Chang, W. Zingg, and A. W. Neumann. "Thermodynamic Model for Cell Spreading," *Colloids Surf.* 42:215–232 (1989).

Van Winkle, W. Jr. "Effect of Environmental Factors on Byssal Thread Formation," *Mar. Biol.* 7:143–148 (1970).

Vargha-Butler, E. I., T. K. Zubovits, H. A. Hamza, and A. W. Neumann. ''Surface Tension Effects in the Sedimentation of Polymer Particles in Various Liquid Mixtures,'' *J. Dispersion Sci. Technol.* 6:357–379 (1985).

Visser, J. ''Measurement of the Force of Adhesion Between Submicron Carbon Black Particles and a Cellulose Film in Aqueous Solution,'' *J. Colloid Interface Sci.* 34:26–31 (1970).

Von Karman, T. ''Über laminare und turbulente reibung,'' *Z. Angew. Math. Mech.* 1:233–252 (1921).

Yonge, C. M. ''On the Primitive Significance of the Byssus in the Bivalvia and Its Effects in Evolution,'' *J. Mar. Biol. Assoc. U.K.* 42:113–125 (1962).

Yonge, C. M., and J. I. Campbell. ''On the Heteromyarian Condition in Bivalvia with Special Reference to *Dreissena polymorpha* and Certain Mytilaceae,'' *Trans. R. Soc. Edinburgh* 68:21–43 (1968).

Waite, J. H. ''Adhesion in Byssally Attached Bivalves,'' *Biol. Rev.* 58:209–231 (1983).

CHAPTER 17

Influence of Chemical Composition of the Water on the Mollusk *Dreissena polymorpha*

Germane A. Vinogradov, Nataliya F. Smirnova, V. A. Sokolov, and A. A. Bruznitsky

On the basis of experiments on the acclimation of *Dreissena* to water of low salinity, it was found that this mollusk is capable of maintaining the balance of sodium and potassium when these ions are at very low levels in the water. The balance concentration (C_b) for potassium is 0.4 mg/L and 0.5–1.0 mg/L for sodium. For *Dreissena polymorpha* to maintain a balance between the metabolic loss and uptake of calcium, concentrations in the water must not be lower than 12–14 mg/L. This concentration is significantly higher than required by other bivalved mollusks living in the same region. pH values less than 6.5 produce a negative effect on the metabolism of sodium, potassium, and calcium. In the process of acclimation of *Dreissena* to acid water, adaptive modifications in sodium and potassium exchange occur allowing this species to maintain a balance of these elements; however, pH values must not be lower than 5.5 for this to occur. A decrease in the sodium concentration reduces resistance of *D. polymorpha* to water acidification. The ionic metabolism in *Dreissena* is more sensitive to decreases in pH than the metabolism of other freshwater bivalved mollusks.

0-87371-696-5/93/$0.00 + $.50

INTRODUCTION

The rate of ion absorption in freshwater mollusks depends upon the ion concentration in the water (Halley and Gibson, 1971; Romanenko et al., 1982). A decrease in ambient sodium, calcium, and potassium concentrations to certain values leads to a reduction in the rate of the influx of these cations into the organism (Vinogradov et al., 1987). Freshwater mollusks are especially sensitive to ambient ion concentrations, in particular, calcium. Species composition, biomass, and distributions are directly related to the amount of calcium in the water; Nduru and Harrison, 1976; Williams, 1970). Calcium is necessary for the growth and building of the shell. Absorption occurs directly from the water and also from food; however the former pathway is far more important, accounting for up to 70–80% of total calcium intake (Romanenko et al., 1982). Calcium is directly absorbed from the water not only through the gills, but also through the mantle (Halley and Gibson, 1971; Pryadko, 1976).

Another important factor influencing distributions of freshwater mollusks is pH. A decrease in pH is known to have a negative impact on the abundance of bivalves and gastropods, with pH values of 5.2–6.0 proving unfavorable, and values of 5.0–5.2 restricting distributions (Roff and Kwaitkowski, 1977; Okland, 1980). Besides, it is known that water acidification causes disturbances in sodium, calcium, and potassium exchange between mollusks and the ambient medium (Vinogradov, 1986; Vinogradov et al., 1987).

Because *Dreissena* is believed to be a relatively recent ''immigrant'' to freshwater (as evidenced by the retention of the veliger stage of marine forms), its response to natural and anthropogenic factors is generally quite different from other freshwater forms (i.e., unionids). For instance, studies have indicated that *Dreissena* has a higher resistance to phenol and salinity when compared to freshwater unionids (Smirnova, 1973a and 1973b). For salinity, Smirnova (1973a) found that concentrations of 5‰ were 100% lethal to *Dreissena* after 168 days. This concentration was slightly lower than the lethal concentration of 5–7‰ found by Karpevich (1955); the difference can be attributed to the fact that Karpevich used Caspian Sea water which has a lower concentration of sodium and chlorine than the ocean water used by Smirnova. When stepwise acclimation was used to determine salinity tolerances, which is the most appropriate method to discern real physiological differences, significant differences were found between *D. polymorpha* (lethal concentration of 10–12‰) and typical freshwater forms (lethal concentration of 2–3‰) (Karpevich, 1955).

Given the unique and peculiar physiology of *Dreissena* and the importance of ion exchange to mollusks in general, a study was undertaken to examine the exchange of calcium, sodium, and potassium in *Dreissena polymorpha* when acclimated to water of low salinity. Other objectives were to determine limiting concentrations of these ions and also to examine the relation between ionic exchange of these minerals and pH.

METHODS

The ion exchange between *Dreissena polymorpha* and the medium was studied by placing 5–10 individuals (total mass of 0.6–1.0 g) into 25–50 mL of aerated water with varying ionic composition for 15–30 min. Water temperature during the experiments was 16–18°C. Each set of experiments consisted of 4–5 replicates. Prior to the experiments, the organisms were held in natural water from the Rybinsk Reservoir. To obtain water of low ionic composition, the water was diluted 5- to 20-fold with distilled water with the necessary amounts of NaCl and KCl added. The ionic composition (mg/L) of water from the Rybinsk Reservoir was: pH = 7.7, Na = 6.7, K = 1.5, Ca = 36.6, Mg = 20.4, HCO_3 = 130.5, Cl = 3.9, and SO_4 = 37.0. For determination of calcium exchange, organisms were placed into distilled water or into solutions of $CaCl_2$ with neutral pH. Experiments were also conducted on individuals acclimated for 28 days in artesian water. This water contained (mg/L): K = 0.2–0.5, Na = 0.8–1.0, Mg = 1.0–1.2, and Ca = 0.8–1.5.

The net flux (NF) of ions was estimated by the change in cation concentrations from the initial values. If the cation concentration in the water decreased, then the NF was considered positive. The rate of the total ion influx (IF) was calculated according to the equation: NF = IF − V. The value V was considered negative, since a loss of salts was always observed during determination of this parameter. A concentration of a cation in the water at which the absorption was balanced by loss (IF = V and NF = O) will further be referred to as the balance concentration (C_b).

Sodium and potassium concentrations were determined with liquid flame analyzer PAZ-2, while Ca^{2+} was determined with atomic absorption spectrophotometer. The necessary pH values in experiments with neutral water were maintained by adding sulfuric acid using an automatic titration block BAT-15 and millivoltmeter pH-340.

RESULTS AND DISCUSSION

Ion Exchange

The kinetics of absorption of sodium, potassium, and calcium from the ambient medium in *D. polymorpha* is described by the Michaelis-Menten equation. The values of the constants for these ions vary essentially (Figures 1, 2, and 3). Sodium and potassium ion-transporting systems possess the greatest affinity for substrates. Ca is intensively absorbed from the water at much higher concentrations. The results indicated rapid change in the equilibrium concentration (C_b) for sodium and potassium, with equilibrium essentially being reached after 5 to 7 days. During this process, a significant decrease in the rate of the total sodium and potassium loss was observed,

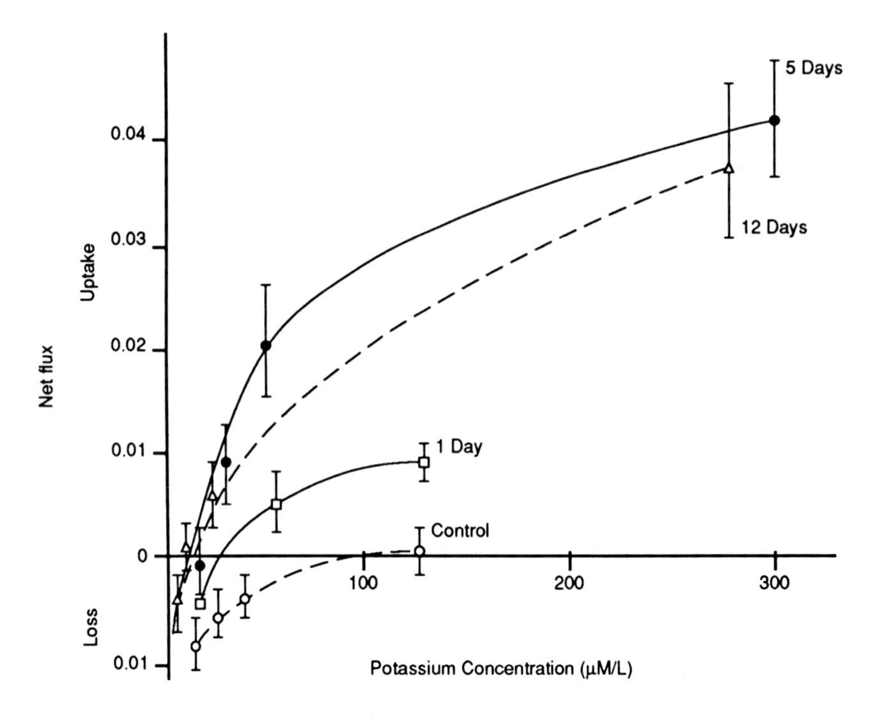

Figure 1. The effect of acclimation to low-salt water on potassium metabolism in *Dreissena*. Uptake and loss given as μmoles/g/hr.

along with a significant increase in the consumption of these ions. Both mechanisms are well-known in freshwater organisms (Lavrova et al., 1979; Vinogradov, 1986). A prolonged exposure (30–36 days) of the mollusks to low-calcium water had no effect on the kinetics of calcium absorption.

These results suggest that *Dreissena* is well adapted to waters with extremely low sodium and potassium levels. In this aspect, the dynamics of sodium and potassium exchange in *Dreissena* differ from those typically found in freshwater mollusks (Vinogradov et al., 1987). However, for calcium, exchange was lower than an earlier study by Vinogradov et al. (1987) in which calcium exchange in *Dreissena* was compared to that of other mollusks from the same habitat. In this earlier study, the minimal concentration at which calcium equilibrium was established was 13–14 mg/L in *Dreissena*, 3 mg/L in *Anodonta cygnea*, and 6–7 mg/L in *Unio pictorum* (Figure 4). This study, however, demonstrates that water with calcium concentrations of less than 10 to 12 mg/L is not suitable for normal calcium metabolism in *D. polymorpha*. The reason why water with low calcium (0.8–1.5 mg/L) had little effect on the kinetics of calcium absorption is not clear. It is possible that the 28-day experiments were not long enough for calcium transport to occur, given the need to adapt to a calcium-free medium. However, a 28-

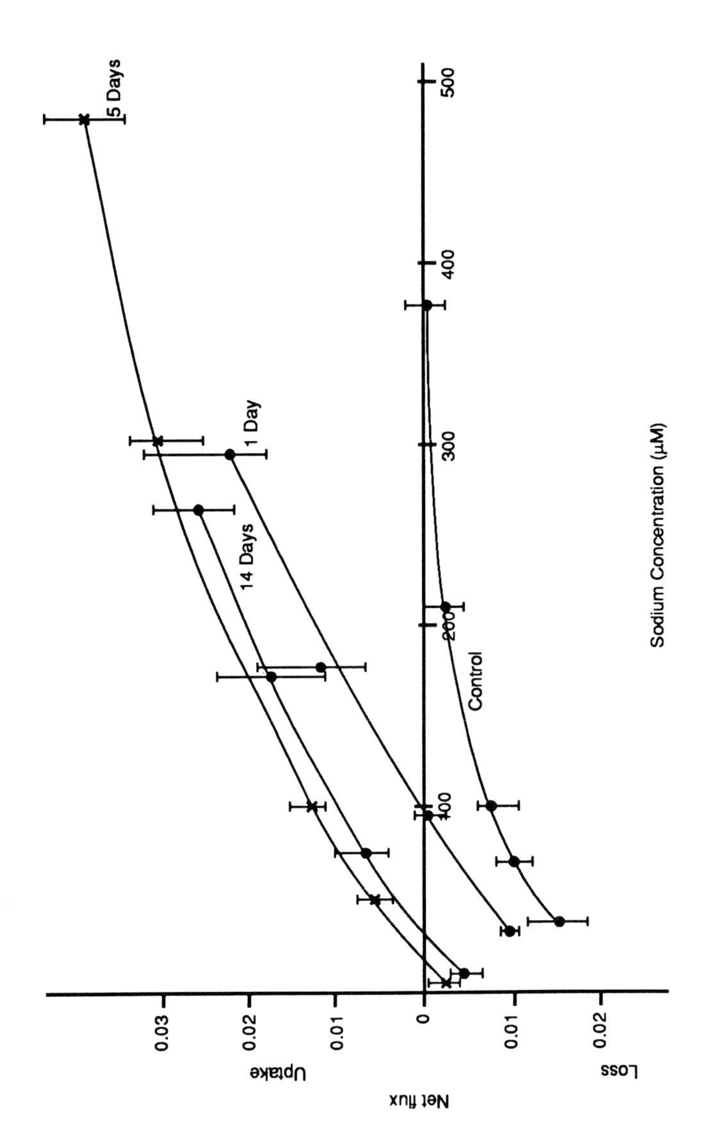

Figure 2. The effect of acclimation to low-salt water on sodium metabolism in *Dreissena*. Uptake and loss given as μM/g/hr.

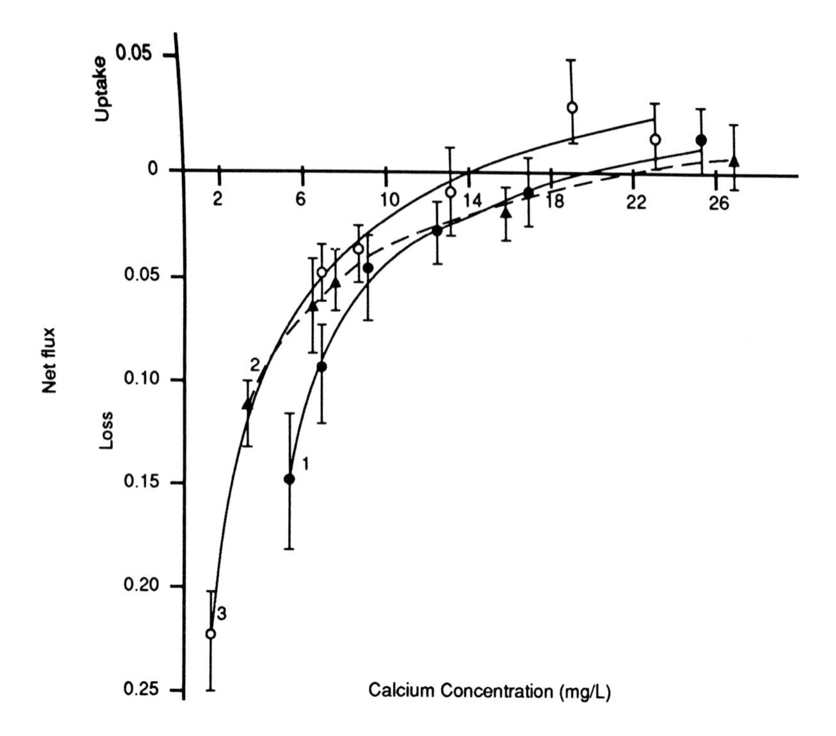

Figure 3. The effect of acclimation to low-salt water on calcium metabolism in *Dreissena*. Uptake and loss given as μM/g/hr. 1 = Individuals in artesian water, 2 = same individuals after 28 days of acclimation in diluted artesian water (Ca = 0.8–1.5 mg/L), 3 = individuals in water from Rybinsk reservoir.

day exposure was sufficient to detect the boundaries of phenotypical variability (Khlebovich, 1981) and should be long enough to detect calcium transport. The calcium requirements of *Dreissena* are obvious when examining distributions in the field. This species is absent from Lake Ladoga in which the calcium content at depths of 0–10 m is 7–12 mg/L, but is found in adjacent lakes having higher calcium levels. *Dreissena* does not occur in the low-salt lakes and rivers of Karelia, while it occurs at the same latitudes in the North Dvina River with calcium levels of about 40 mg/L.

pH

pH values in fresh waters vary to a greater extent than in the shore zone of salt and brackish waters; thus, freshwater animals tolerate a wider range of pH values than marine animals do. It was of special interest then to study reaction of *Dreissena* to decreases in pH since this could determine the potential of the species to settle in continental waters with various degrees of

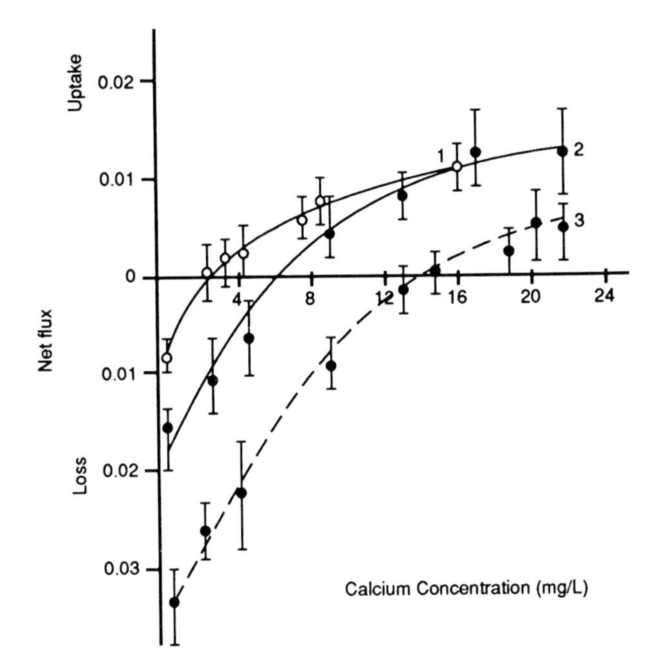

Figure 4. Calcium metabolism in three species of bivalve mollusks from the Volga at different calcium concentrations in water. 1 = *Anodonta piscinalis*, 2 = *Unio pictorum*, 3 = *Dreissena polymorpha*. (Redrawn from Vinogradou, et al., 1987)

fluctuations in pH levels. Several investigations have shown that in the majority of animals, the system of ionic homeostasis is very sensitive to shifts in the pH value (Vinogradov, 1979 and 1986).

Studies of the relation between ion exchange and pH in *Dreissena* showed that a significant decrease in the influx of sodium and calcium occurred when pH values were below 7.0 (Figures 5 and 6). The losses of sodium and calcium exceed the gains when pH values are lower than 6.8–6.9. In similar experiments with other bivalved mollusks, a decrease in the NF was also observed. In these mollusks, however, the losses of sodium and calcium exceeded their influx only at pH values below about 6.0 (Figures 5 and 6). Thus, these results indicate that the metabolism of sodium and calcium in *Dreissena* is much more sensitive to acidification than in other freshwater mollusks. In order to extrapolate the experimental data concerning the relation of *Dreissena* to acidification into nature, we attempted to acclimate *Dreissena* to soft waters that had different pH values. Two differnet experimental waters had a sodium content of 1.0–1.2 mg/L and a calcium content of 2.5 mg/L, but had pH values of 5.0 ± 0.2 and 5.7 ± 0.25, respectively. In the water with pH 5.0, some mortality of the animals was noted in 1 day, and mortality reached 75% after 4 days. Mortality also occurred after 4 days in the water having a pH of 5.7. To prevent further mortality sodium was added to the water to achieve

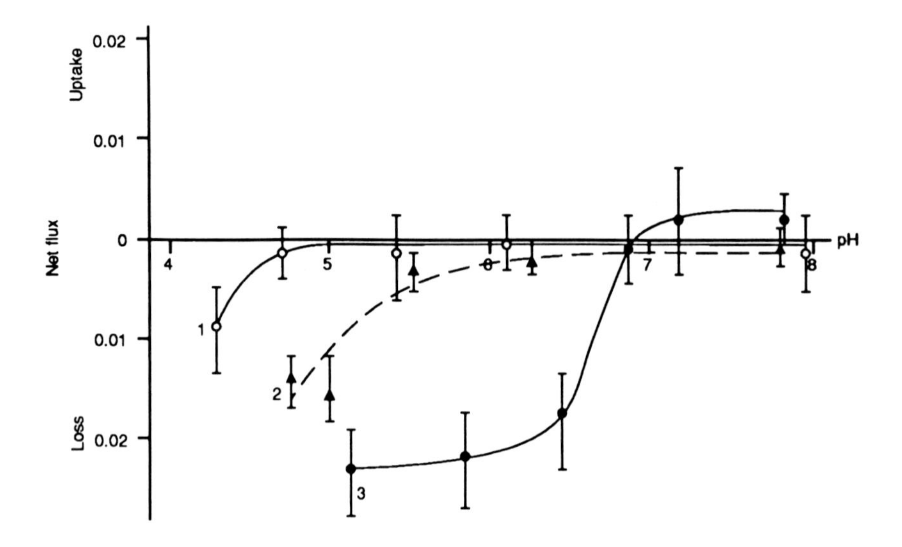

Figure 5. The effect of pH on sodium metabolism in three species of bivalve mollusks. 1 = *Anodonta piscinalis*, 2 = *Unio pictorum*, 3 = *Dreissena polymorpha*.

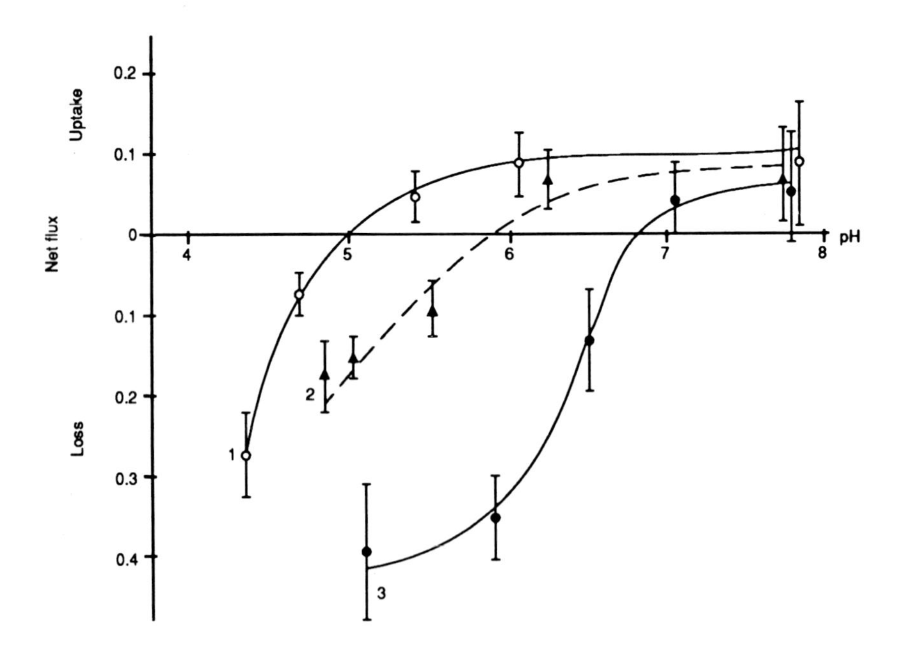

Figure 6. The effect of pH on calcium metabolism in three species of bivalve mollusks. 1 = *Anodonta piscinalis*, 2 = *Unio pictorum*, 3 = *Dreissena polymorpha*.

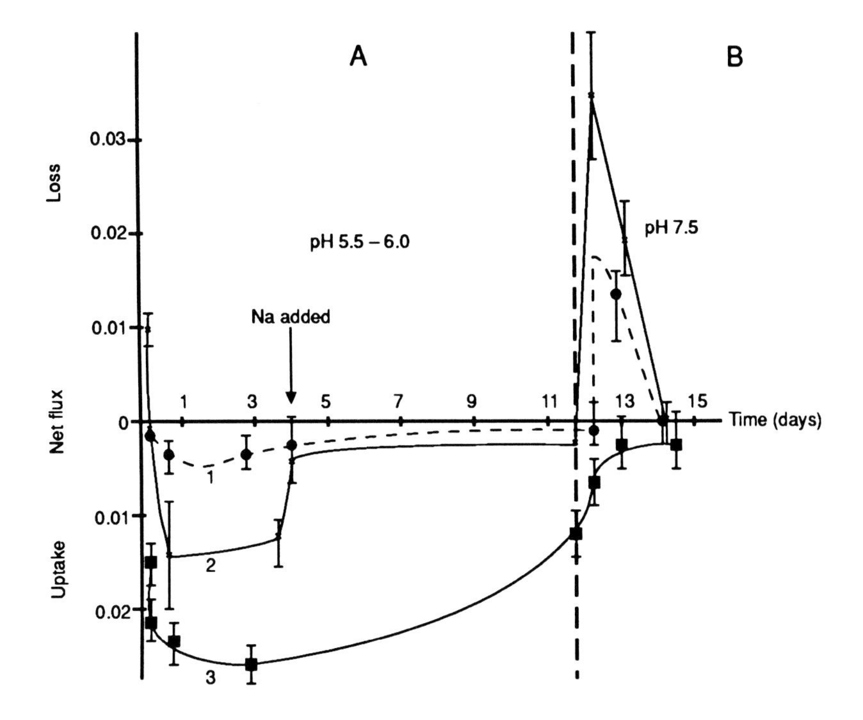

Figure 7. The effect of pH values on the metabolism of sodium, calcium, and potassium in *Dreissena polymorpha*. A = Individuals taken from artesian water (pH 7.8) and placed in low-salt water (pH 5.5–6.0), B = individuals transferred into artesian water after 12 days. Net flux of potassium (1), sodium (2), calcium (3).

a concentration of 2.3 mg/L. It should be emphasized that in the control water (Na = 1.0 mg/L, Ca = 1.6 mg/L, pH = 7.6) all the mollusks were alive during the entire experiment (14 days).

Immediately after the mollusks were placed into acidified water, basic changes in the metabolism of Na, Ca, and K occurred and the ionic balance became negative (Figure 7). The losses of salts from the organisms continued to exceed gains. After 4 days, sodium was added to the water to achieve a concentration of 2.3 mg/L. This additional sodium tended to normalize sodium metabolism, probably by decreasing sodium loss, as found for other freshwater organisms (Vinogradov, 1979 and 1986). As for calcium metabolism, disturbance of its balance was observed during the whole experiment. In the process of acclimation, the extent of calcium imbalance declined. In our view, this may be explained by the fact that calcium concentration in the water gradually increased in the course of the experiment from 2.5 to 14 mg/L due to intensive loss of calcium from the mollusks. After 12 days of acclimation to acid medium, a sharp increase in the rate of influx of salts into the organism occurred with a change in the water pH from 5.5 to 7.5. The net influx of sodium increased many-fold; then after 2 days an equilibrium was established between loss and influx. Similar results were obtained for potassium. Calcium metabolism also quickly recovered with an increase in the pH value; however, net in flux of calcium did not occur since it likely requires a higher calcium concentration.

Based on these findings, a deficit of sodium and potassium is created when *Dreissena* is exposed to an acid medium, which then probably brings about some adaptive compensatory alterations in the ion exchange. These adaptive reactions are clearly displayed when an organism is no longer exposed to low pH and the negative imbalance of the ion exchange is replaced by an influx of salts into the organism. In waters with low pH values, *Dreissena* displayed a high sensitivity and low efficiency of mineral metabolism which may be a factor limiting its distribution, especially in low-salt waters.

REFERENCES*

Halley, G. P., and M. A. Gibson. "Calcium Storage in the Soft Tissues of Freshwater Gastropods: The Influence of Environmental Calcium Concentrations," *Can. J. Zool.*, 49:1002–1004 (1971).

Karpevich, A. F. "Some Data on Formation in the Bivalved Molluscs," *Zool. Zh.* 34:46–67 (1955) (In Russian).

Khlebovich, V. V. "Acclimation of thc Animal Organism," *L. Nauk.* (1981) p. 135.

Lavrova, E. A., L. Ya. Shterman, and YuV. Natochin. "Accumulation of Ions and Their Loss by Rainbow Trout in the Waters with Various Salt Content," *Vopr. Ikhtiol.* 19:148–154 (1979).

Nduru, W. R., and A. D. Harrison. "Calcium as a Limiting Factor in the Biology of Biomphalaria Pfeifferi (Krauss). (Gastropoda:Planorbidae)," *Hydrobiologia* 49:143–170 (1976).

Okland, J. "Acidification in 50 Norwegian Lakes," *Nord. Hydrol.* 11:25–32 (1980).

Pryadko, V. P. "Tissue Characteristics of Calcium Exchange" in *Anodonta cygnea* L. *Dokl. Akad. Nauk. USSR, Ser. Biol.* 9:832–836 (1976).

Roff, H., and R. Kwiatkowski. "Zooplankton and Zoobenthos Communities of Selected Northern Ontario Lakes of Different Acidities," *Can. J. Zool.* 55:899–911 (1977).

Romanenko, V. D., O. M. Arsan, and V. G. Solomatina. "Calcium and Phosphorus in Living Activity of Hydrobionts," *Kiev, Naukova Dumka* (1982) p. 152.

Smirnova, N. F. "Reaction of *Drcissena* to Salinity and Phcnol," *Biologia Vnutr. Vod.* 18:37–39 (1973a).

Smirnova, N. F. "Influence of Some Factors on the Freshwater Bivalved Molluscs," *Tr. IBVV Akad. Nauk. SSSR.* 24(27):90–99 (1973b).

* Note: Citations are presented as provided by the authors. For further information contact: Dr. G. A. Vinogradov, Institute Biology of Inland Waters, Nekovzskiy, Raion, Yaroslavskaya Oblast, Borok, Russia.

Vinogradov, G. A. ''Adaptation of Aquatic Animals with Various Types of Osmoregulation to Low pH in Ambient Medium, in Physiology and Parasitology of Freshwater Animals,'' *L. Nauka* (1979) pp. 17–26.

Vinogradov, G. A. ''Functional Basis for the Effect of Low pH on Fish and Invertebrates,'' in Problems of Aquatic Toxicology and Water Quality Management, Proceedings of the USA-USSR Symposium, EPA-600/9-86/024, Athens, GA (1986) 1–18.

Vinogradov, G. A., A. K. Klerman, and V. T. Komov. ''Peculiarities of Ion Exchange in the Freshwater Molluscs at High Hydrogen Ion Concentrations and Low Salt Content in the Water,'' *Ekologiya* 3:81–84 (1987).

Williams, N. V. ''Studies on Aquatic Pulmonare Snails in Central Africa. I. Field Distribution in Relation to Water Chemistry,'' *Malacologia* 11:153–164 (1970).

CHAPTER 18

Metabolism of the Zebra Mussel *(Dreissena polymorpha)* in Lake St. Clair of the Great Lakes

Michael A. Quigley, Wayne S. Gardner, and Wendy M. Gordon

Laboratory measurements of oxygen consumption and nitrogen (ammonia) excretion rates of zebra mussels, *Dreissena polymorpha*, were obtained from mussels collected at a 5-m site in southwest Lake St. Clair, April to November 1990. Metabolic activity varied widely between spring and fall. Oxygen to nitrogen ratios approached a value of 50 in spring, declined to a minimum of 16 in summer, and increased moderately in fall. The low O:N ratios observed in summer suggest that higher metabolic rates associated with high temperatures and lower supplies of nutritionally rich food may have a negative impact on the condition of Lake St. Clair mussel populations during summer months. Over this interval, mussel populations may be more vulnerable to eradication measures than in spring and fall, and future control strategies might prove more effective if treatment is timed to coincide with such periods of heightened vulnerability.

INTRODUCTION

Although considerable information exists on the biology of the zebra mussel, *Dreissena polymorpha*, the underlying metabolism of the organism is poorly understood. In general, previous studies of European populations have examined changes in metabolic rate (e.g., oxygen consumption rate) relative to ambient temperature and individual size (e.g., dry weight) (Woynarovich, 1961; Dorgelo and Smeenk, 1988; Hamburger and Dall, 1990). While these studies have provided useful data for further description and modeling of zebra mussel energetics, they reveal little about either the chemical identity of metabolic substrates used by the mussel or the implications of seasonal availability of such materials.

The primary objective of our study was to simultaneously measure both oxygen consumption and nitrogen (ammonia) excretion rates of zebra mussels during the ice-free season (April-November). Such data allow the calculation of a corresponding atomic O:N ratio defined as atoms of oxygen consumed per atoms of nitrogen excreted. This ratio has been widely used to examine the metabolism of crustacean zooplankton since it integrates the status of the free amino acid pool relative to tissue growth (anabolism) or tissue degradation (catabolism) of individual organisms (Mayzaud and Conover, 1988). The amount of free amino acids is dependent on an animal's nutritional condition and trophic history, and also on the quantity and quality of available food. In general, animals catabolizing protein exclusively exhibit O:N ratios between 3 and 16. O:N ratios between 50 and 60 indicate use of approximately equal amounts of protein, and either lipids or carbohydrates; and an O:N ratio greater than 60 reflects a primary reliance on lipids or carbohydrates.

At present, no information exists on O:N ratios in zebra mussels and little data are available on molluscan O:N ratios in general. The most comparable results to date have been derived from work with the marine blue mussel (*Mytilus edulis* L.) of the North and Baltic Seas (Bayne et al., 1985). In that study, O:N ratios less than 30 indicated stress in blue mussels, while animals having O:N ratios greater than 50 were considered healthy. In another study, low O:N ratios were associated with stress induced by decreased salinity that resulted in a permanently increased metabolism of free amino acids, and a correspondingly sustained energy loss (Tedengren and Kautsky, 1986). Moreover, this continuous energy loss was manifested in lowered growth rate and smaller size of *M. edulis* in lower saline Baltic waters as compared to North Sea environments. In the Baltic, where blue mussel populations live close to their physiological tolerance limits, exposure to additional stress (diesel oil exposure) slowed metabolic activity and altered catabolic substrate utilization, thus further lowering O:N ratios (Tedengren and Kautsky, 1987).

In addition to measuring seasonal variation in O:N ratios, we examined the effects of animal size, temperature, and oxygen concentration on subsequent respiration rates of zebra mussels. Knowing the effects of these factors on metabolism is useful for energetic modeling and for identifying potential periods of time when mussels are in a lowered state of health and, thus, more vulnerable to control measures.

METHODS

Oxygen consumption and nitrogen excretion rates were determined in adult *Dreissena polymorpha* collected from southwest Lake St. Clair (42° 20′ 00″ N 82° 47′ 30″ W) on a monthly basis from April to November 1990. Animals were held in lake water and in darkness at ambient collection temperatures prior to experiments. Oxygen consumption and nitrogen excretion rates were simultaneously determined on each of 10 medium-sized (i.e., 10–15 mm) individuals. Each animal was placed in a 60-mL BOD bottle filled with filtered (1-μm glass filter) well-aerated lake water. Each bottle was then stoppered and incubated for 5 hr in darkness at ambient collection temperature. In general, oxygen consumption and nitrogen excretion rates of each of the 10 animals were measured per experiment. Oxygen consumption rate was defined as the net decline in dissolved oxygen concentration occurring over the 5-hr incubation interval. Four bottles containing only lake water served as controls. Dissolved oxygen concentration was determined by Winkler titration (Grasshoff, 1983). For measurement of ammonia-nitrogen, 1-mL aliquots were withdrawn from BOD bottles immediately following animal introduction and after the 5-hr incubation period. Concentrations were determined using the microfluorometric method of Gardner (1978). Dry weights of animals in the experiments were determined by drying the shell-free tissue at 60°C for 24 hr.

To determine the relationship between oxygen consumption and animal size, two 24- to 25-mm mussels, four 15- to 16-mm mussels, and ten 5- to 6-mm mussels were placed in 300-mL BOD bottles filled with filtered, aerated lakewater. Four replicates were run for each size class along with three initial and three final animal-free controls. All bottles were incubated in darkness at 10°C for a 5-hr period. Animals were collected in November 1990 and held at 10°C prior to experiments. Net decline in oxygen concentration was monitored using a YSI 5420 self-stirring oxygen probe coupled to a chart recorder. These methods were modeled after those of Dorgelo and Smeenk (1988).

To determine the relationship between oxygen consumption and temperature, the same experiments described above were repeated at 20°C. The animals were acclimated by gradually increasing ambient temperature from 10 to 20°C over a 7-day period prior to initiation of experiments.

Because closed-vessel respirometry was used to estimate oxygen consumption, a progressive decline of available dissolved oxygen occurring over the course of the experiments may have affected mussel oxygen consumption.

To test for such an effect, we measured oxygen consumption rates of groups of mussels exposed to different initial ambient oxygen concentrations ranging nominally from 100 to 25% saturation at a test temperature of 11.3°C.

Adult zebra mussels collected in November 1990 from the vicinity of the Detroit Edison power plant on Lake Erie, Monroe, MI were held for 48 hr in the laboratory darkness and at 11°C prior to experiments. Mussels (15–16 mm in length) were placed individually in 60-mL BOD bottles filled with filtered lake water. Oxygen concentrations were adjusted from 100% saturation by appropriate dilution of highly aerated water with water that had been stripped of oxygen via vigorous bubbling with nitrogen. Oxygen consumption was determined for five individual animals at nominal saturation levels of 100, 75, 50, and 25%. Background changes in oxygen concentration were determined in four initial and four final animal-free controls, and these results were used to correct oxygen consumption rates. Oxygen concentration was measured with the previously described YSI oxygen probe/chart recorder which was calibrated against Winkler titrations at the beginning and end of the experiment.

RESULTS

Seasonal Changes

Seasonal changes in mean oxygen consumption, nitrogen excretion, and O:N ratios are given in Table 1. Highest oxygen consumption rates occurred in spring and early summer, and appeared to be closely related to ambient temperatures. Nitrogen excretion rates were also closely associated with temperature and were relatively low in April, October, and November; and high during the warmer months of May, June, and August. Mean O:N ratios were highest in early spring, declined through summer, and subsequently increased in fall.

Effects of Mussel Size and Temperature on Oxygen Consumption

Results of oxygen consumption measurements on three different zebra mussel size classes, at two temperatures, are listed in Table 2. At 10°C, oxygen consumption rates declined with an increase in size; at 20°C, oxygen consumption of the smallest (5–6 mm) animals was highest, while consumption rates of the two larger (15–16 and 24–25 mm) size classes were similar. There was a significant correlation between dry weight (log) and respiration rate ($p \leq 0.05$) (Zar, 1984). Comparison of regression lines for 10 and 20°C revealed that slopes of the lines (Figure 1) were not significantly different ($p \leq 0.05$).

Table 1. Mean (±S.E.) Oxygen Consumption, Nitrogen Excretion, and O:N Ratios of *Dreissena polymorpha* Collected from Lake St. Clair on 1990 Sampling Dates

Date	Temperature (°C)	Oxygen Consumption[a] ($mgO_2/gDW/d$)	Nitrogen Excretion[a] (ng-atomN/mgDW/hr)	O:N
April 18	7.4	18.6 ± 1.1 (10)	1.1 ± 0.1 (10)	48.3 ± 5.6
May 8	19.0	45.2 ± 4.4 (10)	5.3 ± 0.4 (10)	22.2 ± 1.6
June 22	18.0	32.3 ± 1.4 (10)	4.4 ± 0.4 (10)	20.1 ± 1.3
July 26	21.0	41.0 ± 3.9 (10)	—	—
August 24	21.0	27.4 ± 1.7 (10)	4.8 ± 0.5 (10)	16.0 ± 1.6
October 12	15.5	25.7 ± 0.9 (10)	2.3 ± 0.2 (9)	30.5 ± 3.6
November 15	5.7	11.8 ± 1.4 (9)	1.2 ± 0.1 (10)	29.7 ± 2.3

[a] Number of replicates given in parentheses.

Table 2. Mean (±S.E.) Dry Weight and Oxygen Consumption of Different Size Classes of *Dreissena polymorpha* at Two Different Temperatures

Size Class (mm)	Temperatures (°C)	Dry Weight[a] (mg)	Oxygen Consumption[a] (mgO^2/gDW/d)
5–6	10	0.91 ± 0.11 (39)	87.7 ± 21.0 (4)
	20	1.19 ± 0.14 (40)	82.8 ± 20.2 (4)
15–16	10	10.73 ± 0.80 (16)	27.2 ± 5.4 (4)
	20	13.75 ± 1.03 (16)	32.1 ± 2.1 (4)
24–25	10	28.77 ± 4.12 (8)	15.2 ± 2.4 (4)
	20	17.62 ± 2.90 (8)	35.7 ± 5.1 (4)

[a] Number of replicates in parentheses.

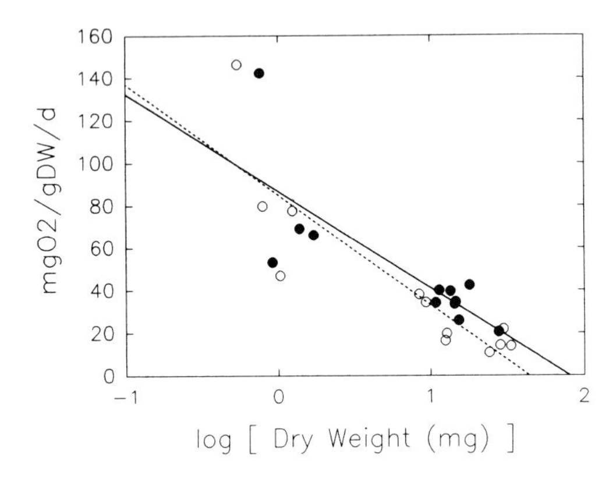

Figure 1. Relationship between dry weight (log) and oxygen consumption in *Dreissena polymorpha*. Open dots and dashed line denote 10°C data points and regression [R = 85.08 − (51.90 log(DW))], respectively; solid dots and solid line denote 20°C data points and regression [R = 83.97 − (37.40 log(DW))], respectively.

Effects of Ambient Oxygen Concentration on Oxygen Consumption

Overall, when oxygen concentrations were decreased in the BOD bottles, corresponding oxygen consumption rates also declined. Although the actual range of percent oxygen saturation was narrower (42–89%) than the 25 100% range we sought to attain, oxygen consumption progressively declined with corresponding decreases in dissolved oxygen. Thus, it appears that *D. polymorpha* has little ability to regulate its respiration rate when encountering decreased ambient oxygen. A subsequent Michaelis-Menten/Lineweaver-Burk double-reciprocal plot (Lehninger, 1975) (Figure 2) and regression analysis indicated that oxygen consumption rate declined significantly ($p \leq 0.05$) relative to ambient oxygen concentration according to the equation:

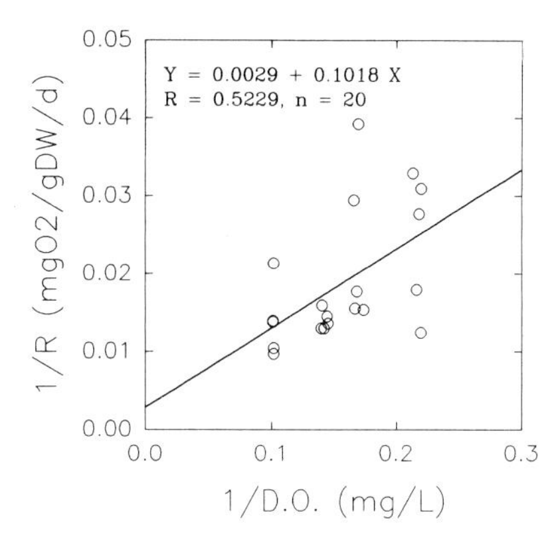

Figure 2. Michaelis-Menten double-reciprocal plot of dissolved oxygen concentration and oxygen consumption in *Dreissena polymorpha*. K_m = 4.8 mgO_2/L; V_{max} = 82.6 mgO_2/gDW/day.

$$V_{max} = 1/[0.0029 + (0.1018/K_m)]$$

where V_{max} (the predicted respiration rate at 100% oxygen saturation; 11.05 mg/L O_2 at 11°C) = 82.6 mgO_2/gDW/day and K_m (the oxygen concentration at which respiration was half that of V_{max}) = 4.8 mgO_2/L.

To examine the possibility that declines in ambient oxygen concentration in closed test vessels may have influenced respiration rates, the determined Michaelis-Menten regression equation was used to predict expected respiration rates for given mean ambient oxygen concentrations existing at the beginning and end of the 5-hr measurement intervals for each sampling date in 1990. Results (Table 3) imply that respiration rates may have declined by as much as 27.6% due to declining ambient dissolved oxygen concentration in test vessels. Actual decreases in respiration should be somewhat less because the final 5-hr oxygen endpoint was used. Higher oxygen concentrations existing prior to the end of the determination interval would presumably have exerted a lower effect on mussel respiration rates.

DISCUSSION

Oxygen Consumption, Nitrogen Excretion, and O:N Ratios

Changes in oxygen consumption, nitrogen excretion, and O:N ratios indicate that conditions in early spring and fall were more favorable to the physiological health of *D. polymorpha* than summer conditions. In April, O:N ratios were near 50, and this value has been identified as the lower limit

Table 3. Oxygen Consumption Rates of *Dreissena polymorpha* Estimated from a Michaelis-Menten Double-Reciprocal Equation for Initial and Final Oxygen Concentrations in Test Vessels on 1990 Sampling Dates

Date	Initial O_2 (mg/L)	Final O_2 (mg/L)	Predicted M-M Oxygen Consumption (mgO_2/gDW/d)		Percent Decline[a]
			Initial	Final	
April 18	12.7	10.9	91.4	81.8	10.5
May 8	8.3	5.6	65.8	47.6	27.7
June 22	8.6	6.6	68.1	54.5	20.0
July 26	8.6	6.5	67.7	54.1	20.2
August 24	10.1	7.8	76.9	62.9	18.2
October 12	9.6	8.3	74.2	65.8	11.3
November 15	11.6	10.8	85.5	81.0	5.3

[a] The percent decline in oxygen consumption predicted as mussels deplete oxygen over the course of each 5-hr experiment.

for physiologically healthy blue mussels (*Mytilus edulis*) (Bayne et al., 1985). Moreover, O:N ratios between 50 and 60 generally indicate that animals are catabolizing roughly equal amounts of protein, and carbohydrates or lipids. Thus, it appears that in spring, mussels are able to rely on nutritionally rich food sources. From April to May, however, O:N ratios decline markedly, indicating greater reliance on protein and lower use of carbohydrate or lipids to support metabolic needs. Over the April-May interval (Table 1), oxygen consumption increased by a factor of 2.4, while nitrogen excretion increased by a factor of 4.8. Increases in both rates may be associated with the rapid temperature increase between April and May. Temperature effects can also be assessed by calculating a Q_{10} (change in reaction velocity accompanying a 10°C temperature rise; Prosser and Brown, 1962). Assuming a linear increase in oxygen consumption and nitrogen excretion rates when the temperature increased from 7.4 (April) to 19.0°C (May), the Q_{10} for oxygen consumption would be 2.3 while the Q_{10} for nitrogen excretion would be 4.3. Most thermochemical (enzymatic) reactions generally range between 2 and 3 (Prosser and Brown, 1962), so the Q_{10} for oxygen consumption would fall within this expected interval. The nitrogen excretion Q_{10} is above this range, implying the presence of an additional source of excretory nitrogen other than that associated with the expected increased excretion corresponding to temperature-related increases in metabolism. This additional output of excretory nitrogen is probably derived either from catabolism of nutritional protein sources, or from catabolism of tissue protein during times of low food availability (i.e., starvation) (Mayzaud and Conover, 1988).

After the April-May period, O:N ratios declined to a minimum value of 16 in August, a value indicating that mainly protein is being catabolized (Mayzaud and Conover, 1988). This low value would thus imply that zebra mussels are either relying on protein as a food source, or catabolizing internal protein under starvation conditions. As such, this period represents a time when zebra mussels are surviving at their physiological limits, and little or no high-quality food may be available. Seasonally low chlorophyll levels in the water column of Lake St. Clair (Nalepa, unpublished data) imply that little food may be available during this time. In addition, the reproductive activity expected during this summer period would impose further energetic demands on mussel metabolism and nutrition (Stanczykowska, Personal communication). This may also explain the fact that O:N ratios increased somewhat after August, but did not return to higher ratios found in the spring which typified physiologically healthier animals.

Effect of Animal Size and Temperature on Oxygen Consumption

Temperature studies did not show any significant differences in respiration rates at 10 and 20°C for animals of comparable dry weight in two of the three size classes examined. These results were surprising since metabolic rate of poikilotherms commonly doubles with an accompanying 10°C rise in temperature (Prosser and Brown, 1962). In the two smaller size classes of mussels, respiration rate changed little between 10 and 20°C (Table 2). A roughly twofold increase in metabolic rate occurred in the largest size class. Since mussels were not fed over the 1-week period when temperature was increased from 10 to 20°C, metabolic rates may have been depressed, thus offsetting metabolic increases expected with increased temperature. If food deprivation did depress metabolic rates, the effect might be greatest in smaller animals due to their proportionally higher metabolic rates and lower reserve energy stores (i.e., lipids). Woynarovich (1961) noted that oxygen consumption in *D. polymorpha* was significantly influenced by the mussel's general condition, fat content, and length of period without food. Seasonal changes noted in this study suggest that temperature, and/or the availability and nutritional value of food strongly influence metabolic rates. Underlying reasons for such different responses are not yet apparent.

The decline in oxygen consumption relative to a decrease in ambient oxygen availability demonstrates the inability of zebra mussels to maintain a constant respiration rate when exposed to relatively low oxygen levels. Inability to adjust metabolically to declining oxygen levels may subsequently affect growth and production. Stanczykowska (1977) noted that *D. polymorpha* requires clean well-oxygenated water, and that growth may sometimes be limited by insufficient amounts of oxygen. These observations suggest that the low oxygen levels may subsequently affect production.

Additionally, the close relationship between ambient dissolved oxygen concentrations and respiration rates suggests that ambient oxygen concentrations during closed-vessel respiration determinations must be considered to obtain valid respiration rates. Because mussel respiration rates are highly influenced by ambient dissolved oxygen concentrations in the water, the

Table 4. Oxygen Consumption Rates of *Dreissena polymorpha* as Determined in Various Studies

Location	Temperature (°C)	Dry Weight (mg)	Oxygen Consumption ($mgO_2/gDW/d$)	Ref.
Lake St. Clair, North America	10	0.9	87.7	Present study
		28.8	15.2	
Lake Maarsseveen, The Netherlands	10	1.3	86.1	Dorgelo and Smeenk (1988)
		27.1	24.7	
Lake St. Clair, North America	20	1.2	82.8	Present study
		13.8	32.1	
Lake Maarsseveen, The Netherlands	20	1.0	106.3	Dorgelo and Smeenk (1988)
		14.4	40.6	
Lake Esrom, Denmark	20	1.5	21.2	Hamburger and Dall (1990)

Note: Comparisons limited to those studies with similar temperatures and mussel dry weights.

decline in available oxygen in closed test vessels occurring over the interval of the respiration determination may bias respiration estimates. Respiration rates thus obtained in the laboratory may underestimate rates found under in situ conditions. In extreme cases, this underestimate could be as high as 28%, assuming that individuals inhabit lake areas where oxygen concentrations are at near-saturation levels.

Comparisons to Previous Estimates of Respiration

Because oxygen consumption was closely related to individual dry weight, comparisons to other studies of respiration rates in *D. polymorpha* were confined to those studies where dry weight and temperature were similar to ours (Table 4). Respiration rates in this study were generally similar to rates reported by Dorgelo and Smeenk (1988) for mussels from Lake Maarsseveen, The Netherlands. However, our rates were much higher than rates reported by Hamburger and Dall (1990) for mussels from Lake Esrom, Denmark; rates at 10°C for similar-sized mussels (1.2–1.5 mg) were 82.2 and 21.2 mgO_2/gDW/day for the two studies, respectively. These different respiration rates may be related to differences in the experimental design. Hamburger and Dall incubated mussels for 20 hr and thus realized a 25% decline in oxygen levels during the experimental period. In contrast, oxygen concentration declined by a mean of 19% over the 5-hr incubation period in this study. Moreover, the methodology of the Lake Esrom study was different from that used in this study; test vessels were gently shaken throughout the 20-hr incubation period in the Lake Esrom study, while the vessels were left undisturbed in this study.

The determined regression equations relating dry weight and oxygen consumption were used to further compare respiration rates. Actual respiration rates from previous studies, all determined on European populations, were

Table 5. Comparison of Actual and Predicted Oxygen Consumption Rates of *Dreissena polymorpha* from Various Studies

Location	Dry Weight (mg)	Temperature (°C)	Oxygen Consumption (mgO_2/gDW/d) Measured Rate	Predicted Rate	Ref.
Lake Balaton,	13.6	10	13.7	26.3	Woynarovich
Hungary		20	31.7	41.6	(1961)
Lake Maarsseveen,	1.3	10	86.1	79.9	Dorgelo and
The Netherlands	15.0		15.8	24.1	Smeenk
	27.1		24.7	10.7	(1988)
Lake Esrom,	4.2	11	15.8	52.7	Hambruger
Denmark	1.5	20	21.2	77.4	and Dall (1990)

Note: Predicted values were calculated from dry weight vs oxygen consumption regression equations determined in this study.

generally less than predicted rates in most cases (Table 5). Thus, respiration rates of mussels from Lake St. Clair (as determined in the present study) were similar to, or slightly higher than, rates observed in European studies for comparable dry weights and temperatures.

REFERENCES

Bayne, B. L., D. A. Brown, K. Burns, D. R. Dixon, A. Ivanovici, D. R. Livingstone, D. M. Lowe, M. N. Moore, A. R. D. Stebbing, and J. Widdows. *The Effects of Stress and Pollution on Marine Animals* (New York: Praeger 1985).

Dorgelo, J., and J. W. Smeenk. ''Contribution to the Ecophysiology of *Dreissena polymorpha* (Pallas) (Mollusca:Bivalvia): Growth, Filtration Rate and Respiration,'' *Verh. Int. Verein. Theor. Angew. Limnol.* 23:2202–2208 (1988).

Gardner, W. S. ''Microfluorometric Method to Measure Ammonium in Natural Waters,'' *Limnol. Oceanogr.* 23:1069–1072 (1978).

Grasshoff, K. ''Determination of Oxygen,'' in *Methods of Seawater Analysis* K. M. Grasshof, M. Ehrhardt, and K. Kremling, Eds. (Weinheim, Germany: Verlag Chemie [now VCH Publishers, Inc.] 1983), pp. 61–72.

Hamburger, K., and P. C. Dall. ''The Respiration of Common Benthic Invertebrate Species from the Shallow Littoral Zone of Lake Esrom, Denmark,'' *Hydrobiologia* 199:117–130 (1990).

Lehninger, A. L. *Biochemistry*. (New York: Worth 1975).

Mayzaud, P., and R. J. Conover. O:N Atomic Ratio as a Tool to Describe Zooplankton Metabolism,'' *Mar. Ecol. Prog. Ser.* 45:289–302 (1988).

Nalepa, T. F. Unpublished data (1990).

Prosser, C. L., and F. A. Brown. *Comparative Animal Physiology* (Philadelphia, PA: W. B. Saunders, 1962).

Stanczykowska, A. ''Ecology of *Dreissena polymorpha* (Pall.) (Bivalvia) in Lakes,'' *Pol. Arch. Hydrobiol.* 24:461–530 (1977).

Stanczykowska, A. Personal communication (1990).

Tedengren, M., and N. Kautsky. ''Comparative Stress Response to Diesel Oil and Salinity Changes of the Blue Mussel, *Mytilus edulis* from the Baltic and North Seas,'' *Ophelia* 28:1–9 (1987).

Tedengren, M., and N. Kautsky. ''Comparison Study of the Physiology and Its Probable Effect on Size in Blue Mussels (*Mytilus edulis* L.) from the North Sea and Northern Baltic Proper,'' *Ophelia* 25:147–155 (1986).

Woynarovich, E. ''The Oxygen Consumption of *Dreissena polymorpha* (Lamellibranchiata) at Different Temperatures,'' *Ann. Biol. Tihany* 28:211–216 (1961).

Zar, J. H. *Biostatistical Analysis* (Englewood Cliffs, NJ: Prentice Hall, 1984).

CHAPTER 19

Chemical Regulation of Spawning in the Zebra Mussel (*Dreissena polymorpha*)

Jeffrey L. Ram and S. Jerrine Nichols

Previous literature suggests that spawning in bivalves is chemically regulated, both by environmental chemical cues and by internal chemical mediators. In a model proposed for zebra mussels, chemicals from phytoplankton initially trigger spawning, and chemicals associated with gametes provide further stimulus for spawning. The response to environmental chemicals is internally mediated by a pathway utilizing serotonin (5-hydroxytryptamine, a neurotransmitter), which acts directly on both male and female gonads. The role of serotonin and most other aspects of the model have been tested only on bivalves other than zebra mussels. The effect of serotonin on zebra mussel spawning was tested. Serotonin (10^{-5} and 10^{-3} M) injected into ripe males induced spawning, but injection of serotonin into females did not. Gametes were not released by 10^{-6} serotonin; in most cases, serotonin injection did not release gametes from immature recipients. Serotonin injection provides a reliable means for identifying ripe male zebra mussels and for obtaining zebra mussel sperm without the need for dissection.

INTRODUCTION

Control methods for zebra mussels usually use biologically nonspecific methods directed at adults (e.g., chlorination) and often require large capital and labor inputs. Problems with these methods include lack of specificity, with consequent danger to other organisms, and the use or production of corrosive chemicals that may damage physical structures.

An alternative to controlling adult zebra mussels is to intervene in the zebra mussel life cycle at an earlier stage, such as when spawn are being produced. Zebra mussels have a high level of spawn production, with a mature female mussel producing more than 30,000 eggs per year (Stanczykowska, 1977). If a method can be developed that disrupts the normal reproductive cycle of the zebra mussel, then population densities may be reduced, thereby decreasing the frequency of application and magnitude of antifouling methods.

Previous experiments with zebra mussels and other bivalves indicate that spawning behavior in bivalves may be under the control of both internal chemicals and environmental chemical cues. The aim of our research is to determine the nature of these chemical cues for spawning in zebra mussels, with the long-term goal of using these chemicals or inhibitors of their receptors to regulate their reproductive success. This chapter first reviews previous evidence for chemical control of spawning in bivalves and then describes recent experiments in our laboratory demonstrating that serotonin (5-hydroxy-tryptamine [5-HT], a neurotransmitter) can elicit spawning in male zebra mussels.

For an organism that has external fertilization and a planktonic larval stage such as the zebra mussel, it is critical that spawning of eggs and sperm occur simultaneously and at a time when food is available for larval growth. Hence, bivalves with pelagic larvae have evolved a signaling system and response pathway for coordinating spawning with the necessary environmental conditions for fertilization and development.

Based on previous literature on bivalves (see as follows), we propose the model in Figure 1 for chemical regulation of zebra mussel spawning. Chemicals produced by algae are known to trigger bivalve spawning (Miyazaki, 1938; Smith and Strehlow, 1983; Smith, 1987; Starr et al., 1990), with males spawning at somewhat shorter latency than females. Therefore, in the model, the first signal for spawning is released by phytoplankton, indicating that appropriate food is available. Following initial activation by phytoplankton, chemical cues from sperm induce females to spawn (Galtsoff, 1938; Starr et al., 1990; Sprung, 1987 and 1989; Walz, 1978), and female chemical cues in turn reactivate the males (Galtsoff, 1940; Stephano and Gould, 1988; Sprung, 1987 and 1989; Walz, 1978), setting up a positive feedback cycle just at the time that phytoplankton for larval development is available.

In both males and females, the environmental spawning cues are detected by specific chemical sensing receptors which, by the way of the nervous

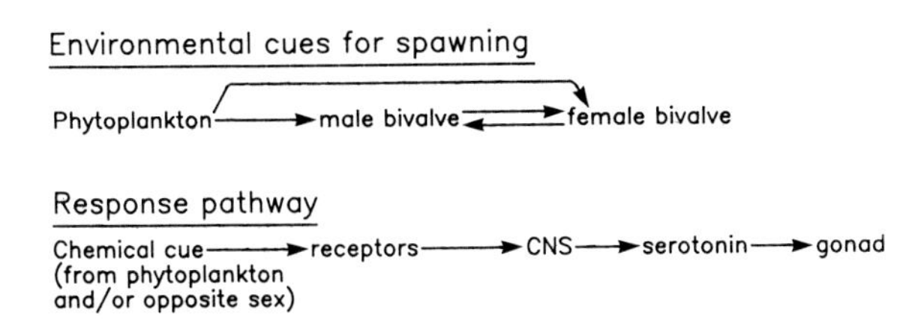

Figure 1. Model of regulation of spawning in *Dreissena polymorpha* and other bivalves.

system, activate the gonad (Smith, 1987). Activation of the gonad results in the mobilization and release of gametes. One of the chemical signals mediating neural activation of the gonads is serotonin (Matsutani and Nomura, 1982, 1986, and 1987; Alcazar et al., 1987; Crawford et al., 1986; Belda and Del Norte, 1988; Gibbons and Castagna, 1984; Braley, 1985). Theoretically, each point along this chemically mediated pathway provides an opportunity for intervention and interface with the reproductive cycle.

Although the published data upon which the model in Figure 1 is based clearly indicate that chemical factors can regulate spawning in bivalves, the only chemical that has been unambiguously identified as having a physiological role in regulating bivalve spawning is serotonin. Since previous experiments had not determined whether serotonin could induce spawning in zebra mussels, our initial experiments to determine the chemical regulators of spawning in zebra mussels began with serotonin.

METHODS

Experiments were conducted on August 21 and 30, and September 7 and 14, 1990. Zebra mussels varied in length from 15 to 25 mm. Several groups of animals were used: one group of 36 animals had been collected in May 1990 from western Lake Erie near Monroe, MI, kept at 4°C, and maintained with light and cleaning until 2 days before the experiment. These animals were transferred to aquarium water (well water) at room temperature and did not spawn during the 2 days prior to the experiment. Another group consisted of 35 animals that had been cultured from the veliger stage in laboratory aquariums at room temperature. Finally, 143 animals were collected within 2 weeks prior to experiments from western Lake Erie, also near Monroe, MI. Mussels in the collection area were still spawning naturally at the time of collection, as evidenced by young veligers in the water column.

Zebra mussels were injected with 0.1 mL of a serotonin solution (10^{-3} M, 10^{-5} M, or 10^{-6} M) or 0.1 mL vehicle with no serotonin. The vehicle for serotonin solutions was artificial freshwater mussel blood, which consisted of 475-mL deionized water, 25-mL Instant Ocean solution (made according to instructions from the manufacturer, Aquarium Systems, Mentor, OH) 40 mg $CaCl_2$-$2H_2O$, and 55 mg KCl. Ion concentrations in this artificial freshwater mussel blood are similar to that used by Motley (1934) in physiological studies concerning the regulation of heartbeat in freshwater mussels. Serotonin (5-hydroxytryptamine-hydrochloride) was obtained from Sigma Chemical. Injections were made between the valves on the ventral side of the animal in or near the hole normally used by the animal's foot.

After injection, animals were placed in individual culture tubes containing 10 mL of well water. Except as noted in results, two drops of water were taken from the bottom of each culture tube between 1 and 4 hr after injection and examined for eggs and sperm under 400 $\times$ power.

After the water was examined for eggs and sperm, squash mounts of the gonad from each animal were examined microscopically (200–400 $\times$) and the sex and gametogenic index were determined. Gametogenic index was scored according to the classification used by Haag and Garton (In press): 0, gonad spent; 1, immature; 2, initial stages of gametogenesis (intermediate immature, first stage at which sex can be identified); 3, late immature (partially ripe) with <50% mature gametes; 4, ripe with >50% mature gametes. The sex and gametogenic index of recipients was, thus, determined only at the end of the experiment; and therefore the number in each experimental and control group varied by chance. Overall, we had 114 female, 72 male, and 28 immature (gametogenic index of 1) zebra mussels.

RESULTS

Serotonin induced spawning in ripe male zebra mussels, but not in females. Figure 2 illustrates the data from all experiments on the response of males to serotonin. Data for 10^{-3} and 10^{-5} M serotonin have been combined since these concentrations gave similar results in all experimental trials. Serotonin induced spawning in 22/23 ripe (Stage 4) males, producing numerous motile sperm, whereas 0 out of 10 ripe males responded to the artificial blood vehicle (difference significant, $p < 0.001$, Fisher exact test). Although we usually waited at least an hour before examining water for spawn, in a few cases spawning was so rapid and copious that the water became cloudy before an hour had passed. In such cases, we examined the water earlier and verified male spawning with latencies as short as 35 min. In partially ripe (Stage 3) males, serotonin induced spawning in 2/8 males vs 0/3 control males (difference not significant). One out of six intermediate immature (Stage 2) males responded to serotonin vs 0/7 controls (not significantly different). In contrast to the high proportion (22/23) of spawning in ripe males with 10^{-3} and 10^{-5} M serotonin, 10^{-6} M serotonin induced spawning in only one out of three ripe males (difference significant, $p < 0.03$, Fisher exact test).

Among female serotonin recipients, small numbers of eggs (usually <10) were found in water surrounding both control and serotonin-injected animals

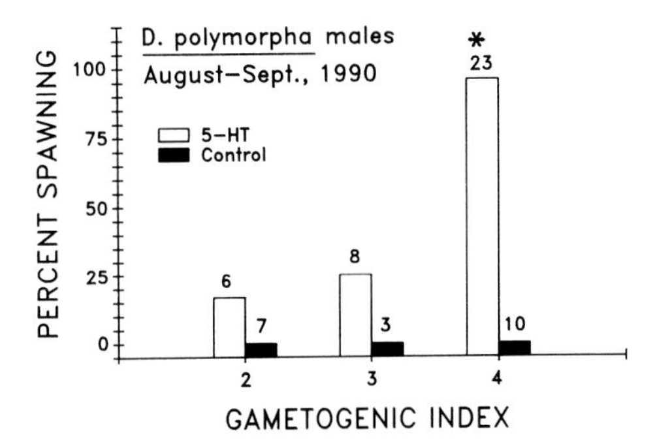

Figure 2. The effect of serotonin (5-HT) on the spawning of male *Dreissena polymorpha*. Methods are described in the text. The numbers above each bar represent the total number of animals having the indicated gametogenic index that were injected with 5-HT (either 10^{-3} or 10^{-5} M) or control solutions. * = The number of animals spawning in response to 5-HT is significantly different from the number spawning in response to control injections ($p < 0.001$, comparing animals with the same gametogenic index, Fisher's exact test; for all nonsignificant comparisons, $p > 0.05$). Data shown are the combined results from experiments conducted on August 21 and 30, and September 7 and 14, 1990.

(e.g., in ripe females, 6/16 experimentals, and 3/7 controls), suggesting perhaps only a mechanical effect of the needle. The quantity of eggs released was always much less than observed with natural zebra mussel spawning.

DISCUSSION

The data reported here are the first demonstration that serotonin can stimulate spawning in zebra mussels, even though only in males under the conditions tested in this study. This is also the first study in which gonadal maturity has been clearly demonstrated to be necessary for serotonin to stimulate spawning in a bivalve (Figure 2). Serotonin injection provides a reliable means of identifying ripe males without the need for dissection. Therefore, this method is useful for obtaining zebra mussel sperm as needed for other experimental studies. For example, studies on control methods that focus on fertilization and development of zebra mussels or sperm motility will be facilitated.

In previous studies of other bivalves (e.g., scallops), both males and females spawned in response to serotonin (Matsutani and Nomura, 1982). The lack of response of female zebra mussels to serotonin in the present study indicates that other factors may be involved in controlling egg release.

In addition to the internal chemical signals such as serotonin, Figure 1 shows that external chemical cues can also control spawning. Which portion of this model would be best for finding a species-specific control method?

Although (as pointed out before) serotonin may be useful in obtaining sperm as needed, serotonin is unlikely to provide a species-specific control method. This is because serotonin is an internal chemical transmitter common to many organisms, including humans. In contrast, the external chemical cues derived from bivalve sperm induce female spawning in a species-specific manner (Galtsoff, 1938). Therefore, a species-specific chemical method for controlling zebra mussel reproduction is more likely to be developed from investigations of the external chemical cues.

How might a species-specific chemical cue for spawning be used in a control strategy for zebra mussels? One method would be to trigger spawning prematurely with phytoplankton or gamete chemical cues. A major spawning episode may normally be triggered by chemicals released by phytoplankton as their density increases after the midsummer clear water phase. If spawning could be artificially triggered earlier (during the clear water phase), then there would be little food available for veligers, and the cohort would die before settling. Sprung (1989) suggests that several natural spawnings in European lakes met just such a fate. Interestingly, if spawning can be activated by local chemical application, positive feedback might trigger a chain reaction between male and female zebra mussels that would spread some distance beyond the site of local chemical application. Alternatively, if inhibitors of the chemical spawning triggers can be developed, another method would be to apply inhibitors in coastal waters at the time that zebra mussels would ordinarily be spawning.

One objection to this scenario is that some investigators have suggested that temperature is the critical trigger for spawning in zebra mussels; however, this is based solely on field observations and not on laboratory tests. While it is clear that below a certain temperature spawning cannot occur, temperature may have mainly a permissive function. That is, it provides conditions for the zebra mussel to spawn but does not directly trigger it. As pointed out by Starr et al. (1990), discussing experiments on *Mytilus*, "Because temperature often varies erratically . . . , the [phytoplankton] bloom may be a more reliable signal of impending favorable [conditions] . . . Phytoplankton as a spawning cue . . . integrates various environmental parameters indicating favorable conditions for larval success." In some bivalve species in which laboratory experiments have demonstrated spawning induced by temperature, the effect is strongly dependent on the amount and rate of heating and the previous history of the animal, and it usually triggers spawning in less than half of the specimens. For example, in oysters, Galtsoff (1938) found chemical stimuli to be far more effective at triggering spawning than temperature.

The use of external species-specific chemicals to disrupt the natural spawning cycle of zebra mussels may have an advantage over the nonspecific toxic chemicals presently used. An important question is whether it would be necessary to treat an entire lake to have an impact on zebra mussel populations at a particular site. Another question is whether these methods may provide a better way to achieve control over zebra mussel populations throughout a lake than would local application of molluscicides. Such questions can only be answered by identification and field testing of chemicals that trigger zebra mussel spawning.

ACKNOWLEDGMENT

The assistance of James Walker is appreciated.

REFERENCES

Alcazar, S. N., E. P. Solis, and A. C. Alcala. ''Serotonin-Induced Spawning and Larval Rearing of the China Clam, *Hippopus porcellanus* Rosewater (Bivalvia:Tridacnidae),'' *Aquaculture* 66:359–368 (1987).

Belda, C. A., and A. G. C. Del Norte. ''Notes on the Induced Spawning and Larval Rearing of the Asian Moon Scallop, *Amusium pleuronectes* (Linne), in the Laboratory,'' *Aquaculture* 72:173–179 (1988).

Braley, R. D. ''Serotonin-Induced Spawning in Giant Clams (bivalvia:tridacnidae),'' *Aquaculture* 47:321–325 (1985).

Crawford, C. M., W. J. Nash, and J. S. Lucas. ''Spawning Induction, and Larval and Juvenile Rearing of the Giant Clam, *Tridacna gigas*,'' *Aquaculture* 58:281–295 (1986).

Galtsoff, P. S. ''Physiology of Reproduction of *Ostrea virginica*. II. Stimulation of Spawning in the Female Oyster,'' *Biol. Bull.* 75:286–307 (1938).

Galtsoff, P. S. ''Physiology of Reproduction of *Ostrea virginica*. III. Stimulation of Spawning in the Male Oyster,'' *Biol. Bull.* 78:117–135 (1940).

Gibbons, M. C., and M. Castagna. ''Serotonin as an Inducer of Spawning in Six Bivalve Species,'' *Aquaculture* 40:189–191 (1984).

Haag, W. R., and D. W. Garton. ''Synchronous Spawning in a Recently Established Population of the European Zebra Mussel, *Dreissena polymorpha*, in Western Lake Erie,'' *Hydrobiologia* (In press).

Matsutani, T., and T. Nomura. ''Induction of Spawning by Serotonin in the Scallop *Patinopecten yessoensis* (Jay),'' *Mar. Biol. Lett.* 3:353–358 (1982).

Matsutani, T., and T. Nomura. ''Pharmacological Observations on the Mechanism of Spawning in the Scallop *Patinopecten yessoensis*,'' *Bull. Jpn. Soc. Sci. Fish.* 52:1589–1594 (1986).

Matsutani, T., and T. Nomura. ''In Vitro Effects of Serotonin and Prostaglandins on Release of Eggs from the Ovary of the Scallop, *Patinopecten yessoensis*,'' *Gen. Comp. Endocrinol.* 67:111–118 (1987).

Miyazaki, I. ''On a Substance Which is Contained in Green Algae and Induces Spawning Action of the Male Oyster,'' *Bull. Jpn. Soc. Sci. Fish.* 7:137–138 (1938).

Motley, H. L. ''Physiological Studies Concerning the Regulation of Heartbeat in Freshwater Mussels,'' *Physiol. Zool.* 7:62–84 (1934).

Smith, J. R. ''The Role of the Nervous System in Algae-Induced Gamete Release by *Mytilus californianus*,'' *Com. Biochem. Physiol.* 86C:215–218 (1987).

Smith, J. R., and D. R. Strehlow. ''Algal-Induced Spawning in the Marine Mussel *Mytilus californianus*,'' *Int. J. Invertebr. Reprod.* 6:129–133 (1983).

Sprung, M. ''Ecological Requirements of Developing *Dreissena polymorpha* Eggs,'' *Arch. Hydrobiol., Suppl.* 79:69–86 (1987).

Sprung, M. ''Field and Laboratory Observations of *Dreissena polymorpha* Larvae: Abundance, Growth, Mortality and Food Demands,'' *Arch. Hydrobiol., Suppl.* 115:537–561 (1989).

Stanczykowska, A. ''Ecology of *Dreissena polymorpha* (Pall.) (Bivalvia) in Lakes,'' *Pol. Arch. Hydrobiol.* 24:461–530 (1977).

Starr, M., J. H. Himmelman, and J.-C. Therriault. ''Direct Coupling of Marine Invertebrate Spawning with Phytoplankton Blooms,'' *Science* 247:1071–1074 (1990).

Stephano, J. L., and M. Gould. ''Avoiding Polyspermy in the Oyster *(Crassostrea gigas)*,'' *Aquaculture* 73:295–307 (1988).

Walz, N. ''The Energy Balance of the Freshwater Mussel *Dreissena polymorpha* Pallas in Laboratory Experiments and in Lake Constance. II. Reproduction,'' *Arch. Hydrobiol., Suppl.* 55:106–119 (1978).

CHAPTER 20

Spawning of Zebra Mussels *(Dreissena polymorpha)* and Rearing of Veligers under Laboratory Conditions

S. Jerrine Nichols

The spawning cycle of the zebra mussel, *Dreissena polymorpha*, is amenable to laboratory manipulations. Techniques are presented that can be used to initiate spawning and rear veligers from fertilized egg to settlement stage. Spawning can be induced in sexually mature mussels by temperature fluctuations or by the addition of ripe gametes. Embryonic survival is excellent until the straight-hinge stage when the first wave of mortality occurs, usually due to improper food. The second critical stage of development occurs just prior to settlement when mortality increases again. Veliger mortality averaged over 90% from egg to settlement. The results indicate that obtaining large numbers of veligers for laboratory experiments to be conducted year-round is difficult.

INTRODUCTION

Spawning and larval development of zebra mussels (*Dreissena polymorpha*) are closely tied to environmental factors such as water temperature and food density (Stanczykowska, 1977). Water temperature controls initiation and duration of spawning, beginning when water temperatures exceed 12–16°C and continuing until temperatures drop below 10°C. In aquatic systems where water temperature remains above 12°C, reproduction continues throughout the year (Stanczykowska, 1977). The influence of food density on spawning has not been established for zebra mussels; however, for some marine mussels and clams, spawning is delayed until the density and type of phytoplankton most acceptable for larval development are present in the water column (Starr et al., 1990). Gamete release in itself is a spawning trigger, as the presence of ripe gametes in the water will initiate spawning in adjacent zebra mussels (Walz, 1978).

The length of time required for development from fertilized egg to settled larvae ranges from 3 days to 3 months depending on water temperature, with growth ceasing below 10°C or above 25°C (Stanczykowska, 1977; Sprung, 1987; Mackie et al., 1989). Food supplies are not considered critical for early developmental stages, as veligers are believed to use yolk material in the egg for sustenance. Larvae begin feeding on algae 1–4 μm in diameter at the straight-hinge stage. Lack of proper food at this stage results in starvation and loss of all larvae. Mortalities as high as 99% are common for veligers in European waters (Morton, 1971; Stanczykowska, 1977).

This information suggests that spawning and veliger development could be manipulated under laboratory conditions. However, with few exceptions, such manipulations have not been published (e.g., Sprung, 1987). Experiments were conducted to determine whether zebra mussels could be induced to spawn at any time of the year and whether veligers thus produced could survive to the settling stage.

METHODS

Spawning

The ability of zebra mussels to spawn under laboratory conditions was tested on two different groups of zebra mussels — newly collected mussels that had been in captivity less than 6 months (referred to as recently caught) and mussels that had been held under laboratory conditions over 6 months (referred to as captive stock). The captive stock was initially brought into the laboratory from Lakes St. Clair and Erie in May 1989 and held until August 1990. Recently caught mussels were collected from these lakes from March to May 1990. Mussels were fed dried algae, *Chlorella* spp., at a rate of 3.2 g of dried algae per 24 hr for each 1000 mussels over 10 mm in length. Water temperature was 20°C, calcium level was at least 125 mg/L, pH was 7.0–7.5, and dissolved oxygen was 9.0 mg/L. All colonies were kept at a photoperiod of 12-hr light, 12-hr dark.

Table 1. Gametogenic Index Used on Captive Stock and Recently Collected *Dreissena polymorpha*

	Female		Male	
Gametogenic Stage	**Size (μm)**	**Features**	**Size (μm)**	**Features**
Stage 1	3–6	Round cells, no nuclei or germinal vesicle, cannot tell from males	3–6	Round cells, no nuclei or germinal vesicle, cannot tell from females
Stage 2	10–20	Visible nuclei but no germinal vesicle	6–9	Round cells, look like stage 1
Stage 3	10–40	<50% Gametes have nuclei and germinal vesicle, >50% still in stage 2	5–8 Not including tail length	<50% Gametes triangle shape with tails, >50% look like stage 2
Stage 4	40–85	>50% Gametes have nuclei and germinal vesicle, <50% still in stage 2	5–8 Not including tail length	>50% Gametes triangle shape with tails, <50% look like stage 2
Stage 5	40–85	Only few stage 4 eggs remain, lots of empty spaces	5–8	Only few stage 4 sperm remain, lots of empty spaces

Sexual maturity was monitored in both captive stock and recently caught mussels. Mussels were dissected; then a portion of the reproductive/digestive tract was removed, placed on a slide with a drop of water, gently squashed with a cover slip, and examined under a compound microscope. The sex and maturity of the gametes were recorded. Characterization of gamete maturity for zebra mussels has been described in studies by Borcherding (1991), Haag and Garton (1992), and Tourari et al. (1988). The system used in this study was different from that in the studies mentioned above because living material was dissected rather than preserved, stained, and sectioned. The staging system used in this study follows more closely that used by King et al. (1989) and Barkati and Ahmed (1989–1990). These two studies compare advantages, as well as disadvantages, of using live gonadal squashes rather than standard histological preparations in *Mytilus edulis* populations.

In this study, gamete development is divided into five stages (Table 1). In the immature or stage 1, males and females are not distinguishable. Only simple round cells, <10 μm in diameter, are visible in the reproductive tissue (Figure 1). Females are described as stage 2 when eggs are visible. Stage 2 eggs have visible nuclei, range in size from 20–40 μm, and have an undeveloped germinal vesicle (Figure 2). By stage 3, some eggs have developed a germinal vesicle, such as seen in Figure 3, although the percentage of such

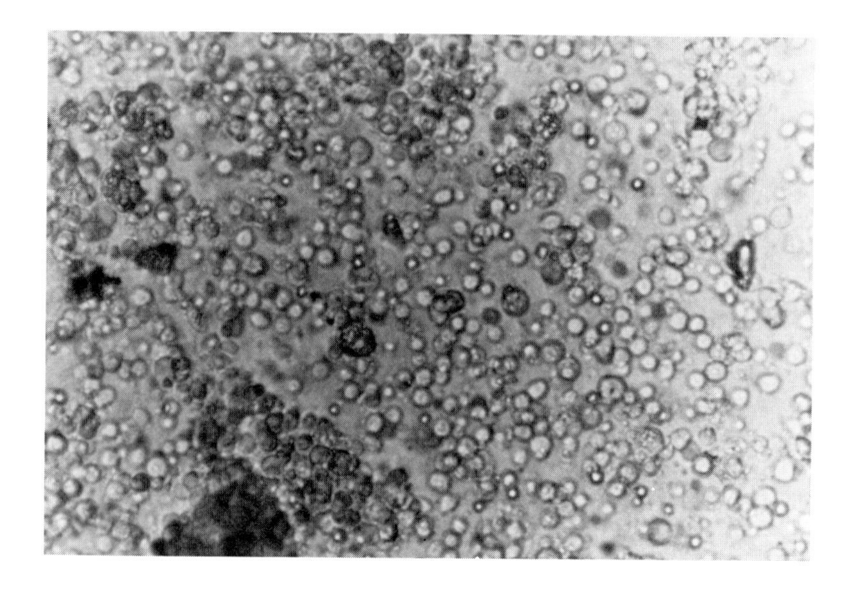

Figure 1. Stage 1 (immature) gametes of *Dreissena polymorpha*. Cells are <10 μm in diameter.

eggs is less than 50%. Eggs with germinal vesicles are usually larger than 40 μm in diameter (maximum size of about 85 μm). In a stage 4 or ripe female, over 50% of the eggs have germinal vesicles. Male gametes are very difficult to identify in the early stages. The only difference between stages 1 and 2 male gametes is size, ranging in size from 6 to 9 μm vs 3 to 6 μm at Stage 1. Mature sperm are triangular in shape, are about 7 μm in length, and have a long tail. In stage 3, mature sperm comprise less than 50% of the cells in the reproductive tissue. In stage 4 or ripe males, mature sperm comprise over 50% of the cells (Figure 4). A stage 5 or spent mussel is characterized by open "spaces" in the reproductive tissues with only one or two residual gametes present.

Spawning under laboratory conditions (without manipulations) was monitored weekly from November 1989 to May 1990. Water flow to aquariums containing culture stock was turned off for 1 hr; and 100 mL of water was removed from the bottom of the aquariums and checked for eggs, sperm, and veligers. The number of eggs and veligers per milliliter was recorded, with sperm listed as either present or absent. In addition, 10 mussels over 12 mm were dissected and checked for sexual maturity.

Two potential spawning triggers where chosen for testing: change in water temperature and presence of ripe gametes. Separate tests were run on both captive stock and recently caught mussels from March through June 1990. For each test, about 200 mussels over 12 mm were placed in each of four

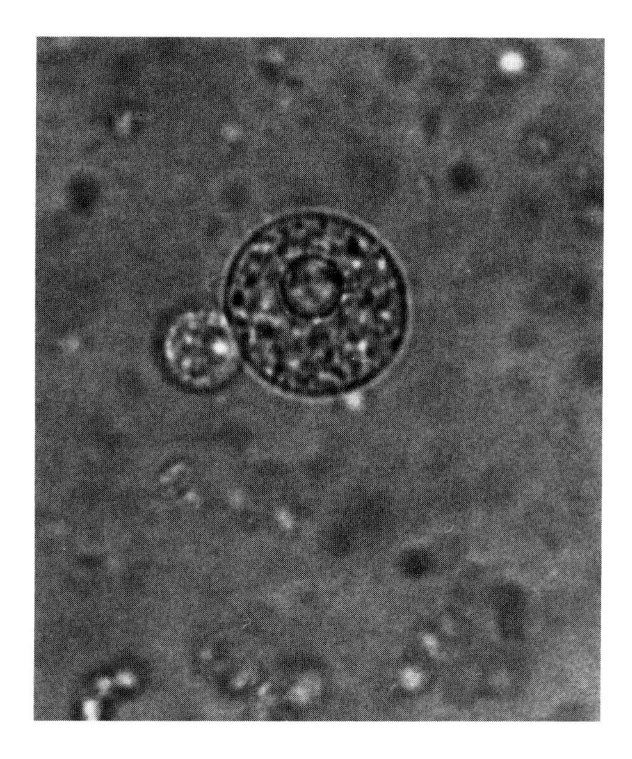

Figure 2. Stage 2 (nonripe) eggs of *Dreissena polymorpha*. No germinal vesicle visible. Eggs are 20–40 μm in diameter.

2-L aquariums. Mussels in one aquarium served as controls (no manipulation), while mussels in the other three aquariums were exposed to a potential spawning trigger. Before the tests, sexual maturity was checked by dissecting 3–4 mussels from each aquarium. If ripe mussels were found in this initial survey, it was assumed that the rest of the sample contained a sufficient number of mature individuals to evaluate the effects of spawning triggers.

There were two types of temperature-induced spawning tests: (1) raising water temperatures to 20°C on recently caught mussels, or (2) subjecting captive stock (which were held at an ambient temperature of 20°C) to water temperature of 4°C for 2 weeks and then returning them to 20°C. The 200 mussels used as controls were not exposed to such temperature manipulations.

Recently caught mussels were exposed to increased temperature merely by transferring them from the lake to the laboratory environment (ambient temperature 20°C). When fresh mussels were collected in 1990, water temperature in the lakes was 11–13°C in March and April, and 15°C in May. This required some modifications in the basic test procedure since no controls were possible, as it was not feasible to maintain mussels at ambient lake

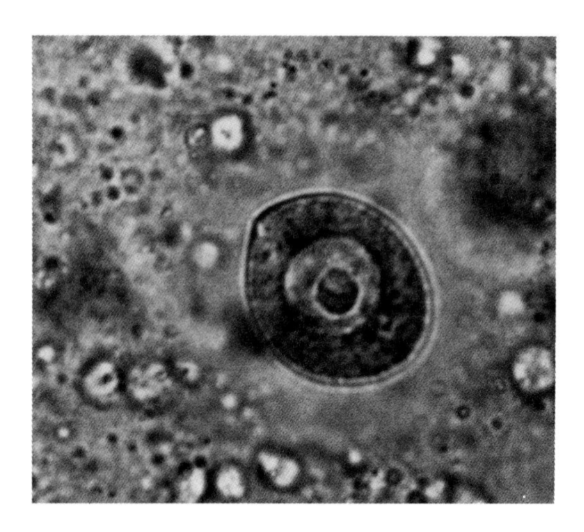

Figure 3. Stage 3 (ripe) eggs of *Dreissena polymorpha*. Germinal vesicle (light gray ring) is visible. Eggs are 40–85 μm in diameter.

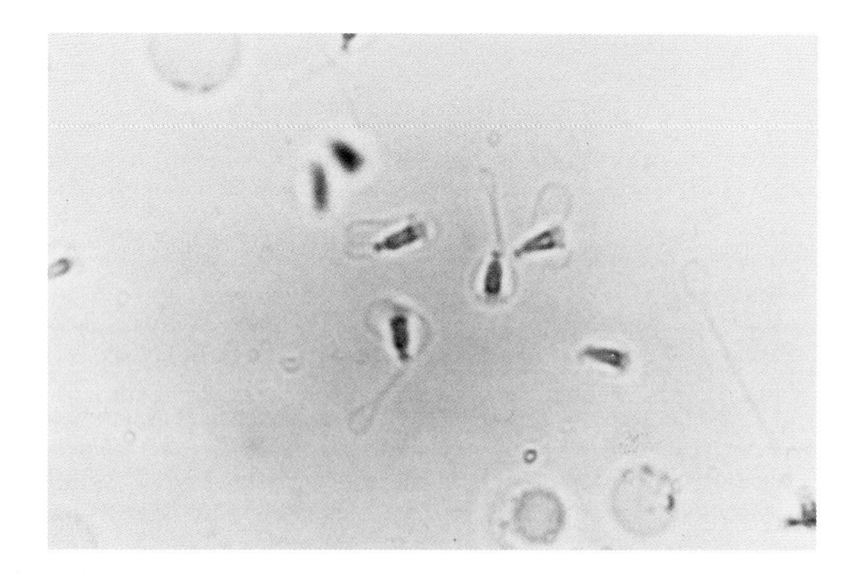

Figure 4. Stage 4 (ripe) sperm of *Dreissena polymorpha*. Triangular cell body is about 7 μm in length.

temperatures in the laboratory. In addition to the four test aquariums, 50 mussels were isolated individually in test tubes and checked daily for spawning. These tests were run longer than 24 hr, continuing for 8 weeks or until spawning occurred. All mussels were dissected as soon as spawning was detected or at the end of the 8-week test period, and gamete maturity was recorded.

Use of ripe gametes as a spawning cue was tested on both recently caught and captive mussels. For recently caught mussels, the influence of temperature change rather than gamete addition could be ruled out as a spawning cue because mussels were used that had been kept in the laboratory for 5 weeks at 20°C, but had not spawned. Ripe gametes were obtained by dissecting a stage 4 mussel, macerating the gonadal tissue in a few drops of water, and adding this suspension to test aquariums. The gonadal tissue from one mussel was used per 2-L aquarium. One set of tests used stage 4 eggs, and another set of tests used stage 4 sperm.

After the application of spawning cues to either recently caught or captive stock, the number of eggs released was monitored by removing 10 mL of water from the bottom of the aquarium (eggs are somewhat adhesive for a short period of time). This aliquot of water was examined for spawn and results listed as number of eggs per milliliter. No effort was made to determine fertilization success. This procedure was repeated every 1 hr for the first 12 hr and then again at 24 hr. Adults were removed after 24 hr and dissected to determine gamete maturity. These specific tests were repeated three times for both temperature changes or ripe gamete additions on captive stock but only once (with four replicates) on recently caught mussels. Results were compared using an analysis of variance or χ^2-test on the number of eggs produced in each replicate for each spawning cue, and compared with the number of stage 4 or ripe females present. Results were considered significant if $p \leq 0.05$.

Rearing

Methods of rearing veligers to the settlement stage were tested on both wild-caught and laboratory-reared individuals. Wild-caught veligers were obtained from plankton tows taken at the Detroit Edison power plant at Monroe, MI (western Lake Erie) from May through August 1990. A number of methods were used to rear veligers, including aquariums with and without aeration, test tubes mounted on a rotating wheel, and a variable speed reciprocating tabletop shaker (Table 2). Types of food tested included bacterial infusions, suspensions of dried *Chlorella* spp., and live green algae (unknown species of 2 μm size) cultured from Lake Michigan water. The medium used to culture these live green algae was autoclaved topsoil (500 mL), mixed with 4 L of well water. Water temperature was maintained at 20°C and all veligers were maintained in well water with a minimum of 125 mg/L dissolved calcium. Guidance for culture water is provided by Sprung (1987) who describes

Table 2. Techniques Used to Rear *Dreissena polymorpha* Veligers in 2-L Aquariums Maintained in the Laboratory; The Mean Number of Veligers at the Beginning and End of a 2-Week Experimental Period also Shown

Technique	Food	Veligers/mL Initial	Veligers/mL Final
With box filter	Dried *Chlorella*	3	0
	Live algae[a]	2	0.3
	Bacteria[b]	1	0
Aeration	Dried *Chlorella*	2	0
	Live algae	2	0
	Bacteria	1	0
No aeration	Dried *Chlorella*	2	0
	Live algae	2	0
	Bacteria	2	0
Shaker table with aeration	Dried *Chlorella*	3	0
	Live algae	2	0
Shaker table without aeration	Dried *Chlorella*	1	0
	Live algae	3	2
Test tubes in ferris wheel	Live algae	0.5	0

[a] Unidentified live green algae, <4 μm in diameter. Density of cells set at 3,000,000/mL.

[b] Bacteria infusions prepared from boiled rice.

reconstituted water that will support all ages of zebra mussels as well as gametes. All rearing chambers, with the exception of test tubes, were provided with either native clam shells (Unionidae) or whole oyster shells as substrate for the veligers. Known numbers of veligers were added to the chambers and survival checked weekly by examining the substrate shells and water. Each type of rearing method (referred to as a test run) was repeated eight times. An analysis of variance was performed on differences in veliger survival using the various rearing techniques, and results were considered significant if $p \leq 0.05$.

RESULTS AND DISCUSSION

Spawning

Zebra mussels spawned readily in the laboratory, even without manipulating environmental factors. Mussels held in captivity since May 1989 spawned almost continually in general culture aquariums from November 1989 to May 1990. Spawning was intermittent, and it was not possible to determine the exact percent of mussels releasing spawn at any given episode. Although 1–2 eggs per milliter could be found weekly, the number increased to 10–15 eggs per milliliter every 6–8 weeks. Live, tailed sperm were never found in general culture aquariums, although clumps of nontailed, inactive sperm were frequently observed. An average of 45% of the mussels dissected during

Table 3. Effect of Temperature Manipulation and Addition of Ripe Gametes on Egg Release by *Dreissena polymorpha* Held in the Laboratory over 6 Months (Captive Stock) or Held Less than 6 Months (Recently Collected Stock)

	Captive Stock		Recently Collected Stock	
	Exp	**Control**	**Exp**	**Control**
Temperature Cue				
Eggs	11	0	13	N/A
Range	(1–16)		(4–15)	
Females	17	27	33	N/A
Sperm Cue				
Eggs	17	0	19	0
Range	(2–25)		(13–20)	
Females	20	22	42	32
Egg Cue				
Eggs	6	0	10	0
Range	(1–13)		(8–11)	
Females	23	24	29	36

Note: N = 1800 for each cue test and 600 for each control. Eggs = mean number/mL released. Range = range of eggs/mL released. Females = percent stage 4 females. N/A = not available.

this time period were stage 4 or ripe (range 30–60%). The percentage for mussels under 24 mm was 44 and 48% for males and females, respectively, and the percentage for mussels over 24 mm was 15 and 85%, respectively. About 8% of the mussels (<24 mm in length) dissected were hermaphrodites. Hermaphroditic zebra mussels have been reported from a few locations in Europe, although histological proof has not been documented (Stanczykowska, 1977; Crochard, 1985; Tourari et al., 1988). Most eggs and embryos produced in general culture aquariums were destroyed during weekly cleaning. However, about 500 veligers survived and settled onto existing colonies. The fact that spawning occurred in the laboratory in the winter and spring is consistent with published literature, indicating that cessation of spawning in the field is, at least in part, a phenomenon related to low temperature (Stanczykowska, 1977).

Increasing temperature and the addition of ripe gametes triggered spawning within 24 hr in every test (and replicate), as long as ripe (stage 4) mussels were present (Table 3). However, the number of eggs released varied significantly in 33–50% of the replicates within each test run using one particular spawning cue as well as between test runs using different spawning cues in both captive stock and recently caught mussels. In captive stock, the largest number of eggs were released using ripe sperm as a spawning cue (mean number of eggs 17/mL), followed by temperature manipulation (11/mL), and finally by the addition of ripe eggs (6/mL).

Regardless of the type of spawning cue, there was no significant relationship between the number of eggs produced and the number of sexually ripe females (stage 4) in captive stock (Table 3). For example, one test run using sperm as a cue resulted in the release of 16 eggs per milliliter (range 9–20) from 210 ripe females in one replicate, and 24 eggs per milliliter (range 22–24) from 50 ripe females in another replicate. As Table 3 shows, colonies exposed to ripe eggs as a spawning cue contained more ripe females (23%) than colonies exposed to either ripe sperm or temperature manipulations, but yielded fewer eggs. Colonies used as controls in captive stock experiments contained more ripe females than colonies used in actual test runs, and yet no spawning occurred. This nonsignificant relationship between number of ripe females present and number of eggs released is a common problem in oyster culture (Dupuy et al., 1977).

Similar results were observed using recently caught stock, except that in general more eggs were released and a greater percentage of females were ripe (Table 3). Spawning occurred in every trial, although the number of eggs released varied from 4 to 20/mL. Addition of ripe sperm stimulated the most release of eggs (mean = 19/mL), followed by temperature manipulations (mean = 13/mL), and addition of ripe eggs (mean = 10/mL). As with captive stock, the number of eggs released showed no significant relationship to the number of ripe females.

Attempts to induce spawning on recently caught mussels ran into immediate difficulties because none of the stock collected in March or early April 1990 were in stage 4 and would not spawn. Water temperature in Lakes St. Clair and Erie had been consistently at or above 12°C for at least 10 days prior to collection and had even reached 15°C for several days. Previous studies have described spawning as occurring in spring immediately after mussels are exposed to 12–15°C water temperature (Stanczykowska, 1977; Walz, 1978; Sprung, 1987). However, spawning did not occur in any of the trial runs using either temperature or the addition of ripe gametes until May 1990, when recently caught stock were sexually mature (stage 4) either because they were ripe when brought in from the lake or had time to ripen under laboratory conditions (Table 4). The length of time required for these mussels to ripen and spawn in the laboratory was related to their collection date rather than laboratory manipulations, such as placing mussels in aquariums or isolating them in test tubes. Colonies in aquariums began to spawn in the following sequence: mussels collected in March were not ripe and took 39 days at 20°C before spawning occurred; mussels collected in April took 23 days; and mussels collected in May took 1 day. The same sequence for the test tube group was 42, 24, and 1 day, respectively. For the test tube stock collected in May 1990, which spawned almost immediately in the laboratory, male mussels consistently spawned up to 12 hr sooner than females. This phenomenon is extremely common in bivalves (Mackie, 1984) and has been recorded before for zebra mussels (Sprung, 1987). By the time spawning occurred for all recently caught stock (after 8 weeks), about 70% of the mussels were in stage 4 compared to about 45% in the captive stock.

Table 4. Spawning Behavior of *Dreissena polymorpha* Collected and Brought into the Laboratory on Three Dates in 1990

	March 28		April 20		May 5	
	A	TT	A	TT	A	TT
Mean number of days to first spawn	39	42	23	24	1	1
% Mussels spawned in 8 wk	*	46	*	60	*	71
% Stage 4 females at end of test	35	54	46	38	32	30

Note: Mussels were held in 2-L aquariums (A) (n = 600) or in test tubes (TT) (n = 50). Results are a mean of all replicates. * = Not determined.

The percent of males and females was 43 and 49%, with 8% hermaphrodites. The immature condition of gonads observed in March and April collections suggests that water temperatures of 12–13°C do not directly cause spawning in spring, as implied in European literature, but act indirectly to facilitate final maturation of gametes, a process which may require 5–6 weeks.

A number of factors indicate that individual zebra mussels in the laboratory spawn at low levels over long periods of time: (1) continual presence of stage 4 (ripe) eggs and sperm in dissected mussels from aquariums where spawning was successful; (2) gametes at all stages of maturity in captive stock; (3) significant variation in egg production; and (4) absence of spent mussels. Low-level spawning by one individual mussel over a long period of time is not uncommon (Mackie, 1984). In a laboratory study on spawning induction, Walz (1978) reported that a single female required 2–6 spawns for all eggs to be released, with the greatest number of eggs produced in the first spawning. In contrast, Haag and Garton (1992) showed a strong degree of synchronization in a wild population of mussels in Lake Erie, with spawning occurring over relatively short period of time in all individuals. Observations on recently caught mussels brought into the laboratory in May 1990 agreed with the findings of Haag and Garton (1992).

Theoretically, the number of ripe eggs (and therefore veligers) released by zebra mussels in the laboratory could be enhanced if gamete ripeness were synchronized, even though there is a great deal of variability between the number of ripe females and number of eggs released. However, recently caught stock (when spawning finally occurred) contained more ripe mussels than captive stock (70% compared to 45%) and, in general, tended to release more eggs. Increasing the number of ripe mussels available for spawning should be possible by following techniques recommended for spawning marine oysters (Dupuy et al., 1977). Marine oysters are collected in early spring when they contain ripe gametes, and held in cool temperatures until spawning is required. Spawning begins after the oysters are warmed and fed cornstarch as a source of easily digested food. Using a similar protocol, ripe zebra mussels can be collected from the field as early as May in most areas. They could be held at less than 10°C until spawning is required. To initiate spawning, zebra mussels could be gradually warmed to 20°C, and fed dried *Chlorella* (cornstarch has not been tried in zebra mussels). However, the ability of these mussels to release ripe gametes decreases after 5–6 months of being held at 4°C even if food is provided (Leitch, Ontario Hydro, Personal communication).

An alternative to this method would be to collect zebra mussels when most individuals contain stage 2 or nonripe gametes. In Lake St. Clair, in 1990, the proper time period was March and early April. These mussels should be held at less than 10°C to prevent gametes from ripening. About 5–6 weeks before spawn is needed, the temperature is increased to at least 15°C, triggering the final maturation of gametes. Live algae could be added at this time to ensure that all potential spawning cues are present, as noted by Starr et al. (1990). This technique provides the greatest degree of synchronization possible in mussels used for spawning, and eliminates any potential decrease in gamete viability caused by long-term storage at low temperatures. Once stage 4 gametes are present, spawning occurs without further manipulation, or can be triggered within 24 hr by adding ripe gametes to the aquariums. As shown, the addition of ripe sperm results in the most egg release. There is also the possibility of triggering spawning in ripe mussels through the addition of serotonin (Ram and Nichols, Chapter 19, this volume).

Regardless of the technique, not all ripe mussels spawn and not all ripe eggs are released (Table 4). Therefore, multiple spawnings can be obtained from the same group of mussels, as has also been noted by Borcherding (1991). This is a common phenomenon in brood stock used for marine shellfish culture. Dupuy et al. (1977) noted that certain female oysters do not spawn readily in capitivity even when full of ripe gametes. In addition, the response of individual oysters to induced spawning efforts in the laboratory is highly variable. For zebra mussels since large numbers are readily available, even if 50% of the females do not spawn, enough eggs are produced with minimal laboratory handling to meet potential experimental needs.

Rearing

Examination of the 10-mL aliquots from spawning aquariums indicated that fertilization of zebra mussel eggs after spawning requires 1–24 hr under laboratory conditions. When eggs are released from females, which are visible as a whitish cloud, they remain in the water column for a short time and rapidly settle to the bottom. Sperm release also appears as a whitish cloud, and the sperm tend to rise to the surface of the water. Although eggs appear to be initially mildly adhesive after fertilization, they do not appear to be injured by strong water flows, or by being rinsed out of the spawning chamber into rearing chambers. The first cell division occurs within 1 hr of fertilization and is distinguished by the appearance of a polar body (see Meisenheimer, 1901 for description of development from fertilization to settling). Growth of the embryo was rapid at 20°C, with straight-hinge larvae appearing in 3–5 days. Previous studies indicate that veligers do not feed prior to the straight-hinge stage (Stanczykowska, 1977; Sprung, 1987). However, I observed material in the intestinal tract of individuals at the straight-hinge stage. The composition of this gut material remains unknown. The first major period of mortality was observed as the veligers reached the straight-hinge stage. Once veligers passed this stage, no significant mortality occurred until just before settling. All rearing chambers contained substrate for settling such as unionid shells and whole oyster shells. No settlement or veliger survival was noted in rearing chambers without substrates. Substrates were placed in the rearing chambers at the same time as the fertilized eggs or young larvae. Veligers as small as the straight-hinge stage were frequently found on the underside of these substrates, and all settling occurred there as well.

The success rate in rearing veligers to settling stage in this study was low (Table 2). Similar experiments conducted at Ontario Hydro also indicate difficulties in rearing veligers from eggs to settling (Leitch, Personal communication). Indeed, most published studies agree that rearing of veligers under laboratory conditions is difficult (Morton, 1971; Sprung, 1987). In contrast, larvae of marine mussels are not as difficult to rear under laboratory conditions, with survival usually about 33% (Dupuy et al., 1977). In the present study, eggs produced from laboratory-induced spawning were easily reared to the straight-hinge stage. Veliger survival to the straight-hinge stage was 13 and 66% in two test runs, but less than 2% of these straight-hinge veligers survived to settling (Table 2). By July 1990, it was determined that the best veliger survival occurred with the use of rearing aquariums placed on a shaker table with live algae at 3 million cells per milliliter. All other methods as described in Table 2 were therefore discontinued. At this time, wild-caught veligers collected with plankton nets from Lake Erie were brought into the laboratory to determine whether the lack of rearing success was due to poor viability of laboratory spawn rather than culture techniques. These wild-caught veligers (straight-hinge or earlier) proved no easier to rear than laboratory stock, with mortality rates averaging nearly 98%. Sixteen test runs with two replicates each were conducted using the live algae/shaker table technique and the veligers from Lake Erie. Of the 16 tests, only 8 had surviving veligers; in these 8 test runs, survival was 22, 15, 10, 3, 2, 1, 0.5, and 0.5%. There was no gradual increase in survival over time; the 22% survival was recorded in the middle of the test run, and in the last test run all the veligers died. Although the number of veligers that survived to final settling was low, several hundred settled larvae were produced. Concentrated efforts using the techniques described here should yield more young.

CONCLUSIONS

All stages of the reproductive cycle of *Dreissena polymorpha* can be induced and, to some degree, maintained under laboratory conditions, enabling experiments on physiology and control mechanisms to be conducted throughout the year. Fertilized eggs can be obtained at any season, as long as the brood stock contains ripe gametes. A high percentage of mussels will remain ripe for months at either 20°C or in cold storage. Veliger stages are not as adaptable to laboratory conditions. Fertilized eggs can be reared to the straight-hinge stage with minimal equipment or food input. However, veliger survival and growth beyond this stage require specific food and substrate conditions that are still poorly understood.

ACKNOWLEDGMENTS

I thank the staff biologists and divers of the Detroit Edison power plant at Monroe, MI for their assistance in obtaining veligers. Contribution No. 812 of the National Fisheries Research Center-Great Lakes, U.S. Fish and Wildlife Service, 1451 Green Road, Ann Arbor, MI 48105.

REFERENCES

Barkati, S., and M. Ahmed. "Reproduction of the Mussel *Mytilus edulis* L. from Lindaspollene, Western Norway," *Oebalia* 16 N.S.:1–14 (1989–1990).

Borcherding, J. "The Annual Reproductive Cycle of the Freshwater Mussel *Dreissena polymorpha* Pallas in Lakes," *Oecologia* (Berlin) 87:208–218 (1991).

Crochard, C. "Essai d'utilisation d'un mollusque bivalve (*Dreissena polymorpha* Pallas), pour estimer le degre de pollution metallique de l'organisme intoxique," Rapport Ministere de l'environment, convention 81.310, Laboratoire d'Ecologie (1985).

Dupuy, J., N. Windsor, and C. Sutton. "Manual for Design and Operation of an Oyster Seed Hatchery," Applied Marine Science and Ocean Engineering, Special Report No. 142, Virginia Institute of Marine Sciences, Glouchester, VA (1977).

Haag, W., and D. Garton. "Synchronous Spawning in a Recently Established Population of the European Zebra Mussel, *Dreissena polymorpha*, in Western Lake Erie," *Hydrobiologia* 243(2):103–110 (1992).

King, P., D. McGrath, and E. Gosling. "Reproduction and Settlement of *Mytilus edulis* on an Exposed Rocky Shore in Galway Bay, West Coast of Ireland," *J. Mar. Biol. Assoc. U.K.* 69:355–365 (1989).

Mackie, G. "Bivalves," in *The Mollusca, Vol. 7, Reproduction,* A. S. Tompa, N. H. Verdonk, and J. A. M. van den Biffelaara, Eds. (New York: Academic Press Inc., 1984), pp. 351–418.

Mackie, G., W. Gibbons, B. W. Munchaster, and I. M. Gray. "The Zebra Mussel, *Dreissena polymorpha*: A Synthesis of European Experiences and a Preview for North America," A report prepared for the Ontario Ministry of the Environment, Water Resources Branch, Great Lakes Section. Queen's Printer, Toronto, Ontario (1989).

Meisenheimer, J. "Entwicklungschichte von *Dreissena polymorpha* Pallas," *Z. Wiss. Zool.* 69:1–137 (1901).

Morton, B. "Studies on the Biology of *Dreissena polymorpha* Pall. V. Some Aspects of Filter-Feeding and the Effect of Micro-Organisms upon the Rate of Filtration," *Proc. Malacol. Soc. London* 39:289–301 (1971).

Sprung, M. "Ecological Requirements of Developing *Dreissena polymorpha* Eggs," *Arch. Hydrobiol., Suppl.* 79:69–86 (1987).

Stanczykowska, A. "Ecology of *Dreissena polymorpha* (Pall.) (Bivalvia) in Lakes," *Pol. Arch. Hydrobiol.* 24:461–530 (1977).

Starr, M., J. Himmelman, and J. Therriault. "Direct Coupling of Marine Invertebrate Spawning with Phytoplankton Blooms," *Science* 247:1071–1074 (1990).

Tourari, A., C. Crochard, and J. Pihan. "Action de la temperature sur le cycle de reproduction de *Dreissena polymorpha* (Pallas) etude "in situ" et au laboratoire," *Haliotis* 18:85–98 (1988).

Walz, N. "The Energy Balance of the Freshwater Mussel *Dreissena polymorpha* Pallas in Laboratory Experiments and in Lake Constance. II. Reproduction," *Arch. Hydrobiol., Suppl.* 55:106–119 (1978).

SECTION III

Effects

CHAPTER 21

The Impact of *Dreissena polymorpha* on Waterworks Operations at Monroe, Michigan: A Case History

Wilfred Laurier LePage

The zebra mussel, *Dreissena polymorpha*, was discovered in the raw water system of the Monroe waterworks in January 1989, when substantial but not alarming numbers of the animal were found. By July 1989, availability of raw water was reduced 20% by the rapid colonization of zebra mussels in the raw water pipelines. This severely strained the ability of the western Lake Erie utility to satisfy the water demands of a population of 45,000. As a result of zebra mussel infestations, several water outages occurred between the fall to early winter of 1989 and spring of 1990. The extent and impact of the mussels on plant operations are reviewed, as well as the remedial measures applied between 1989 and 1991.

INTRODUCTION

Plant Overview

The Monroe plant, an American Water Landmark, is believed to be the longest continually operated water filtration plant in Michigan. It has not missed a day of production since startup March 1, 1924. Extensive modernization of the plant in 1949 initiated the evolution of the modern facilities found at present.

Water enters the system through an underwater crib a mile offshore in the westernmost extremities of Lake Erie (Figure 1). A 76-cm (30-in.) diameter concrete intake siphons water a distance of 1868 m (6100 ft) into the screen chambers of the raw water pump station. Vertical turbine pumps deliver the water to the treatment plant via nearly 9 mi of a 76-cm (30-in.) diameter concrete transmission main. The raw water pump station is unmanned and remotely controlled from the treatment plant.

The entrance to the underwater crib in Lake Erie is 152 cm (60 in.) in diameter and connects to a 76-cm (30-in.) intake pipe situated inside a timber crib with outside dimensions of 9.0 × 9.0 × 2.1 m (29.5 × 29.5 × 7 ft) (Figure 2). The crib rests on the lake bottom 6.7 m (22 ft) beneath the surface of the water at low water datum. The top of the crib is 4.6 m (15 ft) beneath the surface, and the lip of the intake bell is at a depth of 6.1 m (20 ft) below low water datum. Lake elevations during the period covered by this case history were generally about 0.8 m (2.5 ft) above low water datum, thus, adding approximately 0.8 m (2.5 ft) to the depths reported here. The crib is ballasted with stone in each of the four corner chambers and protected externally with rock fenders. The pipe trench inside and outside the crib is also backfilled with stone. The lake floor in the vicinity of the crib is described as "rocky" on navigation charts with rock outcroppings and large boulders littering the bottom. Bedrock in the form of shattered limestone lies beneath silt deposits of varying depth.

The original 1.5×10^4 m^3/day (4 mgd) filtration plant became operational in 1924. It was modernized and expanded to a capacity of 3.0×10^4 m^3/day (8 mgd) in 1949. In 1972, the plant was renovated and enlarged to a capacity of 6.8×10^4 m^3/day (18 mgd) and ozonation was added to the treatment scheme. A standby electric power plant was added in 1978, and an additional 3 million gal finished water reservoir and 9.1×10^4 m^3/day (24 mgd) high service pump station were completed in 1987.

The treatment process consists of preoxidation for taste and odor control with ozone, disinfection with chlorine, clarification with alum, pH correction with hydrated lime, fluoridation, filtration through mixed-media Microfloc filters, and corrosion control with zinc orthophosphate, followed by storage (Figure 3). Stored water is pumped to the system via one of two high-service pump stations, each with a capacity of 9.1×10^4 m^3/day (24 mgd).

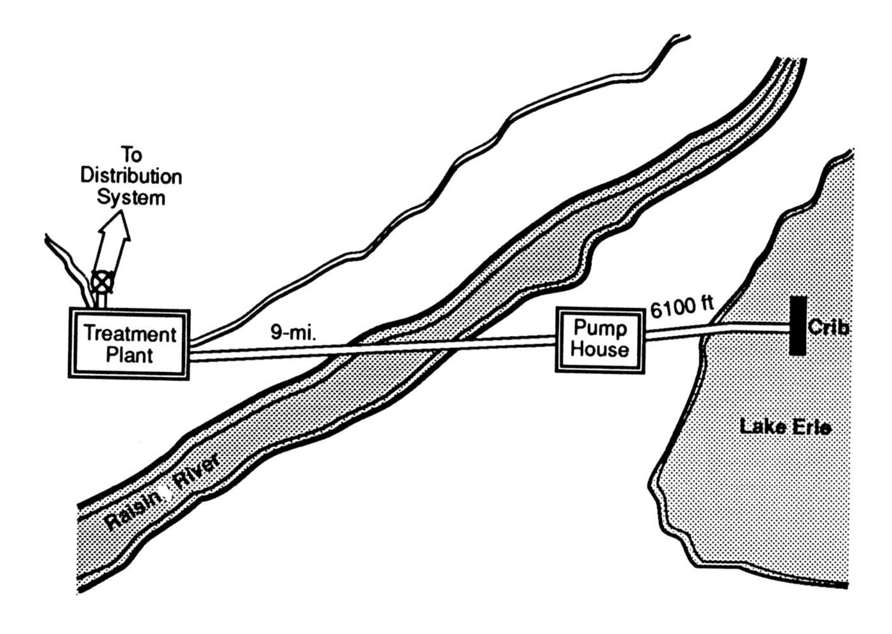

Figure 1. Location of the Monroe waterworks which withdraws water from the western basin of Lake Erie.

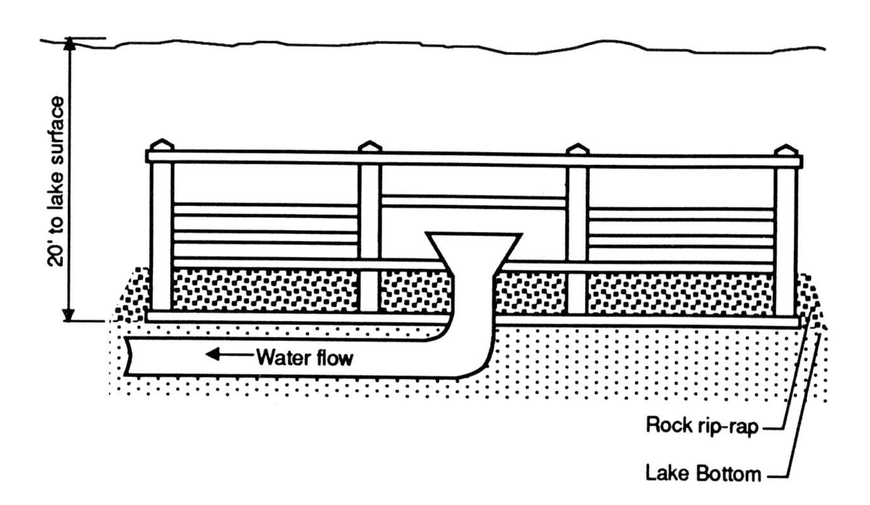

Figure 2. Underwater intake pipe and surrounding, anchoring intake crib located 6,130 ft offshore in the western basin of Lake Erie.

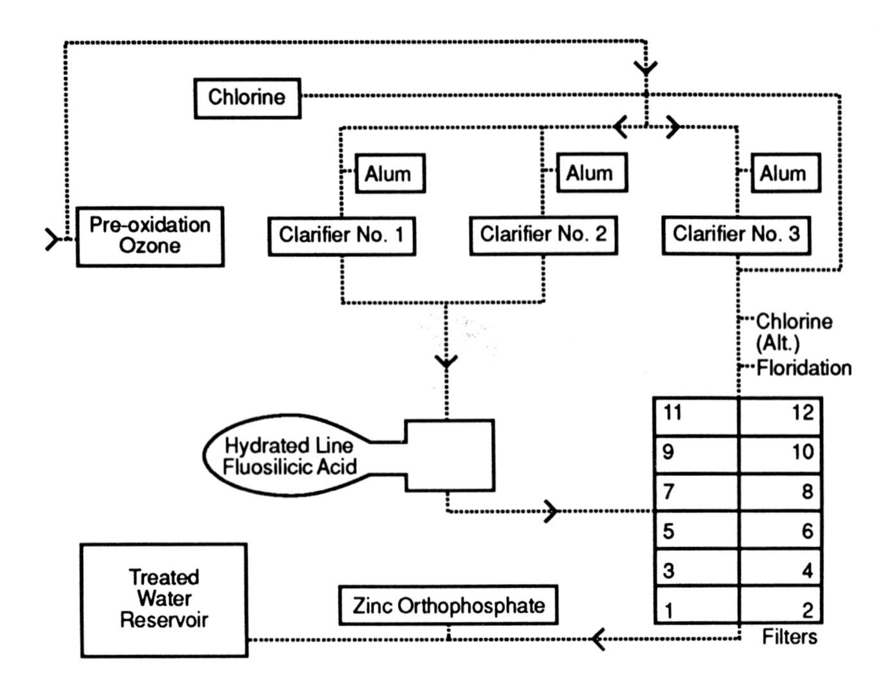

Figure 3. Flow schematics of the water entering the water treatment plant.

The utility serves a combined population of approximately 45,000 people. System demands vary between 2.3 × 10^4 m^3/day (6 mgd) in the winter to more than 4.2 × 10^4 m^3/day (11 mgd) during summer. The treatment plant has a nominal capacity of 6.8 × 10^4 m^3/day (18 mgd) but production was limited (until mid-1990) by the hydraulic capacity of the raw water intake which during 1988 delivered a maximum of 4.4 × 10^4 m^3/day (11.5 mgd).

Long-range plans provided for construction of additional raw water facilities by the mid-1990s. Unanticipated development in the townships, however, advanced the need for additional water by at least 5 years. The severe drought that gripped the area during the summer of 1988 increased water demands as high as 35% above similar periods during the preceding year. This occurred despite a summer-long, odd-even lawn sprinkling ban and numerous short-lived total bans on nonessential water use. Design plans were initiated in 1988 to add an additional intake pipe to meet water demands, but these plans were temporarily postponed when the extent of the zebra mussel problems was realized.

Problems caused by the zebra mussel, *Dreissena polymorpha*, were unknown in the Laurentian Great Lakes prior to 1988 when substantial infestations were discovered in southeastern Lake St. Clair (Hebert et al., 1988). Between 1986 when zebra mussels were introduced into Lake St. Clair, and

PLATE 1. Zebra mussels showing variations in shell color pattern found in a given population in North America. The individual in the lower left is the RR phenotype discussed by Michael Ludyanskiy in this volume. (Photograph courtesy of Ron Peplowski, Detroit Edison Company, Detroit, MI.)

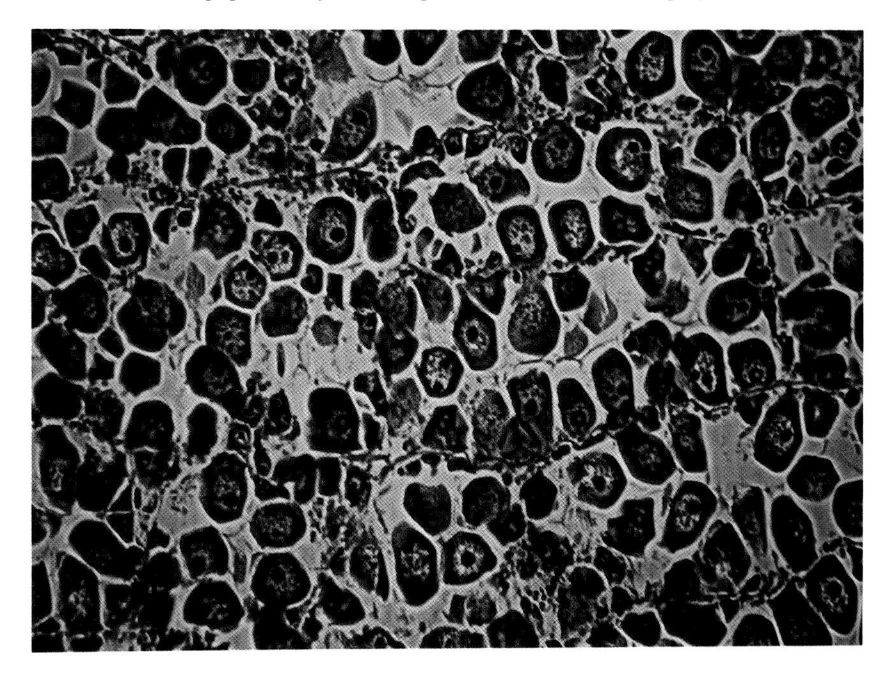

PLATE 2. Section through mature female zebra mussel showing ovarian follicles completely filled with ova prior to spawning. (Photograph courtesy of Dave Garton, Ohio State University, Columbus, OH.)

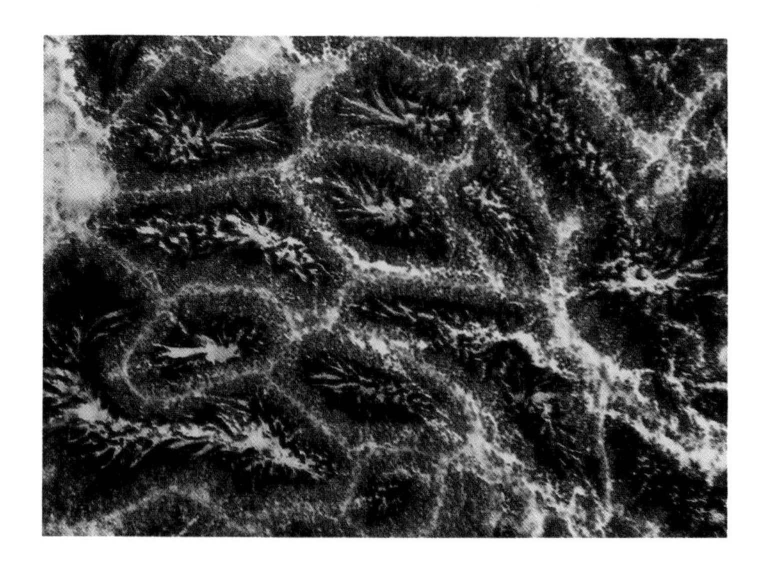

PLATE 3. Section through mature male zebra mussel showing testis with well-developed lobes filled with sperm prior to spawning. Sperm are oriented with tails toward the center of the lobe. (Photograph courtesy of Dave Garton, Ohio State Univeristy, Columbus, OH.)

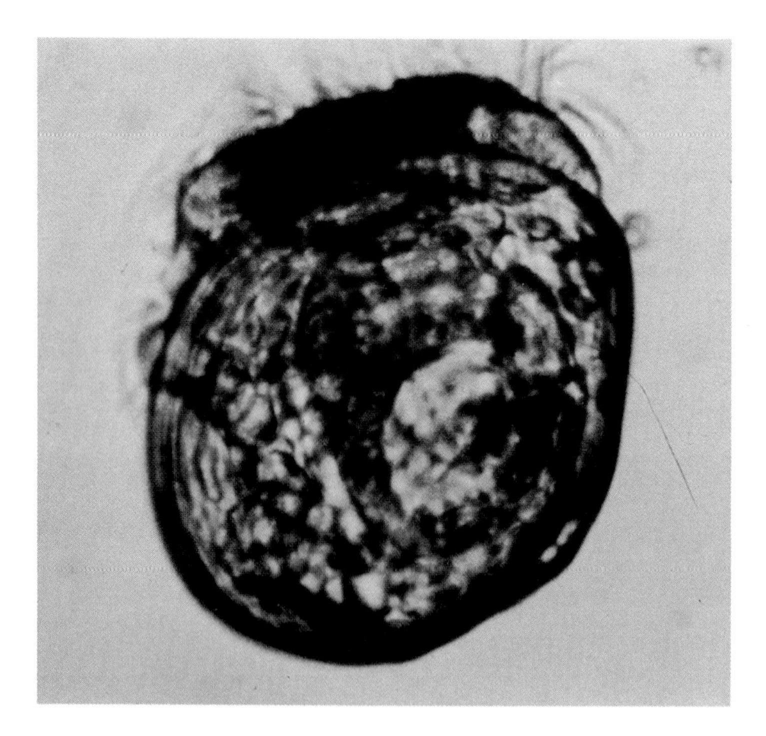

PLATE 4. Lateral view of a live zebra mussel veliger showing rudimentary shell development. Actual size is about 125 μm. (Photograph courtesy of Gordon Hopkins, Ontario Ministry of the Environment, Rexdale, Ontario.)

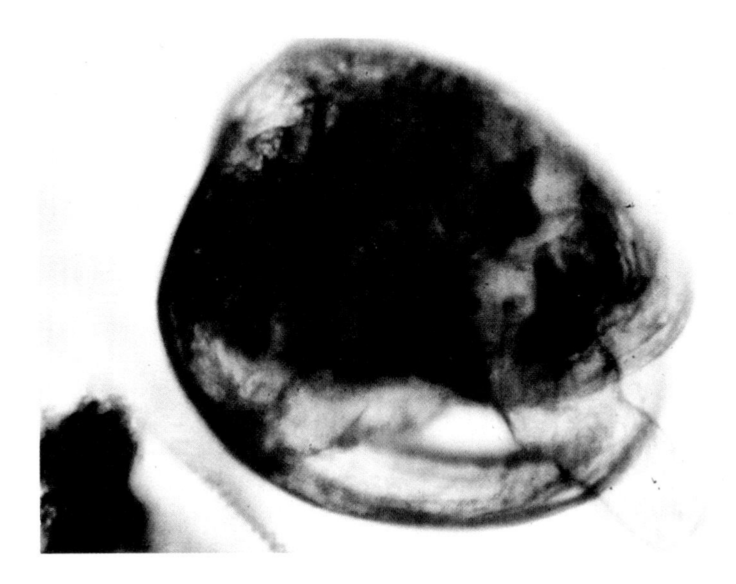

PLATE 5. Live zebra mussel postveliger showing bivalve shell and foot structures. Actual size is about 600 μm. (Photograph courtesy of Gordon Hopkins, Ontario Ministry of the Environment, Rexdale, Ontario.)

PLATE 6. Posterior view of a zebra mussel with inhalant and exhalant siphons actively filtering water. (Photograph courtesy of Wayne Brusate, Commercial Diving and Marine Services, Port Huron, MI.)

PLATE 7. Typical colony of zebra mussels removed from western Lake Erie early in the invasion, April 1989. (Photograph courtesy of Don W. Schloesser, U.S. Fish and Wildlife Service, Ann Arbor, MI.)

PLATE 8. Young-of-the-year zebra mussels attached to submersed aquatic plants (*Vallisneria america*) from Lake St. Clair. (Photograph courtesy of Ron Griffiths, Ontario Ministry of the Environment, London, Ontario.)

PLATE 9. Young-of-the-year zebra mussels attached to a log taken from western Lake Erie. (Photograph courtesy of Ron Peplowski, Detroit Edison Company, Detroit, MI.)

PLATE 10. Windrows of zebra mussels piled on a beach along the Lake Erie shoreline. (Photograph courtesy of Dick Van Nostrand, Bay City Times, Bay City, MI.)

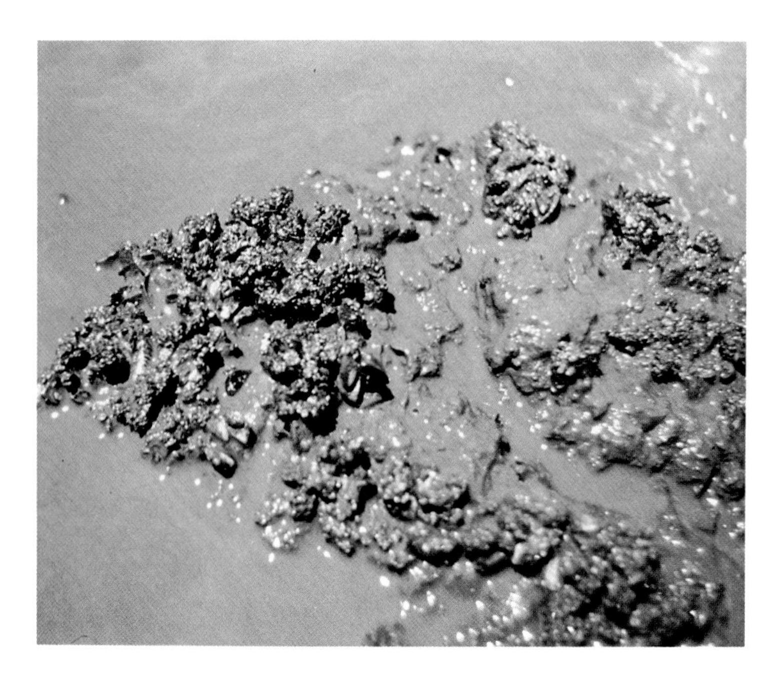

PLATE 11. Clumps of zebra mussels from soft mud substrate in the open waters of western Lake Erie. (Photograph courtesy of Glenn Black, U.S. Fish and Wildlife Service, Ann Arbor, MI.)

PLATE 12. Zebra mussels attached to a crayfish taken from western Lake Erie. (Photograph courtesy of Don W. Schloesser, U.S. Fish and Wildlife Service, Ann Arbor, MI.)

PLATE 13. Zebra mussels attached to a large rock taken from a walleye spawning reef in western Lake Erie. (Photograph courtesy of Joseph Leach, Ontario Ministry of Natural Resources, Wheatley, Ontario.)

PLATE 14. Zebra mussels attached to another non-indigenous bifouling species, *Corbicula fluminea,* from western Lake Erie. (Photograph courtesy of Don W. Schloesser, U.S. Fish and Wildlife Service, Ann Arbor, MI.)

PLATE 15. Zebra mussels attached to a native unionid bivalve collected from Lake Erie. (Photograph courtesy of Peter Yates, Ann Arbor, MI.)

PLATE 16. Dead unionid bivalves on the shore of Lake Erie. This was one of the first indications that zebra mussels were having a negative affect on native bivalves in North America. (Photograph courtesy of Ed Masteller, The Behrend College, Erie, PA.)

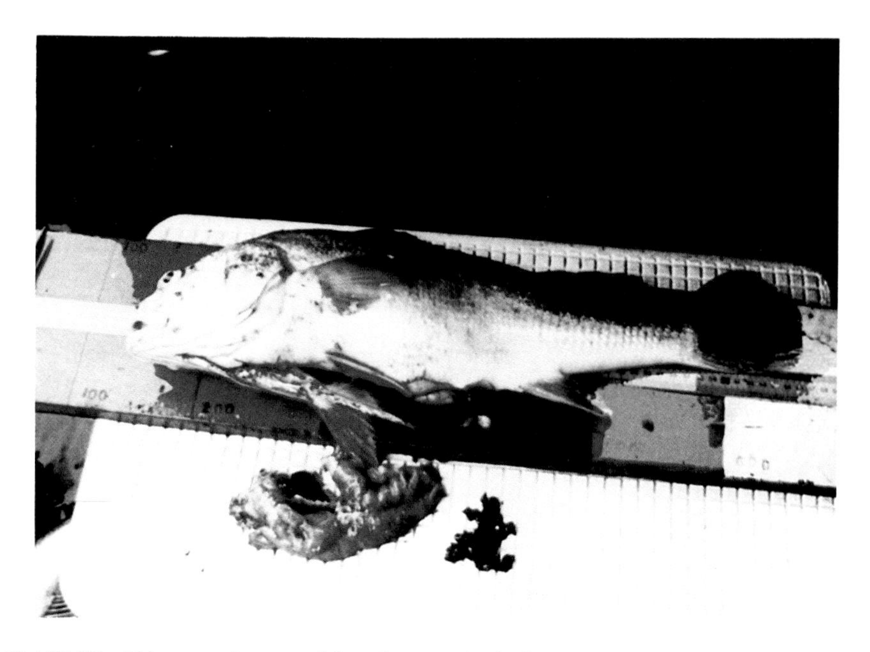

PLATE 17. Zebra mussels removed from the stomach of a freshwater drum from western Lake Erie. (Photograph courtesy of Rich Stickle, U.S. Fish and Wildlife Service, LaCrosse, WI.)

PLATE 18. Zebra mussels and a sponge colony attached to a log from western Lake Erie. Note that both organisms are competing for space. (Photograph courtesy of Ron Peplowski, Detroit Edison Company, Detroit, MI.)

PLATE 19. Zebra mussels attached to a dock post in western Lake Erie. (Photograph courtesy of Ron Griffiths, Ontario Ministry of the Environment, London, Ontario.)

PLATE 20. Zebra mussels attached to an automobile removed from a harbor in northwestern Lake Erie. No mussels were attached to the top portion of the automobile because it was lying upside down when submerged in the water. (Photograph courtesy of Ron Griffiths, Ontario Ministry of the Environment, London, Ontario.)

PLATE 21. Small navigation buoy covered by zebra mussels. The buoy was removed from northwestern Lake Erie where it was observed to be sinking below the water's surface because of the weight of attached mussels. (Photograph courtesy of Joseph Leach, Ontario Ministry of Natural Resources, Wheatley, Ontario.)

PLATE 22. Chain and anchor colonized by zebra mussels. (Photograph courtesy of Ron Griffiths, Ontario Ministry of the Environment, London, Ontario.)

PLATE 23. Zebra mussels attached to the inside of a pipe. (Photograph courtesy of Peter Yates, Ann Arbor, MI.)

PLATE 24. Accumulation of zebra mussels below traveling screens at the Monroe Water Plant. The plant draws water from Lake Erie. (Photograph courtesy of Wil LePage, Monroe Water Works, Monroe, MI.)

PLATE 25. Shells and body tissues of zebra mussels accumulated on traveling screens of the Monroe Water Plant 14 days after treatment with chlorine to remove mussels from the water intake. (Photograph courtesy of Wil LePage, Monroe Water Works, Monroe, MI.)

PLATE 26. Body tissues of zebra mussels clogged in an instrument strainer at the Monroe Water Plant 14 days after chlorination. (Photograph courtesy of Wil LePage, Monroe Water Works, Monroe, MI.)

PLATE 27. Zebra mussels covering a gear of a traveling screen in a power plant along the western shore of Lake Erie. (Photograph courtesy of Ron Peplowski, Detroit Edison Company, Detroit, MI.)

PLATE 28. Clumps of zebra mussels removed from sandy substrates in Lake Erie. At the center of each clump is a pebble which provided the substrate for initial attachment. (Photograph courtesy of Don W. Schloesser, U.S. Fish and Wildlife Service, An Arbor, MI.)

PLATE 29. Zebra mussel dreuses impinged in tubes of a main steam condenser of a power plant along Lake Erie. These dreuses appear to be similar to clumps except they do not have a center substrate for attachment and are believed to have broken off from a larger mussel colony. (Photograph courtesy of Ron Peplowski, Detroit Edison Company, Detroit, MI.)

PLATE 30. Internal strainer plate (41 × 91 cm) of a Coast Guard vessel showing about 50% blockage with zebra mussels. (Photograph courtesy of William Hall, U.S. Coast Guard, Naval Engineering Support, Cleveland, OH.)

PLATE 31. Zebra mussels attached to the steering rudder of a fishing boat removed from western Lake Erie. (Photograph courtesy of James J. Toth, Shrock Marine, Marblehead, OH.)

PLATE 32. Hydroblasting of zebra mussels from screenhouse walls of a power plant along western Lake Erie. (Photograph courtesy of Peter Yates, Ann Arbor, MI.)

1989 when mussels were first found at the Monroe water plant, the mussels spread throughout Lake Erie with phenomenal speed (Griffths et al., 1989). The explosive development of the mussel population in the western basin is highly unusual and has few comparable examples in Europe. The unusual speed and severity of infestations by zebra mussels in Lake Erie are direct causes of the mussel problems at the Monroe waterworks. The current study is a case history of some problems the zebra mussel caused between 1989 and 1991.

MUSSELS IN MONROE WATERWORKS

First Discovery

Zebra mussels were discovered at the Monroe waterworks on January 29, 1989, during routine inspection of the ozonation system. Substantial numbers of then unidentified, and previously unencountered, bivalves were discovered in all three stages of Pass II of the ozone contact chamber where offgases from primary diffusion stages, still containing varying low concentrations of unreacted ozone, are introduced (Figure 4; discussed below). Infestation, which occurred between April 14, 1988, when the systems were previously inspected, and January 29, 1989 was extensive. However, little significance was attached to the animal's presence since it was not uncommon for small gastropods and crustacea to pass through the intake screens and traverse the nearly 10 mi of pipe to the treatment plant. Bivalves were attached to the ozone diffuser header concrete support blocks, the stainless steel diffuser headers, the diffuser end plates, and the floor and lower portion of the walls. Densities varied according to site, but generally decreased in proportion to the distance above the floor and inward from the entrance. Size of the mussels varied, but the majority were in the less than 5-mm range with a maximum of 10–11 mm. It was noted that smaller mussels were not present in any stage of Pass I, where ozone application averaged 1.5 mg/L. Mussels were hand-scraped from the populated surfaces and disposed of in a landfill; the ozone contactor was sealed and returned to service. It was not until months later that the bivalves were identified as zebra mussels and all the ramifications of their presence were revealed.

In order to adequately describe the infestation of zebra mussels at the Monroe waterworks, a brief description of the affected portion of the ozone contactor is necessary (Figure 3) (LePage, 1981).

Pass II of the ozone contact chamber, where the mussels were first discovered in January 1989, is the point at which raw water has traveled 16 km (10 mi) from the underwater crib located in western Lake Erie to the treatment plant (Figure 1). The chamber is divided into three stages, or rooms, which are described as follows:

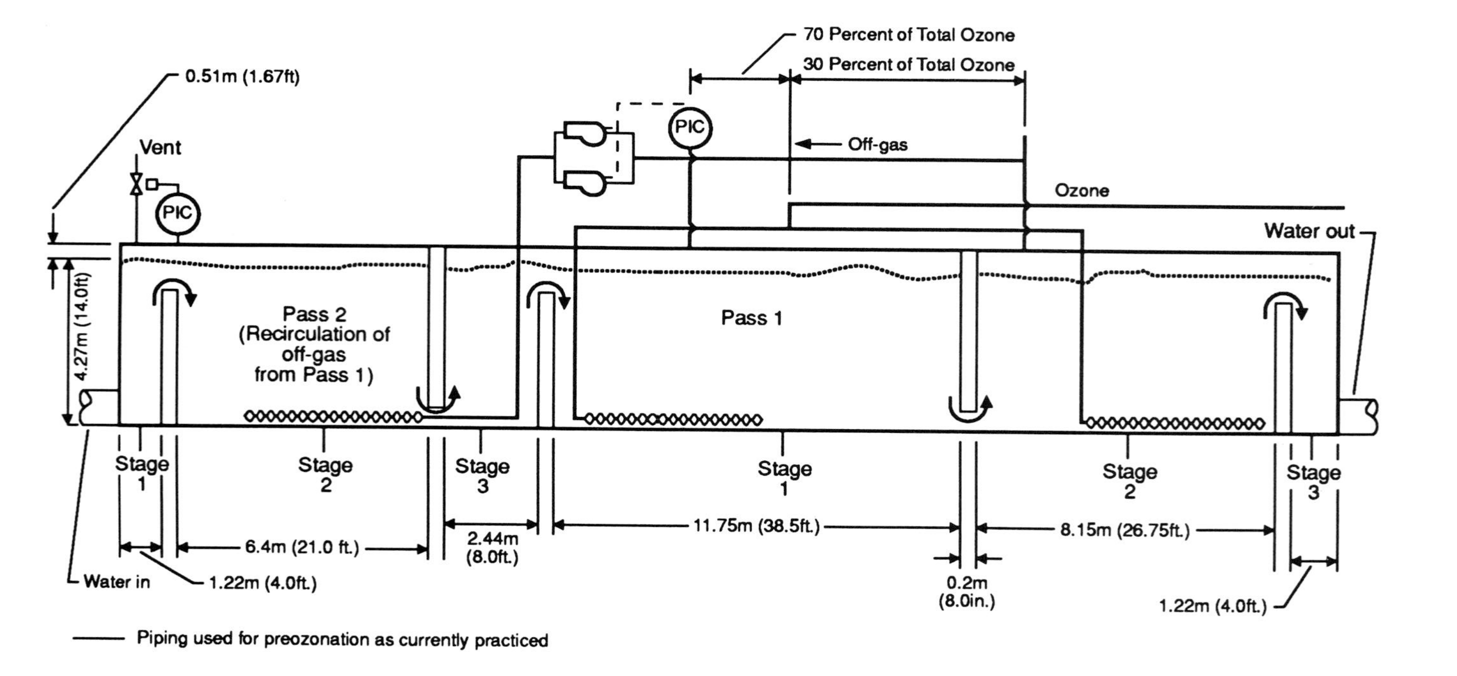

Figure 4. Schematic of the ozone contact chamber at Monroe waterworks.

- Stage 1, the inlet riser stage, is 1.2 m long × 3.0 m wide × 4.9 m high (4 × 10 × 16 ft) and contains no diffusers; hence, no ozone is present. Raw water enters the treatment plant at this point.
- Stage 2, is the offgas recirculation stage, where offgas from Pass I is diffused into raw water entering the ozone contactor. Varying low concentrations of ozone are applied here. This chamber measures 6.4 m long × 3.0 m wide × 4.9 m high (21 × 10 × 16 ft) and contains two horizontal diffuser headers with 70 diffusers attached. The 15-cm (6-in) stainless steel headers are supported on concrete bases to elevate the diffusers approximately 8 in. above the floor.
- Stage 3, is the riser stage between Pass II and Pass I. This stage measures 2.4 m long × 3.0 wide × 4.9 m high (8 × 10 × 16 ft) and contains no diffusers.

On January 29, 1989, before plant operations were impacted by zebra mussels, mussels covered more than 75% of the surface area of concrete supports beneath the headers in Stage 2 of Pass II (Figure 4). Surfaces of the horizontally mounted, 15-cm (6-in.) diameter stainless steel diffuser headers were approximately 20% covered with some animals actually attached to the diffusers. The vertical 15-cm (6-in.) diameter stainless steel pipe at the downstream end of Stage 2 was approximately 20% covered, but only for a distance of 60-100 cm (2-3 ft) above the floor, with only sparse population between the 60- to 100-cm (2-3 ft) mark. A similar pattern was noted on the concrete walls. The extent to which the floor was populated was not ascertained because of mild siltation which also presented a soft and, thereby, probably less desirable site for habitation. Stage 1 had substantially more animals on the walls and of a height equal to the top of the weir separating Stage 1 from Stage 2. While the small amount of ozone present in Pass II appeared to depress the growth of the mussels, it was initially believed that the higher ozone concentrations in the primary diffusion stages of Pass I would be necessary to preclude their survival. This was based on the observation that no mussels were present anywhere in the primary diffusion stages of Pass I.

Underwater Inspections

On June 22, 1989, divers descended to the crib to observe conditions in and surrounding the intake. They noted some mussel attachment to the rock fenders surrounding the crib, but virtually none on the timbers of the crib. The animals encountered were similar in size to those found inside the inlet riser of the ozone contactor.

During the summer of 1989, conditions deteriorated noticeably. In July, unprecedented head losses developed through the gravity-fed intake from Lake Erie. Drawdown in the suction wells of the low service pump station increased almost daily and steadily eroded pump delivery rates due to the decreased suction head. Numerous clusters and individual mussels in the size range of 1–2 cm were collected daily in the traveling screens at the raw water pump station (Figure 5). In 1988, the intake would deliver enough water to sustain

Figure 5. Typical cluster of mussels removed by screens at the Monroe waterworks in the early summer of 1989.

a pumping rate of 4.3×10^4 m^3/day (11.5 mgd). By spring of 1989, the maximum sustainable rate was 4.1×10^4 m^3/day (10.7 mgd), a decrease of 3.0×10^3 m^3/day (0.8 mgd) or 7.0%. By mid-July, the pumps had to be throttled to preclude overpumping the suction wells. On numerous occasions, usually when aggravated by offshore winds which lower lake elevations, pumping rates were limited to about 3.5×10^4 m^3/day (9.2 mgd) down 8.7×10^3 m^3/day (2.3 mgd) or 20% in order to avoid sucking air. Excessive drawdown became an evermore frequent problem. Steady reduction of pumping capacity in the summer of 1989 hastened plans to inspect the intake crib and to penetrate the intake from the lake end. A similar inspection also would originate in the screen chambers and penetrate the intake from the shore end.

On July 15, 1989, divers investigated conditions at the submerged intake crib and inside the 76-cm (30-in.) diameter concrete intake pipe. They reported that the rock fenders and ballast surrounding the timber crib were heavily populated, a lesser population was on the iron work of the access hatch, and only a sparse presence was on the timber crib (mainly in crevices in the wood grain). The interior of the upturned bell of the inlet was covered with mussels between 1 and 20 mm ($^3/_8 \times {}^3/_4$ in.) in length, but not layered. Penetration of the pipeline revealed a less dense but quite uniform pattern of colonization. The bottom one third of the inside circumference of the pipe had an even distribution of approximately 930 animals per square meter ($85/ft^2$). One sixth of the circumference on each side above the bottom one third had approximately 370 mussels per square meter ($35/ft^2$) while the top one third had less than 90 mussels per square meter ($8/ft^2$). At pipe joints, mussels formed clusters at least 5 cm (2-in.) thick and 10 cm (4 in.) wide across the bottom one third of the circumference of the pipe.

On July 30, 1989, divers inspected the traveling screen chambers at the pumping station and penetrated the intake pipe from the shore end. No mussels were found on the screen structures and concrete walls of the screen chamber. Connective piping between the two screen chambers and the intake pipe for a distance of about 10 meters (30 ft) was sparsely populated.

On October 14, 1989, divers inspected the crib and reported that in just three months portions of the intake hatch, latch, and hinges (that were totally cleaned during the July 15 dive) were totally covered to a depth of approximately 10 mm ($^3/_8$ in.) with young-of-the-year mussels. Coverage of the rocks on top of the crib did not appear to have increased substantially; however, new layers of young were forming atop the larger, older mussels.

First Pumping Outage

The first pumping outage attributed to the mussel infestation occurred on September 1, 1989; after pumping at a sustained rate of 3.8×10^4 m^3/day (10 mgd) for more than 16 hr, suction was suddenly and totally lost. The cone check valves on the pumps stalled somewhere between open and shut because of a lack of discharge pressure, and delivery of water to the treatment plant virtually stopped. The pumps would not respond to shutdown commands because the valves were stalled between the limit switches. The 76-cm (30-in.) inlet pipe ran only two thirds full, 7 m (23 ft) below the normal water level in the screen chambers. The pumping rate had to be held at 1.5×10^4 m^3/day (4 mgd) for 2 hr before normal elevation was restored in the suction wells. Thereafter, the rate was raised to 2.5×10^4 m^3/day (6.5 mgd) for an additional 2 hr before head conditions stabilized enough to maintain a rate of 3.1×10^4 m^3/day (8.4 mgd).

This incident prompted measures to establish emergency pumping facilities to obtain water from the Raisin River that flows adjacent to the treatment plant (Figure 1). A series of engine-driven trash pumps situated on the river bank were connected by 15-cm (6-in.) diameter hoses to the inlet of the ozone contactor. This system was successfully tested and kept in readiness in the event of failure of the supply of water from the waterworks.

Treatment of Raw Water Transmission Main

Chlorination to control zebra mussels in the most accessible and longest portion of the transmission main was initiated October 9, 1989. The discovery of mussels at the entrance to the ozonation facility indicated that mussel veligers or postveligers were surviving the trip through more than 10 mi of raw water piping (Figure 1). Since it was not possible to drain the pipe for inspection during summer demands, it was unknown to what extent the pipeline was infested. It was decided that chlorination of the 9 mi of raw water transmission main between the raw water pump station and the treatment plant might be effective in controlling mussels in that vital pipeline. Plans were drawn and a complete chlorination facility was constructed and outfitted to apply chlorine in the discharge header of the raw water pump station.

The literature reports numerous successes using chlorine to control zebra mussels at concentrations ranging from the application of as much as 50 mg/L followed by continuous treatment at 2 mg/L (Clarke, 1952; Greenshields and Ridley, 1957). Others reported successes by periodically producing free chlorine residuals of as little as 0.3 or 0.5 mg/L at the end of the pipe (Jenner, 1983; Hoestlandt, 1968; Jenner, 1984).

Chlorination at Monroe was limited by the equipment immediately available, and by consumer tolerance. It was decided to install a 225 kg/day (500 lb/day) remote vacuum chlorinator and limit the initial application to produce a free residual not to exceed 1.0 mg/L at the end of the pipe. An average chlorine application of 1.86 mg/L produced an average free chlorine residual of 0.36 mg/L at the end of the transmission main. The maximum application was 2.92 mg/L which produced a free chlorine residual of 0.92 mg/L. These data are based on 62 days of continuous chlorination. During the first 23 days of transmission main treatment, the average chlorine application was 2.37 mg/L which produced a free residual of 0.50 mg/L at the end of the pipeline.

Indications that the chlorine was having a marked effect on the mussel population became evident on October 23, 1989 after 14 days of treatment with chlorine. Coincident with the highest free chlorine residual recorded at the end of the pipeline, 0.92 mg/L, troublesome amounts of mussel tissue began to appear in raw water arriving at the treatment plant. The influx of mussel tissue interrupted automatic control of various systems due to clogged strainers (Figure 6). Instruments with small orifices through which raw water had to flow quickly became clogged with decomposing mussel meat and ceased to function. Screens had to be attached to raw water samplers in the laboratory to trap the debris. This condition continued for several days before diminishing.

On October 31, 1989, the ozonation system was removed from service and drained to provide access to the final 15 m (50 ft) of raw water main for visual inspection. Inspection revealed no mussels attached to the wall of the pipe. The only site at which mussels were attached was at the outlet of the

Figure 6. Instrument strainer clogged by mussels.

pipe in the inlet riser of the ozone contactor (Stage 1 of Pass II) and on the walls of the inlet riser. On the floor of the inlet riser, empty mussel shells were piled to a depth of more than 1 m (3 ft) in the corners (Figure 7). Over the baffle, in Stage 2, which is the offgas reinjection stage, empty mussel shells were piled as high as the tops of diffusers throughout approximately half of the floor area (Figure 8). The shells were quite uniform in size with an average length of 2.5 cm (1 in.). All were bleached and totally devoid of meat (Figure 9).

Logic suggests that when the animals first sensed the disagreeable presence of chlorine in their environment, they merely stopped filtering and shut their valves to wait until the foreign substance passed. When they were at last forced to open up and resume filtering or suffocate, they presumably succumbed to the biocide. Following death, the lightweight shell soon separated from the body and was washed, undetected, into the entry stages of the ozone contact chamber. Thereafter, the decomposing remains separated from their byssal attachments and, in suspension, flowed into the treatment plant. This hypothesis might explain the 2-week delay before any effect became noticeable (Figure 6). In all, more than 6 m^3 (8 cu yd) of empty shells were removed from the chamber (Figures 7 and 8). However, after ozonation absolutely no odor was imparted to the drinking water as might be expected from exposure to such quantities of decomposing mussel flesh.

Figure 7. Empty shells in entrance to ozone contact chamber following chlorination of raw water main.

Inspection of the transmission main from each end revealed no remaining mussels attached to the pipe wall. This implied that the infestation was under control all the way back to the raw water pump station, a distance of nearly 14 km (9 mi). Chlorination at the raw water station was discontinued for the winter on December 10, 1989, in the belief that mussels would be inactive and not reproducing during the winter cold water season. This belief was subsequently proved incorrect.

Second and Third Outages

In late 1989, plans were made to clean the raw water intake pipe from the underwater crib in Lake Erie to the pump house prior to winter (Figure 1). However, unfavorable weather and sea conditions forced postponement of the project until the following spring of 1990. On December 14, 1989, the intake pipe was totally blocked by frazzle ice. This was the second and most serious outage. It is believed the mussel population in and around the inlet to the raw water intake contributed heavily to the frazzle ice formation that totally interrupted raw water flow for more than 56 hr. Frazzle ice had

Figure 8. Empty shells in ozone contactor Stage 2, Pass II following chlorination of raw water main.

last been encountered 28 years previously at which time the intake was damaged by ice that separated several joints only 400 m (1300 ft) from shore. The damage was repaired and frazzle ice did not occur again until this incident. The only variables that changed during those 28 years were the repair of the damaged joints and the infestation by zebra mussels. Logic implies that velocity changes and turbulent flows created by the mussel population enhanced the transformation of supercooled water into troublesome shards of frazzle ice. Ordinarily, frazzle ice dissipates as soon as the sun comes out and the supercooling effect is interrupted. Unfortunately, in this instance, heavily overcast skies and frigid temperatures prevailed.

A water emergency was declared and industry and heavy commercial users were asked to shut down. Schools, bars, and restaurants also were required to close and residents were asked to conserve water. The emergency river system was pressed into service and performed well, despite subzero temperatures. Loss of pressure in the distribution system appeared inevitable, prompting the county health officer to issue a boil order. The boil order probably precluded its own necessity, since it served as a final and effective

Figure 9. Empty shells in ozone contact chamber were uniform in length (2–2.5 cm).

deterrent to water use, and no part of the distribution system lost pressure. In the morning of the third day, the sun came out, the ice dissipated, and flow was resumed.

A third, but less severe outage occurred February 16, 1991. Again the emergency water system was utilized and additional water use restrictions were unnecessary. The outage cleared after only 9 hr.

Temporary Auxillary Water Supply

In April 1990, with the advent of improving weather and sea conditions, preparations to clean the intake pipe were initiated. A temporary raw water supply system was erected and tested. The system consisted of two skid-mounted foot valves and suction lines which, in December, had been positioned on the ice about 75 m (250 ft) from shore (Figure 10). When the ice melted, the entire intake assembly sank in about 1.5 m (5 ft) of water. The suction lines were connected to two diesel engine-driven, skid-mounted dredge pumps situated at the lake shore (Figure 11). Flow was conducted 265 m (870 ft) to the pump station via two 40-cm (16-in) diameter seamless plastic pipelines. One pipe discharged directly into the west screen chamber of the pump house, and the second terminated near the cleanout fitting in the yard (Figure 12). The second line would provide wash water for the cleaning operation and also be available to supply additional water to the pump station,

Figure 10. Foot valves at the inlet to the auxiliary raw water supply system.

Figure 11. Auxiliary raw water supply pumps.

Figure 12. Auxiliary raw water supply pipelines.

should it become necessary. A series of float switches installed in the screen chamber indicated suction well elevations within operating tolerance. During tests in conjunction with the river pumping system, sustained rates in excess of 5.7×10^4 m^3/day (15 mgd) were attained.

Cleaning Intake

In late April 1990, divers opened the crib access and released a parachute-like drogue attached to a 10-mm (3/8-in.) diameter rope into the 1.6-km (6100-ft) long pipe from the crib to shore. Water flow through the pipe carried the drogue beyond the cleanout fitting on shore where it was retrieved by a diver. The rope was then used to winch a 10-mm (3/8-in.) diameter wire cable through the cleanout and back to a barge located above the crib. This cable was next used to pull a 13-mm (1/2-in.) diameter steel cable back to the cleanout on shore. This completed arrangements for the actual cleaning operation. At this time mussels within the crib structure were about 15 cm (6 in.) thick on and inside the upturned bell of the intake.

A scrubbing device to be pulled through the intake pipe was designed and built by the cleaning contractor (Figure 13). It consisted of a tubular steel hull with two rows of spring-loaded scraper rings forward, and a combination squeegee/propulsion disk at the stern. Four high-pressure nozzles were designed to wash dislodged debris ahead of the device and back into the lake.

Figure 13. The scrubbing device for the initial pass.

A 13-mm (1/2-in.) diameter steel cable was attached to each end, one to pull the device through the pipeline and the other to retrieve it in the event it became stuck.

In early May 1990, the scrubber was shackled to the cables and inserted into the intake pipe. A wash water pipe was bolted in place and pressurized with water to blow debris away from the scrubber, and the scrubber was pulled through the intake pipe. Observers described the flow of zebra mussels being removed by the scrubber and exiting the crib opening as resembling an eruption of an underwater volcano.

The intake was returned to service. When water pumping began, a tremendous amount of loose debris washed into the screen chambers and less water was obtained than before the cleaning. It required nearly 16 hr of almost continuous operation of the screens before accumulation of debris tapered off and normal flow was restored (Figure 14).

A few days after the scrubber was drawn through the intake pipe a device intended to sweep and squeegee all remaining debris from the pipeline was inserted into the pipe (Figure 15). The new machine was an articulated device consisting of stiff bristle brushes on the leading unit followed by a tight-fitting squeegee on the trailing unit. The squeegee was made of about six

Figure 14. Water screens jammed with mussel debris after cleaning the intake.

layers of 19-mm (3/4-in.) thick reinforced rubber belting bolted together and backed up with steel plates. This tight-fitting device was pulled through the pipe, leaving the pipe free of debris. The pipe was sealed and returned to service. Three hours after returning the intake to service, flow through the west suction well stopped and the operator had to shift to pumps in the east well. By morning, the east well screen also was plugged. Maintenance staff found the screens solidly jammed with zebra mussel debris, the pins sheared in the drive mechanism, and the machinery totally immovable (Figure 14). The screen trash hoppers were so choked with debris that when the access door was opened, the debris surged out on the floor. Apparently, much of the huge pile of loose debris removed from the intake and left on the lake bottom near the intake after the cleaning operation was sucked back into the pipe when flow resumed. It took hours to free the screens and return them to service followed by many hours of screening and shoveling loose shells. Large quantities of loose debris continued to arrive for a week before diminishing. Pulverized shells washed through the screens and onward for a distance of 9 mi to the treatment plant where they settled in the ozone contactor.

Divers went down to airlift the debris from the crib and redistribute it on the lake floor a safe distance away. They reported the debris removed from

Figure 15. The sweeper-squeegee device for the final pass.

the pipeline was stacked up like the cone of a volcano in and around the crib. The divers estimated that they removed the equivalent of "seven or eight large dump truck loads" of shells from the immediate area of the crib.

Treatment of Intake Pipe

Immediately after the pipe was believed to be clean work began to pull a 5-cm (2-in.) diameter high-density polyethylene hose through the interior of the pipe to the intake crib offshore. This pipe was to be used to apply chlorine at the crib 1.6 km (6100 ft) offshore.

The cable from the barge was connected to the end of the 5-cm (2-in) diameter pipe; and the pull proceeded smoothly until, with 0.9 km (3000 ft) in the hole, a thermally fused joint separated and the tightly stretched hose backlashed and broke in a second place. When both ends were withdrawn, about 0.5 km (1800 ft) remained unaccounted for. The next attempt to pull the polyethylene line through the pipe was successful. Attempts to retrieve the missing 0.5 km (1800 ft) of hose were unsuccessful.

Divers installed the equipment on the intake inlet and on June 20, 1990, delivery from the 225 kg/day (500 lb/day) chlorinator was diverted to the offshore point of application. Rates were adjusted to maintain a free chlorine residual of 0.5–1.0 mg/L in water arriving at the treatment plant 10 mi from

Figure 16. Screen-clogging agglomeration following commencement of offshore chlorination.

the point of application. After 3 days of offshore chlorination, the screens again became clogged (Figure 14). The material was unlike anything previously encountered. It consisted of fibrous or threadlike material intermixed with a seemingly gelatinous binder and generously interspersed with whole and broken mussel shells (Figure 16). It formed a mat on the screens and severely impaired flow (Figure 17). The material consisted mainly of pulverized and decomposing remains of zebra mussels left behind during the cleaning operation and others that had washed back in before the crib was vacuumed. The arrival of this material stopped after a few days and did not reappear. Shortly after normal operation was resumed, the Microfloc filters began to air bind. This was an occurrence unheard of in the warm water season. Massive gouts of air were expelled during backwashing which threatened to upset the filter beds despite the greatest caution. It is believed the problem was not caused by air, but instead by methane and carbon dioxide and possibly other gaseous decomposition products of rotting protein resulting from the decay of the animals' byssal threads, which presumably had remained attached to the pipe wall (Greenshields and Ridley, 1957).

On September 12, 1990, after 3 months of chlorination offshore, an underwater inspection of the crib and penetration of the pipe was conducted.

Figure 17. The impervious cake formed by the agglomeration illustrated in Figure 16.

Divers reported a massive accumulation of live mussels on the lake bottom extending more than 15 m (50 ft) in all directions from the crib. Areas of the crib that had been entirely free of mussels during the divers last visit on June 12, were totally covered with 8- to 12-mm (1/4- to 1/2-in.) mussels. Tiny mussels were so entrenched in the gap surrounding the crib door that more than an hour was required to free it. Inside the crib, the rim of the bell of the intake and the ring to which the chlorine diffuser was attached supported a mussel population more than 5 cm (2 in.) thick. From the chlorine diffuser inward, however, the pipe was 100% free of mussels.

An inspection of the ozone contact chambers at the treatment plant revealed a substantial deposit of pulverized zebra mussel shells (Figure 18). These shells were bleached white from exposure to chlorine and ozone and are believed to have been deposited there immediately after the cleaning operation. During this inspection, there was no evidence of animals having attached anywhere in the structure or in the accessible portion of raw water main serving it.

Based on observations, there is good reason to believe that continuous chlorination is essential not only during the warm water months of summer, but during the cold water season, as well.

Figure 18. Bleached shells in the ozone contactor following offshore chlorination.

Chlorination offshore was suspended on November 20, 1990, when the raw lake water temperature dropped to 10.2° C. This was in accordance with the common belief that treatment was necessary only in summer during the reproductive season of the zebra mussels. In January 1991, scarcely 2 months following interruption of continuous treatment, the west suction well at the raw water pump station was drained and opened for routine inspection and cleaning. Approximately 600 live mussels were discovered attached to the wall just above the mud line in an area adjacent to the inlet port from the screen chamber. These observations are supported by the earlier work of Clarke (1952) who found zebra mussels are capable of surviving very high levels of chlorine for as long as 2 weeks. No mussels were found at the treatment plant or in the accessible portion of the raw water transmission main immediately upstream of the ozone contactor.

Offshore chlorination was immediately reinstated, but on an intermittent schedule. Chlorine was applied at the crib for 6 hr each day, Monday through Friday. An average dose of 3.0 mg/L produced a free chlorine residual of 0.50 mg/L in the screen chambers. After 6 weeks of intermittent treatment, the east suction well was dewatered for inspection. Approximately 500 mussels 8–12 mm (1/4–1/2 in.) were discovered attached to the wall at the mud line just as they had been in the west well. The live zebra mussels found in the shorewells are suspected to be migrants that entered the shorewells after continuous chlorination was stopped for the winter.

Permanent Control Strategy

It is apparent by virtue of the September 1990 inspections that continuous chlorination, as practiced at Monroe, is an effective defense against zebra mussels. As a result, future plans include offshore chlorination of the existing intake pipe on a continuous basis and new construction at the raw water pump station to include chemical storage tanks and chemical delivery lines to the new 42-in. diameter intake. In addition, new telemetry and control equipment will provide remote control of chlorine application and continuous monitoring of critical parameters. The current use of liquid chlorine will be discontinued in favor of sodium hypochlorite for reasons of safety at the unmanned facility.

The ease of application and proven effectiveness of offshore chlorination make a strong case for its continued use in mussel management. However, there is uncertainty about its continued use because more stringent regulation of disinfection by-products (DBPs) is expected. The U.S. Environmental Protection Agency (EPA) has advised that the maximum contaminant level (MCL) for total trihalomethanes (TTHM) be lowered from its current 100 μg/L (U.S. Environmental Protection Agency, 1987a and 1987b, 1988). In addition, MCLs are expected to be proposed for a number of other halogenated disinfection by-products, which with TTHM are described as total organic halogen (TOX). This might in the future further impede compliance when using chlorine as a primary disinfectant and mussel control agent.

A lowered MCL for TTHM alone could make compliance difficult while using chlorine in this fashion. Total trihalomethanes increased during chlorination at the shorewell in 1989 (63 μg/L) and at the crib during 1990 (57 μg/L). Variables other than chlorine, such as lake elevation and turdidity are known to have a marked influence on TTHM formation.

During 1990, slightly more than half of all chlorine applied was introduced offshore at the crib, with the remainder applied at the treatment plant following ozonation (Table 1). This translated to an average of 2.94 mg/L applied offshore which produced an average free chlorine residual of 0.59 mg/L in water arriving at the treatment plant. This is believed to be more chlorine than is necessary to achieve our objective. Therefore, if the MCL for TTHM is lowered to 50 μ/L, compliance might be achieved (even though marginally) if chlorination practices are fine-tuned. If the MCL is lower than 50 μg/L, the ability to comply while prechlorinating will be severely compromised.

Table 1. Applied and Residual Chlorine Concentrations Used to Control *Dreissena polymorpha* at the Monroe Waterworks in 1990

	Applied at Crib (mg/L)	Terminal Residual (mg/L)
June	2.49	0.32
July	3.03	0.60
August	3.27	0.73
September	3.05	0.63
October	2.91	0.61
November	2.90	0.63
Average	2.94	0.59

Costs

The financial burden of responding to the mussel invasion has been substantial for the community of 24,000 people in the City of Monroe; costs are estimated at about $300,950 between January 1989 and January 1991 (Table 2). The two largest expenditures to date have been for research and engineering studies and cleaning of raw water intake pipes (about 44% of total expenditures). Many of the other larger expenditures were also associated with these operations (e.g., unbudgeted man-hours, equipment). The actual costs of chlorine control methodology was about $50,000 (about 17% of total costs).

In addition to estimated costs, operating costs also have increased. Offshore chlorination for mussel control has increased annual chlorine consumption 36%. Personnel reassignments to the otherwise unmanned raw water pump station have cut deeply into overtime accounts as well as regular operating and maintenance accounts. The resulting deferred maintenance at the treatment plant also has increased costs due to an undesirable shift from preventive maintenance to response maintenance. Approximately $100,000 more will be spent on new chemical storage and application as well as monitoring and control equipment at the raw water pump station during 1991. If, for any reason, chlorination is discarded in favor of an alternate treatment for mussel control, the cost could soar to as much as an additional $3,000,000 for an ozonation system currently under development (LePage and Bollyky, 1990).

SUMMARY

The threat to the Monroe water supply by the zebra mussel was sudden and serious. On several occasions, zebra mussels nearly brought the water system to a standstill; however, the resourcefullness and cooperation of many

Table 2. Estimated Costs Directly Attributable to the Zebra Mussel at the Monroe Waterworks from January 1989 to January 1991

Education and training	4,500
Congressional testimony	1,000
Research and engineering studies	60,000
Experimental devices	600
Underwater inspections	5,000
Chlorination facility and equipment	34,500
Chlorine delivery line	8,000
Chlorine	7,000
Chlorine solution line entry enclosure	1,500
Contractual service (Vactor rental)	1,000
Vacuum hose fabrication	400
Lost revenue during mussel related outages	20,000
Contractual service during outage	7,000
Purchased water during mussel-related outage	1,200
Operation of emergency river supply	6,000
Installation of discharge check valve at pump station	2,750
Cleaning of raw water intake pipes	73,000
Supervision and technical man-hours	30,000
Unbudgeted man-hours at pump station	37,500
Total	$300,950

people within and outside the water department brought the invaders under control.

Measures to control the mussels at the Monroe waterworks are now recognized as a routine part of the treatment process. The high cost of controlling the animal is passed to the consumer in the form of higher utility rates. However, if new treatments have to be adopted in the future in order to comply with impending changes in water quality standards, the cost of such alternative treatments might substantially increase already high expenses.

The case history of the zebra mussel at the Monroe waterworks demonstrates the extreme problems that can be caused by the mussels when there is little knowledge about them and how fast the animals can affect water supply operations. To this end, the story of the zebra mussel at the Monroe waterworks will serve as a guide and early warning message to water treatment operators throughout North America.

REFERENCES

Clarke, K. B. "The Infestation of Waterworks by *Dreissena polymorpha*, a Freshwater Mussel," *J. Inst. Water Eng.* 6:370–379 (1952).

Greenshields, F., and J. E. Ridley. "Some Researches on the Control of Mussels in Water Pipes," *J. Inst. Water Eng.* 11:300–306 (1957).

Griffiths, R. W., W. P. Kovalak, and D. W. Schloesser. ''The Zebra Mussel, *Dreissena polymorpha* (Pallas, 1971), in North America: Impact on Raw Water Users.'' Nuclear Power Division, Electric Power Research Institute, Palo Alto, CA (1989).

Hebert, P. D. N., B. W. Muncaster, and G. L. Mackie. ''Ecological and Genetic Studies on *Dreissena polymorpha* (Pallas),'' (In press).

Hoestlandt, H. ''Le Probleme du Mollusque *Dreissena polymorpha* dans l'Approvisionnement en Eau Potable de la Region d'Istanbul,'' Organisation Mondiale de la Sante, Bureau Regional de l'Europe (1968).

Jenner, H. A. ''Control of Mussel Fouling in The Netherlands: Experimental and Existing Methods,'' Proceedings Symposium on Condenser Macrofouling Control Technologies, The State-of-the-Art. J. A. Diaz-Tous et al., Eds. (1983).

Jenner, H. A. ''Chlorine Minimization in Macrofouling Control in the Netherlands,'' in *Water Chlorination Vol. 5. Chemistry, Environmental Impact and Health Effects*, R. E. Jolly et al., Eds. (Chelsea, MI: Lewis Publishers, 1984).

The King Company, 13520 Barry Street, Holland, Michigan 49424.

LePage, W. L. ''The Anatomy of an Ozone Plant,'' *J. Am. Water Works Assoc.* 73:2–105 (1981).

LePage, W. L., and L. J. Bollyky. ''A Proposed Treatment for *Dreissena polymorpha*,'' International Ozone Association, Pan American Committee, Norwalk, CT (1990).

U.S. Environmental Protection Agency. The Safe Drinking Water Act as Amended by the Safe Drinking Water Amendments of 1986.

U.S. Environmental Protection Agency. ''National Primary Drinking Water Regulations — Synthetic Organic Chemicals; Monitoring for Unregulated Contaminants; Final Rule,'' *Fed. Reg.* 52(130):25689–25717 (1987a).

U.S. Environmental Protection Agency. ''National Primary Drinking Water Regulations; Filtration and Disinfection; Turbidity, Giardia Lamblia, Viruses, Legionella, and Heterotrophic Bacteria; Proposed Rule,'' *Fed. Reg.* 52(212):42177–42222 (1987b).

U.S. Environmental Protection Agency. ''Drinking Water; Substitution of Contaminants and Drinking Water Priority List of Additional Substances Which May Require Regulation Under the Safe Drinking Water Act,'' *Fed. Reg.* 53(14):1891–1902 (1988).

CHAPTER 22

Infestation of Power Plant Water Systems by the Zebra Mussel (*Dreissena polymorpha* Pallas)

William P. Kovalak, Gary D. Longton, and Richard D. Smithee

Between 1988 and 1990, zebra mussel densities in cooling water intakes were determined at 10 power plants located from southern Lake Huron to western Lake Erie. Mussels were absent at power plants along Lake Huron and the St. Clair River, but occurred at low densities (i.e., $<1000/m^2$) along the Detroit River, and occurred at high densities (i.e., up to $750,000/m^2$) in western Lake Erie. Differences in mussel densities between intakes appear to be related to dispersal rates, water quality, water temperature, and food supply.

High mussel densities and rapid growth rates caused extensive fouling of water systems at the most southern plant, Monroe Power Plant in western Lake Erie. Mussels fouled trash bars, screenhouses, main steam condensers, heat exchangers, and the service water system. Blockage of condensers was up to 20%, with most of the blockage caused by druses (i.e., clusters) sloughed from screenhouses. Screenhouses, where the fouling layer was as much as 12 cm thick, have been routinely dewatered and cleaned mechanically using hydrolasing (21,000 kP; 3000 psi). Low level continuous chlorination (0.5 mg/L total residual chlorine) has been used to control fouling of the service water system.

0-87371-696-5/93/$0.00 + $.50

INTRODUCTION

In Europe, infestation by zebra mussels (*Dreissena polymorpha* Pallas) has caused chronic problems at raw water intakes (e.g., Clarke, 1952; Greenshields and Ridley, 1957; Morton, 1979). Fouling of water systems is facilitated by two characteristics of this species, planktonic veliger larvae and byssal threads. Larvae are readily drawn into water intakes and penetrate most parts of water systems, while byssal threads allow attachment to most surfaces (particularly other mussel shells) resulting in development of thick layers that can block pipes and other conduits. Cases of mussel layers up to 30 cm thick inside pipes have been reported (Clarke, 1952).

Detroit Edison Company operates power plants that are located both upstream and downstream of Lake St. Clair, the focus of zebra mussel introduction in the Laurentian Great Lakes (Hebert et al., 1989; Griffiths et al., 1991). Because of concern about fouling of water systems, studies of zebra mussels were initiated in 1988. This paper describes (1) results of mussel monitoring at cooling water intakes between 1988 and 1990, (2) chronology and extent of infestation at Monroe Power Plant in western Lake Erie, where fouling has been greatest, and (3) control methods presently used to minimize operational impacts.

METHODS

A total of 10 power plants are operated by the Detroit Edison Company; these plants are located between Harbor Beach in southern Lake Huron and Monroe in western Lake Erie (Figure 1). With the exception of Greenwood and Fermi 2, all are coal fired and use once-through cooling. Greenwood burns natural gas and/or residual oil while Fermi 2 is a nuclear power plant; however, both plants use closed-cycle cooling. Marysville and Conners Creek were sampled for zebra mussels even though they were not in operation during this study.

A variety of methods were used to determine the presence of zebra mussels. Typically, inspections consisted of divers examining debris and concrete structures in intake canals, concrete and steel walls of intake structures, and stony substrates adjacent to intakes. Mussel densities were estimated from debris (e.g., wood, concrete, steel) collected by divers from the bottom of intake canals and from standardized substrates. The latter consisted of two 15 × 15 × 2.5 cm concrete plates spaced 15 cm apart on an eyebolt 45 cm long. At each sampling site, substrates were suspended by ropes approximately 1 m below the water surface and 1 m above the canal bottom. The number of substrates varied between power plants, ranging from two in small intakes to six in large intakes.

Quantitative samples were collected from vertical surfaces with metal sleeves of two designs. The sampler used in 1989 was square at one end

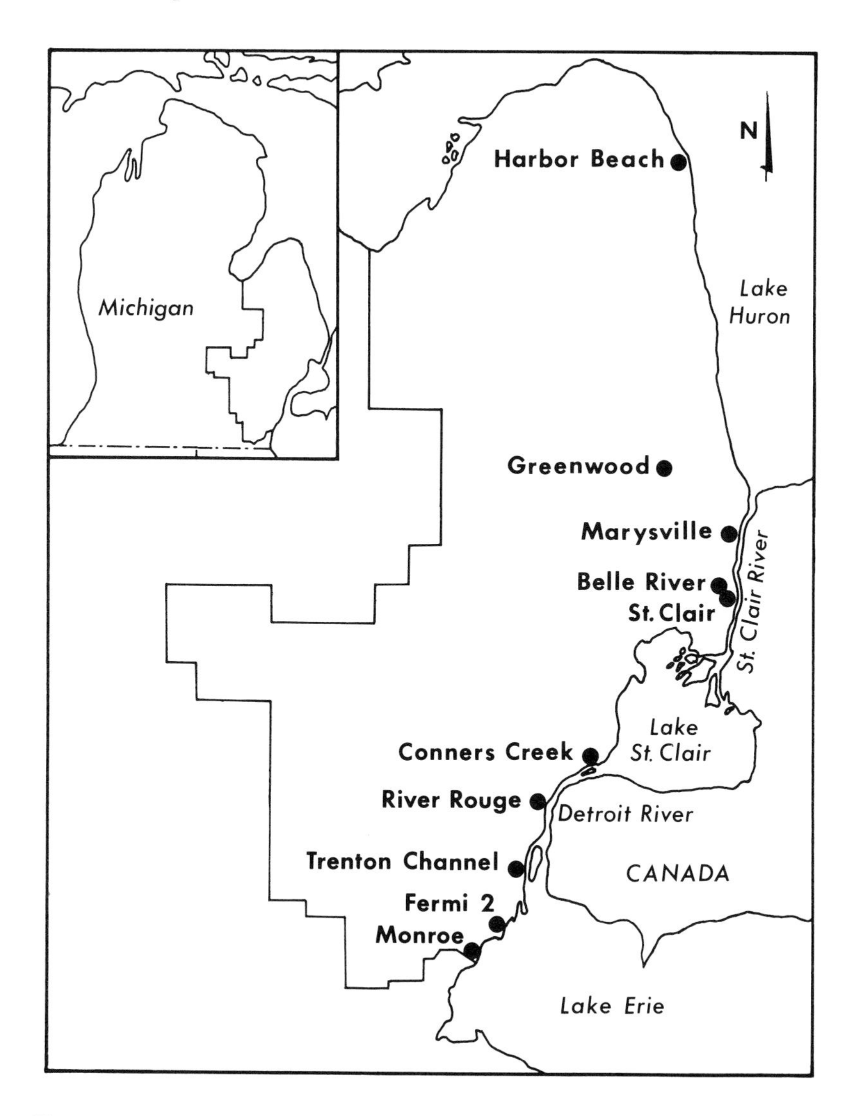

Figure 1. Map of Detroit Edison service area and the locations of 10 power plants examined for *Dreissena polymorpha* in 1989 and 1990.

(10 × 10 cm) and round at the other (10-cm diameter). The round end was fitted with a knee-length nylon stocking. This sampler was used in either of two ways. When densities were low, it was used as a scraper. The square end of the sampler was pushed vertically along a concrete or steel surface over a predetermined distance (typically 20 cm), and dislodged mussels fell into the stocking. When densities were high, the sampler was pushed and held perpendicular to the surface while a spatula was used to remove mussels

Table 1. Generating Capacity, Cooling Water Flow (Summer Maximum at Capacity) and Densities of *Dreissena polymorpha* at the 10 Detroit Edison Power Plants in 1989 and 1990

Plant	Capacity (MWe)	Cooling Flow (m^3/sec)	Densities (number/m^2)	
			1989[b]	1990[c]
Harbor Beach	103	3	0	0
Greenwood	785	0.5	—	0
Marysville[a]	—	<0.1	0	0
Belle River	1280	29	0	0
St. Clair	1361	50	<1	0
Conners Creek[a]	—	<0.1	—	20
River Rouge	516	19	20–50	50–250
Trenton Channel	630	23	0	0
Fermi 2	1093	2	25,000–30,000	15,000–20,000[b]
Monroe	3000	85	700,000–800,000	350,000–500,000[b]

[a] Not operated during 1989 and 1990.
[b] From intake structure and/or debris from intake canal.
[c] From concrete substrates suspended in intake.

around the sampler. The spatula was then used to remove mussels enclosed by the sampler and to direct them into the stocking. In 1990, the sampler was redesigned so that it was round at both ends (10-cm diameter). Rounding the end of the sampler allowed it to be rotated as it was pushed through thicker layers of mussels. Samples were collected with a spatula using the procedure described above.

Measurements of shell length of individuals larger than 5 mm were made with a vernier caliper (to the nearest 0.02 mm), while small individuals (<5 mm) were measured with an ocular micrometer to the nearest 0.1 mm.

RESULTS AND DISCUSSION

Overview of Ten Plants

Small number of mussels (3–11 mm in shell length) were first collected from woody debris in the Monroe intake in early August 1988. These were young of the year spawned in 1988, but subsequent analysis of larger samples (Griffiths et al., 1991) suggested that zebra mussels first arrived in 1987. In late 1988, mussels were not found at intakes of other operating plants.

Densities increased dramatically at Monroe in 1989 and by late summer were 700,000–800,000/m^2 (Table 1). At Fermi 2, located approximately 11 km northeast of Monroe, densities were substantially lower (25,000–30,000/m^2). At other plants, mussels were absent or occurred at low densities. At both Monroe and Fermi 2, mussels declined somewhat in 1990, with densities of 350,000–500,000 and 15,000–20,000/m^2 at the two plants, respectively. At River Rouge, densities increased slightly while at other plants no mussels were found or mussel densities remained low.

Mussel densities at Detroit Edison water intakes appear to be regulated by a number of factors. Intakes on Lake Huron and St. Clair River are upstream of Lake St. Clair, the site of initial introduction. Upstream dispersal, which depends on active transport (e.g., ships and boats), has been slow so that isolated populations in the upper Great Lakes have been found primarily at major ports (Griffiths et al., 1991). Additionally, lower water temperatures and less food probably will discourage substantial population growth in the upper Great Lakes compared to the lower (Stanczykowska, 1977).

Low population densities in the lower Detroit River may be related to low water quality. Both the River Rouge and Trenton Channel plants are downstream of the Rouge River, which historically has been polluted by domestic and industrial waste. Despite recent improvements, this area is still designated a Class A area of concern by the International Joint Commission (Manny et al., 1988). In Europe, zebra mussels are absent from polluted waters (van Urk, 1976).

High mussel densities in western Lake Erie were probably a result of favorable water temperatures and an abundant food supply. Because climatic conditions in the Detroit area are similar to conditions within the mussel's original range along the northern shores of the Black and Caspian Seas (Ruffner and Blair, 1987; cf. Strayer, 1991), the range in annual water temperatures (0–24°C) is also quite similar. In addition, primary production rates are very high in western Lake Erie (Vollenweider et al., 1974). Mussel growth rates are apparently correlated with productivity in the Great Lakes (Nichols, Unpublished data). Similarly, Stanczykowska (1977) reported a correlation between mussel size and growth and seston concentrations in some European lakes.

Decreased mussel densities at Monroe in 1990, as compared to 1989, may have been a result of yearly differences in water temperatures and/or populations exceeding the carrying capacity of the environment. Sudden changes in water temperature observed in May (Nichols, Personal communication) may have desynchronized reproduction since spawning was 3–4 weeks later in 1990 than in 1989, and lower water temperatures during most of the summer may have reduced recruitment and growth. High densities may also have reduced recruitment and growth because of crowding and reduced food supply. Increased water transparency in the western and central basins of Lake Erie following mussel introduction have been attributed to the filtration and biosedimentation of algae by mussels (Leach, Personal communication). However, reductions in spawning and growth do not necessarily translate into reduced fouling of water intakes. Experiences with mussel fouling in Russia have indicated that fouling will remain high even if mussel densities decline (Smirnova, Personal communication). In 1989 when fouling was greatest, mussel densities in Lake Erie were relatively low ($<5000/m^2$ on suitable substrate).

Differences in mussel densities between Monroe and Fermi 2 appeared to be related to differences in cooling water flow. In 1989, mussel densities at Monroe (700,000–800,000/m^2) were about 30 times greater than at Fermi 2 (25,000–30,000/m^2). At maximum flow during the summer (coinciding with the period of mussel spawning) the Monroe plant draws about 85 m^3/sec (1.3 million gpm), whereas Fermi 2 draws about 2 m^3/sec; (30,000 gpm), which is a 40-fold difference. Although many factors may influence mussel densities in water intake lines, the primary factor for power plants with a design intake speed of 15 cm/sec appears to be volume of flow. Because intakes concentrate larvae resulting in mussel densities greater than those in the source water, they are the best place to monitor the early phases of infestation.

Despite moderate mussel densities at the Fermi 2 water intake, no live mussels have been found in the Fermi 2 condenser cooling water system. The Fermi 2 closed-cycle cooling system includes two cooling towers and a small reservoir (pond) which directly supplies water to condensers and other heat exchangers. Between June and October, maximum temperatures in the cooling pond approach 36°C, and exposure of mussels to this high a temperature for less than 1 hr has proven lethal (Jenner, 1983). Thus, high water temperature within Fermi 2 water systems has probably prevented mussel fouling.

Monroe Power Plant — Site Description

The cooling water intake at Monroe is located on the Raisin River approximately 800 m upstream from Lake Erie (Figure 2). At capacity, flow is approximately 85 m^3/sec (1.3 million gpm); a portion of cooling water is drawn from the Raisin River and the other portion from Lake Erie. Most of the flow enters main steam condensers and a variety of smaller heat exchangers. Approximately 2 m^3/sec (30,000 gpm) of water is supplied to the service water system where it is used for the fire protection system and also for ash handling.

The intake canal branches to two screenhouses, each supplying water to two units (Figure 2). Within screenhouses, water supply to steam condensers is separate for the two units, whereas service water is supplied by a common header. Water entering screenhouses first passes through trash bars that exclude large debris (Figure 3). Trash bars are vertical slats set on 7.5-cm (3-in.) centers and organized in 16 racks (8 per screenhouse) which are 3.5 m wide, and 6 m long (submerged portion). Downstream of trash bars are traveling screens (16; 4 per unit) with 9.5-mm (3/8-in.) mesh. They exclude smaller debris that might occlude condenser tubes (19-mm; 3/4-in. diameter). Downstream of traveling screens the screenhouses open into large chambers (i.e., basements) that supply water to service water pumps (three per screenhouse) and to circulating water pumps (three per unit).

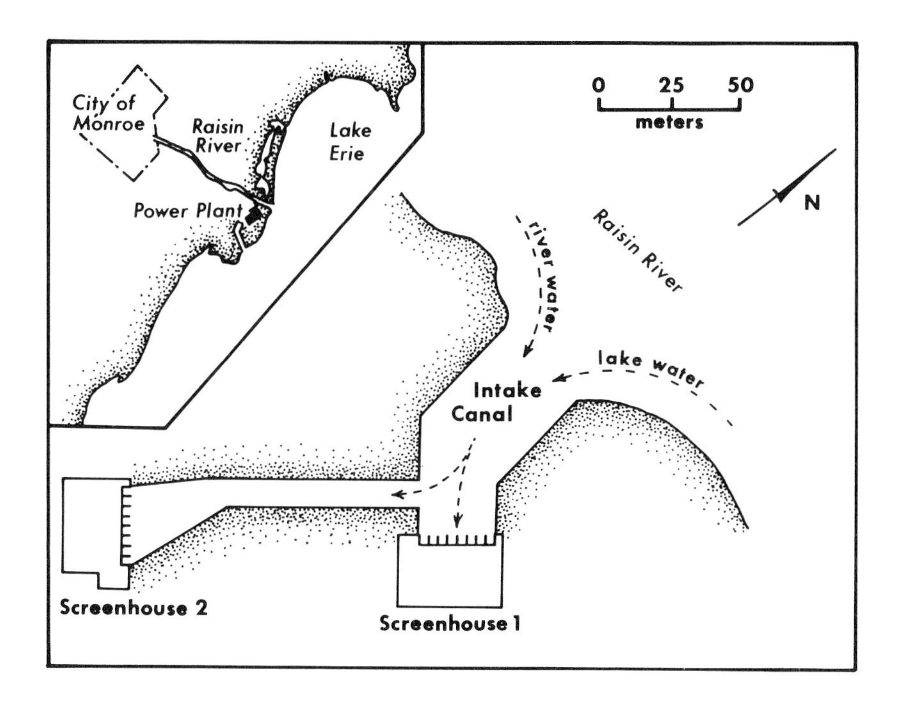

Figure 2. The cooling water intake of the Monroe Power Plant located along the Raisin River in western Lake Erie.

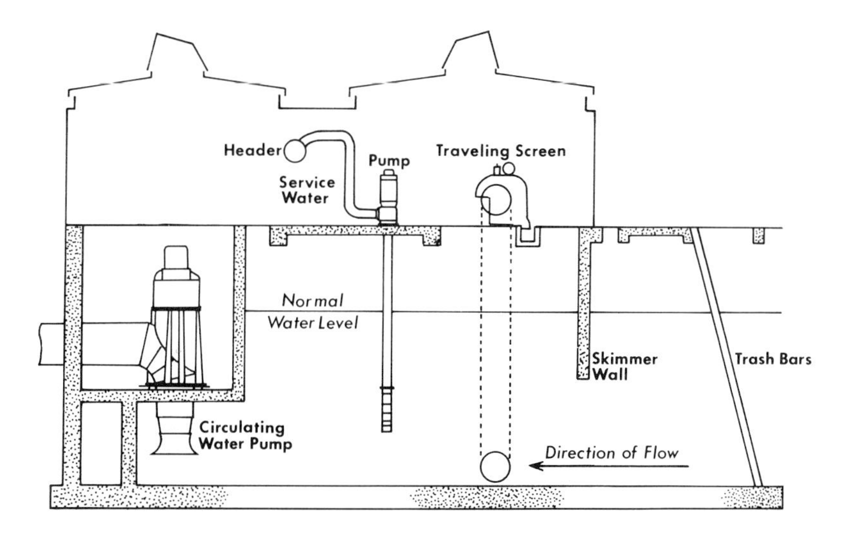

Figure 3. Cross-section view of intake canal and screenhouse at Monroe Power Plant showing internal structures and equipment.

Figure 4. Subsurface view of trash bars occluded by *Dreissena polymorpha*.

Monroe Power Plant — Intake Canal

In 1989, high mussel densities coupled with rapid growth produced an encrusting layer 5–8 cm thick, coating all surfaces forming the intake structure and associated equipment, including trash bars (Figure 4) and circulating water pumps (Figure 5). Densities were similar on all materials (limestone rip-rap, concrete, steel, wood) despite an apparent preference for calcareous substrates observed when densities were low (Kovalak et al., 1990). By late September, this mussel layer was characterized by well-defined druses (i.e., aggregations on flat surfaces; Kirpichenko, 1964).

Blockage of trash bars (Figure 4) was 75% in 1989 but decreased in 1990 because of lower mussel density and also because of earlier evaluation and remediation. Fouling was greater at the bottom and decreased toward the top, possibly because of increased current speeds near the top resulting from complete blockage near the bottom. Mussels not only spanned the space between slats, but also extended as much as 15 cm downstream of the trash bars. Western Lake Erie is subject to large changes in water depth because of wind-driven seiches. Had a major seiche occurred in 1989, the plant probably would have lost water.

Figure 5. Inlet bell of circulating water pump showing characteristic druses (≈clusters) of *Dreissena polymorpha*.

As mussels grew, druses increased in diameter and attachment to the encrusting mass became less secure. Finally, because of mass, gravity and/or current, druses sloughed from the walls and were deposited on the bottom of the screenhouse or became impinged on traveling screens. Between 1989 and 1990, an estimated 500 m^3 (650 yd^3) of mussels were deposited at the

Figure 6. Drive gear on traveling screen covered by *Dreissena polymorpha*. Note that straining surfaces of the screen were free of mussels.

base of sheet piling and concrete in the intake canal; also, an estimated 150 m^3/year (200 yd^3/year) were impinged on the traveling screens in both 1989 and 1990. Druse sloughing and impingement varied seasonally, with a peak of 1.5–2.5 m^3/day (2–3 yd^3/day) of mussels being deposited in September and October. Small peaks of impingement were correlated with strong easterly winds that may have dislodged druses from sheet piling immediately east of the intake canal. Although mussels covered both framing and drive gear on traveling screens (Figure 6), no fouling of strainer surfaces (9-mm mesh) occurred. Traveling screens are cleaned automatically at 8-hr intervals by a backwash spray operating at 550 kP (80 psi).

Monroe Power Plant — Steam Condensers

Some druses, sloughed from surfaces downstream of traveling screens, were entrained by flow through circulating water pumps and transported to main steam condensers (Figure 7) and smaller closed-loop coolers (Figure 8). Druses lodged at the inlet ends of condenser tubes (Figure 9) or were impinged on the tube sheet face. Between 1989 and 1990, such blockage was between 5 and 20%, depending on location in the plant and season. These blockage estimates were based only on druses lodged in condenser tubes (Table 2).

Figure 7. Lower portion of condenser tube sheet showing blockage by druses of *Dreissena polymorpha*. The large number of druses on the floor fell from individual tubes when the water box was drained.

This underestimated blockage because a large number of druses were impinged on the tube sheet but fell to the floor when condensers were drained. If druses on condenser floors are included, estimated blockage would be as high as 35%. Each unit at the Monroe power plant is serviced by two condensers (designated east and west), and the west condensers typically exhibited greater

Figure 8. Tube sheet of a closed-loop cooler that was nearly completely occluded by druses of *Dreissena polymorpha*.

blockage (Table 2). This was attributed to inertial separation of druses in water supply lines. Blockage also was greater in September and October when sloughing was greatest. Blockage decreased during winter months probably because of reduced sloughing of druses and because of fragmentation of druses by turbulence in water boxes.

Figure 9. Condenser tube sheet showing druses of *Dreissena polymorpha* lodged at inlet ends of tubes.

Table 2. Percent Blockage of East and West Main Steam Condensers at Monroe Power Plant by Druses of *Dreissena polymorpha*

		Percent Blockage	
Date	Unit	East Box	West Box
September 1989	2	3	16
September 1989	3	10	4
October 1989	4	—	13
November 1989	3	10	13
April 1990	4	2	8

Initially, blockage by large, mostly dead shells which lodged diagonally and longitudinally in condenser tubes was negligible because large individuals were rare. As the population of mussels grew, shells of large individuals (>25-mm shell length) became common. In early July 1991, blockage by large shells was 10–20% (Figure 10).

Large shells pose potentially more serious long-term operational problems than druses. Individual shells are more easily entrained and transported by

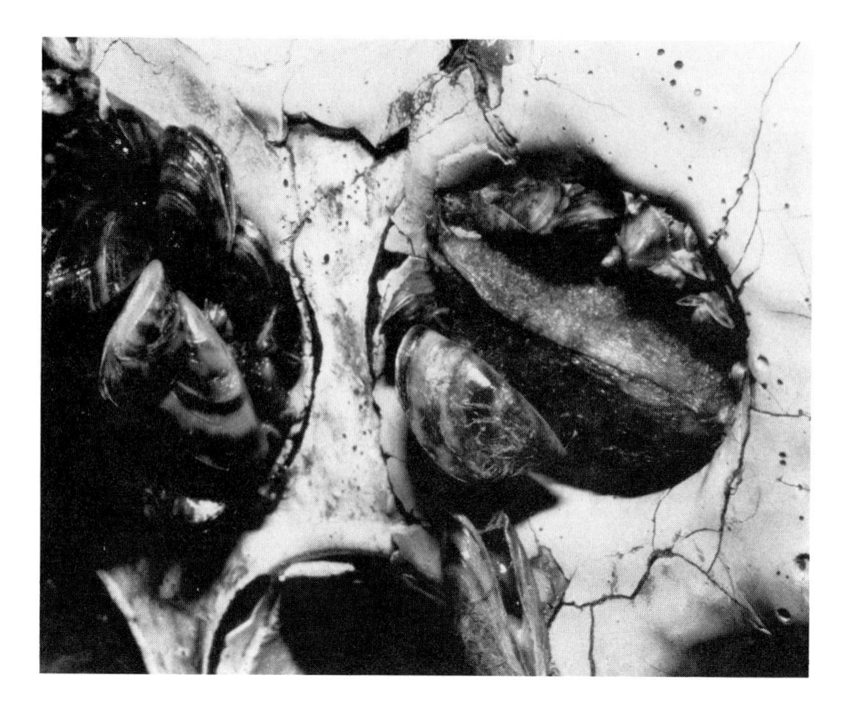

Figure 10. Condenser tubes showing a large dead *Dreissena polymorpha* lodged diagonally and smaller live mussels that subsequently attached at point of blockage.

cooling water flow. Moreover, it takes but a single shell to block a tube which then, in turn, causes small and otherwise innocuous debris to accumulate upstream of the block resulting in complete occlusion of the tube. Small zebra mussels also become attached to the accumulated debris. These small mussels appear to behave normally and although it has not been observed, it is not unreasonable to suggest that uncontrolled they could grow large enough to block tubes as well.

Concern that larval and juvenile zebra mussels will attach and grow large enough to block tubes appeared unfounded. At Monroe, attachment of small mussels occurred only when the tube was first obstructed by some other debris (e.g., dead large mussel shells, sticks, and coal). Zebra mussels are unable to attach when flow rates are greater than 1.5–2.0 m/sec (4–6 ft/sec) (Lyakhov, 1964). Design flows through the Monroe condensers as well as smaller heat exchangers are 2.5 m/sec (7.5 ft/sec). However, changes in water systems with age may reduce flow rates below design specifications making condensers and heat exchangers vulnerable to infestation by larval and juvenile mussels.

Figure 11. Removal of *Dreissena polymorpha* from dewatered screenhouse basement by high-pressure hydrolasing.

Monroe Power Plant — Mechanical Cleaning

Mechanical cleaning was used to control fouling in the inlet canal and the condenser cooling water system. To date, control in the intake canal has been limited to annual cleaning of trash bars by underwater hydrolasing (21,000 kP; 3000 psi). Dislodged mussels drifted onto traveling screens that were operated continuously during cleaning. Each cleaning of trash bars took 8–10 days and cost $25,000–$35,000.

Fouling of steam condensers was controlled by mechanical cleaning of water boxes, and walls and basement floors of screenhouses. Water boxes were cleaned during scheduled maintenance outages and whenever unscheduled outages for other reasons permitted. Mussels were removed from condenser tubes manually using hooked picks. This was cost-effective because druses and shells of large mussels typically were lodged at or near (within 20 cm) inlet ends of tubes. Boroscopic inspection indicated blockage no more than 60 cm into the tubes.

Mussels were removed from walls and basement floors of screenhouses by dewatering and hydrolasing (21,000 kP; 3000 psi) (Figure 11). Hydrolasing was done from inflatable boats that served as floating scaffolds during dewatering. Dislodged mussels accumulated on basement floors and were concentrated, dewatered, and removed by dry vacuum for disposal in a landfill. Between November 1989 and October 1991, four units were cleaned

Table 3. Volume of *Dreissena polymorpha* Removed by Mechanical Cleaning of Screenhouses at the Monroe Power Plant in 1989–1991 (the Number of Spawning Periods Between Cleanings Is Also Given)

Cleaning Date	Unit Number	Number of Spawns Between Cleanings	Volume (m^3)
November 1989	3	1	38
January 1990	1	1	40
April 1990	4	1	44
September 1990	2	2	74
January 1991	3	1	34
October 1991	4	2	80

(Table 3). Cleaning after one spawning season yielded between about 40 m^3 (50 yd^3) of mussels weighing 10 metric tons (15 tons), and cleaning after two spawning seasons (October 1991) yielded about 80 m^3 (100 yd^3) weighing 20 metric tons (30 tons). Dewatering and cleaning of one unit took about 5 days and cost $25,000–$35,000. This included cleaning of 800 m^2 (8500 ft^2) of concrete surface, removal, transport, and disposal of mussels.

Monroe Power Plant — Service Water Systems

Service water systems that supply the fire protection equipment and the ash-handling system were also fouled. In 1989, fouling of low pressure lines (700 kP; 100 psi) was greater than high pressure lines (1700–2100 kP; 250–300 psi), probably because the latter were used less. By late July, some low pressure lines were obstructed by shells lodged in elbows and valves. It was hypothesized that following infestation and a short period of growth (up to 4-mm shell length), the mussels died because of high water temperature attributable to high air temperatures inside the plant. Decomposition of soft tissues coupled with sulfur reducing bacteria in sediment generated large amounts of hydrogen sulfide. Fouled lines were cleared mechanically and were maintained by biweekly flushing with water.

In 1990, mussel densities in the main service water header (low pressure) ranged between 150,000–200,000/m^2 (Figure 12). High densities caused a variety of problems including fouling of valves (particularly disks of butterfly valves) (Figure 13) and occlusion of piping, elbows, and spigots by live and dead mussels (Kovalak et al., 1990).

Chlorination was used to control fouling of the service water system. In 1990, the service water system was chlorinated at 0.5 mg/L total residual chlorine (TRC) (feed rate approximately 2.5 mg/L) for 6 weeks (mid-September-October) to mitigate fouling. Continuous chlorination at 0.5 mg/L TRC will be used to prevent mussel reinfestation during the spawning season (June 1-October 1).

Figure 12. *Dreissena polymorpha* inside a 76-cm (30-in.) diameter pipe of a low-pressure service water header.

Continuous chlorination as a short-term solution to service water treatment is cost-effective because the volume is small (2 m^3/sec; 30,000 gpm) and because there is no need for dechlorination. All of the treated water passes through either the fly ash basin (313-day retention) or bottom ash basin (2-day retention) prior to discharge. Consequently, no residual chlorine has been

Figure 13. A 76-cm (30 in.) diameter butterfly valve that would not close because the disk was fouled by *Dreissena polymorpha*.

detected at basin discharges (Jennings, Personal communication). At this time, it is not known if retention times allow for breakdown and volatilization of trihalomethanes that may have formed.

Other changes to the service water system to minimize operational impacts include increased monitoring of equipment performance coupled with mechanical cleaning where needed, weekly flushing of some low pressure lines, installation of easily cleaned strainers and filters on lines servicing critical equipment (e.g., slurry pumps), and replacement of badly fouled valves. Large butterfly valves have been replaced with self-cleaning sliding gate valves.

FUTURE DIRECTIONS

Although the mussel control program at Monroe achieved the initial goals of insuring unit availability, environmental acceptability, and cost-effectiveness, it will not provide a long-term solution to annual reinfestation. Beginning in 1989, mechanical cleaning was conducted annually during scheduled outages. However, to control maintenance costs, plant strategy has changed so that the period between outages has been extended 18–24 months. In the worst case, this would expose units to three spawning seasons of zebra mussels between cleanings. It is not known whether units will be able to withstand this level of mussel infestation. Consequently, Detroit Edison is exploring other ways to insure reliability of power plant operations.

Chlorination is not a long-term control alternative because of concern about direct (free oxidant) and indirect (THMs) effects on nontarget species, and because of probable damage to piping (carbon steel and aluminum), gaskets, and seals in the service water system. U.S. EPA has encouraged experimentation with less toxic forms of chlorine like chloroamines (Howe, Personal communication) and there has been considerable interest in more benign biocides such as potassium (Fisher, Personal communication). Assessment of the effects of chlorine and other oxidants on water system components has not been undertaken although there is concern that long-term use would be detrimental.

Another potential control method is the use of coatings that prevent or at least minimize fouling and that may also facilitate removal. Currently, several products are being tested, the most promising of which are silicone-based (Figure 14). However, silicone coatings may not be cost-effective because of uncertain life expectancy. Life expectancy is related not only to the nature of the material, but also to the rigorous conditions under which it must be applied. At Monroe outages are scheduled in late fall or late winter to early spring. At low air temperatures it is difficult to dry screenhouse walls to ensure good bonding and curing of coatings.

Alternatively, partial cleanings may be performed underwater during short unscheduled outages or during partial unit outages at night or on weekends. To this end, entrainment of druses is being modeled to determine the source area for most of the druses entering the water boxes. If the source area is small, it might be cleaned quickly using high pressure spray. Another approach is to treat these small areas with an antifouling coating containing copper or zinc (Figure 15). Limited application might control condenser fouling and minimize water quality impacts.

ACKNOWLEDGMENTS

A large number of people inside and outside Detroit Edison made this study possible. We would like to thank the staff at Monroe Power Plant, particularly M. Smolinski, N. Bednar, D. Fahrer, T. Walsh, and C. Jennings, for their assistance and support. R. Peplowski took photographs and B. Kollar and D. Ottey processed quantitative samples of larval and adult zebra mussels. W. Brusate, C. Witherspoon, and the dive crews of Commercial Diving and Marine Services (Port Huron, MI) provided invaluable assistance by collecting samples and conducting visual inspections.

Figure 14. Experimental panels with silicone coating that was effective in controlling fouling by *Dreissena polymorpha*. Note that mussels are attached only to uncoated areas as the metal framing, tabs, and hole in center of concrete plate.

Figure 15. Experimental panels with zinc-epoxy coating that was effective in controlling fouling by *Dreissena polymorpha*.

REFERENCES

Clarke, K. B. "The Infestation of Waterworks by *Dreissena polymorpha*, a Fresh Water Mussel," *J. Inst. Water Eng.* 6:370–379 (1952).

Greenshields, F., and J. E. Ridley. "Some Researches on the Control of Mussels in Water Pipes," *J. Inst. Water Eng.* 11:300–306 (1957).

Griffiths, R. W., Schloesser, D. W., Leach, J. H., and W. P. Kovalak. "Distribution and Dispersal of the Zebra Mussel (*Dreissena polymorpha*) in the Great Lakes region," *Can. J. Fish. Aquat. Sci.* 48:1381–1388 (1991).

Herbert, P. D. N., B. W. Muncaster, and G. L. Mackie. "Ecological and Genetic Studies on *Dreissena polymorpha* (Pallas): A New Mollusc in the Great Lakes," *Can. J. Fish. Aquat. Sci.* 48:1389–1395.

Jenner, H. A. "Control of Mussel Fouling in The Netherlands: Experimental and Existing Methods," in I. A. Diaz-Tous, M. J. Miller, and Y. G. Mussalli, Eds. (Palo Alto, CA: Electric Power Research Institute, 1983).

Kirpichenko, M. Ya. "Phenology, Population Dynamics and Growth of *Dreissena* larvae in the Kuibyshev Reservoir," in Biology and Control of *Dreissena*, B. K. Shtegman, Ed. *Tr. Inst. Biol. Vnut. Vod.* 7(10):15–24 (1964) (Translated 1968 from Russian, Israel Program for Scientific Translations, Jerusalem, Israel).

Kovalak, W. P., Longton, G. D., and R. D. Smithee. "Infestation of Monroe Power Plant by the Zebra Mussel (*Dreissena polymorpha*)," *Proc. Am. Power Conf.* 52:998–1000 (1990).

Lyakhov, S. M. "Work of the Institute of Biology of Inland Waters, Academy of Sciences of the USSR," pp. 55–59, in Biology and Control of *Dreissena*, B. K. Shtegman, Ed. *Tr. Inst. Biol. Vnut. Vod.* 7(10):55–59 (1964) (Translated 1968 from Russian, Israel Program for Scientific Translations, Jerusalem, Israel).

Manny, B. A., Edsall, T. A., and E. Jaworski. "The Detroit River, Michigan: An Ecological Profile," Biological Report 85 (7.17). U.S. Fish and Wildlife Service, Ann Arbor, MI (1988).

Morton, B. S. "Freshwater Fouling Bivalves," in Proceedings of the First International Corbicula Symposium, J. C. Britton, J. Mattice, C. E. Murphy, and L. W. Newland, Eds. Texas Christian University Research Foundation, Fort Worth, TX (1979) 1–14.

Ruffner, J. A., and F. E. Blair. *The Weather Almanac,* 5th ed. (Detroit, MI: Gale Research Company, 1987).

Stanczykowska, A. "Ecology of *Dreissena polymorpha* (Pall.) (Bivalvia) in Lakes," *Pol. Arch. Hydrobiol.* 24:461–530 (1977).

Strayer, D. L. "Projected Distribution of the Zebra Mussel, *Dreissena polymorpha,* in North America," *Can. J. Fish. Aquat. Sci.* 48:1389–1395 (1991).

van Urk, G. "De driehoeksmossel, *Dreissena polymorpha*, in de Rijn," *Water* 9:327–329 (1976).

Vollenweider, R. A., Munawar, M., and P. Stadelmann. "A Comparative Review of Phytoplankton and Primary Production in the Laurentian Great Lakes," *J. Fish. Res. Board Can.*" 31:739–762 (1974).

CHAPTER 23

Impacts of the Zebra Mussel *(Dreissena polymorpha)* on Water Quality and Fish Spawning Reefs in Western Lake Erie

Joseph H. Leach

Patterns in abundance of zebra mussel (*Dreissena polymorpha*) larvae, water transparency, and chlorophyll *a* concentrations were monitored weekly in western Lake Erie from May to November 1988, 1989, and 1990. Zebra mussel colonization of three reef areas that are used by several fish species for spawning was measured in 1989 and 1990 by SCUBA divers. Increases in mean numbers of larvae from 4/L in 1988 to 96/L in 1990 indicate rapid expansion of the population in the western basin. Mean Secchi disc transparencies between 1988 and 1989 increased by 1.24 m (85%) in the western basin and by 1.75 m (52%) in the west-central basin. Mean chlorophyll *a* concentrations between 1988 and 1989 declined 43% in the western basin and 27% in the west-central basin. These changes in quantities of suspended particles are attributed mainly to the filtering activity of settled zebra mussels in western Lake Erie. The clearing of the water may be a positive asthetic effect, but there is concern about the long-term effects of shifting so much organic matter from the pelagic to the benthic components of the ecosystem. Colonization of rocky reefs by zebra mussels proceeded rapidly since the invasion of the area with mean counts of up to 342,000/m^2 found on Sunken Chicken reef in 1990. The populations are skewed toward smaller mussels (<5 mm) with percentage of adults (>10 mm) accounting

0-87371-696-5/93/$0.00 + $.50

for between 5 and 24% of the total mussel population. Zebra mussels have completely altered the character of reef substrates but adverse impacts on walleye (*Stizostedion vitreum vitreum*) spawning have not been observed.

INTRODUCTION

Zebra mussels (*Dreissena polymorpha*) were first found in North America in June 1988 (Hebert et al., 1989). The vector of introduction is believed to be ballast water (Hebert et al., 1989). Although the specific location and date of introduction is unknown, the evidence points to an introduction of veligers in Lake St. Clair, possibly in 1986 (Griffiths et al., 1989; Griffiths et al., 1991). Although the first adult mussels were collected in the western basin of Lake Erie in July 1988, there was an earlier unofficial siting that predates even the Lake St. Clair collections. Four mussels (ca. 10 mm in length) were observed attached to the hull of a commercial fishing tug drydocked at Kingsville, Ontario in December, 1987 (Penner, G., 1988). When the economic and ecological implications of the invader were perceived from European literature (Stanczykowska, 1977), observations on the spread and colonization of zebra mussels in Lake Erie were initiated.

The purpose of this study was to measure the growth and impacts of an expanding population of *Dreissena polymorpha* on water quality and spawning reefs in western Lake Erie.

METHODS

In Europe, the major impact of the zebra mussel is the fouling of underwater structures and devices used by man. Because of its large filtering capacity, the mussel is capable of altering energy and nutrient flow in ecosystems (Stanczykowska, 1977).

A study of biomass and production of zooplankton in western Lake Erie was underway before the zebra mussel was first reported from the lake (unpublished data). This study provided data on zebra mussel larvae abundance, water transparency, and chlorophyll *a* concentrations. Zooplankton and water samples were collected weekly from May to November from four stations in each of the western and west-central basins (Figure 1). Plankton samples were collected by hauling a 50-cm diameter conical net (mesh size 76 μm), equipped with a calibrated flow meter, vertically through the water column. Samples were stored in 4% sugared formalin (Haney and Hall, 1973). Subsampling and counting routines were similar to those given in Evans and Sell (1983). No effort was made to distinguish the various planktonic stages of the mussel. Water samples from 1-m depth were collected in a Van Dorn sampler and filtered through GF/C filters. Chlorophyll *a* was determined by acetone extraction (Strickland and Parsons, 1968). Surface water temperatures and Secchi disc transparencies were measured at each station.

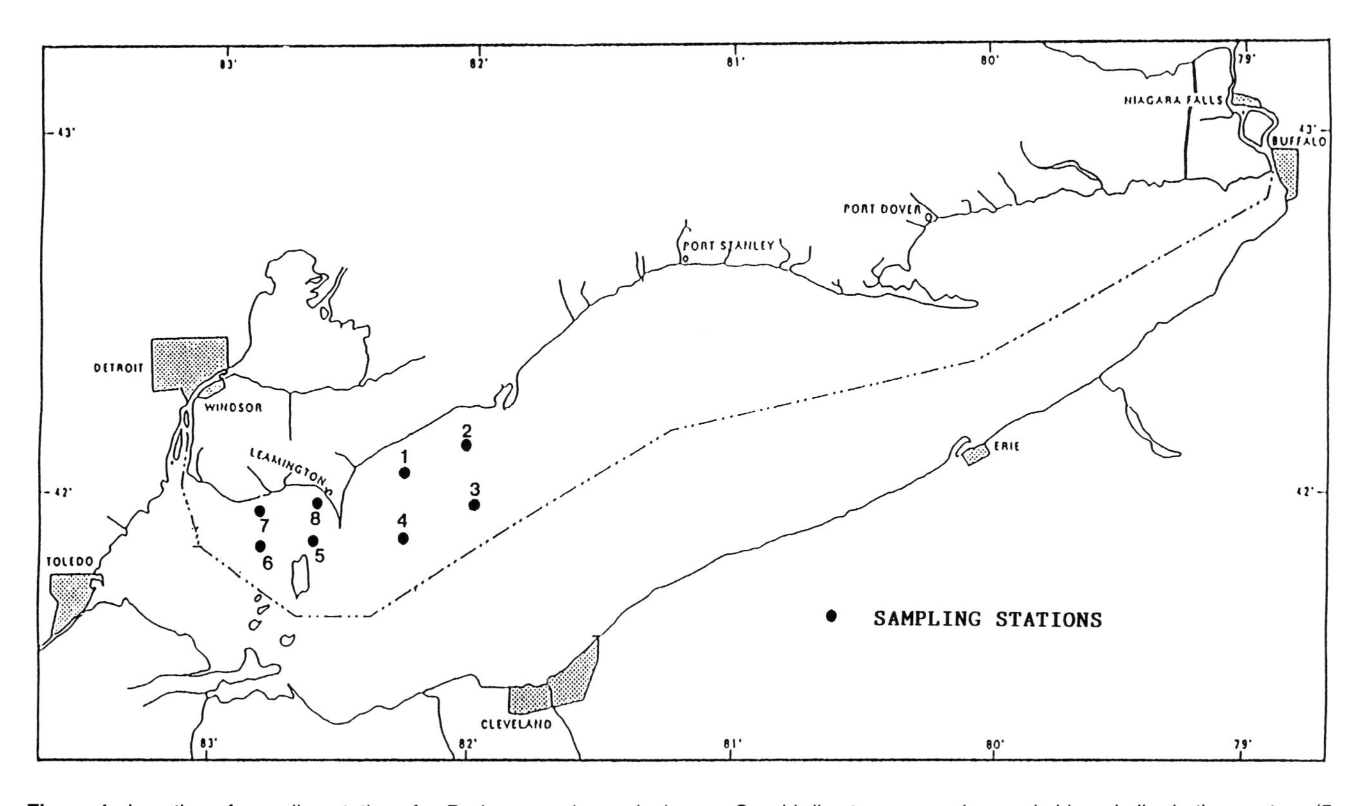

Figure 1. Location of sampling stations for *Dreissena polymorpha* larvae, Secchi disc transparencies, and chlorophyll *a* in the western (5 to 8) and west-central basins (1 to 4) of Lake Erie.

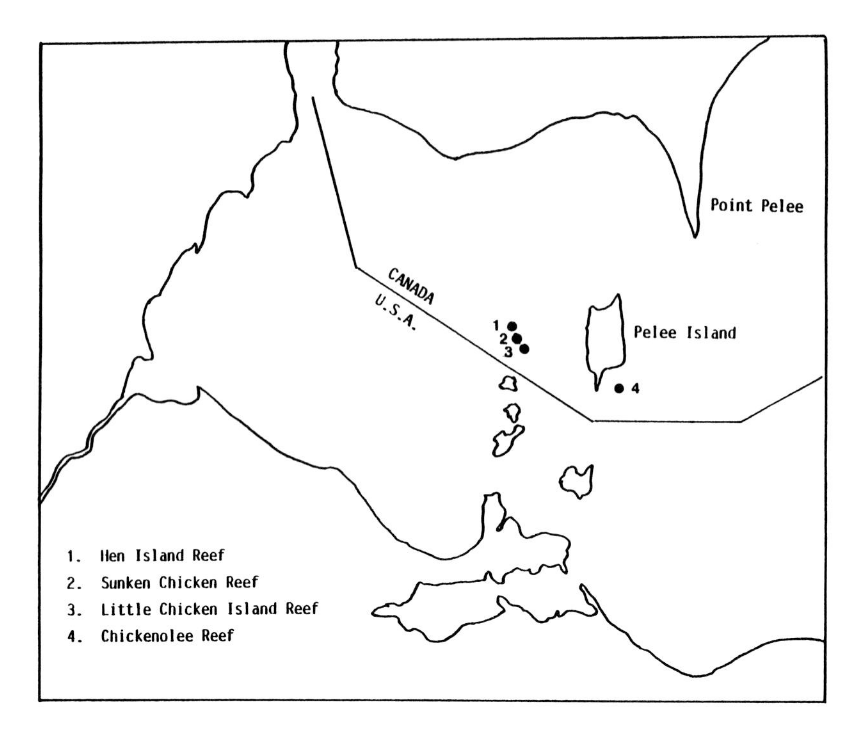

Figure 2. Location of fish spawning reefs used for assessment of *Dreissena polymorpha* colonization in the western basin of Lake Erie.

Colonization of fish spawning reefs was determined by counting zebra mussels attached to rock samples collected randomly by scuba divers (Figure 2). The number of subsamples counted and measured per location varied from 6 to 12 depending on the uniformity of mussel distributions. Individuals were measured to the nearest whole millimeter. Three reef areas (locations 1, 3, and 4) were selected in 1989 for an initial assessment of zebra mussel colonization. An additional reef, Sunken Chicken (location 2), was selected in the spring of 1990 for an in-depth study of possible impact on fish (e.g., walleye) spawning.

RESULTS

Larval Abundance

Dispersion of the zebra mussel in the Great Lakes has been aided by its planktonic larval (i.e., veliger) stage. At western basin stations (Figure 3), larvae were first observed in late June 1988. A peak of 20/L was counted in

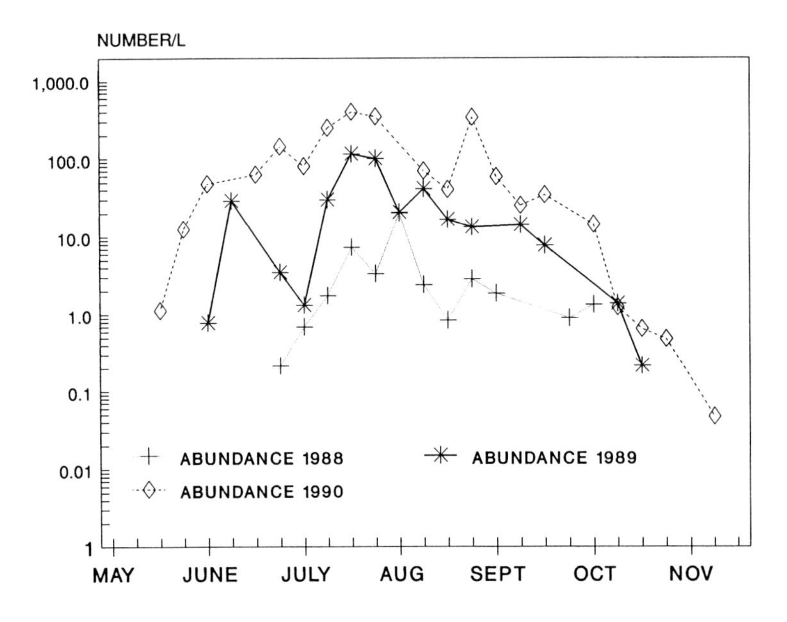

Figure 3. Mean abundance of *Dreissena polymorpha* larvae in the western basin of Lake Erie from May to November 1988–1990.

the first week of August, and larvae were found at much lower numbers through the first week of October. During the 14-week period when larvae were present, the average concentration was about 3.6/L.

In response to an enlarged population of adults which colonized the basin in 1988, the abundance and duration of larvae increased in 1989 (Figure 3). Larvae were found at western basin stations from the first week of June through the third week of October, a period of 18 weeks. Abundance peaked in the third week of July at almost 116/L. Average concentration of veligers and postveligers in the 1989 season was about 30/L.

Abundance of zebra mussel larvae at western basin stations continued to increase in 1990 (Figure 3). Two peaks occurred in 1990, 400/L in the third week of July and 338/L in the last week of August. The highest individual station abundance, 628/L, occurred at station 7 on July 23, 1990. Larvae were found earlier in 1990 (i.e., third week of May) and persisted longer in the plankton (i.e., until the second week of November); the average seasonal concentration (96/L) was much higher than in the previous 2 years.

Abundance of larvae in the west-central basin was less than that measured in the western basin in all 3 years surveyed (Figure 4). In 1988, there was a lag of 23 days between first occurrence of larvae at western and west-central basin stations. There were early and late peaks but abundance never exceeded 4/L. Average abundance of larvae over the 13-week period of occurrence was

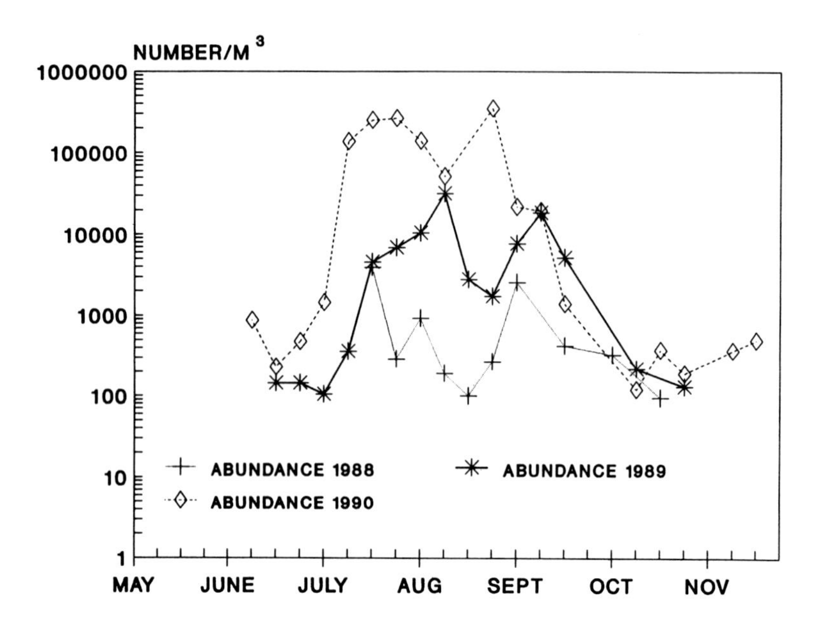

Figure 4. Mean abundance of *Dreissena polymorpha* larvae in the west-central basin of Lake Erie from May to November 1988–1990.

less than 1/L. Average counts increased to 6/L and 68/L in 1989 and 1990, respectively, but these were considerably less than larvae abundances from the western basin in those years. Two peaks of abundance occurred in both 1989 and 1990 but the peaks did not coincide. In 1990, a late July peak of 263/L was followed by a late August peak of 345/L. Maximum abundance (i.e., 1225/L) occurred at station 1 on August 24, 1990. The majority of these larvae were small (<100 μm), and therefore a new brood. Larvae appeared in the 1990 plankton during a 22-week period from mid-June to late November, which is a considerable increase over the mid-July to mid-October period in 1988.

Apparently, the presence of larvae in the water is not regulated solely by temperature. Temperatures at first occurrence of veligers in the western basin declined from 21.7°C in 1988 to 14.5°C in 1990 (Table 1). Temperatures at last occurrence ranged from 18.1°C in 1988 to 7.5°C in 1990. In the west-central basin, surface water temperature at first occurrence of veligers was 24.1°C, 18.3°C, and 17.1°C in 1988, 1989, and 1990, respectively. Corresponding temperatures for last observations of veligers were 14.2°C, 13.5°C, and 9.9°C, respectively.

Table 1. Surface Water Temperatures and Dates of First and Last Occurrences of *Dreissena polymorpha* Larvae in Plankton Samples from the Western and West-Central Basins of Lake Erie from 1988 to 1990

	First Occurrence		Last Occurrence	
	Date	°C	Date	°C
Western Basin				
1988	June 27	21.7	October 3	18.1
1989	June 7	19.2	October 24	11.6
1990	May 22	14.5	November 14	7.5
West-Central Basin				
1988	July 19	24.1	October 19	14.2
1989	June 19	18.3	October 23	13.5
1990	June 12	17.1	November 19	9.9

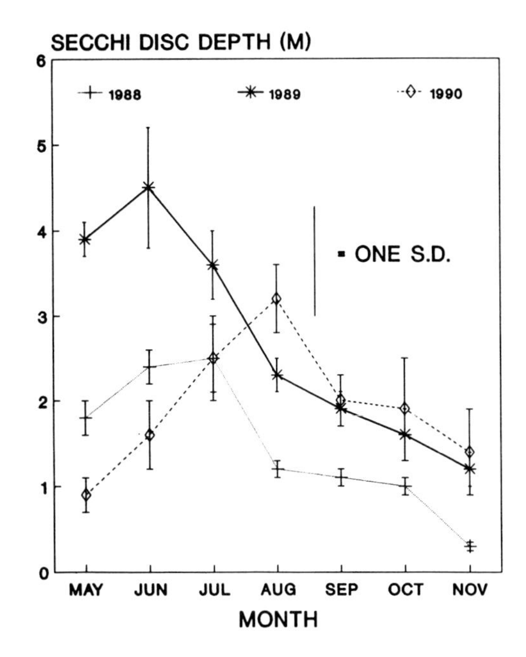

Figure 5. Mean Secchi disc transparencies in the western basin of Lake Erie from May to November 1988–1990.

Water Transparency and Chlorophyll *a*

Over the period of this study water transparency increased substantially. Mean transparency during the May to November period at western basin stations increased 85% between 1988 and 1989 (Figure 5). Windy conditions on sampling dates in the spring and early summer of 1990 caused considerable

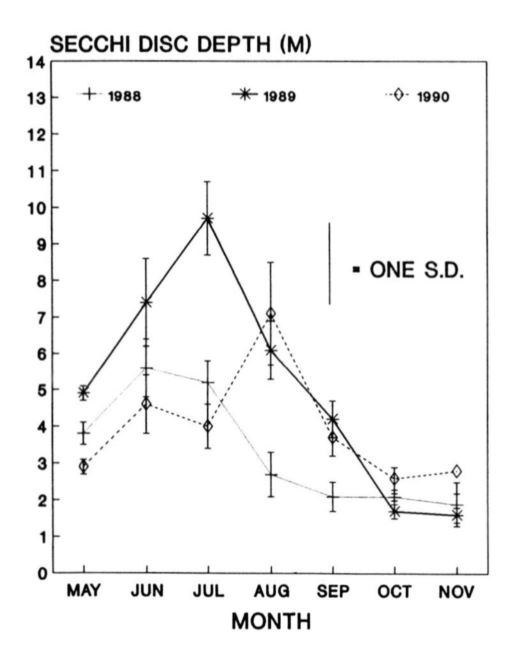

Figure 6. Mean Secchi disc transparencies in the west-central basin of Lake Erie from May to November 1988–1990.

resuspension of bottom sediments and are reflected in lower transparencies. In the August to November period, mean monthly transparencies were greater than those in the corresponding period of 1989. At west-central basin stations mean transparency for the May to November period increased 52% between 1988 and 1989 (Figure 6). No apparent further increase was recorded in 1990.

In general, chlorophyll *a* concentrations declined in Lake Erie between 1988 and 1989 (Figures 7 and 8). At western basin stations, mean chlorophyll *a* concentrations declined 43% between 1988 and 1989 (Figure 7). Although spring concentrations in 1990 were higher than those of 1989, summer and autumn concentrations were lower; over the May to November period, there was an 18% decrease from 1989 and a 54% decrease from 1988. In the west-central basin mean chlorophyll *a* concentrations during the growing season declined 27% between 1988 and 1989 (Figure 8). No further decline in mean chlorophyll *a* was measured in 1990, although autumn concentrations were less than those in 1989.

Densities on Fish Spawning Reefs

Offshore reef areas in the western basin of Lake Erie are utilized by walleye (*Stizostedion vitreum vitreum*), lake whitefish (*Coregonus*

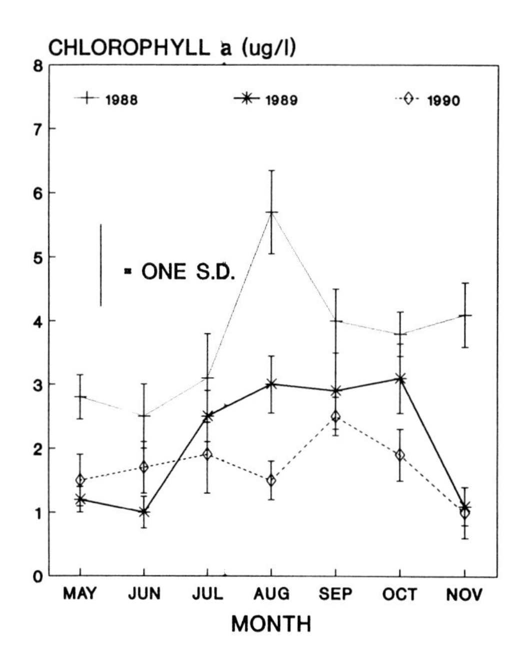

Figure 7. Mean chlorophyll *a* concentrations in the western basin of Lake Erie from May to November 1988–1990.

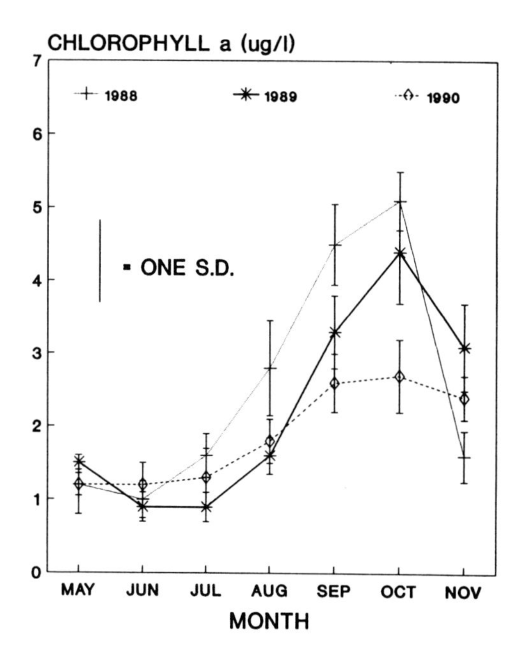

Figure 8. Mean chlorophyll *a* concentrations in the west-central basin of Lake Erie from May to November 1988–1990.

clupeaformis), white bass (*Morone chrysops*), and other species for spawning (Goodyear et al., 1982). The principal reef areas are bedrock which is sedimentary in origin (Herdendorf and Braidech, 1972). The bottom surface adjacent to rocky outcrops consists of unconsolidated sediments including boulders, cobbles, pebbles, sand, and mud (Herdendorf and Braideck, 1972). The bedrock and gravel components of the reef areas were quickly colonized by zebra mussels, which caused concern among scientists that reproduction of walleye and possibly other species could be affected.

The first observation of fish spawning reefs were made by scuba divers at three reef areas on May 16, 1989 (Figure 2). Zebra mussel coverage ranged from about 5% colonization on Little Chicken Island reef and Hen Island reef to 95% coverage on Chickenolee reef. Mean estimates of populations ranged from 1,200/m^2 on Little Chicken Island reef to 3,500/m^2 on Hen Island reef to 30,000/m^2 on Chickenolee reef. Shell lengths up to 12 mm long were found at Chickenolee reef and up to 20 mm long at the other reefs. By October 1989, zebra mussel populations on three reefs had increased to mean abundances of 112,000/m^2, 180,000/m^2, and 233,000/m^2 at Chickenolee, Little Chicken Island, and Hen Island reefs, respectively (Table 2). Maximum shell length ranged from 23 mm at Chickenolee through 24 mm at Hen Island to 30 mm at Little Chicken Island reef. The proportion of the populations with shells longer than 10 mm varied from 5% at Little Chicken Island to 20% at Chickenolee reef.

Sunken Chicken shoal was sampled intensively on April 20, 1990 (Table 2) in connection with a study of walleye spawning (Fitzsimons et al., 1991). The mean abundance at 12 locations on the shoal was 138,000/m^2 with about 10% of the population longer than 10 mm. Abundance was skewed in favor of a large number of small mussels (<5 mm) which represented the late hatch from the previous year (Figure 9).

The October 22, 1990 survey indicated further increases in numbers of zebra mussels colonizing the reefs (Table 2). The annual increase was 15 and 27% for Hen Island and Chickenolee reefs, respectively. Densities of mussels on Sunken Chicken reef increased 148% between April 20, and October 22, 1990. Adult populations (>10 mm) increased, but the main increase in abundance was due to settlement of larvae produced in 1990.

DISCUSSION

The abundance and seasonal cycle of zebra mussel larvae provide an indication of the potential to colonize areas downstream and a measure of the status of the adult population. Stanczykowska (1977) considered zebra mussels to have the highest fecundity of freshwater mollusks. The high fecundity, along with a planktonic larval stage, probably accounts for the rapid dispersal of the mussel downstream from Lake St. Clair. By the end of 1988, mussels

Table 2. Estimated Mean Numbers of *Dreissena polymorpha* (Total and >10 mm in Length) on Western Basin Reef Areas in 1989 and 1990

Date	Reef	Mean Number/m² Total (S.E.)	Mean Number/m² >10 mm	n	Percent >10 mm
1989 (October 13)	Chickenolee	112,000 (19,200)	22,400	6	20
	Little Chicken Island	180,000 (26,900)	9,000	6	5
	Hen Island	233,000 (27,350)	14,000	6	6
1990 (April 20)	Sunken Chicken	138,000 (23,300)	14,000	12	10
1990 (October 22)	Chickenolee	142,000 (19,650)	34,300	8	24
	Sunken Chicken	342,000 (34,020)	18,000	8	5
	Hen Island	268,000 (28,020)	17,500	9	6

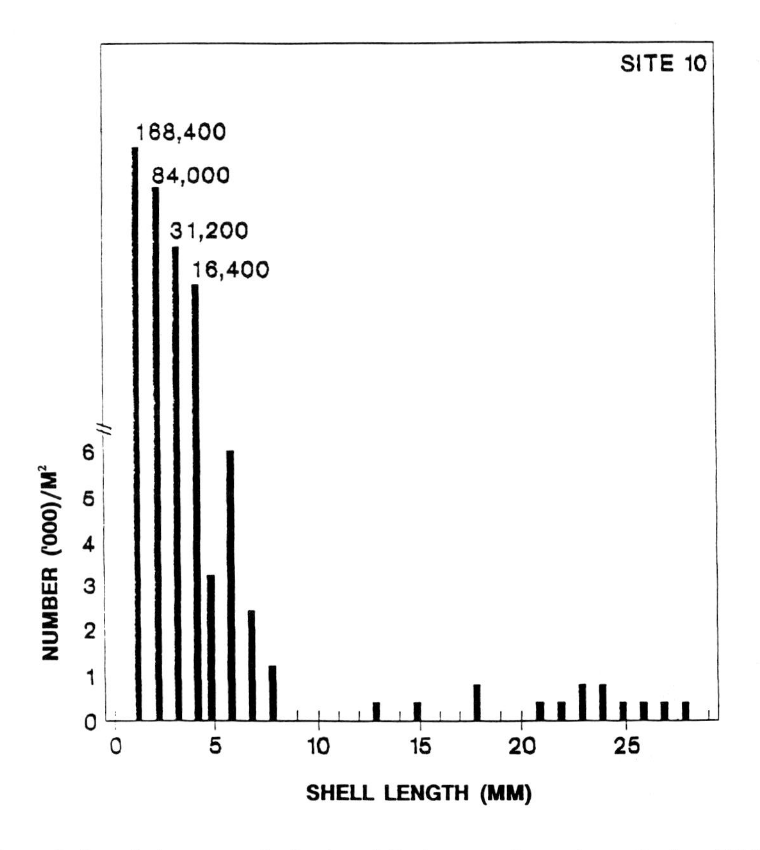

Figure 9. Length frequency distribution of *Dreissena polymorpha* on Sunken Chicken reef of western Lake Erie on April 20, 1990.

were found as far east as Long Point on the north shore of Lake Erie, a distance of 250 km from the mouth of the Detroit River. Mussels colonized all of Lake Erie in fall 1989 and had spread through the Welland Canal and Niagara River into Lake Ontario.

The high fecundity of the zebra mussel is reflected in veliger data; mean abundance of larval mussels between 1988 and 1989 increased 730 and 570% in the western and west-central basins, respectively. The maximum abundance of larvae in both areas in 1990 was 1200/L which is much less than the 9000/L reported from some lakes in Poland (Stanczykowska, 1977). However, the latter abundances were from water samples collected directly above adult mussels, whereas abundances from Lake Erie were from samples collected in the pelagic zone. In the pelagic zone of the Polish lakes, the abundance of larvae never exceeded 500/L.

In European lakes, reproduction commences when water temperatures reach 12 to 16°C (reviewed by Mackie et al., 1989). The lowest temperature (14.5°C) recorded in this study when larvae were first observed in the plankton is in the middle of this range. However, water temperatures at first occurrence of larvae were usually above 16°C, which indicates that other factors may be important in reproduction.

In Europe, the duration of the individual planktonic phase is quite variable and ranges from 8 to 10 days in Polish lakes (Stanczykowska, 1964) to about 5 weeks in Lake Constance (Walz, 1975). Sprung (1989) found a range of from 18 to 33 days in two German lakes over 2 years. The length of the planktonic phase was not accurately measured in the present study; however, in 1990 the lag between first appearance of larvae in the plankton and first settlement of postveligers on settling chambers in the same areas was 44 days (unpublished data). Fraleigh et al. (1990) recorded a period of 45 days between attachment and first observation of veligers near Maumee Bay in western Lake Erie.

The period when larvae are found in the plankton in Europe is usually from 3 to 5 months, but can extend from 2 to 8 months (reviewed by Mackie et al., 1989). Larvae were found in the western basin for 3.3 months in 1988, 4.6 months in 1989, and 5.8 months in 1990. Corresponding lengths of time in the west-central basin were 3.1, 4.2, and 5.3 months, respectively. The increase in abundance of larvae and length of time in the plankton over the 3 years of the present study are indications of an expanding population of zebra mussels.

The western Lake Erie population increased rapidly between 1988 (and possibly earlier) and 1990 but may now be approaching its asymptotic maximum. Mean numbers of larvae at western basin stations increased 733% between 1988 and 1989 but only 220% between 1989 and 1990. Mean numbers of adult zebra mussels on Hen Island reef increased 300% between 1988 and 1989 but only 25% between 1989 and 1990. Similarly, the mean population of adults on Chickenolee reef increased 273% between 1988 and 1989 and 53% between 1989 and 1990.

Abundance of zebra mussel larvae in European lakes is variable between years and, in some cases, is not related to abundance of adults (Stanczykowska, 1977). In this study, peaks in abundance of larvae varied between years and also between basins (Figures 3 and 4). Whereas, two peaks occurred in the west-central basin in all 3 years, only one peak occurred in the western basin in 1988 and 1989 and two peaks occurred in 1990. The double peaks in the west-central basin were 4–6 weeks apart each year, but the timing was not synchronous between years. In 1990, peak abundances of planktonic larvae and recently settled larvae on steel-settling chambers (unpublished data) were bimodal in both basins. Time of peak settlement lagged maximum abundance of larvae by about 1 week for the first peak and about 4 weeks for the second. Garton and Haag (1990) also found temporal variability in peak abundances of veligers between 1989 and 1990 in the vicinity of South Bass Island in the western basin. Differences between spawning activity of local adults and patterns of larvae abundance led these authors to conclude that discrete populations in the western basin were spawning asynchronously.

Filtration rates of zebra mussels have been studied extensively in Europe (Kryger and Riisgård, 1988; Reeders and bij de Vaate, 1990; Stanczykowska, 1977; and others). Water filtration rates of individual mussels (21–22 mm shell length) have been estimated at 8–44 mL/hr in Poland (Stanczykowska, 1975), 135–150 mL/hr in The Netherlands (Reeders et al., 1989) and 260–286 mL/hr in Denmark (Kryger and Riisgård, 1988). In some areas, the bioprocessing of seston by mussels is considered useful in water quality management (Wisniewski, 1990; Reeders and bij de Vaate, 1990). The clearing capacity of zebra mussels in Lake Ijsselmeer and Lake Markermeer in The Netherlands is capable of filtering those lakes at least once or twice a month (Reeders et al., 1989).

Estimations of populations in the entire western basin have not yet been attempted but concentrations on three rocky reefs in the Pelee Island area have been estimated (Table 2). MacIsaac et al. (1992) used equations derived by Kryger and Riisgård (1988) to calculate clearance rates of the western basin reefs populations which ranged from 0.4 to 516 mL/individual/hr depending on shell length. They then applied these clearance rates to population densities of zebra mussels on the reefs to calculate filtering impacts on a well-mixed water column 5 m deep (which approximates mean depths in the western basin). The filtration impact on Hen Island reef in October 1990 was estimated at 132 m^3/m^2/day. MacIsaac et al. (1992) estimated that the filtering impact of the settled mussels greatly exceeded that of the veliger population and the herbivorous zooplankton.

Mussel population estimates used in calculations are limited to areas of suitable substrate. Population estimates for the remaining area of the basin are not available. From the water volume of the basin (i.e., 24.2 km^3) and a filtration rate estimate of 1 L/day per individual (Stanczykowska, 1977), a theoretical population of 7400/m^2 settled adults would be required to filter the basin in 1 day. The higher filtration rates of Reeders et al. (1989) and Kryger and Riisgård (1988) would require theoretical adult populations of $2170/m^2$ and $1135/m^2$, respectively, to filter the basin in 1 day.

The increases in transparency and decreases in chlorophyll *a* noted in the present study are probably due to the enormous filtering capacity of zebra mussels. Concentrations of chlorophyll *a* in the western basin now range between 1 and 2 μg/L, which classifies the basin at the oligotrophic end of the trophic scale. Concentrations in the west-central basin now range between 1 and 3 μg/L. In 1972, chlorophyll *a* from six stations in the same area of the western basin averaged 16.8 μg/L and from four inshore stations in the west-central basin averaged 11.4 μg/L (Leach, 1975). Joint efforts by the United States and Canada to reduce phosphorus loadings to Lake Erie are responsible for much of the decline in chlorophyll *a* between the early 1970s and the mid-1980s (GLWQB, 1987). The decline of chlorophyll *a* since 1988 is due, in part, to the filtering activity of zebra mussels. Plankton diatoms in the Bass Island region of the western basin declined over 80% between the mid-1980s and 1990 (Holland-Beeton, 1990). Impacts of the decline of planktonic algae on the pelagic food web are not yet known. There is concern that the food web in western Lake Erie could be altered by a shift in energy from the pelagic to benthic components of the ecosystem. In California, Cloern (1982) found that suspension-feeding bivalves (mainly *Tapes japonica, Musculus senhousia*, and *Gemma gemma*) were sufficiently abundant to filter the entire volume of South San Francisco Bay in 1 day, and were capable of controlling phytoplankton biomass during summer and fall.

The importance of the western basin reefs for walleye spawning is recognized (Regier et al., 1969; Goodyear et al., 1982); therefore, the rapid colonization of these areas by zebra mussels has caused concern about their future use by this important sport and commercial species. Questions raised by the colonization included: (1) in view of the complete change in the surficial structure of the reefs, would walleye spawn over them, and (2) would dissolved oxygen in the interstitial areas around the mussels be sufficient to permit egg development?

Studies were undertaken in April and May 1990 to determine if use of the reefs by spawning walleye would be effected by the presence of zebra mussels (Fitzsimons et al., 1991). Very briefly, the investigations indicated that: (1) walleyes spawned randomly on the reefs (i.e., there was no relationship between numbers of eggs collected and concentrations of zebra mussels and the presence of zebra mussels had no significant effect on the viability of collected eggs); and (2) dissolved oxygen concentrations measured in interstitial spaces were near saturation levels indicating that the presence of zebra mussels had no apparent effect on dissolved oxygen due primarily to the low water temperature (7°C), wind-generated currents, and shallowness of mussel beds (<2 cm).

In general, there appeared to be no adverse impacts of zebra mussel colonization on walleye reproduction in the western basin in 1990. Bottom trawling for young-of-the-year walleye in the western basin in August 1990 indicated an average index of abundance which was much larger than those recorded for the previous 3 years (OMNR, 1991). These 1990 data indicate that the spawning substrate remained favorable for walleye reproduction. It is possible that spawning conditions were improved by the creation of additional crevices to hold developing eggs.

There is no evidence in European literature of adverse impact of zebra mussel invasions on fish populations. A comparison between the invasion of western Lake Erie and Lake Balaton in Hungary has been noted by Schloesser et al. (1990). An important fish species in Lake Balaton is the pike perch (*Stizostedion lucioperca*), which is similar to the walleye in North America. Birò (1977) reported commercial harvests of pike perch from Lake Balaton between 1902 and 1973. The data (incomplete during World War II) indicate no impacts on the harvest of pike perch which could be attributed to zebra mussels in Lake Balaton in the early 1930s.

CONCLUSIONS

The present study supports the following conclusions: (1) zebra mussel populations in western Lake Erie expanded rapidly between 1988 and 1990; (2) the filtration capacity of settled mussels is large and mainly responsible for recent increases in water transparencies and declines in chlorophyll *a*; and (3) the character of substrates on rocky spawning reefs has been completely altered through colonization by zebra mussels, but walleye spawning has not been adversely affected.

ACKNOWLEDGMENTS

I am grateful to G. Bell, S. Geddes, G. Ives, A. Matthews, R. Sutherland, and A. Wormington for technical assistance and to C. Brousseau, D. Hurley, S. J. Nepszy, K. Stewart, and anonymous reviewers for providing helpful comments and suggestions. Contribution No. 91–10 of the Ontario Ministry of Natural Resources, Research Section, Fisheries Branch, Box 5000, Maple, Ontario, Canada L6A 1S9.

REFERENCES

Birò, P. ''Effects of Exploitation, Introductions, and Eutrophication on Percids in Lake Balaton,'' *J. Fish. Res. Board Can.* 34:1678–1683 (1977).

Cloern, J. E. ''Does the Benthos Control Phytoplankton Biomass in South San Francisco Bay?,'' *Mar. Ecol. Progr. Ser.* 9:191–202 (1982).

Evans, M. S., and D. W. Sell. ''Zooplankton Sampling Strategies for Environmental Studies,'' *Hydrobiologia* 99:215–223 (1983).

Fitzsimons, J. P., J. Leach, S. Nepszy, and V. Cairns. ''Effects of Zebra Mussels on Walleye Reproduction in Western Lake Erie,'' Abstract of a paper presented at the 34th Conference on Great Lakes Research, Buffalo, NY, June 3–6, 1991.

Fraleigh, P. C., P. L. Klerks, and R. C. Stevenson. ''Temporal Changes in Zebra Mussel (*Dreissena polymorpha* Pall.) Veliger Densities and Veliger Settling Rates in Western Lake Erie near Maumee Bay,'' Abstract of a paper presented at the International Zebra Mussel Research Conference, Columbus, OH, December 5–7, 1990.

Garton, D. W., and W. R. Haag. ''Seasonal Patterns of Larval Abundance of *Dreissena* in Western Lake Erie: What a difference a year makes,'' Abstract of a paper presented at the International Zebra Mussel Research Conference, Columbus OH, December 5–7, 1990.

Goodyear, C. D., T. A. Edsall, D. M. Demsey, G. D. Moss, and P. E. Polanski. ''Atlas of Spawning and Nursery Areas of Great Lakes Fishes,'' U.S. Fish Wildlife Service, Ann Arbor, MI, FWS/OBS–82/52 (1982).

GLWQB. ''1987 Report on Great Lakes Water Quality to the International Joint Commission,'' Great Lakes Water Quality Board, IJC Great Lakes Regional Office, Windsor, ON (1987).

Griffiths, R. W., W. P. Kovalak, and D. W. Schloesser. ''The Zebra Mussel, *Dreissena polymorpha*, (Pallas 1771), in North America: Impacts on Raw Water Users,'' in *Symposium: Service Water System Problems Affecting Safety-Related Equipment* (Palo Alto, CA: Electric Power Research Institute, 1989).

Griffiths, R. W., D. W. Schloesser, J. H. Leach, and W. P. Kovalak, ''Distribution and Dispersal of the Zebra Mussel (*Dreissena polymorpha*) in the Great Lakes Region,'' *Can. J. Fish. Aquat. Sci.* 48:1381–1388 (1991).

Haney, J. F., and D. J. Hall. ''Sugar-Coated Daphnia: A Preservation Technique for Cladocera,'' *Limnol. Oceanogr.* 18:331–333 (1973).

Hebert, P. D. N., B. W. Muncaster, and G. L. Mackie. ''Ecological and Genetic Studies on *Dreissena polymorpha* (Pallas): A New Mollusc in the Great Lakes,'' *Can. J. Fish. Aquat. Sci.* 46:1587–1591 (1989).

Herdendorf, C. E., and L. L. Braidech. ''Physical Characteristics of the Reef Area of Western Lake Erie,'' Ohio Division Geological Surveys, Report No. 82 (1972).

Holland-Beeton, R. E. ''Plankton Diatoms in Hatchery Bay, Western Lake Erie, Before and After the Invasion of the Zebra Mussel,'' Abstract of a paper presented at the International Zebra Mussel Research Conference, Columbus, OH, December 5–7, 1990.

Kryger, J., and H. V. Riisgård. ''Filtration Rate Capacities in 6 Species of European Freshwater Bivalves,'' *Oecologia (Berlin)* 77:34–38 (1988).

Leach, J. H. ''Seston Composition in the Point Pelee Area of Lake Erie,'' *Int. Rev. Ges. Hydrobiol.* 60:347–357 (1975).

MacIsaac, H. J., and W. G. Sprules, O. E. Johannsson, and J. H. Leach. "Filtering Impacts of Larval and Sessile Zebra Mussels (*Dreissena polymorpha*) in Western Lake Erie," *Oecologia* (In press).

Mackie, G. L., W. N. Gibbons, B. W. Muncaster, and I. M. Gray. "The Zebra Mussel, *Dreissena polymorpha*: A Synthesis of European Experiences and a Preview for North America," Queens Printer for Ontario (1989).

OMNR. Lake Erie Fisheries Report 1990. Prepared for the Lake Erie Committee Meeting, Great Lakes Fishery Commission, Niagara Falls, NY (March 25–26, 1991).

Penner, G. Kingsville, Ontario, Personal communication (1988).

Reeders, H. H., and A. bij de Vaate. "Zebra Mussels (*Dreissena polymorpha*): A New Perspective for Water Quality Management," *Hydrobiology* 200/201:437–450 (1990).

Reeders, H. H., A. bij de Vaate, and F. J. Slim. "The Filtration Rate of *Dreissena polymorpha* (Bivalvia) in Three Dutch Lakes with Reference to Biological Water Quality Management," *Freshwater Biol.* 22:133–141 (1989).

Regier, H. A., V. C. Applegate, and R. A. Ryder. "The Ecology and Management of the Walleye in Western Lake Erie," Great Lakes Fishery Commission Technical Report 15 (1969), 101 pp.

Schloesser, D. W., W. P. Kovalak, and T. F. Nalepa. "Comparison of the Zebra Mussel Invasion in Western Lake Erie, North America and Lake Balaton, Hungary," Abstract of a paper presented at the International Zebra Mussel Research Conference, Columbus, OH, December 5–7, 1990.

Sprung, M. "Field and Laboratory Observations of *Dreissena polymorpha* Larvae: Abundance, Growth, Mortality and Food Demands," *Arch. Hydrobiol.* 115:537–561 (1989).

Stanczykowska, A. "On the Relationship Between Abundance, Aggregations and "Condition" of *Dreissena polymorpha* Pall. in 36 Masurian Lakes," *Ekol. Pol., Ser. A,* 12:653–690 (1964).

Stanczykowska, A. "Ecosystem of the Mikolajskie Lake. Regularities of the *Dreissena polymorpha* Pall (Bivalvia) Occurrence and Its Function in the Lake," *Pol. Arch. Hydrobiol.* 22:73–78 (1975).

Stanczykowska, A. "Ecology of *Dreissena polymorpha* (Pall.) (Bivalvia) in Lakes," *Pol. Arch. Hydrobiol.* 24:461–530 (1977).

Strickland, J. D. H., and T. R. Parsons. *A Practical Book of Seawater Analysis,* Bulletin of the Fisheries Research Board of Canada No. 167 (1968).

Walz, N. "Die Besiedlung von kunstlichen Substraten durch Larven von *Dreissena polymorpha*," *Arch. Hydrobiol., Suppl.* 47:423–431 (1975).

Wisniewski, R. "Shoals of *Dreissena polymorpha* as Bio-Processor of Seston," *Hydrobiologia* 200/201:451–458 (1990).

CHAPTER 24

Biomass and Production of Zebra Mussels (*Dreissena polymorpha*) in Shallow Waters of Northeastern Lake Erie

Ronald Dermott, Joanne Mitchell, Ian Murray, and Elise Fear

Qualitative observations, biomass, and production of zebra mussels (*Dreissena polymorpha*) were determined in shallow waters of northeastern Lake Erie from 1989 to 1991. In 1990, 1 year after mussel colonization, the dry weight biomass (B) was 72 g/m^2 (shell-free dry weight). This biomass represented 70% of the total macroinvertebrate community biomass. Annual production (P) of mussel tissue between May 1990 and May 1991 was 147 g/m^2 with an annual P/B ratio of 4.7. Growth rate, as measured by changes in shell length, peaked in early summer, while the increase in dry weight peaked in late summer. The percentage of both lipid and soft tissue, relative to total weight, decreased over the summer. The macroinvertebrate population increased by an order of magnitude on bedrock colonized by zebra mussels, as compared to adjacent bedrock not colonized by mussels.

0-87371-696-5/93/$0.00 + $.50

INTRODUCTION

Invasion of the Great Lakes by the zebra mussel, *Dreissena polymorpha* occurred rapidly and with a dramatic population explosion (Griffiths et al., 1989). Drifting veligers spread the species downstream from Lake St. Clair, so that by the fall of 1988 the central basin of Lake Erie had been colonized (Hebert et al., 1991). Settled mussels were first found along the north shore of eastern Lake Erie between Long Point and the entrance to the Welland Canal in early September 1989 (Figure 1). The extensive fissured limestone bedrock and rubble along the northeast shore of the lake provide ideal habitat for zebra mussels, which attach to hard substrates by their byssus. Although poorly studied, this habitat has historically supported populations of filter feeding caddis flies (*Hydropysche* spp.), other invertebrates, and attached algae (Shelford and Boesel, 1942; Barton and Hynes, 1978).

The objective of this study was twofold: (1) to estimate seasonal changes in growth, biomass, and production of zebra mussels during the first year of colonization in northeastern Lake Erie, and (2) to determine any changes in the benthic macroinvertebrate community in response to the increase in zebra mussel abundances.

METHODS

Qualitative observations of zebra mussels were made in the vicinity of Port Colborne in September 1989 (Figure 1). Subsequently, quantitative samples were obtained monthly near Longbeach, approximately 10 km west of the Welland Canal from May to November 1990 and in March and May 1991. Samples were collected from rubble on exposed limestone bedrock that sloped gradually down to a depth of 3 m, about 1 km offshore.

On each quantitative sampling date, rocks up to 35 cm in diameter were brought to the surface from a depth of 1.0–1.5 m using a long-handled dip net. The abundance of attached mussels was estimated from counts within a quadrate frame of 100 cm^2. Counts from 25 to 30 quadrates were made, from which a total of 200 mussels were randomly removed from rocks and returned to the laboratory. A 2-L water sample was collected at a distance of 20 to 30 m offshore, and used to estimate the weight of suspended solid (i.e., seston) by vacuum filtration through a GF/C glass fiber filter of 1.2 μm retention. The organic content (ash-free-dry weight) of seston was estimated as loss of weight following ignition at 500°C for 4 hr.

Mussels were blotted to remove exterior water and dried at 60°C for 24 hr. Shell lengths of mussels greater than 4 mm were measured along their greatest length using a vernier caliper, and individual dry weight including shell was recorded. During May, July, and October 1990, approximately 30 live mussels were measured, weighed, dried, and weighed to estimate the

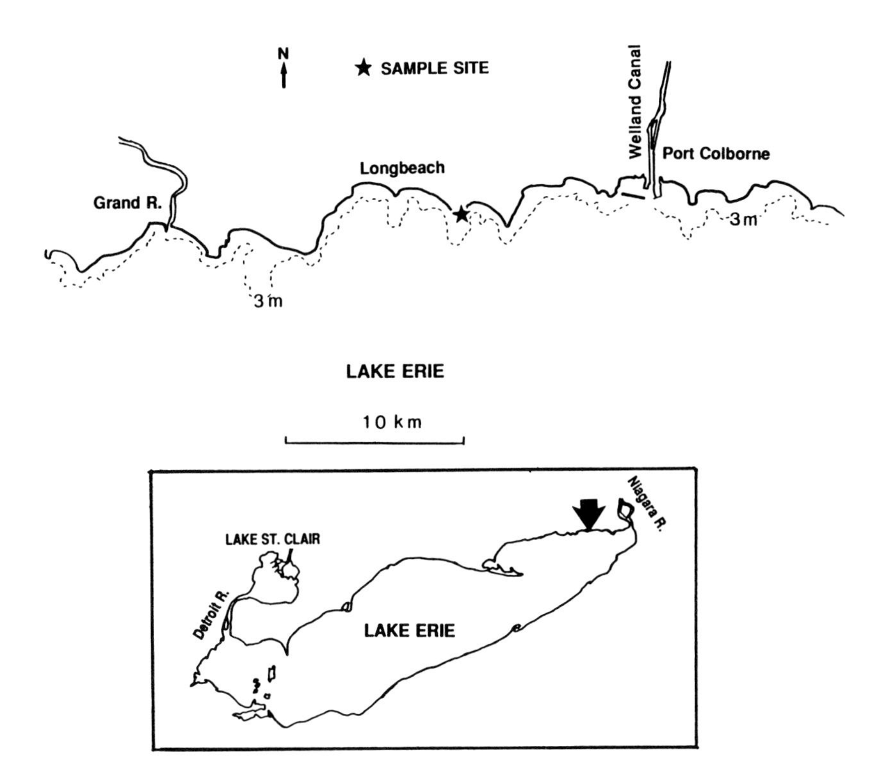

Figure 1. Location of study site and the 3-m depth contour along northeastern Lake Erie.

water content of fresh individuals. On sample dates between August 1990 and May 1991, four to six rocks were taken to the laboratory to estimate the number of small, newly settled mussels. The number of individuals within a quadrate area of 8.5 cm^2 were counted using a magnifying lens. A minimum of nine quadrates were examined from the surface and two from the sides of each rock. Newly settled mussels (<4 mm) were measured using a stereomicroscope and measured with a microlens fitted micrometer. The average dry weight of newly settled mussels was obtained by weighing groups of 10 individuals in each 1-mm size class.

Length-frequency histograms were plotted as the percentage of total animals in 1-mm size classes from 1 to 20 mm. Average length and dry weight of mussels were determined for each sample date and length-weight regressions, and the average weight of each size class was then calculated. Production over a period of 366 days was calculated for the population using the size frequency method based on the abundance of mussels per square meter in each size class on each sample date and the average dry weight per individual for each size class (Krueger and Martin, 1980). The generation time was assumed to be 1 year.

Lipid and Soft Tissue

Lipid content was measured as loss in weight during solvent extraction over 3 weeks in a 2:1 mixture of chloroform and methanol, followed by rinsing in ethanol and redrying (Costopoulos and Fonds, 1989; Smith, 1990). In June and October, mussels were maintained for 1 month without food at 20°C in order to estimate the lipids levels after starvation. Soft tissue of weighed individuals was digested from shells using 0.1 g/L pancreatin for 3 days, in order to measure shell weight (Dingerkus and Uhler, 1977). Dry weight of the soft tissue was calculated as the difference in total weight minus shell weight.

Associated Studies

Macroinvertebrates were collected from areas of hard substrates in May, July, August, and October 1990. The frame of an Ekman grab (240 cm^2), without doors or springs, was placed onto sampling areas in 1 m of water. Loose rocks and pebbles within the frame were removed and placed in a bucket. A hose attached to a bilge pump was used to extract the water, loosened debris, and organisms on the substrate within the frame. Debris in the outflow water was collected in a bag made from 210-μm nylon mesh. This material and the invertebrates attached to the removed rocks were preserved in 10% formalin for later analysis. During July and August, samples from bare bedrock, within 10 m of the areas sampled in May, July, August, and October were also collected to estimate the abundance of macroinvertebrates on bedrock colonized by mussels, as well as adjacent areas having no visible mussels. The preserved, wet weights of the various types of macroinvertebrates in the analyzed samples were converted to dry shell-free weights using correction factors for individual taxa.

In November 1990, scuba divers collected samples at a depth of 2 m from about 500 m offshore to compare nearshore and offshore mussel abundances. Offshore samples were collected with a larger frame (0.1 m^2) than that used nearshore to allow divers to scrape mussels off the rocks. Mussels, macroinvertebrates, and debris within the frame were sucked up into numbered 210-μm mesh bags, preserved and taken to the laboratory for analysis. Additional rocks were collected, to estimate the abundance of small, newly settled mussels living offshore.

RESULTS AND DISCUSSION

Abundance

Although mussels were observed in the area in September 1989, ice scour during winter storms removed most of the mussels from the exposed bedrock prior to the first sampling in 1990. At that time, mussels were only observed within bedrock fissures and under large rocks. The abundance of mussels increased from 384/m^2 (S.E. = 144) in May 1990 to 2551/m^2 (S.E. = 544) in July, probably as a result of the migration of small mussels up from within the rubble and rock fissures (Figure 2). By November, the total population had reached 54,317/m^2 (S.E. = 7,963) of which, 2,335 (S.E. = 340) were the large mussels of the 1989 year-class which had shells larger than 8 mm. A t-test indicated that there was no significant difference in the abundance of adult-sized mussels inshore vs offshore (t = 0.6, df = 27). However, the abundance of young-of-the-year mussels was significantly less offshore (t = -4.5, df = 17) where they averaged only 12,500/m^2 (S.E. = 3,626). High mortality in this area over the winter, most likely due to ice scour and heavy surf, reduced the mussel population to 20,479/m^2 (S.E. = 4,153) of which only 1,160/m^2 were larger mussels of the 1989 year-class.

The abundance of the 1989 year-class in northeastern Lake Erie was similar to that of the long-established, multiple year-class populations in European lakes, which may range from 30/m^2 (Walz, 1978) to 4500 (Kajak and Dusoge, 1975), but generally are less than 2000/m^2 (Stanczykowska, 1976 and 1978). In September 1989, young-of-the-year mussels increased the population to the dramatic populations of 30,000–200,000/m^2 on rocky shoals of western Lake Erie in 1990 (Leach, Personal communication) where the species has been present for at least two generations.

Growth Rate

In early September 1989, newly settled mussels in northeastern Lake Erie averaged 1.1 mm (S.E. = 0.1). By late September, mussels near Port Colborne had grown to 3.1 mm (S.E. = 0.5). In early July 1990, only 1.5% of the population was smaller than 3 mm and by August no mussels less than 3 mm long were collected. This indicates that little reproduction prior to August was successful. The 1990 year-class settled between August 31 and September 11, 1990 when the settled young averaged 1.15 mm (S.E. = 0.04) with an average dry weight of 0.02 mg (Table 1). Average length of the 1990 year-class reached only 2.21 mm (S.E. = 0.12) by November 8, 1990 when their average dry weight was 0.18 mg. Growth of newly settled mussels in the autumn of 1990 was considerably less than that during 1989, probably due to competition with large mussels for both food and settling sites. By October 1990, the average length of the 1989 year-class had reached 15.00 mm (S.E. = 0.3), and ranged between 9 and 21 mm (Table 1, Figure 3). Growth over the winter was negligible. By May 3, the average length of the 1989 year-class was 15.73 mm and the average length of the 1990 year-class had reached 2.49 (S.E. = 0.12 mm).

Because there was no overlap in size of the two generations (Figure 3), average growth in 1990 was calculated from average length. The growth rate

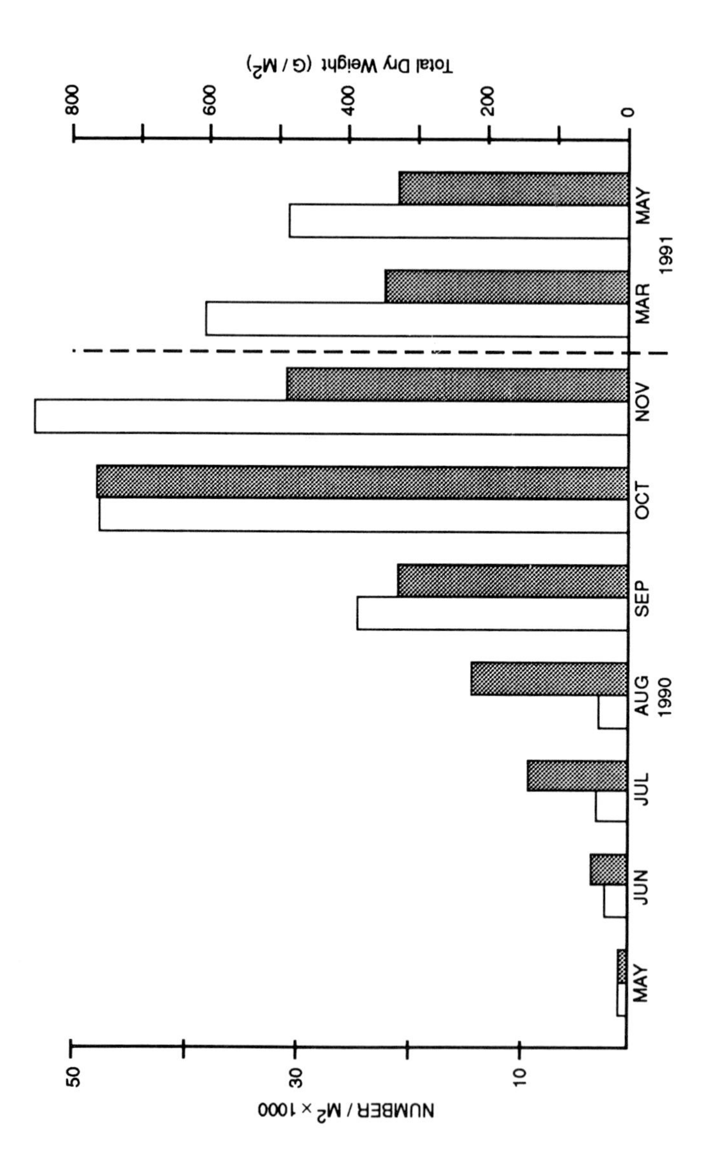

Figure 2. Abundance (clear bars) and total dry weight biomass of *Dreissena polymorpha* (hatched bars) at the nearshore study site in northeastern Lake Erie 1990–1991.

TABLE 1. Seasonal Changes in Shell Length and Dry Weight of the *Dreissena* Population in Northeastern Lake Erie from 1990 to 1991

Date	Mean Length (mm)	Increased Length per Interval	Dry Weight with Shell (mg)	Percent Soft Tissue	Dry Weight Shell-Free (mg)	Increased Tissue Weight	Water Temp (°C)	Seston AFDW (mg/L)
1989 Year-Class								
May 2	4.73		9.49	17.6	1.68		13	2.30
		2.70				2.64		
Jun. 8	7.43		27.37	15.8	4.32		14	6.39
		2.66				2.81		
Jul. 6	10.09		56.42	12.6	7.12		21	1.87
		1.85				3.80		
Aug. 13	11.94		91.47	11.9	10.92		23	2.61
		2.25				5.48		
Sept. 11	14.19		154.97	10.6	16.40		21	2.49
		0.81				1.39		
Oct. 10	15.00		193.01	9.2	17.79		15	2.56
		−0.38				−0.78		
Nov. 8	14.62		184.55	9.2	17.01		8	5.74
		0.41				2.38		
Mar. 22	15.03		194.15	10.8	19.39		8	—
		0.70				1.88		
May 3	15.73		248.87	9.8	21.27		12	3.35
1990 Year-Class								
Sept. 11	1.15		0.124	17.9	0.022			
		0.68				0.084		
Oct. 10	1.83		0.596	17.9	0.107			
		0.38				0.072		
Nov. 8	2.21		1.006	17.9	0.179			
		−0.02						
Mar. 22	2.19		1.199	*	—			
		0.30				0.044		
May 3	2.49		1.771	12.5	0.223			

Note: Percent soft tissue not available for March 1991 samples (*).

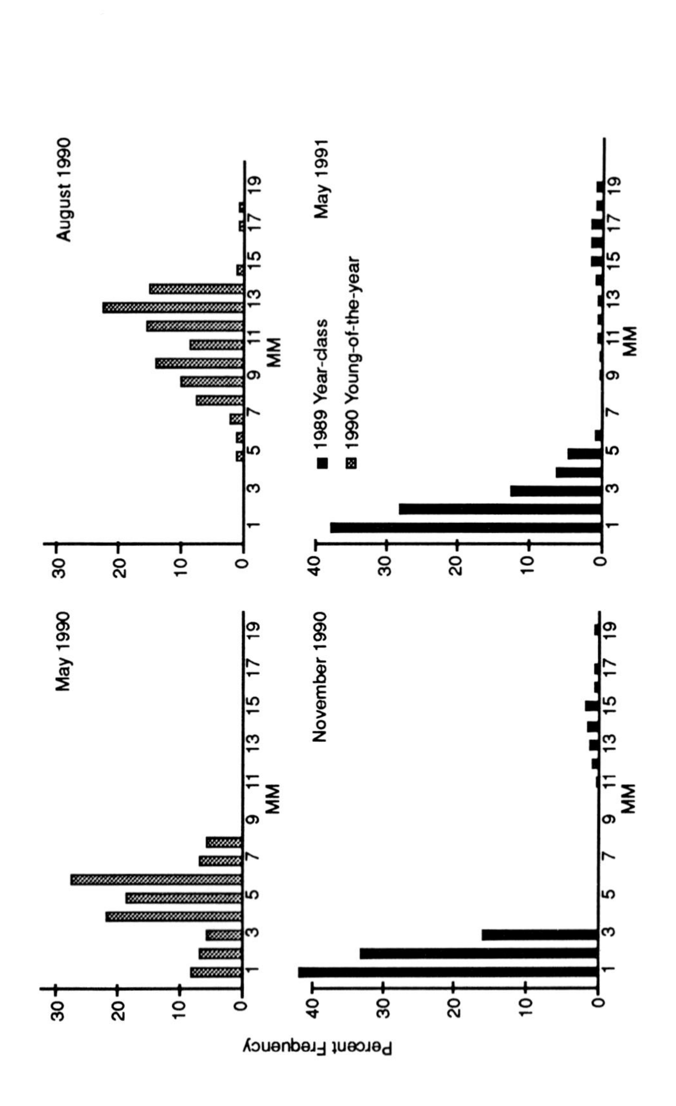

Figure 3. Size frequency distributions of 2-year-classes of *Dreissena polymorpha* at the nearshore study site in May, August, and November 1990 and May 1991.

TABLE 2. Length-Weight Regressions for *Dreissena polymorpha* Collected in Northeastern Lake Erie from May to October 1990

Total wet weight (with shell)	mg = exp (2.663 * $\log_n$(X mm) − 1.410)	n = 110, r^2 = 0.97
Wet tissue (shell free)	mg = exp (2.695 * $\log_n$(X mm) − 2.066)	n = 110, r^2 = 0.96
Total dry weight[a] (with shell)	mg = exp (2.612 * $\log_n$(X mm) − 2.019)	n = 178, r^2 = 0.96
Dry tissue (shell free)	mg = exp (2.198 * $\log_n$(X mm) − 3.224)	n = 178, r^2 = 0.90
Percent soft tissue	percent = (−0.775 * (X mm) + 20.941)	n = 178, r^2 = 0.43

[a] Mean % dry weight = 50.65% shell included.

of the 1989 year-class declined from 2.70 mm/month in May 1990 to 0.81 mm/month in October (Table 1). No growth was measured after October 10. Shell growth was less in July and August than during May and June, when the higher water temperatures in midsummer may have been above the optimum for growth (Walz, 1979). This reduced growth occurred during the single breeding period of the eastern Lake Erie mussel population in 1990 (Riessen et al., 1990). At this time, energy may have been transferred from growth into the production of eggs and sperm. Morton (1969) had measured a cessation in growth during the time of breeding in European populations.

Changes in Soft Tissue

Although the increase in shell length slowed in midsummer, weight of soft tissue continued to increase over the summer, reaching a maximum growth rate in September (Table 1). The proportion of soft tissue to total weight had a negative correlation with length as a result of seasonal changes and allometric growth (Table 2). Reduced shell growth during the breeding period was also reflected by a drop in lipid levels of mussels from 4.1% (S.E. = 0.5) of total dry weight in July to 2.0% (S.E. = 0.3) by August (Figure 4). Changes in lipid levels reflect both the amount of food available (Walz, 1979) and loss of lipid reserves during reproduction. High seston levels during June was a result of phytoplankton and cladophora blooms and may have provided a high food supply during the time of maximum shell growth and high lipid reserves (Table 1). Starvation of 7-mm mussels in June reduced their lipid reserves to levels similar to that of the larger 15-mm mussels in October. Mussels starved during October had a proportional reduction in lipid levels (65% loss) but the initial levels of 1.57% (S.E. = 0.04) were much lower.

Biomass and Production

Average total dry weight biomass of mussels was 0.28 kg/m^2 during 1990. The numerous 1990 year-class composed less than 4% of this total. This

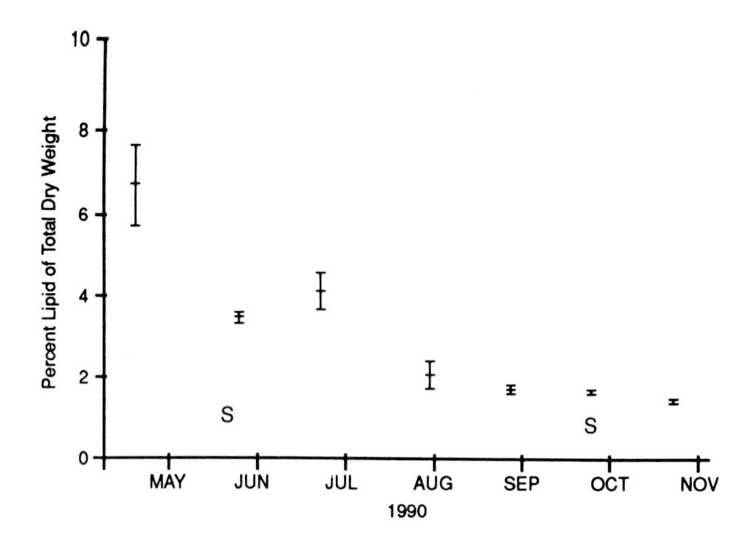

Figure 4. Average lipid content of the 1989 year-class of *Dreissena polymorpha* at the nearshore study site in northeastern Lake Erie. The lipid levels marked S in June and October are from individuals starved for 30 days at 20°C.

biomass value was similar to the 0.2 to 0.8 kg/m^2 (including dry shell) in European lakes (Stanczykowska et al., 1983). However, the biomass of mussels in recently invaded Volga reservoirs had reached 6 kg/m^2 when populations exceeded 10,000 per square meter (Mordukhai-Boltovskoi, 1979).

Production of soft tissue, over the 366-day sample period (i.e., May 2, 1990 to May 3, 1991) amounted to 242.9 g/m^2 dry weight (S.E. = 51.6). As large mussels of 1989 year-class were about 605 days old on May 3, 1990, correcting for the cohort production interval of 365/605 (Benke, 1979) the annual production by zebra mussels was 146.5 g/m^2 (dry shell-free). This production does not include the organic matrix of shells or the byssal threads produced by mussels. Walz (1979) calculated that the byssus may represent up to 5% of the carbon content of soft tissues. Average annual biomass of soft tissue during 1990 and 1991 was 30.84 g/m^2 at the study site. The annual corrected production (CPI) biomass ratio (P/B) was 4.7. European values of P/B for established *Dreissena polymorpha* populations normally range between 0.4 and 1.0 (Stanczykowska, 1976). However, high production of young *Dreissena* in Lake Constance resulted in a very high annual P/B ratio of 6.8 for a newly established population (Walz, 1978). P/B ratios for most freshwater bivalves range between a maximum of 3.5 for *Sphaerium* (Holopainen, 1979) to a minimum of 0.2 for slow growing unionid clams (Edmondson and Winberg, 1970; Lewandowski and Stanczykowska, 1975).

In comparison, the annual production of all macroinvertebrates amounted to only 50 g/m^2 (dry weight) in the midsection of the Bay of Quinte, Lake

Ontario (Johnson and Brinkhurst, 1971) under water quality and temperature conditions similar to that in northeastern Lake Erie. Macroinvertebrate biomass values are not available for the inshore region of Lake Erie, thus comparisons to the current biomass of zebra mussels are not possible. Wood (1963) found that total macroinvertebrate biomass in the shallow western basin of Lake Erie averaged 99 g/m^2 wet weight (about 14 g/m^2 dry weight) where native mollusks comprised 83% of this weight.

Lacking data on the macroinvertebrate community prior to the arrival of zebra mussels, it is not possible to separate seasonal effects from effects related to the presence of zebra mussels. By June 1990, all areas of fissured substrate below a depth of 0.5 m were colonized by zebra mussels. On rocks and gravel, the seasonal increase in abundance of *Gammarus fasciatus*, small chironomids, (mostly *Crictopus* spp., *Psectrocladius* sp., and *Microtendipes* spp.), and oligochaetes (mainly *Nais* spp.) was due to summer reproduction (Table 3). Conversely, mayflies (*Heptagenia* sp., *Stenonema* spp.), and caddis flies (mostly *Hydropsyche recurvata, Cheumatopsyche* sp., and *Helicopsyche borealis*) are more common during the fall and spring (Barton and Griffiths, 1984). The proportion of the total macroinvertebrate biomass (shell-free) represented by *Dreissena polymorpha* in the littoral of northeastern Lake Erie increased from 5% in May to 68% in October (Table 3).

Attached zebra mussels increase the number of crevices on the otherwise smooth bedrock, thus creating habitat for the amphipod *Gammarus* sp., the mayfly *Caenis* sp. and the leeches *Erpobdella punctata* and *Helobdella stagnalis*. During midsummer, total invertebrate abundance increased on bedrock colonized by mussels, as compared to adjacent bedrock that was uncolonized by *Dreissena polymorpha* (Table 4). Most of this increase was due to large populations of the amphipod *G. fasciatus*. However, the increase was statistically significant only for *Gammarus, Dreissena*, and gastropods. Species composition was similar on rocks, gravel, bare rock, and bedrock. From the data available during the first year of colonization, there was no evidence that zebra mussels had a negative impact on any of the major macroinvertebrate species present.

CONCLUSION

Zebra mussels colonizing the rock outcrops in northeastern Lake Erie represent a large pool of organic matter, and provide interstitial habitat which was formerly absent. Between May 1990 and 1991, the zebra mussel population averaged 21,272 per squared meter of which 2,003 were large mussels of the 1989 year-class. This population represented a shell-free dry biomass of 30.8 g/m^2 and had an annual production of 146 g/m^2 of soft tissue that potentially may not be passed on to the higher trophic levels of the lake. The macroinvertebrate biomass present in the colonies of zebra mussels has also

TABLE 3. Seasonal Changes in the Density and Shell-Free Dry Weight Biomass of Macroinvertebrates on Hard, Nearshore Substrates along the Northeast Shore of Lake Erie

Taxa	May 2, 1990 Density (number/m²)		May 2, 1990 Biomass (g/m²)		July 6, 1990 Density (number/m²)		July 6, 1990 Biomass (g/m²)		October 10, 1990 Density (number/m²)		October 10, 1990 Biomass (g/m²)	
Total	10420	(3198)	4.872	(0.892)	47573	(7620)	34.373	(9.560)	121373	(59496)	64.621	(25.606)
Oligochaetes	1173	(460)	0.173	(0.084)	13040	(5247)	3.119	(1.269)	3200	(1170)	1.004	(0.894)
Leeches	7	(7)	0.034	(0.034)	412	(96)	1.085	(0.283)	467	(267)	0.210	(0.114)
Gammarus	1860	(439)	3.219	(0.599)	15026	(2080)	25.377	(6.892)	25773	(9241)	18.824	(4.853)
Dreissena	1220	(652)	0.255	(0.127)	1080	(266)	2.044	(0.503)	85146	(48989)	43.656	(25.115)
Gastropods	66	(66)	0.007	(0.007)	507	(346)	0.802	(0.722)	653	(163)	0.308	(0.076)
Pisidium	80	(61)	0.009	(0.007)	666	(255)	0.090	(0.024)	160	(69)	0.042	(0.017)
Mayflies	66	(66)	0.040	(0.040)	53	(53)	0.007	(0.007)	733	(393)	0.031	(0.016)
Caddisflies	1527	(381)	0.741	(0.184)	504	(208)	0.297	(0.124)	333	(224)	0.142	(0.093)
Chironomids	2706	(869)	0.155	(0.052)	13640	(642)	0.603	(0.028)	4027	(1067)	0.124	(0.028)

Note: Mean and standard error (in parentheses) of triplicate samples collected from 1-m depth with a modified Eckman grab. Totals include minor unlisted taxa.

TABLE 4. Macroinvertebrate Densities and Shell-Free Dry Biomass on Exposed Limestone Bedrock in Northeastern Lake Erie, with and without *Dreissena polymorpha* Present

	Exposed Limestone Bedrock							
	Without *Dreissena*				With *Dreissena*			
Taxa	Density (number/m²)		Biomass (g/m²)		Density (number/m²)		Biomass (g/m²)	
Total	4558	(2464)	0.491	(0.265)	44636	(8108)	22.529	(4.092)
Oligochaetes	2200	(1586)	0.132	(0.095)	7612	(5296)	0.314	(0.217)
Leeches	13	(13)	0.035	(0.035)	200	(61)	0.522	(0.158)
Gammarus	1146	(807)	0.208	(0.146)	24144	(4760)	6.256	(1.233)
Dreissena	0	(0)	0.0	(0)	904	(181)	12.733	(2.557)
Gastropods	146	(58)	0.025	(0.005)	600	(69)	1.983	(1.818)
Pisidium	13	(13)	0.003	(0.003)	306	(109)	0.038	(0.013)
Mayflies	26	(13)	0.008	(0.004)	784	(420)	0.030	(0.016)
Caddis flies	53	(35)	0.031	(0.020)	226	(104)	0.127	(0.058)
Chironomids	747	(301)	0.020	(0.008)	9120	(7236)	0.406	(0.290)

Note: Means and standard errors (in parentheses) for three samples collected in July and August 1990. Totals include minor unlisted taxa.

increased, mostly due to large populations of the amphipod *Gammarus fasciatus*.

ACKNOWLEDGMENTS

We thank Mr. Honsberger for allowing access to the rocky shoals for sampling and the dive unit of NWRI for collecting comparable offshore samples.

REFERENCES

Barton, D. R., and M. Griffiths. "Benthic Invertebrates of the Nearshore Zone of Eastern Lake Huron, Georgian Bay, and North Channel," *J. Great Lakes Res.* 10:407–416 (1984).

Barton, D. R., and H. B. N. Hynes. "Wave Zone Macrobenthos of the Exposed Canadian Shores of the St. Lawrence Great Lakes," *J. Great Lakes Res.* 4:26–45 (1978).

Benke, A. C. "A modification of the Hynes Method for Estimating Secondary Production with particular significance for multivoltine Species," *Limnol. Oceanogr.* 24:168–171 (1979).

Costopoulos, C. G., and M. Fonds. "Proximate Body Composition and Energy Content of Plaice (*Pleuronectes platessa*) in Relation to the Condition Factor," *Neth. J. Sea Res.* 24:45–55 (1989).

Dingerkus, G., and L. D. Uhler. "Enzyme Clearing of Alcian Blue Stained Whole Small Vertebrates for Demonstration of Cartilage," *Stain Technol.* 52:229–232 (1977).

Edmondson, W. T., and G. G. Winberg. *A Manual on Methods for the Assessment of Secondary Productivity in Fresh Waters, IBP Handbook No. 17.* (Oxford, England: Blackwell Scientific Publications, 1971).

Griffiths, R. W., W. P. Kovalak, and D. W. Schloesser. "The Zebra Mussel, *Dreissena polymorpha*, (Pallas 1971) in North America: Impacts on Raw Water Users," in Symposium: Service Water System Problems Affecting Safety-Related Equipment, (Palo Alto, CA. Electric Power Research Institute, Nuclear Power Division, 1989), 11–26.

Hebert, P. D. N., C. C. Wilson, M. H. Murdoch, and R. Lazar. "Demography and Ecological Impacts of the Invading Mollusc *Dreissena polymorpha,*" *Can. J. Zool.* 69:405–409 (1991).

Holopainen, I.J. "Population Dynamics and Production *Pisidium* species (Bivalvia, Sphaeriidae) in the Oligotrophic and Mesohumic Lake Paajarvi, Southern Finland," *Arch. Hydrobiol. Suppl.* 54:466–508 (1979).

Johnson, M. G., and R. O. Brinkhurst. "Production of Benthic Macroinvertebrates of Bay of Quinte and Lake Ontario," *J. Fish. Res. Board Can.* 28:1699–1714 (1971).

Kajak, Z., and K. Dusoge. "Macrobenthos of Lake Taltowisko," *Ekol. Pol.* 23:295–316 (1975).

Krueger, C. C., and F. B. Martin. "Computation of Confidence Intervals for the Size-Frequency (Hynes) Method of Estimating Secondary Production," *Limnol. Oceanogr.* 25:773–777 (1980).

Leach, J. Ontario Ministry of Natural Resources, Personal communication (1990).

Lewandowski, K., and A. Stanczykowska. "The Occurrence and Role of Bivalves of the Family Unionidae in Mikolajski Lake," *Ekol. Pol.* 23:317–334 (1975).

Mordukhai-Boltovskoi. "The Volga River and Its Life," *Monographie Biologicae 33* The Hague, The Netherlands: Junk Publishers, Inc., (1979).

Morton, B. S. "Studies on the Biology of *Dreissena polymorpha* Pall. III. Population Dynamics," Proceedings of the Malacological Society of London 38:471–481 (1969).

Riessen, H. P., T. A. Ferro, and R. A. Kamman. "Distribution of Zebra Mussel Veligers in Eastern Lake Erie During the First Year of Colonization," International Zebra Mussel Research Conference, Columbus, OH, December 1990.

Shelford, V. E., and M. W. Boesel. "Bottom Animal Communities of the Island Area of Western Lake Erie in the summer of 1937. *Ohio J. Sci.* 42:179–190 (1942).

Smith, S. L. "Egg Production and Feeding by Copepods Prior to the Spring Bloom of Phytoplankton in Fram Strait, Greenland Sea. *Mar. Biol.* 106:59–69 (1990).

Stanczykowska, A. "Biomass and Production of *Dreissena polymorpha* (Pall.) in Some Masurian lakes." *Ekol. Pol.* 24:103–112 (1976).

Stanczykowska, A. "Occurrence and Dynamics of *Dreissena polymorpha* (Pall.) (Bivalvia)," *Verh. Int. Verein. Theor. Angew. Limnol.* 20:2431–2434 (1978).

Stanczykowska, A., E. Jurkiewicz-Karnikowska, and K. Lewandowski. "Ecological Characteristics of Lakes in Northeastern Poland Vs Their Trophic Gradient. 10. Occurrence of Mollusks in 42 Lakes," *Ekol. Pol.* 31:459–476 (1983).

Walz, N. "The Production of the *Dreissena* — Population and Their Significance in the Nutrient Cycle of Lake Constance," *Arch. Hydrobiol.* 82:482–499 (1978).

Walz, N. ''The Energy Balance of the Freshwater Mussel *Dreissena polymorpha* Pallas in Laboratory Experiments and in Lake Constance. V. Seasonal and Nutritional Changes in the Biochemical Composition,'' *Arch. Hydrobiol. Suppl.* 55:235–254 (1979).

Wood, K. G. ''The Bottom Fauna of Western Lake Erie, 1951–1952,'' Proceedings of the 6th Conference on Great Lakes Research, June 13–15, 1963 University of Michigan-Greak Lakes Res. Div.; (1963), Pub. No. 10: 258–265.

CHAPTER 25

Effects of Zebra Mussels (*Dreissena polymorpha*) on the Benthic Fauna of Lake St. Clair

Ronald W. Griffiths

Zebra mussels, *Dreissena polymorpha*, invaded Lake St. Clair in late 1986. By September 1990, they occurred throughout the southern and eastern areas of the lake at densities of up to 10,000/m^2. Changes in species richness and benthic invertebrate abundance in southeastern Lake St. Clair following the establishment of zebra mussels were similar to those noted in northwestern Lake St. Clair where zebra mussels were not present. Several unique changes, however, were noted in benthic faunal composition in the southeastern part of the lake. The relative abundance of *Gammarus*, Tricladida, snails, *Helobdella, Polycentropus*, and *Spirosperma ferox* (Eisen) increased from 0-2.5% (median 1.2%) prior to the appearance of the mussels to 10-29% (median 27%) of the fauna in September 1990 after the establishment of zebra mussels, a value approaching the relative abundance of these taxa in the northwestern area of the lake. The removal of seston by filtering mussels has promoted a shift in habitat structure from a relatively homogeneous environment of turbid water and silty sand substrata to an environment of clearer water with patches of macrophytes, mussel colonies, and silty sand spaces. The altered habitat structure probably accounts for the increased abundance of amphipods, flatworms, and snails, whereas

0-87371-696-5/93/$0.00 + $.50

the deposition of feces and pseudofeces by the mussels is likely the reason for the increased abundance and richness of worms. The increased water clarity and changes in benthic fauna suggest that the mussels have had an oligotrophication effect on water quality conditions in the southeastern area of the lake. Future water quality assessments must take this biotic-induced change into account, so that improved water quality conditions are not mistakenly attributed to anthropogenic abatement programs.

INTRODUCTION

Zebra mussels, *Dreissena polymorpha* (Pallas), have the potential to alter the ecology of aquatic systems by their ability to rapidly establish large populations. In European lakes, settled postveligers (mussels <1 mm in length) have reached 1,000,000/m^2 (Lewandowski, 1982 and 1983; Afanas'yev and Protasov, 1987), while juveniles (mussels >2 mm) and adults (mussels >10 mm) have approached 30,000/m^2 (Walz, 1973; Burla and Lubini-Ferlin, 1976; Franchini, 1978; Suter, 1982). In the Great Lakes, populations of juveniles and adults approached 400,000/m^2 in the western basin of Lake Erie in the autumn of 1990 (MacIsaac et al., 1991). These population densities generally develop quickly, usually within a few years after their introduction, principally because female fecundity increases exponentially with age from 10,000 to 1,000,000 eggs per year (Walz, 1978).

Large populations of zebra mussels affect the ecology of aquatic ecosystems by altering the flow of energy through the food web via their filter-feeding activity. Moderate densities of adult zebra mussels (i.e., 300–1000/m^2) have been estimated to consume 4–18% of the net phytoplankton production (Hamburger et al., 1990; Stanczykowska, 1977), and settled postveligers probably consume at least an equal quantity of phytoplankton (Lewandowski, 1983). Furthermore, the removal of seston by mussels can increase the depth of the photic zone, thus promoting the growth of aquatic macrophytes in nearshore areas and thereby initiating a cascade of effects throughout the aquatic system (Reeders and bij de Vaate, 1990).

Unfortunately, few Europeans documented the impact that zebra mussels had on the benthic biota of lakes. Zebra mussels may benefit some macroinvertebrate by increasing the organic content of sediments via the deposition of pseudofaeces and faeces. This waste matter can contribute a substantial amount to the annual tripton settling to the bottom (Hamburger et al., 1990; Walz, 1978; Stanczykowska, 1977) and is an excellent source of food for specific bottom organisms such as midges (Izvekova and L'ova-Katchanova, 1972). This additional organic matter and the mechanical retention of deposited material by mussel colonies may account for the observations of Wiktor (1969) — who noted that benthic macroinvertebrate biomass near zebra mussel colonies was twice that found elsewhere — and by Dusoge (1966) — who found a direct relationship between the abundance of zebra mussels and the abundance of other invertebrates. In contrast, zebra mussels may negatively affect some benthic species. Hebert et al. (1991) found that the lipid content of unionids, a good substrate for zebra mussels (Lewandowski, 1976), decreased as the biomass of attached zebra mussels increased, suggesting that the attached zebra mussels reduced the feeding efficiency of their unionid hosts. Sebestyen (1938) noted that the unionid population in Lake Balaton, Hungary, declined 4 years after zebra mussels were first found, while Arter (1989) concluded that the decline of the unionid *Anodonta cygea,* in Lake Hallwil, Switzerland, was probably a result of zebra mussels, which invaded the lake in the 1970s. Snails may have difficulty obtaining food because mussels attached to their shells may restrict their mobility, whereas hydropsychid and polycentropodid caddis flies may be competitively excluded because of mussels interfering with the construction of their nets.

Effects that zebra mussels will have on benthic fauna in North America are not clear. From an applied point of view, an understanding of such effects is important because water quality evaluations are primarily based on the structure and composition of the aquatic fauna, especially benthic macroinvertebrates (e.g., Griffiths, 1991 and 1987; Johnson et al., 1987; Johnson and McNeil, 1986; Veal and Osmond, 1968; Brinkhurst et al., 1968). The effect of zebra mussels on macroinvertebrates needs to be understood, so that future water quality assessments will not improperly attribute macroinvertebrate changes to anthropogenic causes (i.e., abatement programs). The purpose of this study was to document the changes in the benthic macroinvertebrate fauna of Lake St. Clair induced by the invasion and establishment of zebra mussels and to assess the water quality implications of these changes.

METHODS

Effects of the zebra mussel, *Dreissena polymorpha*, on the native macroinvertebrate fauna of Lake St. Clair were examined by comparing the biota at six sites before and after the invasion of zebra mussels. Three sites (72, 83, 88) were located in the northwestern Lake St. Clair, where zebra mussels were absent in 1990, and three sites (5, 15, 63) in southeastern Lake St. Clair, were zebra mussels were abundant in 1990 (Figure 1). Changes in the benthic fauna at the sites in the northwestern area of the lake were attributed to annual and seasonal variations, while changes at the sites in the southeastern area that differed from the sites in the northwestern area were considered to be directly or indirectly caused by zebra mussels. Preinvasion macroinvertebrate data were available from surveys in April and July 1977 (Hiltunen and Manny, 1982) and May 1983 (Griffiths, 1987), while postinvasion data were available from surveys in September 1990 and June 1991 (Griffiths, unpublished data). Site numbers and locations correspond to those of Pugsley et al. (1985) and Nalepa and Gauvin (1988).

Sites in the main lake were located at depths between 4.5 to 6.0 m, while sites in Anchor Bay were located at depths of 3.5 to 4.0 m. Sediments at the

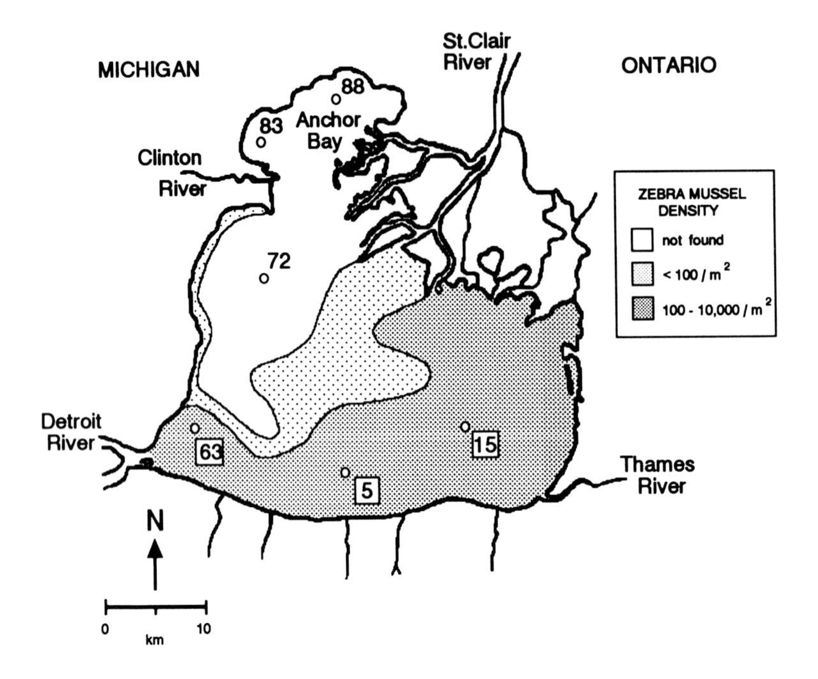

Figure 1. Location of sampling sites and abundance of *Dreissena polymorpha* in Lake St. Clair, September 1990. Abundance estimates derived from 29 sampling sites. (From Griffiths, R.W. Unpublished data, 1991.)

six sites were similar, chiefly composed of silty sand (Table 1). Nutrient and organic carbon concentrations in the sediments were also comparable, although the sites in the southeastern area were slightly more enriched, probably as a result of agricultural inputs primarily from the Thames River (Leach, 1972 and 1980). Sediment concentrations of metal and organic contaminants at these sites were low and probably have little effect on the macroinvertebrate fauna (Griffiths, 1987).

Sampling methods used to collect benthic macroinvertebrates were similar in all surveys. The benthic fauna was sampled with a Ponar grab (area of about 0.05 m^2), with three samples taken at each site. Each sample was washed through a No. 30 (U.S. Standard Sieve Series) mesh screen (aperture 0.60 mm); the retained sediment, debris, and organisms were then placed into labeled jars. In 1977, these samples were preserved with 10% formalin, and at a later date the organisms were sorted from debris using forceps with the aid of a dissecting microscope. In 1983, samples were transported to a field laboratory where the live organisms were sorted from sediment and debris in white enamel trays using forceps. In 1990, the samples were preserved with 20% buffered formalin, and at a later date the organisms were sorted from sediment and debris as described above. Invertebrates were placed in 30-mL bottles and preserved with ethanol or 10% formalin.

TABLE 1. Sediment and Depth Characteristics of Sampling Sites in Southeastern and Northwestern Lake St. Clair, May 1983

	Southeastern Area			Northwestern Area		
Sampling sites[a]	**63**	**5**	**15**	**72**	**83**	**88**
Total organic carbon (mg/g)	16.0	13.0	24.0	9.0	9.6	18.0
Total phosphorus (mg/g)	0.3	0.5	0.6	0.2	0.3	0.2
Total Kjeldahl nitrogen (mg/g)	0.6	0.8	0.9	0.5	0.7	0.6
Gravel (%)	2.0	1.4	0.0	0.2	0.4	2.2
Sand (%)	63.0	45.2	62.7	60.4	58.7	57.1
Silt and clay (%)	34.6	53.3	36.3	39.3	40.8	40.5
Depth (m)	5.0	6.0	6.0	5.5	4.0	4.0

Source: Griffiths, Unpublished data, 1991.

[a] See Figure 1 for location of sampling sites.

Benthic invertebrates from the 1977 study were identified by Schloesser and Hiltunen, while those from the later studies were identified by the author with the assistance of Sardella (1983 study) and Gutschi (1990 study). For the 1977 study, organisms were generally identified to genus, except midges and worms which were identified only to family. For the 1983 and 1990 studies, all organisms were generally identified to genus, with annelids identified to species.

Profile analysis (i.e., a Model II, repeated-measures ANOVA; Morrison, 1976) was used to determine if changes in species richness or macroinvertebrate abundance between surveys differed significantly between areas of the lake with (i.e., southeastern area) and without (i.e., northwestern area) zebra mussels. Samples were nested within sampling sites (subject factor) and sites were nested within areas of the lake (grouping factor). Number of taxa per sample and ln-transformed density were used as dependent variables; their value for each survey composed the trial factor (multivariate dependent variable). Systat (Wilkinson 1987) was used to conduct these analyses.

RESULTS

Zebra mussels, absent in the 1978 and 1983 surveys, were a major component of the benthic community in southeastern Lake St. Clair in September 1990; mussel densities averaged from 2500 to 7500/m^2 (Figure 1), and mussels generally accounted for 60–90% of the macroinvertebrate fauna. In June 1991, zebra mussels similarly dominated the benthic fauna at site 63, with a mean density of about 20,000/m^2. Single individuals were found on aquatic vegetation, especially *Vallisneria americana* and *Elodea canadensis*, whereas a layer of mussels (up to 4 cm thick) was found on rocks and unionid shells (e.g., *Lampsilis, Ligumia*). In addition, many clumps of zebra mussels up to 10 cm in diameter were collected, which consisted of numerous small animals attached to a few larger mussels.

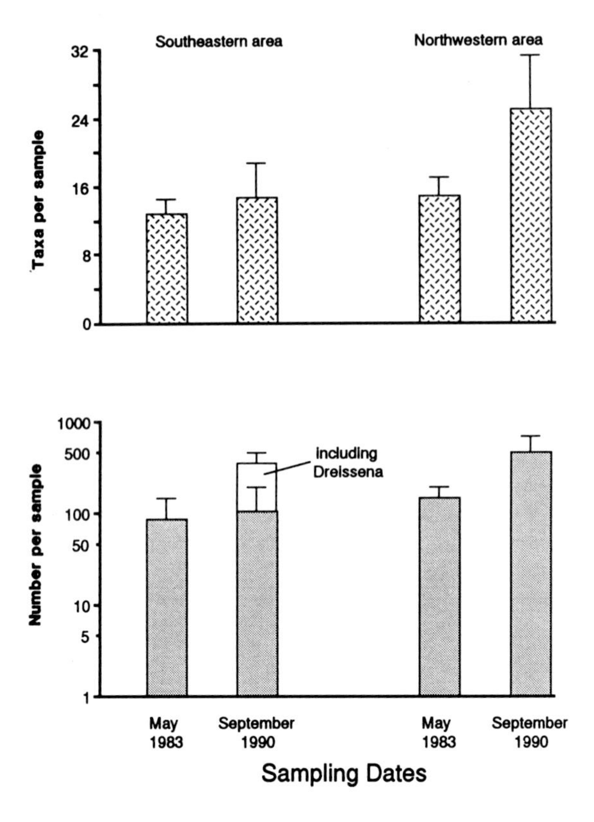

Figure 2. Comparisons of mean species richness (number of taxa/0.05 m^2) and macroinvertebrate density (number/0.05 m^2) between May 1983 and September 1990 in northwestern and southeastern Lake St. Clair. Error bars represent 1 S.D. *Dreissena polymorpha* excluded from values, except where noted.

Changes in benthic community structure in southeastern Lake St. Clair following the appearance of zebra mussels were similar to those in the northwestern area where zebra mussels were not found (Figure 2). Species richness, in terms of the number of taxa per sample, was greater in September 1990 than in May 1983 in both areas of the lake. The increase in richness in the southeastern area was not significantly different from that observed in the northwestern area between surveys ($p \geq 0.22$; Table 2). Similarly, the abundance of macroinvertebrates was greater in September 1990 than in May 1983 in both southeastern and northwestern Lake St. Clair (Figure 2). The increase in abundance of macroinvertebrates in the southeastern area was considerably less than that observed in the northwestern area (20 vs 300, respectively), although these values were not significantly different from one another ($p \geq 0.07$; Table 2). Interestingly, the mean density of zebra mussels in September 1990 was about 290 individuals per sample, making the difference in total

TABLE 2. Summary of Repeated-Measure ANOVAs that Tested Two Hypotheses

	Hypothesis 1[a]				Hypothesis 2[b]			
Source	**MS**	**df**	**F**	**p**	**MS**	**df**	**F**	**p**
Between lake areas	249	1	2.14	0.22	3.38	1	5.82	0.07
Sites within areas	116	4			0.58	4		
Sites within southeastern area	105	2	8.86	0.02	1.03	2	4.56	0.06
Sites within northwestern area	127	2	3.71	0.09	0.13	2	0.77	0.50
Samples within sites within areas	23.1	12			0.20	12		
Samples within sites within southeastern area	11.9	6			0.23	6		
Samples within sites within northwestern area	34.3	6			0.17	6		

Note: Density values were ln-transformed before analyses. Zebra mussels were not included in the analyses.

[a] H1: The difference in species richness between the 1983 and 1990 surveys in the southeastern area of Lake St. Clair (mean difference = 1.9 taxa) equaled that in the northwestern area (mean difference = 10.1 taxa).

[b] H2: The difference in invertebrate abundance between the 1983 and 1990 surveys in the southeastern area of Lake St. Clair (mean difference = 19) equaled that in the northwestern area (mean difference = 298).

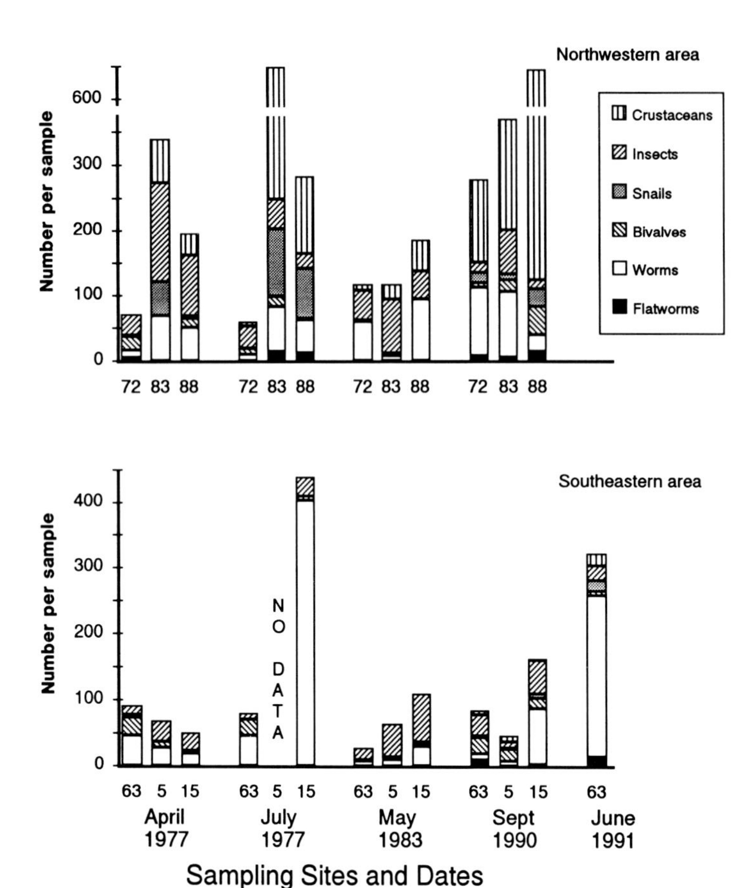

Figure 3. Taxonomic composition (number/0.05 m^2) of macroinvertebrate fauna, excluding *Dreissena polymorpha*, at sites in northwestern (top panel) and southeastern (lower panel) Lake St. Clair, before (1977 and 1983) and after (1990 and 1991) the appearance of *Dreissena polymorpha*.

invertebrate abundance between surveys virtually the same in the two areas of the lake (i.e., 310 vs 300, respectively) (Figure 2).

In contrast, several unique changes were noted in benthic faunal composition (relative abundance) of southeastern Lake St. Clair following the establishment of zebra mussels (Figure 3). These changes made the community more closely resemble that in northwestern Lake St. Clair. For instance, in 1983 and 1977, *Gammarus*, Tricladida, snails, *Helobdella, Polycentropus*, and *Spirosperma ferox* (Eisen) were rare (<4 individuals per 0.05 m^2) in the southeastern area of the lake, accounting for 0–2.5% (median 1.2%) of the total fauna (Table 3). In northwestern Lake St. Clair, meanwhile, this group of taxa accounted for 3.2–53% (median 16%) of the benthic fauna at a site

TABLE 3. Mean Abundance (Number/0.05 m^2) of Benthic Macroinvertebrates at Three Sites in Southeastern Lake St. Clair Before and After the Establishment of Zebra Mussels

Sampling Date	April (1)[a] 1977			July (1)[a] 1977		May (2)[b] 1983			September (3)[c] 1990			June (3)[c] 1991
Sampling Site	**63**	**5**	**15**	**63**	**15**	**63**	**5**	**15**	**63**	**5**	**15**	**63**
Dreissena polymorpha	**0**	**0**	**0**	**0**	**0**	**0**	**0**	**0**	**376**	**340**	**143**	**1026**
Insects												
Bugs												
Sigara lineata										<1		<1
Caddis flies												
Cheumatopsyche				1					<1			
Oecetis			3		1		2		<1		9	
Phylocentropus												3
Polycentropus									<1			<1
Mayflies												
Hexagenia		18	4	1	2	6	20	23	7	<1	13	
True flies												
Chironomidae[d]	13	12	19	7	27							
Chironomus								10	<1		1	<1
Cryptochironomus						2	3	<1	<1		1	
Cryptotendipes									<1			
Demicryptochironomus						1			<1			
Hamischia									<1			<1
Polypedilum (Tripodura)												3
Pseudochironomus									<1	<1	10	
Stictochironomus							2	<1	13	<1	3	8
Tanytarsus								2	3		5	
Tribelos								3				
Potthastia								2				
Epicocladius						<1	1	<1				
Parakiefferiella						1						
Ablabesmyia						1	6	9			<1	

TABLE 3 (continued). Mean Abundance (Number/0.05 m²) of Benthic Macroinvertebrates at Three Sites in Southeastern Lake St. Clair Before and After the Establishment of Zebra Mussels

Sampling Date	April (1)[a] 1977			July (1)[a] 1977		May (2)[b] 1983			September (3)[c] 1990			June (3)[c] 1991
Sampling Site	**63**	**5**	**15**	**63**	**15**	**63**	**5**	**15**	**63**	**5**	**15**	**63**
Dreissena polymorpha	**0**	**0**	**0**	**0**	**0**	**0**	**0**	**0**	**376**	**340**	**143**	**1026**
Clinotanypus						<1		<1		<1		
Coelotanypus						4	12	13	4	8	8	1
Procladius						3	5	7		<1	<1	4
Crustaceans												
Amphipods												
Gammarus					<1				7	9	2	20
Mollusca												
Clams												
Pisidium	23	7	4	22	5	<1	2	3	6	18	15	2
Sphaerium	3	<1				1	3	1	18			5
Mussels												
Unionidae	1	<1		<1	<1				<1			
Snails												
Ferrissia												1
Bithynia tentaculata	<1	<1							1			
Amnicola	2		<1						1			9
Probythinella lacustris						<1		1	<1		5	
Physella gyrinus										1		3
Gyraulus								<1				3
Elimia livescens	1	<1										<1
Valvata			<1								2	
Annelida												
Leeches												
Erpobdellidae		<1	3		7							
Helobdella		<1				<1		<1		4	6	<1
Polychaetes												
Manayunkia speciosa	106	<1	10	41	<1	1			<1			

Worms												
Lumbricidae									<1			
Stylodrilus heringianus									<1			
Tubificidae[d]	46	27	19	47	403							
Aulodrilus americanus							4	<1	1			
Branchiura sowerbyi							3		<1	4		
Limnodrilus hoffmeisteri						2	1	6	<1	<1	27	65
Limnodrilus udekemianus												5
Potamothrix moldaviensis												71
Quistadrilus multisetosus												4
Spirosperma ferox							<1		4			18
Immature Tubificidae						3	1	23	3	2	58	79
Nematoda												
Roundworms	6	16	74	9	96	13	15	17	<1	<1	<1	<1
Nemertea												
Proboscis Worms	<1								<1			2
Platyhelminthes												
Flatworms												
Tricladida		<1			<1				9	<1	3	15
Neorhabdocoela	<1							<1		<1		
Mean number of taxa	—	—	—	—	—	11.0	12.3	14.7	18.3	10.3	15.3	19.3
Mean abundance	203	85	136	129	544	40	79	126	85	50	168	325

Note: Taxa represented by a single specimen were not included.

[a] Hiltunen and Manny, 1982.
[b] Griffiths, 1987.
[c] This study.
[d] Not identified beyond family in 1977.

(Table 4). In September 1990, the abundance of these taxa in the southeastern area was noticeably greater (14–23 individuals per 0.05 m^2) and accounted for 10–29% (median 27%) of the fauna, values approaching the 29–58% (median 33%) observed in the northwestern area of the lake. In June 1991, these taxa remained a major component of the fauna at site 63 in southeastern Lake St. Clair (70 individuals per 0.05 m^2) accounting for about 22% of the total fauna, a value similar to that previously noted in northwestern Lake St. Clair during the same time of year.

Several additional taxa appeared in southeastern Lake St. Clair following the establishment of zebra mussels. *Pseudochironomus* and *Stylodrilus heringianus* Claparede accounted for 1–6% of the benthic fauna in September 1990, while *Polypedilum* accounted for 1% of the fauna at site 63 in June 1991 (Table 3). Since these taxa were part of the fauna in northwestern Lake St. Clair (Table 4), their appearance again suggests that the faunal composition of southeastern Lake St. Clair shifted toward that observed in the northwestern area of the lake following the mussel invasion. Additionally, *Limnodrilus udekemianus* Claparede and *Quistradrilus multisetosus* (Smith) accounted for more than 2% of the fauna at site 63 in June 1991, while *Potamothrix moldaviensis* Vejdovsky & Mrazek accounted for 22% of the fauna. These taxa, however, were not observed in northwestern Lake St. Clair prior to the mussel invasion, although *L. udekemianus* and *Q. multisetosus* were present in low numbers in September 1990 (Table 4).

Five taxa noted in southeastern Lake St. Clair in 1977 or 1983 were not re-collected in 1990 or 1991 (Table 3). Four of these taxa were chironomids: *Tribelos, Potthastia, Epicocladius*, and *Parakiefferiella*, while the remaining taxon was a leech, Erpobdellidae. *Tribelos, Epicocladius*, and Erpobdellidae were present in northwestern Lake St. Clair before and after the mussel invasion (Table 4), suggesting that the disappearance of these taxa in the southeastern area of the lake may be related to the occurrence of zebra mussels.

DISCUSSION

Since its introduction in late 1986 (Griffiths et al., 1991), the distribution and abundance of *Dreissena polymorpha* in Lake St. Clair have increased. Shortly after zebra mussels were first discovered, Hebert et al. (1989) reported that the mussels were confined to the south-central area of the lake in August 1988, with a maximum density of macroscopic individuals of about 200/m^2. In 1989, zebra mussels occurred throughout southeastern Lake St. Clair with a maximum density of about 6000/m^2 (Griffiths et al., 1991; Hebert et al., 1991). This study found that in 1990, the mussels had spread throughout the lake except for the northwestern corner, with a maximum density about 10,000/m^2. By early 1991, the maximum density increased to 20,000/m^2, a 100-fold increase between 1988 and 1991.

TABLE 4. Mean Abundance (Number/0.05 m²) of Benthic Macroinvertebrates at Three Sites in Northwestern Lake St. Clair

Sampling Date	April (1)[a] 1977			July (1)[a] 1977			May (2)[b] 1983			September (3)[c] 1990		
Sampling Site[d]	72	83	88	72	83	88	72	83	88	72	83	88
Insects												
Bugs												
Sigara lineata											<1	
Caddis flies												
Ceraclea			2		1	5					2	
Mystacides		<1	1	<1	2	1				<1	1	3
Oecetis	<1	2	4	<1	1	3	<1	<1	<1	<1	8	<1
Setodes	1	<1		3							<1	
Triaenodes			<1								1	
Molanna										1		
Phylocentropus											<1	
Polycentropus	<1		<1								4	
Mayflies												
Hexagenia	6	2	6	<1		2	19	8	17	5	3	2
True flies												
Ceratopogonidae		<1	<1		<1				<1			
Chironomidae[e]	23	149	81	30	41	12						
Chironomus							1	18	10		6	<1
Cryptochironomus									2	<1	1	<1
Demicryptochironomus							1				<1	
Dicrotendipes								1				
Microtendipes								4				
Polypedilum									7			
Pseudochironomus							<1	3			9	<1
Stictochironomus							5	20		1	12	
Tanytarsus										<1	4	
Tribelos								19	6		3	4

TABLE 4 (continued). Mean Abundance (Number/0.05 m^2) of Benthic Macroinvertebrates at Three Sites in Northwestern Lake St. Clair

Sampling Date	April (1)[a] 1977			July (1)[a] 1977			May (2)[b] 1983			September (3)[c] 1990		
Sampling Site[d]	**72**	**83**	**88**	**72**	**83**	**88**	**72**	**83**	**88**	**72**	**83**	**88**
Potthastia								<1				
Epicocladius							<1			<1		1
Heterotrissocladius							1					
Hydrobaenus								5				
Ablabesmyia								<1				<1
Clinotanypus										<1		
Coelotanypus							4	<1		5	8	
Procladius							13	4		<1	2	1
Crustaceans												
Amphipods												
Gammarus		10	26	5	60	92	10	15	24	87	17	179
Hyalella azteca		33	<1		68	<1		4	<1	3	59	16
Isopods												
Caecidotea		5	1		85	19		2	24	38	3	327
Lirceus		16	7		187	8		3			89	
Mollusca												
Clams												
Pisidium	19		11	9	12	3	<1	<1	<1	2	1	3
Sphaerium		<1	<1		3	<1		2		4	16	41
Mussels												
Unionidae		<1			1			<1		<1	<1	
Snails												
Bithynia tentaculata		<1			2					2	<1	
Amnicola	<1	41	3	<1	65	28				2	2	<1
Probythinella lacustris											<1	<1
Lymnaea					<1	1						
Physella	<1	2	1	<1	17	32		<1		3	3	3
Gyraulus		5	<1		10	6	<1				3	

Elimia livescens	<1	<1	<1	<1	3	1	1	<1		<1		
Valvata	2	2	<1	<1	8	8		<1	<1	10		23
Annelida												
Leeches												
Erpobdellidae	<1	<1		<1		<1			2			3
Glossiphonia complanata					1	<1						
Helobdella		1			15	<1	<1	<1	<1	4	3	4
Polychaetes												
Manayunkia speciosa	14	25		1	35		1			12	3	
Worms												
Lumbricidae										2	1	8
Stylodrilus heringianus							<1	3	2			<1
Chaetogaster					2							
Specaria josinae											2	
Stylaria lacustris					2	6					2	
Tubificidae[e]	11	68	50	8	63	43						
Aulodrilus americanus										<1		<1
Aulodrilus pluriseta											1	
Branchiura sowerbyi									5			
L. angustipenis									2			
L. claparedianus									21			
L. hoffmeisteri							9		25	3		6
L. udekemianus										<1	<1	
Quistadrilus multisetosus										<1	<1	
Spirosperma ferox							47	3	10	60	79	5
Immature Tubificidae							2	<1	27	25	12	6
Nematoda												
Roundworms	27	53	33	7	31	35	10		1	6	3	<1
Nemertea												
Proboscis worms		<1									<1	

TABLE 4 (continued). Mean Abundance (Number/0.05 m²) of Benthic Macroinvertebrates at Three Sites in Northwestern Lake St. Clair

Sampling Date	April (1)[a] 1977			July (1)[a] 1977			May (2)[b] 1983			September (3)[c] 1990		
Sampling Site[d]	**72**	**83**	**88**	**72**	**83**	**88**	**72**	**83**	**88**	**72**	**83**	**88**
Platyhelminthes												
Flatworms												
Tricladida			<1	<1	10	12		<1		5	3	12
Neorhabdocoela	5	<1	<1		4	<1				2	1	2
Mean number of taxa	—	—	—	—	—	—	12.3	15.7	16.3	23.0	31.7	20.0
Mean abundance	111	420	230	67	730	319	127	117	189	287	386	652

Note: Taxa represented by a single specimen were not included.

[a] Hiltunen and Manny, 1982.
[b] Griffiths, 1987.
[c] This study.
[d] Zebra mussels were absent from all sites in each study.
[e] Chironomidae and Tubificidae were not identified beyond family in 1977.

TABLE 5. Summary of Repeated-Measure ANOVAs that Tested the Hypothesis: H1: Mean Abundance of Taxon (i) in May Was Equal to that in October

Taxon	May Mean Density (No. 0.05/m²)	Test Result	October Mean Density (No. 0.05/m²)	F (df = 1, 8)	*p*
Gammarus	9.5	<	51.9	19.90	0.002
Hexagenia	34.4	<	90.3	13.00	0.007
Oecetis	1.5	=	3.0	6.65	0.033
Chironomidae	44.6	=	32.0	5.62	0.045
Amnicola	1.2	=	2.5	3.19	0.112
Elimia livescens	0.1	=	0.4	3.57	0.096
Valvata tricarnata	0.1	=	1.0	2.60	0.146
Pisidium	22.4	=	34.0	1.47	0.260
Sphaerium	1.6	=	1.8	1.14	0.316
Helobdella elongata	0.3	=	0.5	2.80	0.133
Helobdella stagnalis	0.3	=	0.1	2.62	0.144
Manayunkia speciosa	29.0	=	35.0	1.24	0.297
Spirosperma ferox	0.6	=	4.7	4.16	0.076
Tubificidae	121.9	=	229.3	1.75	0.222
Nematodes	70.3	=	55.0	2.07	0.188
Flatworms	1.3	=	1.8	0.01	0.918

Source: Hudson et al., 1986.

Note: Three samples collected from nine sites in central Lake St. Clair in May and October of 1983. Abundance values were in-transformed for analyses. For 16 tests and alpha = 0.05, *p* (2 significant values by chance) = 0.146; *p* (3 significant values by chance) = 0.036; *p* (4 significant values by chance) = 0.006.

Changes noted in macroinvertebrate abundance and composition in southeastern Lake St. Clair may in part be a function of season, since the 1990 study was conducted in late summer, while the preinvasion studies were conducted in spring and midsummer. The total density of macroinvertebrates in September 1990, however, was not substantially different from that found in preinvasion studies (Figure 3). Furthermore, an assessment of benthic data reported by Hudson et al. (1986) from nine sites in central Lake St. Clair, using essentially the same methods as used in the present studies, showed that density estimates of 14 of 16 taxa did not differ significantly between May and October of 1983 (Table 5). Only the abundance of *Gammarus* and *Hexagenia* was significantly greater in October than May, probably as a result of recruitment. These data, therefore, suggest that the September 1990 data reflect spring 1990 abundance estimates (i.e., within ±1 ln-unit or x/÷ 2.7) for many of the taxa.

To evaluate whether changes in abundances of *Gammarus*, Tricladida, snails, *Helobdella, Polycentropus*, and *S. ferox* in the southeastern area of the lake were related to the introduction of zebra mussels or simply related to seasonal and annual variations, a repeated-measures ANOVA (see Methods) was conducted contrasting the sum of the ln (abundance) of these taxa before (mean of April and July 1977 and May 1983) and after (September 1990) the appearance of zebra mussels between southeastern and northwestern Lake St.

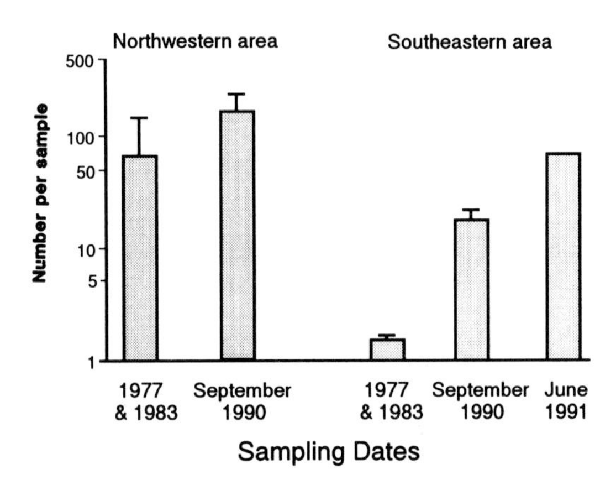

Figure 4. Mean density (number/0.05 m^2) of *Gammarus*, Tricladida, snails, *Helobdella, Polycentropus*, and *Spirosperma ferox* in northwestern and southeastern Lake St. Clair, before (1977 and 1983) and after (1990 and 1991) the appearance of *Dreissena polymorpha* Error bans represent 1 S. D.

Clair. The abundance of *S. ferox* at each site in 1977 was estimated using its relative proportion of the total worm density in May 1983. The average 11.6-fold increase in abundance of these taxa in southeastern Lake St. Clair following the mussel invasion was significantly greater ($F = 7.7$; $df = 1, 4$; $p < 0.05$) than the average 2.4-fold increase in northwestern Lake St. Clair (Figure 4). About 32% of the variance in abundance resulted from sampling error, annual and seasonal factors, etc. while 68% of the variance was unique to the southeastern area of the lake and, possibly, related to the mussel invasion. The high abundance of these taxa at site 63 in June 1991 provides supporting evidence that these taxa have established viable populations in the southeastern area of the lake.

Changes in the benthic macroinvertebrate composition in southeastern Lake St. Clair correspond to observed changes in habitat structure that zebra mussels are known to induce in aquatic systems (Reeders and bij de Vaate, 1990; Reeders et al., 1989). In September 1990, water transparency (i.e., 1.8–2.8 m Secchi depth) in the southeastern area of the lake was about double that reported prior to the mussel invasion (0.5–1.5 m Secchi depth; Leach, 1972 and 1980). This suggests that the filtering activity of the zebra mussel population was sufficient to remove seston faster than algal production and resuspension replaced it. Greater water clarity probably accounts for: (1) the heavy growth of macrophytes at site 63, which consisted mainly of *Potamogeton richardsonii* and *Vallisneria americana*; (2) the sparse growth of macrophytes at site 5, which consisted largely of *Elodea canadensis* and *Najas*; and (3) the growth of filamentous green algae on the sediments at site 15. Prior to the mussel invasion, submergent vegetation was absent at these sites (Schloesser and Manny, 1982). This shift in habitat characteristics from a relatively homogeneous environment of turbid water and silty sand substrata to a clearer water environment of macrophytes, shells, and silty sand patches possibly accounts for the increased abundance of *Gammarus*, Tricladida, snails, *Polycentropus*, and *Polypedilum*. Meanwhile, the deposition of feces and pseudofeces by the mussels is probably responsible for the higher abundance and richness of worms, such as *S. ferox, P. moldaviensis, L. udekemianus*, and *Q. multisetosus*.

Leach (1980) showed that two distinct water masses occur in Lake St. Clair: the water mass in the northwestern area is more oligotrophic, consisting mainly of water from Lake Huron; whereas the water mass in the southeastern area is more eutrophic, consisting of water from streams draining agricultural areas of Ontario. This difference in trophic conditions probably accounts for the observation reported by Griffiths (1987) that the benthic macroinvertebrate fauna in the northwestern area of the lake differed from that in the south-central area. The shift in benthic faunal composition in the southeastern area of the lake toward that found in the northwestern area following the invasion of zebra mussels suggests that trophic conditions in southeastern Lake St. Clair may have similarly shifted toward those in the northwestern area. *S. heringianus, Pseudochironomus, Polycentropus, Tricladida, Amnicola, Probythinella lacustris* (Baker), and flatworms are indicative of mesooligotrophic conditions; while *S. ferox, P. moldaviensis, Polypedilum, Gammarus, Valvata, Elimia livescens* (Menke), and *Ferrissia* are characteristic of mesotrophic conditions (Roback, 1965; Cook and Johnson, 1974; Lauritsen et al., 1985; Rabeni et al., 1985). The occurrence of *Q. multisetosus* and *L. udekemianus*, taxa characteristic of eutrophic conditions, possibly reflects localized enrichment resulting from the accumulation of pseudofeces. This shift in the faunal composition, coupled with the increase in water clarity, thus implies that zebra mussels have had an oligotrophication effect on water quality conditions. Zebra mussels have had a similar effect in the western basin of Lake Erie, as reflected by the 54% decline in mean chlorophyll *a* concentrations from 1988 to 1990 (Leach, Personal communication).

Zebra mussels have been shown to play a substantial role in the nutrient cycling of some European lakes (Stanczykowska and Planter, 1985; Stanczykowska, 1984). They can accumulate and store nutrients in amounts similar to those found in emergent and submergent macrophytes or fish populations. In addition, they can annually cycle 3–6 times their standing stock of nutrients via their filtering-feeding activity, depositing about 50% of the nitrogen and 40% of the phosphorus directly to the sediments as feces and pseudofeces. In effect, zebra mussel populations act as large nutrient pumps, removing nutrients from the limnetic waters and transferring them to benthic sediments.

The oligotrophication of Lake St. Clair will probably continue until the mussel population in the lake stabilizes or declines. Consequently, changes in the benthic biota will also continue, as species tolerant of increasingly oligotrophic conditions become more dominant and those intolerant of these conditions diminish. If the phosphorus pumped into the sediments by the mussels is not readily recycled back into the food web, the overall productivity of the lake may decline.

In summary, these data indicate that the zebra mussel population in southeastern Lake St. Clair (about 5000/m^2) was sufficient to effect physical, chemical, and biological relationships within the lake. The shift in the benthic composition toward species less tolerant of nutrient and organic enrichment reflects more oligotrophic water quality conditions that were induced by the mussel population. Future water quality assessments must take this biotic-induced change into account, so that improved water quality conditions are not mistakenly attributed to anthropogenic abatement programs.

ACKNOWLEDGMENTS

John Westwood, Sonya Gutschi, Don Schloesser, and Tom Nalepa assisted with the collection of the 1990 benthic samples. I thank Bruce Hawkins for collection of the 1991 samples.

REFERENCES

Afanas'yev, S. A., and A. A. Protasov. "Characteristics of *Dreissena* Populations in the Periphyton of Nuclear Power Plant Cooling Ponds," *Hydrobiol. J.* 23:42–49 (1987).

Arter, H. E. "Effect of Eutrophication on Species Composition and Growth of Freshwater Mussels (Mollusca, Unionidae) in Lake Hallwil (Aargau, Switzerland)," *Aquat. Sci.* 51:87–99 (1989).

Brinkhurst, R. O., A. L. Hamilton, and H. B. Herrington. "Components of the Bottom Fauna of the St. Lawrence, Great Lakes," Great Lakes Institute No. PR-33, University of Toronto, Toronto, Ontario, Canada (1968).

Burla, H., and V. Lubini-Ferlin. "Bestandesdichte und Verbreitungsmuster von Wandermuscheln im Zurichsee," *Vierteljahrschr. Naturforsch. Ges. Zurich* 121:187–199 (1976).

Cook, D. G., and M. G. Johnson. "Benthic Macroinvertebrates of the St. Lawrence Great Lakes," *J. Fish. Res. Board Can.* 31:763–782 (1974).

Dusoge, K. "Composition and Interrelationships Between Macrofauna Living on Stones in the Littoral of Mikolajskie Lake," *Ekol. Pol.* 14:755–762 (1966).

Franchini, D. A. "Distribuzione verticale di *Dreissena polymorpha* (Pallas) nel lago di Garda," *Boll. Zool.* 45:257–260 (1978).

Griffiths, R. W. Unpublished data (1991).

Griffiths, R. W. "Environmental Quality Assessment of Lake St. Clair in 1983 as Reflected by the Distribution of Benthic Invertebrate Communities," Ontario Ministry of the Environment, Water Resources Assessment Unit, London, Ontario, Canada (1987).

Griffiths, R. W. "Environmental Quality Assessment of the St. Clair River as Reflected by the Distribution of Benthic Invertebrates in 1985," Environmental Assessment and Habitat Evaluation of the Upper Great Lakes Connecting Channels, M. Munawar and T. Edsall, Eds. *Hydrobiologia* 219:143–164 (1991).

Griffiths, R. W., D. W. Schloesser, J. H. Leach, and W. P. Kovalak. "Distribution and Dispersal of the Zebra Mussel (*Dreissena polymorpha*) in the Great Lakes Region," *Can. J. Fish. Aquat. Sci.* 48:1381–1388 (1991).

Hamburger, K., P. C. Dall, and P. M. Jonasson. "The Role of *Dreissena polymorpha* Pallas (Mollusca) in the Energy Budget of Lake Esrom, Denmark," *Verh. Int. Verein. Theor. Angew. Limnol.* 24:621–625 (1990).

Hebert, P. D. N., B. W. Muncaster, and G. L. Mackie. "Ecological and Genetic Studies on *Dreissena polymorpha* (Pallas): A New Mollusc in the Great Lakes," *Can. J. Fish. Aquat. Sci.* 46:1587–1591 (1989).

Hebert, P. D. N., C. C. Wilson, M. H. Murdoch, and R. Lazar. "Demography and Ecological Impacts of the Invading Mollusc, *Dreissena polymorpha*," *Can. J. Zool.* 69:405–409 (1991).

Hiltunen, J. K., and B. A. Manny. "Distribution and Abundance of Macrozoobenthos in the Detroit River and Lake St. Clair, 1977," Administrative Report No. 82-2, National Fisheries Research Center — Great Lakes, U.S. Fish and Wildlife Service, Ann Arbor, MI (1982).

Hudson, P. L., B. M. Davis, S. J. Nichols, and C. M. Tomcko. "Environmental Studies of Macrozoobenthos, Aquatic Macrophytes and Juvenile Fishes in the St. Clair-Detroit River System, 1983–1984," National Fisheries Research Center — Great Lakes, U.S. Fish and Wildlife Service, Ann Arbor, MI (1986).

Izvekova, E. I., and A. A. L'vova-Katchanova. "Sedimentation of Suspended Matter by *Dreissena polymorpha* Pallas and Its Subsequent Utilization by Chironomidae Larvae," *Pol. Arch. Hydrobiol.* 19:203–210 (1972).

Johnson, M. G., O. C. McNeil, and S. E. George. "Benthic Macroinvertebrate Associations in Relation to Environmental Factors in Georgian Bay," *J. Great Lakes Res.* 13:310–327 (1987).

Johnson, M. G., and O. C. McNeil. "Changes in Abundance and Species Composition in Benthic Macroinvertebrate Communities of the Bay of Quinte, 1966–84," in Project Quinte: Point-Source Phosphorus Control and Ecosystem Response in the Bay of Quinte, Lake Ontario, C. K. Minns, D. A. Hurley, and K. H. Nicholls, Eds. Canadian Special Publication of Fisheries and Aquatic Sciences 86 (1986) pp. 177–189.

Lauritsen, D. D., S. C. Mozley, and D. S. White. "Distribution of Oligochaetes in lake Michigan and Comments on Their Use as Indices of Pollution," *J. Great Lakes Res.* 11:67–76 (1985).

Leach, J. H. Personal communication (1991).

Leach, J. H. "Distribution of Chlorophyll *a* and Related Variables in Ontario Waters of Lake St. Clair," in Proceedings: 16th Conference on Great Lakes Research, International Association for Great Lakes Research (1972), 80–82.

Leach, J. H. "Limnological Sampling Intensity in Lake St. Clair in Relation to Distribution of Water Masses," *J. Great Lakes Res.* 6:141–145 (1980).

Lewandowski, K. ''Unionidae as a Substratum for *Dreissena polymorpha* Pall.,'' *Pol. Arch. Hydrobiol.* 23:409–420 (1976).

Lewandowski, K. ''The Role of Early Development Stages in the Dynamics of *Dreissena polymorpha* (Pall.) (Bivalvia) Populations in Lakes. II. Settling of Larvae and the Dynamics of Numbers of Settled Individuals,'' *Ekol. Pol.* 30:223–286 (1982).

Lewandowski, K. ''Formation of Annuli on Shells of Young *Dreissena polymorpha* (Pall.),'' *Pol. Arch. Hydrobiol.* 30:343–351 (1983).

MacIsaac, H. J., W. G. Sprules, and J. H. Leach. ''Ingestion of Small-Bodied Zooplankton by Zebra Mussels (*Dreissena polymorpha*): Can Cannibalism on Larvae Influence Population Dynamics?,'' *Can. J. Fish. Aquat. Sci.* 48:2051–2060 (1991).

Morrison, D. F. *Multivariate Statistical Methods,* (New York: McGraw-Hill Book Company, 1976).

Nalepa, T. F., and J. M. Gauvin. ''Distribution, Abundance and Biomass of Freshwater Mussels (Bivalvia:Unionidae) in Lake St. Clair,'' *J. Great Lakes Res.* 14:411–419 (1988).

Pugsley, C. W., P. D. N. Hubert, G. W. Wood, G. Brotea, and T. W. Obal. ''Distribution of Contaminants in Clams and Sediments from the Huron-Erie Corridor. I. PCBs and Octachlorostyrene,'' *J. Great Lakes Res.* 11:275–289 (1985).

Rabeni, C. F., S. P. Davis, and K. E. Gibbs. ''Benthic Invertebrate Response to Pollution Abatement: Structural Changes and Functional Implications,'' *Water Resour. Bull.* 21:489–497 (1985).

Reeders, H. H., A. Bij de Vaate, and F. J. Slim. ''The Filtration Rate of *Dreissena polymorpha* (Bivalvia) in Three Dutch Lakes with Reference to Biological Water Quality Management,'' *Freshwater Biol.* 22:133–141 (1989).

Reeders, H. H., and A. Bij de Vaate. ''Zebra Mussels (*Dreissena polymorpha*): A New Perspective for Water Quality Management,'' in Biomanipulation — Tool for Water Management, R. D. Gulati, E. H. R. R. Lammens, M. L. Meijer, and E. van Donk, Eds. *Hydrobiologia* 200/201:437–450 (1990).

Roback, S. S. ''Environmental Requirements of Trichoptera,'' in Biological Problems in Water Pollution, Third Seminar, 1962, C. M. Tarzwell, Ed., U.S. Department of Health, Education and Welfare, Division of Water Supply and Pollution Control, Cincinnati, OH (1965), 118–126.

Schloesser, D. W., and B. A. Manny. ''Distribution and Relative Abundance of Submerged Aquatic Macrophytes in the St. Clair-Detroit River ecosystem,'' National Fisheries Research Center-Great Lakes, U.S. Fish and Wildlife Service, Ann Arbor, MI (1982).

Sebestyen, O. ''Colonization of Two New Fauna-Elements of Pontus-Origin *Dreissena polymorpha* Pall. and *Corophium curvisinum* (G. O. Sars forma devium Wundsch) in Lake Balaton,'' *Verh. Int. Verein. Theor. Angew. Limnol.* 8:169–181 (1938).

Stanczykowska, A. ''Ecology of *Dreissena polymorpha* (Pallas) (Bivalvia) in Lakes,'' *Pol. Arch. Hydrobiol.* 24:461–530 (1977).

Stanczykowska, A. ''Role of Bivalves in the Phosphorus and Nitrogen Budgets in Lakes,'' *Verh. Int. Verein. Theor. Angew. Limnol.* 22:982–985 (1984).

Stanczykowska, A., and M. Planter. ''Factors Affecting Nutrient Budget in Lakes of the R. Jorka Watershed (Masurian Lakeland, Poland). X. Role of the Mussel, *Dreissena polymorpha* (Pall.) in N and P cycles in a Lake ecosystem,'' *Ekol. Pol.* 33:345–356 (1985).

Suter, W. "Der Einflub von Wasservogeln auf Populationen der Wandermuschel (*Dreissena polymorpha* Pall) am Untersee/Hochrhein (Bodensee)," *Schweiz. Z. Hydrol.* 44:149–161 (1982).

Veal, D. M., and D. S. Osmond. "Bottom Fauna of the Western Basin and Near-Shore Canadian Waters of Lake Erie," in Proceedings: 11th Conference on Great Lakes Research, International Association for Great Lakes Research (1968) 151–160.

Walz, N. "Studies on the Biology of *Dreissena polymorpha* Pallas in the Lake of Constance," *Arch. Hydrobiol., Suppl.* 42:452–482 (1973).

Walz, N. "The Production and Significance of the *Dreissena* Population in the Nutrient Cycle in Lake Constance," *Arch. Hydrobiol.* 82:482–499 (1978).

Wiktor, J. "The Biology of *Dreissena polymorpha* (Pall.) and Its Ecological Importance in the Firth of Szczecin," *Stud. Mat. Morck. Inst. Ryb. Gdynia, Ser. A,* 5:1–88 (1969).

Wilkinson, L. "Systat: the System for Statistics," Systat, Inc., Evanston, IL (1987).

CHAPTER 26

Potential of the Zebra Mussel (*Dreissena polymorpha*) for Water Quality Management

Harro H. Reeders, Abraham bij de Vaate, and Ruurd Noordhuis

Techniques are being developed in The Netherlands to use the zebra mussel, *Dreissena polymorpha* (Pallas), in water quality management. Zebra mussels can contribute in the control of eutrophication by filtering phytoplankton out of the water column and by biodeposition of polluted suspended matter via feces and pseudofeces. In this regard, in situ measurements of zebra mussel filtration rates have been performed in the eutrophic Lake Wolderwijd, whereas biodeposition was measured under nearly natural conditions in the Meuse River. Filtration rate and pseudofeces production are mainly determined by the concentration of suspended matter, while water temperature is of minor importance in the range of 5–20°C. The relation between filtration rate and pseudofeces production with shell length is sigmoid in shape. Pollutant concentrations are higher in biodeposits than in naturally suspended matter. Preliminary results of field experiments in a small eutrophic pond show that zebra mussels are able to control phytoplankton growth.

0-87371-696-5/93/$0.00 + $.50

INTRODUCTION

In general, phytoplankton communities in shallow Dutch lakes are characterized by water blooms of cyanobacteria (e.g., *Oscillatoria agardhii*) (Berger, 1987). To control these incidences of water blooms, several measures are being considered by management authorities. One option is the reduction of the density of zooplankton predators, which has been successfully applied in some smaller lakes (Meijer et al., 1989; Meijer et al., 1990; Van Donk et al., 1989). Another option is to increase the density of zebra mussels and allow these filter feeders to decrease phytoplankton densities (Reeders and bij de Vaate, 1990). So far practical applications of zebra mussels in eutrophication control have been examined only by Piesik (1983), who studied the elimination of seston and nutrients on stylon nets in a canal in Poland. On the other hand, the Asian clam (*Corbicula fluminea*) is used in North America to reduce turbidity in fish ponds (Buttner, 1986).

Another application of zebra mussels in water quality management is to increase the sedimentation of polluted suspended matter (Reeders and bij de Vaate, 1992). Zebra mussels filter suspended particles from the water column and deposit a rejected fraction, including the adsorbed pollutants, as biodeposits (i.e., feces and pseudofeces). Studies are presently being conducted to examine the possibility of the construction of a "biological filter," consisting of hanging cultures of zebra mussels at the freshwater inlet of Lake Volkerak-Zoommeer (Figure 1, location A) (Reeders, 1990). The aim is to situate the hanging culture just outside the lake over a depression in the inlet bottom so the pseudofeces can settle and accumulate without being resuspended by water currents or waves.

This chapter reviews the results of various studies we conducted on filtration rate and pseudofecal production of zebra mussels and presents some preliminary data on their use to control the effects of eutrophication.

FILTRATION RATE AND PSEUDOFECES PRODUCTION

Previous studies on the filtration rate of *D. polymorpha* have been performed with different "food" types and under different conditions. An overview of filtration rates found in the literature and recalculated for zebra mussels with a shell length of 22 mm is given in Table 1. The range in values is considerable. Apart from differences in the experimental conditions such as temperature, food type, and food concentration, possible disturbance of zebra mussels under artificial conditions (i.e., the period before and during the execution of the measurements) may be an important cause of this variation. In 1985, a 24-hr measurement of the filtration rate of *D. polymorpha* (shell length 20–22 mm) was conducted *in situ* in Lake Veluwemeer and in the laboratory, using Lake Veluwemeer water (Reeders et al., 1989). The results

Figure 1. A map of The Netherlands showing the locations on which experiments with *Dreissena polymorpha* have been performed. (A) Location for the biological filter in the inlet of Lake Volkerak-Zoommeer; (B) field laboratory at the Meuse River (measurements of the pseudofeces production); (C) "Roggebotsluis" (field experiment). Measurements of the filtration rate were conducted in Lake Wolderwijd.

of these experiments are shown in Figure 2. In Lake Veluwemeer, a maximum and stable level in the filtration rate was reached after 6 hr. In the laboratory, such a stable level was not reached, even by the end of the experiment. Recovery of the mussels after manipulation before the beginning of the experiment took longer under laboratory conditions than in the field. Hence, measurements of filtration rate as well as pseudofeces production should be conducted under conditions as natural as possible to prevent artifacts due to disturbance of the animals.

In 1988, filtration rates of zebra mussels were determined in Lake Wolderwijd, a hypertrophic lake dominated by cyanobacteria (*Oscillatoria agardhii*), and pseudofeces production was determined in a field laboratory at the

TABLE 1. A Review of Filtration Rates of *Dreissena polymorpha* (Shell Length 22 mm) from the Literature

Author	Food	Temperature (°C)	Filtration Rate (mL/mussel/hr)[a]
Kondratev (1963)	Clay suspension	16–17	(133)
Mikheev (1966)	*In situ*	20–22	(52)
Mikheev (1967)	*In situ*	—	2–50
Mikheev and Sorokin (1966)	C^{14}	—	45.5
Stanczykowska (1968)	*In situ*	—	35
Stanczykowska et al. (1975)	*In situ*	22–25	8–44
Alimov (1969)	Clay suspension	18–20	(159)
L'vova-Katchanova (1971)	—	—	43–56.3
Morton (1971)	Graphite	5–30	6.9
Hinz and Scheil (1972)	Suspension	5	13
Walz (1978)	*Nitzschia*	15	(63)
Benedens and Hinz (1980)	Graphite	17.7	4
Kryger and Riisgård (1988)	Suspension	20	286.8
Reeders et al. (1989)	*Chlorella*	13–17.7	40–75
Reeders and bij de Vaate (1990)	*In situ*	10–21	40–75
	In situ	3–5	18–51

[a] Numbers between parentheses have been recalculated for mussels with a shell length of 22 mm after Kryger and Riisgård (1988).

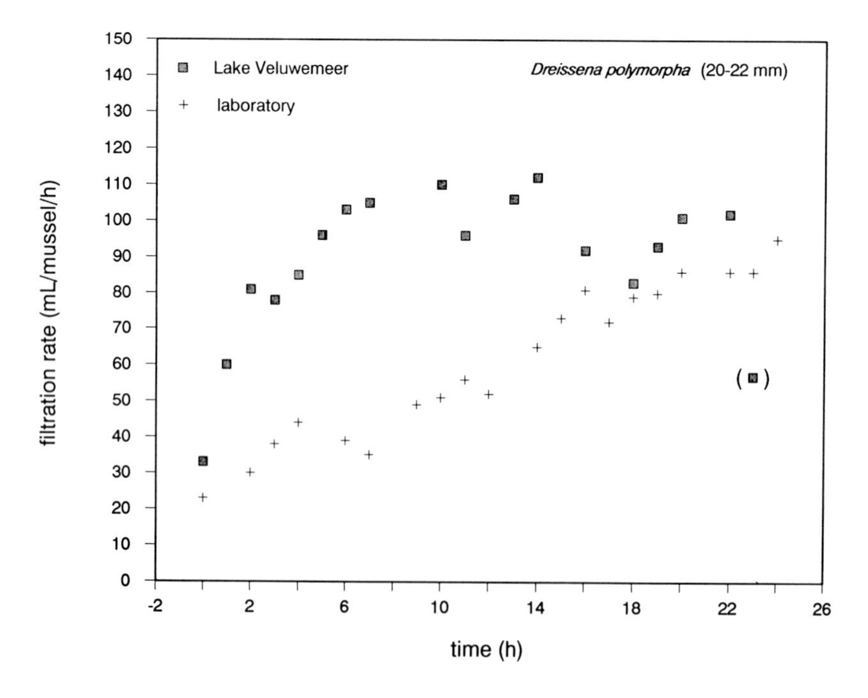

Figure 2. Comparative filtration rate of *Dreissena polymorpha* measured *in situ* in Lake Veluwemeer and in the laboratory.

Meuse River (Figure 1). The methods used are described by Reeders and bij de Vaate (1992) and were based on the desire to keep disturbance of the mussels before and during the experiments to a minimum. From observations of biodeposit production of mussels in these relatively turbid eutrophic and shallow water bodies, it was concluded that over 90% of the excreted solids were pseudofeces.

The relationship of both filtration rates and pseudofeces production to temperature, suspended matter, and mussel size is given in Table 2. The filtration rate decreased exponentially with suspended matter content, while the relationship between pseudofeces production and suspended matter content was linear. A model for the relation between filtration rate and temperature shows that the filtration rate is independent of temperature between 7 and 20°C, but decreases below 5°C and above 20°C (Reeders and bij de Vaate, 1989). For pseudofeces production (PSF, milligram per mussel per day), a two-factor regression model could be fitted with suspended matter content (C, mg/L) and temperature (T, °C):

$$PSF = -23.01 + 1.21{*}C + 1.90{*}T \qquad (p < 0.001;\ R^2 = 0.92)$$

In this regression model suspended matter accounts for 80% of the variation and temperature (>6.4°C) only for 12%.

FIELD EXPERIMENTS

Real-scale experiments which measure changes in water pollutant levels before and after passage through a biological filter system have not been performed to date. However, in 1988 pseudofeces were collected from zebra mussels suspended in cages at the inlet of Lake Volkerak-Zoommeer, the location of the intended biological filter (Figure 1, location A). The pseudofeces were analyzed for organic and inorganic pollutants. In general, concentrations of cadmium, zinc, PCB-153, pp-DDE, and PAHs were higher in pseudofeces than in suspended matter (Figure 3). The exact reason for this difference is not entirely clear, but food selection by the mussel may offer an explanation. Zebra mussels prefer particles in the 15–40 μm size range (Ten Winkel and Davids, 1982). Thus smaller particles, to which greater concentrations of pollutants are adsorbed than to larger particles (Salomons and Förstner, 1984) would be more subject to rejection as pseudofeces. Concentrations of pollutants in the pseudofeces would, therefore, be higher than in suspended material. In these experiments, particles less than 10 μm comprised 82.3% of the collected pseudofeces but comprised 71% of suspended matter.

In 1990, a field experiment for eutrophication control was started in two hypertrophic ponds at Roggebotsluis (Figure 1, location C). The ponds have

TABLE 2. The Relation Between Filtration Rate and Pseudofeces Production of *Dreissena polymorpha* with Water Temperature (T) and Suspended Matter Content (C, Dry Weight) for Mussels with a Shell Length of 22 mm and 19 mm, Respectively, and the Relation with Size (mm Shell Length L; mg Dry Body Weight W)

	Filtration Rate (mL/mussel/hr)	Pseudofeces Production (mg DW/mussel/day)
Temperature (°C)	No significant relation	$PSF = 6.6^{*}\ 0.78^{*}T$ ($p < 0.05$; $R^2 = 0.35$)
Suspended matter content (mg/L)	$FR = 187.1^{*}e^{-0.037^{*}c}$ ($p < 0.001$; $R^2 = 0.70$)	$PSF = 5.54 + 0.97^{*}C$ ($p < 0.001$; $R^2 = 0.80$)
Shell length (mm)	$FR = 15.43/(0.293 + 52.38^{*}e^{-0.367^{*}L})$ ($p < 0.001$; $R^2 = 0.59$)	$PSF = 34.87/(1 + 34.87^{*}0.83^{L})$ ($p < 0.001$; $R^2 = 0.56$)
Body weight (mg DW)	$FR = 5.132^{*}W^{0.608}$ ($p < 0.001$; $R^2 = 0.60$)	$PSF = 2.513^{*}W^{0.659}$ ($p < 0.001$; $R^2 = 0.60$)

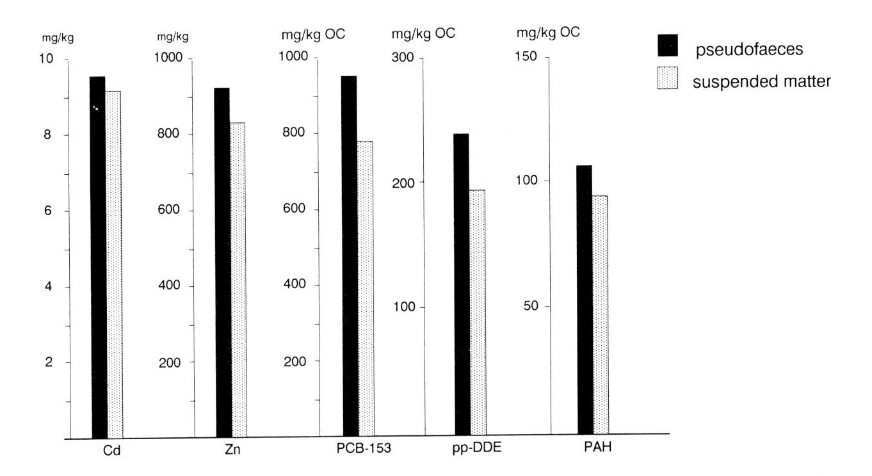

Figure 3. Concentrations of various pollutants in the pseudofeces of *Dreissena polymorpha* and in suspended matter at the inlet of Lake Volkerak-Zoommeer, the location of the intended biological filter. Concentrations given as milligram per kilogram dry weight or milligram per kilogram organic carbon (OC).

a surface area of 0.72 and 0.39 ha and a mean depth of 1.5 and 1.0 m, respectively, and are connected by means of culverts. Zebra mussels were not found in the ponds, and each summer a severe bloom of cyanobacteria developed. On April 3 the ponds were isolated by closing the culverts. The larger of the two ponds was stocked with zebra mussels: on March 20 about 1.9×10^6 mussels (larger than 4 mm) were added, and on April 5 another 2.0×10^6 were added. Thus a density of 540 animals per square meter (360 animals per cubic meter) was realized, which is the density required to accomplish sufficient reduction of phytoplankton growth in shallow, eutrophic Dutch lakes (Reeders and bij de Vaate, 1990). Both ponds have a mud bottom so wire netting (about 2000 m^2) was placed in the larger pond about 10–50 cm above the bottom. This prevented the added mussel aggregations from sinking into the mud. Figure 4 shows the length-frequency distribution of the stocked population along with the calculated filtration capacity (L/day) and pseudofeces production (g/day) per 1000 individuals of this population in relation to the suspended matter content. Filtration rates and pseudofeces production were determined from equations given by Reeders and bij de Vaate (1990). At the start of the experiment, at a suspended matter content of 13 mg/L, the calculated filtration capacity and pseudofeces production of the entire population of zebra mussels was 5100 m^3/day and 39.4 kg/day, respectively.

Figures 5–7 show some preliminary results of the experiment. On March 12, before stocking the mussels, transparency in both ponds (expressed in Secchi disk depth) was 60 cm. On March 26, when only a portion of the

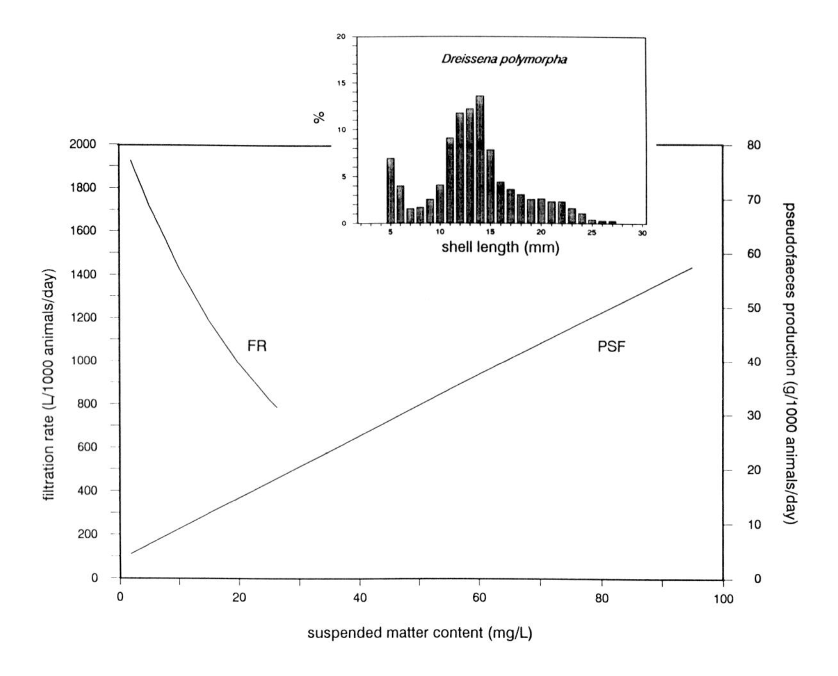

Figure 4. Length-frequency distribution of the *Dreissena polymorpha* population stocked in the experimental pond at Roggebotsluis and also the calculated filtration capacity (FC) and pseudofeces production (PSF) of this population in relation to suspended matter. The equations are (DMC = dry weight suspended matter content):

$$FC = 2074.9 \star e^{-0.037 \cdot DMC} \text{ L/1000 animals/day}$$
$$PSF = 3.27 + 0.572 \star DMC \text{g/1000 animals/day}$$

zebra mussels was added to the larger pond and the two ponds had not yet been isolated from each other, Secchi disk depth was reduced to 50 cm in the treated pond and to 40 cm in the reference pond. This decline in Secchi disk depth was caused by a bloom of the diatom *Diatoma elongatum*. At this time, total phytoplankton biovolume in the treated pond and the reference pond was 65 and 72 mm^3/mL, respectively. Diatoms accounted for 53 and 63% of these biovolumes, respectively, while cyanobacteria accounted for 41 and 32%, respectively. On the first sampling date (April 9) after stocking the second part of the mussels and closing the culverts, algal biovolume had decreased to 39 mm^3/mL and 8 mm^3/mL in the reference and treated pond, respectively; and Secchi disk depth increased to 55 and 110 cm, respectively. In both ponds, the *Diatoma* population had declined considerably, but the difference in the biovolume and Secchi disk depth in the two ponds was a result of the virtual disappearance of cyanobacteria from the treated pond. From then on, cyanobacteria remained absent in the treated pond, while in the reference pond they (mainly *Aphanizomenon flos-aquae* and *Oscillatoria* spp.) developed a heavy bloom.

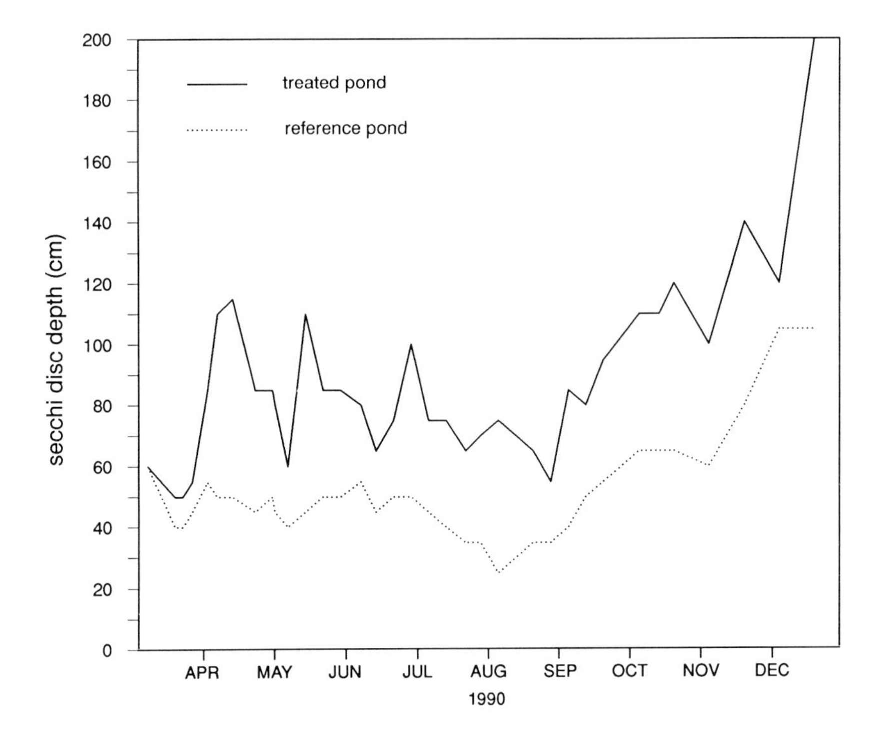

Figure 5. Changes in the Secchi disk depth over a 1-year period in experimental ponds with (treated) and without (reference) *Dreissena polymorpha*. The experimental pond had mussel densities of 540 per square meter. The experiments were conducted at Roggebotsluis in 1990.

Throughout the rest of the year, the suspended matter content was much lower in the treated pond and the water remained much clearer. A minimal Secchi disk depth of 25 cm and an extremely high suspended matter content in the reference pond on July 30 were connected to bloom of the cyanobacterium *Anabaena* sp. which was totally absent in the treated pond at that time. From August until December, Secchi disk depth steadily increased in both ponds to reach a maximum of 200 cm in the treated pond and 105 cm in the reference pond. Filamentous cyanobacteria such as *Aphanizomenon flos-aquae* and *Oscillatoria* spp. will often be too large to be selected as suitable food by zebra mussels, and their disappearance from the treated pond may partly have been a secondary effect of zebra mussel filtration. For example, as a result of zebra mussels filtering out particles, both organic and inorganic suspended matter contents were lower in the treated pond and the amount of light entering the water column increased. As cyanobacteria are relatively fast growers under poor light conditions, the conditions created by the zebra mussels will have weakened their competitive advantage toward other groups

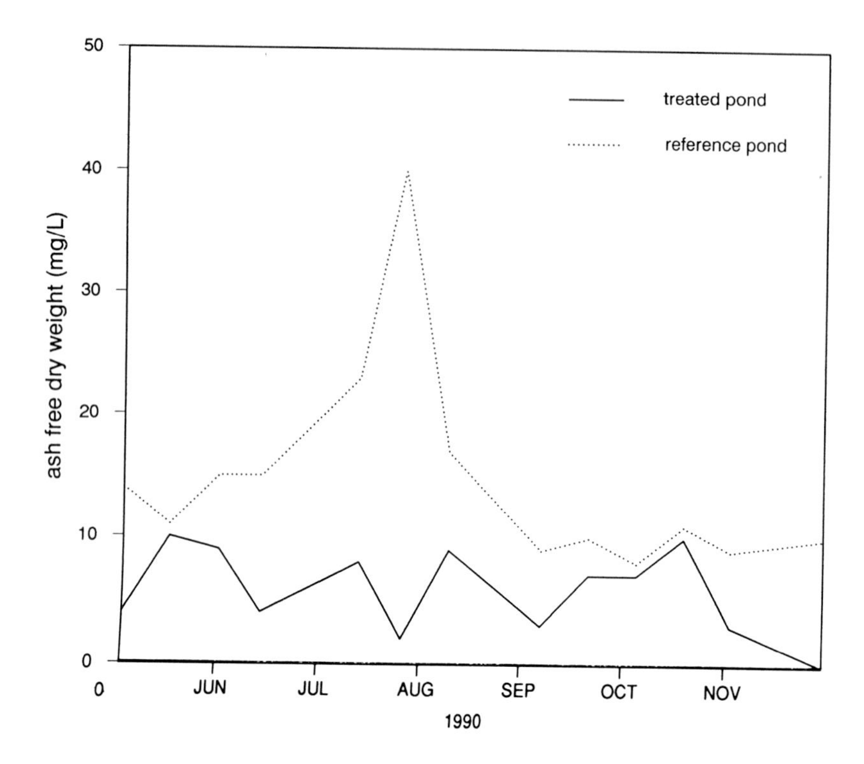

Figure 6. Changes in the ash-free dry weight of suspended matter over a 1-year period in experimental ponds with (treated) and without (reference) *Dreissena polymorpha*. The experimental pond had mussel densities of 540 per square meter. The experiments were conducted at Roggebotsluis in 1990.

of phytoplankton. Further improvement of water transparency is probably prevented by the presence of fish species such as bream *Abramis brama* and roach *Rutilus rutilus*, both being benthic feeders. The larger fish stir up bottom material in searching for chironomid larvae, while smaller fish often prey on zooplankton, resulting in reduced numbers of phytoplankton consumers. Nevertheless, the presented results of this field experiment demonstrate promising potentials of zebra mussels for programs of water quality management.

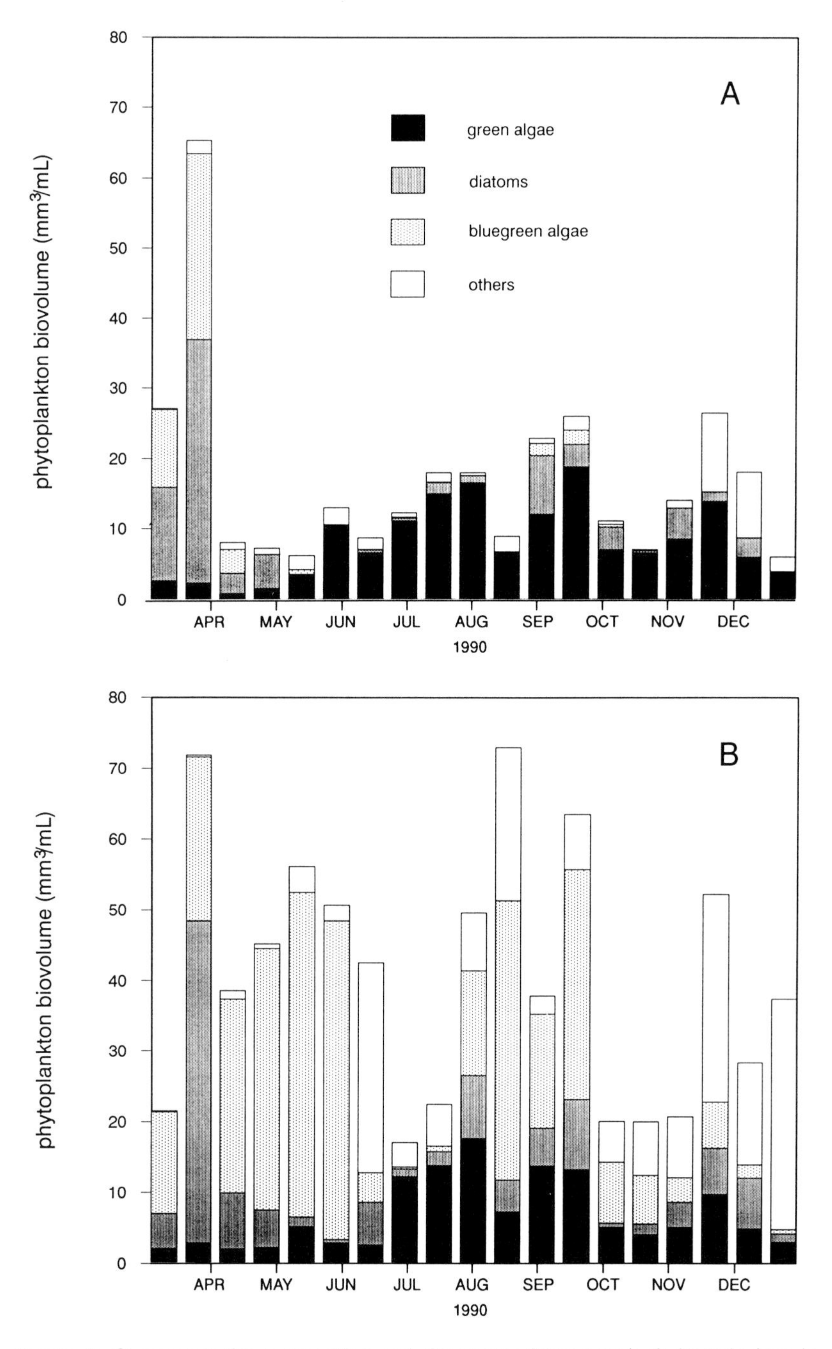

Figure 7. Changes in the composition and abundance (biovolume) of phytoplankton in experimental ponds with (treated) and without (reference) *Dreissena polymorpha*. The experimental pond had mussel densities of 540 per square meter. The experiments were conducted at Roggebotsluis in 1990. (A) Treated pond; (B) reference pond.

REFERENCES

Alimov, A. F. "Nekotorye obscie zakonomernosti processa filtracii u dvustvorcatych molljuskov," *Zh. Obshch. Biol.* 30:621–631 (1969).

Benedens, H.-G., and W. Hinz. "Zur Tagesperiodizitat der Filtrationsleistung von *Dreissena polymorpha* und *Sphaerium corneum* (Bivalvia)." *Hydrobiologia* 69:45–48 (1980).

Berger, C. "Habitat en ecologie van *Oscillatoria agardhii* Gomont." Ph.D. Thesis, State University Groningen, Groningen, The Netherlands (1987).

Buttner, J. K. "*Corbicula* as a Biological Filter and Polyculture Organism in Catfish Rearing Ponds," *Prog. Fish-Cult.* 48:136–139 (1986).

Hinz, W., and H.-G. Scheil. "Zur Filtrationsleistung von *Dreissena, Sphaerium* und *Pisidium* (Eulamellibranchiata)," *Oecologia (Berlin)* 11:45–54 (1972).

Kondratev, G. P. "O nekotoryh osobennostjah filtracii u presnovodnyh molljuskov. Naucnye doklady vyssej skoly," *Biol. Nauk. (Moscow)* 1:13–16 (1963).

Kryger, J. and H. U. Riisgård. "Filtration Rate Capacities in 6 Species of European Freshwater Bivalves," *Oecologia (Berlin)* 77:34–38 (1988).

L'vova-Katchanova, A. A. "On the Role of *Dreissena polymorpha* Pallas in Processes of Self-Purification of the Water of Uchinskoye Reservoir," in *Kompleksnye Issledovanija Vodoemoov* (Moskwa: Izdat. Moskovskogo University, 1971), pp. 196–203.

Meijer, M. L., A. J. P. Raat, and R. W. Doef. "Restoration by Biomanipulation of Lake Bleiswijkse Zoom (The Netherlands): First Results," *Hydrobiol. Bull.* 23:49–57 (1989).

Meijer, M. L., M. W. de Haan, A. W. Breukelaar, and H. Buiteveld. "Is Reduction of the Benthivorous Fish an Important Cause of High Transparency Following Biomanipulation in Shallow Lakes?," *Hydrobiologia* 200/201:303–315 (1990).

Mikheev, V. P. "O skorosti filtracii vody Drejssenoj," *Tr. Inst. Biol. Vod. Akad. Nauk. (USSR)* 12:134–138 (1966).

Mikheev, V. P. "Filtration Nutrition of the *Dreissena*," *Tr. Vses. Nauchno-Issled. Inst. Prudov. Ryb.* 15:117–129 (1967).

Mikheev, V. P., and Yu. I. Sorokin. "Quantitative Studies of *Dreissena* Feeding Habits by the Radiocarbon Method," *Zh. Obshch. Biol.* 27:463–472 (1966).

Morton, B. S. "Studies on the Biology of *Dreissena polymorpha* Pall. V. Some Aspects of Filter-Feeding and the Effect of Micro-organisms upon the Rate of Filtration," Proc. *Malacol. Soc. London* 39:289–301 (1971).

Piesik, Z. "Biology of *Dreissena polymorpha* (Pall.) Settling on Stylon Nets and the Role of this Mollusc in Eliminating the Seston and the Nutrients from the Water-Course," *Pol. Arch. Hydrobiol.* 30:353–361 (1983).

Reeders, H. H. "Hangcultures driehoeksmosselen (*Dreissena polymorpha*), resultaten van onderzoek in 1989 [Suspended Cultures of Zebra Mussels (*Dreissena polymorpha*), Results of Studies Made in 1989]," Institute for Inland Water Management and Waste Water Treatment, Report 90.030, Lelystad, The Netherlands (1990).

Reeders, H. H., and A. bij de Vaate. "Zebra Mussels (*Dreissena polymorpha*): A Perspective for Water Quality Management," *Hydrobiologia* 200/201:437–450 (1990).

Reeders, H. H., and A. bij de Vaate. "Bioprocessing of Polluted Suspended Matter from the Water Column by the Zebra Mussel (*Dreissena polymorpha* Pall.)," *Hydrobiologia* 239:53–63 (1992).

Reeders, H. H., A. bij de Vaate, and F. J. Slim. "The Filtration Rate of *Dreissena polymorpha* (Bivalvia) in Three Dutch Lakes with Reference to Biological Water Quality Management," *Freshwater Biol.* 22:133–141 (1989).

Salomons, W., and U. Förstner. *Metals in the Hydrocycle* (Berlin, Germany: Springer-Verlag, 1984).

Stanczykowska, A. "The Filtration Capacity of populations of *Dreissena polymorpha* Pall. in Different Lakes, as a Factor Affecting Circulation of Matter in the Lake," *Ekol. Pol. Ser. B* 14:265–270 (1968).

Stanczykowska, A., W. Lawacz, and J. Mattice. "Use of Field Measurements of Consumption and Assimilation in Evaluation of the Role of *Dreissena polymorpha* Pall. in a Lake Ecosystem," *Pol. Arch. Hydrobiol.* 22:509–520 (1975).

Ten Winkel, E. H., and C. Davids. "Food Selection of *Dreissena polymorpha* Pallas (Mollusca:Bivalvia)," *Freshwater Biol.* 12:553–558 (1982).

Van Donk, E., R. D. Gulati, and M. P. Grimm. "Food Web Manipulation in Lake Zwemlust: Positive and Negative Effects During the First Two Years," *Hydrobiol. Bull.* 23:19–34 (1989).

Walz, N. "The Energy Balance of the Freshwater Mussel *Dreissena polymorpha* Pallas in Laboratory Experiments and in Lake Constance. I. Pattern of Activity, Feeding and Assimilation Efficiency," *Arch. Hydrobiol., Suppl.* 55:83–105 (1978).

CHAPTER 27

Predation of the Zebra Mussel (*Dreissena polymorpha*) by Freshwater Drum in Western Lake Erie

John R. P. French III and Michael T. Bur

Environmental and economic problems associated with the colonization of zebra mussels (*Dreissena polymorpha*) in western Lake Erie created a need to investigate control mechanisms. Predation by fishes is one potential means of control, but predation on zebra mussels by native fishes in Lake Erie is unknown. The freshwater drum (*Aplodinotus grunniens*) is the most likely fish predator since it is the only fish with pharyngeal teeth capable of crushing mollusk shells. In 1990, freshwater drum were collected in western Lake Erie from 9 sites near rocky reefs and 13 sites with silt or sand bottoms, and gut contents were examined. Predation on zebra mussels increased as drum size increased. Small drum (200–249 mm in length) fed mainly on dipterans, amphipods, and small fish; small zebra mussels (<2.0 mm in length) comprised only a trace of gut contents. Medium-sized drum (250–374 mm in length) at reef sites fed predominantly on dipterans in May and on zebra mussels in July and September. The diet of medium-sized drum

Figure 1. Freshwater drum (*Aplodinotus grunniens* Rafinesque).

in locations with silt or sand bottoms shifted from zebra mussels in July to small fish in September. Large drum (>375 mm in length) fed almost exclusively on zebra mussels (seasons and locations combined). The smallest drum capable of crushing zebra mussel shells was 265 mm. Since freshwater drum over 375 mm feed heavily on zebra mussels, they may become a possible biological control mechanism for mussels in portions of North America.

INTRODUCTION

The zebra mussel (*Dreissena polymorpha* Pallas) has spread throughout the Great Lakes and has become the dominant species of the benthic community on hard substrates (Hebert et al., 1989; Griffiths et al., 1989). Zebra mussels have heavily encrusted fish spawning reefs in western Lake Erie and have created other environmental and economic problems.

The roach (*Rutilus rutilus* Linnaeus), a cyprinid fish that reaches a length of 350 mm, is a successful predator of zebra mussels in Europe because it has pharyngeal teeth to crush shells (Stanczykowska, 1977; Prejs et al., 1990). However, the roach is not found in Lake Erie. A review of the existing fishes in Lake Erie revealed that the freshwater drum (*Aplodinotus grunniens* Rafinesque) (Figure 1) is a potential predator of zebra mussels. This species is a large fish of the family Sciaenidae with strong molariform pharyngeal teeth (Krumholz and Cavanah, 1968) (Figure 2). This chapter describes predation by the freshwater drum on zebra mussels in order to evaluate this fish as a potential biological control for mussels in western Lake Erie and possibly other areas in North America.

Figure 2. Upper and lower molariform pharnygeal teeth of a freshwater drum from western Lake Erie.

METHODS

Freshwater drums were collected at 22 sites throughout western Lake Erie (Figure 3); 9 sites were near rocky reefs in the southeastern portion of the western basin, while the other 13 sites were located throughout the western basin. The latter sites had sediments ranging from silt to sand and water depths between 3 and 12 m. Drum were collected during daylight in May and September 1990. Because few drum were collected at reefs in May, all reef sites and two additional sites near reefs were sampled during daylight in July. Drum were collected with a bottom trawl (7.9-m headdrop) or gill nets (each having five panels of 64-, 76-, 89-, 102-, and 104-mm stretch mesh) to collect samples. Trawls were towed for 10–30 min periods and gill nets were set for 2 to 2.5 hr at each site. Drum were removed from the trawl and gill nets, measured (total length) to the nearest 1 mm, and weighed to the nearest 5 g.

A maximum of 15 drum (minimum total length of 200 mm) were selected randomly for each length group of 25-mm intervals. Guts of these fish were removed, placed in labeled jars with 10% formalin, and taken to the laboratory. Prey in the gut between the esophagus and anus were examined under a dissecting microscope (10×), identified to the lowest feasible taxonomic level, and counted. Crushed zebra mussel shells were also recorded to determine the ability of various length groups of drum to crush shells. Internal

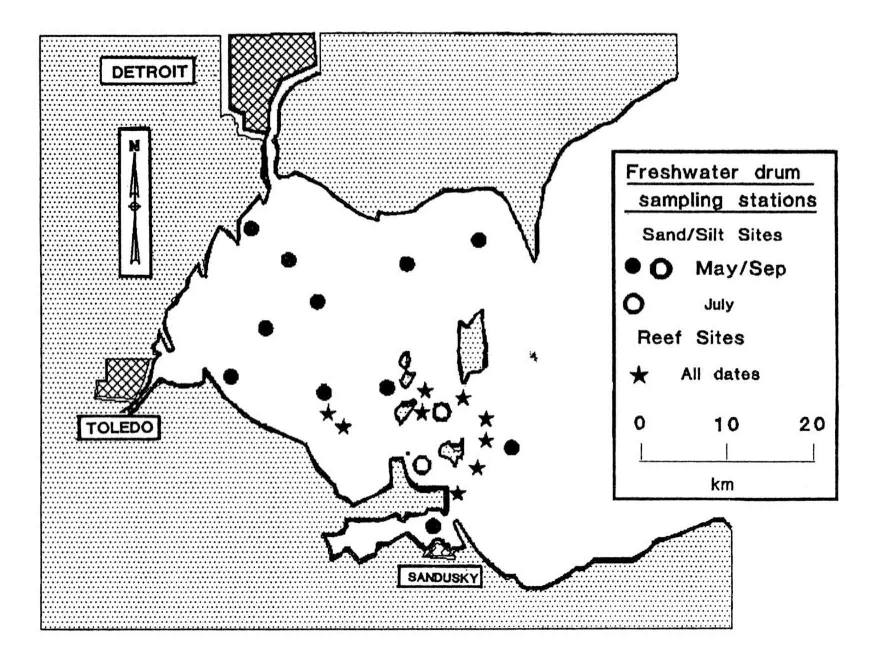

Figure 3. Map of western Lake Erie where freshwater drum were collected in 1990.

septa of only crushed right shells were counted to avoid a possible overestimation of mussels in guts (Prejs et al., 1990). Volumes of prey were determined by water displacement. Three prey groups (zebra mussels with shells, other invertebrates, and fish) were dried at 105°C to constant total dry weight (g) and then heated to 525°C to obtain ash weight (g). Ash-free body weight was the difference between total dry weight and ash weight.

For analyses, drum were separated into three size classes based on total length: small (200–249 mm), medium (250–374 mm), and large (375–574 mm). Three percentages — frequency of occurrence (F), number (N), and volume (V) — were used to indicate the proportion of each kind of prey in fish guts. Index of relative importance (IRI) of each prey type was calculated as IRI = (N = V)F (Pinkas et al., 1971).

RESULTS

Zebra mussels were present in guts of 57 (20.9%) of 273 freshwater drum; 19 (6.9%) guts were empty (Table 1). Numbers and volumes of zebra mussels in guts increased with size class of drum; medium and large-sized drum fed primarily on zebra mussels. Dipterans, amphipods, and small fish were also important food of small and medium-sized drum less than 375 mm in length.

TABLE 1. Percent Frequency (F), Percent Number (N), and Percent Volume (V) of Prey in Guts of Freshwater Drum from Western Lake Erie, 1990 (Number of Drum Given in Parentheses)

Length (mm)	Nonreef Sites									Reef Sites					
	200–249			250–374			375–574			200–249			250–374		
	F	N	V	F	N	V	F	N	V	F	N	V	F	N	V
May 8–15		(30)			(61)			(3)			(0)			(3)	
Zebra mussels	27	1	t	11	t	t	67	99	100	—	—	—	—	—	—
Amphipoda	59	11	1	41	2	3	33	1	t	—	—	—	33	100	100
Trichoptera	37	9	5	13	1	t	—	—	—	—	—	—	—	—	—
Diptera	100	79	93	82	93	95	—	—	—	—	—	—	—	—	—
Mollusca	—	—	—	—	—	—	—	—	—	—	—	—	—	—	—
Fish	—	—	—	—	—	—	—	—	—	—	—	—	—	—	—
Miscellaneous	41	t	t	32	4	1	33	1	t	—	—	—	—	—	—
July 16–18		(12)			(18)			(1)			(18)			(32)	
Zebra mussels	—	—	—	28	47	91	100	88	99	17	t	t	22	15	68
Amphipoda	75	4	12	44	1	t	100	t	t	89	5	34	69	7	15
Trichoptera	92	4	7	72	2	t	100	1	t	67	3	3	66	3	t
Diptera	100	30	79	78	36	6	100	1	t	100	64	7	87	73	13
Mollusca	8	t	t	28	9	1	100	10	t	—	—	—	9	t	t
Fish	—	—	—	—	—	—	—	—	—	22	t	48	13	t	1
Miscellaneous	50	62	1	39	4	1	33	1	t	44	28	8	46	2	t
Sept 26–28		(16)			(33)			(8)			(10)			(23)	
Zebra mussels	13	1	t	3	72	19	100	100	100	30	t	t	30	88	53
Amphipoda	19	t	1	6	t	t	—	—	—	60	40	29	13	5	1
Trichoptera	38	4	10	15	3	t	—	—	—	40	t	t	9	t	t
Diptera	88	94	32	45	16	1	—	—	—	70	56	12	26	6	t
Mollusca	—	—	—	—	—	—	38	t	t	20	t	t	4	1	t
Fish	10	t	57	61	5	80	—	—	—	10	t	59	22	t	44
Miscellaneous	19	1	1	9	4	t	—	—	—	60	4	t	4	t	1

t = trace.

Sample size was consistently greater at sites with silt or sand substrates than at sites over rocky reefs. In May, almost all drum were from nonreef sites. Mean catch per unit effort (CPUE) in May, July, and September was 0.5, 17.6, 6.4 at rocky reefs; and 7.1, 22.0, and 8.6 at sites with silt or sand substrates.

Small Drum (200–249 mm in Length)

Zebra mussels were consumed by small drum, but were not a prevailing food item (Table 1). Although zebra mussels were in 16.3% of guts of small drum throughout the sampling season, they constituted only a small percentage of the volume (Figure 4). More drum fed on zebra mussels in May and September than in July. Traces of uncrushed 1–3 mm mussels were in 14 of 86 guts of small drum. From 10 to 80 trichopterans (primarily *Oecetis*) were found in 11 of 14 guts of small drum containing zebra mussels. An *Oecetis* larva found in a drum collected in May had several 2-mm mussels attached to its case, suggesting that some drum may consume zebra mussels incidentally while feeding on trichopterans. At all sampling sites, dipterans, amphipods, trichopterans, and fish had higher IRI values than zebra mussels (Figure 5).

Medium Drum (250–374 mm in Length)

Zebra mussels were the prevailing food by volume of medium-sized drum at reefs in July and September and at sites with silt or sand substrates in July (Table 1). Zebra mussels comprised 68 and 91% of gut volumes of drum at reef and nonreef sites in July and 53 and 19% in September. At all sites in July, they were in less than 30% of drum 250–324 mm and in 75% of drum over 324 mm (Figure 4). Fewer medium drum ate mussels in September than in July. In July, the volume of mussels increased with total length of drum, but fairly large volumes (46–62%) were in drum 300–374 mm in September. The smallest drum that crushed zebra mussel shells was 265 mm in length. We could not determine whether zebra mussels or fish dominated biomass of food in guts because soft bodies of mussels and fish were completely digested before they passed through the pyloric valve. Intestines were fully packed with shells and shell fragments when drum fed heavily on mussels; however, there was no evidence that the digestive tract was damaged by sharp shell fragments. Dipterans were the prevailing food by volume of medium-sized drum (n = 61) from nonreef sampling sites in May, while fish were the dominant food in September. For all sampling sites, dipterans had the highest IRI value, followed by zebra mussels (Figure 5). However, at nonreef sites, zebra mussels had the highest IRI value.

Although zebra mussels were the most abundant item in the guts of drum by volume and dry weight (Table 2), most of the biomass was composed of

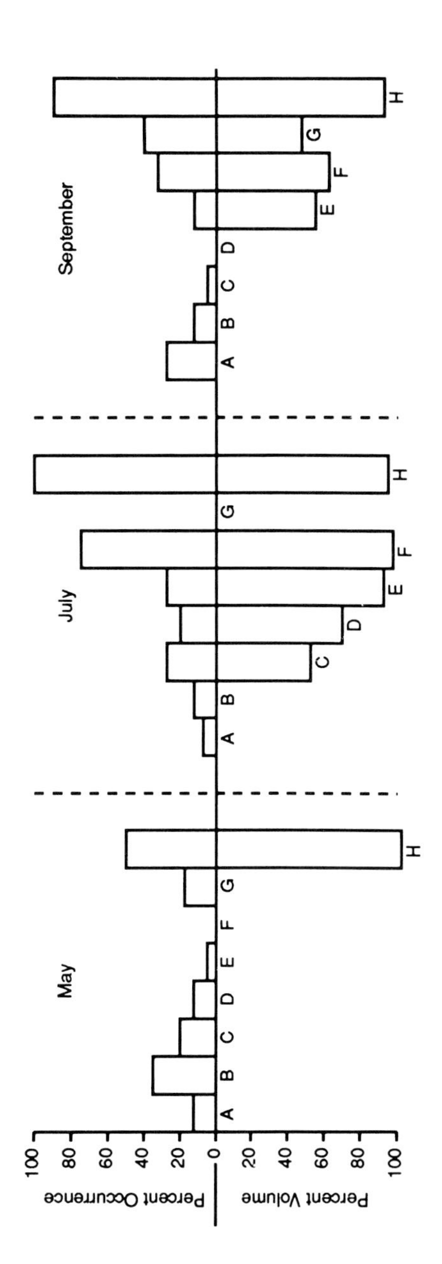

Figure 4. Percentage occurrence and volume of *Dreissena polymorpha* in freshwater drum collected from western Lake Erie in May, July, and September 1990. Percentages are given for each of eight length (mm) groups of drum: A = 200–224, B = 225–249, C = 250–274, D = 275–299, E = 300–324, F = 325–349, G = 350–374, and H = >375.

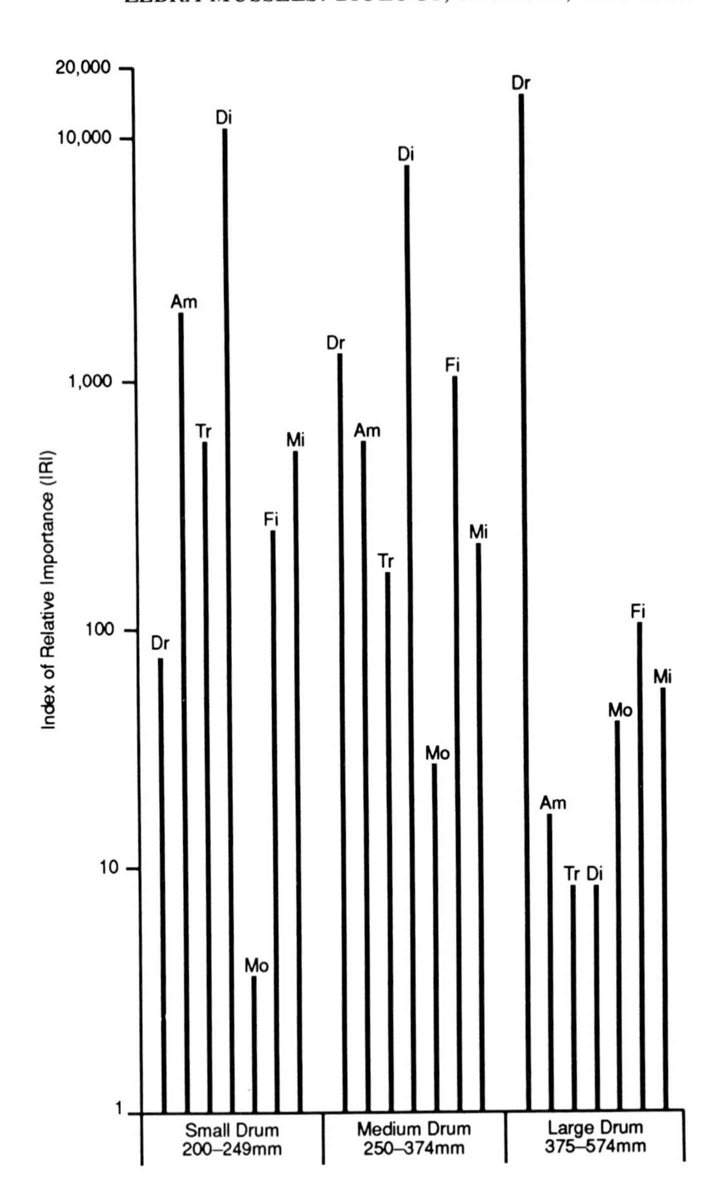

Figure 5. Index of relative importance (IRI) of prey in guts of freshwater drum from western Lake Erie, 1990. Dr = *Dreissena*, Am = Amphipoda, Tr = Trichoptera, Di = Diptera, Mo = other Mollusca, Fi = fish, Mi = miscellaneous.

TABLE 2. Mean Biomass of *Dreissena polymorpha* (D), Other Invertebrates (I), and Fish (F) in Guts of Freshwater Drum of Three Size Classes from Western Lake Erie, 1990

Dry Weight (g)		Ash Weight (g)			Ash-Free Weight (g)					
Length (mm)	Prey	May	July	Sept.	May	July	Sept.	May	July	Sept.
200–249	D	t	t	t	t	t	t	t	t	t
	I	0.17	0.03	0.03	0.07	0.02	0.02	0.10	0.01	0.01
	F	—	t	0.05	—	t	t	—	t	0.05
250–374	D	t	0.50	0.54	t	0.47	0.52	t	0.03	0.02
	I	0.20	0.04	0.01	0.06	0.03	0.01	0.14	0.01	t
	F	—	t	0.22	—	t	0.06	—	t	0.16
375–574	D	4.46	15.80	10.30	4.19	14.90	9.89	0.27	0.90	0.41
	I	t	0.04	t	t	0.03	t	t	0.01	t
	F	—	0.01	0.28	—	t	0.08	—	0.01	0.20

t = Trace < 0.005 g.

shells and shell fragments. Ash weights were 92–97% of whole dry mussels weights and only 22–32% of fish weights.

Large Drum (375–574 mm in Length)

Although sample size was small (total catch = 17), zebra mussels were the prevailing food in guts of 12 large drum on all three sampling dates at sites with silt or sand substrates (Table 1). Over 70% of zebra mussels in the guts were crushed, and nearly all uncrushed mussels were juveniles (mostly less than 3 mm) that had settled on and were still attached to adult shells. Three large drum captured near reefs in July had eaten zebra mussels almost exclusively. Zebra mussels averaged 97–99% of the total dry weight of food ingested by large drum (Table 2). The IRI values of zebra mussels were the highest in all prey in the guts of large drum (Figure 5).

DISCUSSION

Since the invasion of Lake Erie by *Dreissena polymorpha*, food habits of the freshwater drum have changed; and this species has now become an important component of drum diets. Prior to the invasion of zebra mussels, crustaceans and larval and pupal stages of aquatic insects dominated the diet of freshwater drum. In the late 1940s, mayflies (*Hexagenia*) and amphipods (*Gammarus*) were the principal prey items (Daiber, 1952); however, after the *Hexagenia* population declined in the 1950s (Carr and Hiltunen, 1965), larval midges (Chironomidae) and zooplankton (mostly Cladocera) became the dominant prey of drum (Griswold and Tubb, 1977; Bur, 1982).

More large drum were caught at sampling sites with silt or sand substrates than sites at rocky reefs, and large drum seemed to better forage for zebra mussels on soft compared to hard substrates. Heavy predation of zebra mussels by drum at sites that had mostly silt or sand bottoms suggests that drum picked zebra mussels from shells of native bivalves (Hebert et al., 1989), submersed logs and twigs, or mussel colonies that became dislodged. Because drum collected near reefs had a lower volume of zebra mussels in their guts, other foods (such as amphipods, dipterans, and fish) were either more abundant or selectively preyed upon. We do not know whether these food items were preferred or simply more abundant because no benthic samples were collected concurrently to determine abundance. Furthermore, the possibility exists that mussels were attached firmly to reef structures and within interstices, and could not be easily removed by the fish.

Freshwater drum, especially large adults, may reduce zebra mussel populations in some habitats of Lake Erie. However, the high fecundity of zebra mussels probably precludes their extermination by predation, as was shown by Robinson and Wellborn (1988) for *Corbicula fluminea* Muller. The smallest drum (265 mm) in which more than one half of the food bulk was zebra mussels is slightly larger than the European roach (Prejs et al., 1990). Although drum over 265 mm crushed and consumed mussels, drum over 375 mm are the most efficient predators. Drum over 250 mm comprised 19.5% of the total sample (n = 883, including drum under 200 mm). Drum over 350 mm were not common. Most drum reach a length of 250 mm at age 4+ years and >375 mm at age 7 years (Bur, 1984). Therefore, if the population of freshwater drum in western lake Erie responds to this new, abundant food source by increasing in numbers (in the form of mussels), it will probably take several years to increase sufficiently to suppress zebra mussel populations. Resource management agencies of the United States and Canada should use regulations and public education programs to encourage the release of all drum to increase biological control of *Dreissena polymorpha* in Lake Erie.

ACKNOWLEDGMENTS

We thank Michael McCann and Richard Stickel of the National Fisheries Research Center—Great Lakes for aiding in collections of freshwater drum; and Douglas Wilcox, John Gannon, Bruce Davis, and Randy Owens of the National Fisheries Research Center—Great Lakes for reviewing this manuscript. Contribution 807 was granted by the National Fisheries Research Center—Great Lakes, U.S. Fish and Wildlife Service, Ann Arbor, MI.

REFERENCES

Bur, M. T. "Food of Freshwater Drum in Western Lake Erie," *J. Great Lakes Res.* 8:672–675 (1982).

Bur, M. T. "Growth, Reproduction, Mortality, Distribution, and Biomass of Freshwater Drum in Lake Erie," *J. Great Lakes Res.* 10:48–58 (1984).

Carr, J. F., and J. K. Hiltunen. "Changes in the Bottom Fauna of Western Lake Erie from 1930 to 1961," *Limnol. Oceanogr.,* 10:551–569 (1965).

Daiber, F. C. "The Food and Feeding Relationships of the Freshwater Drum, *Aplodinotus grunniens* Rafinesque in Western Lake Erie," *Ohio J. Sci.* 52:35–46 (1952).

Griffiths, R. W., W. P. Kovalak, and D. W. Schloesser. "The Zebra Mussel, *Dreissena polymorpha* (Pallas, 1771), in North America: Impact on Raw Water Users," in Symposium: Service Water System Problems Affecting Safety-Related Equipment, (Palo Alto, CA: Nuclear Power Division, Electric Power Research Institute, 1989), 11–26.

Griswold, B. L., and R. A. Tubb. ''Food of Yellow Perch, White Bass, Freshwater Drum, and Channel Catfish in Sandusky Bay, Lake Erie,'' *Ohio J. Sci.* 77:43–47 (1977).

Hebert, P. D. N., B. W. Muncaster, and G. L. Mackie. ''Ecological and Genetic Studies on *Dreissena polymorpha* (Pallas): A New Mollusc in the Great Lakes,'' *Can. J. Fish. Aquat. Sci.* 46:1587–1591 (1989).

Hyslop, E. J. ''Stomach Contents Analysis — A Review of Methods and Their Application,'' *J. Fish Biol.* 17:411–429 (1980).

Krumholz, L. A., and H. S. Cavanah. ''Comparative Morphometry of Freshwater Drum from Two Midwestern Localities,'' *Trans. Am. Fish. Soc.* 97:429–441 (1968).

Pinkas, L., M. S. Oliphant, and I. L. K. Iverson. ''Food Habits of Albacore, Bluefin Tuna, and Bonito in California Waters,'' *Calif. Fish Game* 152:1–105 (1971).

Prejs, A., K. Lewandowski, and A. Stanczykowska-Piotrowska. ''Size-Selective Predation by Roach (*Rutilus rutilus*) on Zebra Mussel (*Dreissena polymorpha*): Field Studies,'' *Oecologia (Berlin)* 83:378–384 (1990).

Robinson, J. V., and G. A. Wellborn. ''Ecological Resistance to the Invasion of a Freshwater Clam, *Corbicula fluminea*: Fish Predation Effects,'' *Oecologia (Berlin)* 77:445–452 (1988).

Stanczykowska, A. ''Ecology of *Dreissena polymorpha* (Pall.) (Bivalvia) in Lakes,'' *Pol. Arch. Hydrobiol.* 24:461–530 (1977).

CHAPTER 28

Investigations of the Toxicokinetics of Hydrophobic Contaminants in the Zebra Mussel (*Dreissena polymorpha*)

Susan W. Fisher, Duane C. Gossiaux, Kathleen A. Bruner, and Peter F. Landrum

Physiological and toxicokinetic parameters for the zebra mussel, *Dreissena polymorpha*, were examined along with effects of feeding and temperature on selected measurements. Filtration rate for *Dreissena* was related to the algal concentration and ranged from 352 to 2651 mL/gDW/hr. There was a trend toward higher filtration rates in smaller mussels but the trend was insignificant. Oxygen consumption was inversely proportional to mussel size and directly proportional to temperature. Oxygen consumption ranged from 6.9 mgO_2/gDW/day at 4°C to 60.8 mgO_2/gDW/day at 23°C. Uptake clearance rates for contaminants from water exhibited similar relationships with temperature and mussel size and an additional direct proportionality with the lipophilicity of contaminants as represented by log K_{ow}. At 20°C, mean uptake clearances ranged from 428 to 1073 mL/gDW/hr across a range of compounds with log Kow values of 5.2–6.7. Efficiency for oxygen accumulation was much lower than that for contaminants, while filtration rate for a wide range of particle sizes was similar to uptake clearances for contaminants. Thus, it appears that high filtration rates are not a result of oxygen requirements but rather food requirements, and thus dissolved contaminants are effectively accumulated. Elimination of contaminants was relatively slow with half-lives ranging from 41 to 173 hr for the range of contaminants studied. The presence of food complicates contaminant accumulation by sorbing contaminants and reducing their availability on a whole-water concentration basis while also increasing rates of contaminant elimination. Overall, high filtration rates, relatively high bioconcentration potential, and high fecundity of zebra mussels will probably affect cycling of contaminants in the Great Lakes.

INTRODUCTION

Polychlorinated biphenyls (PCBs) and polynuclear aromatic hydrocarbons (PAHs) are consistently identified as Great Lakes priority pollutants because of their presence in detectable concentrations in water, biota, sediment, and suspended particles of the Great Lakes (Fitchko, 1986). The high lipid solubility of these chemicals facilitates sorption to particulate organic matter in water and sediment, and they also partition readily into the lipid tissue of aquatic organisms. Although sorption to sediment is generally thought to reduce biological availability and thus decrease exposure to organisms in the water column, it is clear that benthic invertebrates — which ingest and process contaminated sediment — may have a major role in transferring sorbed contaminants to pelagic food chains (Neff, 1979; D'Itri and Kamrin, 1983; Landrum, 1988).

Less studied, but potentially as important, is the role of zebra mussels in contaminant cycling in the Great Lakes. High filtering rates increase the probability that populations of zebra mussels will be exposed to a wide range of pollutants including hydrophobic contaminants such as PCBs and PAHs. In addition, mussels actively filter contaminated particles and algae. If algae are not ingested by mussels, they are eliminated as pseudofeces and deposited on the bottom. The biological availability of contaminants sorbed to pseudofeces has not been investigated but may serve as a reservoir from which feeding and burrowing activities of benthic invertebrates can recycle PCBs and PAHs. In order to accurately assess the impact of zebra mussels on the Great Lakes, it is imperative to understand the extent to which such contaminants can be accumulated in mussels.

The purpose of the current study was to examine the basic toxicokinetics of zebra mussels in the Great Lakes. Specific objectives were to: (1) determine accumulation and loss kinetics of selected environmental pollutants; (2) determine filtering rate as a function of particle size and respiration rates as a function of animal mass and environmental temperature; (3) examine the influence of size and temperature on toxicokinetics; (4) correlate kinetic parameters with log Kow; and, (5) perform an interlaboratory comparison of toxicokinetic parameters in two mussel populations. This information will increase our knowledge of the potential impact of zebra mussels on contaminant cycling in the Great Lakes and North America.

METHODS

Zebra mussels used in these studies were collected from two geographically distinct locations. Individuals used in experiments at the Great Lakes Environmental Research Laboratory (GLERL) were collected at a site in Lake St. Clair (42°20′00″ N and 82°47′30″ W) with a water depth of 5 m. Mussels were collected using an epibenthic sled, cleaned with lake water, and placed in a cooler containing lake water and transported to the laboratory. In the laboratory, mussels were transferred to aerated aquariums. Individuals used in experiments at Ohio State University (OSU) were collected at a site in Lake Erie with a water depth of 3 m, located 15 m offshore from Catawba Point, Marblehead, OH. In general, mussels were removed from rocks, rinsed with lake water, and placed in coolers with aeration for transport to holding facilities. Adult mussels were maintained in 208-L Plexiglas(TM) aquaria filled with Lake St. Clair water (GLERL stock) or aged, aerated tap water (OSU stock). Initial collections of both experimental stocks of mussels were made in July 1990 when water temperature was between 21 and 23°C. Mussels were maintained in the laboratory at room temperature and fed a diet of pelleted *Chlorella* at the rate of approximately 3.3 g *Chlorella* per 1000 mussels per day. Water was monitored daily for ammonia buildup and changed completely every 2–3 days. Mean (±S.D.) wet weight of the Lake St. Clair mussels used in experiments was 78.7 ± 29.6 mg wet tissue (n = 74, range 30.1–156.7 mg) and mean length was 17.5 ± 2.9 mm (n = 74, range 11–23 mm); mean (±S.D.) wet weight of Lake Erie mussels used in the experiments was 96.8 ± 21.6 mg tissue (n = 80, range 61.1–137.0 mg) and mean length was 21.2 ± 1.51 mm (n = 80, range 20.0–25.0 mm).

Chemicals used in kinetic experiments were ^{14}C-labeled *p,p′*-DDT (12.2 mCi/mMol, OSU; 13.5 mCi/mMol, GLERL); 2,2′,4,4′,5,5′-hexachlorobiphenyl (HCBP, 20.0 mCi/mMol, OSU; 17.6 mCi/mMol, GLERL); [^3H]benzo(*a*)pyrene (BaP, 33.1 Ci/mMol); 3,4,3′,4′-tetrachlorobiphenyl (TCBP, 15.7 mCi/mMol); and [^3H]pyrene (Pyr, 34 Ci/mMol). Radiopurity of compounds was >98%, as confirmed by thin-layer chromatography (TLC) on silica gel plates using hexane:benzene (8:2 V:V) and liquid scintillation counting (LSC) prior to use in kinetic experiments (Landrum, 1988).

Uptake Clearance and Elimination Rate Constants

Several days prior to initiation of experiments, adult zebra mussels were removed from stock cultures by severing their byssal threads with a razor blade. Individual mussels were then allowed to reattach to glass microscope slides (OSU) or glass petri dishes (GLERL) in culture water contained in 38-L aquariums. Only mussels which secreted new byssal threads and reattached within 24–48 hr were used in kinetic experiments.

To initiate uptake clearance experiments, ^{14}C- or ^{3}H-labeled compounds were added in bulk to 15–20 L of experimental water (Lake St. Clair water filtered through Gelman AE glass fiber filters was used in GLERL experiments, unfiltered Lake Erie water was used in OSU experiments) using 100 μL of acetone or methanol carrier. Bulk experimental water was allowed to equilibrate for 1 hr (GLERL) or 24 hr (OSU). Contaminated water was then divided among treatment vessels. Contaminant concentrations varied from 4 pg/L to 1 ng/L. Mussels in GLERL experiments were individually exposed to each contaminant in 600-mL beakers filled with 500 mL of treated lake water to afford a mass-balance analysis. Mussels used in OSU experiments were exposed to contaminants in one of three 40-L aquariums at densities of 3/L during the uptake phase. In both laboratories, uptake experiments were conducted for 6 hr and samples of zebra mussels were withdrawn for determination of total radioactivity after 0.5, 1.5, 2.0, 3.0, and 6.0 hr of exposure. Water samples of 2 mL (GLERL) or 1 mL (OSU) were withdrawn initially and at each sampling time. A minimum of three replicates were analyzed at each sampling time.

Mussels which had been simultaneously exposed to contaminants along with mussels that were used in uptake experiments were transferred to clean, uncontaminated water in order to measure elimination of each compound. In GLERL experiments, elimination was measured in static systems filled with lake water. The elimination water was renewed daily. In OSU experiments, mussels were placed in flow-through systems through which Lake Erie water was added with a gravity feed at a flow rate of 500 mL/hr. This flow rate was sufficient to keep eliminated contaminants below detectable limits in the water. Mussels were withdrawn from elimination experiments at specific time periods for liquid scintillation counting along with water samples to confirm that secondary uptake was not occurring. For OSU experiments, sampling times during the elimination phase were 0, 1, 3, 7, 17, 24, and 488 hr. For GLERL experiments, samples were withdrawn at 0, 12, 24, 96, 144, 192, 264, and 360 hr. A minimum of three replicates were used for each sample time.

Mussels were analyzed for total radioactivity immediately after removal from uptake and elimination experiments. Individuals were blotted dry, weighted, and measured (i.e., shell length). Soft tissue was then dissected from the shell and each component weighed separately. Shells and soft tissues were placed separately in 20-mL glass scintillation vials to which 5 mL of scintillation cocktail (dioxane:naphthalene: PPO, 1000/100/5; OSU) or 12 mL of RPI 3a70b scintillation cocktail (GLERL) were added. Water samples were pipetted directly into vials containing scintillation fluid. Total radioactivity was measured by liquid scintillation counting. Each vial was counted for a minimum of 5 min in a Beckman LS 6000IC scintillation counter with automatic quench control (OSU) or 10 min on an LKB 1217 Rack Beta scintillation counter (GLERL). Samples were corrected for quench using the external standards ratio method after subtracting background. The efficiency for measuring radiolabeled compounds after direct extraction by scintillation cocktail was examined by comparing organisms with the same exposure where the tissue had been predigested with Protosol. Both methods yielded equivalent tissue concentrations.

The standard temperature at which uptake and elimination experiments were run was 20°C. However, temperature was altered in a series of experiments conducted at GLERL to address effects of changing temperature on toxicokinetic parameters. In these experiments, uptake and elimination experiments for HCBP and BaP were conducted at 4, 12, and 20°C. Organisms examined at lower temperatures were acclimated from 20°C to lower temperatures at 2°C/day and held at experimental temperatures for a minimum of 24 hr prior to use. In addition to temperature experiments, effects of feeding on accumulation and elimination were evaluated for HCBP at OSU. Accumulation of HCBP was measured when mussels were exposed to contaminated water without food, to contaminated water in the presence of uncontaminated food, and to clean water with [^{14}C]HCBP-contaminated *Chlorella*. The alga was contaminated by adding radiolabeled HCBP to a slurry consisting of 12 g dried *Chlorella* in 120 mL of distilled water. The slurry was mixed on a rotary shaker for 48 hr after which a subsample was centrifuged at high speed for 10 min in a clinical centrifuge. The supernatant and pellet were then analyzed for total radioactivity with the finding that less than 1% of total HCBP was in the aqueous phase. Zebra mussels were exposed to contaminated *Chlorella* in static systems using a constant *Chlorella* drip. Elimination was measured in flow-through systems as previously described.

Respiration Rate

Oxygen consumption and clearance of oxygen from water was measured at GLERL as a function of temperature. In these experiments, one mussel (in replicates of three) was placed in a 60-mL BOD bottle with filtered lake water under yellow light (lambda > 500 nm) for 6 hr. Initial and final samples of oxygen content of the water were determined as described by Grasshoff (1983). Oxygen consumption was measured for mussels held at 4, 10, 15, and 23°C. For each temperature, oxygen consumption was measured both in the absence and presence of contaminants.

Filtration Rate

The ability of adult zebra mussels to remove particles from the water was assessed as a function of particle size. To measure filtration rate, a stock solution of 8 mg, Sun *Chlorella* in 50-mL lake water was prepared in a 100-mL beaker. The solution was sonicated to break up any large clumps of *Chlorella*. This solution was added to 1.5 L of lake water in a volumetric flask and inverted several times to suspend fine material. The average concentration of *Chlorella* particles used in filtration studies was 45,110 particles per milliliter with a size range of 2- to 20-μm diameter. A portion of the solution (600 mL) was poured into each of two 1-L beakers. Wire screen supports were suspended approximately one quarter of the depth of the beaker, and the solution was stirred on a water-driven magnetic stirring plate with a magnetic stir bar. A 20-mL water sample was withdrawn from each of the duplicate beakers for determination of particle size and concentration. One zebra mussel was then placed on each wire screen and monitored for filtering activity. After 3 hr from initial opening, the mussel was removed and a second 20-mL sample was withdrawn. Mussels were measured for length, and the tissue removed and weighed. The size and concentration of particles were determined with a TA II Coulter Counter using the method of Sheldon and Parsons (1967). The aperture setting used for particle size determination was 140 μm. Filtration rates (particle clearance) were calculated from the following equation (Vanderpleog et al., 1982):

$F = (V \ln(C/Z))/(t\ n)$
F = filtering rate
V = volume of experimental container
C = concentration of algae at the end of experiment
Z = concentration of algae at the beginning of experiment
t = experimental duration
n = number of animals in experimental chamber

Clearance was calculated for 11 particle size classes ranging from mean spherical diameters of 2.0–20.0 μm.

Wet vs Dry Weight Ratio

Wet and dry weights were measured by first weighing the soft tissues wet and then drying them to a constant dry weight at 60°C. Data from the respiration and uptake experiments were thus expressed as a function of total weight, wet weight, and dry weight.

Mass Balance Model

Kinetic processes of accumulation and loss can be modeled through either concentration-based or mass balance models depending on experimental configurations. In the case of studies performed at GLERL, the experimental design dictates that mass balance models be employed. The basic assumptions employed in this model are that the mass of contaminant does not change in the system and that no biotransformation of contaminants takes place. Biotransformation potential of these organisms was examined once and no biotransformation was found for a 6-hr exposure (Landrum, Unpublished data). In most uses of the mass balance model, contaminants are assumed to partition between the organism and the water. This produces the following equation:

$$\frac{dQ_a}{dt} = k_1 Q_w - k_d Q_a \quad (1)$$

Then, assuming mass balance in the system:

$$A = Q_w + Q_a \quad (2)$$

Where Q_a is the quantity of contaminant in the animal (ng), k_1 is the conditional uptake rate constant (per hr), Q_w is the quantity of contaminant in the water (ng), k_d is the conditional elimination rate constant (per hr), t is time (hr), and A is the total amount of compound in the system. In the present study, sorption to walls of experimental vessels and to shells may have contributed to overall mass balance. However, sorption to these two items was small, being only 1.6 ± 0.2% for the vessel and 0.6 ± 0.5% for shells of the total BaP in the system. Similar fractions were found for the other contaminants examined. Further, the amount of sorption was essentially constant throughout all experiments. Thus, sorption was assumed to occur with the addition of water to the beaker and mussels to the water, and would not contribute significantly to the mass balance. As a result, both sorption to glassware and shells were not incorporated into the model equation. Because elimination of compounds is not significant over the course of uptake experiments, the elimination term can be removed from the equation and the calculation simplified to the following integrated initial rates equation:

$$k_1 = (-\ln(1 - Q_a/A))/t \quad (3)$$

This conditional rate constant is a system dependent value and must be converted to a system independent clearance (k_u) by the following equation:

$$k_u = k_1 \text{ (Volume of water/Wet mass of tissue)} \quad (4)$$

k_u describes the volume of water scavenged of contaminant per amount of organism per time and has units of milliliters per gram per hour. This coefficient is conditional on the enviro.1mental conditions under which measurements are made. This calculation ignores the shell and water inside the shell as being part of the organism.

The model for determining accumulation in OSU experiments was simplified by the experimental design chosen. Because the volume of water employed for exposure was large compared to the total mass of organisms, the mass balance assumption was not required since the concentration of contaminants in the water did not change significantly over the period of the experiment ($p < 0.05$). Thus, the following equation could be employed:

$$\frac{dC_a}{dt} = k_u\, C_w - k_d\, C_a \tag{5}$$

where C_a is the concentration in the animal (ng/g), k_u is the uptake clearance (mL/g/hr), C_w is the concentration in the water (ng/mL), and all the remainder of the terms are the same as defined previously. When the initial rates assumptions of no biotransformation and no elimination over the course of the uptake exposure are employed, the equation simplifies:

$$\frac{dC_a}{dt} = k_u\, C_w \tag{6}$$

which integrates to:

$$C_a = k_u\, C_w\, t \tag{7}$$

The elimination data were fit to a first order elimination model for both GLERL and OSU experiments.

$$\frac{dC_a}{dt} = k_d\, C_a \tag{8}$$

Integration of the above yields:

$$\ln C_a = \ln C_a^o - k_d\, t \tag{9}$$

where C_a^o is the initial organism concentration (ng/g) at the beginning of the elimination experiment, and all the rest of the terms are as previously defined.

Bioconcentration factors were calculated from uptake and elimination rate constants:

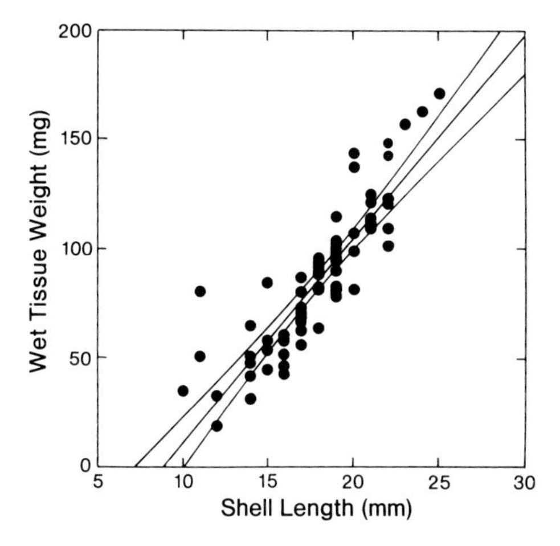

Figure 1. Relationship between wet tissue weight (WW in mg) and shell length (L in mm) in *Dreissena polymorpha*. WW = 9.39(±0.57) L − 85.5, $r^2 = 0.764$, $p < 0.001$, n = 85 (value in parentheses is ±S.E.).

$$BCF = k_u/k_d \tag{10}$$

Elimination half-lives were calculated directly from k_d values:

$$t_{(1/2)} = \frac{0.693}{k_d} \tag{11}$$

Multiple regressions were conducted for various toxicokinetic parameters as a function of temperature, organism mass, and/or physical parameters such as log K_{ow} using the Systat multiple regression program (Wilkinson, 1988). Analysis of variance was performed on regression parameters.

RESULTS

Physiological Characteristics

Regression analysis of shell length and wet tissue weight of zebra mussels in the GLERL experiments indicated that length was positively correlated with weight (Figure 1). In addition, the dry to wet tissue ratio was 0.131 ± 0.025 (mean ± S.D., n = 12) and wet tissue weight accounted for 16.2 ± 2.6% (mean ± S.D., n = 18) of the total (i.e., tissue and shell) weight, while 54.5 ± 2.6% (mean ± S.D., n = 18) of the total weight was water.

The mean filtering rate for all particle sizes depended linearly on particle concentration: filtering rate (mL/gWW/hr) = 21.8 (±6.1) Cp (mgDW/L) +

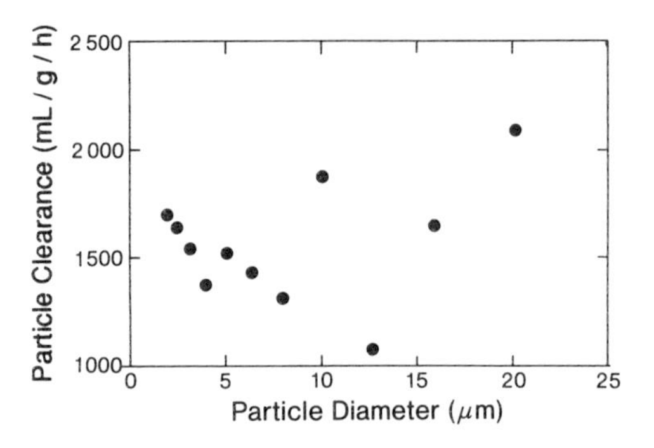

Figure 2. Clearance of particulate material from water by actively filtering adult *Dreissena polymorpha* as a function of mean particle size (μm).

TABLE 1. Mean (±S.E.) Oxygen Consumption and Clearance Rates of *Dreissena polymorpha* as a Function of Temperature

Temperature (°C)	Oxygen Consumption[a] ($mgO_2/g/d$)	Clearance of O_2[b] (mL/g/h)
23	60.8 ± 6.1 (n = 14)	40.8 ± 4.1
15	39.6 ± 4.9 (n = 29)	21.4 ± 2.6
10	19.5 ± 0.8 (n = 14)	9.5 ± 0.4
4	6.9 ± 0.7 (n = 15)	3.1 ± 0.3

[a] A Q_{10} value of 2.14 was calculated on a dry weight basis.
[b] Oxygen clearance was calculated on a wet weight basis.

288.7 (±237.8) ($r^2 = 0.52$, $p < 0.01$, n = 12), where Cp is the particle concentration. The range of filtering rates was 352–2651 mL/gWW/hr. Smaller mussels tended to have higher filtering rates than larger mussels. Filtration rates were not related to particle size over a size range of 2.0–20.0 μm in mean spherical diameter (Figure 2).

Oxygen consumption varied significantly ($p < 0.001$) with temperature from a low of 6.9 mgO_2/gDW/day at 4°C to a high of 60.0 mgO_2/gDW/day at 23°C (Table 1). From these data, a Q_{10} value of 2.14 of oxygen consumption was determined. When the weight of mussels was factored into the relationship between temperature and oxygen consumption in a multiple regression analysis, these two variables together explained 84.6% of the variation in oxygen consumption (Figure 3). As the weight of mussels increased, the rate of oxygen consumption declined significantly. Thus, the highest rate of oxygen uptake (i.e., clearance from the water) was measured in relatively small mussels (5–10 mgDW per mussel) at 23°C, and the lowest clearance of oxygen from water occurred in larger mussels (15–30 mgDW per mussel) at 4°C.

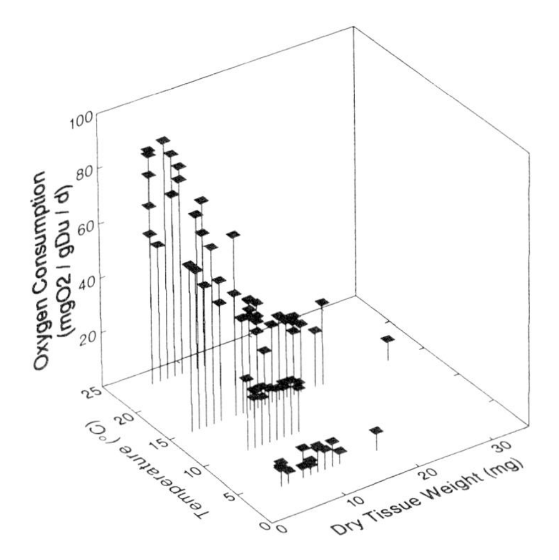

Figure 3. Oxygen consumption (R in mgO^2/gDW/day) as a function of temperature (T in °C) and dry weight (DW in mg) in *Dreissena polymorpha*. R = 9.5 (±3.7) − 1.71 (±0.24) gDW + 3.2 (±0.2) T; r^2 = 0.841, p <0.001, n = 72 (value in parentheses is ±S.E.).

TABLE 2. Toxicokinetic Parameters for Selected Organic Contaminants for *Dreissena polymorpha* in GLERL and OSU Experiments

Experiment	Compound	k_u[a] (mL/g/hr)	k_d[a] (/hr)	BCF	$t_{1/2}$ (hr)	K_{ow}
GLERL	B(a)P	838	0.009	76,182	63	5.98
	(n = 36)	(348)	(0.002)			
GLERL	HCBP	1073	0.004	268,250	173	6.77
	(n = 22)	(407)	(0.002)			
GLERL	DDT	736	0.0068	108,235	102	6.19
	(n = 16)	(354)	(0.001)			
GLERL	TCBP	796	0.0169	47,100	41	5.95
	(n = 14)	(317)	(0.002)			
GLERL	Pyrene	428	0.0096	44,583	72	5.2
	(n = 16)	(202)	(0.001)			
OSU	HCBP	167	0.004	41,750	173	6.77
	(n = 30)	(81)	(0.0001)			
OSU	DDT	124	0.008	53,000	87	6.19
	(n = 30)	(126)	(0.0002)			

[a] Values given are means with standard deviations in parentheses.

Uptake and Elimination of Xenobiotics

For all contaminants studied, accumulation was rapid with significant uptake of each contaminant taking place within the uptake clearance phase of the experiment (Table 2). For the GLERL experiments, uptake clearance rates varied from a high of 1073 mL/gWW/hr for HCBP to a low of 428 mL/gWW/hr for relatively water soluble pyrene. Elimination rate constants, in

TABLE 3. Effects of Temperature on Accumulation and Elimination of BaP and HCBP in *Dreissena polymorpha* in GLERL experiments

Compound	T (°C)	k_u[a] (mL/g/hr)	k_d[a] (/hr)	BCF	$t_{1/2}$ (hr)
BaP	4 (n = 12)	415 (149)	0.0021 (0.001)	197,619	330
	12 (n = 16)	514 (225)	0.006 (0.001)	85,666	115
	20 (n = 12)	882 (403)	0.009 (0.002)	98,000	77
HCBP	4 (n = 15)	564 (280)	0.001 (0.0005)	564,000	693
	12 (n = 16)	715 (280)	0.004 (0.001)	178,750	138
	20 (n = 12)	1048 (459)	0.0039 (0.002)	268,718	175

[a] Values given are means with standard deviations in parentheses.

contrast, were low, indicating preferential retention by tissues and were monophasic for all compounds. BCF values calculated from GLERL experiments demonstrated that substantial residues of hydrophobic xenobiotics can be expected in environmental settings. Furthermore, kinetic parameters provide insight into the processes which determine body burden and which appear to vary by compound. For example, the high uptake clearance of TCBP (796 mL/gWW/hr) was compensated by a comparatively high elimination rate constant (0.0169/hr). Thus, the BCF value for TCBP was relatively low at 47,100. In contrast, the uptake clearance rate for pyrene was the lowest of all compounds examined in GLERL experiments (428 mL/gWW/hr), but the low elimination rate constant (0.0096/hr) rendered a BCF value of 44,583, which was comparable to that of TCBP.

The uptake clearance rate for HCBP in OSU experiments was very low compared to that determined in GLERL experiments (Table 2). However, uptake clearance rates for DDT were similar between the two sets of experiments as were elimination rate constants for both chemicals. The low k_u found in OSU experiments for HCBP resulted in a depressed BCF value relative to GLERL results. However, because elimination rate constants for both DDT and HCBP were similar for both data sets, the $t_{1/2}$ determined for these chemicals were comparable for both OSU and GLERL experiments.

For two chemicals, HCBP and BaP, effects of changing temperature were measured on toxicokinetic parameters in GLERL experiments. Uptake clearance rates for both HCBP and BaP increased significantly as temperature increased from 4 to 23°C (Table 3). Elimination rate constants likewise increased in response to increasing temperature. The net result of both changes was to cause a significant reduction in the projected BCF values and a similar depression in $t_{1/2}$ values as temperature increased. These trends were clearly evident for BaP. However, the elimination rate constant for HCBP did not

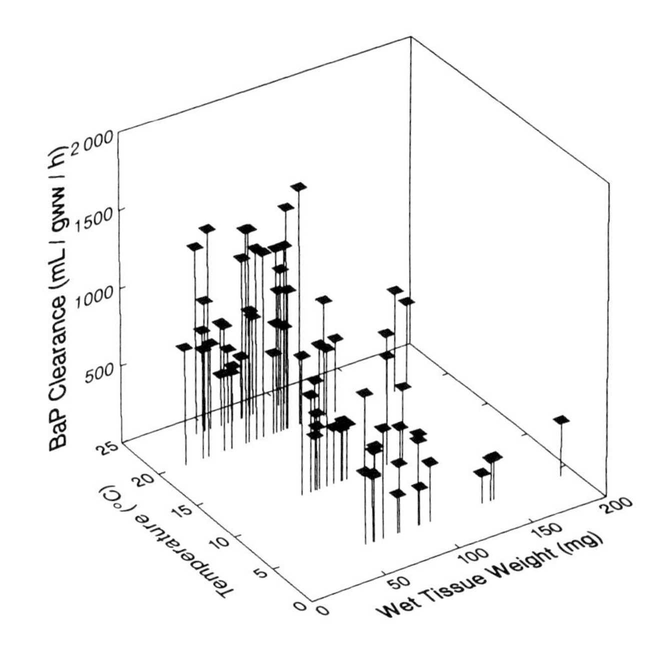

Figure 4. Relationship between uptake clearance rate for BaP (k_{uBaP} in mL/gWW/hr), temperature (T in °C), and wet tissue biomass (WW in mg) in *Dreissena polymorpha*. $k_{uBaP} = 532\ (\pm 141) - 2.2\ (\pm 1.1)\ WW + 20.9\ (\pm 5.2)\ T$; $r^2 = 0.264$, $p < 0.001$, n = 68 (value in parentheses is ±S.E.).

change appreciably when water temperature was increased from 12 to 20°C. Thus, BCF values and $t_{1/2}$ did not increase as a function of elevated temperature for HCBP.

Multiple Regression Analyses

When the mass of a mussel was introduced as a variable in addition to temperature in kinetic experiments, a negative relationship occurred between mass and uptake clearance rate of BaP (Figure 4) and HCBP (Figure 5). In the case of BaP, temperature and mass accounted for only 26.4% of the total variation in k_u with temperature being the more important of the two variables. For HCBP, the combination of temperature and mass accounted for 38% of the variation in k_u. The rather low r^2 values indicate that other factors are important in determining uptake clearance rates in addition to temperature and mass.

For the GLERL experiments, uptake clearance rates (Figure 6), elimination rate constants (Figure 7), and BCF values (Figure 8) were examined as a function of both temperature and log K_{ow} in multiple regression analyses. Uptake clearance was significantly correlated with temperature and K_{ow} (Figure 6); the two independent variables contributed equally to changes in uptake

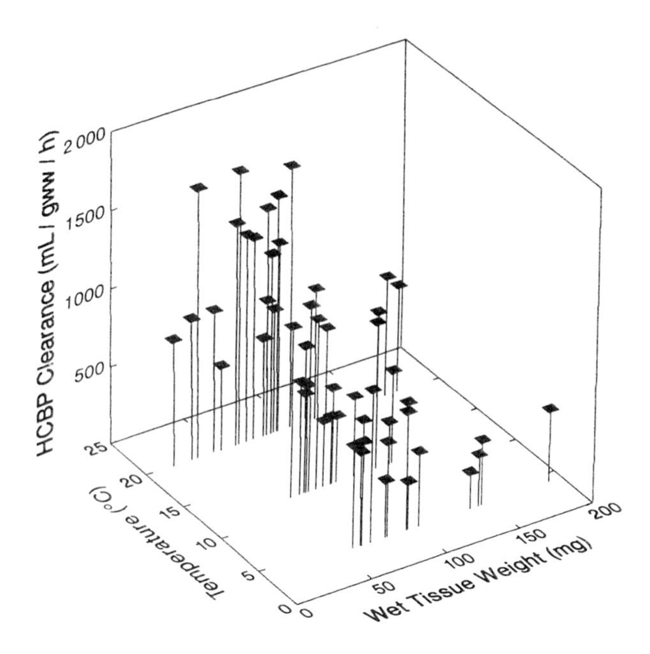

Figure 5. Relationship between uptake clearance rate for HCBP (k_{UHCBP} in mL/gWW/hr), temperature (T in °C), and wet tissue biomass (WW in mg) in *Dreissena polymorpha*. k_{UHCBP} = 738 (±153) − 3.6 (±1.2) WW + 29.7 (±6.5) T; r^2 = 0.38, p <0.001, n = 54 (value in parentheses is ±S.E.).

clearance rates and explained 75.8% of the variability in k_u. Uptake clearance rates increased significantly as temperature and contaminant lipophilicity increased. Elimination rate constants were likewise responsive to changes in temperature and K_{ow} although the latter was negatively correlated with elimination (Figure 7). Finally, BCF, which is a function of both uptake clearance and elimination rate constants, showed a significant negative relationship with temperature and a positive relationship with log K_{ow} (Figure 8). Both of these independent variables contributed equally to variation in the BCF.

Influence of Feeding and Exposure Route on Kinetics

Uptake clearance of HCBP in OSU experiments declined by a factor of 3 when uncontaminated *Chlorella* was added to the water as a food source (Table 4), while the elimination rate constant increased by a factor of 13. The accumulation of HCBP directly from algae in uncontaminated water was also measured for HCBP. In this case, K_u and K_d values of 7 mL/g/hr and 0.030/hr, respectively, were estimated. Bioconcentration factors and $t_{1/2}$ values varied as the route of exposure changed, with highest values occurring when chemicals were accumulated directly from water and lowest values occurring when assimilation from algae was required.

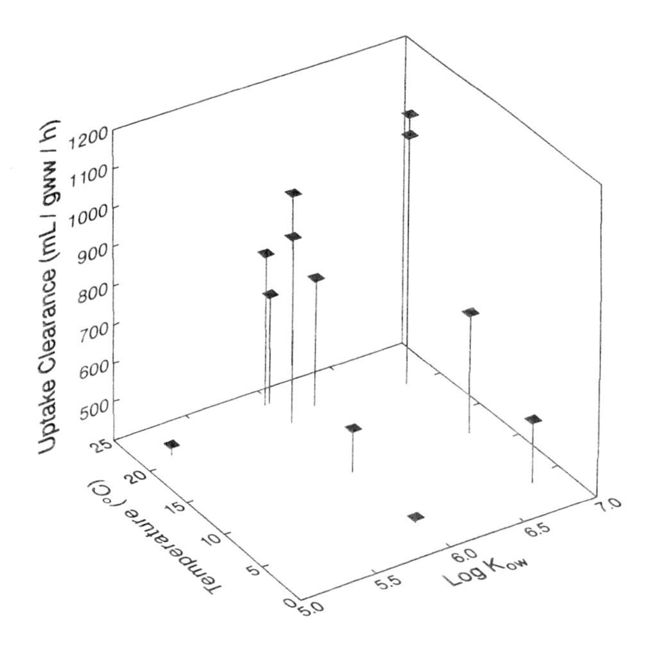

Figure 6. Relationship between mean uptake clearance rate (k_u in mL/gWW/hr) for a series of hydrophobic contaminants, temperature (T in °C), and log octanol:water partition coefficient (log K_{ow}) in *Dreissena polymorpha*. $k_u = -1819 (\pm 543) + 344 (\pm 81) \log K_{ow} + 25.7 (\pm 5.7) T$; $r^2 = 0.71$, $p = 0.002$, $n = 12$ (value in parentheses is ±S.E.).

DISCUSSION

Physiological Parameters

Results of filtering experiments suggest that small zebra mussels may have higher filtering rates than large mussels. Indeed, although the regression between mussel size and filtering rates was not significant (probably because of the small sample size), both respiration and contaminant clearance were correlated with size. For the marine mussel *Mytilus*, filtering rates are reported to be inversely, correlated with size (Vahl, 1973; Bayne and Widdows, 1978).

Although there are numerous reports of selective filtering of zebra mussels in response to food quality and particle size (Lee et al., 1972; Morton, 1971), selective filtering was not apparent in the present studies (Figure 2). However, the range of particle size employed in other studies was wider than the present study, ranging from a few to several hundred microns. An optimum particle size for ingestion by zebra mussels has been identified in the range of 15–40 μm (Ten Winkel and Davids, 1982). Thus, the particle size range used in this study was probably too narrow to permit evaluation of the relationship between particle size and filtering rate.

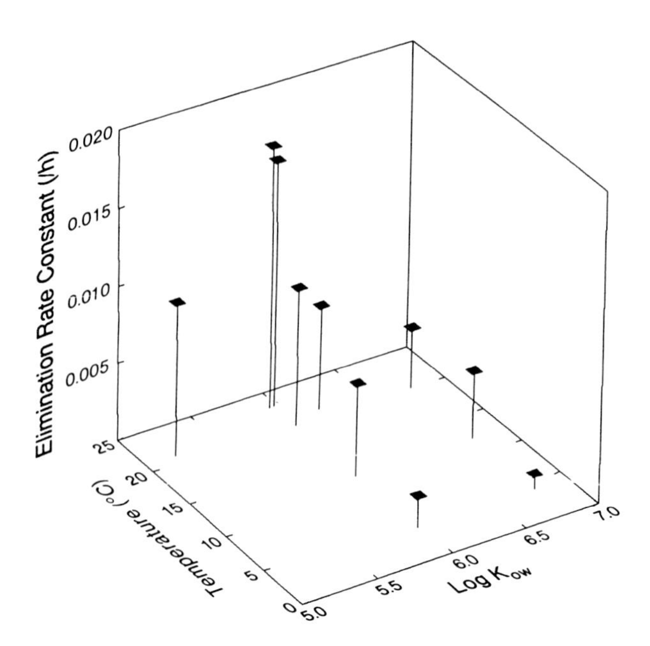

Figure 7. Changes in elimination rate constant ($k_{d \text{ in } h}$) for a series of hydrophobic xenobiotics as a function of temperature (T in °C) and log octanol:water partition coefficient (log K_{ow}) in *Dreissena polymorpha*. $k_d = 0.025 (\pm 0.14) - 0.004 (\pm 0.002) \log K_{ow} + 0.00045 (\pm 0.00015)$ T; $r^2 = 0.059$, $p < 0.01$, n = 12 (value in parentheses is ±S.E.).

Direct correlation of filtering rate and particle concentration as seen in this study may be due, in part, to the condition of the organisms. Experiments were performed after mussels had been in the laboratory for several weeks. When animals were preexposed to *Chlorella* concentrations of 50 mg/L for 24 hr prior to initiating the filtration study, the filtration rate declined. However, the rate was still higher than previously found for zebra mussels (Ten Winkel and Davids, 1982). A second potential reason for the high filtration rates is that *Chlorella* is not the preferred food and may be deficient in some required nutrients.

The influence of temperature on filtering rate in zebra mussels has been carefully studied with two contrasting findings. Several authors report that filtering rate increases linearly with a rise in temperature within the 8–25°C range in both *Dreissena* and *Mytilus* (Schulte, 1975; Ali, 1970; Stanczykowska, 1977). Others report a temperature optimum of 12.5–15.0°C for filtering with a marked decline in filtering capacity on either side of the optimum (Theede, 1963; Walz, 1978). Although filtering rate was not measured as a function of temperature in the current study, two closely related functions — oxygen consumption and contaminant uptake — showed a steady increase in response to increasing temperature (Tables 1 and 3). These results

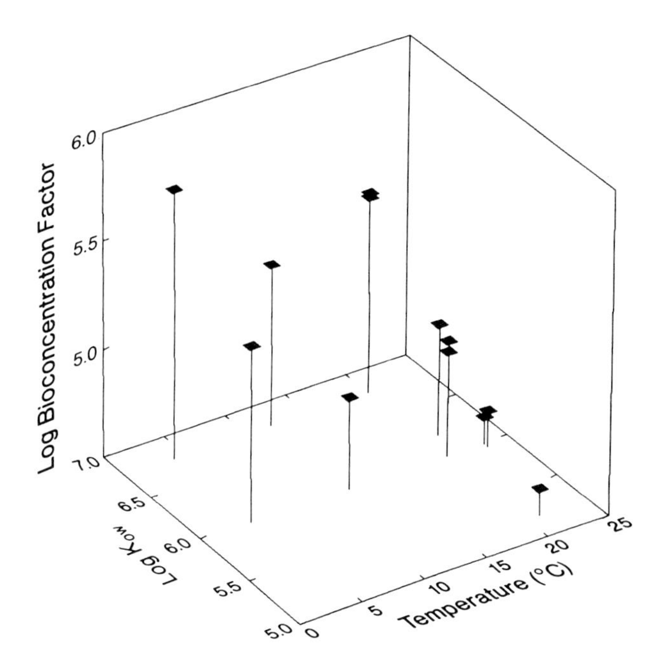

Figure 8. Dependence of log bioconcentration factor (log BCF) on temperature (T in °C) and log octanol:water partition coefficient (log K_{ow}) in *Dreissena polymorpha*. Log BCF = 2.4 ± 0.7 + 0.5 ± 0.1 log K_{ow} − 0.023 ± 0.007 T; r^2 = 0.79, p <0.001, n = 12 (value in parentheses is ±S.E.).

agree with studies on *Mytilus* which show that increased oxygen consumption and other physiological functions were related to increases in temperature. For example, Boryslawskj et al. (1985) state that the initial response of *Mytilus* to an increase in temperature is an observable elevation in the rate of ciliary beating. This activity increases the rate at which water is transported across gills, and thereby accentuates exposure of gill tissue to dissolved oxygen. If an increase in ciliary beating is the initial response to elevated temperature, this would also provide a mechanistic explanation for the observed increase in filtering activity with temperature: namely, cilia which line the incurrent siphon may beat more rapidly as temperature goes up; this, in turn, raises the amount of particulate material coming into the digestive system. The inverse relationship between mussel size and oxygen consumption agree with studies on *Mytilus* (Figure 3). In *Mytilus*, smaller mussels cleared more oxygen from water per milligram of tissue than did larger animals (Boryslawskj et al., 1985).

Comparing clearances for oxygen and contaminants, oxygen is accumulated much less efficiently than organic contaminants (Tables 1 and 3). The uptake rate of contaminants is similar to the filtration rate. Thus, one can speculate that high filtration rates were not related to oxygen requirements, but to obtain adequate food resources. Because of high filtration rates, the accumulation of nonpolar organic contaminants is rapid as described as follows.

Toxicokinetics of Xenobiotics Compared to *Mytilus*

Accumulation of neutral lipophilic chemicals (log K_{ow} 3–6) in *Mytilus* is generally described as rapid (Lee et al., 1972; Clark and Finley, 1975; Nunes and Benville, 1979; Geyer et al., 1982; Widdows et al., 1983; Broman and Ganning, 1986) with significant body burdens being achieved within a few hours of exposure. For the most part, accumulation studies performed with *Mytilus* have involved steady-state analyses; that is, BCF or BAF values were determined from concentrations in mussels at steady state divided by the average concentration of chemical in the water. These studies indicate there is a fairly rapid approach to steady state, frequently within 10–30 days (Geyer et al., 1982; Hartley and Johnston, 1983).

In uptake experiments with zebra mussels, clearance of hydrophobic contaminants from water was rapid and linear throughout the 6-hr uptake phase (Table 2). There was no evidence of an approach to steady state or a leveling off of the uptake curve as has been reported for experiments conducted for similar duration with *Mytilus* (Ernst, 1977; Geyer et al., 1982). This may be attributable to the fact that the soft tissue of *Dreissena* consists of about 11.4% lipid (Walz, 1979) while the lipid content of *Mytilus* is less than 2% (Renberg et al., 1986). Since these materials are preferentially deposited in lipid-rich tissues, zebra mussels should accumulate significantly greater levels of contaminants.

Elimination of xenobiotics from contaminated mussels was slow and monophasic in nature (Table 2). Relatively slow k_d values have been reported for *Mytilus* and other bivalves when exposed to PCBs and PAHs (Kannan et al., 1989; Solbakken et al., 1982; Doherty, 1990; Ingebrigsten et al., 1988; Pruell et al., 1986); however, quantitative comparisons of k_d values cannot be made because of the different methods used. *Mytilus*, which has been most intensively studied, shows considerable variation in depuration activity among different studies. Elimination is frequently biphasic with an initial rapid phase followed by a second slower phase. This pattern has been found for a variety of chemicals including PCBs and PAHs (Broman and Ganning, 1986; Adema and Compaan, 1975; Hansen et al., 1978; Widdows et al., 1983). On the other hand, most investigators have found that depuration is monophasic and that the rate depends on the lipophilicity of the chemical (Clark and Finley, 1975; Pruell et al., 1986; Lee et al., 1972; Dunn and Stich, 1976; Hawker and Connel, 1986) and the lipid content of the organism (Landrum, 1988). When rapid depuration was seen, a low-level but persistent residue was retained following elimination of 75–90% of the maximum body burden (Clark and Finley, 1975; Lee et al., 1972). Depuration in these cases may thus actually be biphasic in nature with the apparent retention of low-level residues representing the second, slower phase. This viewpoint is substantiated by studies in which the distribution of contaminants among different tissues was analyzed in *Mytilus*. Elimination varied considerably by tissue type; tissues with relatively high lipid levels eliminated xenobiotics at lower rates than lipid-poor tissues (Widdows et al. 1983). The initial rapid elimination noted in *Mytilus*, when up to 90% of contaminants is eliminated, represents elimination from organs with low lipid contents while the slower, long-term depuration reflects elimination from lipid reserves. The route of exposure may also play an important role in this determination. For example, Clark and Finley (1975) found that when petroleum hydrocarbons were absorbed directly across the gill, elimination occurred relatively quickly when animals were placed in clean water. However, when the same compounds were assimilated from food, absorbed across the gut, and stored in the hepatopancreas, elimination was much slower.

Although it is difficult to compare kinetic parameters, the potential for bioconcentration appears to be greater for *Dreissena* than for *Mytilus*. Renberg et al. (1986) reported a concentration factor of 13,000 for trichlorobiphenyl in *Mytilus*. Further, BAFs for a series of chemicals with log K_{ow} values between 5 and 6 ranged from 2,940 to 49,600 (Geyer et al., 1982). For a series of compounds with similar log K_{ow} values, BCFs in *Dreissena* varied from a low of 41,750–268,250 (Table 2). Thus, the potential for concentration of hydrophobic compounds in zebra mussels appears to be about an order of magnitude higher than for *Mytilus*. Again, this may be directly related to the difference in lipid content of the two species. As noted earlier, the lipid content of *Dreissena* is nearly an order of magnitude higher than the lipid content of *Mytilus*. As found in *Mytilus* (Widdows et al., 1983), a linear relationship was apparent between log K_{ow} and BCF in *Dreissena* (Figure 8).

Previous studies with *Mytilus* clearly indicate the importance of lipid content when determining kinetic rate constants as well as projected steady-state values. Lipid levels and, therefore, accumulation rates of individual chemicals vary significantly over a seasonal period. Hansen et al. (1978) estimated that changes in lipid content accounted for 18–32% of the variation in accumulation between individual mussels. Similarly, PCB content in *Mytilus* exposed to environmental PCBs varied in response to changes in lipid levels; in addition, the attainment of steady-state widely reported in other studies with *Mytilus* were only apparent since BCFs would change as a function of lipid metabolism (McDowell-Capuzzo et al., 1989). Contaminants accumulated by *Mytilus* were preferentially retained in lipid-rich tissues (Lee et al., 1972; Renberg et al., 1986; Mattson et al., 1988; Ingebrigsten et al., 1988), thus substantiating the hypothesis that lipid content is an important parameter of accumulation in *Mytilus*.

When the kinetic parameters from GLERL and OSU experiments are compared to each other, the uptake clearance and elimination rate constants were consistently lower in OSU experiments (Table 2). This was particularly noticeable for HCBP where the k_u values measured were 167 and 1073 mL/gWW/hr for OSU and GLERL experiments, respectively. The k_u for HCBP is believed to be artificially low in OSU experiments. Mussels used in the OSU-HCBP uptake experiment had been maintained in culture for nearly 2 months, during which time a protocol for culturing the mussels was still being developed. The feeding regime being used at the time was inadequate to maintain the mussels, and weight loss was observed among stock culture. Since much of the weight loss probably involved lipid mobilization, it is not surprising that k_u for OSU experiments was very low. For DDT, the k_u value found in OSU experiments was lower than found in GLERL experiments. Since these experiments were all performed with freshly collected mussels, the difference may be a function of size (Table 2). Mussels in OSU experiments were approximately 5–10 mm larger than those used in GLERL experiments. Larger animals can be expected to show lower uptake clearance rates than smaller animals (Figures 4 and 5). The use of different kinetic models may also account for part of the difference between kinetic measurements for the two experimental populations.

In zebra mussels, an increase in temperature had the effect of increasing uptake clearance and elimination rates for BaP and HCBP (Table 3, Figures 4 and 5). The increase in k_d was proportionately greater than the increase in k_u so that overall, BCF values declined with an increase in temperature. In multiple regressions of temperature and K_{ow}, and BCF, a significant negative correlation was demonstrated between temperature and BCF with the series of compounds used in regression analyses (Figure 8). Thus, the effect of temperature on toxicokinetic parameters does not appear to be limited to specific compounds. The greater effect of temperature upon elimination rate in comparison to uptake clearance has been demonstrated in other freshwater aquatic organisms such as the midge, *Chironomus riparius* (McIntyre, 1988; Lydy et al., 1990).

In *Mytilus*, temperature exerts a clear influence on contaminant accumulation; an increase in temperature leads to an increase in accumulation rates (Boryslawskj et al., 1985). In this study, when environmental temperature was increased from 5 to 15°C, accumulation of petroleum hydrocarbons increased significantly. Elimination rate constants were also seen to be responsive to changing temperature although the two phases of the biphasic elimination curve were affected differently. The rate constant for the initial rapid phase of depuration increased linearly with temperature. However, the second phase of depuration was temperature independent.

Higher filtration and respiratory rates were measured for smaller than large zebra mussels (Figure 3). Smaller mussels also showed relatively greater uptake clearance rates for BaP and HCBP (Figures 4 and 5). Because k_d was not analyzed as a function of mass, it is not clear whether projected steady-state values would increase or decrease in smaller mussels. Animals which feed selectively on zebra mussels within a smaller size range may be exposed to increased levels of contaminants if accumulation is skewed toward smaller mass mussels.

Muncaster et al. (1990) reported that size was negatively correlated with levels of accumulation in two freshwater mollusks. Melaouah (1990) demonstrated that accumulation of amino acids from water in mollusks was dependent upon larval size for which a negative correlation with size was observed. However, accumulation of the latter compounds may not be passive and, therefore, involve a different uptake mechanism (Swinehart and Cheng, 1987; Widdows et al., 1983). Several authors have argued that the higher filtration and respiratory rates measured in small mussels should also lead to increased accumulation of contaminants (Vahl, 1973; Bayne and Widdows, 1978).

Reasonably good correlations were found for kinetic parameters when regressed against temperature and K_{ow} (Figures 6, 7, and 8). The lipophilicity of the chemical, as expressed by log K_{ow}, is clearly an important determinant of uptake clearance rates (Figure 6), elimination rate constants (Figure 7), and BCF (Figure 8). Of particular note was the negative relationship of elimination rate constant to K_{ow} (Figure 7). This has also been reported for *Mytilus* (Hawker and Connel, 1986). Likewise, there are several reports of BCF values for neutral lipophilic contaminants increasing as a function of log K_{ow} (Ernst, 1977; Geyer et al., 1982), although Pruell et al. (1986) found that BCFs for PCBs were significantly higher than for PAHs possessing identical log K_{ow} values. In general, results of the current study on *Dreissena* and the literature reports of *Mytilus* confirm the predictive value of log K_{ow} in forecasting levels of bioaccumulation under laboratory conditions.

Impact on Contaminant Cycling

Kinetic experiments conducted in these studies have used water as the primary medium from which contaminants were sorbed. However, studies with *Mytilus* have shown that significant accumulation of hydrophobic xenobiotics can also occur from food (Clark and Finley, 1975; Arapis et al., 1984) and sediment (Dame and Dankers, 1988; Pruell et al., 1986; Doherty, 1990). In OSU experiments, preliminary attempts were made to measure the relative rates of accumulation of HCBP from water, rates of accumulation from water in the presence of an algal food source, and rates from the assimilation of contaminated algae. When uncontaminated algae were fed to zebra mussels in contaminated water, k_u was reduced threefold while elimination was increased by a factor of 12 (Table 4). From this result, it can be inferred

TABLE 4. Influence of Route of Exposure in HCBP Clearance and Elimination in *Dreissena polymorpha* in OSU Experiments

Exposure Route	k_u[a] (mL/g/hr)	k_d[a] (/hr)	BCF	$t_{1/2}$ (hr)
Water (no food)	167	0.004	41,750	173
(n = 30)	(81)	(0.001)		
Water (with food)	59	0.055	1,072	13
(n = 30)	(16)	(0.009)		
Algae (clean water)	7	0.03	233	23
(n = 60)	(3)	(0.02)		

[a] Values given are means with standard deviations in parentheses.

that the uptake of HCBP may have been impeded by sorption of HCBP to algae, thereby reducing its availability for uptake. Uptake measured in this experiment potentially reflects accumulation from both food and water but, since the binding of HCBP to algal cells was not measured in this experiment, definitive conclusions are precluded. The order of magnitude increase in k_d when uncontaminated algae was present in the water suggests that elimination of HCBP from zebra mussel tissues is enhanced by the presence of a sorbent in the gut of the mussel. The combination of lower uptake and increased elimination led to a marked reduction in BCF and $t_{1/2}$ for HCBP (Table 4). Our data also indicate that contaminants can be readily assimilated from food into mussels, although uptake clearance rates for this route were significantly lower than for the other two routes. Because assimilation from contaminated food is slow, it is likely that this route did not contribute significantly to the body burden of mussels exposed for 6 hr to contaminated water in the presence of *Chlorella*. Although accumulation from food is comparatively slow, it may be a significant route of accumulation in nature especially when mussels settle in benthic environments. Hydrophobic contaminants which reach the benthic zone will most likely be sorbed to organic materials including algae, dissolved organic carbon, and sediment. The ability of zebra mussels to assimilate contaminants from these media must be considered in order to fully understand contaminant accumulation. The relative affinity of compounds for each medium as well as dynamic processes such as lipid metabolism, temperature, and phytoplankton abundance will ultimately determine the complex equilibria between all compartments.

One potentially important issue which must be considered is whether zebra mussels can restructure contaminant cycling by removing contaminants from the pelagic zone and concentrating them in the benthic environment. Filtered material which is not assimilated is compacted into a pellet, wrapped with a mucous coating, and deposited on the bottom as pseudofeces. Benthic organisms may be exposed to highly concentrated hydrophobic xenobiotics if these organically rich pseudofeces are subsequently consumed.

The concern for benthos is not academic. Based on some rough calculations, the collective filtering capacity of the mussels is relatively large. If it is assumed that a single mussel can filter 61 mL/hr, then each animal will filter a volume of 1.46 L/day. If it is further assumed that zebra mussels are present in densities of 10,000/m^2 (which is conservative) and that they cover 1% of the lake, it would take 24 and 52 days for the mussels to completely filter Lake St. Clair and the western basin of Lake Erie, respectively. If a more realistic mussel density of 50,000/m^2 was assumed, the filtering time is reduced to 10 days for both lakes. Clearly, the mussels can potentially exert a major influence on phytoplankton distributions in these lakes. If mussels filter contaminated material (either water or food), then the opportunity for redirecting contaminant distributions in these lakes also exists. Furthermore, the finding that assimilation of HCBP from contaminated *Chlorella* was slow implies that HCBP may remain in unassimilated algal cells in significant amounts. The deposition of this pollutant in pseudofeces or feces is highly probable. Gauging the impact of the zebra mussels on the Great Lakes must therefore include an assessment of where these compounds are going and whether alterations in patterns of contaminant cycling are occurring.

SUMMARY

Using both mass-based and concentration-based kinetic models, *Dreissena polymorpha* was shown to rapidly accumulate substantial body burdens of PCBs and PAHs. Differences in kinetic parameters between two experimental populations of mussels were attributable primarily to mussel length. Smaller mussels showed much higher uptake clearance rates than did larger mussels. Environmental temperature profoundly altered uptake clearance and elimination rates such that total accumulation was significantly greater at lower temperatures. The relatively high lipid content of zebra mussels probably accounts for accentuated accumulation of contaminants relative to *Mytilus*.

It is clear that zebra mussels can accumulate hydrophobic chemicals from several sources and that the levels of accumulation will be comparatively high. In feeding experiments, zebra mussels assimilated HCBP from contaminated algae even though at a much lower rate than absorption from water. These preliminary experiments showed that accumulation from food as well as from water will occur in nature. The impact of zebra mussels on contaminant cycling is believed to be potentially important in the Great Lakes.

ACKNOWLEDGMENTS

GLERL contribution number 748 is appreciated. The OSU component of this research was supported in part by the Ohio Sea Grant College Program, project R/MR-4 under grant NA89AA-D-SG B2 with additional support from the Ohio State University and the Ohio Board of Regents.

REFERENCES

Adema, D. M. M., and H. Compaan. "Accumulation and Elimination of Dieldrin by Mussels, Shrimps and Guppies," Central Laboratory TNO Report (1975).

Ali, R. M. "The Influence of Suspension Density and Temperature on the Filtration Rate of *Hiatella arctica*," *Mar. Biol.* 6:291–302 (1970).

Arapis, G., S. Bonotto, G. Nuyts, A. Bossus, and R. Kirchman. *Soc. Physiol. Veg. Abstr.* (1984), pp. 33–34.

Bayne, B. L., and J. Widdows. "The Physiological Ecology of Two Populations of *Mytilus edulis* L." *Oecologia (Berlin)* 37:137–162 (1978).

Boryslawsky, M., A. C. Garrod, J. T. Pearson, and D. Woodhead. "Processes Involved in the Uptake and Elimination of Organic Micropollutants in Bivalve Molluscs," *Mar. Environ. Res.* 17:310 (1985).

Broman, D., and B. Ganning. "Uptake and Release of Petroleum Hydrocarbons by Two Brackish Water Bivalves, *Mytilus edulis* and *Macoma baltica* (L.)" *Ophelia* 25:49–57 (1986).

Clark, R. C., and J. S. Finley. "Uptake and Loss of Petroleum, Hydrocarbons by the Mussel, *Mytilus edulis* in Laboratory Experiments," *Fish. Bull.* 73:508–515 (1975).

Dame, R. F., and N. Dankers. "Uptake and Release of Materials by a Wadden Sea Mussel Bed," *J. Exp. Mar. Biol. Ecol.* 108:207–216 (1988).

D'Itri, F. M., and M. A. Kamrin. *PCBs: Human and Environmental Hazards* (Woburn, MA: Butterworth Publishers, 1983).

Doherty, F. G. "The Asiatic Clam, *Corbicula* spp. as a Biological Monitor in Freshwater Environments," *Environ. Monitor. Assess.* 15:143–181 (1990).

Dunn, B. P., and H. F. Stich. "Release of the Carcinogen Benzo(a)pyrene from Environmentally Contaminated Mussels," *Bull. Environ. Contam. Toxicol.* 14:398–401 (1976).

Ernst, W. "Determination of the Bioconcentration Potential of Marine Organisms — A Steady-State Approach," *Chemosphere* 11:731–740 (1977).

Fitchko, J. "Literature Review of the Effects of Persistent Substances in Great Lakes biota," Report of the Health of Aquatic Communities Task Force, International Joint Commission, Windsor, Ontario (1986).

Geyer, H., P. Sheehau, D. Kotzias, D. Freitag, and F. Korte. "Prediction of Ecological Behavior of Chemicals: Relationship Between Physicochemcial Properties and Bioaccumulation of Organic Chemicals in the Mussel, *Mytilus edulis*," *Chemosphere* 11:1121–1134 (1982).

Grasshoff, K. "Determination of Oxygen," in *Methods of Seawater Analysis,* K. Grasshoff, M. Ehrhardt, and K. Kremling, Eds. (Weinheim, Germany: Verlag Chemie, 1983), pp. 203–229.

Hansen, N., V. P. Jensen, H. Appelquist, and E. Morch. "The Uptake and Release of Petroleum Hydrocarbons by the Marine Mussel, *Mytilus edulis,*" *Prog. Water Technol.* 10:351–359 (1978).

Hartley, D., and J. B. Johnston. "Use of the Freshwater Clam *Corbicula manilensis* as a Monitor for Organochlorine Pesticides," *Bull. Environ. Contam. Toxicol.* 31:33–40 (1983).

Hawker, D. W., and D. W. Connel. "Bioconcentration of Lipophilic Compounds by Some Aquatic Organisms," *Ecotoxicol. Environ. Saf.* 11:184–197 (1986).

Ingebrigsten, K., J. E. Solbakken, G. Norheim, and I. Natstad. "Distribution and Elimination of ^{14}C-octachlorostyrene in Cod (*Gadus morhua*), Rainbow Trout (*Salmo gairderi*), and Blue Mussel (*Mytilus edulis*)," *J. Toxicol. Environ. Health* 25:361–372 (1988).

Kannan, N., S. Tanabe, R. Tatsukawa, and D. J. H. Phillips. "Persistency of Highly Toxic Coplanar PCBs in Green-Lipped Mussels (*Perna viridis* L.). *Environ. Pollut.* 56:65–76 (1989).

Landrum, P. F. "Bioavailability and Toxicokinetics of Polycyclic Aromatic Hydrocarbons Sorbed to Sediment for the Amphipod, *Pontoporeia hoyi,*" *Environ. Sci. Technol.* 23:588–595 (1988).

Lee, R. F., R. Sanerherber, and A. A. Benson. "Petroleum Hydrocarbons: Uptake and Discharge by the Marine Mussel *Mytilus edulis,*" *Science* 177:343–346 (1972).

Lydy, M. J., T. W. Lohner, and S. W. Fisher. "Influence of pH, Temperature and Sediment Type on the Toxicity, Accumulation and Degradation of Parathion in Aquatic Systems," *Aquat. Toxicol.* 17:27–44 (1990).

Mattson, N. S., E. E. Gidius, and J. E. Solbakken. "Uptake and Elimination of (Methyl-^{14}C) Trichlorfon in Blue Mussel (*Mytilus edulis*) and European Oyster (*Ostrea edulis*)-Impact of Neguvon Disposal on Mollusc Farming," *Aquaculture* 71:9–14 (1988).

McDowell-Capuzzo, J., J. W. Farrington, P. Rantamaki, C. Hovey Cifford, B. A. Lancaster, D. F. Leavitt, and X. Jia. "The Relationship Between Lipid Composition and Seasonal Differences in the Distribution of PCBs in *Mytilus edulis* L. *Mar. Environ. Res.* 28:259–264 (1989).

McIntyre, D. O. "The Effect of Temperature on Uptake Rate Constants and Bioconcentration Factors (BCFs) for Six Organochlorines in the Aquatic Insect, *Chironomus riparius,*" PhD Thesis, Ohio State University, Columbus, OH (1988).

Melaouah, N. "Absorption et metabolisation de substances organiques dissoutes au cours du developpemant larvaire de *Mytilus edulis* L. (Bivalves)," *Oceanol. Acta* 13:245–255 (1990).

Morton, B. S. "Studies on the Biology of *Dreissena polymorpha* Pall. V. Some Aspects of Filter Feeding and the Effect of Micro-organisms upon the Rate of Filtration," *Proc. Malacol. Soc. London* 39:289–301 (1971).

Muncaster, B. W., P. D. N. Herbert, and R. Lazar. "Biological and Physical Factors Affecting the Body Burden of Organic Contaminants in Freshwater Mussels," *Arch. Environ. Contam. Toxicol.* 19:25–34 (1990).

Neff, J. M. *Polycyclic Aromatic Hydrocarbons in the Aquatic Environment* (Essex, England: Applied Science Publishers Ltd., 1979).

Nunes, P., and P. E. Benville. "Uptake and Depuration of Petroleum Hydrocarbons in the Manila Clam, *Tapes semidecussata* Reeve," *Bull. Environ. Contam. Toxicol.* 21:719–726 (1979).

Pruell, R. J., J. L. Lake, W. R. Davis, and J. G. Quinn. "Uptake and Depuration of Organic Contaminants by Blue Mussels (*Mytilus edulis*) Exposed to Environmentally Contaminated Sediment," *Mar. Biol.* 91:497–507 (1986).

Renberg, L., M. Tarkpea, and G. Sundstrum. "The Use of the Bivalve *Mytilus edulis* as a Test Organism for Bioconcentration Studies. II. The Bioconcentration of Two ^{14}C-labeled Chlorinated Parafins," *Ecotoxicol. Environ. Saf.* 11:361–372 (1986).

Schulte, E. H. "Influence of Algal Concentration and Temperature in the Filtration Rate of *Mytilus edulis*," *Mar. Biol.* 30:331–341 (1975).

Sheldon, R. W., and T. R. Parsons. "A Practical Manual on the Uses of the Coulter Counter in Marine Science," Coulter Electronics, Inc. Toronto, Ontario (1967).

Solbakken, J. E., F. M. H. Jeffrey, A. H. Knap, and K. H. Palmork. "Accumulation and Elimination of 9-^{14}C Phenathrene in the Calico Clam (*Macrocallista maculata*)," *Bull. Environ. Contam. Toxicol.* 28:530–534 (1982).

Stanczykowska, A. "Ecology of *Dreissena polymorpha* (Pall.). (Bivalvia) in Lakes," *Pol. Arch. Hydrobiol.* 24:401–530 (1977).

Swinehart, J. H., and M. A. Cheng. "Interactions of Organic Pollutants with Gills of the Bivalve Molluscs *Anodonta califoriensis* and *Mytilus californianus*: Uptake and Effect on Membrane Fluxes. II, *Comp. Biochem. Physiol.* 88C:293–299 (1987).

Ten Winkel, E. H., and C. Davids. "Food Selection by *Dreissena polymorpha* Pallas (Mollusca:Bivalvia)," *Freshwater Biol.* 12:553–558 (1982).

Theede, H. "Experimentelle unterschungen uber die filtrierleistung der mies-muschel *Mytilus edulis* L." *Kiel. Meeresforsch.* 19:20–41 (1963).

Vahl, O. "Pumping and Oxygen Consumption Rates of *Mytilus edulis* L. of Different Sizes," *Ophelia* 12:45–52 (1973).

Vanderpleog, H. A., J. A. Bowers, O. Chapelski, and H. K. Soo. "Measuring in situ Predation by *Mysis relicta* and Observations on Underdispersed Microdistributions of Zooplankton," *Hydrobiologia* 93:109–119 (1982).

Walz, N. "The Energy Balance of the Freshwater Mussel *Dreissena polymorpha* Pallas in Laboratory Experiments and in Lake Constance. I. Pattern of Activity, Feeding and Assimilation Efficiency," *Arch. Hydrobiol. Suppl.* 55(1):83–105 (1978).

Walz, N. "The Energy Balance of the Freshwater Mussel *Dreissena polymorpha* Pallas in Laboratory Experiments and in Lake Constance. V. Seasonal and Nutritional Changes in the Biochemical Composition," *Arch. Hydrobiol. Suppl.* 314:235–254 (1979).

Widdows, J., S. L. Moore, K. R. Clarke, and P. Donkin. "Uptake, Tissue Distribution and Elimination of [1-^{14}C-] Naphthalene in the Mussel *Mytilus edulis*," *Mar. Biol.* 76:109–114 (1983).

Wilkinson, L. "Systat: The system for Statistics," Systat Inc., Evanston, IL (1988).

CHAPTER 29

Toxicity of Heavy Metals to the Zebra Mussel (*Dreissena polymorpha*)

Michiel H. S. Kraak, Daphna Lavy, Merel Toussaint, Hans Schoon, Wilma H. M. Peeters, and Cees Davids

The zebra mussel, *Dreissena polymorpha*, was chosen as a test organism for toxicological laboratory experiments since it plays an important role in the aquatic food chain. The sublethal effects of Cu, Zn, Cd, and combinations of these metals on this species were determined after acute and chronic exposures. *Dreissena* appears to be a suitable test organism for toxicological laboratory experiments because it provided reproducible results, was easy to collect and handle, and had low control mortality. The EC_{50} filtration rate (48 hr) for Cu, Cd, and Zn were 41, 388, and 1350 μg/L, respectively. *Dreissena* was able to regulate Zn at higher external concentrations than Cu, while Cd could not be regulated. When metals were added as equitoxic mixtures in acute experiments, several significant interactions between the metals occurred. These results indicate that the effects of mixtures cannot be predicted from the effects of the metals tested individually. It appears that metals can contribute to the toxicity of a mixture at or below the no observed effect concentrations (NOEC) for the metals tested individually. Concentrations of Cu and Cd, which had an equal negative effect after 48 hr, produced completely different impacts on *Dreissena* during chronic exposure, Cd being far more toxic than Cu.

0-87371-696-5/93/$0.00 + $.50

INTRODUCTION

The zebra mussel, *Dreissena polymorpha*, plays an important beneficial role in various freshwater ecosystems in Europe. This mussel is able to reduce high abundances of phytoplankton by its high filtration activity (Stanczykowska et al., 1975; Lewandowski, 1983; Reeders et al., 1989). It is the main source of food for diving ducks (Stanczykowska, 1977; Suter, 1982) which winter in large numbers in The Netherlands, and for benthivorous fish such as roach (*Rutilus rutilus*) and carp (*Cyprinus carpio*). The zebra mussel was chosen as a test organism for toxicological laboratory experiments because adverse effects of toxicants on this mussel may affect the aquatic food chain (Scholten et al., 1989). It has also been used as an early warning system (Kramer et al., 1989) and a biomonitoring organism (Karbe et al., 1975).

Heavy metals are known to reduce the performance of bivalve mollusks, including *Dreissena*; effects include shell closure for long periods of time (Slooff et al., 1983; Doherty et al., 1987; Salinki and V.-Balogh, 1989; Kramer et al., 1989), production of fewer byssus threads (Martin et al., 1975), and reduced heart rate (Davenport, 1977; Akberali and Black, 1980; Grace and Gainey, 1987). In addition, it has been demonstrated that heavy metals reduce the filtration rate of some marine bivalves (Watling, 1981; Manley, 1983; Grace and Gainey, 1987; Redpath and Davenport, 1988; Krishnakumar et al., 1990).

In many toxicological laboratory experiments, acute LC_{50} values for one toxicant are determined. This is unrealistic because contaminated lakes and rivers contain a great variety of toxicants. In our experiments, we determined the acute, sublethal effects of Cu, Zn, Cd, and equitoxic mixtures of these metals on the filtration rate of zebra mussels. Filtration rate is a more realistic parameter when compared to mortality (Abel, 1976) and it is an important aspect of the ecological role of zebra mussels. In addition, chronic toxicity experiments were conducted with Cu and Cd. These experiments reflect the field situation better than acute experiments because in ecosystems animals are often exposed to low concentrations of toxicants throughout their lifetime.

MATERIALS AND METHODS

Acute Toxicity Experiments

Mussels and water were collected from Lake Markermeer, a relatively unpolluted lake in the The Netherlands. The experiments were conducted in temperature-controlled aquariums (15°C) containing aerated, oxygen-saturated water. pH was 7.9 and hardness was 15D (= 107 mg Ca/L). Each experimental treatment consisted of 25 mussels of 16–20 mm, placed in a plastic aquarium containing 3 L of filtered (0.45 μm filter pore size) lake water. Experiments were carried out in duplicate for each treatment. One dose-response relationship is based on the activity of 350 mussels. The following concentrations of heavy metals were tested: Zn: 0, 0.2, 0.5, 1.0, 2.0, 5.0, and 10.0 mg/L; Cd: 0, 0.1, 0.2, 0.5, 0.75, 1.0, 2.0, and 5.0 mg/L; and Cu: 0, 10, 20, 50, 100, 200, 500, and 1000 μg/L. Concentrations were made from stock solutions of 1000 mg/L $CuCl_2$, $ZnCl_2$, and $CdCl_2$.

Metals were added to water in the aquariums, within 1 day of when the mussels were collected. Water was renewed and metals were added again after 24 and 48 hr. Water samples were taken at 1, 24, 25, and 48 hr and analyzed for added metals by flame or furnace atomic absorption spectrophotometry (AAS). The average concentrations of metals to which animals were exposed during experiments were determined from these values using integral calculation. Filtration rates were measured after 48 hr.

To determine filtration rates, mussels were fed either *Chlamydomonas eugametos* or *Scenedesmus acuminatus* (30,000 cells per milliliter), and filtration rates were calculated from the decrease in algal concentration over time, according to Coughlan's formula (1969):

$$m = \frac{M}{nt} \, nl \, \frac{Co}{ct}$$

in which:
- m = filtration rate in milliliter per mussel per hour
- M = volume of the test solution (3,000 ml)
- n = number of animals per aquarium
- t = duration of the experiment in hours
- Co = algal concentration at the beginning of determination of the filtration rate
- Ct = algal concentration at the end of determination of the filtration rate

Filtration rates of experimental treatments were expressed as a percentage of controls. Results are given as dose-response relationships, from which the EC_{50} filtration values were calculated by probit analysis (Finney, 1971) and the no observed effect concentration (NOEC) filtration values using Williams' test (Williams, 1971).

After experiments were finished, five mussels from each aquarium were removed from their shells and byssus, freeze-dried, weighed, dissolved by wet destruction with HNO_3 and H_2O_2, and analyzed for heavy metals by flame or furnace AAS following Timmermans et al. (1989). Results were tested successively using Bartlett's test for homogeneity of variances, analysis of variance (one-way ANOVA), and *Scheffe's* test for posterior comparison of means.

Acute Mixture Toxicity Experiments

The only difference in methods between the acute toxicity experiments and the acute mixture toxicity experiments was that in the latter equitoxic mixtures of the metals were applied. In an equitoxic mixture the metals are added as fractions of their EC_{50}:

$$fi = Ci/EC_{50}\ i \qquad (1)$$

in which: fi = fraction of metal i
Ci = concentration of metal i
EC_{50} i = EC_{50} of metal i

The toxic units (TU) are the sum of the added fractions:

$$TU = \Sigma\ fi \qquad (2)$$

in which: TU = toxic units
Σ fi = the sum of the added fractions

Equitoxicity is achieved by adding equal fractions of the metals (Sprague, 1970).

In these experiments a range of toxic units was applied. Results were plotted as dose-response relationships, and EC_{50} filtration values were determined by probit analysis (Finney, 1971). If the EC_{50} of a mixture is below 1 TU, equal to 1 TU, or above 1 TU — then the effects of the metals are more than additive (synergism), additive, and less than additive (antagonism), respectively. The following combinations were testing: Cu + Zn, Cu + Cd, Zn + CD, and Cu + Zn + Cd; and the following concentrations were added: 0, 0.2, 0.5, 1.0, 1.5, 2.0, and 5.0 TU.

Chronic Toxicity Experiments

In the chronic toxicity experiments, only 48-hr EC_{50} filtration values for Cu and Cd were studied. Experiments lasted 9 weeks for Cu and 11 weeks for Cd. The first 48 hr of the chronic experiments were conducted in the same manner as the acute toxicity experiments. Between 48 hr and 11 weeks water samples were taken, and new water and metals were added each working day. Mussels were fed *Chlamydomonas eugametos* each Monday and Friday and *Scenedesmus acuminatus* each Wednesday. Filtration rates were determined each Wednesday. For Cd, in contrast to Cu, many animals died during the experiments; thus a Lethal $Time_{50}$ (LT_{50}) could be determined by probit analysis (Finney, 1971). After the experiments were finished, soft tissues of

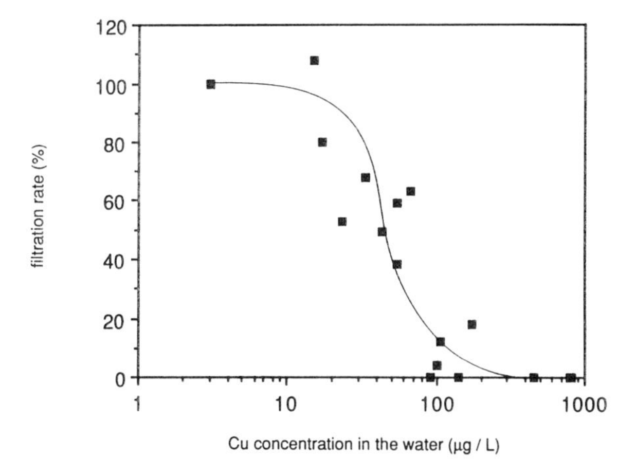

Figure 1. Filtration rate (expressed as a percentage of controls) of *Dreissena polymorpha* after 48-hr exposure plotted against Cu concentration in water.

mussels were analyzed for heavy metals by flame or furnace AAS following Timmermans et al. (1989).

RESULTS AND DISCUSSION

Dreissena polymorpha appeared to be a suitable test organism for toxicological laboratory experiments. This mussel was easy to collect and handle, and survived well in chronic experiments. Filtration rates in the controls were among the higher values found in the literature (Morton, 1971; Benedens and Hinz, 1980; Sprung and Rose, 1988).

Acute Toxicity Experiments

Figure 1 gives the effect of Cu on the filtration rate of *Dreissena* after 48 hr. Reproducible results were obtained, since the variation between the filtration rate of animals exposed to equal Cu concentrations in the two experiments was usually below 15%. The NOEC filtration can be defined as the highest metal concentration in the water which did not lead to a significant ($p < 0.05$) decrease in filtration rate, and the EC_{50} filtration can be defined as the metal concentration in the water which led to a 50% decrease in the filtration rate. NOEC filtration values and EC_{50} filtration values for Cu, Zn, and Cd indicated that Cu was very toxic to zebra mussels (Table 1), as it is to many marine bivalves (Abel, 1976; Watling, 1981; Redpath and Davenport, 1988). Cd was about 10 times less toxic than Cu, and Zn was the least toxic. In contrast to these results, Zn is often more toxic than Cd for marine bivalves (Watling, 1981; Redpath and Davenport, 1988).

TABLE 1. No Observed Effect Concentration (NOEC) Values and EC_{50} Values for Filtration Rate of *Dreissena polymorpha* after 48-hr Exposure to Cu, Zn, or Cd

	NOEC	EC_{50} (μg/L)
Cu	16	41
Cd	175	388
Zn	191	1350

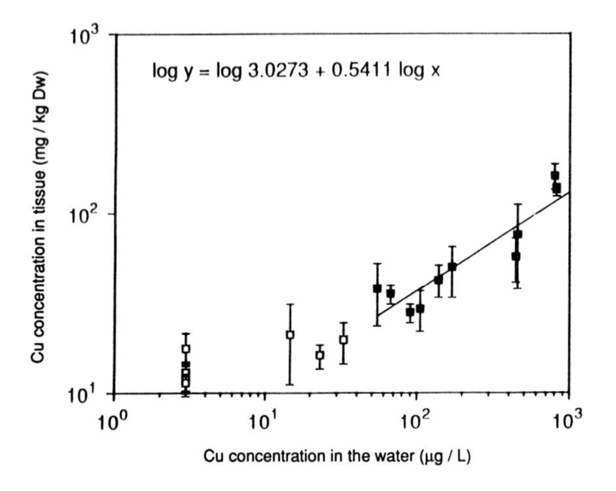

Figure 2. Cu concentration in soft tissue of *Dreissena polymorpha* after 48-hr exposure plotted against Cu concentration in water. Open squares are control mussels and exposed mussels that did not differ significantly ($p < 0.05$) from control mussels. Filled squares indicate Cu concentrations in exposed mussels which differed significantly ($p < 0.05$) from the controls.

Our experiments showed that filtration rate is a sensitive parameter when compared to mortality. For example, in the case of Cu, even at the highest concentration tested (i.e., 1000 μg/L), no mortality occurred after a 48-hr exposure. Thus, the acute LC_{50} for Cu would have been higher than 25 times the acute EC_{50} filtration (41 μg/L). It is, therefore, concluded that this sublethal parameter is a far more realistic endpoint for assessing the toxicity of heavy metals than mortality.

In Figure 2, the Cu concentration in mussel soft tissue is plotted against metal concentration in the water. The NOEC accumulation can be defined as the highest metal concentration in the water which did not lead to a significant ($p < 0.05$) increase of metal concentration in mussels. Above this concentration, accumulation of metal in the animals took place. The NOEC accumulation for Cu and Zn were 28 and 191 μg/L, respectively, suggesting that Zn could be regulated up to higher external concentrations than Cu. In contrast to Cu and Zn, any elevated Cd concentration in the water led to a significant ($p < 0.05$) increase in the Cd concentration in mussels. Thus, a NOEC accumulation could not be determined for Cd, suggesting that Cd cannot be regulated by *Dreissena*. This agrees with the results of Amiard et al. (1987) for the marine mussel *Mytilus edulis*. The explanation for these results is that Cu and Zn are essential elements, whereas Cd is a nonessential element. From these experiments it can be concluded that background values for heavy metals in *Dreissena* are 10–20 μg/gDW for Cu, 90–130 μg/gDW for Zn, and up to 1.5 μg/gDW for Cd.

In comparing concentrations that had an effect in our experiments to concentrations in water quality criteria for Dutch waters (2.5 μg/L Cd, 50 μg/L Cu, and 200 μg/L Zn), it appears that concentrations for the NOEC filtration for Cu and Zn, concentrations for the NOEC accumulation for Cu and Zn, and concentrations for the EC_{50} filtration for Cu were all below these water quality standards.

Acute Mixture Toxicity

For a combination of two metals, several significant ($p < 0.05$) interactions were apparent: Cu + Zn had a less than additive effect, and Cu + Cd had a more than additive effect. EC_{50} values were 1.63 and 0.64 TU, respectively. EC_{50} values for Zn + Cd and Cu + Zn + Cd were not significantly different from 1.0 TU (0.94 and 0.84 TU, respectively); thus these metal combinations had an additive effect. The additive effect of Cu + Zn + Cd may be the result of the more than additive effect of Cu + Cd compensated for by the less than additive effect of Cu + Zn. One must conclude, however, that the effects of mixtures could not always be predicted from the effects of the metals tested individually.

There is little agreement in the literature concerning the toxicity of mixtures of heavy metals. An extensive overview is given by the EIFAC (1987). Test organisms varied from bivalves to frogs, with emphasis on fish. Both mortality and sublethal parameters were chosen as endpoints. Different exposure times were studied and a variety of metal mixtures was tested. The results of Spehar and Fiandt (1986) and of Enserink et al. (1991) suggested that when a mixture consists of a large number of metals the effect will be additive, but concerning mixtures of two or three metals no general trend is to be seen in the literature.

The usefulness of mixture toxicity experiments has been clearly demonstrated in the present study. The EC_{50} filtration of an equitoxic mixture of Cu + Zn + Cd did not significantly differ from 1 TU (0.84), being 14 μg/L Cu, 130 μg/L Cd, and 450 μg/L Zn. However, the NOEC filtration values for

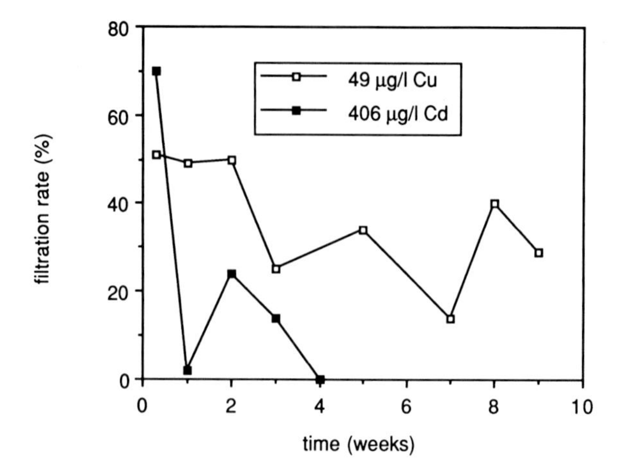

Figure 3. Filtration rate (expressed as a percentage of controls) of *Dreissena polymorpha* exposed to 49 µg/L Cu and 406 µg/L Cd.

these metals tested individually were 16 µg/L Cu, 175 µg/L Cd, and 191 µg/L Zn (Table 1). Thus, the EC_{50} filtration of the mixture of Cu + Zn + Cd was already found at Cu and Cd concentrations at or below the NOEC filtration values for these metals tested individually. It is remarkable that a metal could already contribute to the toxicity of a mixture below its NOEC. These results are supported by Könemann (1981). We conclude from our study that toxicity experiments with metal mixtures often reflect actual pollution exposure conditions, and resulting responses in aquatic ecosystems are more realistic than responses with single toxicant experiments.

Chronic Toxicity

Filtration rate and mortality were examined during and after exposure to 49 µg/L Cu or 406 µg/L Cd, which is approximately the EC_{50} filtration 48 hr for these metals (Table 1). For Cu, mussels filtered continuously but at a reduced level throughout the entire experiment. For Cd, in contrast, there was a total cessation of filtration rate after 4 weeks (Figure 3). Mortality for Cu was low, with only 6 of 25 mussels dying. However, mortality for Cd was much higher, with only one mussel surviving the 11-week exposure period (Figure 4). The LT_{50} (the time at which 50% of the mussels died) for 406 µg/L Cd was highly reproducible, being 28 days for both experiments. Mortality in controls during these chronic exposure experiments was always below 20%, and in several experiments mortality was zero.

It can be concluded that concentrations of Cu and Cd, which have an equally negative effect after 48 hr, have completely different impacts on *Dreissena* during long-term exposure; Cd is far more toxic than Cu. This again

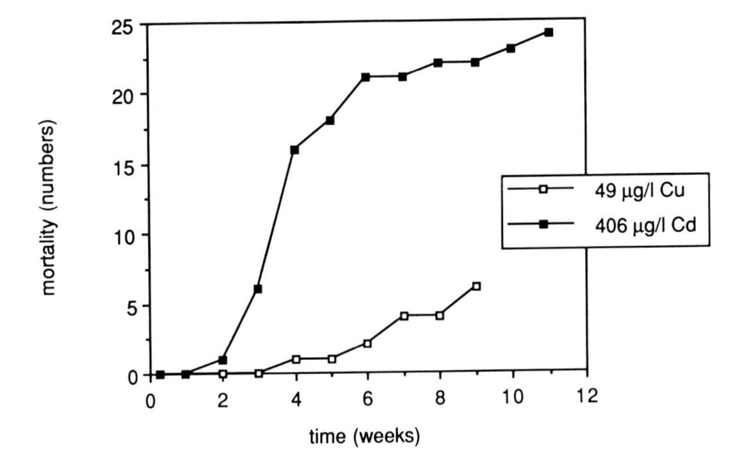

Figure 4. Mortality (numbers) of *Dreissena polymorpha* exposed to 49 µg/L Cu and 406 µg/L Cd.

TABLE 2. Cu and Cd Concentrations (Mean ± S.D.) in *Dreissena polymorpha* Exposed for 48 hr to 53 µg/L Cu and 460 µg/L Cd (Acute), and for 9 weeks to 49 µg/L Cu, and for 11 weeks to 406 µg/L Cd (Chronic)

	Blank	Acute	Chronic
Cu	13 ± 2	35 ± 4	199 ± 47
Cd	1.3 ± 0.3	74 ± 19	540 ± 288

Note: Values given in mg/kgDW.

can be explained by differences between essential (Cu) and nonessential elements (Cd). Of course, much more Cd than Cu was added, but it was our intention to compare equal toxicities and not equal concentrations.

Metal concentrations in mussel tissues at the end of chronic experiments were significantly elevated compared to concentrations in tissues of control animals and tissues of animals exposed for 48 hr (Table 2). For Cu, however, mussels kept filtering throughout the experiments. It does not appear that the Cu concentration in tissue, after acute exposure, caused the decrease in filtration rate. These results suggest that, at least in the acute experiments, filtration rates of zebra mussels are mainly decreased by detection of metals in the water and not by accumulation in the tissue. This idea is supported by results showing that there is no significant ($p < 0.05$) difference between toxicity of Zn and Cu immediately after addition of the metals, when accumulation could not have taken place yet, and toxicity after a 48-hr exposure. Moreover, the NOEC filtration values after 48 hr are in agreement with the detection limits of an early warning system, using the immediate valve-movement response of *Dreissena* as a parameter (Slooff et al., 1983; Kramer et al., 1989).

CONCLUSIONS

The present study of metal toxicity in *Dreissena polymorpha* demonstrated that: (1) Cu was very toxic in acute experiments, Cd was about 10 times less toxic than Cu, and Zn was the least toxic metal tested; (2) Zn could be regulated up to higher threshold concentrations than Cu, but Cd could not be regulated; (3) effects of equitoxic mixtures could not be predicted from the effects of the metals tested individually; (4) metals could contribute to the acute toxicity of a mixture at or below the NOEC values for the metals tested individually; (5) during chronic exposure, Cd had far more deleterious effects than Cu; (6) filtration rates in acute experiments probably were reduced because of metal detection in the water and not because of accumulation in the tissue.

ACKNOWLEDGMENTS

We thank Professor Dr. Nico M. van Straalen for his comments and Miranda Aldham-Breary, M.Sc. for improving the English text.

REFERENCES

Abel, P. D. ''Effect of Some Pollutants on the Filtration Rate of *Mytilus*,'' *Mar. Pollut. Bull.* 7:228–231 (1976).

Akberali, H. B., and J. E. Black. ''Behavioural Responses of the Bivalve *Scrobicularia plana* (Da Costa) Subjected to Short-Term Copper (CuII) Concentrations,'' *Mar. Environ. Resour.* 4:97–107 (1980).

Amiard, J. C., C. Amiard-Triquet, B. Berthet, and C. Metayer. ''Comparative Study of the Patterns of Bioaccumulation of Essential (Cu, Zn) and Non-essential (Cd, Pb) Trace Metals in Various Estuarine and Coastal Organisms,'' *J. Exp. Mar. Biol. Ecol.* 106:73–89 (1987).

Benedens, H. G., and W. Hinz. ''Zur Tagesperiodizität der Filtrationsleistung von *Dreissena polymorpha* und *Sphaerium corneum* (Bivalvia),'' *Hydrobiologia* 69:45–48 (1980).

Coughlan, J. ''The Estimation of Filtering Rate from the Clearance of Suspensions,'' *Mar. Biol.* 2:356–358 (1969).

Davenport, J. ''A study of the Effects of Copper Applied Continuously and Discontinuously to Specimens of *Mytilus edulis* (L.) Exposed to Steady and Fluctuating Salinity Levels,'' *J. Mar. Biol. Assoc. U.K.* 57:63–74 (1977).

Doherty, F. G., D. S. Cherry, and J. Cairns, Jr. ''Valve Closure Responses of the Asiatic clam *Corbicula fluminea* Exposed to Cadmium and Zinc,'' *Hydrobiologia* 153:159–167 (1987).

EIFAC. ''Water Quality Criteria for European Freshwater Fish,'' Revised Report on Combined Effects on Freshwater Fish and Other Aquatic Life of Mixtures of Toxicants in Water,'' EIFAC Technical Paper (37), Revue 1 (1987).

Enserink, E. L., J. L. Mass-Diepeveen, and C. J. van Leeuwen. ''Combined Toxicity of Metals; an Ecotoxicological Evaluation,'' *Water Resour.* 25:679–687 (1991).

Finney, D. J. *Probit Analysis,* 3rd ed. (London, England: Cambridge University Press, 1971).

Grace, A. L., and L. F. Gainey, Jr. ''The Effects of Copper on the Heart Rate and Filtration Rate of *Mytilus edulis*,'' *Mar. Pollut. Bull.* 18:87–91 (1987).

Karbe, L., N. Antonacopoulos, and C. Schnier. ''The Influence of Water Quality on Accumulation of Heavy Metals in Aquatic Organisms,'' *Verh. Int. Verein. Theor. Angew. Limnol.* 19:2094–2101 (1975).

Könemann, H. ''Fish Toxicity Tests with Mixtures of More than Two Chemicals: A Proposal for a Quantitative Approach and Experimental Results,'' *Toxicology* 19:229–238 (1981).

Kramer, K. J. M., H. A. Jenner, and D. de Zwart. ''The Valve Movement Response of Mussels: A Tool in Biological Monitoring,'' *Hydrobiologia* 188/189:433–443 (1989).

Krishnakumar, P. K., P. K. Asokan, and V. K. Pillai. ''Physiological and Cellular Responses to Copper and Mercury in the Green Mussel *Perna viridis* (Linnaeus),'' *Aquat. Toxicol.* 18:163–174 (1990).

Lewandowski, K. ''Occurrence and Filtration Capacity of Plant-Dwelling *Dreissena polymorpha* (Pall.) in Majca Wielki Lake,'' *Pol. Arch. Hydrobiol.* 30:255–262 (1983).

Manley, A. R. ''The Effects of Copper on the Behavior, Respiration, Filtration and Ventilation Activity of *Mytilus edulis*,'' *J. Mar. Biol. Assoc. U.K.* 63:205–222 (1983).

Martin, J. M., F. M. Piltz, and D. J. Reish. ''Studies on *Mytilus edulis* Community in Alamitos Bay, California. V. The Effects of Heavy Metals on Byssal Thread Production,'' *Veliger* 18:183–188 (1975).

Morton, B. ''Studies on the Biology of *Dreissena polymorpha* Pall. V. Some Aspects of Filter Feeding and the Effect of Micro-organisms upon the Rate of Filtration,'' *Proc. Malacol. Soc. London* 39:289–301 (1971).

Redpath, K. J., and J. Davenport. ''The Effect of Copper, Zinc and Cadmium on the Pumping Rate of *Mytilus edulis* L.,'' *Aquat. Toxicol.* 13:217–226 (1988).

Reeders, H. H., A. bij de Vaate, and F. J. Slim. ''The Filtration Rate of *Dreissena polymorpha* (Bivalvia) in Three Dutch Lakes with Reference to Biological Water Quality Management,'' *Freshwater Biol.* 22:133–141 (1989).

Salinki, J., and K. V.-Balogh K. ''Physiological Background for Using Freshwater Mussels in Monitoring Copper and Lead Pollution,'' *Hydrobiol.* 188/189:445–454 (1989).

Scholten, M. C. T., E. Foekema, W. C. de Kock, and J. M. Marquenie. ''Reproduction Failure in Tufted Ducks Feeding on Mussels from Polluted Lakes,'' Proceedings of the Second European Symposium on Avian Medicine and Surgery, Utrecht, The Netherlands (March 1989).

Slooff, W., D. de Zwart, and J. M. Marquenie. ''Detection Limits of a Biological Monitoring System for Chemical Water Pollution Based on Mussel Activity,'' *Bull. Environ. Contam. Toxicol.* 30:400–405 (1983).

Spehar, R. L., and J. T. Fiandt. ''Acute and Chronic Effects of Water Quality Criteria-Based Metal Mixtures on Three Aquatic Species,'' *Environ. Toxicol. Chem.* 5:917–931 (1986).

Sprague, J. B. ''Measurement of Pollutant Toxicity to Fish. II. Utilizing and Applying Bioassay Results,'' *Water Resour.* 4:3–32 (1970).

Sprung, M., and U. Rose. ''Influence of Food Size and Food Quantity on the Feeding of the Mussel *Dreissena polymorpha*,'' *Oecologia (Berlin)* 77:526–532 (1988).

Stanczykowska, A., W. Lawacz, and J. Mattice. ''Use of Field Measurements of Consumption and Assimilation in Evaluating the Role of *Dreissena polymorpha* Pall. in a Lake Ecosystem,'' *Pol. Arch. Hydrobiol.* 22:509–520 (1975).

Stanczykowska, A. ''Ecology of *Dreissena polymorpha* (Pall.) (Bivalvia) in lakes,'' *Pol. Arch. Hydrobiol.* 24:461–530 (1977).

Suter, W. ''Der Einfluss van Wasservogeln auf Populationen der Wandermuschel (*Dreissena polymorpha* Pall.) am Untersee/Hochrhein (Bodensee),'' *Schweiz. Z. Hydrol.* 44:149–161 (1982).

Timmermans, K. R., B. van Hattum, M. H. S. Kraak, and C. Davids. ''Trace Metals in a Littoral Foodweb: Concentrations in Organisms, Sediment and Water,'' *Sci. Total Environ.* 87/88:477–494 (1989).

Watling, H. ''The Effects of Metals on Mollusc Filtering Rates,'' Trans. R. Soc. S. Afr. 44:441–451 (1981).

Williams, D. A. ''A Test for Differences Between Treatment Means When Several Dose Levels are Compared with a Zero Dose Control,'' *Biometrics* 27:103–117 (1971).

CHAPTER 30

Bioaccumulation, Biological Effects, and Food Chain Transfer of Contaminants in the Zebra Mussel (*Dreissena polymorpha*)

W. Chr. de Kock and C. T. Bowmer

A series of toxicological field studies were conducted between 1976 and 1990 using the zebra mussel, *Dreissena polymorpha*. This species has become an important biomonitoring organism in The Netherlands, accumulating contaminants rapidly at rates specific to a given compound until steady-state concentrations are attained. The pattern of controlled, diked, inland waterways in the low-lying Netherlands is briefly discussed in terms of contaminant transport and environmental pollution problems. Accumulation of cadmium (Cd) to steady state in the zebra mussel (40–60 days) is demonstrated in a field experiment. The apparent effect of Cd on the actual growth of transplanted *D. polymorpha* in active biomonitoring studies is presented, and the concentration at which growth ceases is determined. A case study of the effects of an industrial Cd spill in the Maas River in 1988 is reviewed. The accumulation of cadmium, copper, and zinc in *D. polymorpha* was monitored before and after the spill, and physiological measurements were made to assess the resultant strain on organisms. A reduced scope for growth (SFG) was confirmed at a known elevated internal (tissue) concentration of Cd. Additionally,

0-87371-696-5/93/$0.00 + $.50

zebra mussels from several sites were histopathologically examined and, although no evidence of the Cd accident could be demonstrated, some populations (both natural and transplanted) were observed to be in a generally poor state of health. Finally, a study focusing on the transfer of cadmium and organochlorine contaminants from zebra mussels to the tufted duck (*Aythya fuligula*) held in experimental ponds is discussed. The subsequent transfer to the duck eggs is highlighted, and the resulting teratogenic effects are described. The results of the biomonitoring studies using *D. polymorpha* and the complementary investigation of contaminant transfer to *A. fuligula* are briefly discussed in relation to the recent colonization of zebra mussels in North America.

INTRODUCTION

The zebra mussel, *Dreissena polymorpha*, demonstrated its ability to invade and colonize large, new areas when it entered the Rhine Basin at the beginning of the 19th century (Ellis, 1978). The repetition of this invasion, this time on the North American continent, has been a primary motivation in the compilation of the information presented here. This chapter examines: (1) the susceptibility of this species to contamination (mostly to Cd), (2) looks at the toxicological consequences that may be expected from the sudden filling of a sessile epibenthic niche, and (3) the introduction of a new food source for wading birds and ducks, resulting in a new link in the food chain.

The release of xenobiotic substances into the aquatic environment has traditionally been controlled by the setting of discharge standards based on chemical analysis and toxicological testing. Often, however, the only means of relating such data to field situations is either by crudely applying arbitrary dilution factors (as in the case of actual discharges) or by applying statistical extrapolation techniques (as in the case of impact studies). Currently available methods to predict the ecological consequences of discharges to the environment are generally inadequate, yet feedback from the receiving system is essential. Chemical stress resulting in strain on aquatic organisms is difficult to realistically assess when testing is conducted on laboratory-exposed populations. Strain can be defined as the endogenous condition of the organism as expressed in a disturbance of biochemical, cytological, and physiological mechanisms. Indirect effects at higher levels of biological organization (i.e., populations and communities) are even more difficult to assess because of the absence of ecological realism in laboratory test procedures (de Kock and Kuiper, 1981; Kuiper, 1984). Regrettably, comprehensive biomonitoring programs in the field (defined here as the repeated measurement of variables so that trends may be determined and compared to a standard; Holdgate, 1976) are often considered too expensive and lacking in focus.

Many ecologists have long felt that the possibility of unseen, long-term effects to biota as a result of contamination is a real threat. Catastrophic effects are generally easy to recognize in the field, requiring no impact assessment but rather a cleanup operation. There has always been an underlying fear that the most important effects of contamination in the aquatic environment may be extremely subtle, occurring at concentrations far below those measurable in acute tests, and over far longer time spans. Effects, such as a reduced chance of survival, are of crucial importance in the long term because many aquatic organisms, including macroinvertebrates, are long-lived.

We will always need inexpensive, straightforward, and sensitive bioassays as a screening mechanism, preferably ones with a high predictive ability and a higher degree of ecological realism than those currently used. At the same time, more comprehensive and well-directed field monitoring programs are equally necessary to provide a connection between impact on receiving water bodies, standards, and point source-related research. The prospective information on toxicological effects deemed necessary for setting emission standards in The Netherlands is, at present, invariably based on tiered testing schemes under laboratory conditions. Field studies are not included.

Just as the marine mussel *Mytilus edulis* has proven in the marine environment (de Kock, 1986), *D. polymorpha* to be an excellent indicator of bioavailable pollutants in the freshwater environment. This is of vital importance, because long-term toxic effects are primarily related to the internal concentrations of the compound and not to some ambient external level.

This chapter describes the use of *D. polymorpha* in field monitoring studies between 1976 and 1990 to assess bioavailable pollutant concentrations, and long-term sublethal toxic effects. We show how *D. polymorpha* can survive adverse environmental conditions, and also play a key role in the food chain transfer of contaminants. A number of topics have been extracted from a large body of (largely unpublished) data to demonstrate the usefulness of *D. polymorpha* as an ecotoxicological test species.

METHODS

Study Site

The pathway of contaminants in the aquatic environment considered here begins with industrial releases into the Rhine and Maas Rivers in Switzerland, Germany, France, Belgium, and The Netherlands. Following downstream transport, contaminants are retained in the sediment of eutrophic freshwater lakes in The Netherlands (Figure 1). Contaminant accumulation is the combined result of particle sedimentation at low current velocities (including adsorbed trace metals and organic micropollutants), enhanced adsorption of dissolved contaminants onto suspended matter at increasing pH values, and accumulation by freshwater phytoplankton (Salomons and Mook, 1980). These

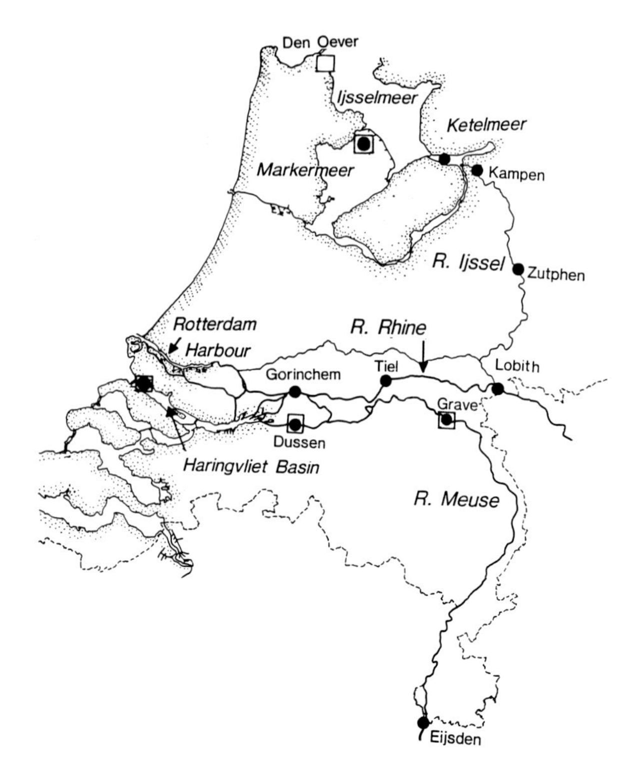

Figure 1. A map of The Netherlands showing major rivers, lakes, and basins where field samples were collected and exposure experiments were conducted. Open square = field samples; filled circle = exposure experiments conducted; filled circle within a square = both field samples and exposure experiments.

factors ultimately lead to the differential exposure of the resident fauna to contamination within their particular ecosystem microniche or subcompartment.

Freshwater contaminant sinks in The Netherlands were created when man-made protective barriers were constructed during the 20th century. Of particular importance are Lake Ijsselmeer, created in 1932, and the Haringvliet Basin, separated from the North Sea in 1970 (Figure 1). Table 1 shows data on the percent retention of dissolved and adsorbed fractions of different heavy metals in these areas, as determined by filtration using a 0.45-μm membrane filter (Rijkswaterstaat, 1984). Clearly, the majority of adsorbed metals are retained in the system, as is a substantial part of the dissolved fractions of Zn, Cr, and Cd transported by the rivers. Estimates are based on a single mass balance inventory conducted in 1978. The underlying analytical data are based on measurements by atomic absorption spectrometry of the water or suspended sediment sample.

TABLE 1. Estimated Percent Retention of Riverine Input of Heavy Metal Fractions in Several Freshwater Lakes in The Netherlands

	Metal									
Location	**Zn**		**Cu**		**Cr**		**Cd**		**Ni**	
Hollands Diep/Haringvliet	18	86	3	88	30	84	58	88	4	85
Lakes Ketelmeer/IJsselmeer	90	75	47	84	87	71	80	91	48	55

Source: Rijkswaterstaat, 1984.

Note: The first value of each metal is the dissolved fraction and the second value is the adsorbed fraction. The source of metals for Lakes Hollands Diep/Haringvliet is the Rhine and Maas Rivers, while the source for Lakes Ketelmeer/IJsselmeer is the Rhine River.

Biomonitoring

Passive and active biomonitoring activities using the zebra mussel, *Dreissena polymorpha*, have been carried out in the Rhine, Maas, and IJssel Rivers since 1976. Passive biomonitoring is when native populations are directly sampled and analyzed for their chemical body burden. Active biomonitoring is defined as the translocation and exposure of standardized groups of individuals at selected locations. This type of monitoring has been used to assess the bioavailability of waterborne contaminants, and is central to the monitoring program. Active biomonitoring allows the standardization of methodology and the subsequent direct intercomparison of locations, populations, accumulated body burdens, and toxic effects. An example of the methods employed in a typical active biomonitoring study follows.

An active biomonitoring experiment with zebra mussels was started on October 31, 1988 as part of a long-term program aimed at gathering biological samples for storage in a chemical and pathological specimen bank. A total of 2200 of that were 15–25 mm in length were collected by hand from a population living on dike stones at 0.5-m depth in Lake Markermeer, a lake considered to be relatively unpolluted.

It should be mentioned that in The Netherlands the majority of inland waterways and lakes are surrounded by retaining dikes (as well as are parts of the maritime coastline) where stones serve as a substrate for mussel attachment. Samples of 135 mussels each were selected randomly from this Lake Markermeer stock. The samples were immediately placed in rigid polyethylene baskets (average mesh size, ca. 1.5 cm^2) and suspended in the water column for 3 days at the site of collection. During this acclimation period, byssus attachment to the baskets and other mussels occurred, and clumps began to form. The samples were then transported in moist polyethylene bags to the exposure stations at the Dutch-German frontier (Lobith, Rhine River) and the Dutch-Belgian frontier (Eijsden, Maas River), among other locations. The samples were suspended from floating monitoring platforms where possible. Reference samples were left suspended at the Lake Markermeer location.

Following an exposure period of 6 weeks, samples were retrieved and thoroughly rinsed with ambient water at the exposure location to remove associated silt, pseudofeces, feces, and small organisms. A depuration period was considered to be of no special advantage, in agreement with Boalch et al. (1981); and samples were immediately deep-frozen in polyethylene bags at −20°C in the laboratory, with the exception of individuals which were used for scope for growth analysis and histopathological screening. Subsequent treatment of the mussel samples involved thawing at room temperature, careful draining of mantle water (20 min), removal of all soft tissues with a titanium scalpel, and mixing with a homogenizer modified with a titanium rotor shaft and cutting blades. Homogenates thus consisted of the pooled tissues of approximately 100 individuals. Subsamples of homogenates were stored in acid-cleaned glassware for metal analysis. Cd and Cu were determined by electrothermal atomic absorption spectrometry (ET-AAS) and Zn by flame atomic absorption spectrometry (F-AAS) using approximately 2-g homogenate, digested in H_2SO_4 and H_2O_2 and diluted with distilled water. Separate subsamples were used to determine dry weight (after 16 hr at 105°C) and ash content (after 4 hr at 600°C). This was done to express analytical results on an ash-free dry weight basis. Using merely a dry weight basis may introduce bias due to salinity variations in the tissue fluids. Pearson's coefficients of variation for Cd, Cu, and Zn analyses — determined by using a standard mussel homogenate — were 8.6, 12.2, and 8.2, respectively.

Scope for Growth (SFG)

SFG was used in the 1988 active biomonitoring study to determine physiological impacts of the contaminants. This measure is defined as the amount of energy per unit time available to an organism (or population) for somatic growth and gamete production under specific experimental circumstances. Details on the methodology for measuring SFG in mussels can be found in Bayne et al. (1985). Caged mussels were transported to the laboratory in a cool box, cleaned of epibiotic growth, placed in 90 L of continuously aerated tap water (not chlorinated) and fed with the green alga *Chlorella pyrenoidosa*. The mussels were allowed to acclimate to these conditions for 24 hr before SFG measurements were initiated. Feeding rates were determined by measuring the volume of water cleared of algal cells per hour (clearance rate). The concentration of *C. pyrenoidosa* used was about 20,000 cells per milliliter, and flow rates through the measurement chambers were set at 20 mL/min. Algal concentrations were determined with a particle data counter from samples recovered by peristaltic pumping from the inflow and outflow tubes of the measurement cells during 12 consecutive hours. For the clearance rate measurements, 10 mussels were placed in each of 8 identical 500-mL experimental chambers. Each chamber had a magnetic stirrer to keep the algal cells from settling. These same chambers were used immediately afterward for determining oxygen consumption and ammonia excretion rates. Experiments lasted for 4 hr under dark conditions with the chambers sealed off from the atmosphere and with no water flow. Oxygen concentrations were determined by micro-Winkler titration with spectrophotometric endpoint determination. Ammonia (NH_4-N) concentrations were measured according to Grashoff and Ehrhardt (1976). At the end of each 16-hr measurement run, fecal material was collected for the purpose of measuring absorption efficiency. Algae and feces were separately filtered over glowed and preweighed Whatman GF/C filters. After drying (2 hr, 90°C) and ashing (2 hr, 450°C) the absorption efficiency (e) was determined (Conover, 1966). Scope for growth was calculated (in J/hr) as follows:

$$SFG = E_A - (E_R + E_U) \quad (1)$$

$$E_A = eE_C \quad (2)$$

where: SGF = scope for growth
E_A = absorbed energy from algae in J/hr
E_C = energy consumed (filtered from ambient water) in J/hr
e = absorption efficiency (ability to absorb food energy)
E_R = energy respired in J/hr
E_U = energy excreted in J/hr

The absorption efficiency e, known as the Conover ratio, was calculated as follows:

$$e = \frac{F - E}{(1 - E)F} \quad (3)$$

where: F = ash-free dry weight:dry weight ratio for algae
E = ash-free dry weight:dry weight ratio for feces

Histopathological Screening

The general histological techniques employed to prepare the mussels used in the 1988 active biomonitoring study for histological sectioning can be found in Bancroft and Stevens (1977). All duplicate slides containing sections of two whole, individual zebra mussels were stained with Ehrlichs hematoxylin and eosin. The slides were examined microscopically and about 20 tissues, each from a different organ, were screened and allocated a score on a scale of 1 to 5. The score is a numerical value depending primarily on the cytological state of a tissue, including the general morphology of the relevant organ. The highest score is given for normal tissue, using reference material as a guide. Tissue/organ scores were averaged for each group of individuals to create a clear picture of condition in that organ within a population. The tissue quality scoring system is supported by a classified microphotographic library, which contains examples of various pathological conditions, ranked according to their apparent severity as reported in the literature for bivalve mollusks. The following tissues were examined: pharynx, stomach, digestive gland, style-sac, intestine, rectum/anus, gonad sacs, gametes and blood spaces, a range of muscle tissue (including the adductors and foot muscles), kidney, nerve tissue, byssal gland, gills, mantle epithelia, and short siphons. The criteria used for this cytological score are given below, while general background information can be found in Bayne et al. (1980).

Score	Description
5	This indicates a normal condition (i.e., epithelia, basement membranes, brush borders, and supporting connective tissue are regular in appearance; no inflammation is present and the whole area shows no unusual cytology.)
4	A slightly abnormal condition is present (i.e., a localized light inflammation, or small irregularity)
3	A moderately abnormal condition is present (i.e., inflammation; hyperplasia; or other irregularity covering a substantial area of tissue, such as a reduction in the area normally occupied by the tissue, atrophy, etc.)
2	A severely abnormal condition is present; the majority of the tissue is inflamed, atrophied, hyper/neoplastic or otherwise abnormal; a substantial part of the tissue or organ may show degeneration or atrophy
1	No normal tissue remains

A more general histopathological condition (health) index of individuals was also used for *D. polymorpha*. This index is based on the extent and severity of cytological abnormalities or lesions in the whole individual. This type of general index has been used by a number of authors in recent years (e.g., Peters and Yevich, 1989) and summarizes the health of invertebrates in a very general way according to the following criteria.

Index	Description
5.0	An individual showing no apparent tissue abnormalities; well nourished
4.5	An individual showing no obvious tissue abnormalities but where feeding condition is suboptimal (winter/postspawning)
4.0	An individual showing slight tissue abnormalities in one organ or part of a system (e.g., abnormal vacuolation or slight/local inflammation)
3.5	An individual showing moderate abnormalities/lesions in one organ or part of a system (e.g., inflammation, necrosis, etc.)
3.0	A definite pathological condition is recognizable, i.e., serious abnormalities in one organ, or a more diffuse general condition (e.g., lesions, inflammation, evidence of parasitation, starvation, etc.)
2.0	Individual exhibiting extensive pathology (e.g., lesions, neoplasia, extensive parasite infestation/infection); condition apparently critical, involving several organs and sometimes coupled with signs of starvation/wasting
1.0	Terminal case, substantial wasting of key tissues, little apparent function remains

Food Chain Experiment

Two groups of eight tufted ducks (*Aythya fuligula*) (four males and four females per group) were each held in cages of 45 m^2 and provided with a central pond of 8-m^2 surface area and 1-m depth. The pond was continuously replenished with filtered water from a neighboring canal at a rate of 200 L/hr, giving a water residence time of about 9.4 days. The ducks were fed daily with 500 g of zebra mussels each (wet weight, shells included), supplemented seasonally with grain products. The mussels were collected from Lake Markermeer (less contaminated populations) and from Haringvliet Basin (highly contaminated populations). Mussels were cleaned of debris and deep frozen for future use. A portion was defrosted daily for feeding to the ducks. The experiment started in summer 1982 and lasted until autumn 1985.

RESULTS

Cadmium Bioavailability

Figure 2 shows how pollutants may reach *Dreissena polymorpha* from different geochemical compartments (the water column, suspended particles, sediment, and interstitial water) and through different pathways in the ecosystem. *D. polymorpha*, being an epifaunal, sessile organism, is exposed to the direct influence of river/lake water, and in areas of deposition is subject to a constant rain of particles with adsorbed contaminants. Mussels may be found attached to hard objects projecting out of soft bottoms and as a result may be exposed to contaminated, fine-grained sediments. Furthermore, its food source — phytoplankton — may contain contaminants. The relative contribution of dissolved or particle-associated pollutants to the accumulation of such substances in zebra mussels remains, however, largely unknown. In an attempt to examine the relative contribution of various compartments to the body burden in *D. polymorpha* and other aquatic organisms, de Kock and Marquenie (1982) calculated correlation coefficients for pollutant concentrations in abiotic compartments and concentrations in organisms (Table 2).

The data were taken from a biogeochemical field study conducted in 1977, in which biotic and abiotic samples were simultaneously collected at about 50 locations in the Rhine and Maas Rivers and their downstream lakes (Rijkswaterstaat, 1984). As shown in Table 2, correlations between Cd in *D. polymorpha* and Cd in the four geochemical compartments are generally lacking. This led to the hypothesis that only specific Cd species are biologically available to the zebra mussel.

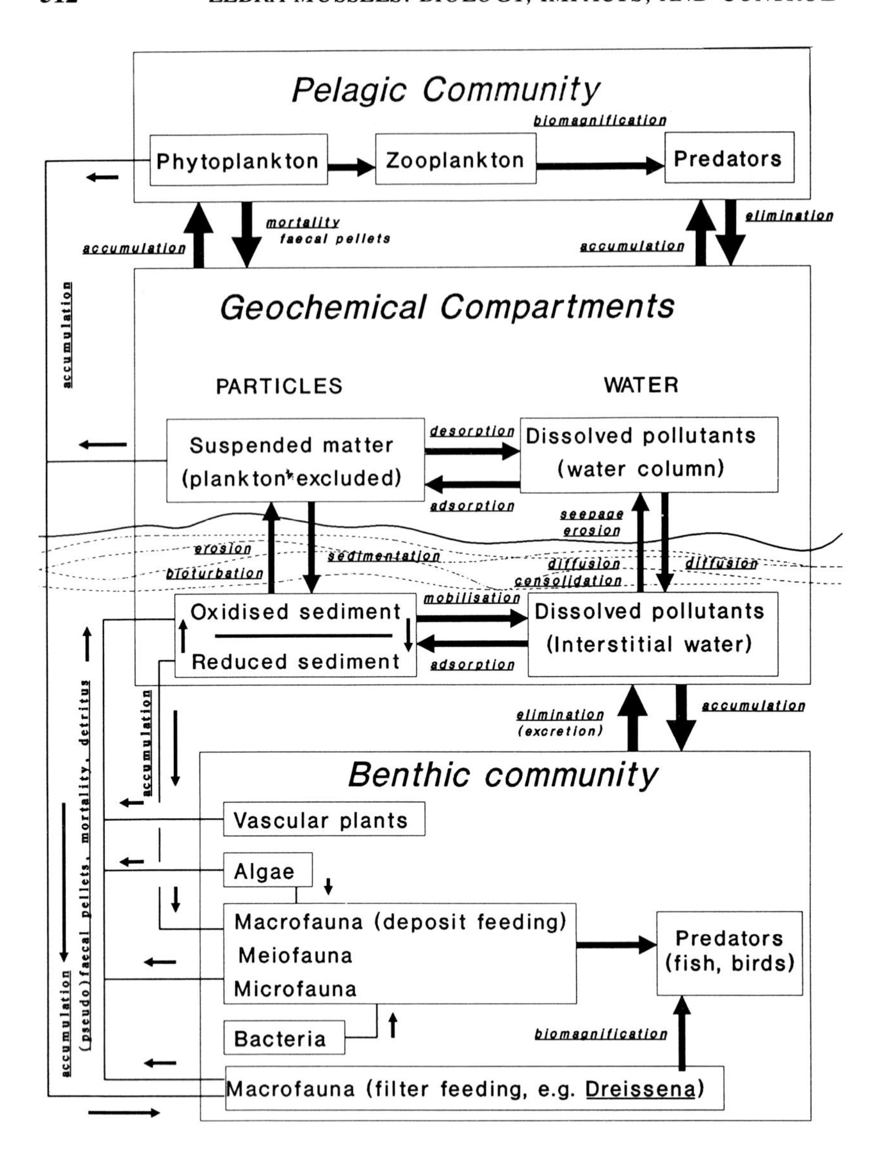

Figure 2. Schematic representation of the interactions between the biotic and abiotic compartments of aquatic systems, mediating the transfer of pollutants.

The idea of differential bioavailability of various Cd species has been tested in the field (Castilho et al., 1984) by using dialysis membranes which were exposed for a number (usually five or six) of 7-day periods, along with zebra mussels that were exposed for about 40 days (to reach steady state) at river and lake locations. Electroactive cadmium (Cd^{2+} ions or labile dissociating complexes), more resistant ozone-oxidizable organic Cd complexes,

TABLE 2. Coefficients of Linear Correlation for Cadmium Concentrations in Biotic and Abiotic Freshwater Compartments in The Netherlands, 1977

	Zp	D	A	G	Fw	Sm	Iw
Zp	—						
D	0.84* (37)	—					
A	0.54* (26)	0.97* (20)	—				
G	0.36 (19)	0.75* (16)	0.63* (13)	—			
Fw	0.18 (37)	0.31 (36)	0.37 (19)	0.59* (9)	—		
Sm	0.44 (17)	0.30 (10)	0.28 (11)	(2)	0.62* (18)	—	
Iw	−0.06 (46)	−0.01 (33)	0.01 (26)	−0.18 (18)	−0.18 (35)	−0.08 (18)	—
S	0.18 (48)	0.05 (35)	0.05 (26)	−0.01 (16)	0.45* (36)	0.63* (16)	−0.21 (45)

Source: de Kock and Marquenie, 1982, revised.

Note: Zp = zooplankton (pooled species); D = *Dreissena polymorpha* (bivalve mollusk filter feeder); A = *Anodonta anatina* (bivlave mollusk filter feeder); G = *Gymnocephalus cernua* (fish); Fw = filtrate water; Sm = suspended matter; Iw = interstitial water; S = sediment; * = significant at $p = 0.05$; number of paired observations in parentheses.

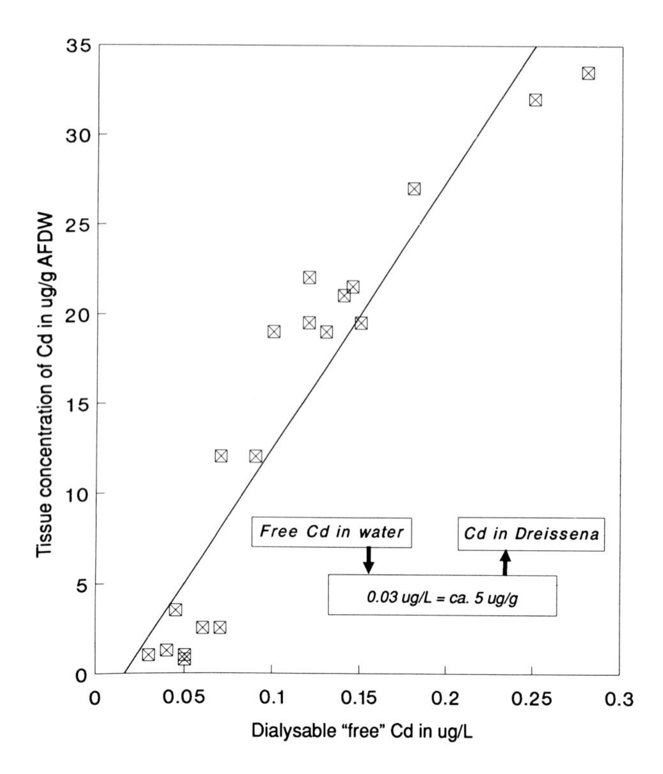

Figure 3. A linear regression showing the relationship between "free" ionic Cd in water, and bioaccumulation in *Dreissena polymorpha*. Body burden on the basis of ash-free dry weight.

and refractory Cd were separately determined by chemical fractionation of the dialysates by using differential pulse anodic stripping voltametry in addition to AAS analysis. It was found that the concentration of free electroactive Cd in the water column correlated well with levels found in *D. polymorpha* after about a 6-week exposure in an active biomonitoring experiment. In this experiment, caged samples were simultaneously transplanted from a comparatively unpolluted area to a site on the Rhine-IJssel Rivers-Ketelmeer-IJsselmeer Lakes trajectory (Figure 1). The relationship between electrochemically free Cd in the water column and steady-state concentrations found in *D. polymorpha* proved to be nearly linear (Figure 3). The study also showed that the electroactive fraction in the downstream Lake IJsselmeer was relatively small when compared to upstream river locations. In the lake a larger percentage of Cd appeared to be incorporated into stable complexes, a factor presumably related to longer residence times, the production of extracellular compounds by algal blooms, and higher pH values. This example of a biogeochemical study for Cd serves to emphasize that various coexisting pro-

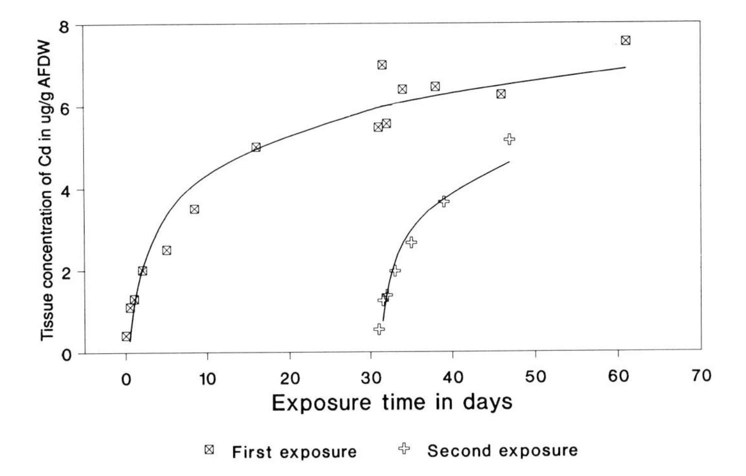

Figure 4. Accumulation of Cd in *Dreissena polymorpha* exposed at Kampen in the River IJssel, a branch of the Rhine River. The first group of mussel samples was exposed in September and was followed 31 days later by a second mussel group to determine the course of bioaccumulation over time (solid lines = log regression curve). Concentrations on the basis of ash-free dry weight.

cesses in the aquatic ecosystem will determine the actual toxic stress factor in the target organism (e.g., *D. polymorpha*). This stress factor is not some ambient external concentration, but rather the bioavailable internal load of toxic elements or compounds accumulated by the organism from its immediate surroundings over a period of time.

Cadmium Accumulation and Growth

To estimate the time required to achieve steady-state tissue levels of Cd, a batch of zebra mussel samples was exposed as a reference in an active biomonitoring study. After 31 days, a second batch of mussel samples was exposed to see if the accumulation rate and steady-state values would be similar. Each sample was randomly composed of 100–120 individuals of similar length (15–20 mm) and taken from a population living in the northern Lake IJsselmeer (Breezanddijk). At a location near Kampen on the IJssel River, Cd stress was found to be relatively low compared to more upstream locations on the Rhine River (e.g., at Lobith near the German-Dutch frontier). The concentration at steady state was about 6.5 μg/g ash-free dry weight. The accumulation occurred over a period of 45–60 days. As Figure 4 shows, both the first and the second test groups had similar accumulation patterns (with a rapid initial phase) followed by a gradual reduction in rate until steady state was reached. As a result of these experiments, a period of 40- to 60-day exposure was considered to be of sufficient length for routine biomoni-

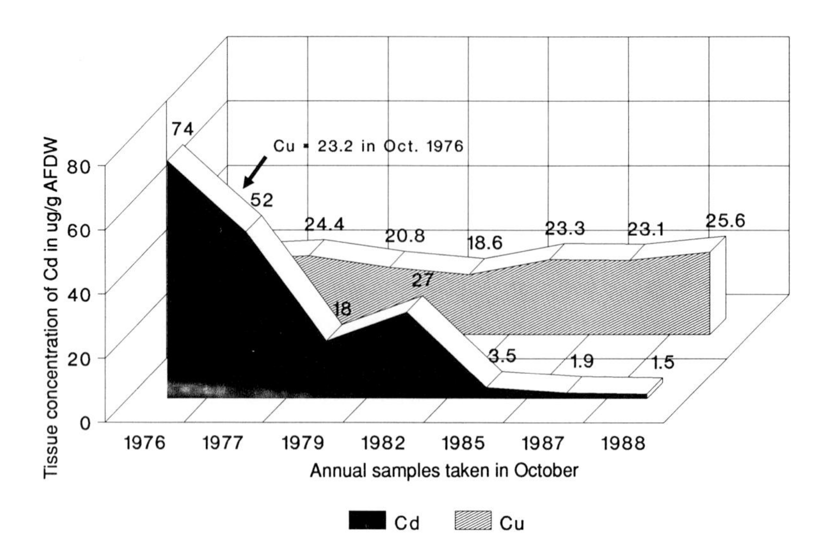

Figure 5. The bioavailability of cadmium and copper in Rhine River water at Lobith, as reflected by concentrations in tissues of natural and transplanted *Dreissena polymorpha* from 1976 to 1988. Concentrations are expressed in μg/g on an ash-free dry weight basis.

toring of Cd concentrations with zebra mussels, and has been adopted as a (6 week) standard ever since. Other heavy metals (Cu, Cr, Zn) usually take less time to accumulate to steady state. Similarly, the process of accumulation to steady state for some PCB congeners (PCB 101, -138, and -153) in comparable experiments with *Mytilus edulis* in the Scheldt estuary took about 45 days (de Kock, Unpublished data).

In practice, many such biomonitoring exercises have been conducted down through the years. This is illustrated in Figure 5, in which data on natural and transplanted populations of *D. polymorpha* have been combined to give an overview of Cd and Cu steady-state tissue concentrations measured close to the Dutch-German border at Lobith. It is immediately obvious that Cd inputs to the Rhine River have been drastically reduced since the 1970s, while the steady-state levels of Cu accumulated by *D. polymorpha* have remained essentially the same for more than 15 years. In 1976, Cd concentrations in mussels were in excess of 70 μg/g ash-free dry weight, dropping to about 20 μg/g in the early 1980s, and leveling off at 2–4 μg/g since 1985. What the situation was like before 1976 is not clear; however, it can be assumed that for a number of years at least, it was as bad, if not worse, in terms of general contamination.

Of a somewhat anecdotal nature is the observation that in 1976, when high Cd levels were present in the Rhine, numbers of *D. polymorpha* were apparently far lower. It was extremely difficult to find enough specimens at Lobith upon which to perform a chemical analysis (for which 2- to 5-g wet tissue weight was needed). This tends to agree with other measurements of the effect of high Cd levels on *D. polymorpha* presented below. The assumption being made here is that at internal Cd concentrations of about 70 μg/g ash-free dry weight (in the presence of other bioaccumulated elements and compounds), the survival chance of this species was severely reduced. This illustrates just how resistant this species is to environmental contamination.

An advantage of field exposure experiments is the feasibility of using growth as a meaningful toxicological stress parameter. Mean ash-free dry weight per individual (obtained as part of the chemical analysis of exposed mussel samples) provides a good measure of stress. In fact, the technical monitoring principles applied, i.e., the use of uniform, statistically similar mussel samples (age, size, random sample composition) at the outset of exposure experiments ensures ideal test populations for biological stress measurements. Figure 6 gives weight increments (growth factors) for zebra mussels exposed for 60 days at five river locations and one lake location in 1979. As the inset in Figure 6 shows, there was little difference in the phytoplankton food supply at the river locations, as measured by chlorophyll *a* levels. The Lake Ketelmeer location generally had higher phytoplankton concentrations despite the missing autumn peak value. As noted earlier, however, the bioavailability of Cd decreases during transport in the downstream direction toward the basins and lakes (Figure 1). This was found to be the case in the River IJssel (northern Rhine branch), with availability decreasing in the order: Lobith > Zutphen > Kampen and in the Waal River (western Rhine branch): Lobith > Tiel > Gorinchem. Weight increments of *D. polymorpha* showed a straight line relationship with Cd concentrations in soft tissues. Extrapolating the regression line through growth factor = 1 reveals a concentration of 42.5 μg/g at which growth would theoretically cease (see Figure 6). Although a direct dose-effect relationship for Cd was apparent, it may not be the only metal causing the reduction in growth; other pollutants in the Rhine system may have varied similarly to the Cd concentration in the animals, thereby adding up to an overall effect. The growth factor is defined as the ratio between the mean ash-free dry weight per mussel of the exposed sample and the mean ash-free dry weight of mussels collected randomly from the same natural stock at the place of origin when the exposure experiment was initiated (in this case 60 days earlier).

Cadmium and Scope for Growth

A regular, annual biomonitoring exercise commenced in 1987, which has been repeated annually and will continue for the forseeable future, to provide a specimen bank of *D. polymorpha* tissue homogenates reflecting water quality in the Rhine and Maas Rivers and their downstream basins. This specimen bank is available to third parties for research purposes.

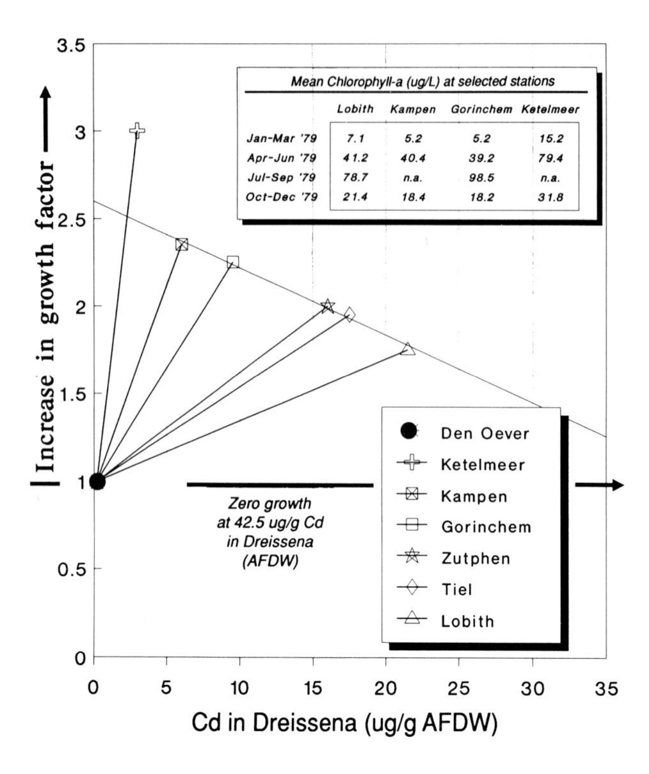

Figure 6. Growth of *Dreissena polymorpha* after 60-day exposure (expressed as a growth factor) is shown in relation to accumulated Cd burdens, observed at various Rhine River and IJssel River locations, and at the Den Oever collection site in 1979. Mean chlorophyll *a* concentrations (μg/L), indicative of the available food resources for *D. polymorpha* at selected locations, are also presented. n.a. = Not available. (Chlorophylla data from Ministry of Transport and Public Works, Rijkswaterstaat, quarterly water quality reports).

During the 6-week exposure of zebra mussels in the winter of 1988–1989, a dramatic increase of Cd concentrations in water samples was observed at the Eijsden monitoring station on November 23, 1988. This was later shown to be the result of an accidental Cd discharge 2 days earlier upstream in the Maas River. A maximum Cd concentration in the Dutch sector of the river of 100 μg/L was reported by van Vuuren (1989) in relation to this accident. Subsequently, the zebra mussel samples — which by pure coincidence were being exposed at Eysden (Dutch-Belgian border), Grave, Dussen, Haringvliet and Lake Markermeer reference site — were recovered on December 16, 1988; and the soft tissues were analyzed for Cd, Cu, and Zn (Table 3). In addition to the routine analysis of Cd, Cu, and Zn, the accident offered an opportunity to further investigate the possible relationship between growth and Cd concentration observed in earlier years (i.e., Figure 6). On the basis

TABLE 3. The Concentration of Metals (μg/g) in *Dreissena polymorpha* Exposed for 6 Weeks During Active Biomonitoring Experiments in the Rhine and Maas Rivers and in Lake Markermeer

	Metal		
Location	**Cadmium**	**Copper**	**Zinc**
Eijsden (Maas River)	21.80	20.8	386
Lobith (Rhine River)	1.51	25.6	298
Lake Markermeer	0.54	14.3	158

Note: Concentration given on an ash-free dry weight basis.

of experience with the blue mussel (*Mytilus edulis*) in estuarine areas of The Netherlands (de Kock, Unpublished data), part of the exposed zebra mussel samples were used for the determination of their scope for growth.

Results presented in Figure 7 are limited to zebra mussels, which were retrieved after a 43-day exposure from stations at Eijsden (Maas R.), Lobith (Rhine R.), and Lake Markermeer (control and place of origin). Cd levels in zebra mussels exposed in 1987 at Eijsden were less than 5 μg/g but increased to over 20 μg/g as a result of the spill in 1988 (Table 3). It is evident that clearance rates of zebra mussels (in triplicate groups of 10 mussels each) were all much lower than at the unaffected Lobith site on the Rhine, and the reference site in Lake Markermeer (Figure 7). Food adsorption efficiency, respiration, and excretion rate appeared to be similar in all three exposure groups. Scope for growth values of the Eijsden samples were all negative, indicating a lack of available energy for somatic or reproductive growth. In this study, an apparently short exposure (a few days at most) to about 100 μg/L of cadmium resulted in a residual tissue concentration of 21.8 μg/g some 25 days after the accident, coupled with a negative potential for growth. It should, however, be noted that the zinc concentration in mussels at Eijsden was also somewhat higher than in mussels at Markermeer and the Rhine River and may also have affected the SFG.

Histopathology

Mussels from five of the 1988 active biomonitoring locations were histologically sectioned to assess their general health and to carry out a histopathological screening of their tissues. The primary purpose was to look at any evidence of tissue damage which would verify the SFG results mentioned above, at least for the Eijsden individuals. The resulting data have been presented in Bowmer et al. (1991), and are briefly summarized below.

The histopathological condition or health index varied from 2.73 to 3.72 on a scale of 1 to 5 (Table 4). A population in winter condition is expected

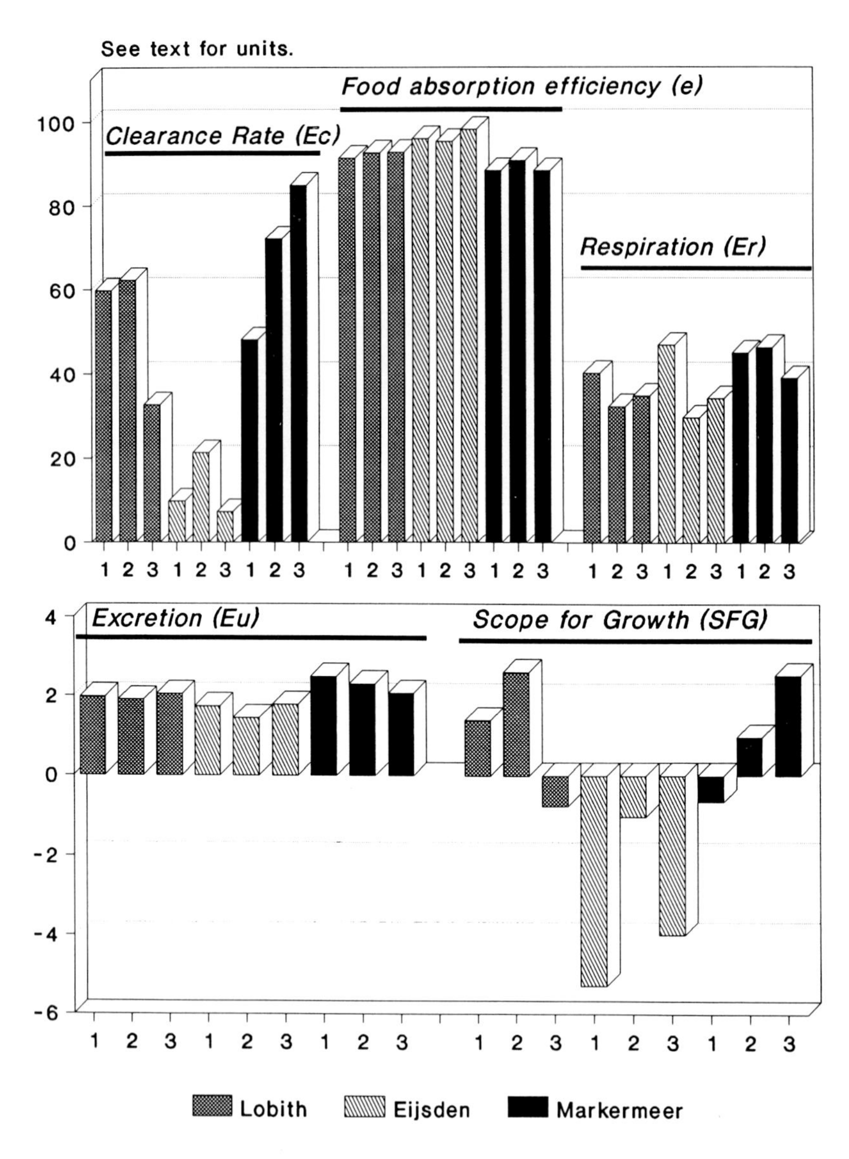

Figure 7. Physiological parameters measured in scope for growth experiments on *Dreissena polymorpha* exposed to an industrial Cd spill in the Maas River. Given are the clearance rate (in mL/hr per individual), food absorption efficiency (Conover ratio × 100), respiration ($\mu g\ O_2$/hr per individual), excretion ($\mu g\ NH_4$-N/hr per individual) rates, and the calculated scope for growth (J/hr per 10 individuals). The numbers on the x-axis indicate each of the 3 replicates, with each replicate consisting of 10 adult zebra mussels. The Eijsden site where the Maas enters The Netherlands was affected by the Cd spill, while the Lobith site where the Rhine enters The Netherlands was not. The Lake Markermeer site was used as a clean reference.

TABLE 4. Mean (±S.D.; n = 15–30) Histopathological Condition (Health Index) of *Dreissena polymorpha* Observed at Five Locations in The Netherlands During the Winter Season 1988–1989

Location	Passive Biomonitoring (Natural Population)	Active Biomonitoring (Transplanted Population)
Markermeer	3.72 ± 0.68	3.58 ± 0.33
Eijsden	—	3.47 ± 0.46
Grave	2.80 ± 0.44	3.13 ± 0.70
Dussen	2.73 ± 0.49	3.08 ± 0.43
Haringvliet	—	2.88 ± 0.54

Note: See text for derivation of values.

to score at least 3.5–4, based on previous experience with *D. polymorpha* (Bowmer, Personal observation). Mussels sampled at the control site in Lake Markermeer and mussels exposed at this site in a polyethylene basket for 6 weeks showed little difference in their health index scores (3.72 ± 0.68 vs 3.58 ± 0.33, respectively). Lowest mean scores were recorded among the natural mussels at Dussen (2.73) and Grave (2.80). These latter scores must be considered very low, reflecting a less than healthy condition. The active biomonitoring mussels exposed at these two sites showed reduced health index scores (3.08 and 3.13, respectively), relative to the Lake Markermeer active biomonitoring samples (3.58). In contrast, the active biomonitoring mussels from Eijsden, which contained the highest Cd concentrations, scored considerably better (3.47) than the rest of the exposed populations. The group exposed in Haringvliet Basin at the former mouth of the Maas and Rhine Rivers, now a freshwater lake sealed off by a dam, scored only 2.88. These results bear no relationship to the accumulated Cd concentrations in the same zebra mussel populations which in decreasing order were Eijsden > Grave > Dussen > Haringvliet (Bowmer et al., 1991). This general health index scale is nonlinear, because surviving individuals with a lower score are relatively scarce and presumably die off rapidly. Thus, the majority of populations examined to date generally have not had scores below 2.5.

The specific cytological conditions underlying these general health index scores have been treated elsewhere (see Methods), and Figure 8 shows the resulting mean cytological scores for the tissues of six key organs in each of 15–30 mussels from every exposure. It is apparent that the digestive gland or hepatopancreas had consistently lower scores than most of the other tissues. However, in winter this organ never appears in optimum cytological condition, often showing signs of degeneration of the smaller lobules and branches, with a generally ragged appearance of the digestive epithelium. Individual specimens in all the low-scoring populations, whether natural or active biomonitoring exposed, showed additional frequent lesions of the digestive gland; these ranged from degenerating lobules to entire branches, coupled with

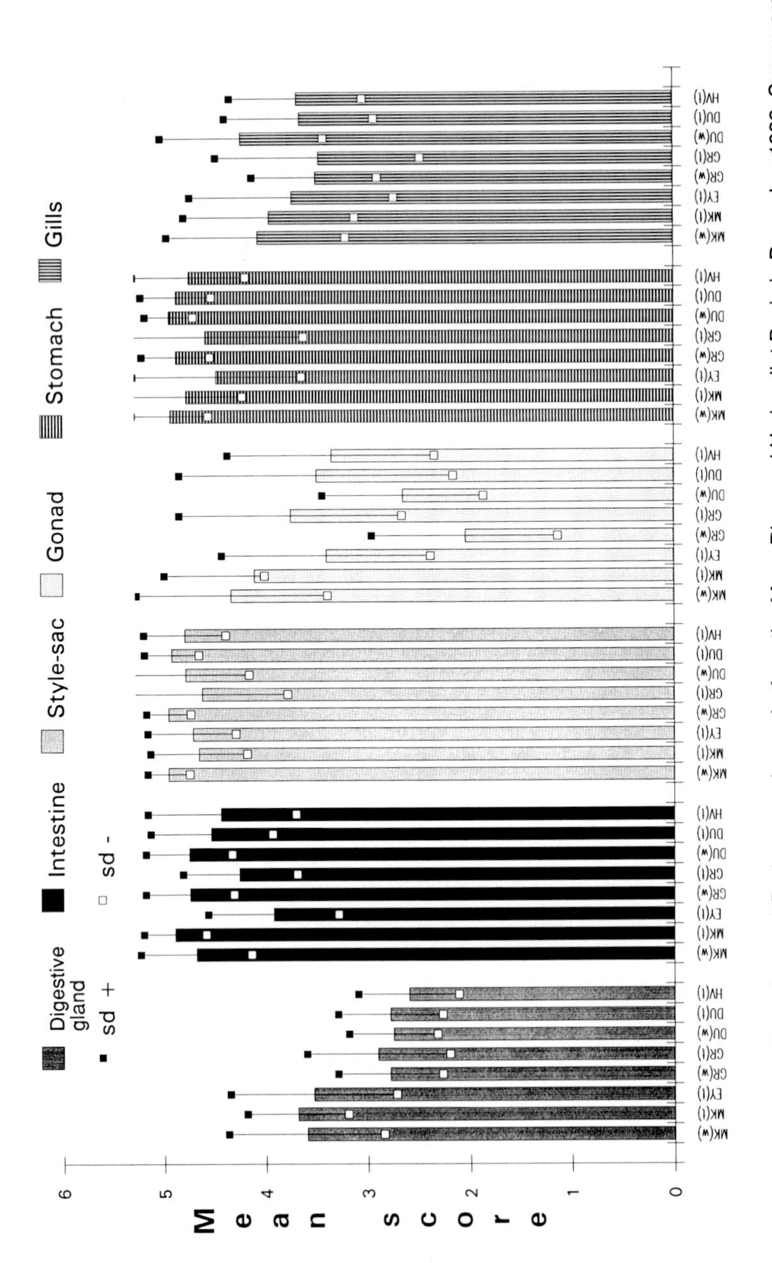

Figure 8. Cytological condition of natural *Dreissena polymorpha* from the Maas River and Haringvliet Basin in December 1988. Scores are based on a scale of 1 to 5, with a score of 5 being considered cytologically normal. The standard deviation is given by squares and vertical lines. MK = Markermeer, EY = Eijsden, GR = Grave, DU = Dussen, HV = Haringvliet; w = natural, t = transplanted.

extensive granulocyte and lymphocyte infiltration/inflammation, often of the entire organ. Only in a limited number of cases were clear signs of tissue repair and recovery seen. In the most severe cases (i.e., individual scores of 1 or 2), the entire digestive gland showed inflammatory lesions and large patches of wasted tissue, often involving adjacent organs. A mean score of ca. 3.5, as displayed by both Lake Markermeer and Eijsden populations, thus is not unusual and indicates no apparent cytological strain (at the light microscopy level) in this organ in winter.

Cytological scores of <3.0 at the remaining downstream stations (Dussen, Grave, Haringvliet) indicate a general trend toward loss of functionality of the digestive gland and indicate cause for concern. The gonad varied in cytological quality from relatively normal at the Lake Markermeer reference site to extremely poor among the Grave and Dussen natural populations, indicating a poor state of readiness for reproductive maturation during the following spring at these Maas River areas. The active biomonitoring populations exposed at these sites appeared to retain the cytological integrity of the gonads throughout their 6-week exposure. The gills are another variable organ in terms of cytological condition, and one which may be subject to direct damage by waterborne agents. Gill condition tended to be better in the Lake Markermeer populations and also in the Dussen natural population. Damage to this organ, which had been expected in the Eijsden population in the wake of the Cd spill in the Maas River, was not found. The intestine, stomach, and style-sac generally showed a healthy cytological condition in all the populations examined. Regrettably, the histological sections of whole mussels did not allow an adequate view of the kidney, nervous system, or heart area for assessment.

The most consistently occurring parasite infestation in all groups was tentatively described as an ''unidentified protozoan'' which occurred in small numbers (one to six individuals per section) in the digestive gland, and was largely restricted to the tubes leading into the smaller lobules. These large elongate cells (ca. 50–350 μm long) appeared to possess a large dark-staining nucleus and a highly vacuolar cytoplasm. No immediate damage was seen in the immediate vicinity of these parasites, and their presence did not correlate with the observed lesions of the digestive gland. The control group had a near 90% infestation rate, and the rest varied from 60 to 95%. The only group which was free of this parasite were those exposed at Eijsden. However, in a later study of young zebra mussels from the Haringvliet area, considerable digestive gland degeneration in the presence of this parasite was noted (Bowmer et al., In preparation). A broad infestation of one individual with trematode larvae was noted; and a single incidence of an organism that consisted of spherical groups of 10–15 dark-staining rounded cells (reminiscent of an Ascetosporian protozoan), which had invaded the entire blood system, was also observed.

Contaminant Transfer to Diving Ducks

Historically, the interest of water management authorities in the environmental risks posed by toxic pollutants arose from alarm signals among higher trophic levels (endotherms) (Carson, 1962; Risebrough et al., 1970). The results of research by Koeman and van Genderen (1972) became well-known in The Netherlands, demonstrating the relationship between the industrial point source of a number of chlorinated hydrocarbons (particularly the insecticide telodrin), and populations of the sandwich tern (*Sterna sandvicensis*, a fish predator) and the eider duck (*Somateria mollissima*, a shellfish predator) in the Dutch Wadden Sea some 90 nautical miles from Rotterdam Harbor and ca. 40 transport days further along the coastal residual currents.

Freshwater systems such as Lake IJsselmeer and Haringvliet Basin (Figure 1) which also act as sediment traps (Table 1) received little attention until about 1980, although these systems are internationally recognized as important waterfowl areas for resident as well as migratory birds. The production capacity of these hypertrophic waters has become beneficial to large numbers of migratory freshwater bird species since the separation of these basins from the sea in previous decades. According to current local estimates, the maximum numbers of diving ducks overwintering on open waters in the Lake IJsselmeer area are about 30,000–50,000 *Aythya ferina* (pochard); 85,000–115,000 *Aythya marila* (scaup); and 85,000–115,000 *Aythya fuligula* (tufted duck). This accounts for about 15–20% of the total migratory pochard and tufted duck populations in northwest Europe, and more than 50% of the scaup population (data from government reports 1978, 1981). Scaups and tufted ducks are voracious predators of zebra mussels in the shallow (usually <5 m) downstream freshwater lakes of The Netherlands. Between 80 and 95% of the diet of adult tufted ducks in Lake IJsselmeer consists of zebra mussels.

Given the known ability of *D. polymorpha* to accumulate contaminants, food chain transfer of metals and organic contaminants could constitute a significant risk to predatory duck species that depend on zebra mussels for their nourishment. Indeed a decade ago, elevated concentrations of Cd and higher chlorinated PCB (polychlorinated bipheryl) congeners (PCB 138, -153, and -180) were found in the liver tissue of adult tufted ducks ensnared in commercial fishing nets in Lakes IJsselmeer and Markermeer (Table 5). The levels found were indicative of polluted environments, although these locations are regarded as being relatively clean. This set of data, however, did not allow conclusions to be drawn regarding dose-effect relationships.

Haringvliet Basin, which as already mentioned acts as a sediment trap and is immediately influenced by the Rhine-Maas pollutant input, is also likely to have contaminant inputs. Haringvliet Basin also contains a large biomass of *D. polymorpha* which could be used as an experimental food source.

TABLE 5. Mean (±S.D.) Concentrations of Cadmium and Some Higher Chlorinated PCB Congeners in Liver Homogenates of Three Diving Duck Species in the Lake IJsselmeer-Lake Markermeer area, December 1979 to March 1981

Species	Cd (μg/g)	PCB-138 (μg/kg)	PCB-153 (μg/kg)	PCB-180 (μg/kg)
Aythya fuligula (tufted duck) n = 47	2.5 ± 1.8	199 ± 128	310 ± 227	167 ± 128
Aythya ferina (pochard) n = 9	1.2 ± 1.2	122 ± 47	166 ± 50	109 ± 38
Aythya marila (scaup) n = 9	2.7 ± 2.3	196 ± 121	354 ± 390	173 ± 190

Source: Original data (de Kock), not previously published.

Note: Concentrations given on an ash-free dry weight basis.

A modified field experiment was set up in summer 1982 in order to investigate the extent of pollutant transfer from zebra mussels to tufted ducks and associated biological effects. The experiment started with juvenile birds (age 3 months) and lasted until autumn 1985 when the animals were 3 years of age. The Haringvliet (H) group received zebra mussels that were fished by beam trawl from Haringvliet Basin, the other group (M) was fed with mussels caught in the lesser polluted Lake Markermeer (Figure 1), which is protected from contaminated river water by dikes. The difference in pollution degree between the two areas is reflected in the relative concentrations of PCB congeners, DDE (1,1 dichloro-2,2 dichlorophenyl ethene, the principal metabolite of DDT), and hexachlorobenzene (HCB) in the total soft tissue homogenates of *D. polymorpha* (Figure 9). Cd concentrations in the soft tissue homogenates were 0.47 and 5.00 μg/g in the Markermeer and Haringvliet populations, repectively.

One half of the duck group (four individuals) in each cage was sacrificed in July 1983 after a 10-month exposure; and the other half, in October 1985 after a 37-month exposure to a mussel diet. Animals were dissected for general histopathological screening and analysis of liver tissue for heavy metal (including Cd), and organochlorine compounds (PCB, DDE, HCB). No gross histopathological abnormalities were observed, but differences in chemical stress were clearly demonstrated (Table 6 and Figure 9). Higher pollutant concentrations were found in livers of ducks fed on zebra mussels from Haringvliet Basin. Coincident with the higher liver pollutant load were somewhat lower mean weights of body, liver, and (especially) kidney (Table 7).

Organochlorine contaminant concentrations in H group livers (female) after a 37-month exposure, along with concentrations of the same compounds detected in the eggs, are shown in Figure 10. These eggs were laid in 1985 after a 34-month exposure of the mother ducks to contaminated zebra mussels.

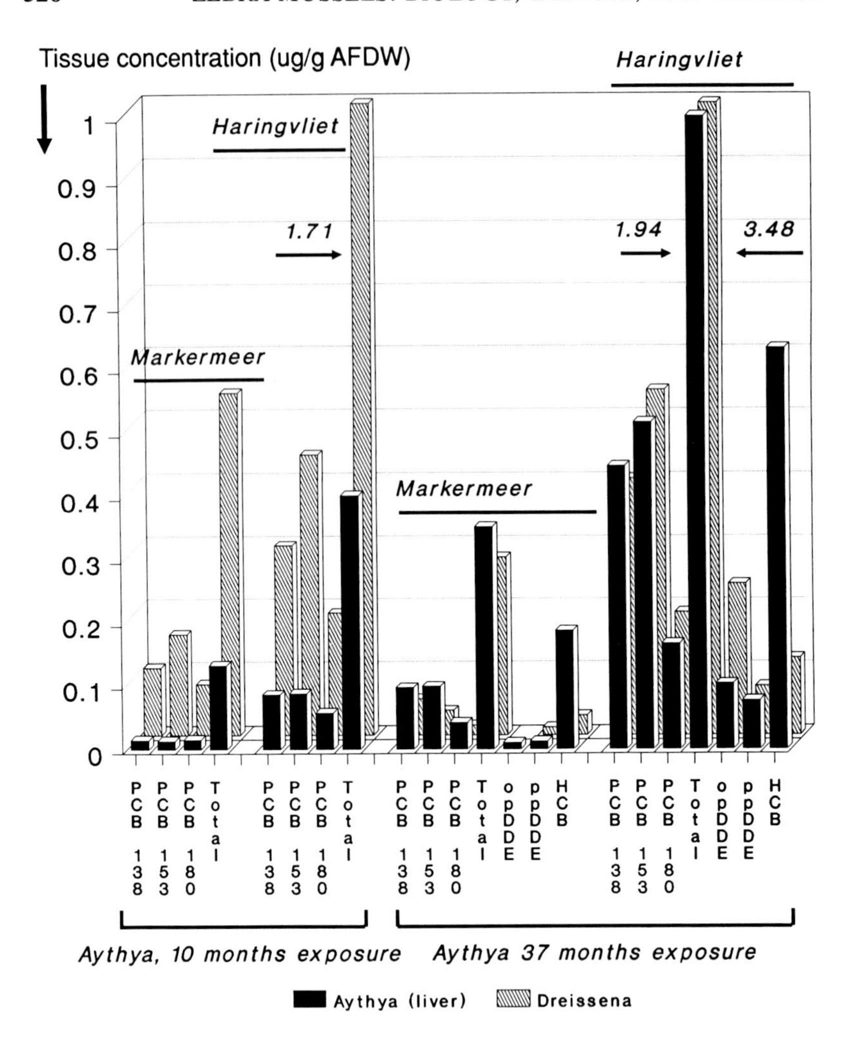

Figure 9. Concentrations (in μg/g) of PCB-138, -153, -180, total PCBs, o,p-DDE, p,p′-DDE, and HCB bioaccumulated in *Dreissena polymorpha* and the resultant bioaccumulation in the tufted duck *Aythya fuligula* fed on a diet of these mussels. Two groups of ducks were examined: one fed on mussels from the relatively unpolluted Lake Markermeer and the other fed on mussels from Haringvliet Basin. Note levels (right) accumulated in the ducks after 37 months on a diet of Haringvliet mussels. Concentrations in *Dreissena polymorpha* are based on total body burden (ash-free dry weight), and those in ducks are based on liver homogenates (dry weight).

Whereas the Cd concentrations in the eggs were low, indicating little or no maternal transfer (notwithstanding the absence of M group control data, Table 6), those of the PCB congeners in the H group were higher by a factor of 2.1–4.2 when compared to levels in the mother's liver. The same result applied to o,p- and p,p-DDE (factor 4.3–4.8), but the result for HCB was

TABLE 6. Mean (±S.D.) Cadmium Concentrations in Maternal Liver Tissues and Eggs of the Tufted Duck *Aythya fuliga* after 10- and 37-month Exposures to Zebra Mussels as a Food Source from Lake Markermeer and Haringvliet Basin

Location	Maternal Livers	Eggs
Markermeer	0.39 ± 0.17 (at 10 mo)	n.a.
	0.73 ± 0.18 (at 37 mo)	n.a.
Haringvliet	1.46 ± 0.26 (at 10 mo)	n.d.; <0.025 (at 22 mo)
	1.94 ± 0.37 (at 37 mo)	n.d.; <0.030 (at 34 mo)

Note: Concentrations given in μg/g on a dry weight basis; n.a. = not analyzed; n.d. = not detectable.

TABLE 7. Mean (±S.D.) Weight Characteristics of *Aythya fuliga* After 37-month Exposure Time to Their Respective Food Sources from Lake Markermeer and Haringvliet Basin

Location	Total Wet Weight[a]	Liver Wet Weight[a]	Kidney Wet Weight[a]
Markermeer	582 ± 33	13.5 ± 0.6	4.7 ± 0.5
Haringvliet	547 ± 38	12.7 ± 0.5	3.6 ± 0.4

[a] Weights given in grams.

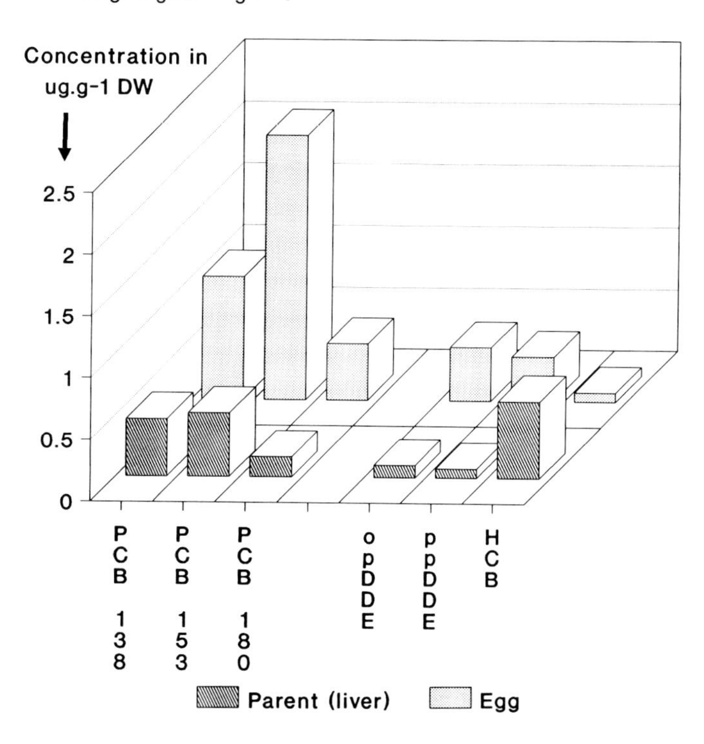

Figure 10. Concentrations (μg/g) of PCBs, DDE, and HCB in liver homogenates and eggs of the tufted duck (*Aythya fuligula*) fed with contaminated zebra mussels from Haringvliet Basin during the period from 1982 to 1985. Concentrations given on a dry weight basis.

TABLE 8. Overview of the Reproductive History of Groups of the Tufted Duck *Aythya fuliga* Fed with Zebra Mussels from Lake Markermeer and Haringvliet Basin During 1982–1985

	Location			
	Markermeer		Haringvliet	
Parameter	**22 mo**	**34 mo**	**22 mo**	**34 mo**
Number of egg-laying females	1	2	2	2
Number of eggs	7	17 (8 + 9)	11 (5 + 6)	11 (5 + 6)
Mean egg length (mm)	n.a.	60.12	n.a.	57.54
Mean egg weight (g)	n.a.	50.11	n.a.	46.96
Number of eggs naturally reared by parents	7	8	None	None
Number of eggs in incubator (abandoned nests)	0	8	10	7
Embryo mortality	0	0	9	1
Number of eggs for chemical analysis	0	1	9	4
Number of chicks hatched	7	16	1	6

Note: Exposure time given in months (mo); n.a. = not analyzed.

less conclusive. Furthermore in the eggs of the M group, all organochlorine compounds were below the detection limit of the GC-MS (gas chromatography-mass spectrometry) mode used. The combined results strongly suggest that the maternal transfer of chlorinated hydrocarbons to the eggs led to a reduced reproductive success resulting from a high embryonal mortality rate. PCBs are especially suspect in this regard, considering the results of Reijnders (1986), who conducted a similar experiment with common seals fed in captivity with fish from polluted coastal waters. More research is needed with respect to the allocation of hydrophobic organochlorine compounds to the developing egg and the dynamics of such processes.

Ducks became fertile after 2 years in captivity and showed pair forming, courtship behavior, and nest-building activities. However, in contrast to the M group (control), the H group usually showed more nervous behavior, built untidy nests, and neglected their eggs once they were laid.

Table 8 shows that two females in the H group produced a total of 22 eggs (5–6 per nest) in 1984 and 1985. Of these, 17 fertilized eggs were artificially incubated with a breeding success of 7 hatched chicks (about 40%), and an embryo mortality of 10 (about 60%). In contrast, three females in the M group produced a total of 24 eggs, of which 23 were either naturally reared by the mother ducks or incubated, leading to a final breeding success of 100%. One egg from this group was purposely used for chemical analysis. After 10 months, the levels of chlorinated hydrocarbons (Figure 9) and Cd (Table 6) in the duck livers were elevated in the H group as compared to the M group. After a 37-month exposure, concentrations of Cd and higher chlorinated PCB congeners were higher than found after 10 months. These contaminants therefore seemed to be slowly accumulating over time.

DISCUSSION

Dreissena polymorpha was scarce in the Rhine River at Lobith in the mid-1970s, presumably due to environmental pollution (possibly acting in concert with lowered oxygen levels; de Kock, Personal observations). Bioaccumulated cadmium concentrations in excess of 70 μg/g were recorded in 1976 at Lobith, reflecting part of a body burden which included a wide spectrum of organic and inorganic contaminants. The fact that some individuals could still be found suggests just how resilient this species can be. The studies reported here show that at an internal body burden of only 21.8 μg/g of Cd alone, the calculated scope for growth was negative, predicting that there was no energy available for somatic or reproductive growth. Scope for growth has been used in connection with fish production studies, but has recently received attention as a general physiological parameter with respect to the use of invertebrates in pollution biomonitoring. In marine mussels, SFG measurements have been shown to correlate with bioaccumulated pollutant levels in field studies (Widdows et al., 1981a). However, natural variables (e.g., spawning condition, concentrations of food, suspended matter, salinity, etc.) may interact unfavorably and mask the effects of contaminants. We do know that the accumulated Cd concentration of 21.8 μg/g occurred during an exposure period of a few days duration (at the most), at concentrations of 100 μg/L Cd in water (van Vuuren, 1989); the short exposure may explain the lack of histopathological effects.

Tending to confirm the scope for growth data for *D. polymorpha* are the actual growth studies carried out in the Rhine and IJssel Rivers in 1979, which show a real reduction in somatic growth of some 30% at bioaccumulated levels of ca. 20 μg/g Cd and the presence of other contaminants. Extrapolation of the data shows a total inability to grow at ca. 40 μg/g Cd. Other contaminants present at the time were metals such as Cu, Zn, and Hg. However, these other metals were never identified as a particular problem in the same series of studies. Although concentrations at river locations were generally somewhat higher than in downstream reservoirs, the levels of Cu and Zn accumulated in *D. polymorpha* at Lobith (where the Rhine enters The Netherlands) remained roughly similar from 1976 to 1988. The role of organic contaminants in the observed stress and the resulting strain remains an uncertain factor.

In a laboratory study, Hemelraad et al. (1986a and 1986b) showed that the freshwater clam *Anodonta cygnea* bioaccumulated Cd to steady state at a level of ca. 60 μg/g after about a 70-day exposure to 25 μg/L dissolved Cd. This bioaccumulation was accompanied by gradual reduction in body dry weight. Although the animals were not fed, it seemed quite possible that a gradual breakdown of the carbohydrate metabolism contributed to the body weight reduction. The authors further noted that mortality was low up to 100 days, but "strongly increased" thereafter. They also showed similar general patterns of Cd accumulation in *Anodonta anatina* and *Unio pictorum*. A comparison between the studies of Hemelraad et al. (1986a and 1986b) on *Anodonta cygnea* and the field data for *D. polymorpha* indicates that the zebra mussel does not have any special tolerance to modern pollution levels, but rather has tolerance levels to Cd (one of the most common and toxic metals) that are similar to other widely distributed and highly successful exploiters of soft substrates. Levels of PCBs, DDE, and HCB in the tissues of *D. polymorpha* from Haringvliet Basin, measured during the tufted duck experiments, were considerably elevated. The reason for such levels is that this is a sedimentation area and organic particles with adsorbed organic contaminants tend to settle out, providing a local source of pollution. In turn, significant histopathological abnormalities in zebra mussels were found in this area (Bowmer et al., 1991).

What began as a local study of contaminant tranfer in a food chain is perhaps of wider importance in the context of the zebra mussel problem in North America. The results of trophic transfer studies with the tufted duck *Aythya fuligula* (e.g., behavioral disturbances of mature adults, growth retardation, and embryonal mortality as a consequence of pollutant transfer from zebra mussels and the reallocation of hydrophobic chlorinated hydrocarbons to developing eggs) should be taken seriously, since such effects may indeed occur in the field. It is quite conceivable that the European population of diving ducks feeding in wintertime on zebra mussels (which occur abundantly in large polluted downstream freshwater lakes in The Netherlands) have become subject to anthropogenic chemical stress, with potential reproductive disturbance and weakened population dynamics as a result. More data are required, especially on PCB uptake and allocation dynamics in the female birds. We tentatively suggest that zebra mussels in The Netherlands may play a role in the subsequent breeding success of migratory *Aythya* species in the Holarctic and Palearctic environment. Resident *Aytha* species in polluted sedimentary areas of The Netherlands (about 8000 breeding pairs of the tufted duck) run an even larger risk. In the absence of field data, this must remain largely speculative in nature. However, the arrival of *Dreissena polymorpha* in the Great Lakes and inland waterways of North America is probably permanent, given the European experience of the last two centuries. It is logical to assume that some part of the waterfowl community will begin to exploit this new resource. Given the capacity of *D. polymorpha* to accumulate contaminants (in general) and its ideal position as a prey item along the banks of water bodies (a role not filled by the larger burrowing clams which are substantially protected by the sediment in shallow water), it would be wise to carefully examine the pathways and flux of organic contaminants in this (new) part of the ecosystem.

This chapter was meant to convey some or our experiences with *D. polymorpha* as a tool in toxicological field studies. This species is hardy enough to survive moderate pollutant stress and experimental treatment, but at the same time reacts sensitively enough to be used as a biomonitoring species for studying the effects of chemical stress at the biochemical, cytological, histopathological, and physiological level of biological organization. It is exactly this ability to survive which allows us to measure the degree of strain caused by pollutants in the ecosystem. Furthermore, the combination of its dominant filter-feeding role in various freshwater ecosystems and its handiness in experimental toxicological field studies, including the trophic transfer of pollutants to waterfowl, deserves further attention because of the large spatial and temporal scales of pollution effects that may be involved.

ACKNOWLEDGMENTS

The authors acknowledge their deep debt of gratitude for the help and advice of Dr. G. van Urk (recently deceased) as biological coordinator of the Dutch ZMAS heavy metal program in the 1970s and early 1980s. We thank colleagues old and new, whose work we have used: especially Drs. J. M. Marquenie, who participated in many of the ZMAS experiments; Dr. Zhengqiang Shao, who more recently carried out the SFG measurements; and Drs. M. C. Th. Scholten and Mr. E. Foekema, who freely allowed us to use the tufted duck material. Furthermore, we give our thanks to the team in Den Helder, especially Mr. G. Hoornsman, Mr. H. van het Groenewoud, Mrs. M. van der Meer, Mr. P. Roele, and Mr. B. Schrieken. Thanks are also due to Marie Cecile Hendriks for typing the manuscript.

REFERENCES

Bancroft, J. D., and A. Stevens. Theory and Practise of Histological Technique (Edinburgh, England: Churchill Livingstone, 1977).

Bayne, B. L., D. A. Brown, K. Burns, D. R. Dixon, A. Ivanovici, D. R. Livingstone, D. M. Lowe, N. M. Moore, A. R. D. Stebbing, and J. Widdows. *The Effects of Stress and Pollution on Marine Animals* (New York: Preager Publishers, 1985).

Bayne, B. L., D. A. Brown, F. Harrison, and P. D. Yevich. "Mussel Health," in *The International Mussel Watch* (Washington, D.C.: National Academy of Sciences, 1980), pp. 163–236.

Boalch, R., S. Chan, and D. Taylor. "Seasonal Variation in the Trace Metal Content of *Mytilus edulis*," *Mar. Pollut. Bull.* 12:276–280 (1981).

Bowmer, C. T., M van der Meer, and M. C. Th. Scholten. "A Histopathological Analysis of Wild and Transplanted *Dreissena polymorpha* from the Dutch Sector of the River Maas," *Comp. Biochem. Physiol.* 100:225–229 (1991).

Carson, R. *Silent Spring* (Greenwich, CT: Fawcett Publications, 1962) (Crestbook 681, Reprinted 1982, London, England: Pelican Books Publishers, 1982).

Castilho, P. del, R. G. Gerritse, J. M. Marquenie, and W. Salomons. "Speciation of Heavy Metals and the In-situ Accumulation by *Dreissena polymorpha*: A New Method," in *Complexation of Trace Metals in Natural Waters*, C. J. M. Kramer and J. C. Duinker, Eds., (The Hague, The Netherlands: Martinus Nijhoff/Dr. W. Junk Publishers, 1984), pp. 445–448.

Conover, R. J. "Assimilation of Organic Matter by Zooplankton," *Limnol. Oceanogr.* 11:338–354 (1966).

de Kock, W. Chr. "Monitoring Bio-available Marine Contaminants with Mussels (*Mytilus edulis* L.) in The Netherlands," *Environ. Monitor. Assess.* 7:209–220 (1986).

de Kock, W. Chr., and J. Kuiper. "Possibilities for Marine Pollution Research at the Ecosystem Level," *Chemosphere* 10:575–603 (1981).

de Kock, W. Chr., and J. M. Marquenie. "The Effects of Discharges of Certain Metals and Organochlorine Compounds such as PCBs and Pesticides in Marine Ecosystems," Commission of The European Communities, EEG 4/81/81 Final Report, TNO Report CL82/97 (1982).

Ellis, A. E. "British Freshwater Bivalve Mollusca," Synopses of the British Fauna, No. 11, published for the Linnaean Society of London, (London, England: Academic Press, 1978).

Grasshoff, K., and M. Ehrhardt. "Automated Chemical Analysis," in *Methods of Seawater Analysis*. K. Grasshoff, Ed. (Weinheim, Germany: Verlag Chemie, 1976), pp. 263–298.

Hemelraad, J., D. A. Holwerda, and D. I. Zandee. "The Pattern of Cd Accumulation in *Anodonta cygnea*," *Arch. Environ. Contamin. Toxicol.* 15:1–7 (1986a).

Hemelraad, J., D. A. Holwerda, K. J. Teerds, H. J. Herwig, and D. I. Zandee. "A Comparative Study of Cadmium Uptake and Cellular Distribution in the Unionidae *Anodonta cygnea, Anodonta anatina* and *Unio pictorum*," *Arch. Environ. Contam. Toxicol.* 15:9–21 (1986b).

Holdgate, M. W. "Closing Summary," in *Marine Ecology and Oil Pollution,* J. M. Baker, Ed. (Limited, England: Applied Science Publishers, 1976), pp. 525–535.

Koeman, J. H., and H. van Genderen. "Tissue Levels in Animals and Effects Caused by Chlorinated Hydrocarbon Insecticides, Chlorinated Biphenyls and Mercury in the Marine Environment Along The Netherlands Coast," in *Marine Pollution and Sea Life,* M. Ruivo, Ed. FAO (1972) (England: Publ. Fishing News (Books) Limited, 1972), pp. 527–588.

Kuiper, J. "Marine Ecotoxicological Tests: Multispecies and Model Ecosystem Experiments," in *Ecotoxicological Testing for the Marine Environment, Vol. 1.* G. Persoone, E. Jaspers, and C. Claus, Eds. (Bredene, Belgium: State University Ghent and Institute of Marine Scientific Research, 1984).

Peters, E. C., and P. P. Yevich. "Histopathology of *Ceriantheopsis americanus* (Cnidaria:Ceriantharia) Exposed to Black Rock Harbour Dredge Spoils in Long Island Sound," *Dis. Aquat. Organisms* 7:137–148 (1989).

Reijnders, P. J. H. "Reproductive Failure in Common Seals Feeding on Fish from Polluted Coastal Waters," *Nature (London)* 324:456–457 (1986).

Risebrough, R. W., J. Davis, and D. W. Anderson. "Effects of Various Chlorinated Hydrocarbons," *Oreg. State Univ. Environ. Health Sci. Ser* 1:40–53 (1970).

Rijkswaterstaat. ''Zware metalen in aquatische systemen, geochemisch en biologisch onderzoek in Rijn en Maas en de daardoor gevoede bekkens,'' *Overzicht van het onderzoek en conclusies, 6 bijlagen,* E. H. Hueck-van der Plas, Ed. Rijkswaterstaat: RIZA en Deltadienst i.s.m. MT-TNO, IB en WL. (1984) (In Dutch).

Salomons, W., and W. G. Mook. ''Biogeochemical Processes Affecting Trace Metal Concentrations in Lake Sediments (IJsselmeer, The Netherlands),'' *Sci. Total Environ.* 16:217–229 (1980).

Scholten, M. C. Th., E. Foekema, W. Chr. de Kock, and J. M. Marquenie. ''Reproduction Failure in Tufted Ducks Feeding on Mussels from Polluted Lakes,'' in Proceedings 2nd European Symposium on Avian Medicine and Surgery, Dutch Association of Avian Veterianarians, Eds. Utrecht, The Netherlands (1989) 365–371.

van Vuuren, W. E. ''Cadmium calamiteien op de Maas: een triest record in November 1988,'' *Water* 22:23–25, 35 (1989) (in Dutch).

Widdows, J., D. K. Phelps, and W. Galloway, ''Measurement of Physiological Condition in Mussels Transplanted Along a Pollution Gradient in Narragansett Bay,'' *Mar. Environ. Res.,* 4:181–194 (1981a).

SECTION IV

Mitigation

CHAPTER 31

Monitoring and Control of *Dreissena polymorpha* and Other Macrofouling Bivalves in The Netherlands

Henk A. Jenner and Joke P. M. Janssen-Mommen

To date, three fouling bivalves (the zebra mussel, *Dreissena polymorpha*; the Asian clam, *Corbicula fluminea*; and the brackish water mussel, *Mytilopsis leucophaeta*) are present in The Netherlands. Of these three species, the zebra mussel is the most troublesome and control is often necessary. Settlement, growth, and effectiveness of control of the zebra mussel within water systems can be detected and followed with a settlement monitor. In addition, the behavior of zebra mussel valve movements is a useful monitoring method that shows toxic effects during chemical control. At present, there are six primary methods used to control macrofouling bivalves in The Netherlands; these include: chlorination, surface coatings, heat treatment, drying, water velocity, and microsieves. In most instances, control of bivalves at a particular water pumping facility relies primarily on one of these six methodologies. However, the primary method of control may be and often is supplemented with additional control methods that are more efficient based on the facility, time of year, and mussel life stage than the primary control method. This use of integrated control technology has subtantially reduced macrofouling problems caused by bivalves in The Netherlands.

0-87371-696-5/93/$0.00 + $.50

INTRODUCTION

Zebra Mussel

The introduction of the freshwater zebra mussel, *Dreissena polymorpha*, in the Great Lakes of North America will have severe consequences for raw water users (Mackie et al., 1989; Griffiths et al., 1989). In Europe, zebra mussels have been severe biofoulers for many years (Wilhelmi, 1923; Mikheev, 1961; Kirpichenko et al., 1962; Gillet and Micha, 1985). These problems originate from the mussel attachment to any solid substrate by byssal threads (Mattice, 1984). In natural waters, the amount of suitable substrate for settlement of mussels is an important factor regulating population size (bij de Vaate, 1991). Unfortunately, cooling systems with concrete walls form excellent substrate for mussel to attachment. Growth of mussels in pipes and canals is rapid because the environment is protected against storm and ice and has a continuous flow of water supplying food and oxygen, and mussels are free of predation. Growth in rivers can be twice as high as in open lakes. This is reflected in the size of first year mussels, with those from rivers reaching a maximum length of 16 mm and means of 6.1–9.0 mm by mid-September, while those from lakes reach a maximum of only 12 mm and means of 2.6–4.1 mm (Table 1).

Fouling problems in Europe caused by the zebra mussel have existed for about 100 years (Clarke, 1952). Drinking water plants often have problems with fouling mussels in their extended pipeline network where mussels reduce flow by increasing wall resistance, thereby causing higher costs for pump energy (Schalenkamp, 1971a and 1971b). In power plants, mussels block condensor tubes, sometimes in combination with the hydroid *Cordylophora caspia*, causing reduced cooling efficiency. The obstruction of flow in power plants increases corrosion of tubes by silt and bacterial slime formation (Dexter, 1985; Characklis et al., 1985). In addition fouling in most industries (for instance, the chemical industry) occurs more often where smaller instead of a few large coolers are used. Most of the smaller coolers have tubes and sheets of common iron alloys which are highly susceptible to damage by corrosion and erosion. Fouling also causes condenser backpressure in cooling operations (Sneek and Jenner, 1985). At present, the fouling problems with zebra mussels in The Netherlands are less severe than those in the Great Lakes (Griffiths et al., 1989).

Asian Clam

A second macrofouling freshwater species, recently found in The Netherlands, is the Asian clam, *Corbicula fluminea*, (Aldrige and McMahon, 1978; McMahon, 1982; Doherty et al., 1986). The Asian clam originated from

TABLE 1. Number of *Dreissena polymorpha* Settled on Acrylic Cylinders in Lake and River Environments (September 1982–1985)

Location	Number of spat/m² (>750 μm long)	Length (mm)	
		Mean	Maximum
Lake Environment			
IJsselmeer			
1982	45,000	3.6	9
1983	29,000	2.6	4
1984	16,000	3.0	7
1985	36,000	4.1	12
River Environment			
Waal			
1984	420	8.8	16
1985	560	9.0	14
Hollandsch Diep			
1984	60,000	6.1	13
1985	16,000	6.8	15

(sub)tropical areas, and in northern temperate regions and is normally restricted to thermal plume areas of power plants. In winter at temperatures below 2°C, populations often experience high mortality in temperate regions (Graney et al., 1980). In early summer, however, the clam population can grow rapidly from clams that survived in warm water of power plants, and these areas serve as reproductive sources for range extension in the summer. Only juvenile *C. fluminea* produce a fragile byssus which is not suitable for anchoring on concrete walls (Mattice, 1983). According Belanger, et al. (1985) the clam prefers coarse sediments of sand and pebbles. In The Netherlands, Asian clams are now found in the Rhine and Meuse Rivers, especially in thermally elevated outlet canals of power and chemical plants (Jenner and bij de Vaate, 1991). This new species is believed to have been introduced in 1986 (bij de Vaate and Greijdanus-Klaas, 1990). The Asian clam is not expected to be serious in The Netherlands because the thick muddy bottoms of industrial inlet areas are not suitable substrate and winter mortality should be high. In the United States, high numbers are often found in cooling towers (Cherry et al., 1990). The use of cooling towers in The Netherlands is restricted to a few months of the year when outlet water temperatures exceed limits allowed by water authorities and duration of use is too short for any fouling problem.

Brackish Water Mussel

A third macrofouling species found in The Netherlands is the brackish water mussel, *Mytilopsis leucophaeta* (i.e., *Concheria cochleata*), which produces a byssus strong enough to foul intake systems. This species originates from (sub)tropical areas of the Caribbean (Marelli and Gray, 1983).

Reproduction of the brackish water mussel usually starts at a water temperature greater than 15°C (Schütz, 1969). However, it is still not clear whether there is one or two distinct reproduction periods and how old the mussels are before they can spawn. It is currently thought that mussels reach sexual maturity when about 11 mm in length and that young of the year can spawn within the same season (Vorstmann, 1933; Schütz, 1969). Wintertime mortality is high in the 1- and 2-year-old mussels.

In The Netherlands, zebra and brackish water mussels are found next to each other in a canal connecting Amsterdam with the coast (Noordzee Kanaal). The population of brackish water mussels is growing fast in this canal due to increasing chloride concentrations. Consequently, the population of zebra mussels is decreasing. Chloride concentrations in the canal range from 3500 mg/L chloride on the seashore side to 1000 mg/L chloride in Amsterdam. The upper chloride limit for zebra mussels in brackish water is 500–1100 mg/L of chloride (Wolff, 1969).

The brackish water mussel alone is not a serious threat to power plants and industries in The Netherlands. Sometimes the explosive growth in brackish water of the hydroid *Cordylophora caspia* in combination with mussel fouling makes control necessary, for which hypochlorite is used in a discontinuous regime of 4 hr with and 4 hr without chlorine applied several weeks a year.

So far, no fouling problems have been attributed to *Corbicula fluminea*, but problems with *Dreissena polymorpha* and *Mytilopsis leucophaeta* do occur and are increasing in frequency and severity. This is illustrated by the forced shutdown of a nuclear power plant in 1985 at the River Waal (Rhine branch) due to severe fouling of inlet water boxes and condenser tubes.

MONITORING

Settlement and Growth

To enable plant operators to follow the process of bivalve settlement, growth, and effectiveness of control procedures a settlement monitor has been developed for byssus-producing bivalves (Figure 1). The principle of the monitor is that bivalve mussel spat enters and populates the settlement monitor, thereby allowing easy inspection and determination of time of settlement, growth, and control effectiveness (Figures 2 and 3). Zebra mussel spat (i.e., pediveliger) has a length of about 0.3 mm and exhibits valves and a foot at which byssus threads are produced for attachment. The settlement monitor is a closed cylindrical container (35-cm diameter and 50-cm height) made of PVC, with a vertical water flow from top to bottom. Special care should be taken in selecting the location and design of the water withdrawal point. The entrance should preferably be isokinetic to ensure that spat concentrations in the monitor correspond to those in the source water flow. The mussel monitor mimics a section of a cooling system (Jenner, 1983). To enable an optimal

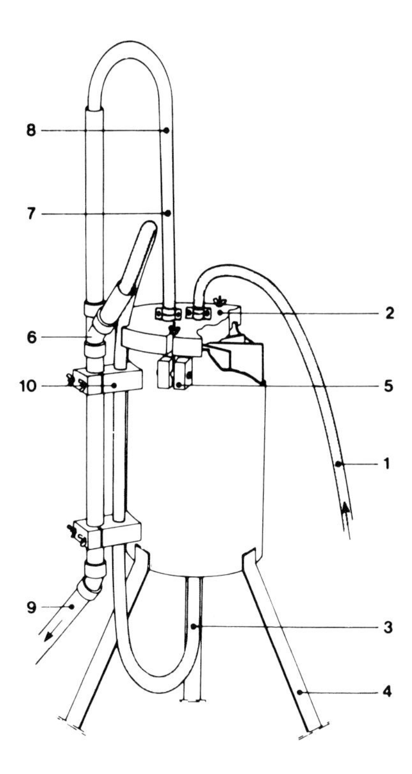

Figure 1. Monitoring apparatus for biofouling bivalves: (1) raw water supply, (2) detachable cover, (3) drain hose, (4) stand, (5) cover mounting clamps, (6) coupling tee, (7) overflow hose, (8) water level during operation, (9) outlet end of PVC tube, and (10) holder for PVC tube and drain hose.

environment for settlement the water velocity in the monitor is kept low at about 1 cm/sec.

The monitor can be used for spat settlement of bivalves such as *Dreissena polymorpha* (Figures 2 and 3) and *Mytilopsis leucophaeta*, but not for *Corbicula fluminea* since *Corbicula* forms no byssus for attachment on solid substrates. Settlement numbers and growth can be investigated by inserted panels of PVC. In the field ropes can be used, with the spat settling preferentially in the crevices between the strands of the ropes, or acrylic tubes/plates of PVC.

Valve Movement

A new monitor system, the "early warning monitor," uses the valve movement pattern of bivalves for detection of pollutants in the aquatic environment (Slooff et al., 1983; Jenner et al., 1989; Kramer et al., 1989). Besides application in water quality surveillance, in general, the monitor is

A

B

Figure 2. Interior of biofouling monitor (A) and a PVC panel removed from the monitor (B) showing zebra mussel colonization after one summer.

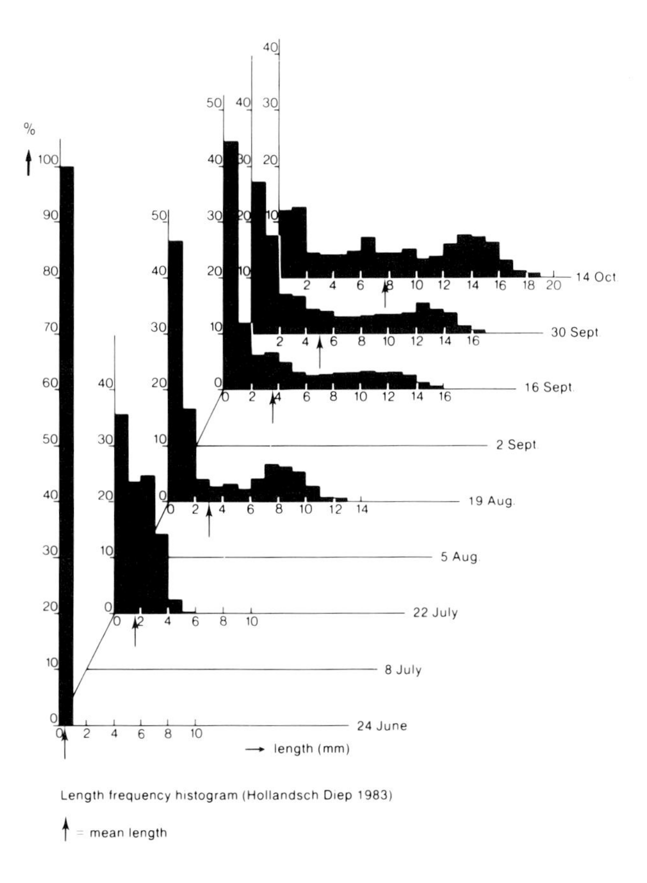

Figure 3. Spat settlement in a biofouling monitor and growth of *Dreissena polymorpha* in the lower part of the Rhine River (Hollandsch Diep) in 1983.

used to study the behavior of mussels and clams exposed to different chlorine dosing during control applications. Normal valve movement behavior of the zebra mussel can be described as continuously open with low and regular periods of activity. Activity is defined as the number of counts calculated by a program for passing a voltage level, corresponding with a movement (opening/closure) of the valves (Jenner et al., 1989). Normally closure is incomplete, about 50%, and the valves open again within a few seconds. Periods of high activity are caused by byssus formation and feces and pseudofeces production. High water turbidity, for instance, leads to high activity due to pseudofeces production.

The effect of chlorination on valve closure pattern of the zebra mussel is presented in Figure 4. A chlorine concentration of 40 μg/L total residual oxidants (TRO) can be detected by the mussel. Increasing concentrations lead to increasing periods of closure, alternated with short periods of high activity.

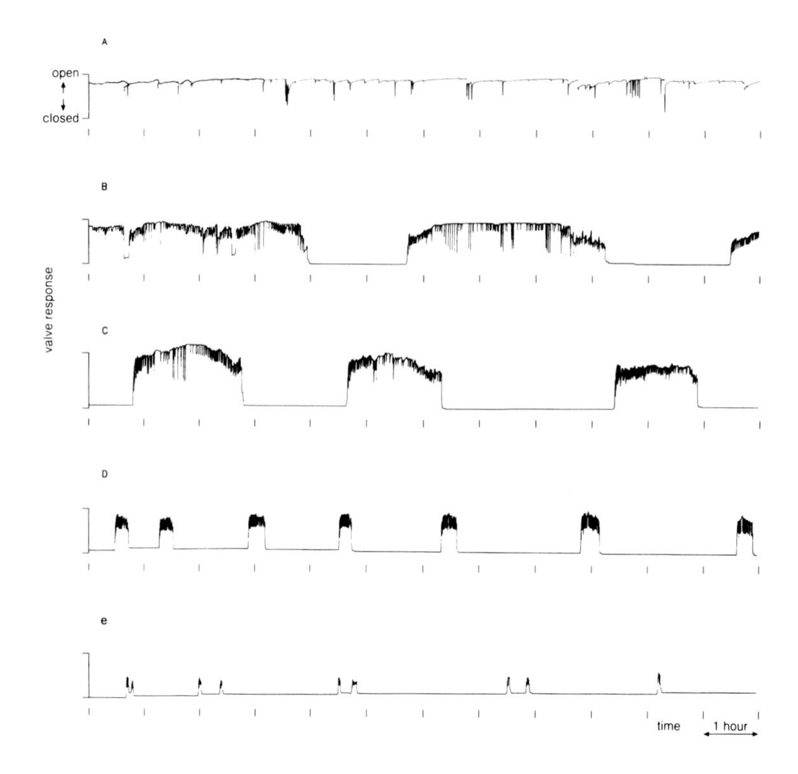

Figure 4. Valve movement of *Dreissena polymorpha* at varying chlorination control concentrations (total residual oxidants): (A) control, (B) 37 µg/L, (C) 55 µg/L, (D) 180 µg/L, and (E) 550 µg/L.

At 500 µg/L TRO, the zebra mussel is closed nearly all the time due to the toxic action of chlorine. Higher concentrations of chlorine seem no more effective, at least for the situation in The Netherlands.

Figure 5 presents closure time, in percentage of a 3-week chlorination application for 11 mussels with their date of mortality. Mussels that close immediately withstand a period of 3 weeks before they are killed, but mussels that show activity (= open) are killed more rapidly. At the end of the period all mussels were dead.

Fouling Areas in Cooling Systems

The main distinction in types of cooling water systems is in offshore vs onshore intakes. Usually offshore intakes consist of long intake pipes of several hundred meters to even a few kilometers. The cooling water pumps are situated on shore in front of condenser systems (e.g., power plants) or filter basins (e.g., drinking water plants). These pipes have an intake dome or crib in front but they are not designed for regular drying for manual

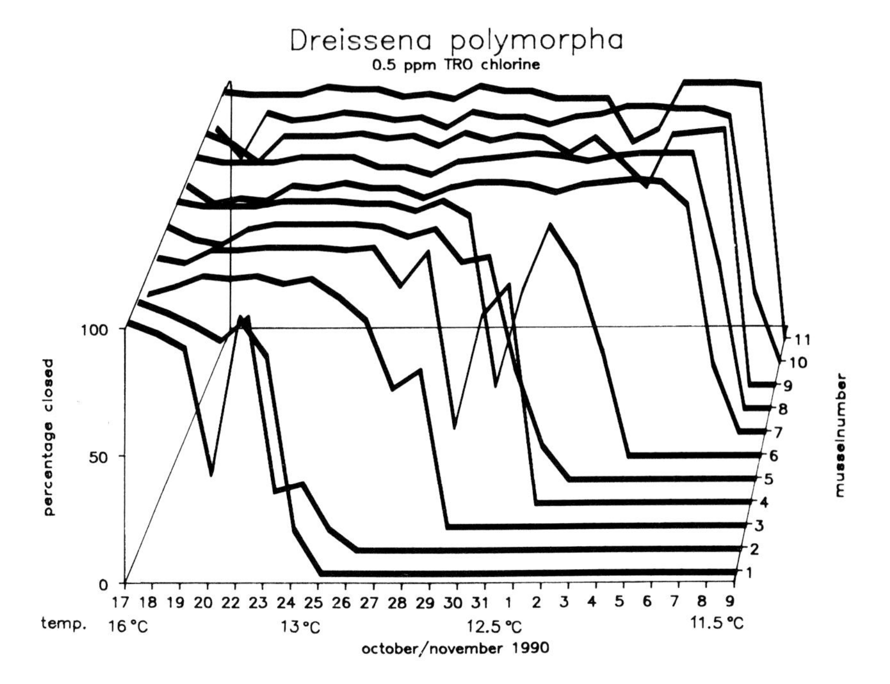

Figure 5. Valve movement pattern (closure) of *Dreissena polymorpha* during exposure to 0.5 mg/L total residual oxidants (TRO) of chlorine. Chlorination period was 3 weeks from October 17 to November 8, 1990. Mortality is expressed as 0% closure (gaping) of the valves.

cleaning. The onshore intake system has forebays, trash racks, and pumps next to each other. The water is normally pumped through condensers with a mean water velocity of 1.5–2.5 m/sec.

In long offshore intake pipes fouling is often restricted to the first several hundred meters. This is caused by sedimentation of spat due to lowered turbulence of the water, as was determined for *Mytilus edulis* at a small coupled desalting/power plant (25 MW) in The Netherlands. The cooling process water pipe (1200 m length, 1.5 m diameter) with a water velocity of 0.5 m/sec acts as a sedimentation environment for the mussel spat. In the first part of the intake, mussel fouling was heavy; this diminishes to zero within 200 m of the pipe inlet. The sedimentated spat seems to smother in silt within the pipe beyond 200 m of the inlet. This phenomenon was tested in an experiment with dosed glass beads (330 μm diameter), having the same sinking rate as alive mussel spat. By sieving the water at the end of the pipe, glass beads could be detected under the microscope with skimming light. The result was 1 bead per cubic meter water out of the 10^7 beads per cubic meter dosed in the intake. The sinking rate for *M. edulis* was experimentally found to be linear between 10 mm/sec at a length of 0.5 and 40 mm/sec at a length of 2 mm. In freshwater with a lower density, a slightly higher sinking rate can be expected for *D. polymorpha*; thus the sinking rate will be about 7.5 mm/sec for spat of 0.3-mm length. At the floor of the culvert other physical transport mechanisms (sliding, saltation, and rolling) will occur. With a culvert of 200-cm diameter and a water velocity of 100 cm/sec at a sinking rate of 7.5 mm/sec, the bulk of the fouling will be found within the first 200–400 meters; in theory the rest of the culvert will show little or no mussel fouling.

Onshore intake systems consist of trash racks with openings between the bars of approximately 10 cm and rotating screens with mesh diameters of between 4 and 10 mm. Hydroids and mussels grow on trash racks and screens. In most cases the growth of hydroids is more of a nuisance than that of mussels. However, if the growth of *D. polymorpha* and *M. leucophaeta* goes unnoticed, fouling will occur, but it usually takes several years before operational problems are noticed. For example, a shutdown of a nuclear plant for manual cleaning of mussels from the intake system has occurred. The fouling is now controlled by regular manual cleaning during normal maintenance outages, and the amount of debris and shells is limited to a few cubic meters. At another plant, zebra mussel fouling is controlled by means of thermal treatment. Fouling only occurs in the forebays into which heated water is pumped (about 38°C) into the intake basins once a year.

CONTROL

Chlorination

Chlorination with hypochlorite is the most commonly used method of controlling macrofouling bivalves in cooling water systems in Europe. In The Netherlands, the emphasis on large-scale use of chlorine is changing because of concern about environmental damage to aquatic life caused by chlorine and its resulting halogenated by-products. As outlined in the Federal Register of November 1982, the U.S. EPA gave an effluent limitation guideline for chlorine of 0.2 mg/L total residual chlorine (TRC) for power plants greater than 25 MW. Discharge concentrations are limited to 2 hr/day, unless the need in combating macrofouling is demonstrated. A 2-hr regime will be ineffective in the control of macrofouling by mussels and clams according to the experience in Europe (Whitehouse et al., 1985).

In The Netherlands, water authorities have growing concern about the impact of by-products from antifouling chlorination, i.e., chloro-bromoform, halogenated benzenes, and phenols. At KEMA, the Dutch Electricity Research Company, a research program has been started to update toxicity data of seawater and freshwater chlorination and chlorinated by-products. This study is part of a larger program for the study of chlorinated organics and their

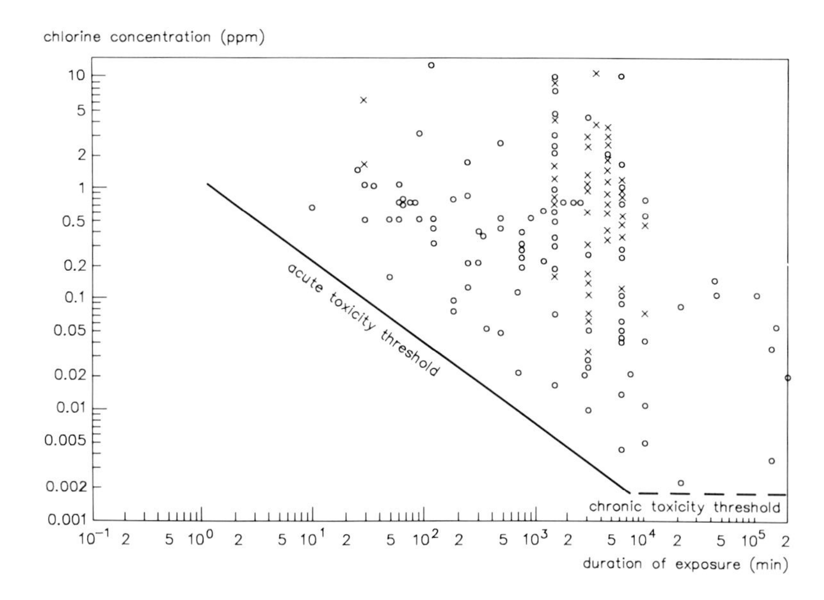

Figure 6. Toxicity of chlorine with the expected chronic and acute toxicity threshold for freshwater organisms (0), worldwide data is completed to 1990 (x). (Modified from Mattice, J. S., and H. E. Zittel. *Water Pollut. Control Fed.* 48:2284–2308 [1976]. With permission.)

effects on hormonal control mechanisms of thyroid hormones and shifts in enzyme systems in aquatic organisms. The literature research started with data published originally by Mattice and Zittel (1976) and updated by Lewis (1983) on chronic and acute toxicity thresholds. The study is completed from 1976 up to 1990 (excluding a lot of problematic data from the available literature caused by insufficient analytical uniformity in measuring). The result (presented in Figure 6) for freshwater agrees well with the original figure of Mattice and Zittel (1976) indicating that their thresholds are correct and useful.

Dreissena polymorpha and *M. leucophaeta*, depending on water temperature and condition of the mussels, can be killed with low levels of chlorine in a relatively short period of 2–4 weeks. The chlorination period should be planned at the end or after the reproduction period to ensure that no "new" spat will enter the system. Otherwise, water temperature should not be dropped below 15°C because it will enlarge disproportionately the duration for complete kill of mussels. In The Netherlands, this program is called "autumn chlorination." Killed mussels are small enough to pass through the cooling system without blocking condenser tubes. It can be necessary to chlorinate twice a year if reproduction success of the mussel is so high that mussels form clusters and when chlorination is performed, fall from pipes and block power plant condenser tubes.

If the water temperature is low (i.e., <15°C) the time required by effective chlorination will be prolonged. In Figure 7, results of mussel toxicity at chlorine concentration of 0.5 and 1.0 mg/L total residual oxidants (TRO) are given. A concentration of 1.0 mg/L initiated a more rapid beginning of mortality compared with 0.5 mg/L. A chlorine concentration of 0.5 mg/L TRO measured in the vicinity of the spot where fouling occurs — normally the inlet of the condensers — is sufficient. According to Doherty et al. (1986) and Ramsay et al. (1988) *Corbicula fluminea* could be killed by continuous chlorination for 2–3 weeks at concentrations of 0.5 mg/L total residual chlorine (TRC) at water temperatures of 20–25°C. The chlorine concentrations as used in the present study are valid within the framework of chemical water parameters (i.e., pH, water temperature, and chlorine demand) of rivers and lakes in The Netherlands. At other locations, differences in physical parameters will cause variation in chlorine concentrations needed for effective control.

Surface Coatings

In the past, the use of toxic paints containing copper was a common method to control mussel fouling in The Netherlands. Paints based on tributyl tin oxide were introduced as a substitute for copper and widely used. The environmental problems with tributyl tin oxide are so extensive that its use is now restricted and a complete ban is being considered. In power plants, its application has already been banned.

New nontoxic coatings based on low surface tension create extreme smooth surfaces that are now available and are being tested as macrofouling control agents. Results are promising; depending on costs and technical feasibility, these coatings may be an effective and economical control methodology in water systems, particularly to prevent mussel settlement.

Heat Treatment

Heat treatment to control macrofouling is a commonly used control method (Jenner and Janssen-Mommen, 1990). This type of treatment, however, is restricted to a few power plants. This is due to the fact that a special design is required for the cooling water system at an early stage of plant building. Adaptations afterward are often expensive or technically difficult.

The mortality curve of *D. polymorpha* (Figure 8) is about 4°C lower than that of marine mussels. The mortality of *M. leucophaeta* is in the same range as marine mussels (Van der Kolk, 1990). It has to be noted that our results are valid for the climatic conditions as found in the inland and seashore waters of The Netherlands. Mussel populations from different geographical locations with different water temperatures can have higher or lower lethal temperatures during heat treatment, as was shown by Graham et al. (1975) in California.

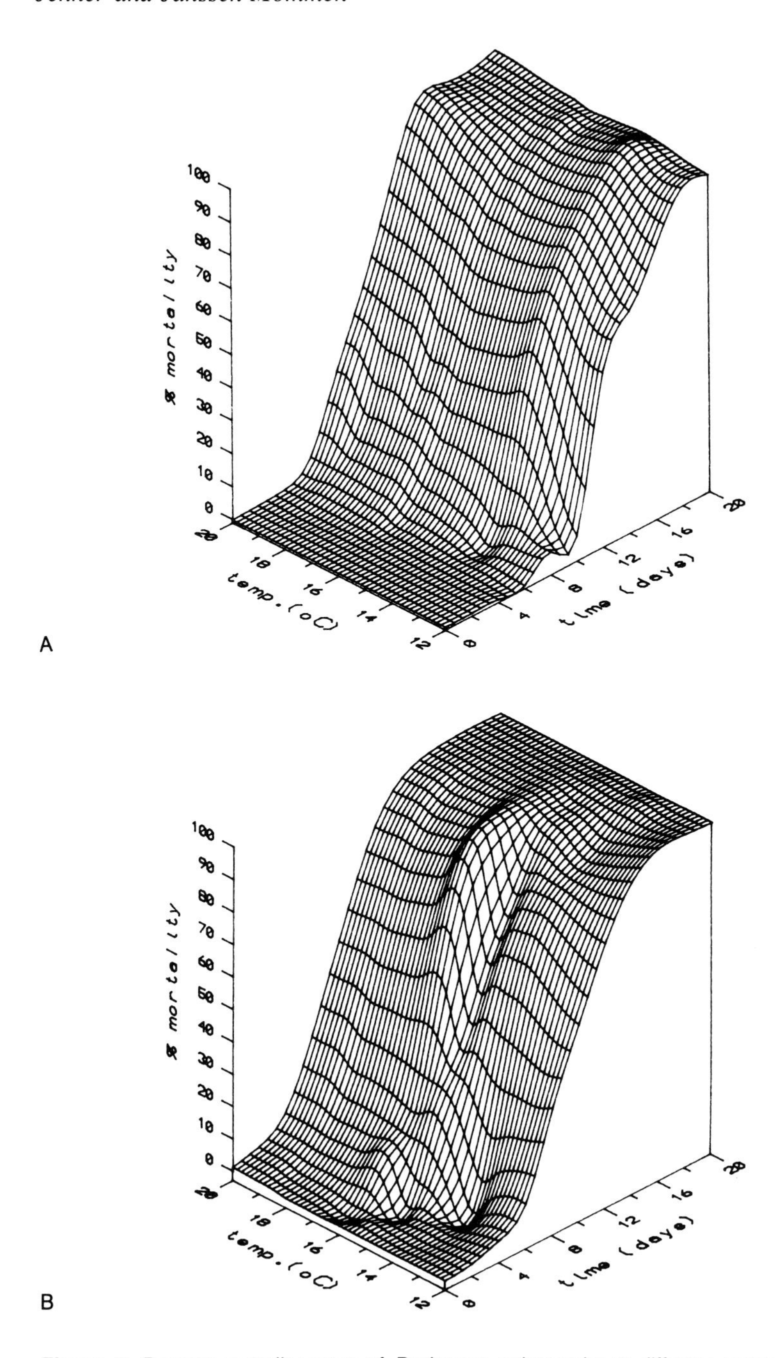

Figure 7. Percent mortality rates of *Dreissena polymorpha* at different temperatures during chlorination of 0.5 mg/L (A) and 1.0 mg/L (B) total residual oxidants (TRO).

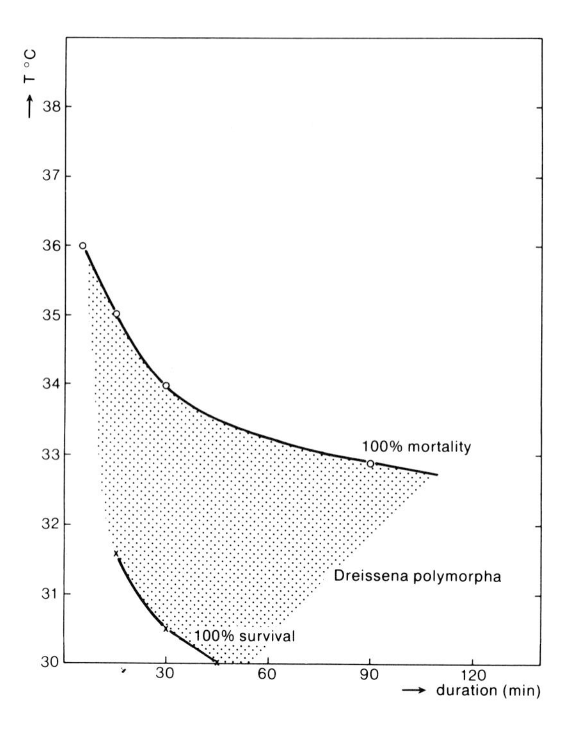

Figure 8. Relationship between survival and 100% mortality of *Dreissena polymorpha* during exposure to different temperatures.

Lethal temperatures were found to be 2–3°C higher for mussels acclimated at higher water temperatures. In general, heat treatment has to be applied for freshwater and brackish water only once or twice a year.

Drying

Sometimes it is possible to drain and dry a pipe causing macrofouling bivalves to experience lethal levels of desiccation. An advantage of this option is that simple exposure is the effective control method, but it can be accelerated by air heating. This method has been used at a drinking water plant in two pipelines transporting raw lake water to sand filter beds.

Water Velocity

In cooling water pipes, the settlement of zebra mussel spat is suppressed at water velocities >1.5 m/sec. However, mussels that settle at reduced velocities will stay and grow when velocities are increased to 1.5 m/sec. The expansion joints of concrete pipes are a favorable spot for settlement. In freshwater, the growth of mussels in condenser water boxes is normally restricted to the edges and fouling in these spots rarely presents a problem. These conclusions are based on the experience of power plant operators at inspections of pipes during maintenance periods. For barnacles and marine mussels a velocity of 3 m/sec is necessary to prevent settling according to our knowledge based on 15 years of experience. Lewis (1964) has summarized the results from literature on water velocity vs settlement success. A critical velocity for preventing settlement of barnacles and marine mussels is 0.5 m/sec for smooth surfaces of pipe walls. To remove attached mussels, water velocities of at least 5 m/sec are necessary. In the design of new plants a combination of high water velocities, smooth finishing of the walls, and rounding off of all sharp edges in intake and culverts can be utilized to greatly reduce the surfaces susceptible to mussel fouling.

Microsieves

At small water withdrawal plants, such as service water plants with low water flows of 1–2 m^3/sec, the installation of microsieves can be an effective control method. The required mesh size is about 100 μm. In the case of power plants, the method will be less feasible due to the high number of needed installations. Shephard et al. (1971) discussed methods of controlling fouling at desalting plants and determined that filtering was impractical due to sieve clogging by microbial material.

CONCLUSIONS

Fouling problems caused by the zebra mussel, *Dreissena polymorpha*, in The Netherlands are not generally solved by one control methodology. For each location or situation, the best solution is a combination of mechanical, physical, and chemical applications in that declining order of acceptance. However, most existing power plants in Europe have no other alternative than to chlorinate. The conclusion of Mattice and Zittel (1976) is followed extensively: use of chlorination has to be of brief duration, of minimum effective concentration, and in a timely fashion. To this end, monitoring of macrofouling bivalves via an in-line monitoring box and a mussel valve monitor aids in the efficient and economical application of a selected control methodology in The Netherlands.

ACKNOWLEDGMENTS

Comments and suggestions of two anonymous reviewers were quite useful in the revision of the manuscript.

REFERENCES

Aldridge, D. W., and R. F. McMahon. "Growth, Fecundity, and Bioenergetics in a Natural Population of the Asiatic Freshwater Clam, *Corbicula manilansis* Philippi, from North Central Texas," *J. Moll. Stud.* 44:49–70 (1978).

Belanger, S. E., J. L. Farris, D. S. Cherry, and J. Cairns, Jr. "Sediment Preference of the Freshwater Asiatic clam, *Corbicula fluminea*," *Nautilus* 99:66–72 (1985).

bij de Vaate, A. "Occurrence and Aspects of Population Dynamics of the Zebra Mussel, *Dreissena polymorpha* (Pallas, 1771), in the Lake IJsselmeer area (The Netherlands)," *Oecologia (Berlin)* 86:40–50 (1991).

bij de Vaate, A., and M. Greijdanus-Klaas. "The Asiatic Clam, *Corbicula fluminea* (Müller, 1774) (Pelecypoda, Corbiculidae), a New Immigrant in The Netherlands," *Bull. Zoöl. Mus. Univ. Amsterdam* 12(12):173–177 (1990).

Characklis, W. G., R. Bakke, and A.-I. Yeh. "Microbial Fouling and Its Control: a Phenomenological Approach," in EPRI Condenser Biofouling Control Symposium: The State-of-the-Art, W. Chow and Y. Mussalli, Eds. Lake Buena Vista, FL (June 18–20, 1985).

Cherry, D. S., J. Farris, J. R. Bidwell, A. Mikailoff, R. L. Shema, and J. W. McIntire. "Application of a Molluscicide and Environmental Fate and Effects at the Beaver Valley Power Station, Duquesne Light Company," EPRI International Macrofouling Symposium, Orlando, FL (December 4–6, 1990), 355–385.

Clarke, K. B. "The Infestation of Waterworks by *Dreissena polymorpha*, a Freshwater Mussel," *J. Inst. Water Eng.* 6:370–379 (1952).

Dexter, S. C. "Fouling and Corrosion," in EPRI Condenser Biofouling Control Symposium: The State-of-the-Art, W. Chow and Y. Mussalli, Eds. Lake Buena Vista, FL (June 18–20, 1985).

Doherty, F. G., J. L. Farris, D. S. Cherry, and J. Cairns, Jr. "Control of the Freshwater Fouling Bivalve *Corbicula fluminea* by Halogenation," *Arch. Environ. Contam. Toxicol.* 15:535–542 (1986).

Gillet, A., and J. C. Micha. "Etude de la biologie d'un mollusque bivalve, *Dreissena polymorpha* P., en vue de son èlimination dans le circuit "Eau brute" de la centrale nuclèaire de Tihange," *Trib. CEBEDEAU* 504:3–28 (1985).

Graham, J. W., R. W. Moncreiff, and P. H. Benson. "Heat Treatment for the Control of Marine Fouling at Coastal Electric Generating Stations," Proceedings of the Conference of the Institute of Electrical and Electronics Engineers, *Ocean* (1975), pp. 926–930.

Graney, R. L., D. S. Cherry, J. H. Rodgers, Jr., and J. Cairns, Jr. "The Influence of Thermal Discharges and Substrate Composition on the Population Structure and Distribution of the Asiatic Clam, *Corbicula fluminea*, in the New River, Virginia," *Nautilus* 94:130–135 (1980).

Griffiths, R. W., W. P. Kovalak, and D. W. Schloesser. "The Zebra Mussel, *Dreissena polymorpha* (Pallas, 1771) in North America: Impact on Raw Water Users," Service Water System Problems Affecting Safety-Related Equipment, (Palo Alto, CA: Nuclear Power Division, Electric Power Research Institute, 1989).

Jenner, H. A. "A Microcosm Monitoring Mussel Fouling," in EPRI Symposium on Condenser Macrofouling Control Technologies: The State-of-the Art, I. A. Diaz-Tous, M. J. Miller, and Y. G. Mussalli, Eds. Hyannis, MA (June 1–3, 1983).

Jenner, H. A., F. Noppert, and T. Sikking. "A New System for the Detection of Valve Movement Response of Bivalves," *KEMA Sci. Tech. Rep.* 7:91–98 (1989).

Jenner, H. A., and J. P. M. Janssen-Mommen. "Control of Macrofouling in Cooling Water Systems," EPRI International Macrofouling Symposium, Orlando, FL (December 4–6, 1990).

Jenner, H. A., and A. bij de Vaate. "Wordt de Aziatische mossel, *Corbicula fluminea* een probleem in Nederland?," *Water* 24(4):101–103 (1991).

Kirpichenko, M. Ya., V. P. Mikheev, and E. P. Stern. "Battling over Growth of *Dreissena* at Hydroelectric Power Plants," Oak Ridge National Laboratory, Oak Ridge, TN (1962) (Translated from the Russian *Electr. Stan.* 5:30–32, ORNL-tr-4705).

Kramer, K. J. M., H. A. Jenner, and D. de Zwart. "The Valve Movement Response of Mussels: a Tool in Biological Monitoring," *Hydrobiologia* 188/189:433–443 (1989).

Lewis, B. G. "Water Flow and Marine Fouling in Culverts: A Review of Literature up to 1962," CERL Note No. RD/L/M 60 (1964).

Lewis, B. G. "Effects of Continuous Chlorination on Mussels and Validation of Preliminary Model," CEGB Report No. TPRD/L/2594/N83 (1983).

Mackie, G. L., W. N. Gibbons, B. W. Muncaster, and I. M. Gray. "The Zebra Mussel, *Dreissena polymorpha*: a Synthesis of European Experiences and a Preview for North America," Report prepared for: Water Resources Branch Great Lakes Section (1989).

Marelli, D. C., and S. Gray. "Conchological Redescriptions of *Mytilopsis sallei* and *Mytilopsis leucophaeta* of the Brackish Western Atlantic," *Veliger* 25:185–193 (1983).

Mattice, J. S., and H. E. Zittel. "Site-Specific Evaluation of Power Plant Chlorination," *J. Water Pollut. Control Fed.* 48:2284–2308 (1976).

Mattice, J. S. "Freshwater Macrofouling and Control with Emphasis on *Corbicula*," in EPRI Symposium on Condenser Macrofouling Control Technologies: The State-of-the-Art, A. I. Diaz-Tous, M. J. Miller and Y. G. Mussalli, Eds. (June 1–3, 1983).

Mattice, J. S. "Chlorination of Power Plant Cooling Waters," in Water Chlorination: Chemistry, Environmental Impact and Health Effects, Vol. 5, Jolley, R. L., R. J. Bull, W. P. Davis, S. Katz, M. H. Roberts, Jr., and V. A. Jacobs, Eds. Proceedings of the Fifth Conference on Water Chlorination, Williamsburg, VA (June 3–8, 1984) 39–62.

McMahon, R. F. "The Occurrence and Spread of the Introduced Asiatic Freshwater Clam *Corbicula fluminea* in North America 1924–1982," *Nautilus* 96:134–141 (1982).

Mikheev, V. P. "Experiments on Destroying *Dreissena polymorpha* by Heating the Water," *Bjul. Inst. Biol. Vodochr. Moskwa* 11:10–12 (1961).

Ramsay, G. G., J. H. Tackett, and D. W. Morris. "Effect of Low-Level Continuous Chlorination on *Corbicula fluminea*," *Environ. Toxicol. Chem.* 7:855–856 (1988).

Schalenkamp, M. "Warnung vor der wandermuschel *Dreissena polymorpha* Pallas und bekämpfung derselben," Sonderdruck aus GWA 1971/3 des Schweiz, Vereins von Gas- und Wasserfachmännern (1971a).

Schalenkamp, M. "Neueste erkenntnisse über die wandermuschel *Dreissena polymorpha* Pallas (DPP) und ihre bekämpfung," Sonderdruck aus GWA 1971/11 des Schweiz, Vereins von Gas- und Wasserfachmännern, (1971b).

Schütz, L. "Ökologische untersuchungen über die benthos fauna im Nordostseekanal. III. Autoecologie der vagilen und hemisessilen arten im bewuchs der pfähle: makrofauna," *Int. Rev. Ges. Hydrobiol.* 54:553–588 (1969).

Shepherd, B. P., P. G. LeGros, J. C. Williams, D. C. Mangum, and W. F. Mcllhenny. "Intake Systems for Desalting Plants," Research and Development Progress Report No. 678, U.S. Department of the Interior (1971).

Slooff, W., D. de Zwart, and J. M. Marquenie. "Detection Limits of a Biological Monitoring System for Chemical Water Pollution Based on Mussel Activity," *Bull. Environ. Contam. Toxicol.* 30:400–405 (1983).

Sneek, E. J., and H. A. Jenner. "Dutch Experience with Condenser Maintenance," in EPRI Condenser Biofouling Control Symposium: The State-of-the-Art, W. Chow and Y. Mussalli, Eds. Lake Buena Vista, FL (June 18–20, 1985).

Van der Kolk, A. J. "Oecologie en oecofysiologie van de brakwater mossel *Mytilopsis leucophaeta*," Report University of Nijmegen, The Netherlands, No. 288 (1990).

Vorstmann, A. G. "Zur biologie der brackwassermuschel *Congeria cochleata*," *Nyst. Verh. Int. Ver. Limnol.* 6:182–186 (1933).

Whitehouse, J. W., M. Khalanski, M. G. Saroglia, and H. A. Jenner. "The Control of Biofouling in Marine and Estuarine Power Stations: A Collaborative Research Working Group Report for Use by Station Designers and Station Managers," CEGB, EdF, ENEL and KEMA, CEGB NW-Region-191-9-85 (1985).

Wilhelmi, J. "Beitrage zur praktische biologie der wandermuschel *Dreissenia polymorpha* und ihrer bekämpfung in wasserwerken," *Dtsch. Wasserwirtsch.* 7:107–109; 9:125–126 (1923).

Wolff, W. J. "The Mollusca of the Estuarine Region of the Rivers Rhine, Meuse and Scheldt in Relation to the Hydrography of the Area. II. The Dreissenidae," *Basteria* 33:93–103 (1969).

CHAPTER 32

Control Program for Zebra Mussels (*Dreissena polymorpha*) at the Perry Nuclear Power Plant, Lake Erie

Louise K. Barton

The zebra mussel (*Dreissena polymorpha*) presents a substantial threat to reduce or block water flow in power plant systems. Mussels were first discovered at the Perry Nuclear Power Plant in September 1988. In response, a three-part zebra mussel program was developed that includes monitoring, chemical treatment, and research studies. Monitoring consists of using artificial substrates, sidestream monitors, plankton nets, and diver observations in addition to direct visual inspections of raw water system components during maintenance and repair. A proactive approach using chemical treatments was adopted to control mussel fouling. In addition to using the plant chlorination system, chemical control methods included the use of commercial molluscicides to treat areas upstream of the chlorination system. Three studies were conducted at Perry nuclear power plant to determine the effectiveness of various applications of chlorine using the plant chlorination and dechlorination systems. Results indicate that continuous chlorination is required for adult mussel control, and that intermittent chlorination is effective for veliger control.

0-87371-696-5/93/$0.00 + $.50

INTRODUCTION

The zebra mussel (*Dreissena polymorpha*) has several characteristics which enhance its ability to disrupt water flow in power plants. First, microscopic veligers are easily entrained into raw water systems. Second, mussels are capable of attaching in any system with water velocities less than 2 m/sec (Lyakhov, 1968). Third, mussels can accumulate in layers up to 30 cm thick in large pipes (Clarke, 1952). Several types of problems have been identified with zebra mussel fouling in power plant raw water systems. Initially, layers of attached mussels reduce or block flow through large diameter piping, intake trash racks, and traveling screens. Eventually, shells or clumps of shells breaking free of their attachments block openings in downstream piping, heat exchangers, strainers, and traveling screens. Finally, attachment points accumulate other debris and serve as sites for corrosion.

With these problems in mind, personnel at the Perry Nuclear Power Plant developed a three-part program to minimize the fouling of water systems by the zebra mussel. This chapter documents this program to serve as a reference for other power plants that may encounter zebra mussels in the future.

STUDY SITE

The Perry Nuclear Power Plant is located on the south shore of the central basin of Lake Erie, approximately 56 km east of Cleveland, OH (Figure 1). It is a boiling water nuclear reactor with 1250-MWe gross electrical output. Cooling water is withdrawn from Lake Erie through two submerged intake structures located about 780 m from shore at a water depth of 6 m. A single 3-m diameter concrete intake tunnel connects the two intake structures with the onshore service water pump house. These intake structures provide once-through service water and makeup water for the closed-cycle condenser cooling system, which uses a closed-loop, natural draft cooling tower. Average water withdrawal is 331,114 L/min.

MONITORING

Early detection of zebra mussel infestations is crucial since mussels have the ability to rapidly colonize plant raw water systems. For example, the presence of mussels was detected in September 1988 at the Eastlake Power Plant of the Cleveland Electric Illuminating Company. A treatment program was not immediately initiated at this facility; and during a 4-week period, the population density of zebra mussels increased to a point that caused in-plant clogging of strainers, heat exchangers, and small diameter pipes.

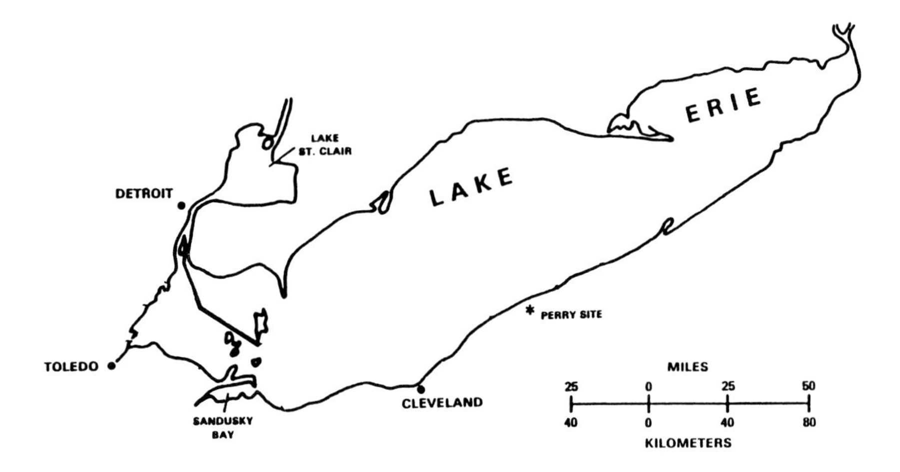

Figure 1. Location of the Perry Nuclear Power Plant (*) on the South Shore of the Central Basin of Lake Erie.

In addition to visually inspecting raw water systems when they are opened for maintenance and repair, monitoring methods used at Perry Nuclear Power Plant include the use of divers, artificial substrates, sidestream monitors, and plankton nets. Divers are currently used to collect mussel samples and monitor infestations. They have also been used to take underwater videotapes of the water basins and intake tunnel. Artificial substrates include concrete blocks and plastic baskets suspended by rope into intakes and water basins. Substrates are kept in place all year and are designed so that they can be easily removed weekly for inspection of any settling mussels. Sidestream monitors are flow-through containers that receive water diverted from plant systems. They are fitted with slides and inspected weekly for veliger settlement. Monitoring for veligers in incoming service water consists of taking weekly samples with an 80-μm mesh plankton net.

Zebra mussels were first discovered at Perry Nuclear Power Plant in September 1988. The initial collection consisted of 19 mussels that were found during a routine monitoring program for another fouling bivalve, the Asian clam (*Corbicula fluminea*). In May 1989, zebra mussels were found attached to samplers placed in the plant service water pump house. By fall 1989, the population in the pump house forebays had increased to about 1100/m^2. However, no mussels were found on traveling screens and no operational problems were experienced. In February 1990, about 30 m of the inside of one of the intake tunnels was inspected using a video camera. Mussel coverage was 100%, and the mussel layer was about 2.5–3.8 cm thick. Subsequently, a zebra mussel program was implemented that included monitoring, treatment, and several research projects.

TABLE 1. Densities (number/L) of Veligers of *Dreissena polymorpha* in Incoming Service Water at the Perry Nuclear Power Plant on Lake Erie Between June 1990 and September 1991

1990			1991		
Date	Number/L	Temp. (°C)	Date	Number/L	Temp. (°C)
June 18	0	20.0	May 20	0	13.3
June 25	0	18.9	May 29	<1	17.8
July 03	3	20.6	June 06	<1	10.6
July 18	<1	19.4	June 12	<1	21.1
August 02	35	20.6	June 19	<1	16.1
August 07	98	20.6	June 26	<1	12.2
August 14	40	21.7	July 03	<1	13.9
August 21	13	21.7	July 10	1	22.8
August 28	15	23.3	July 17	6	23.4
September 04	3	22.2	July 24	180	25.6
September 11	3	22.2	July 31	140	22.8
September 17	6	19.4	August 07	14	23.3
September 25	3	16.1	August 14	15	23.3
October 02	3	17.2	August 21	16	22.2
October 09	3	17.2	August 28	23	23.3
October 16	1	15.6	September 04	8	20.6
			September 11	5	22.8
			September 18	<1	22.8
			September 25	0	18.9

Results of veliger monitoring in 1990–1991 are shown in Table 1. Veligers were first detected on July 3, 1990 and on May 29, 1991. Although maximum densities were found in late July/early August in both years, maximum densities were about twice as high in 1991 as compared to 1990. In 1990, newly settled mussels were first detected in the power plant in late August, at which time densities were about 130,000/m^2; in 1991, newly settled mussels were first found in the power plant in late July.

CHEMICAL TREATMENT

Once zebra mussels were detected in the water supply of the power plant in 1990, immediate attention was given to potential control methods. Physical, mechanical, and chemical alternatives were evaluated in order to select and implement the most effective treatment strategy. Although mechanical methods such as scraping and hydroblasting were available to be used on system components if required, a chemical method to control mussels was selected.

Chemicals used for mussel control included chlorine as well as two commercial molluscicides. Chlorine was selected over other chemical methods such as bromination and chlorobromine combinations primarily because the Perry Nuclear Power Plant has a chlorination and dechlorination system already in place. However, the chlorination system does not treat plant components upstream of the service water pumps, which includes offshore intake structures, the intake tunnel, service water, and emergency service water forebays and basins.

The chlorination system provides 0.8% sodium hypochlorite to plant service water, emergency service water, and circulating water systems. Delivery mechanisms are operated by timers; sodium hypochlorite is injected for 30 min into the suction bells of the service water and into emergency service water pumps once every 12 hr. The timer is connected to a similar mechanism on the dechlorination system. When the timer activates sodium hypochlorite addition, it also activates dechlorination pumps; these pumps deliver sodium sulfite to the service water and emergency service water systems prior to discharge into Lake Erie. The concentration is adjusted to respond to system demand; however, the end of pipe concentration prior to dechlorination is 0.5 mg/L Total Residual Chlorine, as analyzed by amperometric titration.

The effectiveness of this treatment was determined in two ways. First, over 40 visual inspections of raw water system components were conducted in 1990 and 1991. Second, substrate monitors were inspected weekly for new settlement. No mussels or newly settled veligers were found on any plant components and substrate monitors after chemical treatment.

Commercial molluscicides selected for use at the Perry Nuclear Power Plant included a blend of alkyl dimethyl benzyl ammonium chloride and dodecylguanidine hydrochloride, and didecyl dimethyl ammonium chloride. Depending on the product and the water temperature, applications lasted 6–12 hr at concentrations of 2.5–15.0 mg/L. Only one treatment was applied annually, and that was at the end of the settlement period. The chemical was injected into the intake cribs and allowed to travel through plant water systems. Active ingredients were detoxified by adsorption onto bentonite clay prior to discharge into Lake Erie.

The effectiveness of each application was first measured by observing mortality of mussels placed in a flow-through container of plant service water and subject to the chemical treatment. Two to three weeks after each treatment, divers inspected service water basins and the intake tunnel. Both chemicals were successful in killing mussels; mortality observed both in the flow-through containers and in plant systems was over 90%.

RESEARCH STUDIES

In addition to implementing a treatment program, studies to evaluate the effectiveness of chlorination for mussel control were conducted using existing systems. Chlorine is a well-documented biocide and has been used extensively for control of slime and algae. Perry Nuclear Power Plant has a regulatory discharge limit for chlorine (maximum daily discharge time is 2 hr, with concentration limitations of 0.2 mg/L for a 30-day average and 0.5 mg/L on a daily basis), but the presence of a dechlorination system allowed the testing of a variety of chlorine applications. Two tests were completed in 1989 and a third was conducted during summer 1990.

The purpose of the first test was to determine the effectiveness of intermittent chlorination for control of mussels that were 8–15 mm in length. Although a prior study indicated that intermittent chlorination would not provide effective treatment (Clarke, 1952), the test was conducted for two reasons. First, mussels had not been found in plant systems beyond the point of chlorination injection. This indicated that the treatment might be providing control. Second, the chlorination and dechlorination systems normally operate intermittently, injecting sodium hypochlorite once every 12 hr for a duration of 30 min.

A flow-through test basket constructed of plastic and fine-mesh screening was placed in a large, open, service water weir structure at the discharge end of the service water system on September 29, 1989. Mussels placed in this test basket were taken from a similar basket previously placed in the service water pump house forebay. Mussels were transferred without being detached from their substrate and were kept continuously in service water. Mussels in the basket that remained in the service water pump house forebay served as a control for the test since the forebay is not chlorinated. Sodium hypochlorite was injected via normal system operation resulting in an average concentration of 0.5 mg/L (Total Residual Chlorine) at the basket placed in the service water weir structure. Water temperature and pH during the test fluctuated with lake conditions, with temperature varying between 11 and 17°C and pH varying between 7.8 and 8.2. After 28 days of intermittent chlorine treatment, no mussel mortality occurred in the basket located in the service water weir structure; in fact, mussels grew in this time period since the size range of mussels at the end of the test was 9–18 mm. Also, no mussel mortality occurred in the control basket in the pump house forebay. These results are similar to other studies which demonstrate that zebra mussels simply close their shells or "clam up" when they sense certain chemicals in the water (Jenner and Janssen-Mommen, 1989). Clark (1952) reported that zebra mussels could easily remain closed for over 1 week. Thus, under the conditions of this experiment, intermittent chlorination for zebra mussels proved ineffective as a control measure.

A second test was conducted to determine the ability of the existing chlorination and dechlorination systems to operate continuously and reliably, and also to determine the effectiveness of continuous chlorination for zebra mussel control. A flow-through test box constructed of plastic was placed in the service water pump house and connected to a service water header on December 1, 1989. Water was diverted from the service water header through the test box. Mussels were again taken from the service water pump house forebay and placed in the test box. Mussels left in the forebay served as controls. Sodium hypochlorite was injected continuously into the service water

TABLE 2. Number of Live and Dead *Dreissena polymorpha* Observed During Continuous Chlorination Test Conducted in December 1–28, 1989

	Live	Dead	Total
Attached	1211	471	1682
Detached	401	230	631
Total	1612	701	2313

system to maintain a total residual chlorine concentration of 0.3–0.5 mg/L in the test box. The test was originally designed to continue until 100% mortality was achieved; however, after 27 days a chemical feed-pump for the dechlorination system malfunctioned and the experiment was terminated. The 27-day treatment resulted in a total mortality of 30% in the test box (mortality was determined by lack of valve closure response). In addition, 17% of the remaining live mussels detached from the substrate (Table 2). Based on the assumption that dead mussels would eventually drop from the substrate, a total removal of 47% was achieved. Again, no mussel mortality was observed in the service water pump house forebay. The cold water temperatures at which this test was conducted (0–4°C) probably had a significant impact on mortality rates in the test box. Mussel filtration rates are approximately five times lower in cold water (Kornobis, 1977); and since reduced filtration results in less chemical ingestion, chlorine treatments require longer duration in cold water (Jenner and Janssen-Mommen, 1989; Greenshields and Ridley, 1957). According to Jenner (1985), the concentration of chlorine used in this test, if conducted at higher temperatures, should have achieved 100% mortality in 14–21 days. Although the test ended prematurely, it demonstrated that continuous chlorination could be effective for mussel control. Even under conservative conditions of cold water, 47% removal is significant over a relatively short test period.

A third test was conducted during summer 1990 to verify the effectiveness of intermittent chlorination to control veligers. Because zebra mussels had not been found in the plant beyond the injection point of the chlorination system, it appeared that the existing system was providing control. However, based on the results of the first test, the existing control method of intermittent chlorination did not appear to be effective. Four flow-through chambers were placed as settlement monitors in the plant. Sodium hypochlorite was injected intermittently via normal system operation. The chambers were examined weekly for any settled veligers. One live veliger was found in the settlement monitor off the main service water header, but no other live veligers were found at any other time during the season (May through October). These observations, along with inspections conducted on raw water system components when open for maintenance or repair, verified the effectiveness of intermittent chlorination for control of zebra mussel veligers. Thus, veligers appear to be more susceptible to chemical treatment than adults, probably because the shell has not yet fully developed.

CONCLUSIONS

Perry Nuclear Power Plant has taken the approach that the best method for avoiding problems with zebra mussels is early detection followed by preventative treatment of plant water systems. Four recommendations for an optimal zebra mussel control program have been developed: (1) monitor veliger entrainment and settlement to define peak periods so that treatment programs can be timed accordingly; (2) apply treatment prior to the development of a large infestation of adult mussels because it is a better strategy to maintain clean systems rather than to mitigate problems once they occur; (3) optimal treatment should target veligers because they appear to be more susceptible to chemical treatment than adults and controlling mussels before they attach will reduce the potential to develop sites suitable for development of corrosion; and (4) continuous treatment offers the best control strategy but, if impractical, intermittent treatment should be applied at intervals which are as short as practical.

REFERENCES

Clarke, K. B. "The Infestation of Waterworks by *Dreissena polymorpha*, a Freshwater Mussel," *J. Inst. Water Works Eng.* 6:370–378 (1952).

Greenshields, F., and J. E. Ridley. "Some Researchs on the Control of Mussels in Water Pipes," *J. Inst. Water Works Eng.* 11:300–306 (1957).

Jenner, H. A. "Chlorine Minimization in Macrofouling Control in The Netherlands," in *Water Chlorination: Chemistry, Health Effects, and Environmental Impact, Vol. 5*, R. L. Jolly et al., Eds. (Chelsea, MI: Lewis Publishers, Inc., 1985), pp. 1425–1433.

Jenner, H. A., and J. P. M. Janssen-Mommen. "Control of the Zebra Mussel in Power Plants and Industrial Settings," Abstract, Second International Conference on the Zebra Mussel in the Great Lakes, New York Sea Grant and U.S. Fish and Wildlife Service, Rochester, NY (November 1989).

Kornobis, S. "Ecology of *Dreissenaa polymorpha* (Pall.) (Dreissenidae, Bivalva) in Lakes Receiving Heated Water Discharges," *Pol. Arch. Hydrobiol.* 24:531–545 (1977).

Lyakhov, S. M., "Work of the Institute of Biology of Inland Waters, Academy of Sciences of the USSR," pp. 55–59. in Biology and Control of Dreissena, B. K. Shtegman, Ed. *Tr. Inst. Biol. Vnut.* 7(10):55–59 (1968) (Translated 1968, Israel Program for Scientific Translations, Jerusalem).

CHAPTER 33

Chemical Addition Strategies for Zebra Mussel (*Dreissena polymorpha*) Control in Once-Through Service Water Systems

Renata Claudi and David W. Evans

In dealing with the threat of the zebra mussel invasion at its numerous sites across the province, Ontario Hydro has evaluated a number of possible strategies (both chemical and nonchemical) for short- and long-term control of zebra mussels. For the short term, chemical mitigation appears to be the only feasible option. This chapter examines possible chemical control strategies and the factors to be considered in selecting a particular treatment or combination of treatments.

The chemical addition strategies can essentially be divided into treatments aimed at established adult mussels and treatments targeted at preventing settlement of viable postveligers. Hydro has focused principally on preventing settlement by frequent, intermittent chlorination or by low level continuous chlorination of critical water systems.

0-87371-696-5/93/$0.00 + $.50

INTRODUCTION

Since the initial reports of zebra mussels being found in the Great Lakes, there has been a great deal of uncertainty regarding the selection of an appropriate chemical control strategy. Operators were frequently unclear as to their goals; should they control adult mussels or settling postveligers? Will 100% mortality be achieved or is it even necessary? What chemicals should they use? At what concentration and how often should they apply the chemical? The intention of this chapter is to answer these and similar questions, and to clarify the aims and means of achieving a specific control strategy. Specific examples are presented, based on our experience at Ontario Hydro.

As Ontario's principal supplier of electricity (serving 3.6 million customers) with an in-service capacity of 28,200 MW, Ontario Hydro has a large number of nuclear, fossil, and hydraulic generating sites and transmission facilities dispersed over a wide area (Figure 1). Most of them are in the Great Lakes basin and surrounding watershed. Since zebra mussels first appeared at the Nanticoke Thermal Generating Station (TGS) on Lake Erie in 1989, plans for protection of the Hydro facilities on the Great Lakes and connecting waterways have been implemented. To date, chemical treatment had proved to be the only preventative tool in the fight against the zebra mussel, while hot water flushing and mechanical removal (scraping, water jetting) have been used after the fact to deal with established colonies of adult mussels. As hot water flushing is not always practical, chemicals remain the principal method of control at Ontario Hydro, and will remain so for the next few years.

Field experience in control strategies at Ontario Hydro has focused on: (1) preventing mussel infestations by intermittent or continuous hypochlorination and (2) mechanical removal of established adults. The only pesticide currently approved in Canada for control of zebra mussels is chlorine (as gaseous chlorine or sodium hypochlorite). By law, chlorine levels in the combined station discharge are limited to less than 10 μg/L.

TREATMENT STRATEGIES

The four basic ways to apply a chemical treatment are once at the end of the season, periodically, intermittently, and continuously (Figure 2). Characteristics of each treatment should be considered before a decision on control strategy is made.

In the end-of-season treatment enough oxidizing or nonoxidizing chemical is applied for a period sufficient to kill all adults established in the system at the end of the breeding season. This is the approach of many European facilities and has been used successfully in North America. The end-of-season treatment presupposes that the targeted system can tolerate one season of zebra mussel fouling.

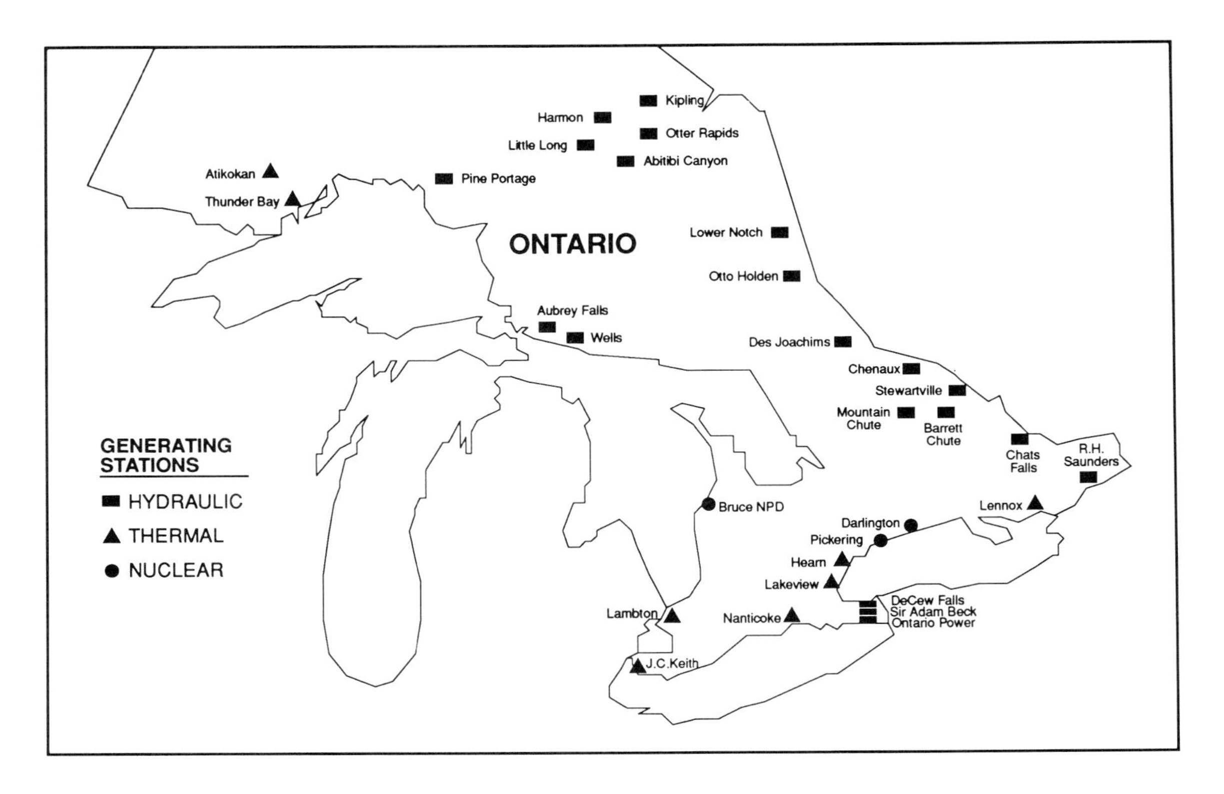

Figure 1. Generating stations of Ontario Hydro.

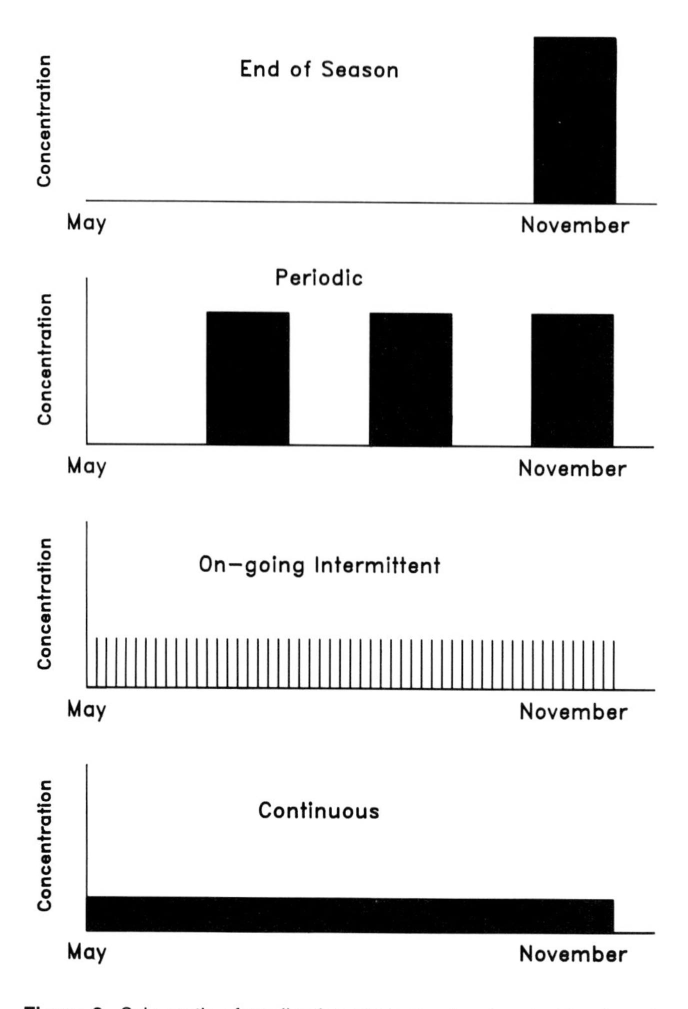

Figure 2. Schematic of application strategies for chemical treatment.

Periodic treatment is only a variant of the end-of-season treatment. Chemical treatment more often is applied during the periods when densities and size of the adult mussels remain fairly low. As in the previous treatment, adult mussels are the target. The chemical concentration needed and the duration of application should be similar to that used for the end-of-season treatment.

Intermittent dosing at frequent intervals (every 12–24 hr) is aimed at preventing zebra mussel infestations in the first place. By killing postveligers that have settled since the previous treatment, 12 or 24 hr earlier, infestation of the system is avoided. At the postveliger stage, the zebra mussel is assumed to be far more susceptible than in its later, adult phase. Although quantitative data on the relative susceptibility of postveliger and adult mussels are currently lacking, field observations at Nanticoke TGS appear to confirm this assumption. Intermittent treatment is used extensively by Ontario Hydro. A standard treatment regime consists of dosing with relatively high levels of chlorine: 2 mg/L for $^1/_2$ hr and at 12-hr intervals throughout the course of the zebra mussel breeding and settlement season. This treatment has proved effective on the six Naticoke units dosed in this fashion in 1990. No new settlement was observed in these systems, but adult mussels that settled there in 1989 survived. As the postveligers seem to be more susceptible, the concentration of chemical used and the duration of application will be significantly less than if adults were the target.

Continuous treatment is designed to discourage all settling in the system. Concentrations of chemical needed can be very low, but the application has to be constant. Typically, continuous treatment is chosen for systems which cannot tolerate any amount of fouling, for example, fire protection water systems. Continuous treatment at 0.5 mg/L total residual chlorine has been used on two units at Nanticoke TGS since the entire fire protection water supply of the station originates from these units. This type of treatment not only prevented any new settlement, but a continuous application of 0.5 mg/L over 90 days also caused 100% mortality of the adult mussels which had settled in the system during the previous season (1989).

FACTORS IN DECISION MAKING

Before selecting a chemical and the appropriate application strategy, many different and often competing factors need to be considered. Basically, these factors can be categorized into biological, physical, and chemical (as affecting operations) and economic components (including safety and regulatory restrictions).

Biological Factors

The extent of the potential mussel invasion at a particular site will depend on the proximity and abundance of the nearest population of adult mussels. When there are reports of only small, isolated populations of adult mussels near the site, veliger densities during the breeding season will probably be too low to cause a major infestation. In such a case, an end-of-season treatment may be all that is required. This would allow a deferral of major capital expenditure on an intermittent or continuous addition system for another year.

The greater concern is when sizable breeding populations of adult mussels, with densities greater than 50,000/m^2, are located within 200 km of the site. If the prevailing winds and currents are coming from the direction of these breeding colonies, a high influx of veligers during the upcoming breeding season can be anticipated. The conservative approach, in this case, is to assume the worst and to prepare for a full-scale invasion before the next breeding season.

Before choosing a chemical treatment strategy, the ecological impact on receiving waters must be considered. For example, if fish spawning beds are located near the water discharge channel of the plant, a treatment regime that would not adversely affect fish eggs or fry would be needed. This might involve using a chemical that is known not to harm the eggs or fry, or it might mean avoiding chemical treatment for the duration of the local fish spawning season. Site-specific testing, such as has been done in the United States with commercial molluscicides (Cherry et al., 1990) may very well be required. Some chemicals or their by-products may be excessively persistent.

Features of the incoming water (such as temperature, pH, and particulate load) may make some chemicals more appropriate than others. Water temperatures can change the efficacy of treatment chemicals in potentially complex ways. For example, the fraction of undissociated hypochlorous acid, the principal biocidal chlorine species in fresh water, will increase at low temperature. However, as a result of the lower metabolic activity of the mussels at low temperature, the net effect will be a reduction in the efficacy of a given chlorine dose as the water temperature decreases. This might make chlorine perhaps less appropriate for end-of-season treatment at some locations.

Physical and Chemical Factors Affecting Operations

Once the probability and size of the invasion has been evaluated, physical and chemical factors affecting operations need to be examined. Characteristics of the water system, such as pipe size, system volume, linear and volumetric flow rates, and materials of construction need to be examined. If the system contains many components susceptible to plugging (such as the spray nozzles on fire protection water systems) or small diameter pipework (such as sampling or instrumentation and control tubing), then excluding adult mussels by intermittent or continuous addition would be advisable. Conversely, if the system is tolerant of plugging (large bore pipework; strainers to facilitate shell removal; redundant systems, e.g., condenser water boxes, which can be sequentially isolated for cleaning and debris removal) then periodic or end-of-season treatment may afford adequate protection.

Systems that flow only infrequently, or remain virtually stagnant, may require only occasional, manual dosing because very long contact times with the chemical can be assured. Systems which have to be available 100% of the time should be kept as clean as possible. In such a situation, either continuous or intermittent chemical addition is advised.

The materials of construction also need to be considered since the choice of treatment strategy and chemical agent may affect these materials both directly and indirectly. For example, the effect of strong oxidizing agents, such as ozone, on the elastomers used in many plants is largely unknown. To requalify elastomers used in nuclear plants would be a very expensive and time-consuming process. Many materials are prone to underdeposit corrosion. Underdeposit corrosion by sulfate-reducing bacteria is common in infestations of the marine blue mussel (*Mytilus edulis*). Similarly, in the experience of Ontario Hydro, underdeposit corrosion by sulfate-reducing bacteria is also a serious problem in freshwater systems. This suggests that heavy colonization by zebra mussels will cause corrosion similar to that observed with marine macrofouling. Even if the deposits of adult mussels are removed mechanically or by thermal or chemical treatment, the byssal threads usually remain attached, with the continued potential for underdeposit corrosion. Concern for underdeposit corrosion is one of the major reasons why Ontario Hydro has tried to avert the colonization of systems through the use of intermittent or continuous chemical treatment. If the water is also used later for other processes downstream, the treatment chemicals have to be compatible with this downstream process. This is generally not an issue in once-through cooling water applications. All of these material constraints will determine whether the proposed treatment should be aimed at preventing the settlement of viable postveligers or if it is sufficient to deal with established colonies of adults after the fact.

Other Factors

In addition to the biological, physical, and chemical aspects of the treatment, economics play a major role in choosing a treatment strategy. Elements to be considered in the economic analysis are operating costs (both labor and consumables) and cost and availability of the application equipment. In the case of Ontario Hydro, economic considerations had some impact on choosing NaOCl over gaseous Cl_2 as the chlorinating agent; gaseous Cl_2 is cheaper for large-scale systems, but the lower capital cost and shorter delivery times favored use of sodium hypochlorite. Another economic consideration is the cost of meeting regulatory requirements, including the cost of monitoring, control, and reporting. In Canada the only pesticide approved for control of zebra mussels has been chlorine; the cost of licensing alternative chemicals would be significant. No economic analysis would be complete without assessing the cost of down time due to fouling of water systems. In the utility sector this can be very significant: the replacement energy cost for a single 900-MW nuclear unit is $250,000/day.

In addition to economics, issues related to both public and worker safety must also be considered when choosing a treatment strategy. Safety concerns were a major factor in the decision by Ontario Hydro to use liquid sodium hypochlorite, rather than gaseous chlorine. It was judged that shipping and handling chlorine gas was inherently more hazardous than liquid hypochlorite. More elaborate safety training of site personnel and local emergency response services (fire fighting, police, medical) are required with the gaseous chlorine option.

REVIEW OF TREATMENT STRATEGIES

After outlining the factors, which are likely to play a key role in selection of a treatment strategy, we will examine the impact of these factors on decision-making.

End-of-Season Treatment

End-of-season treatment is acceptable if the mussel infestation is not likely to be massive and the affected system will tolerate a certain amount of macrofouling. Some means of dealing with the shell and soft tissue debris downstream must be in place, since there will be a large release of shells and soft body parts. A route for disposing of this debris, such as sanitary landfill, must also be available. This has been a problem in some jurisdictions, mostly from the aesthetic standpoint. Concerns about zebra mussels accumulating enough toxic materials to require disposal at a hazardous waste site so far have been groundless. Analysis of year-old mussels (10–15 mm in length) from the Nanticoke intake indicated no significant bioaccumulation of toxic substances such as heavy metals, pesticides, or herbicides.

End-of-season treatment can be implemented with either oxidizing or nonoxidizing chemicals. The required length of application is a function of the chemical used, ambient temperature of the water, and local chemistry (pH, suspended solids, organic loading). Generally, fairly high doses of the chemical are required. The treatment relies directly on the toxic effects of the chemical being added.

Shorter treatment times (6–24 hr) are generally required for nonoxidizing biocides than for oxidizing agents, such as hypochlorite. Apparently mussels do not perceive most of the nonoxidizing biocides as noxious, and continue to filter the biocide treated water. Since oxidizing biocides such as chlorine are recognized as noxious at concentrations in the range of 1–2 mg/L, the adult mussels will close and remain closed for up to 2 weeks. Thus, successful end-of-season treatment with oxidizing chemicals is expected to require continuous dosing at high levels for at least this length of time.

Since the adults in the system will range from 1 to 10 mm, large volumes of organic and inorganic debris will be generated. Close monitoring of the treatment is required to ascertain that maximum mortality was achieved in all infested areas. Experience has shown that 10 to 20% of the mussels will survive the treatment (probably as a result of inadequate dosing in low flow areas within the system). It is generally not feasible to continue treating until 100% mortality is achieved. Corrosion problems could become severe beneath the colonies of attached adults, depending on the materials of construction. Byssal threads tend to remain even after the adult mussel has died and sloughed off. These remaining byssal threads can promote underdeposit corrosion and can provide points of attachment for incoming veligers during the next breeding season.

Periodic Treatment

Periodic treatment during the breeding season is aimed at regular elimination of all adult mussels present in the system. Periodic treatment is appropriate if the system can tolerate some degree of macrofouling. As with the end-of-season treatment, either oxidizing or nonoxidizing chemicals can be used, although nonoxidizing biocides are expected to give quicker results and will therefore minimize operational upsets or interruptions. With periodic treatment, adults in the system will be smaller (generally less than 5 mm) so the resulting volume of debris will be less than in the end-of-season treatment. This represents the major advantage of this strategy over the previous one. Also, 100% mortality is desirable; however, as these treatments are repeated periodically, it is not essential. The amount of chemical needed will be as high for each periodic treatment as that for the end-of-season treatment, making this an expensive option.

Ongoing Intermittent Treatment

Intermittent treatment is aimed at eliminating freshly settled postveligers in the system. The goal is to treat the settled postveligers while their shells are still fragile and permeable and while the byssal attachment is minimal. As noted earlier, intermittent treatment with NaOCl has been used extensively by Ontario Hydro. By minimizing the frequency of intermittent treatment, environmental impact on the receiving waters is kept to a minimum. Also, operating costs are likely to be lower than those for continuous treatment. Intermittent treatment is recommended if the system cannot tolerate the macrofouling of adult mussels, or if the operational preference is to exclude them. There are several advantages to this approach. With fragile shells only about 250 μm long, postveligers will easily pass out of the system. This eliminates downstream plugging of system component and debris disposal problems. Underdeposit corrosion that can result from colonization by adults is avoided.

Both intermittent and continuous treatment (discussed later) will also control microfouling to some extent. By limiting biofilm formation, the treatment may also play a key direct role in preventing settlement of postveligers. Surface preconditioning by macromolecule sorption followed by biofilm formation is thought to be a prerequisite to freshwater and marine macrofouling. Disruption of this biofilm will inhibit subsequent postveliger attachment.

Unlike continuous treatment, the chemical addition with its monitoring hardware does not need to be available 100% of the time, so it is more readily maintained. However, an outage of several days could result in newly settled veligers growing sufficiently to tolerate the intermittent treatment, thus leading to colonization of the pipe work.

Continuous Treatment

Continuous, low-level treatment is designed to prevent or discourage any settlement of postveligers. One possibly important distinction between continuous and intermittent treatment is that it may not be necessary to achieve 100% mortality of the postveligers; merely discouraging settlement within the system is sufficient. Continuous treatment at concentrations currently used will not kill established adults unless applied for a long period of time (i.e., an entire breeding season). Periodic or end-of-season treatments are the preferred methods of eliminating established adult mussels. Ongoing chlorine minimization studies suggest that, with continuous treatment, much lower levels of total residual chlorine (well below 0.5 mg/L) may be sufficient to prevent postveliger settlement, but the long-term effect of these levels on adult zebra mussels is uncertain.

This treatment, to our knowledge, has only been done with oxidizing chemicals to date. Lower concentrations needed to discourage settlement may be lower than concentrations needed to actually kill freshly settled postveligers; thus lower residual oxidant levels may be achieved in the receiving waters, compared with intermittent treatment. However, because it is continuous addition, the total chemical loading to the receiving waters may be higher than with intermittent dosing. Continuous treatment is probably most suited to relatively small, critical subsystems (e.g., safety-related systems), where macrofouling cannot be tolerated and where a very high level of availability is required. The fire water protection systems at Ontario Hydro facilities are treated in this fashion. As with intermittent treatment, underdeposit corrosion is avoided and microfouling is limited. However, chemical addition and monitoring equipment will have to be available at all times during the breeding season.

SUMMARY

When all factors are considered, it is unlikely that there would be a single best strategy for all raw water users. Many users will find that a combination of approaches will serve them best. One combination may be continuous treatment of low volume but critical systems throughout the mussel season, coupled with end-of-season treatment for large unobstructed, high-flow cooling systems.

To date, only a limited number of chemicals have been approved in North America for the treatment of zebra mussels. With more chemicals of varying efficacy coming on the market, the operator's choice will grow. Hopefully, this brief outline will clarify some of the factors to consider in the choice of treatment strategy and chemical.

Although chemicals are currently a principal tool for controlling the zebra mussel invasion, in the long run (especially for new facilities) nonchemical options (such as thermal flushing capabilities or exclusion by design) will be preferred. This is especially important considering the ever-decreasing limits on the discharge of treatment chemicals.

REFERENCE

Cherry, D. W., J. Farris, J. R. Bidwell, A. Mikailoff, R. L. Shema, and J. W. McIntire. in Proceedings of the EPRI International Macrofouling Symposium, Orlando, FL (December 1990), 355–385.

CHAPTER 34

Laboratory Efficacies of Nonoxidizing Molluscicides on the Zebra Mussel (*Dreissena polymorpha*) and the Asian Clam (*Corbicula fluminea*)

Robert F. McMahon, Bradley N. Shipman, and David P. Long

The efficacies of three nonoxidizing, molluscicidal agents against zebra mussels (*Dreissena polymorpha*), were investigated including: an aromatic hydrocarbon — 2-(thiocyanomethyl-thio)benzothiazole (TCMTB), a cationic polyquaternary ammonium compound — poly-[oxyethylene(dimethyliminio)ethylene(dimethyliminio)ethylene dichloride] (PQ1), and a cationic polyquaternary ammonium compound — 1,1′-(methyliminio)*bis*(3-chloro-2-propanol), polymer cross-linked with *N,N,N′,N′*tetramethyl-1,2-ethanediamine (PQ2). TCMTB and PQ1 were also tested against Asian clams, *Corbicula fluminea*. Samples (n ≈ 25) of adult zebra mussels were statically exposed to TCMTB (sample shell length range = 13.5–22.8 mm) and PQ1 (shell length range = 12.8–23.1 mm) at 20°C; and samples of adult Asian clams (n = 65–75), to TCMTB (sample shell length range = 12.2–33.8 mm) and PQ1 (shell length range = 11.5–35.3 mm) at 25°C in continually aerated medium. Juvenile (shell length = 4–11 mm) and adult zebra mussels (shell length = 13–27 mm) (n = 50) were exposed to PQ2 at 20°C. TCMTB

0-87371-696-5/93/$0.00 + $.50

was lethal to zebra mussels at concentrations ≥0.5 mg manufactured product per kilogram and to Asian clams at concentrations ≥0.125 mg manufactured product per kilogram. PQ1 was lethal to zebra mussels and Asian clams at concentrations ≥0.5 and 0.25 mg manufactured product per kilogram, respectively. Lethal concentrations of PQ2 were ≥3 mg manufactured product per kilogram for both juvenile and adult zebra mussels. Juvenile zebra mussels were significantly less tolerant of PQ2 than adults ($p < 0.05$) at 5, 7, and 9 mg/kg. As TCMTB ≥1 mg/kg or PQ1 ≥2 mg/kg induced 100% mortality in zebra mussels and Asian clams more rapidly than exposure to 0.3–0.5 mg/kg residual chlorine (336–505 hr), they may be effective molluscicides for control of bivalve macrofouling in raw water systems. PQ2 proved less toxic to zebra mussels than either TCMTB or PQ1. However, as PQ2 is a registered flocculent in potable water treatment systems at ≤5 mg manufactured product per kilogram, it has potential as a control agent for zebra mussels and Asian clams in municipal water treatment facilities.

INTRODUCTION

Ability to attach to hard surfaces with byssal threads makes the zebra mussel, *Dreissena polymorpha* (Pallas), a major macrofouler of industrial, agricultural, municipal, and power station raw water systems (Clarke, 1952; Greenshields and Ridley, 1957; Griffiths et al., 1989; Mackie et al., 1989; Hebert et al., 1989; McMahon, 1990). The postveliger larvae of this species, when entrained on intake waters, can settle in raw water handling systems in extraordinary numbers, particularly in areas immeditately downstream from intakes where loss of natural turbulence induces settlement (Greenberg et al., 1991). Following dense settlement, mussels grow to sizes that seriously impair and block flow (Clarke 1952; Greenshields and Ridley, 1957; LePage and Bollyky, 1989; McMahon, 1990). Fouling by zebra mussels will become an increasingly important economic problem as this species continues to disperse throughout North America. While a number of control technologies are presently available, development of new, innovative, cost-effective, and environmentally acceptable controls (including new molluscicides) will be required as macrofouling problems become more widespread.

Chemical control of zebra mussel fouling is necessary since mitigation and prevention is not always practical by nonchemical means, particularly in preexisting facilities not readily amenable to retrofitting of nonchemical control technologies. In such facilities fouled piping can be of small diameter, inaccessible, not readily dewatered, and extend for great distances, making manual cleaning difficult (Clarke, 1952; Greenshields and Ridley, 1957; LePage and Bollyky, 1989; McMahon, 1990). The small size and near neutral buoyancy of the zebra mussel postveliger settlement stage (180–290 µm, Mackie et al., 1989) makes straining them from intake water impractical for high volume raw water facilities. Use of chemical agents may also provide a short-term option for zebra mussel control, prior to development and implementation of long-term nonchemical solutions. Molluscicides could also act as fail-safe backups to nonchemical technologies.

Oxidizing agents, such as chlorine, have a long history of use for control of zebra mussels in Europe (Jenner, 1984 and 1990; Mackie et al., 1989 for review). However, chlorination of raw source waters can produce carcinogenic trihalomethane compounds (LePage and Bollyky, 1989); and oxidizing biocides can exacerbate metal corrosion rates, increasing maintenance costs and reducing system performance. Among oxidizing chemicals alternative to chlorine, ozone shows promise for control of zebra mussels. Ozonation is toxic to zebra mussels at relatively low concentrations (<1.5 mg/kg) and does not produce trihalomethanes; however, ozonation may require installation of complex, expensive generation and application systems, and may not be compatible with system materials, particularly rubber gaskets and seals (LePage and Bollyky, 1989).

Nonoxidizing molluscicides have advantages over oxidizing molluscicides because they can be cost-effective, relatively inert to system metalurgies and materials, environmentally acceptable, and incapable of producing deleterious or carcinogenic by-products. In addition, nonoxidizing biocides are toxic to bivalves at low concentrations; are readily inactivated; require little special storage and handling facilities; and can be applied with simple pumps, piping, and sparger lines; and do not produce carcinogenic by-products (McMahon, 1990; McMahon and Tsou, 1990).

The objective of this study was to test the efficacy of three nonoxidizing chemical agents as molluscicides for control of zebra mussels. The agents studied included: an aromatic hydrocarbon, a cationic polyquaternary ammonium compound, and a cross-linked cationic polyquaternary ammonium compound. Comparative testing of the first two of these compounds was also carried out on the freshwater macrofouling bivalve, *Corbicula fluminea*, the Asian clam (for reviews of Asian clam biofouling see, McMahon, 1983; Henager et al., 1985; Isom, 1986).

METHODS

Collections

Zebra mussels were collected from Lake Erie near the mouth of the Raisen River downstream from Monroe, MI and were flown to Arlington, TX overnight in a cooled, insulated container. Asian clams were collected from the Clear Fork of the Trinity River, 1 km below the outfall of Benbrook reservoir, Tarrant County, TX. Specimens of both species were held in the laboratory in continuously aerated, aged, dechlorinated tap water until utilized in experiments. Zebra mussels were maintained at 10°C in a 284-L refrigerated holding tank, while Asian clams were held in 57-L aquariums at 22–25°C. Experiments of molluscicide efficacy were initiated within 30 days of

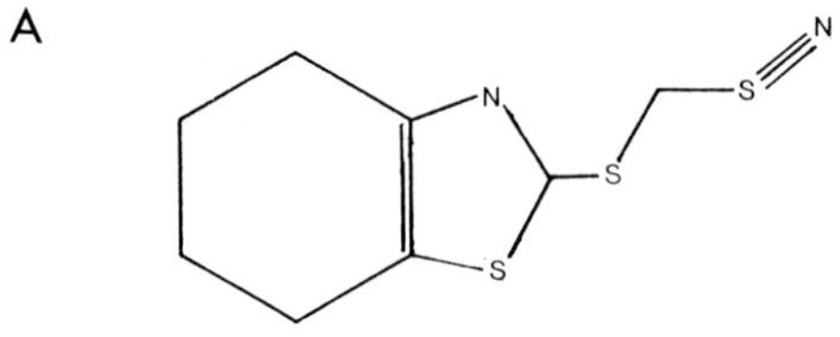

2-(Thiocyanomethylthio)benzothiazole

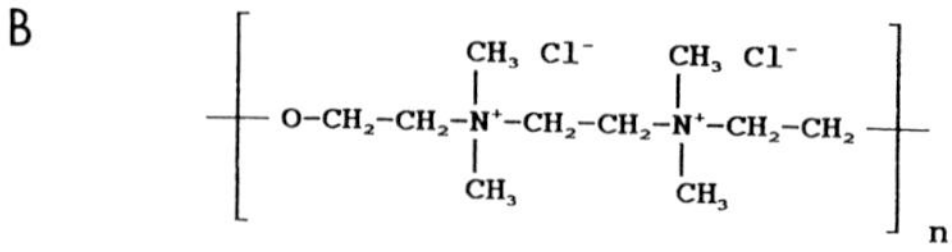

Poly[oxyethylene(dimethyliminio)ethylene (dimethyliminio)ethylene dichloride]

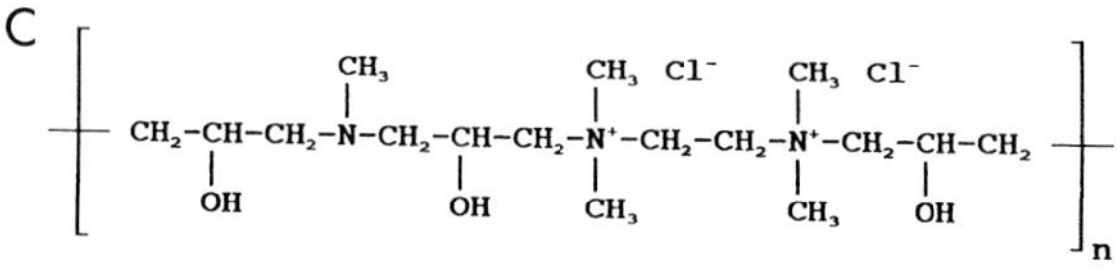

1,1′-(Methyliminio)bis(3-chloro-2-propanol),polymer with N,N,N′,N′-tetramethyl-1,2-ethanediamine

Figure 1. Chemical structures of nonoxidizing molluscicides tested for efficacy in control of *Dreissena polymorpha* and *Corbicula fluminea* macrofouling: (A) TCMTB; (B) PQ1; (C) PQ2.

collection for either species. Specimens of both species did not spawn and remained in good condition during the holding period.

Molluscicides

Nonoxidizing molluscicides tested included an aromatic hydrocarbon, 2-(thiocyanomethylthio)benzothiazole (TCMTB) which as concentrated manufactured material was 30% by weight active product in a solvent system with a specific gravity of 1.08 g/mL (Figure 1A). The second tested product was a cationic polyquaternary ammonium compound, poly[oxyethylene (dimethyliminio)ethylene(dimethyliminio)ethylene dichloride] (PQ1). PQ1 was a straight-chain ionene polymer with positively charged nitrogen atoms in the backbone of its polymeric chain (Figure 1B). As the concentrated manufactured material, PQ1 was 60% active product in aqueous solution with a specific gravity of 1.15 g/mL. The third chemical agent was also a cationic polyquaternary ammonium compound, 1,1′-(methyliminio)*bis*(3-chloro-2-propanol), polymer cross-linked with *N,N,N′,N′*-tetramethyl-1,2-ethanediamine

(PQ2). As concentrated manufactured material, PQ2 was 25% by weight active product in aqueous solution with a specific gravity of 1.09 g/mL. PQ2 was a cross-linked ionene polymer with positively charged nitrogen atoms in the backbone of the polymeric chain (Figure 1C). All three compounds have relatively high flash points (TCMTB = 50°C; PQ1 and PQ2 >100°C), were stable in solution, and were nonfoaming. Throughout this paper, unless explicitly stated otherwise, all test concentrations of these agents are expressed as milligrams of the concentrated manufactured product (as delivered from the supplier) per kilogram of solution with water (mg/kg, equivalent to ppm).

Efficacy Testing

The method for chemical efficacy testing was that described by McMahon and Lutey (1988) and McMahon et al. (1989 and 1990). Zebra mussels and Asian clams were statically exposed to a range of chemical concentrations in 17 L of continually aerated, dechlorinated tap water in glass covered 19-L perspex tanks. Tanks were maintained at 20°C for zebra mussels and 25° C for Asian clams under a 12 hr dark-12 hr light cycle. Test temperatures approximated the modal midsummer ambient water temperatures for both species. The 5°C elevation in test temperature for Asian clams reflected its tropical range (McMahon, 1991). In each test tank, approximately 25 zebra mussels or Asian clams were exposed to a specific chemical concentration. The efficacies of TCMTB and PQ1 were tested against adult specimens of both zebra mussels and Asian clams while that of PQ2 was tested against juvenile and adult zebra mussels.

Concentrations of TCMTB tested against zebra mussels were 0.0-, 0.25-, 0.5-, 1.0-, 2.0-, and 4.0-mg manufactured product per kilogram, while concentrations tested against Asian clams were 0.0, 0.125, 0.25, 0.5, 1.0, 2.0, and 4.0 mg/kg. Zebra mussels were exposed to 0.0, 0.5, 1.0, 2.0, 4.0, and 8.0 mg/kg of PQ1; and Asian clams were exposed to 0.0, 0.25, 0.5, 0.1, 2.0, 4.0, and 8.0 mg/kg. Juvenile and adult zebra mussels were exposed to manufactured product PQ2 concentrations of 0, 0.5, 1.0, 3.0, 5.0, 7.0, and 9.0 mg/kg. In all cases, the 0.0 mg/kg medium was a control, consisting of only dechlorinated tap water.

Three replicate samples of Asian clams (n = 25) were exposed to each test concentration of TCMTB or PQ1. All Asian clams utilized in testing were adults with shell lengths ranging from 12.2 to 33.8 and 11.5 to 35.3 mm in efficacy tests of TCMTB and PQ1, respectively. Limited availability of zebra mussels allowed only 1 sample of 25 to be exposed to each test concentration of these compounds. All zebra mussels utilized were adults with shell lengths ranging from 13.5 to 22.8 and 12.2 to 23.1 mm in efficacy tests of TCMTB and PQ1, respectively. The efficacy of PQ2 was tested against samples (n = 25) of both juvenile and adult specimens of zebra mussels. Juveniles were considered to be less than 1 year old with an anterior-posterior shell length ranging from 4 to 11 mm, while adult mussels were considered to be greater than 1 year old with shell length ranging from 13 to 27 mm (Mackie et al., 1989). For both juvenile and adult mussels, 2 replicate samples (n = 25) were utilized at each test concentration of PQ2.

In all tests, specimens were acclimated to test tanks and media for 24 hr. Zebra mussels were also acclimated to the test temperature of 20°C for an additional 6 days. In efficacy tests of TCMTB and PQ1 against zebra mussels, specimens were allowed to attach to walls and floors of test tanks during the acclimation period and were not removed until after molluscicide exposure had killed them. In tests of PQ2 efficacy, samples of juvenile and adult zebra mussels were first allowed to attach to walls and floors of glass crystallization dishes (9 cm in diameter by 5 cm in height) before being placed in test tanks. The dishes were covered with 1-mm nylon mesh held in place by an elastic band to prevent mussel escape. Confinement of test specimens in this manner allowed rapid inspection of specimens for mortality with minimum disturbance because dishes could be removed from holding tanks. After the acclimation period, tanks were dosed with appropriate amounts of chemicals to achieve test concentrations. Test concentrations were achieved by adding an appropriate volume of an 8,000 or 16,000 mg/kg stock solution to tanks after removing an equivalent volume of water. Stock solutions were freshly made from concentrated agents immediately prior to addition to test tanks. After addition of chemicals to test tanks, survivorship of individuals was tested at least daily throughout exposure periods. Exposure to test concentrations continued until all specimens died, or for a long enough period to assure determination of treatment efficacy.

Behavior of test specimens was observed during each determination of sample survivorship. For both species, the proportion of sample individuals with parted shell valves and extended siphons or closed valves was recorded. Parted valves and extended siphons are characteristic of normal bivalve siphoning behavior, while withdrawn siphons and closed shell valves are a common response to waterborne irritants (Mattice, 1979; McMahon et al., 1989).

Survivorship was tested by observing the capacity of individuals to close shell valves in response to an external tactile stimulus. The posterior mantle margins of zebra mussels were gently touched near the siphons with a blunted dissecting needle. Living individuals responded to this stimulus by withdrawing siphons and rapidly adducting valves. If this response was not observed, the individual was considered to be dead. Asian clams die without valve gaping (McMahon, 1979a). Thus, in order to test their survivorship, individuals were removed from tanks and a dissecting needle tip forced to a depth of 2–3 mm between the shell valves near the siphon (McMahon, 1979a;

McMahon and Lutey, 1988). This probing did not damage clams as indicated by near 100% survivorship of control specimens. Valves of living individuals resisted probe entry and reclosed rapidly on its removal, while valves of dead specimens offered little resistance to probing and remained open after probe withdrawal. After death, individuals of both species were removed from tanks and their shell lengths (measured as the greatest linear distance from the anterior to posterior shell margin) measured to the nearest 0.1 mm with dial calipers.

Tank medium was replaced every 3–4 days to maintain molluscicide concentration and to prevent accumulation of deleterious metabolic wastes. At each replacement, tanks were thoroughly rinsed. Specimens were not fed during experiments because the demand of suspended particulate food material for test chemicals could have reduced exposure concentrations. Lack of food did not appear to deleteriously affect specimens as evidenced by high control survivorship.

Tank pH and water temperature were monitored at each survivorship observation. In all tests, tank water temperature did not range more than 1.5°C from the experimental temperature. Range of pH across all tests combined was 6.5–7.5. Throughout exposure durations, there were no significant differences in tank pH or temperature across concentrations of an agent ($p > 0.05$) or between control and test tanks. The pH range encountered fell within that experienced by zebra mussels (Stanczykowska, 1977) and Asian clams (Kat, 1982; McMahon, 1991) in their natural habitats. Therefore, pH and temperature variation are believed to have had no influence on experimental outcomes.

RESULTS

Mortality of control individuals over exposure periods (range of duration for all tests = 378–1411 hr) was extremely low, varying between 0 and 3.1% (Table 1). Extremely high control survivorship indicates that mortality in treated specimens resulted from the toxic effects of chemical exposure and not from uncontrolled extrinsic factors.

Efficacy of TCMTB

TCMTB was toxic to zebra mussels at ≥0.5-mg manufactured product per kilogram (Table 1, Figure 2). It was not toxic at the lowest concentration tested (0.25 mg/kg) after 826-hr exposure. Mean time to death (MTD) decreased significantly ($p < 0.05$, ANOVA) with increased TCMTB dosage, ranging from 189.4 hr for 4.0 mg/kg to 659.0 hr for 0.5 mg/kg. Corresponding LT_{50} values (LT_{50} = estimated time for 50% sample death [Bliss, 1936])

TABLE 1. Summary of Results of Efficacy Tests of Various Chemical Agents for the Control of Zebra Mussel (*Dreissena polymorpha*) and Asian Clam (*Corbicula fluminea*) Macrofouling of Raw Water Systems

Treatment Level mg/kg Manufactured Product	Treatment Level mg/kg Active Chemical	Sample Size	Mean Time to Death (hr)	Standard Deviation	LT_{50} (hr)	Time for 100% Mortality of Sample (hr)
2-(Thiocyanomethylthio)benzothiazole (TCMTB)						
Dreissena polymorpha						
0.00	0.000	50 (2% dead after 826 hr)				
0.25	0.075	25 (0% dead after 826 hr)				
0.50	0.150	25	659.0	±61.9	652.5	758
1.00	0.300	25	335.2	±48.4	336.0	485
2.00	0.600	25	228.5	±51.5	221.6	313
4.00	1.200	25	189.4	±14.5	183.7	260
Corbicula fluminea						
0.000	0.0000	75 (1.3% dead after 1411 hr)				
0.125	0.0375	75	604.2	±181.9	661.1	1411
0.250	0.0750	75	484.6	±71.8	448.9	735
0.500	0.1500	75	422.8	±80.9	401.4	566
0.000	0.0000	65 (3.1% dead after 160 hr)				
1.000	0.3000	65	102.1	±22.7	96.1	160
2.000	0.6000	65	97.2	±16.8	90.8	127
4.000	1.2000	65	89.9	±18.9	103.5	120
Poly[oxyethylene(dimethyliminio)ethylene(dimethyliminio)ethylene dichloride] (PQ1)						
Dreissena polymorpha						
0.0	0.0	50 (2% dead after 826 hr of exposure)				
0.5	0.3	25	714.3	±135.7	700.2	826
1.0	0.6	25	514.0	±112.6	498.7	680
2.0	1.2	25	231.4	±48.9	215.7	313
4.0	2.4	25	189.3	±31.8	173.7	244
8.0	4.8	25	146.9	±36.9	124.3	197

		Corbicula fluminea				
0.00	0.00	75 (0% dead after 446 hr)				
0.25	0.15	75 (21.3% dead after 446 hr)			556.1	—
0.50	0.30	75	278.9	±71.5	255.7	378
1.00	0.60	75	211.7	±36.0	208.6	283
0.00	0.00	75 (3.1% dead after 160 hr)				
2.00	1.20	75	64.0	±23.8	54.3	113
4.00	2.40	75	58.2	±20.8	49.5	101
8.00	4.80	75	59.5	±22.8	44.8	101
		1,1′-(Methyliminio)*bis*(3-chloro-2-propanol),polymer with *N,N′N′,N′*,-Tetramethyl-1,2 Ethanediamine (PQ2)				
		***Dreissena polymorpha* Juveniles (SL = 4–11 mm)**				
0.0	0.000	50 (2% dead after 1690 hr)				
0.5	0.125	50 (0% dead after 1690 hr)				
1.0	0.250	50 (2% dead after 1690 hr)				
3.0	0.750	50	867.9	±265.8	874.8	1295
5.0	1.250	50	390.9	±65.6	374.2	502
7.0	1.750	50	312.2	±57.8	289.6	408
9.0	2.250	50	205.9	±61.6	197.2	346
		***Dreissena polymorpha* Adults (SL = 13–27 mm)**				
0.0	0.000	50 (0% dead after 1690 hr)				
0.5	0.125	50 (0% dead after 1690 hr)				
1.0	0.250	50 (0% dead after 1690 hr)				
3.0	0.750	50	954.9	±180.1	946.4	1295
5.0	1.250	50	636.2	±129.6	608.9	899
7.0	1.750	50	532.3	±87.0	516.3	669
9.0	2.250	50	502.2	±99.6	531.9	633

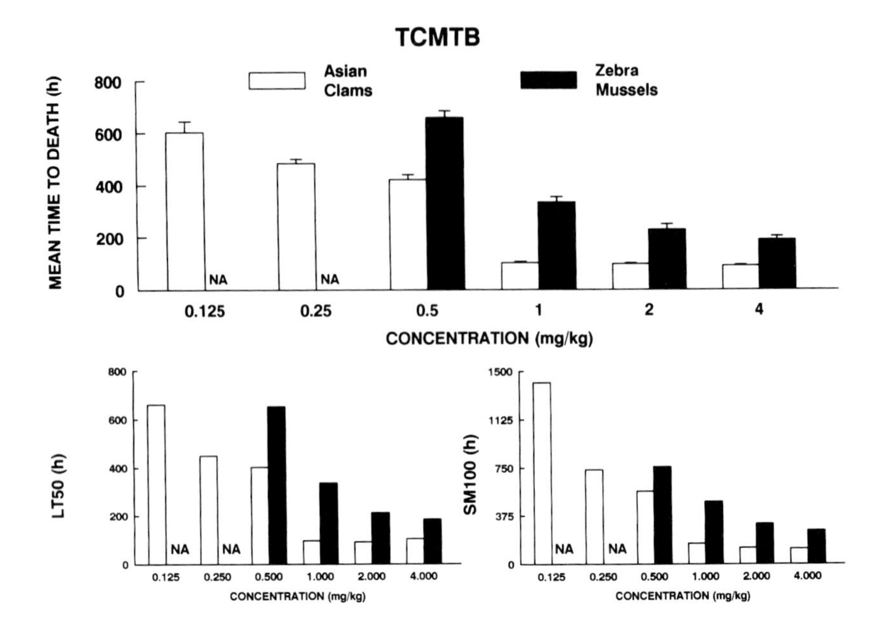

Figure 2. Mean time to death, LT_{50} (estimated hours for 50% mortality of sample), and SM_{100} values (hours for complete mortality of the sample) (Y axes) in samples of *Dreissena polymorpha* and *Corbicula fluminea* statically exposed to different concentrations (as milligram manufactured product per kilogram of solution, X axes) of TCMTB.

were 183.7 and 652.5 hr, respectively. Corresponding SM_{100} times (SM_{100} = time for 100% sample mortality) ranged from 260 to 758 hr, respectively.

TCMTB was also toxic to Asian clams. No significant differences occurred among the mean time-to-death values of replicate samples at any one concentration ($p > 0.05$, ANOVA); therefore, combined data from replicate samples were utilized in all further analyses (combined sample size = 65–75) (Table 1, Figure 2). Complete sample mortality was recorded in all tested concentrations (0.125–4.0 mg/kg). Mean time-to-death values ranged from 604.2 hr at 0.125 mg/kg to 89.9 hr at 4.0 mg/kg; increased dosage significantly reduced mean tolerance time ($p < 0.05$, ANOVA). Corresponding LT_{50} values ranged from 103.5 hr in 4.0 mg/kg to 661.1 hr in 0.125 mg/kg, while SM_{100} values ranged from 120 to 1411 hr over the same range of concentrations (Table 1, Figure 2).

Least squares linear regression analysis relating shell length to individual time to death indicated that time to death was significantly ($p < 0.05$) reduced in larger zebra mussels only at TCMTB concentrations of 0.5 mg/kg. In all other lethal TCMTB concentrations, size did not affect time to death. Similarly, larger Asian clams displayed a significantly reduced tolerance to TCMTB at 0.5 mg/kg and 2.0 mg/kg, but not at other lethal concentrations.

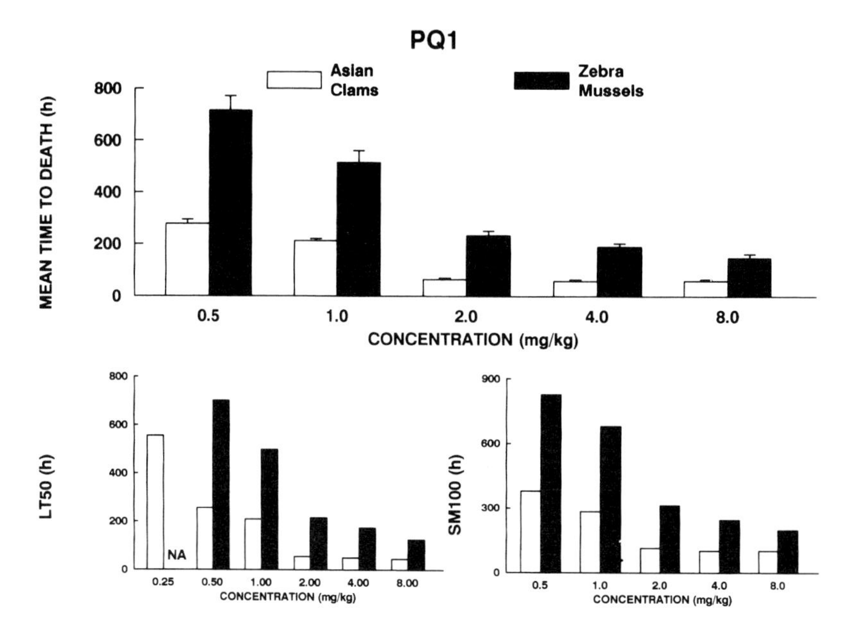

Figure 3. Mean time to death, LT_{50} (estimated hours for 50% mortality of sample), and SM_{100} values (hours for complete sample mortality) (Y axes) in samples of *Dreissena polymorpha* and *Corbicula fluminea* statically exposed to different concentrations (as milligram manufactured product per kilogram of solution, X axes) of PQ1.

Efficacy of PQ1

PQ1 proved toxic to zebra mussels at all tested concentrations (0.5- to 8.0-mg manufactured product per kilogram) (Table 1, Figure 3). Mean time to death decreased significantly ($p < 0.05$, ANOVA) with increased dosage, ranging from 146.8 hr in 8 mg/kg to 714.3 hr in 0.5 mg/kg. Corresponding LT_{50} values ranged from 124.3 to 770.2 hr and SM_{100} values, from 197 to 826 hr, respectively.

Since there were no significant differences in the mean time-to-death values among replicate Asian clam samples exposed to equivalent concentrations of PQ1 ($p > 0.05$, ANOVA), replicate data were combined for further analysis (combined sample size = 75). PQ1 killed Asian clams at all tested concentrations ranging from 0.25-mg manufactured product per kilogram to 8.0 mg/kg (Table 1, Figure 3). In 0.25 mg/kg, 21.3% of the sample had succumbed to PQ1 exposure by experiment termination at 446 hr, yielding an LT_{50} value of 556.1 hr. Sample mean time-to-death values ranged from 59.5 hr in 8.0 mg/kg to 278.9 hr in 0.5 mg/kg. LT_{50} values ranged from 44.8 hr in 8.0 mg/kg to 255.7 hr in 0.5 mg/kg with corresponding SM_{100} values ranging from 101 to 378 hr.

Least squares linear regression analysis relating shell length to individual time-to-death values indicated that time to death was significantly ($p < 0.05$) reduced in larger zebra mussels only at 1.0 mg/kg PQ1, and significantly increased in larger specimens only at 8.0 mg/kg. In all other lethal concentrations, specimen size was not related to time to death. In contrast, time to death was related to shell length in Asian clams at all PQ1 test concentrations other than 0.25 mg/kg, larger individuals being significantly more tolerant of PQ1 in 2.0, 4.0, and 8.0 mg/kg, but significantly less tolerant in 0.5 and 1.0 mg/kg.

Efficacy of PQ2

PQ2 at 3.0- to 9.0-mg manufactured product per kilogram proved toxic to both juvenile and adult zebra mussels (Table 1). No significant differences ($p = 0.25–0.85$; *t*-tests) were recorded between mean time-to-death values in paired replicate samples at equivalent test concentrations for either juvenile or adult mussels, with the exception of juveniles in 9.0 mg/kg (T = 29.3, $p > 0.0001$). As only a single significant difference occurred in eight replicate mean time-to-death comparisons, data for replicates were combined in further analyses.

On exposure to PQ2, mean time to death for juvenile zebra mussels (shell length = 4–11 mm) ranged from 205.9 hr in 9.0 mg/kg to 867.9 hr in 3.0 mg/kg, corresponding to an LT_{50} range of 197.2–874.8 hr and an SM_{100} range of 346–1295 hr, respectively (Table 1). In adult zebra mussels (shell length = 13–27 mm), mean time-to-death values ranged from 502.2 hr in 9.0 mg/kg to 954.9 hr in 3.0 mg/kg, corresponding to an LT_{50} range of 531.9–946.4 hr and an SM_{100} range of 633–1295 hr, respectively.

A comparison of juvenile and adult mean time-to-death values showed that juvenile mussels had significantly reduced survivorship ($p < 0.05$, *t*-tests) in all lethal concentrations of PQ2 with the exception of 3.0 mg/kg (Table 1). At 5.0, 7.0, and 9.0 mg/kg, juvenile values were 61, 59, and 41% of adult values, respectively (Figure 4). This pattern of reduced juvenile mussel tolerance to PQ2 was also reflected in juvenile LT_{50} and SM_{100} values.

A highly significant PQ2 dose effect ($p < 0.05$) was evident for both juvenile and adult zebra mussels ($p < 0.05$, ANOVA). Increased dosage resulted in progressive mean time-to-death reduction in both groups (Table 1, Figure 4). Values for juvenile mussel at all lethal PQ2 concentrations tested were significantly different from each other ($p < 0.05$, Tukey a posteriori test), while for adults, values were not significantly different at 3.0 and 5.0 mg/kg ($p > 0.05$), but were different at 7.0 and 9.0 mg/kg ($p < 0.05$). Values recorded at 7.0 and 9.0 mg/kg were also different from those recorded at 3.0 and 5.0 mg/kg. The pattern of increased mortality rate with increased PQ2 dosage was also reflected in the LT_{50} and SM_{100} values of both juvenile and adult mussels.

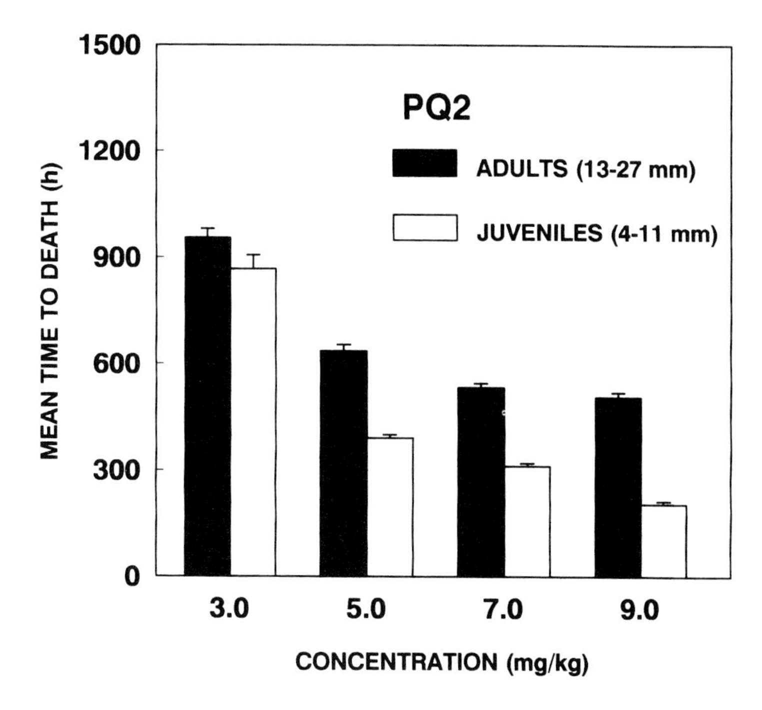

Figure 4. Mean time to death (MTD) in hours (Y axis) for juvenile (shell length = 4–11 mm) and adult (shell length = 13–27 mm) *Dreissena polymorpha* statically exposed to PQ2.

Least squares linear regression analysis indicated that shell length in adult zebra mussels was not significantly related to time to death in any lethal PQ2 concentration (p = 0.76–0.99). Time to death in juvenile mussels was significantly correlated with shell length at 3.0, 5.0, and 7.0 mg/kg ($p < 0.005$). The negative slope values of these regressions indicated that smaller juveniles had significantly elevated tolerance to PQ2.

Behavioral Responses

Behavioral responses of individuals to the three chemicals varied between species, within species, and between concentrations. Asian clams, initially responded to all lethal concentrations of TCMTB by closing shell valves. Initial closure was followed by periodic valve opening and siphoning. Only 10–20% of the test individuals were observed to maintain open valves and extended siphons at any one time during exposure to TCMTB concentrations between 0.5- and 4.0-mg manufactured product per kilogram. At 0.125 and 0.25 mg/kg, 30–50% of individuals opened valves and extended siphons. Throughout exposure, a greater proportion of control individuals had open valves and extended siphons (60–100%), than did Asian clams exposed to any test concentration of TCMTB.

Some individuals of zebra mussels were observed to have opened valves and extended siphons in all tested TCMTB concentrations. The proportion of siphoning individuals in control tanks remained relatively high throughout the exposure period at 85–100%. Similar to Asian clams, the percentage of siphoning mussels was reduced when exposed to TCMTB, ranging from 60 to 70% in 0.25 mg/kg to 20–30% in 4.0 mg/kg. Partial inhibition of siphoning by both Asian clams and zebra mussels on exposure to TCMTB suggests that both species had a limited capacity to detect and respond to the presence of this chemical in inhalant water.

Zebra mussels did not appear to avoid exposure to PQ1. Throughout the exposure period, similar proportions of individuals in control and treated samples had open valves and extended siphons regardless of PQ1 concentration (0–8.0 ppm). In contrast, Asian clams displayed some sensitivity to this chemical. At all tested PQ1 concentrations (0.25–8.0 mg manufactured product per kilogram), proportions of sampled individuals with open valves and extended siphons were similar to those of controls for the first 2–5 days exposure. Thereafter, the proportion of siphoning individuals in treated samples declined. The onset of reduction in siphoning behavior occurred more rapidly in samples exposed to elevated PQ1 concentrations. In the latter stages of exposure, all individuals exposed to 1.0–8.0 mg/kg kept the valves continuously shut, and 50–70% of individuals in 0.25 and 0.50 mg/kg had closed valves. These data suggest that like zebra mussels this species did not initially recognize the presence of PQ1 in inhalant water, but unlike zebra mussels responded to its presence in the latter stages of exposure.

As with PQ1, zebra mussels did not display extensive avoidance of PQ2, particularly at lower test concentrations. Throughout the 1690-hr exposure period, 70–100% of control individuals had open valves and extended siphons. Similar levels of siphoning occurred in juvenile and adult mussels exposed to PQ1 at 0.5–5.0 mg/kg. In 7.0 and 9.0 mg/kg, the percentage of actively siphoning individuals declined to approximately 20–40% of the sample, suggesting that PQ2 may have been irritating at these concentrations.

After 14 days exposure to PQ2, both juvenile and adult zebra mussels in 5.0, 7.0, and 9.0 mg/kg were observed to detach from the byssus while alive. In 3.0 mg/kg, byssal detachment of juveniles and adults was delayed, being first observed after 28–32 days exposure. After detachment, byssal attachment was not reformed. Byssal detachment generally closely preceded death. Byssal detachment preceding death was also observed in zebra mussels exposed to TCMTB and PQ1. Thus, byssal detachment appears to be a response to chronic lethal stress in zebra mussels. No byssal detachment was observed in juveniles or adults exposed to PQ2 concentrations of 0.0, 0.5, and 1.0 mg/kg in which mussel mortality was negligible.

Efficacy Comparisons

Zebra mussels and Asian clams varied in their susceptibility to TCMTB and PQ1. Both chemicals appeared more toxic to Asian clams than to zebra mussels. Static exposure to TCMTB proved lethal to Asian clams at concentrations as low as 0.125-mg manufactured product per kilogram, while it was not lethal to zebra mussels below 0.5 mg/kg (Table 1). At concentrations lethal to both species (0.5–4.0 mg/kg), mean time to death, LT_{50}, and SM_{100} values for zebra mussels where 25–200% greater than for Asian clams (Figure 2). Similarly, zebra mussels appeared more tolerant of PQ1 than Asian clams. While this compound was lethal to both species at the lowest test concentrations (0.25 mg/kg for Asian clams and 0.5 mg/kg for zebra mussels), mean time to death, LT_{50}, and SM_{100} values for Asian clams were 30–50% those recorded in equivalent PQ1 concentrations for zebra mussels (Figure 3).

While there were intraspecific differences between the tolerances of zebra mussels and Asian clams to TCMTB and PQ1, there appeared to be little within species differences in the toxicity of these two chemicals to either zebra mussels or Asian clams at equivalent manufactured product concentrations (Table 1). For either zebra mussels or Asian clams, similar concentrations of the two materials produced very similar mortality times. In contrast, PQ2 at equivalent manufactured product concentrations was less toxic to adult zebra mussels than TCMTB and PQ1. Its mean time to death, LT_{50}, and SM_{100} values were two- to fivefold greater than those recorded at similar manufactured product concentrations of TCMTB or PQ1.

In order to determine whether some of the intraspecific and interspecific differences in the efficacies of the three tested chemicals resulted from differences in the percentage of active product in the manufactured materials (TCMTB = 30%, PQ1 = 60%, and PQ2 = 25% active product in inert solvent solutions), the mean time-to-death values for samples of both species were plotted against the actual concentration of active chemical for each agent tested (Figure 5). Analyzed in this manner, TCMTB appears to be the most toxic of the agents tested against either species. At equivalent active product concentrations (0.3, 0.6, and 1.2 mg/kg), mean time-to-death values for zebra mussels exposed to TCMTB were 45–82% those of specimens exposed to PQ1. Corresponding reductions in mean time-to-death values for Asian clams exposed to TCMTB relative to PQ1 were 37–71%.

When compared on the basis of active product concentration, PQ2 was less toxic to zebra mussels than TCMTB and PQ1 (Figure 5). At an equivalent active product concentration of 1.2 mg/kg, mean time-to-death values for mussels exposed to TCMTB were 29.8% of those of adult mussels and 48.5% of those of juvenile mussels exposed to PQ2. Corresponding values for mussels exposed to PQ1 were 36.4 and 59.2% of the mean time-to-death values recorded for adult and juvenile mussels exposed to PQ2, respectively.

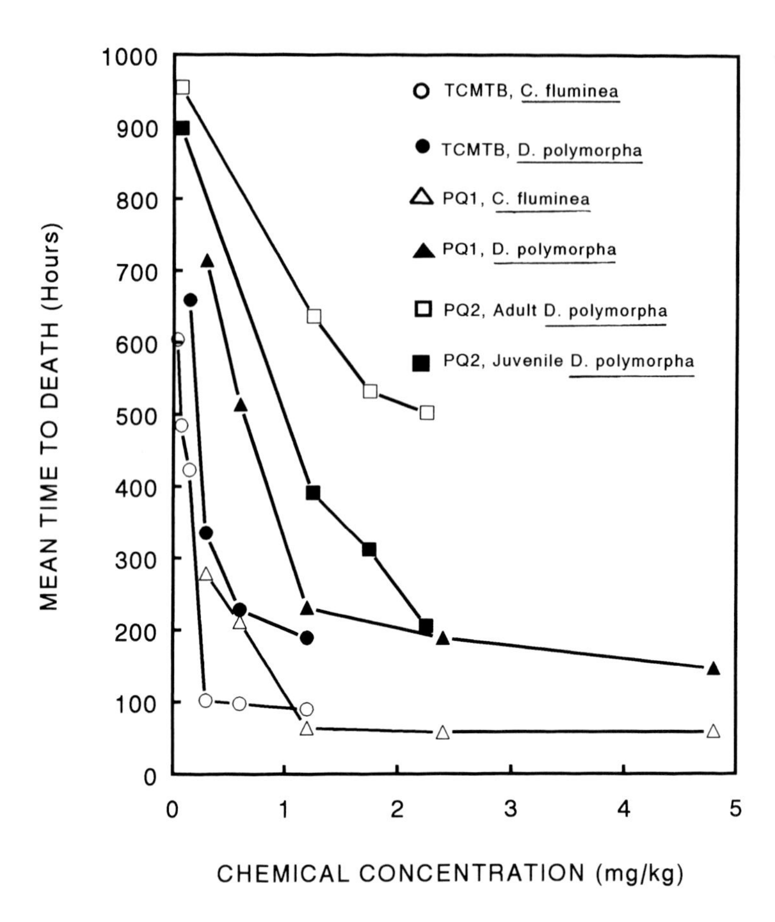

Figure 5. A comparison of mean time to death (MTD) (Y axis) of samples of *Dreissena polymorpha* and *Corbicula fluminea* statically exposed to varying concentrations of the active products of TCMTB, PQ1, and PQ2 at 20°C (*D. polymorpha*) or 25°C (*C. fluminea*).

DISCUSSION

Static laboratory tests indicated that all three tested chemicals had molluscicidal properties for zebra mussels and two tested chemicals, TCMTB and PQ1, had molluscicidal properties for Asian clams. Thus, these test compounds appeared to have potential application for control of fouling by zebra mussels and Asian clams. At higher concentrations, exposure to all three chemicals resulted in kill rates of zebra mussels equivalent to or greater than those produced by exposure to oxidizing biocides. Continuous exposure to PQ2 at 7.0–9.0 mg/kg (juvenile SM_{100} = 346–408 hr; adult SM_{100} = 633–669 hr) resulted in 100% sample kill times similar to those achieved by continuous exposure of zebra mussels to the 0.3–0.5 mg/kg residual chlorine

(SM_{100} > 336–504 hr) recommended for mitigation and control of macrofouling by zebra mussels (Mackie et al., 1989; Jenner, 1984). Continuous exposure of zebra mussels to 1.0–4.0 mg/kg TCMTB (SM_{100} = 260–485 hr) or 2.0–8.0 mg/kg of PQ1 (SM_{100} = 197–313 hr) resulted in equivalent or more rapid mortality than can be achieved by continuous exposure to 0.3–0.5 mg/kg residual chlorine. Exposure of Asian clams to PQ1 and TCMTB also resulted in mortality rates greater than those produced by chlorination. At 2.0–8.0 mg PQ1/kg and 0.5–4.0 mg TCMTB/kg, Asian clam SM_{100} values (range = 101–113 hr and 120–566 hr, respectively) and were considerably lower than the 288–672 hr required for 100% kills of this species by continuous chlorine exposure (Tilly, 1976; Mattice, et al., 1982; Ramsey et al., 1988) at the ≤0.5 mg/kg total free residual chlorine concentrations allowable by the U.S. EPA (1982).

The tolerance of zebra mussels to both TCMTB and PQ1 was considerably greater than that of Asian clams. At equivalent lethal concentrations of TCMTB and PQ1, mean time-to-death values for zebra mussels were 1.6–3.3 and 2.4–3.6 times greater than that of Asian clams, respectively (Figures 2 and 3). In addition, TCMTB induced Asian clam mortality at lower concentrations than it did for zebra mussels. While the reduced tolerance of Asian clams to these biocides may have partially resulted from the 5°C higher test temperature (25 vs 20°C), these results are consistent with observations that zebra mussels have 2–3 times higher tolerance of continuous chlorination than Asian clams (Mattice, 1979; Mattice, et al., 1982; Jenner, 1984; Ramsey, et al., 1988; Mackie, et al., 1989). The elevated molluscicide tolerance of zebra mussels requires further confirmation, but probably should be a consideration in the development of any chemical control program for this species that is based primarily on previous experience with Asian clams.

The elevated tolerance of zebra mussels to TCMTB and PQ1 is somewhat surprising because zebra mussels displayed less avoidance of both compounds than did Asian clams. A greater proportion of zebra mussels maintained siphoning behavior than did individuals of Asian clams in equivalent concentrations of both chemicals; therefore, they should have experienced a greater exposure to the toxic effects of the chemicals. The reduced tolerance of Asian clams to tested molluscicides may lie in this species elevated filtration and metabolic rates. Asian clams have one of the highest filtration rates recorded among freshwater bivalves. An average sized individual (shell length = 25 mm) siphons 750–1000 mL/hr at 20–25°C (Mattice, 1979; Lauritsen, 1986). In contrast, the filtration rate of a zebra mussel with approximately equivalent tissue mass appears much lower, being 50–275 mL/hr (Sprung and Rose, 1988). Elevation of siphoning rates in Asian clams are associated with elevated metabolic rates. Thus, Asian clams maintain higher relative activity levels and oxygen consumption rates than do most other freshwater bivalve species including zebra mussels (McMahon, 1979b and 1983; Alexander and

McMahon, 1991). Elevated filtering rates of Asian clams may have resulted in greater relative water flow over gill and mantle tissues; and thus may have increased molluscicide exposure relative to zebra mussels even if they did siphon less frequently. In addition, elevated metabolic rates of Asian clams could make them more susceptible to molluscicides (such as those examined in this study, and oxidizing molluscicides such as chlorine, ozone, and bromine) which damage epithelial gas exchange surfaces, potentially reducing the capacity to maintain adequate rates of oxygen consumption (Sprague, 1971).

At lethal concentrations, whether as active or manufactured product, longer exposures to PQ2 were required to kill zebra mussels than to TCMTB and PQ1. Mortality rates indicate that TCMTB appeared to be the most toxic of tested compounds to both zebra mussels and Asian clams at the active product concentrations tested (Figure 5). In contrast, PQ2 was the least toxic of tested chemicals to zebra mussels and perhaps also to Asian clams, although this was not tested in this study. PQ1 was only slightly less toxic than TCMTB to either species. Although there were differences in relative efficacies, all three chemicals were able to induce similar or higher mortality rates in both species than continuous exposure to 0.3–0.5 mg/kg free residual chlorine concentrations (Jenner, 1984; Tilly, 1976; Ramsey et al., 1988). Another nonoxidizing molluscicide consisting of the two active compounds, dimethylbenzylammonium chloride and dodecylguanidine hydrochloride, is also reported to induce mortality in both species at higher rates than oxidizing biocides (Lyons et al., 1988 and 1990).

The relatively high efficacy of tested chemicals against zebra mussels and Asian clams compared to chlorination may result from reduced sensitivity of both species to these compounds. Both zebra mussels (Jenner, 1990) and Asian clams (Mattice et al., 1982) detect and respond to low concentrations of chlorine by closing shell valves, preventing exposure of their tissues to its toxic effects. Both zebra mussels (Mikheev, 1968) and Asian clams (Mattice et al., 1982) have a relatively high tolerance of anoxia and can avoid exposure to molluscicides (such as chlorine) for many days while remaining anaerobic with continuously closed valves (duration of tolerated valve closure is reduced at increased temperatures). Only after normal siphoning behavior is resumed are tissues exposed to the effects of waterborne molluscicides. In contrast to avoidance of oxidizing molluscicides by continuous valve closure, zebra mussels and Asian clams opened shell valves and extended siphons a portion of the time during exposure to lethal concentrations of all three test compounds. During initial exposure, siphoning activity generally remained at levels similar to controls, but then declined with increasing exposure duration. The extent of decline in siphoning activity varied inversely with molluscicide concentration. Lack of continuous valve closure may have resulted in both species receiving greater and more immediate contact with tested nonoxidizing molluscicides than initially occurs when valves are continuously closed in response to irritating oxidizing biocides. Similar lack of reactivity by zebra mussels and Asian clams has been reported for another nonoxidizing, surfactant molluscicide (Lyons et al. 1988 and 1990).

Exposure of zebra mussels to increasing concentrations of the three tested biocides resulted in progressive increases in mortality rates reflected in reductions in mean time to death, LT_{50} and SM_{100} values (Table 1). In contrast, static exposures of Asian clams to PQ1 concentrations above 2.0 mg/kg and TCMTB concentrations above 1.0 mg/kg resulted in little appreciable increase in mean time to death, LT_{50}, and SM_{100} values (Table 1, Figures 2 and 3). Thus, while application of increased dosages of TCMTB, PQ1, and PQ2 reduce exposure times necessary to achieve 100% mitigation of zebra mussels, application of TCMTB above 1.0 mg/kg or PQ1 above 2.0 mg/kg may not significantly decrease the exposure times necessary to achieve 100% mitigation of Asian clam macrofouling.

Based on our results, complete (100%) control of zebra mussel macrofouling can be achieved by application of 1.0 mg/kg TCMTB as manufactured product for 20 days, 2.0 mg/kg for 14 days, or 4.0 mg/kg for 12 days at water temperatures above 20°C; for PQ1, corresponding values are 2.0 mg/kg applied for 14 days, 4.0 mg/kg for 11 days, or 8.0 mg/kg for 8 days; and for PQ2 applications, corresponding values for adult mussels greater than 1 year in age are 3.0 mg/kg for 54 days, 5.0 mg/kg for 38 days, 7.0 mg/kg for 28 days, and 9.0 mg/kg for 27 days. Due to their reduced tolerance to PQ2, corresponding mitigation times for juvenile mussels less than 1 year old with PQ2 would be 54 days at 3.0 mg/kg, 21 days at 5.0 mg/kg, 17 days at 7.0 mg/kg, and 15 days at 9.0 mg/kg.

After mitigation of zebra mussel or Asian clam macrofouling by short-term application of chemicals at relatively high concentrations, any of the tested nonoxidizing molluscicides could potentially be utilized for "biostatic" control of these species through continuous application at low concentrations to prevent reoccurrence of juvenile settlement and subsequent system fouling. Continuous application of TCMTB and PQ1 at 0.5 mg/kg and PQ2 at 3.0 mg/kg (manufactured product) would prevent settlement and survival of zebra mussels. For zebra mussels, biostatic applications would be required only when monitoring revealed the presence of postveligers in intake waters (shell length >180 μm) (Mackie et al., 1989; McMahon, 1990; McMahon and Tsou, 1990).

There was little correlation between specimen size and time to death for Asian clams exposed to TCMTB and for zebra mussels exposed to TCMTB and PQ1, indicating that the size and age distributions of fouling populations would probably not need to be a consideration in the development of control programs with these chemicals. In contrast, when exposed to 2.0, 4.0, or 8.0 mg/kg of PQ1, larger Asian clams displayed a marked increase in time to death; while at 0.5 and 1.0 mg PQ1 per kilogram, larger individuals had significantly reduced survival times. Thus, development of control programs for Asian clams with PQ1 may require evaluation of the size distribution of fouling Asian clam populations.

Mortality rates of adult zebra mussels in lethal concentrations of PQ2 were independent of mussel size. In contrast, although survival times of juveniles were reduced relative to adults, smaller juvenile mussels were more resistant to toxic effects of PQ2 than were larger juveniles, as indicated by a significant negative correlation between shell length and time to death in juveniles exposed to 3.0, 5.0, and 7.0 mg/kg. This result was somewhat unexpected based on the general expectation that smaller, younger bivalves would be more susceptible to molluscicide treatment than larger, older individuals.

Juvenile mussels were more susceptible to PQ2 than adults at all lethal concentrations except 3.0 mg/kg. Juvenile mean time-to-death values over 5.0–9.0 mg/kg were 41–61% those of adult mussels. Similarly, smaller Asian clams were more susceptible to toxic effects of PQ1 above 1.0 mg/kg. These results suggest that utilization of these and other molluscicides to control zebra mussel and Asian clam infestations will be more cost-effective if directed against smaller, younger individuals rather than against more resistant older, established infestations with a high proportion of adults. If an intermittent control strategy is chosen, application of molluscicides should be frequent enough (i.e., at least annually) to prevent fouling individuals from reaching adult sizes more resistant to molluscicide exposure.

CONCLUSIONS

Rapid mortalities induced in zebra mussels and Asian clams by TCMTB and PQ1 compared to those achieved by environmentally acceptable limits of continuous chlorination suggest that they hold promise for use as molluscicides to control macrofouling by these species in raw water systems. As continuous exposure to low concentrations of both chemicals also induced 100% mortality in zebra mussels and Asian clams, they could be used as biostatic chemicals for prevention of bivalve macrofouling. The ability of TCMTB and PQ1 to induce mortality in clams and mussels at low concentrations reduces environmental risk because it precludes the necessity for discharge of high biocide concentrations to natural source waters.

Periodic shock application with TCMTB, PQ1, or other oxidizing and nonoxidizing molluscicides may not completely alleviate flow restriction problems associated with zebra mussel macrofouling. Zebra mussels appear to sustain much greater rates of growth and mat formation in North American habitats than reported in Europe (Griffiths et al., 1989; McMahon, 1990; McMahon and Tsou, 1990). In Lake Erie, mussels have caused severe fouling in power station raw water piping within periods of 1–2 months (McMahon,

1990). Thus, even frequent shock application of molluscicides may not prevent water blockage in fouling prone components. In such cases, continuous, low level, biostatic application of control chemicals may be the most efficacious strategy.

PQ2 was less toxic than TCMTB and PQ1 to zebra mussels, making it less efficacious as a molluscicide in most raw water systems. However, annual to twice yearly application of PQ2 to water treatment plant intakes at 5.0 mg/kg to kill recently settled juvenile mussels could prevent mussel fouling from reaching levels that restrict flow below demand. PQ2 is registered for use as a flocculent in potable water systems at manufactured product concentrations ≤5.0 mg/kg. As PQ2 concentrations of ≥3.0 mg/kg induce 100% mortality in zebra mussels, this compound may be efficacious for control of zebra mussel fouling in raw water intakes and transmission lines of municipal potable water treatment systems.

Based on a minimum temperature for successful reproduction of 12°C and a maximal long-term upper thermal limit of 28°C, zebra mussels will eventually range throughout the majority of fresh waters in southern Canada and all but the most southern portions of the United States (McMahon, 1990). The zebra mussel could potentially foul major industrial, municipal, and power generating facilities, as well as navigation locks and water control structures along the length of U.S. navigable inland waterways. Capacity for byssal attachment makes zebra mussels a potentially more damaging macrofouler than Asian clams which have been estimated to cost the U.S. power industry over $1 billion per annum in control, and equipment repair and replacement (Isom, 1986). The eventual cost of zebra mussel macrofouling in the Great Lakes has been estimated to be $2–4 billion (Roberts, 1990). As the zebra mussel eventually spreads throughout the United States and southern Canada, the annual combined cost of Asian clam and zebra mussel macrofouling is likely to increase at least 5- to 10-fold present levels if more effective and environmentally acceptable control measures are not developed.

The nonoxidizing molluscicides described here increase the arsenal of potential molluscicides for control of freshwater bivalve macrofouling of raw water systems. Integration of improved chemical and nonchemical bivalve macrofouling control technologies and development of technologies specifically tailored to control such fouling in specific raw water facilities will be required to achieve cost-effective, environmentally acceptable bivalve macrofouling control for North American raw water facilities.

ACKNOWLEDGMENTS

Dr. William P. Kovalak, Principal Biologist for Detroit Edison, generously collected the zebra mussels utilized in this research. Randal T. Melton assisted with experimentation and data analysis. This research was supported by a grant from Buckman Laboratories International, Inc.

REFERENCES

Alexander, J. E., Jr., and R. F. McMahon. "Temperature and Oxygen Concentration Limits for Zebra Mussels Determined by Respirometry," Abstr. Zebra Mussels: Mitigation Options for Industry, Toronto, Ontario, February 1991.

Bliss, C. I. "The Calculation of the Time-Mortality Curve," *Ann. Appl. Biol.* 24:815–852 (1936).

Clarke, K. B. "The Infestation of Waterworks by *Dreissena polymorpha*, a Freshwater Mussel," *J. Inst. Water Works Eng.* 6:370–378 (1952).

Greenberg, A. B., G. Gubanich, J. Ciaccia, Jr., W. Mucci, and D. Garton. "Investigation and Impact of Zebra Mussels at Cleveland Division of Water," Abstr. Zebra Mussels: Mitigation Options for Industry, Toronto, Ontario, February 1991.

Greenshields, F., and J. E. Ridley. "Some Researches on the Control of Mussels in Water Pipes," *J. Inst. Water Works Eng.* 11:300–306 (1957).

Griffiths, R. W., W. P. Kovalak, and D. W. Schloesser. "The Zebra Mussel, *Dreissena polymorpha* (Pallas, 1777), in North America: Impact on Raw Water Users," in Symposium: *Service Water System Problems Affecting Safety-Related Equipment* (Palo Alto, CA: Nuclear Power Division, Electric Power Research Institute, 1989), 11–27.

Hebert, P. D. N., B. W. Muncaster, and G. L. Mackie. "Ecological and Genetic Studies on *Dreissena polymorpha* (Pallas): A New Mollusc in the Great Lakes," *Can. J. Fish. Aquat. Sci.* 46:1587–1591 (1989).

Henager, C. H., Sr., P. M. Daling, and K. I. Johnson. "Bivalve Fouling of Nuclear Power Plant Service-Water Systems," NUREG/CR-4070, PNL-5300, Vol. 3., U.S. Nuclear Regulatory Commission, Office of Nuclear Regulatory Research, Division of Radiation Programs and Earth Sciences (1985).

Isom, B. G. "Historical Review of Asiatic Clam (*Corbicula*) Invasion and Biofouling of Waters and Industries in the Americas," *Am. Malacol. Bull. Spec. Ed.* No. 2:1–5 (1986).

Jenner, H. A. "Chlorine Minimization in Macrofouling Control in The Netherlands," *Water Chlorination* 5:1425–1433 (1984).

Jenner, H. A. "Biomonitoring in Chlorination Anti-fouling Procedures to Achieve Discharge Concentrations as Low as Possible," in *International Macrofouling Symposium: Symposium Notebook* (Palo Alto, CA: Electric Power Research Institute, 1990), 9–11.

Kat, P. W. "Shell Dissolution as a Significant Cause of Mortality for *Corbicula fluminea* (Bivalvia:Corbiculidae) Inhabiting Acidic Waters," *Malacol. Rev.* 15:129–134 (1982).

Lauritsen, D. D. "Filter-Feeding in *Corbicula fluminea* and Its Effect on Seston Removal," *J. N. Am. Benthol. Soc.* 5:165–172 (1986).

LePage, W., and L. J. Bollyky. ''The Impact of *Dreissena polymorpha* on Water Works Operations at Monroe, Michigan (USA),'' Abstr., Zebra Mussel (*Dreissena polymorpha*) in the Great Lakes, Second International Conference, New York Sea Grant Extension and U.S. Fish and Wildlife Service, Rochester, NY, November 1989.

Lyons, L. A., O. Codina, R. M. Post, and D. E. Rutledge. ''Evaluation of a New Molluscicide for Alleviating Macrofouling by Asiatic Clams,'' *Proc. Am. Power Conf.* 50:1–8 (1988).

Lyons, L. A., J. C. Petrille, S. P. Donner, R. L. Fobes, F. Lehmann, P. W. Althouse, L. T. Wall, R. M. Post, and W. F. Buerger. ''New Treatment Employing a Molluscicide for Macrofouling Control of Zebra Mussels in Cooling Systems,'' *Proc. Am. Power Conf.* 52:1012–1021 (1990).

Mackie, G. L., W. N. Gibbons, B. W. Muncaster, and I. M. Gray. ''The Zebra Mussel, *Dreissena polymorpha*: A Synthesis of European Experiences and a Preview for North America,'' Ontario Ministry of the Environment, Water Resources Branch, Great Lakes Section, Queen's Printer, Toronto, Ontario, 1989.

Mattice, J. S. ''Interactions of *Corbicula* sp. with power plants,'' in *Proceedings, First International Corbicula Symposium,* J. C. Britton, Ed. (Fort Worth, TX: Texas Christian University Research Foundation, 1979), 119–138.

Mattice, J. S., R. B. McLean, and M. B. Burch. ''Evaluation of Short-Term Exposure to Heated Water and Chlorine for the Control of the Asiatic Clam,'' Publication No. 1748, Oak Ridge National Laboratory, Environmental Sciences Division, U.S. Technical Information Service, Department of Commerce, Springfield, VA (1982).

McMahon, R. F. ''Tolerance of Aerial Exposure in the Asiatic Clam, *Corbicula fluminea* (Müller),'' in Proceedings, First International Corbicula Symposium, J. C. Britton, Ed. (Fort Worth, TX: Texas Christian University Research Foundation, 1979a) 227–241.

McMahon, R. F. ''Response to Temperature and Hypoxia in the Oxygen Consumption of the Introduced Asiatic Clam, *Corbicula fluminea* (Müler),'' *Comp. Biochem. and Physiol. Ser. A* 63:383–388 (1979b).

McMahon, R. F. ''Ecology of an Invasive Pest Bivalve, *Corbicula*,'' in *The Mollusca, Vol. 6, Ecology,* W. D. Russel-Hunter, Ed. (New York: Academic Press, 1983) pp. 505–561.

McMahon, R. F. *The Zebra Mussel — U.S. Utility Implications* (Palo Alto, CA: Electric Power Research Institute, 1990).

McMahon, R. F. ''Mollusca: Bivalvia,'' in *Ecology and Classification of North American Freshwater Invertebrates,* J. H. Thorp and A. P. Covich, Eds. (New York: Academic Press, 1991), pp. 315–399.

McMahon, R. F., and R. W. Lutey. ''Field and Laboratory Studies of the Efficacy of Poly[oxyethylene(dimethyliminio)Ethylene(dimethyliminio)ethylene Dichloride] as a Biocide Against the Asian Clam, *Corbicula fluminea,*'' in *Proceedings: of the Service Water Reliability Improvement Seminar* (Palo Alto, CA: Electric Power Research Institute, 1988), 11–27.

McMahon, R. F., and J. L. Tsou. ''Impact of European Zebra Mussel Infestation to the Electric Power Industry,'' *Proc. Am. Power Conf.* 52:988–997 (1990).

McMahon, R. F., B. N. Shipman, and J. A. Ollech. ''Effects of Two Molluscicides on the Freshwater Bivalves, *Corbicula fluminea* and *Dreissena polymorpha*,'' in *Service Water System Reliability Improvement Seminar: Addendum* (Palo Alto, CA: Electric Power Research Institute, 1989), 55–81.

McMahon, R. F., B. N. Shipman, and D. E. Erck. ''Effects of Two Molluscicides on the Freshwater Macrofouling Bivalve, *Dreissena polymorpha*, the Zebra Mussel,'' *Proc. Am. Power Conf.* 51:1006–1011 (1990).

Mikheev, V. P. ''Mortality Rate of *Dreissena polymorpha* in Anaerobic Conditions,'' in *Biology and Control of Dreissena,* B. K. Shtegman, Ed. (Jerusalem, Isreal: Isreal Program for Scientific Translations Ltd., 1968), pp. 65–68.

Ramsey, G. G., J. H. Tackett, and D. W. Morris. ''Effect of Low-Level Continuous Chlorination of *Corbicula fluminea*,'' *Environ. Toxicol. Chem.* 7:855–856 (1988).

Roberts, L. ''Zebra Mussel Invasion Threatens U.S. Waters,'' *Science* 249:1370–1372 (1990).

Sprague, J. B. ''Measurement of Pollutant Toxicity to Fish. III. Sublethal Effects and Safe Concentrations,'' *Water Res.* 5:245–266 (1971).

Sprung, M., and U. Rose. ''Influence of Food Size and Food Quantity on the Feeding of the Mussel *Dreissena polymorpha*,'' *Oecologia (Berlin)* 77:526–532 (1988).

Stanczykowska, A. ''Ecology of *Dreissena polymorpha* (Pall.) (Bivalvia) in Lakes,'' *Pol. Arch. Hydrobiol.* 24:461–530 (1977).

Tilly, L. J. ''Clam Survival in Chlorinated Water,'' Report No. DP-1393, United States Energy Research and Development Administration, Washington, DC (1976).

United States Environmental Protection Agency (1982); Effluent Limitation Guidelines, Pretreatment Standards and Source Performance Standards Under Clean Water Act, Steam Electric Power Generating Point Source Category, 40 CFR, Parts 125 and 423, *Fed. Reg.* 45:68328–68337 (1982).

CHAPTER 35

Chemical Oxidants for Controlling Zebra Mussels (*Dreissena polymorpha*): A Synthesis of Recent Laboratory and Field Studies

John E. Van Benschoten, James N. Jensen, Donald Lewis, and Thomas J. Brady

Chemical oxidants have been used for decades for disinfection and for controlling biofouling. The use of oxidants for controlling zebra mussels is but a recent application of this class of chemicals for biofouling control. Recent laboratory and field study data for chlorine, ozone, potassium permanganate, chloramines, and hydrogen peroxide are presented and discussed. The rate of mortality of zebra mussels depends on a number of factors, including the oxidant type, oxidant concentration, contact time, and temperature. Plots of the log of oxidant concentration vs the log of contact time required for 50% mortality are generally linear. The ability of chemical oxidants to kill adult zebra mussels follows the order: chlorine ≈ ozone > potassium permanganate. The ability of chemical oxidants to reduce the number of zebra mussel veligers follows the order: chlorine ≈ ozone > chloramines > potassium permanganate ≈ chlorine dioxide > hydrogen peroxide. Mortality experiments with adult mussels performed under controlled laboratory conditions yielded similar mortality data as field studies under similar conditions. Mechanisms of zebra mortality are hypothesized and strategies for controlling mussels are examined.

0-87371-696-5/93/$0.00 + $.50

INTRODUCTION

In engineered structures such as water treatment plants, power plants, and water intake pipes, zebra mussel colonization can result in losses in hydraulic capacity, clogging of strainers and filters, obstruction of valves, and nuisance problems associated with the decay of proteinaceous flesh and removal of shells. Chemical oxidants, particularly chlorine, have been used widely for control of microorganisms, slimes, and attached biota. In virtually all water treatment plants, for example, chlorine is added to the treated water for disinfection. Many power plants have used chlorine for control of biofouling, although in recent years the use of chlorine has been reduced greatly due to environmental concerns. The combination of proven effectiveness with familiarity in applying chlorine both in water treatment plants and power installations has resulted in the widespread use of this chemical for zebra mussel control. However, strategies for applying chlorine to optimize treatment effectiveness are not well developed.

While chlorine has become the short-term method of choice for zebra mussel control in North America, other oxidants are also of interest. It is important to identify oxidants that may function as well as chlorine for zebra mussel control but do not result in adverse environmental impacts. Oxidants such as ozone, potassium permanganate, chloramines, and hydrogen peroxide may offer advantages over chlorine in terms of environmental impact (e.g., lower production of currently regulated disinfection by-products), but their effectiveness for controlling zebra mussels has not yet been fully demonstrated.

For chemical oxidants, the mechanisms of zebra mussel mortality are not well understood. Consequently, current application strategies may not be optimal in terms of effectiveness or in minimzing oxidant use. To illustrate the uncertainty in control mechanisms, at least four different chemical addition strategies are being used in once-through cooling systems using chlorine (Evans and Claudi, 1991). These include: (1) intermittent dosing at high concentrations during the spawning season; (2) low-level continuous dosing during the spawning season; (3) short-term (2–3 weeks) continuous, high-level dosing several times during the spawning season; and (4) short-term (2–3 weeks) continuous high-level treatment at the end of the season.

To develop efficient and effective control strategies for zebra mussels, a more fundamental understanding of the factors affecting mortality of zebra mussels by chemical oxidants is needed. The objectives of this chapter are to: (1) summarize laboratory and field results for controlling zebra mussels by chemical oxidants; (2) contrast and compare laboratory and field data by means of simple disinfection models; and (3) assess current control strategies based on available laboratory and field data.

BACKGROUND

Chemical Oxidants for Disinfection and Biofouling Control

The most common disinfectants and biofouling control chemicals also are fairly strong oxidants (i.e., electron acceptors). Frequently used oxidants include chlorine, monochloramine (NH_2Cl), ozone (O_3), chlorine dioxide (ClO_2), hydrogen peroxide (H_2O_2), and permanganate (MnO_4^-). In general, disinfection rates increase with oxidant strength (Rosenblatt, 1975), although some strong oxidants (e.g., peroidate) are weak disinfectants. Oxidants typically are added to contact chambers in drinking water and wastewater treatment facilities to reduce the numbers of pathogenic microorganisms (i.e., disinfection) and to remove slimes. Oxidants are also frequently added to power plant cooling waters to reduce buildup of organisms on condenser walls (i.e., biofouling control). In power plants, biofilms and other biological growths on condenser walls reduce heat transfer rates and can diminish significantly the efficiency of power production. Biofouling can arise from microbiologically produced biofilms (microfouling) or from the growth of larger organisms such as clams and mussels (macrofouling).

Chlorine has been used in North America for drinking water disinfection since 1908 (White, 1986) and for wastewater disinfection since 1892 (EPA, 1986). It is the most common disinfectant and biofouling control agent in North America. Chlorine is added to water as a gas (Cl_2) or salt (e.g., calcium hypochlorite, $Ca(OCl)_2$, and sodium hypochlorite, NaOCl). In either case, the added chlorine reacts rapidly with water to form hypochlorous acid (HOCl) and hypochlorite (OCl^-). The distribution of chlorine between these two chemical species depends on the pH and, to a smaller extent, temperature. The sum of the concentrations of HOCl and OCl^- is called free available chlorine (FAC). FAC reacts rapidly and completely with nitrogen-containing compounds (such as ammonia and amino acids) to form chloramines (also called combined available chlorine). The sum of FAC and chloramine concentrations is called total chlorine. Chloramines are weaker disinfectants and biofouling control agents than FAC (National Research Council, 1987). One of the FAC-ammonia by-products, monochloramine (NH_2Cl), has been found to be effective in controlling the Asiatic clam (*Corbicula fluminea*) and sometimes is used as a disinfectant in drinking water treatment (Cameron et al.,

1989). Chloramines are generally the predominant disinfectants in wastewater treatment and may be the primary biofouling control agents in chlorinated cooling waters containing significant quantities of ammonia and organic nitrogen (Jensen et al., 1990). The chemistry and applications of chloramines have been reviewed elsewhere (Wolfe et al., 1984).

Ozone has been used as a drinking water disinfectant/oxidant for many years in Europe and is becoming increasingly popular in North America. It is an extremely strong oxidant, ranking behind only fluorine in strength among commonly available chemicals. Ozone is also a very good disinfectant. For example, the same level of mortality of the bacterium *Escherichia coli* over a given time can be achieved by using 67 times less ozone than chlorine (National Research Council, 1987). Ozone decomposes rapidly in water. The decomposition rate is greater at high pH and low alkalinity. The high cost of ozone production and relatively rapid rate of ozone self-destruction has limited the use of ozone as a disinfectant.

Potassium permanganate is used widely in the drinking water industry for oxidation of iron and manganese and for control of taste and odor problems. Hydrogen peroxide has not proven to be as practical as other oxidants for use in water and wastewater treatment.

Disinfection Kinetics

The effectiveness of disinfection usually depends on the rate at which disinfection process occurs. Many parameters have been identified as having a significant impact on the rate of disinfection, including: disinfectant type, disinfectant concentration, organism type (including strain and metabolic state), organism density, contact time, and temperature. For process design and operation, it is desirable to develop models to quantitatively describe the effects of these parameters on disinfection kinetics. Disinfection models vary greatly in their complexity, empiricism, and number of adjustable constants. The most sophisticated models explicitly account for the effects of disinfectant concentration, organism density, and contact time. For these models, the effects of disinfectant type, organism type, and temperature are included in the adjustable constant(s). Thus, one would calculate different rate constants for different disinfectants. However, one rate constant would describe the disinfection kinetics of a single organism with a given disinfectant at a specified temperature.

Several disinfection models are listed in Table 1. The simplest model, Chick's law, describes only the effects of contact time on organism density. The Chick-Watson law extends Chick's law to take into account the depend-

TABLE 1. Summary of Disinfection Model

Name	ln $(N/N_0)^a$	Ref.
Chick's law	$-kt$	Chick, 1908
Chick-Watson law	$-kC^nt$	Montgomery, 1985
Generalized model	$-kC^nt^m$	Hom, 1972
Series event model	$-kC^nt + \Sigma_i(kC^nt^i/i!)$	Severin et al., 1984
Collins-Selleck model	$-k\ln(Ct/\tau)$	Selleck et al., 1978

[a] Shown in column 2 are the variables used to describe organism survivorship, (ln (N/N_0). N = organism density at any time, N_0 = initial organism density, C = disinfectant concentration, t = contact time, m, n = constants, τ = Ct at the end of the initial lag.

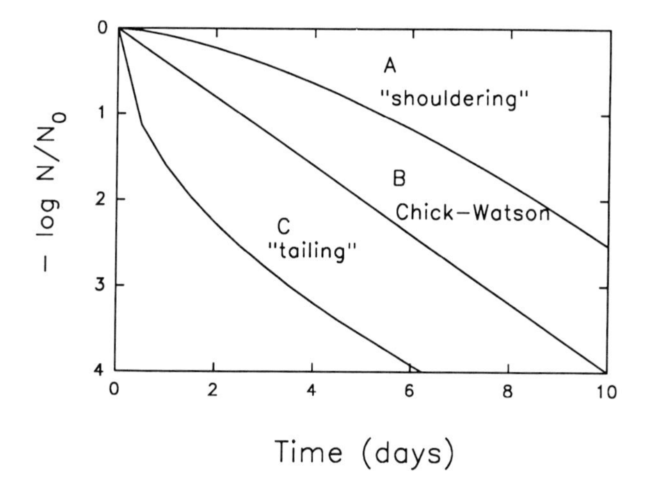

Figure 1. Examples of three generalized disinfection survival curves.

ency of the disinfection rate on the disinfectant concentration. Thus, if disinfection follows the Chick-Watson law, a plot of $-\log (N/N_0)$ vs time should be a straight line (Figure 1, line B). In practice, plots of $-\log (N/N_0)$ against contact time yield curvilinear lines. The generalized model is an empirical attempt to describe both shouldering and tailing phenomena (Figure 1, lines A and C). Note that from the generalized model (Table 1), a plot of the log (disinfectant concentration) vs the log (contact time) at a given percent mortality yields a linear relationship with a slope of m/n. The series event model, nominally based on the Poisson distribution (Wei and Chang, 1975), is used to describe shouldering. The shouldering or initial lag phenomenon is attributed to the necessity of multiple disinfection "hits" prior to mortality. This phenomenon is observed in the mortality of microbial clumps (Wei and Chang, 1975) and multicellular organisms (Montgomery, 1985). The Collins-Selleck model was developed to describe both the initial lag and tailing (Selleck et al., 1978). Both the Chick-Watson law and Collins-Selleck model point to the importance of the parameter Ct, that is, disinfectant concentration multiplied by the contact time. The Ct concept has been used to develop minimum disinfection criteria for both drinking water and wastewater treatment.

SYNTHESIS OF DATA

Over the past several years, numerous laboratory scale studies as well as field scale applications of oxidants for control of zebra mussels have been conducted. A compilation of study results for control of adult and veliger zebra mussels by chemical oxidants is shown in Tables 2–8. Oxidant concentrations listed in column 1 of these tables represent, in most cases, a mean value of the total residual concentration. Variability in the oxidant residual concentrations during some studies shown in the tables was considerable. In the static tests, oxidant residuals decreased with time and oxidant doses are given in the tables.

The highest degree of mortality observed and the corresponding contact time are reported in column 3 of the tables. Methods for assessing mortality in zebra mussel studies vary. In studies of adult mussels conducted at the University at Buffalo, for example, mortality was determined by observing mussel activity after a 24-hr recovery period in untreated water (DeGirolamo et al., 1991). Mussels exhibiting shell movement, siphoning, or motility were scored as alive. In other studies, mortality was evaluated by observations of mussels with ''gaping'' shells (Lewis, 1991). Still other investigators gently probe mussels to test for valve adduction or other movement (Martin et al., 1990). Comparisons of study results would be aided by the use of standardized protocols for assessing mortality.

Because zebra mussels are extremely sensitive to the presence of chemical oxidants, a lag period usually was observed between the time of introduction of an oxidant and the time when mortality was first observed. The time lag is presumably a result of mussels closing their valves to avoid exposure to the oxidant. Under almost all study conditions of adult mortality, a considerable lag period (2–18 days) occurred.

Adult Mussels — Chlorine

Three observations are apparent in the studies examining the response of adult mussels to chlorine (Table 2). First, the range of chlorine concentrations

TABLE 2. A Summary of Selected Data for Chlorination of Adult Zebra Mussels

Conc (mg/L)	Temp (°C)	Mortality[a] (%)	Time Lag[b] (d)	Comments	Ref.
0.5	18–21	100(8–9)	4	Flow through	DeGirolamo et al., 1991
0.2	9–15	50(27)	12	Niagara River	
0.5	9–15	50–70(27)	6		
1.0	9–15	100(27)	4		
0.25	12–15	90(21)	7	Flow through	Jenner and Janssen-Mommen, 1989
0.5	12–15	93(16–17)	7		
1.0	12–15	95(14–15)	2–3		
0.5	17–27	100(9)	5	Static	Klerks and Fraleigh, 1991
0.9	17–27	100(5)	4	Size: 10–15 mm	
3.0	17–27	100(6)	3		
0.32	7.2–17.5	50(53.5)		Flow through	Kerks and Fraleigh, 1991
0.62	7.2–17.5	50(32)		Size: 14–16 mm	
1.74	7.2–17.5	50(16.5)			
4.3	4.8–9.5	50(25)			
9.02	4.8–9.5	50(20)			
0.5	20–22	7(20)		Size 2–6 mm	Martin et al., 1990
1.0	20–22	100(13)			
2.5	20–22	100(9)			
5.0	20–22	100(7)			
1.0	8–12	100(25)	<8	Size: 0.75–2 mm	Van Benschoten (Unpublished)
1.0	8–12	70(28)	~18	Size: >2–5 mm	Lewis
0.60 ± 0.23	10–11.5	40(15)	7	Field	(Unpublished)
0.57 ± 0.20	21.5–24	100(8)	4	Field	

[a] Shown in parenthesis is the time (in days) required for the stated percent mortality.
[b] The time lag is defined as the number of days between the time of chemical introduction and the first observed mortality.

TABLE 3. A Summary of Selected Data for Ozonation of Adult Zebra Mussels

Conc (mg/L)	Temp (°C)	Mortality[a] (%)	Time Lag[b] (d)	Comments	Ref.
0.28	20	40	6	Size: 16.4 ± 2.4 mm	Lewis, 1991
0.54	20	50	5	Size: 16.2 ± 2.5 mm	
1.1	20	100 (14)	3	Size: 16.0 ± 2.1 mm	
1.3	20	100 (11)	4	Size: 12.4 ± 1.3 mm	Lewis, 1991
2.2	20	100 (7)	3	Size: 12.3 ± 1.3 mm	
4.6	20	100 (4)	3	Size: 12.1 ± 1.2 mm	

[a] In parenthesis is the time (in days) required for the stated percent mortality.
[b] The time lag is defined as the number of days between the time of chemical introduction and the first observed mortality.

TABLE 4. A Summary of Selected Data for Control of Adult Zebra Mussels Using Potassium Permanganate

Conc (mg/L)	Temp (°C)	Mortality[a] (%)	Time Lag[b] (d)	Comments	Ref.
0.3		50(26–28)	10–11	Static	Klerks and Fraleigh, 1991
0.6		50(12–28)	9		
2.1		100(9)	6		
0.5	Cold	100(56)		Flow through	Klerks and Fraleigh, 1991
2.5		100(12)			

[a] In parenthesis is the time (in days) required for the stated percent mortality.
[b] The time lag is defined as the number of days between the time of chemical introduction and the first observed mortality.

TABLE 5. A Summary of Selected Data for Chlorination of Zebra Mussel Veligers

Conc (mg/L)	Temp (°C)	Mortality (%)	Comments	Ref.
0		10	Static	Matisoff et al., 1990
0.5		~90	[Cl_2] not constant	
1.0		~90	Contact time = 180 min	
2.5		~50		
0.2	21–22	99	Flow through	Neuhauser et al., 1991
0.5	21–22	99	Residence time = 30 min	
1.0	21–22	99.9		
0.5	Cold	76	Flow through	Van Cott et al., 1991
1.0		85		
2.5		91		
0.65	19–22	100	Field	Lewis (unpublished)

TABLE 6. A Summary of Selected Data for Ozonation of Zebra Mussel Veligers

Conc (mg/L)	Temp (°C)	Mortality (%)	Comments	Ref.
0.1	16	52	Flow through	Neuhauser et al., 1991
0.2	16	98	Residence time = 30 min	
0.5	16	99		
1.0	16	100		

TABLE 7. A Summary of Selected Data for Control of Zebra Mussel Veligers by Chloramines

Conc (mg/L)	Temp (°C)	Mortality (%)	Comments	Ref.
Control		59	Static tests	Matisoff et al., 1990
0.1		49		
1.0		69	60 min exposure	
1.5		75	Mortality corrected	
2.0		66	For controls	
5.0		100		
0.25–3		79–91	Flow through	
1.5		90 (60 min)	Static	
5		95 (60 min)	Static	Van Cott et al., 1991

TABLE 8. A Summary of Control of Zebra Mussel Veligers by Other Oxidants

Conc (mg/L)	Temp (°C)	Mortality (%)	Comments	Ref.
			Hydrogen Peroxide	
3.0	22	89	Flow through	Neuhauser et al., 1991
6.0	22	94	Pulse 30 min/12 hr	
9.0	22	95.5		
0.25	Cold	30	Flow through	Van Cott et al., 1991
0.5		30	H_2O_2 with iron	
1.0		<Control		
			Permanganate	
0.5	Cold	~0	Flow through	Van Cott et al., 1991
1.0		60		
2.5		64		
0		10	Static	Matisoff et al., 1990
0.5		30 (3 min)	Oxidant conc decreased	
1.0		40 (3 min)		
2.5		60 (3 min)		
			Chlorine Dioxide	
0.5		46 [46]	Flow through	Van Cott et al., 1991
1.0		39 [23]	Control mortality in []	
2.0		40 [23]		
3.0		60 [40]		
5.0		70 [10]		

is relatively small, ranging generally from 0.2 to 2.5 mg/L. Second, chlorine is quite effective for controlling zebra mussels; however, contact time can be considerable, especially as temperature decreased. Third, the lag time generally decreased as the chlorine dose increased.

Adult Mussels — Ozone

While considerable interest exists in the use of ozone for zebra mussel control, few studies have been conducted (Table 3). Given that ozone is a stronger oxidant and a better bactericide than chlorine, it is surprising to note that ozone and chlorine are remarkably similar in terms of the contact time required for 100% mortality and the time lag observed prior to first mortality. Although mortality results are similar for ozone and chlorine, byssal threads of zebra mussels were destroyed by ozone within approximately the first 24 hr of exposure. When placed in fresh water after ozonation, mussels were able to regenerate their byssal threads within a day or two. Byssal thread destruction was not observed in the chlorination studies of DeGirolamo et al. (1991) and has not been reported for other studies using chlorine. If detachment of mussels is sufficient so that they can be removed from a structure and collected, ozonation may prove to be an effective zebra mussel control method for some facilities.

Adult Mussels — Potassium Permanganate

In the treatment of potable water, potassium permanganate is used widely for control of foul tastes and odors. Consequently, the use of this oxidant for zebra mussel control is of considerable interest; yet few studies have been conducted. Results indicate that this oxidant is effective for control of zebra mussels, but that the concentrations and contact times required for 100% mortality appear to be greater than for chlorine or ozone (Table 4). For potable water treatment, an upper limit on the concentration of potassium permanganate is imposed by the discoloration of water caused by this oxidant.

Veligers — Chlorine

The results of studies to assess the response of veligers to chlorine are more difficult to interpret than results for adult mussels. Some problems include large fluctuations in veliger number concentrations, high natural mortalities, and difficulties in assessing mortality for these microscopic organisms.

In some studies, the percent loss of veligers from the water column rather than mortality was reported (Neuhauser et al., 1991). The actual time required for veliger mortality is uncertain. Difficulties in data interpretation aside, the results shown in Table 5 suggest that chlorination may be quite effective for controlling zebra mussel veligers at low chlorine concentrations.

Veligers — Ozone

Only one study has examined the response of veligers to ozonation (Table 1). The response shown is for loss of veligers from the water column, not mortality. Based on this criterion, ozone is effective at concentrations above 0.1 mg/L. The degree of removal from the water column increased with increased ozone concentration. In this same study, analysis of sediments after 3 hr showed that about 50% of the veligers exposed to 1 mg/L ozone (0.6 mg/L ozone residual at sediment sampling location) survived. This rather surprising result indicates that even as veligers, zebra mussels are resistant to strong oxidants such as ozone over short time periods.

Note that ozone and chlorine are equally effective (on a weight basis) for the removal of veligers. Thus, ozone and chlorine show comparable efficiencies against both adult mussels (Tables 2 and 3) and veligers (Tables 5 and 6).

Veligers — Chloramines

Recent studies at a water treatment plant in Cleveland have shown that chloramines were quite effective in controlling veliger larvae (Table 7). A mortality of over 90% was found in both static and continuous flow tests at concentrations above 1.5 mg/L. The effectiveness of chloramines at lower concentrations is not known at present.

Veligers — Other Oxidants

Shown in Table 8 is a summary of results for veliger studies using hydrogen peroxide, potassium permanganate, and chlorine dioxide. All of these oxidants appear to be less effective for veliger control than chlorine, ozone, and chloramines. For hydrogen peroxide, for example, 95% removal was observed only at a relatively high concentration of 9 mg/L. It should be noted, however, that mortality may not be required for an oxidant to be effective as a control agent. If the presence of the oxidant prevents settlement but does not cause mortality, the oxidant would be considered successful. Additional research is needed to determine how oxidants affect veliger settlement.

DISCUSSION

Factors Influencing Mortality Kinetics

It is possible to identify several important factors affecting the kinetics of mortality of zebra mussels by chemical oxidants. These factors include oxidant type and concentration, contact time, and temperature.

The effects of concentration and contact time for mortality of adult mussels by chlorine are illustrated by the data of Jenner and Janssen-Mommen (1989) (Figure 2). The time required for a given mortality level decreases with increasing concentration. Also shown in Figure 2 are the lag times prior to the first observations of mussel mortality. At concentrations of 0.25 and 0.5

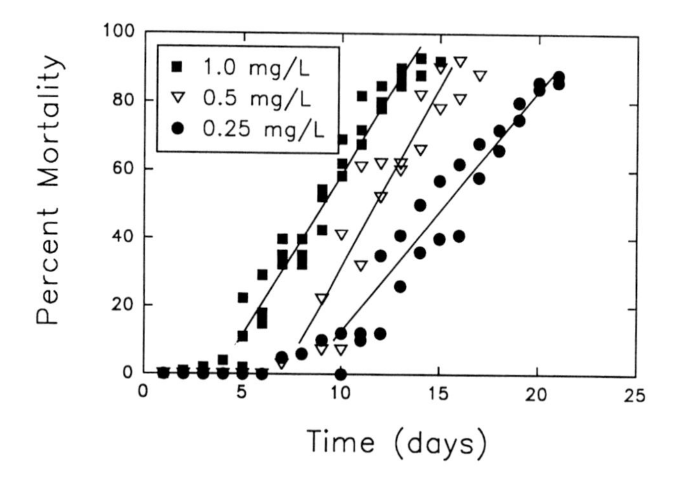

Figure 2. The relationship between zebra mussel mortality and time of exposure for three concentrations of chlorine. Studies conducted at 12–15°C. (Redrawn from Jenner, H. A., and J. P. M. Janssen-Mommen. Abstr., First International Conference on the Zebra Mussel in the Great Lakes, Rochester, NY, November 1989. With permission.)

mg/L, no mortality was observed for approximately 6 days after introduction of the oxidant. The rate of mortality appears to be affected only slightly by the oxidant concentration. Best fit lines through the linear portions of the mortality curves in Figure 2 yield slopes of 9.1, 9.9, and 6.9 for concentrations of 1.0, 0.5, and 0.25 mg/L, respectively.

Data for the ozonation of adult mussels at 20°C are illustrated in Figure 3. As with chlorine, a lag period was observed and the contact time for a given degree of mussel mortality is related inversely to ozone concentration.

Mussel mortality upon exposure to chemical oxidants may be caused by asphyxiation or limited glycolysis because of prolonged shell closure, or by accumulation of toxicants because of continued periodic siphoning during treatment. Although the exact mechanism of mortality is unknown at present, the importance of asphyxiation in zebra mussel mortality may be assessed by an examination of the lag period before death. If the lag period is independent of the oxidant concentration, then mortality simply may be caused by asphyxiation induced by attempts of the mussel to limit exposure to the perceived toxin. Data for chlorination (Figure 2) and ozonation of mussels (Figure 3) suggest this is not occurring. Higher oxidant concentrations appear to result in shorter lag periods.

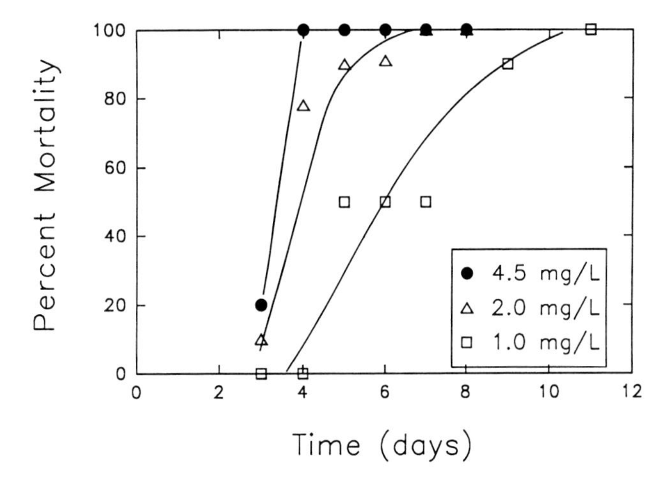

Figure 3. The relationship between zebra mussel mortality and time of exposure for three concentrations of ozone. Studies conducted at 20°C. (From Lewis, D. Abstr., Zebra Mussels Mitigation Options for Industries, Toronto, Ontario, February 1991. With permission.)

Zebra mussels apparently do not completely limit metabolic activity for a time proportional to the perceived toxicity of the oxidant. These observations are consistent with valve movement studies of zebra mussels conducted by Kramer et al. (1989). The presence of chlorine was found to limit the time period that valves were open, but even at 0.5 mg/L chlorine, valve movement was detected during a 12-hr exposure period. Thus, asphyxiation alone cannot explain zebra mussel mortality in the presence of chemical oxidants.

Temperature has a very strong effect on chlorine-induced adult mussel mortality (Figure 4). For the warm water tests shown in Figure 4, 100% mortality was achieved in 9 days while at 9–15°C, 100% mortality was not reached even after 3 weeks of continuous chlorination. From the data in Table 2, the time to 50% mortality decreases by 2.8 to 4.7 times when normalized to a 10°C reduction in water temperature. A 10°C decrease in temperature reduces the disinfection strength of chlorine by 1.6–1.7 times for *Escherchia coli* (Haas, 1990) and 3.0–3.3 times for *Giardia lamblia* cysts (National Research Council, 1987). Temperature appears to have a slightly stronger effect on zebra mussel mortality than on the inactivation of microorganisms. This difference may be attributable to the effect of temperature on zebra mussel metabolism. It is known that mussel respiration increases up to about 30°C, and thereafter decreases (Alexander and McMahon, 1991). Additional data are needed to verify this hypothesis, but many existing studies indicate higher mortality rates with increased temperature.

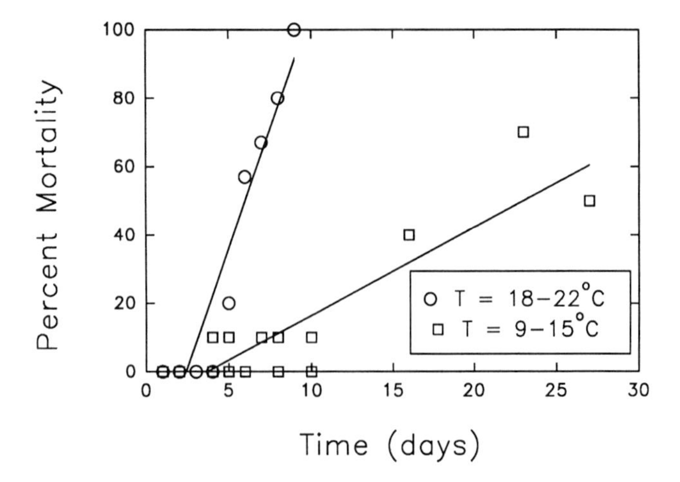

Figure 4. The relationship between zebra mussel mortality and temperature for a chlorine concentration of 0.5 mg/L. (From DeGirolamo, D. J., J. N. Jensen, and J. Van Benschoten. Abstr., American Water Works Association Annual Conference, Philadelphia, PA, June 1991. With permission.)

Disinfection Models

The observations that concentration and contact time are important variables influencing zebra mussel mortality by chemical oxidants suggest that traditional disinfection type models may be useful in describing observed mortality data. The type of disinfection model that should be used for this purpose is uncertain. However, insights to the type of disinfection phenomena occurring can be gained by examination of $-\log (N/N_0)$ vs time plots (Figure 1). Shown in Figure 5 is a plot of $-\log (N/N_0)$ vs time for zebra mussel mortality and chlorine. The curvilinear plots exhibit both shouldering and an initial lag characteristic of multicellular organisms. If one uses the generalized model (Hom, 1972) to describe the shouldering and initial lag, then plots of log (oxidant concentration) vs log (contact time) for a given percent mortality should yield linear relationships useful for comparative purposes.

A plot of log (chlorine concentration) vs the log of time required for 50% mortality of adult zebra mussels is shown in Figure 6. Data for studies conducted at five temperature ranges are presented. The plots appear to be linear and the effects of temperature are well differentiated. Interestingly, the slopes of these lines are all less than -1 (-1.4 to -3.2), implying that n is <1 for a Chick-Watson type of disinfection model. From a practical standpoint, this result means that a reduction in chlorine concentration by one half results in less than a doubling in the required contact time for a given mortality level. This response, also reported by Jenner (1985), implies that the use of lower concentrations minimizes the total mass of chlorine required.

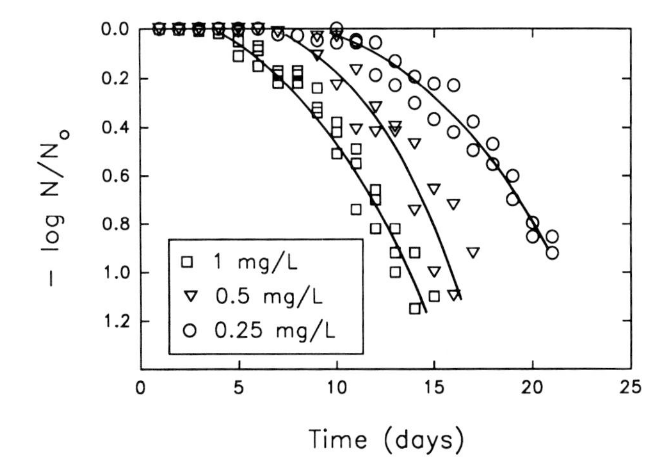

Figure 5. The relationship between survivorship ($-\log N/N_0$) and time of exposure for three concentrations of chlorine at 12–15°C. (From Jenner, H. A., and J. P. M. Janssen-Mommen. Abstr., First International Conference on the Zebra Mussel in the Great Lakes, Rochester, NY, November 1989.)

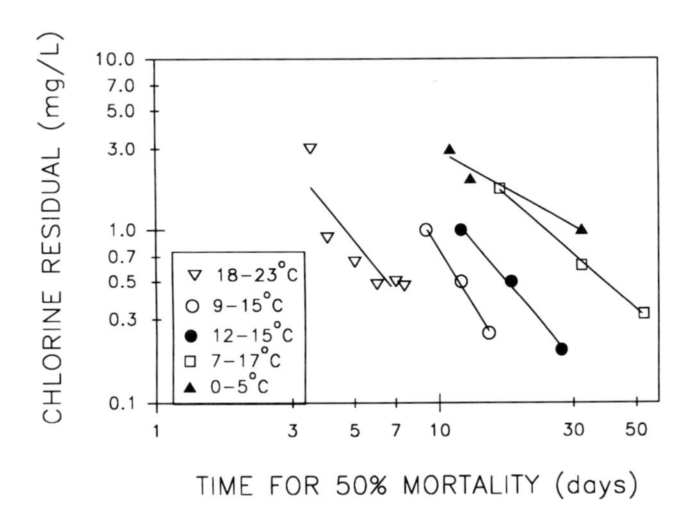

Figure 6. Logarithmic plots of chlorine concentration vs time for 50% mussel mortality as a function of water temperature. Data at 18–23°C from Klerks and Fraleigh (1991), Lewis (unpublished), and DeGirolamo et al. (1991); data at 9–15°C from DeGirolamo et al. (1991); data at 12–15°C from Jenner and Janssen-Mommen (1989); data at 7–17°C from Klerks and Fraleigh (1991); and data from 0–5°C from DeGirolamo et al. (1991).

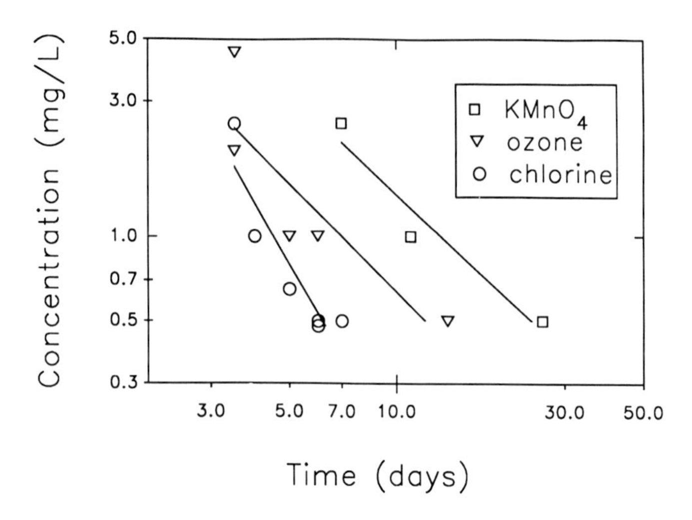

Figure 7. Logarithmic plots of oxidant concentration vs time for 50% mussel mortality for three oxidants. Data for ozone from Lewis (1991); data for KMnO from Van Cott et al. (1991); data for chlorine from Klerks and Fraleigh (1991), Lewis (unpublished), and DeGirolamo et al. (1991).

Plots of the log of oxidant concentration vs the log of contact time (50% mortality) for several oxidants are shown in Figure 7. Chlorine is the most effective of the three oxidants on a weight basis, followed by ozone and potassium permanganate. Although chlorine appears to be more effective than ozone, in fact, most of the data points for the two oxidants are quite close suggesting that they are of similar effectiveness.

Control Strategies

Selection of an optimal strategy for control of the zebra mussel by chemical oxidants is problematic. First, relationships between zebra mussel mortality and oxidant type, dose, and contact time have not been fully developed. The preceding discussion is a first attempt to quantify these relationships. Second, it is uncertain what life stage of the zebra mussel should be the target of the control strategy. It is not apparent from laboratory or field data that one life stage or size class of the zebra mussel is more susceptible to chemical oxidants than others. If periodic treatments are undertaken, the time interval between treatments defines the maximum age of the mussels that are treated. For example, if 3-month-old mussels were more susceptible to a certain oxidant than older mussels, it may then be more effective to implement treatment at 3-month intervals rather than for longer periods. Third, water users must balance the potential need for increased oxidant concentrations and, possibly, increased contact times with limitations on effluent water quality. In controlling zebra mussels, water users may be forced to add chemical oxidants further upstream in their process and/or increase oxidant concentrations. However, water users generally have some upper limit on the amount of oxidant they may add. In drinking water treatment, the upper limit on chlorine may be determined by a maximum contaminant level (MCL) on the trihalomethanes (THMs), a class of chlorination by-products. THM production increases with increasing chlorine concentration and contact time (Fleischaker and Randtke, 1983). (For chlorine dioxide users, there is an additional MCL on the sum of residual chlorine dioxide, chlorite, and chlorate.) Other water users (e.g., municipal wastewater treatment facilities, cooling water users, and industrial processes) have limitations on the amount of total chlorine that can be discharged. Fourth, water users seek to minimize costs. Minimizing the amount of chemical used to control zebra mussels will result in cost savings in two ways. By limiting halogen or other disinfecting chemical use, there is a direct savings in chemical costs. In addition, reduced chemical usage minimizes the costs associated with the use of taste and odor controls such as dechlorination processes or activated carbon.

Treatment Variables

To implement a zebra mussel control strategy with chemical oxidants, treatment personnel must select the oxidant type and concentration, contact time, and time of year at which treatment occurs. In terms of concentration and contact time, two strategies generally are employed. In the first strategy, one may add oxidant doses for short periods with the objective of killing all mussels that have settled in the water system. The second strategy involves continuous addition of oxidant to prevent settlement.

Intermittent treatment is appealing since chemical costs can be less than for continuous treatment and the likelihood of equipment failure or other operational problems is reduced. Because treatment is intermittent, settlement between treatments is expected and water systems for which this type of control is designed must be able to tolerate some infestation. Intermittent treatment has three disadvantages. First, there is an inherent risk in any strategy that allows mussels to periodically enter a system and settle. This is particularly true for industrial users who often have complex service water pipes. Unless it is known that sufficient residual oxidant is reaching each area of the plant during treatment, there will remain uncertainty about potential infestation. Second, environmental regulations are set at extremely low limits for oxidant residuals in industrial effluents. To effectively control mussels at the far reaches of a water system using intermittent treatment, high oxidant doses may be required. In the event that residual oxidant concentrations may violate regulations for residual oxidant, addition of reducing agents to the effluent is necessary. Third, the use of high oxidant concentrations may not represent the best treatment strategy in terms of minimizing the total amount of chemical applied. In developing a zebra mussel control strategy, it is important to assess whether there is an equal tradeoff between oxidant concentration and contact time for a given mussel mortality level. In terms of a Chick-Watson type of disinfection model, such a tradeoff is to ask whether the exponent of concentration term is equal to one. Data for chlorination (Figure 6) indicate that dependence of mortality time on concentration is higher than first order in concentration. The management implication of this result is that lower concentrations of chlorine minimize the total amount of chlorine needed for a given mortality level.

The second common strategy, continuous treatment, usually involves continuous dosing at lower oxidant levels. In terms of treatment costs, continuous treatment may be more costly than intermittent treatment, even if somewhat higher residual oxidant concentations are required on an intermittent basis. For example, continuous dosing at 0.2 mg/L for 4 months would require about 70% more oxidant than 2 weeks of dosing at 1.0 mg/L.

In terms of the time for implementing control strategies, water users have the choice of adding chemical oxidants in the summer or winter months. Clearly, mortality of adult mussels would be faster in the summer when water temperatures are higher. However, to minimize operational problems during treatment for zebra mussels, it may be advantageous to implement control strategies during winter, even though zebra mussel mortality rates are slower. In water treatment plants, for example, winter chlorination for zebra mussel control may minimize potential taste and odor complaints and THM formation. Industries often schedule maintenance during winter. Thus, a longer duration of treatment and associated extra costs may not be of great concern compared to the potential disruption of normal operations if treatment is undertaken during summer. However, additional research is needed to define concentration, contact time, and mortality relationships at cold temperatures.

SUMMARY AND CONCLUSIONS

In controlling the fouling of water systems by zebra mussels, mussel mortality depends on a number of factors including oxidant type, oxidant concentration, contact time, and temperature. The ability of chemical oxidants to kill adult zebra mussels follows the order: chlorine ≈ ozone > potassium permanganate. The ability of chemical oxidants to remove zebra mussel veligers from the water column follows the order: chlorine ≈ ozone > chloramines > potassium permanganate ≈ chlorine dioxide > hydrogen peroxide. In the few studies to date, mortality experiments on adult mussels performed under controlled conditions yielded mortality data similar to that from field studies under similar conditions.

Disinfection models developed for microorganisms have been shown to be useful for describing the impact of oxidant concentration and contact time on mortality kinetics. Plots of the log of oxidant concentration vs the log of contact time required for 50% mortality are generally linear. The log-log plots also illustrate the importance of temperature on mortality kinetics for chlorine. For example, the time to achieve 50% mortality at a chlorine residual of 1 mg/L at water temperatures of 18–23°C and 7–17°C, was about 5 and 30 days, respectively. Most investigators also observe an initial lag in mortality, a response characteristic of multicellular organisms.

The results of this survey may be used to direct control strategies. Two examples are noteworthy. First, mortality kinetics are expected to be much slower at the lower water temperatures experienced in the winter months. Second, reducing chlorine concentrations by one half results in 50% mortality in less than twice the contact time. Thus, overall chemical costs may be reduced at lower oxidant doses. Other factors such as effluent limitations, maintenance schedules, costs, and water quality tissues may dominate the choice of an oxidant application schedule.

REFERENCES

Alexander, J. E., and R. F. McMahon. ''Temperature and Oxygen Concentration Limits for Zebra Mussels — Determined by Respirometry,'' Abstr., Zebra Mussels Mitigation Options for Industries, Toronto, Ontario, February 1991.

Cameron, G. N., J. M. Symons, D. Bushek, and R. Kulkarni. ''Minimizing THM Formation During Control of the Asiatic Clam: a Comparison of Biocides,'' *J. Am. Water Works Assoc.* 81:10:53–61 (1989).

Chick, H. ''An Investigation of the Laws of Disinfection,'' *J. Hygiene* 8:92–158 (1908).

DeGirolamo, D. J., J. N. Jensen, and J. Van Benschoten. ''Inactivation of adult zebra mussels by chlorine,'' Abstr. American Water Works Association Annual Conference, Philadelphia, PA, June 1991.

Environmental Protection Agency (EPA) ''Design Manual: Municipal Wastewater Disinfection,'' Office of Research and Development, Water Engineering Research Laboratory, United States Protection Agency Report 625/1-86/021 (1986).

Evans, D., and R. Claudi. ''Chemical Addition Strategies for Zebra Mussel Control in Once Through Cooling Systems,'' Abstr., Zebra Mussels Mitigation Options for Industries, Toronto, Ontario, February 1991.

Fleischacker, S. J., and S. J. Randtke. ''Formation of Organic Chlorine in Public Water Supplies,'' *J. Am. Water Works Assoc.* 75(3):132–138 (1983).

Haas, C. N. ''Disinfection,'' in *Water Quality and Treatment*. F. W. Pontius, Ed. American Water Works Association, (New York: McGraw-Hill Book Company, 1990) 877–932.

Hom, L. W. ''Kinetics of Chlorine Disinfection in an Ecosystem,'' *ASCE J. Sanit. Eng. Div.* 98:183–194 (1972).

Jenner, H. A. "Chlorine Minimization in Macrofouling Control in The Netherlands," in *Water Chlorination: Chemistry, Health Effects and Environmental Impact, Vol. 5*, R. L. Jolly et al., Eds. (Chelsea, MI: Lewis Publishers, 1985), pp. 1423–1433.

Jenner, H. A., and J. P. M. Janssen-Mommen. "Control of Zebra Mussel in Power Plants and Industrial Settings," Abstr., First International Conference on the Zebra Mussel in the Great Lakes, Rochester, NY, November 1989 (1990).

Jensen, J. N., C. LeCloirec, and J. D. Johnson. "Measurement of Chlorine Residuals in Chlorinated Cooling Waters: Effect of Organic Nitrogen. in *Water Chlorination: Chemistry, Health Effects and Environmental Impact, Vol. 6* R. L. Jolly et al., Eds. (Chelsea, MI: Lewis Publishers, 1990), pp. 535–544.

Klerks, P. L., and P. C. Fraleigh. "Control of Adult Zebra Mussels with Sodium Hypochlorite, Potassium Permanganate, and Hydrogen Peroxide with Iron," Abstr., Zebra Mussels Mitigation Options for Industries, Toronto, Ontario, February 1991.

Kramer, J. M., H. A. Jenner, and D. de Zwart. "The Valve Movement Response of Mussels: A Tool in Biological Monitoring," *Hydrobiologia* 188/189:433–443 (1989).

Lewis, D. "Effect of Ozone on Adult Zebra Mussels Under Different Temperature Regimes," Abstr., Zebra Mussels Mitigation Options for Industries, Toronto, Ontario, February 1991.

Martin, I. D., M. A. Baker, and G. L. Mackie. "Comparative Efficacies of Sodium Hypochlorite and Registered Biocides for Controlling the Zebra Mussel, *Dreissena polymorpha* (Bivalve Heterodonta)," Abstr. Zebra Mussels: The Great Lakes Experience, University of Guelph, Guelph, Ontario, February 1990.

Matisoff, G., P. Fraleigh, A. B. Greenberg, G. Gubanich, G. L. Hoffman, P. L. Klerks, P. L. McCall, R. C. Stevenson, W. Van Cott, and M. E. Wenning. "Controlling Zebra Mussels at Water Treatment Plant Intakes — Part II. Veliger Dose/Response Static Tests," Abstr., International Macrofouling Symposium, Electric Power Research Institute. Orlando, FL, December 1990.

Montgomery, J. M., Consulting Engineers. *Water Treatment Principles and Design* (New York: John Wiley & Sons, Inc., 1985).

National Research Council. "Disinfectants and Disinfectant By-Products," *Drinking Water and Health, Vol. 7* Subcommittee on Disinfectants and Disinfection By-Products, Safe Drinking Water Committee, Board on Environmental Studies and Toxicology, Commission on Life Sciences (Washington, DC: National Academy Press, 1987).

Neuhauser, E. F., J. E. Van Benschoten, and J. N. Jensen. "Effect of Selected Ovidants on Zebra Mussel Veligers — Part III," Abstract, Zebra Mussels Mitigation Options for Industries, Toronto, Ontario. February 1991.

Rosenblatt, D. H. "Chlorine and Oxychlorine Species Reactivity with Organic Substances," in *Disinfection: Water and Wastewater,* J. D. Johnson, Ed. (Ann Arbor, MI: Ann Arbor Science Publishers, Inc., 1975), pp. 249–276.

Selleck, R. E., B. M. Saunier, and H. F. Collins. "Kinetics of Bacterial Deactivation with Chlorine," *ASCE J. Sanit. Eng. Div.* 104:1197–1212 (1978).

Severin, B. F., M. T. Suidan, and R. S. Engelbrecht. "Series Event Kinetic Model for Chemical Disinfection," *ASCE J. Sanit. Eng. Div.* 110:430–439 (1984).

Van Cott, W., R. C. Stenenson, P. Fraleigh, G. Matisoff, and P. L. Klerks. "Controlling Zebra Mussels at Water Treatment Plant intakes. III. Preliminary overview," (Unpublished manuscript, 1991).

Wei, J. H., and S. L. Chang. ''A Multi-Poisson Distribution Model for Treating Disinfection Data,'' in *Disinfection: Water and Wastewater,* J. D. Johnson et al. Eds. (Ann Arbor, MI: Ann Arbor Science Publishers, Inc., 1975) pp. 11–48.

White, G. C. *Handbook of Chlorination,* 2nd ed. (New York: Van Nostrand Reinhold Company, 1986).

Wolfe, R. L., N. R. Ward, and B. H. Olsen. ''Inorganic Chloramines as Drinking Water Disinfectants: A Review,'' *J. Am. Water Works Assoc.* 76:74–88 (1984).

CHAPTER 36

Controlling Zebra Mussel (*Dreissena polymorpha*) Veligers with Three Oxidizing Chemicals: Chlorine, Permanganate, and Peroxide + Iron

Paul L. Klerks, Peter C. Fraleigh, and Robert C. Stevenson

This study investigated the control of zebra mussel (*Dreissena polymorpha*) veligers in water intakes with three oxidizing control chemicals: sodium hypochlorite, potassium permanganate, and hydrogen peroxide with iron (subsequently referred to as "peroxide"). Experiments consisted of short-term static exposures of veligers to chlorine and permanganate, and flow-through exposures to chlorine, permanganate, and peroxide. Results of static experiments showed that application concentrations ≥0.5 mg/L chlorine, resulted in 100% mortality after 2 hr. Static permanganate applications of 0.5–2.5 mg/L had inconsistent effects. Flow-through exposures to all oxidants resulted in increased percentages of dead veligers among veligers in the outflows of bioboxes. This effect was much stronger for chlorine than for permanganate or peroxide. Flow-through exposures also resulted in an inactivation effect where a reduction in the number of veligers between the inflow and outflow of bioboxes occurred. The inactivation effect was stronger for permanganate than for peroxide, with the effect for chlorine being intermediate. These two effects combined resulted in a drastic reduction of the number of live veligers as the water passed through the bioboxes. However, the inactivation component may be specific to

0-87371-696-5/93/$0.00 + $.50

low-flow conditions of bioboxes because some inactivation also occurred in control bioboxes. The relative effectiveness of the three oxidants in killing veligers (chlorine $>$ permanganate $=$ peroxide) differed from the situation with older settled mussels in that permanganate is relatively more effective against settled mussels. Permanganate will be as effective against veligers as chlorine if the inactivation occurs in water intake systems as well, because inactivation will lead to a secondary kill through a prolongation of the exposure time. Taking into account that planktonic veligers in an intake pipe will only be exposed for the time it takes the water to pass through the pipe, control strategies aimed at the veliger stage appear only effective with water residence times in excess of 1 hr. Control aimed at settled mussels will be more effective in situations with short water residence times.

INTRODUCTION

Water intakes are ideally suited for zebra mussels (*Dreissena polymorpha*), as the walls provide a substrate for attachment while the water flow ensures a continuous food supply. Thick crusts of zebra mussels have been found on the insides of water intake pipes (Clarke, 1952; Kovalak et al., 1990). To prevent biofouling-related problems, a control program is often needed when raw water is taken in from an area with zebra mussels.

The control of zebra mussels can potentially be achieved by several strategies. Control can be accomplished by killing veligers before they settle. Alternatively, biofouling can also be prevented if zebra mussels are kept from settling by the creation of an environment in which settlement is postponed. Finally, control can be exerted by killing the mussels after they have settled, by either continuous or periodic control measures. An important factor for determining which strategy is optimal for a certain situation is the relative effectiveness of antifoulants on veliger survival, veliger settlement, and survival of attached mussels.

Oxidants are often used as chemical control agents, especially in drinking water intakes where the choice of alternatives is limited by the need to supply a nontoxic end product. Traditionally, chlorine has been the oxidant of choice, and its effectiveness against settled zebra mussels is well established (Greenshields and Ridley, 1957; Jenner, 1985). However, effects of chlorine on the veliger stage of the zebra mussel have not been widely investigated. In one study, veligers were still alive after being exposed to 7.5 mg/L residual chlorine for 1 hr (Wilhelmi — cited in Clarke, 1952). This low sensitivity of the larvae to chlorine is consistent with a median lethal concentration of 6.7 mg/L chlorine reported for an 8-hr exposure of larvae of the Asiatic clam, *Corbicula fluminea* (Sickel, 1976 — cited in Doherty and Cherry, 1988). Thus, chlorine might not be the best antifoulant for use against veligers. Moreover, concerns about ecological effects of chlorine itself and public health effects of chlorination-derived trihalomethanes have spurred interest in the use of alternatives to chlorine (Cameron et al., 1989; Jenner, 1983). Potassium permanganate and hydrogen peroxide are strong oxidizers that can potentially be used for controlling zebra mussels. However, their effectiveness against zebra mussels has not been established, though data obtained for other organisms are not very encouraging. Potassium permanganate is neither very effective as a bacterial disinfectant (Cleasby et al., 1964; Waite and Fagan, 1980) nor in controlling juveniles and adults of the Asiatic clam (Cameron et al., 1989; Chandler and Marking, 1979). Similarly, hydrogen peroxide is not a very strong bactericide (Baldry, 1983; Toledo et al., 1973; Von Bockelmann and Von Bockelmann, 1972; Waite and Fagan, 1980). However, it has been reported that addition of iron to hydrogen peroxide greatly improves its efficiency in controlling biofouling (Ikuta et al., 1988).

This study addressed the possibility of controlling zebra mussel veligers with various oxidants. Initially, field-collected veligers were exposed to sodium hypochlorite and potassium permanganate in static systems. Subsequently, hydrogen peroxide + iron (hereafter referred to as "peroxide") was added to the list of oxidants under investigation, while a switch was made to flow-through exposure systems. In these experiments, veligers naturally present in the water were sampled before and after they passed through bioboxes in which they were exposed to chlorine, permanganate, and peroxide. During these same experiments, veliger settlement was monitored on settling plates inside the bioboxes. The effects of chlorine, permanganate, and peroxide on veligers are compared with results obtained for adult mussels (Klerks and Fraleigh, 1991).

METHODS

Static Exposures

Veligers were collected by repeated tows with an 80-μm mesh, Wisconsin-style plankton net in western Lake Erie near the Toledo water intake (41° 42.0′ N, 83° 15.5′ W) during June and July 1990. Predatory crustaceans were eliminated by sieving the plankton samples through a 250-μm screen, in both the field and laboratory. The less than 250-μm fraction, containing veligers, was diluted with lake water from the collection site and stored overnight in the dark. Within 24 hr, lake water containing veligers was divided among several 100-mL beakers (16–80 mL volume in the first experiment and 14–41 mL volume in the second experiment). Experiments were started by adding stock solutions of tested oxidant chemicals. Exposures were run in duplicate with a control and applied concentrations of 0.5, 1.0, and 2.5 mg/L, free chlorine (Cl_2) as sodium hypochlorite, and potassium permanganate. In the first experiment, chlorine and permanganate were added only at the beginning of the experiment. In the second experiment, oxidants were added at the start of exposures and at 5, 60, and 120 min into the experiment in amounts needed to reach starting concentrations.

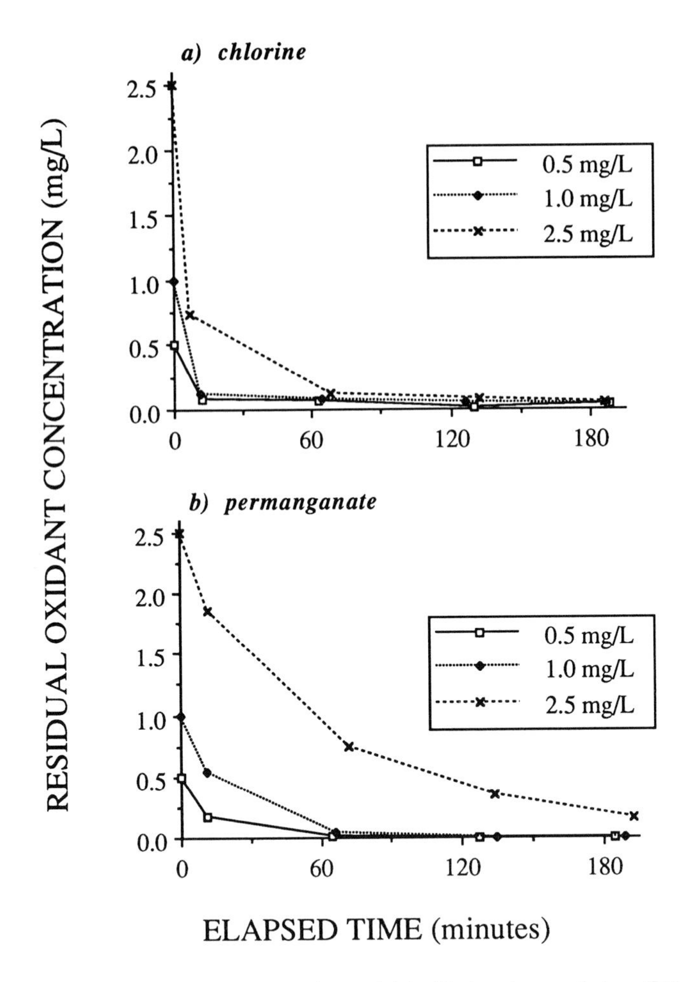

Figure 1. Residual oxidant concentrations of (a) chlorine (as mg/L free Cl_2) and (b) permanganate (as mg/L $KMnO_4$) after one-time applications of 0.5, 1.0, and 2.5 mg/L in Lake Erie water containing veligers of *Dreissena polymorpha.*

Concentrations of free chlorine and permanganate were determined on 10-mL samples using the *N,N*-diethyl-p-phenylenediamine (DPD) colorimetric method (APHA et al., 1985), with a Bausch & Lomb Spectronic 20 spectrophotometer'* at 515 nm, and standard curves generated with stock solutions of each oxidant. In the first experiment, chlorine and permanganate concentrations decreased rapidly after the initial additions of the oxidants (Figure 1). The static renewal procedure used in the second experiment compensated for the rapid loss of oxidant; and it resulted in mean residual oxidant

* Use of name brand manufacturer does not imply U.S. Government endorsement.

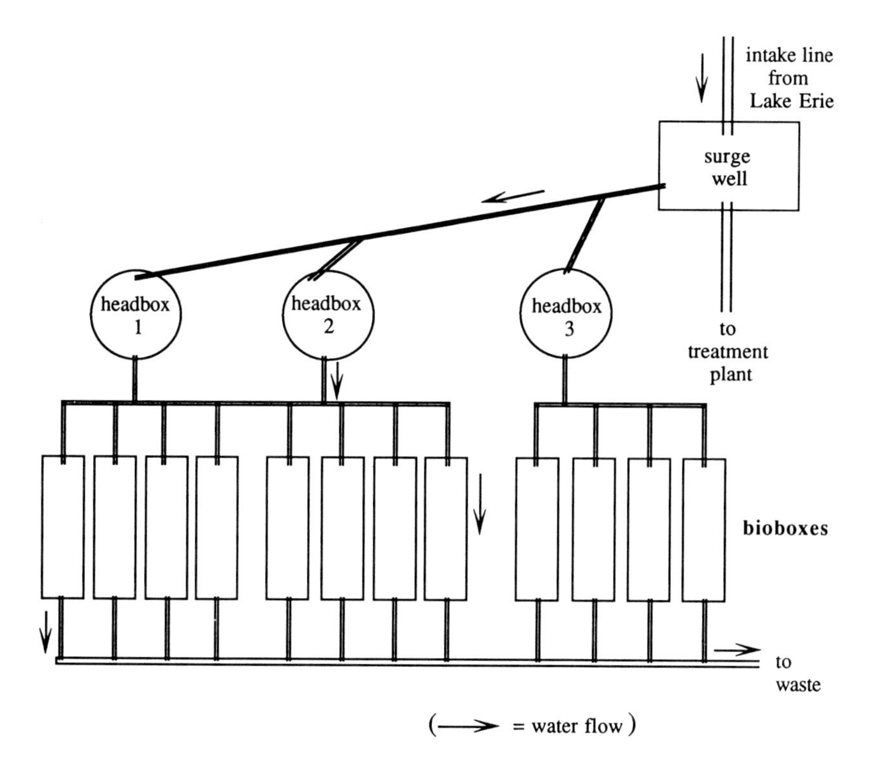

Figure 2. Diagram (not to scale) of experimental design with bioboxes used in flow-through exposures of *Dreissena polymorpha.*

concentrations of 0.23, 0.43, and 1.83 mg/L free chlorine and 0.38, 0.78, and 2.37 mg/L $KMnO_4$, respectively, for the calculated concentrations of 0.5, 1.0, and 2.5 mg/L.

Densities of live and dead veligers were determined at specific times during experiments. Duplicate veliger subsamples were taken after stirring beaker contents, and volumes of 1-mL were counted in Sedgewick-Rafter counting cells at magnification × 40. The distinction between live and dead veligers was based upon the presence or absence of ciliary movement, either inside the translucent shell or on an extended velum.

Flow-Through Exposures

Flow-through exposures used the natural complement of veligers present in untreated water at the low service pump station of the drinking water supply system for the City of Oregon (Ohio), located on western Lake Erie. Exposure systems consisted of Plexiglass™ bioboxes placed in a side-arm of the intake system (Figures 2 and 3). Bioboxes (inside dimensions: 99 cm long × 22

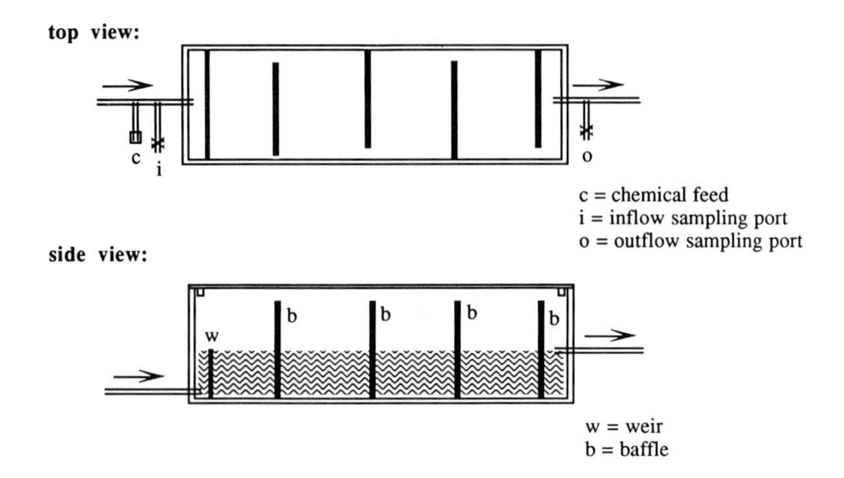

Figure 3. Detailed views (not to scale) of a biobox used in the flow-through exposures of *Dreissena polymorpha* veligers.

cm wide × 30 cm high) were arranged in groups of four, including one group for each oxidant and one for control. Flow rates of raw water into bioboxes were controlled by in-line valves and kept at 1 L/min ± 5%. Water levels in the boxes, controlled by the height of the outflow opening, were approximately 13 cm. The residence time of the water in the bioboxes, and thus the expected exposure time of veligers moving passively with the water, was approximately 27 min.

Chemicals used for treatments were sodium hypochlorite (Jones Chemicals: industrial grade, min. 12.5% NaOCl by weight), potassium permanganate (Carox: Carus®, free-flowing grade, typical purity 98%), and hydrogen peroxide (Mallinckrodt: 30% solution, stabilized) applied together with ferrous sulfate (Sherman Research Laboratories: technical grade). The experiments with chlorine and permanganate were run at applied concentrations of 0.5, 1.0, and 2.5 mg/L (free Cl_2 and $KMnO_4$ concentrations, respectively). Hydrogen peroxide and ferrous iron were applied together in the following combinations in the first two experimental periods: 0.25/0.25, 0.5/0.25, and 1.0/0.25 mg/L (H_2O_2 and Fe concentrations, respectively) and 1.0/0.25, 2.0/0.50, and 5.0/1.25 mg/L in later experimental periods. Stock solutions of treatment chemicals were made every other day, at a concentration of 500 times the exposure concentration. These stock solutions were kept in plastic carboys (Nalgene, LDPE) and pumped by Liquid Metronics Series A3 metering pumps at a rate of 2.0 mL/min. ±10% into the raw water supply line into bioboxes. Mixing of the raw water and treatment chemicals was facilitated by static mixers built into the line between the injection point and bioboxes, and a 13 cm high weir just inside the biobox (Figure 3). Chemical flow rates were checked daily, by determining the amount of stock solution pumped out of each bottle. Treatments were stopped once every 2 weeks. Bioboxes were then cleaned and specific treatments reassigned to different boxes within each set of four. This resulted in the exposures being done over four 2-week periods.

Samples from inflow and outflow sampling ports of each biobox were analyzed once or twice daily for each treatment chemical. Either free chlorine or both free and total chlorine were analyzed on 25-mL samples using the DPD colorimetric method and appropriate built-in programs of the Hach DR-2000 spectrophotometer. Permanganate concentrations were determined using the same free chlorine method with a correction factor of 0.891 (APHA et al., 1985). Total iron concentrations were determined by the 2,4,6-tri-(2-pyridyl)-1,3,5-triazine (TPTZ) colorimetric method, using the iron-TPTZ program of the Hach DR-2000 spectrophotometer. Hydrogen peroxide concentrations were determined by thiosulfate titration, using a Hach hydrogen peroxide test kit and a Gilmont micrometer buret. Standard solutions of all treatment chemicals were employed to verify method accuracy. Precision was assessed by analyzing duplicate samples approximately biweekly. The following coefficients of variation were obtained by replicate analyses during another experiment in bioboxes that followed the same procedures as used in this study: 9.8% for free chlorine, 2.5% for total chlorine, 2.3% for potassium permanganate, 3.7% for hydrogen peroxide, and 3.6% for total iron (values averaged over different days and treatment concentrations). Treatment concentrations were determined by subtracting concentrations measured at inflow and outflow sampling ports of the control box. These analytical blanks averaged 0.01 and 0.01 mg/L free Cl_2, 0.02 and 0.01 mg/L total Cl_2, 0.01 and 0.01 mg/L $KMnO_4$, 0.23 and 0.24 mg/L H_2O_2, and 0.29 and 0.27 mg/L Fe, at biobox inflows and outflows, respectively. Concentrations of treatment chemicals measured in inflows and outflows of bioboxes are summarized in Table 1. Concentrations of free chlorine, total chlorine, and permanganate in outflow samples were below the applied concentrations, indicating average demands of approximately 0.7 mg/L free chlorine, 0.3 mg/L total chlorine, and 0.4 mg/L permanganate. Free chlorine and permanganate concentrations were already well below the applied levels as the oxidant-spiked water entered bioboxes. This might have been due to a rapid use immediately after mixing raw water and test oxidant, or in the period between the start of sampling and the start of DPD reaction in the analyses (i.e., about 3 min). Concentrations of hydrogen peroxide and total iron were close to applied concentrations, both in inflows and outflows. The possibility of a small demand for peroxide is suggested by the consistency of the concentrations in outflows to be somewhat lower than those in inflows.

The following water quality characteristics were determined once or twice daily on water samples from the three headboxes: water temperature, dissolved oxygen (Winkler method), turbidity (Hach, model 16800 turbidimeter), conductivity (Hach model 44600 conductivity/TDS meter), and pH (Hach ONE

TABLE 1. Concentrations of Treatment Chemicals (in mg/L, with $25 \leq n \leq 88$) in Inflows and Outflows of Bioboxes, During Exposures of Veligers to Oxidants

Applied Conc		Residual Conc (mean ± S.D.) In Inflows	In Outflows	
	0.5	0.162 ± 0.110	0.051 ± 0.043	Free Cl_2
		0.437 ± 0.082	0.296 ± 0.061	Total Cl_2
Cl_2	1.0	0.705 ± 0.252	0.329 ± 0.187	Free Cl_2
		0.871 ± 0.140	0.609 ± 0.131	Total Cl_2
	2.5	2.233 ± 0.460	1.803 ± 0.442	Free Cl_2
		2.651 ± 0.489	2.215 ± 0.466	Total Cl_2
	0.5	0.365 ± 0.082	0.275 ± 0.076	$KMnO_4$
$KMnO_4$	1.0	0.849 ± 0.114	0.699 ± 0.097	$KMnO_4$
	2.5	2.021 ± 0.370	1.809 ± 0.357	$KMnO_4$
	0.25	0.241 ± 0.062	0.200 ± 0.081	H_2O_2
	0.5	0.566 ± 0.086	0.518 ± 0.089	H_2O_2
H_2O_2	1.0	1.122 ± 0.278	1.048 ± 0.250	H_2O_2
	2.0	1.885 ± 0.294	1.903 ± 0.267	H_2O_2
	5.0	5.053 ± 0.774	4.856 ± 0.653	H_2O_2
	0.25	0.272 ± 0.053	0.264 ± 0.055	Total Fe
Fe	0.50	0.547 ± 0.088	0.526 ± 0.076	Total Fe
	1.25	1.324 ± 0.185	1.300 ± 0.158	Total Fe

TABLE 2. Water Quality Characteristics During Flow-Through Exposures of Veligers to Oxidants

	Mean ± S.D.	n	Range
Water temperature (°C)	22.53 ± 2.73	83	15.67–26.27
pH	7.94 ± 0.14	82	7.62–8.30
Dissolved oxygen (mg/L O_2)	8.76 ± 1.45	82	5.13–11.65
Conductivity (μS/cm)	301.0 ± 35.8	83	235.0–382.0
Turbidity (NTU)	17.97 ± 16.04	82	3.76–80.47
Alkalinity (mg/L $CaCO_3$)	88.10 ± 5.94	55	79.33–104.33
Hardness (mg/L $CaCO_3$)	122.67 ± 13.63	55	101.00–153.67
Calcium (mg/L Ca)	34.71 ± 4.30	55	28.40–42.00
Magnesium (mg/L Mg)	8.70 ± 1.40	55	4.90–12.47

pH meter). In addition, separate water samples from headboxes were analyzed daily by the chemistry laboratory of the Division of Water of the City of Toledo for total alkalinity, total hardness, and calcium and magnesium concentrations. Water quality characteristics during the four 2-week experiments are summarized in Table 2. Water temperatures were fairly constant during the first three periods and then declined during the last period from late September to early October. Dissolved oxygen concentrations generally showed a 100% saturation.

Samples from the outflow of bioboxes used for chlorine exposures, including controls, were collected once per week for determination of trihalomethanes. The following four trihalomethanes were determined: chloroform, dichlorobromomethane, chlorodibromomethane, and bromoform. Detection limits of these trihalomethanes were 0.5, 0.5, 1.0, and 1.0 μg/L, respectively.

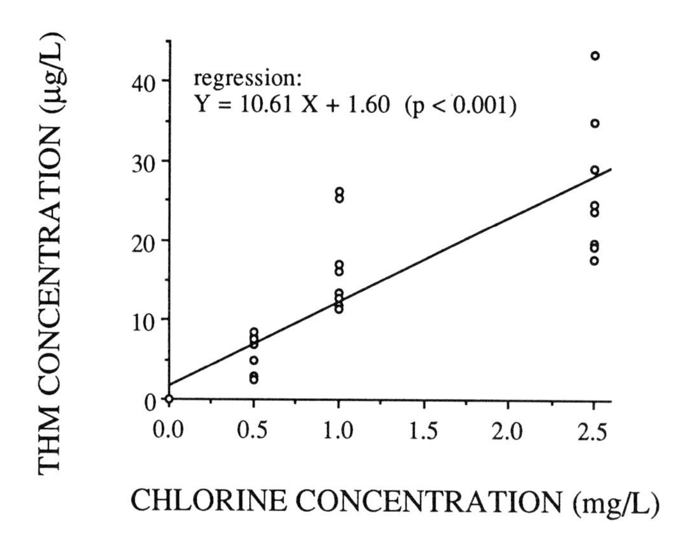

Figure 4. Trihalomethane (THM) concentrations in weekly samples from biobox outflows treated with chlorine concentrations ranging from 0 to 2.5 mg/L free Cl_2. Values are sums of chloroform, bromoform, bromodichloromethane, and dibromochloromethane concentrations.

Concentrations of trihalomethanes in outflows of the chlorine-treated bioboxes were generally higher at higher treatment concentration (Figure 4). The maximum trihalomethane level encountered was 43.5 µg/L. Levels of individual trihalomethanes occurred as chloroform > bromodichloromethane > chlorodibromomethane > bromoform (note: no bromoform was detected in samples). Concentrations in the control were below detection limits.

Effects of the oxidants on postveliger mussel settling were studied with the use of various substrates placed in the bioboxes. Plastic, glass, and minerit plates (19 × 30 cm) 2, 4, and 1 plates per biobox, respectively, were placed vertically against the baffles inside bioboxes. Plates were inspected weekly for the presence of settled mussels.

Effects of chlorine, permanganate, and peroxide on veligers were determined in the same experimental design and simultaneously with the assay of settling effects. During the first experimental period, veligers were sampled twice daily from biobox outflow sampling points. A subsequent comparison of veliger densities in the water going into bioboxes revealed significant differences among some of the bioboxes (Figure 5). The protocol was therefore changed so that these differences could be taken into account. For the subsequent experimental periods, both biobox inflows and outflows were sampled. Inflow samples were collected after turning off dosing pumps and flushing the sampling port with untreated water. To obtain reasonable numbers of veligers, 10- or 20-L water samples were collected from the sampling ports (in replicates of 2 L). Each 2-L batch was filtered through a 65-µm plankton

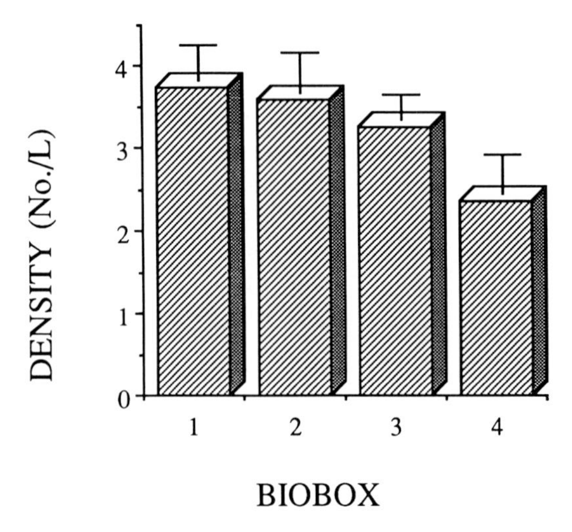

Figure 5. Total mean densities (mean + S.E., n = 14) of *Dreissena polymorpha* veligers entering bioboxes (i.e., untreated veligers). This set of boxes was used for chlorine exposures.

net to concentrate the veligers. The bucket was rinsed with untreated water to prevent any further exposure to oxidants. The 5 or 10 replicates were combined in the same plankton bucket, after which veligers were transferred into a 10-mL beaker using a small volume of water (i.e., 2–5 mL). Densities of live and dead veligers were then determined by counts in a 1-mL Sedgewick-Rafter cell at magnification × 40, on either three subsamples or the total sample using the same counting procedure and live/dead criterion as were used in static experiments.

Data Analyses

Results from the first static experiment were analyzed by analysis of variance on ln-transformed values and results from the second static exposure were analyzed by χ^2-test with Yates' adjustment for continuity (Sokal and Rohlf, 1981). For the flow-through exposures, data on percent of veligers dead in outflow samples were combined within a treatment when densities of veligers (live + dead) were below 2/L. Data below 2/L were combined for subsequent days with low densities, until cumulative densities reached 2/L. All data for the last two periods, when densities were consistently low, were combined into one data point for each treatment. Analyses of percent dead among veligers in the outflows were done by regression analyses using arcsine transformed data (arcsine $\sqrt{p}$). One data point for the permanganate control was deleted after Grubbs' test (Sokal and Rohlf, 1981) showed it to be an outlier ($p < 0.005$). Data for the three control groups were not pooled, because the percent dead was significantly different among the three groups in an analysis of variance

(p = 0.031). For analyses on differences in densities between inflow and outflow samples, data for a biobox were combined when live densities in inflows were below 2/L, while data for the last two periods were again combined into one data point per treatment. Statistical analyses of outflow veliger densities expressed as a percentage of inflow densities were done by regression analysis on log-transformed data. Densities of total and live veligers did not differ among the three control bioboxes (p = 0.809 and 0.815, respectively, in analysis of variance) so that these data were pooled in subsequent analyses. Total chlorine concentrations were estimated from free chlorine concentrations for days that total chlorine concentrations were not measured. This allowed the inclusion of all observations in regressions of oxidant effects on total chlorine concentrations, and was made possible by a tight relationship between free and total chlorine concentrations ($p = 0.0001$, $r^2 = 0.973$). For summary of oxidant effects on veligers in the flow-through system, effects were corrected for control effects (APHA et al., 1985).

RESULTS

Static Exposures

In static exposures veliger survival was determined just after the initial dosing with chlorine and permanganate and after 1, 2, and 3 hr. It appeared that a high oxidant demand in the test system, indicated by rapidly declining oxidant concentrations, resulted in an absence of further increases in mortality after 3 min. Observations made at different times were therefore pooled. Both chemical treatments showed statistically significant mortalities (Figure 6). Chlorine had a stronger effect than permanganate (overall mortalities 70 and 27%, respectively, and 9% in the control). In contrast to permanganate, mortalities among the three chlorine concentrations did not exhibit a clear increase in effect with increasing chlorine level.

In the second static exposure, where concentrations of chlorine and permanganate were repeatedly increased to initial concentration levels, an increase in mortality with exposure time (at $p < 0.001$) was observed for all three chlorine levels (Figure 7a). Again, no differences were evident among the three chlorine treatments. Mortalities in permanganate exposures (Figure 7b) averaged only 10% and were not discernable from those in controls where an average mortality of 5% occurred.

Flow-Through Exposures

In flow-through exposures where veligers were collected as they came out of the bioboxes, percentages of dead veligers were higher with increasing exposure concentrations. Regressions of percent dead on midbox residual

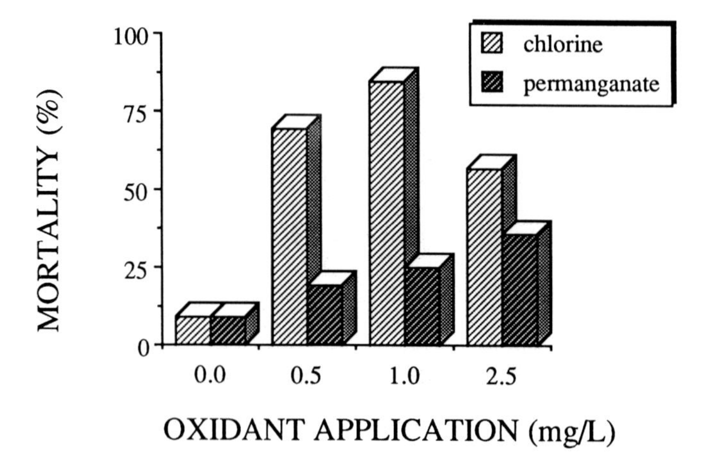

Figure 6. Percent mortality of *Dreissena polymorpha* veligers after one-time applications of 0.5, 1.0, and 2.5 mg/L chlorine and permanganate. Values are the averages for two replicates per treatment and for observations made just after the oxidant additions and at 60, 120, and 180 min after initial exposures.

oxidant concentration were found for all three oxidants (Figure 8). Average mortalities for the various oxidants were 37% for chlorine, 17% for permanganate, and 11% for peroxide. However, this increased proportion of dead veligers among total veligers in outflows was not the only effect of oxidants.

Densities of all veligers in biobox outflows were drastically below those in the inflows. This also occurred in control boxes, where outflow densities of total veligers were on average 56% below those in inflows, and became even stronger as a function of exposure concentration (Figure 9). This effect was observed for all three oxidants. Densities in the outflows of treatment bioboxes were reduced from inflow densities by an average of 86% for the various chlorine concentrations, 90% for permanganate, and 73% for the peroxide treatments. The reduction in veligers between inflows and outflows, combined with the increased proportion of dead veligers among veligers coming out of bioboxes, resulted in an even stronger reduction densities densities of live veligers in outflows relative to those in the inflows (Figure 10). For all chlorine treatments combined, densities of live veligers in outflows were on average 91% below inflow densities, compared to 90% for permanganate, 72% for peroxide, and 53% in control bioboxes.

DISCUSSION

Short-term treatments of veligers with oxidants usually resulted in veliger mortality. This effect was found for all three oxidants. In static chlorine exposures, mortalities exceeding 50% were achieved within 1 hr, while 100%

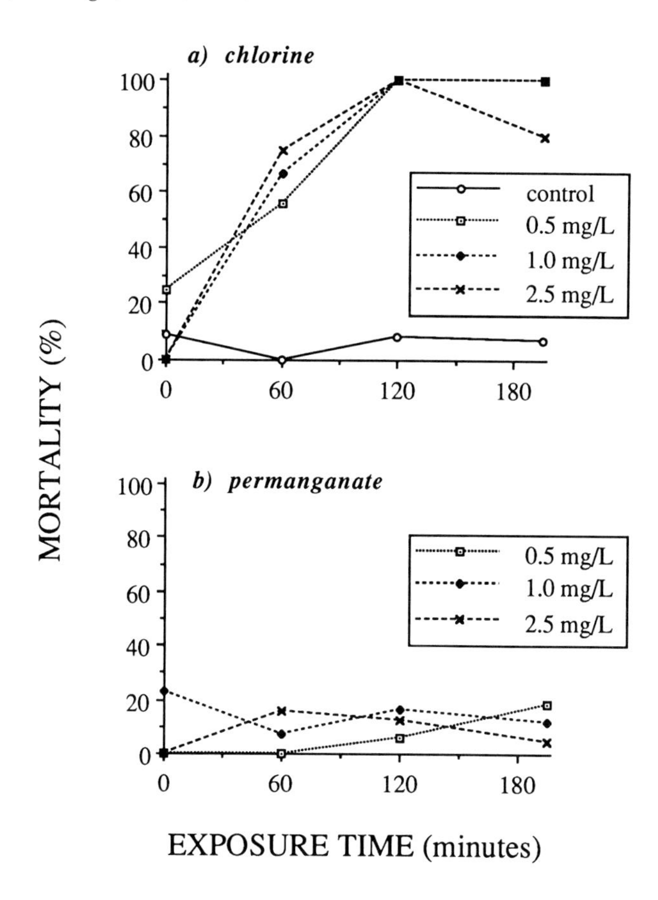

Figure 7. Percent mortality of *Dreissena polymorpha* veligers after treatments with applied concentrations of 0.5, 1.0, and 2.5 mg/L (a) chlorine and (b) permanganate. Following hourly measurements, additional oxidant was added to maintain the initial concentrations. Percentages are means (n = 2).

mortality was reached after 2 hr. These high mortalities occurred with chlorine treatments as low as 0.5 mg/L. Similar results were obtained for the flow-through chlorine exposures, where 27–61% mortalities were achieved in the approximately 27-min exposures. Treatments with chlorine did not result in levels of trihalomethanes above the current U.S. drinking water limit of 100 μg/L. However, our data showed that compliance with regulations could become problematic after the anticipated reduction of this legal limit, if raw water is treated with chlorine. Permanganate treatments had a small effect on survival. Mortality was observed in one of the two static exposures, while significant but low mortality rates were evident in the flow-through exposures. Peroxide was not tested in the static exposures, and showed a small effect on survival in flow-through systems.

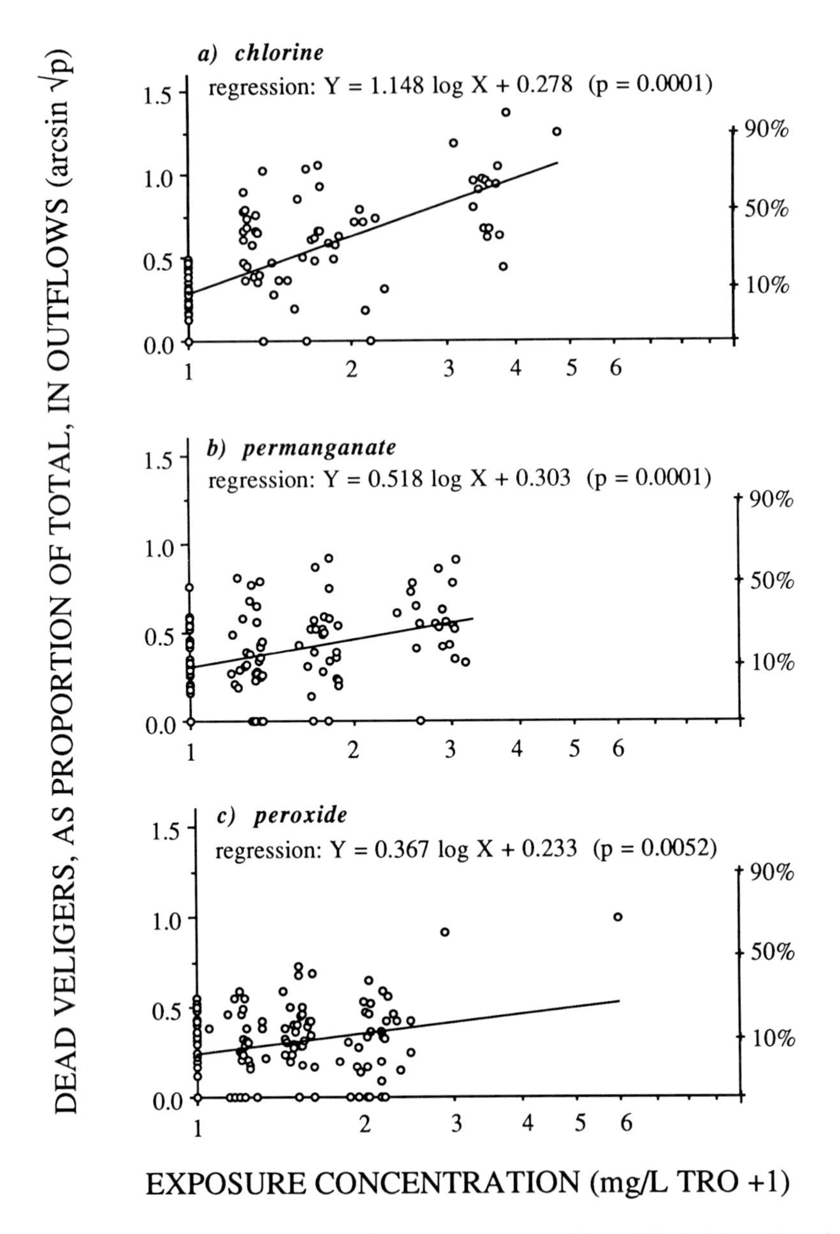

Figure 8. Dead *Dreissena polymorpha* veligers, as proportions of the total number of veligers, in biobox outflows in which veligers were exposed to (a) chlorine, (b) permanganate, or (c) peroxide. Arcsine transformed proportions are plotted against the average oxidant concentration measured in inflow and outflow (on a logarithmic scale). Scale with actual percentages shown on right.

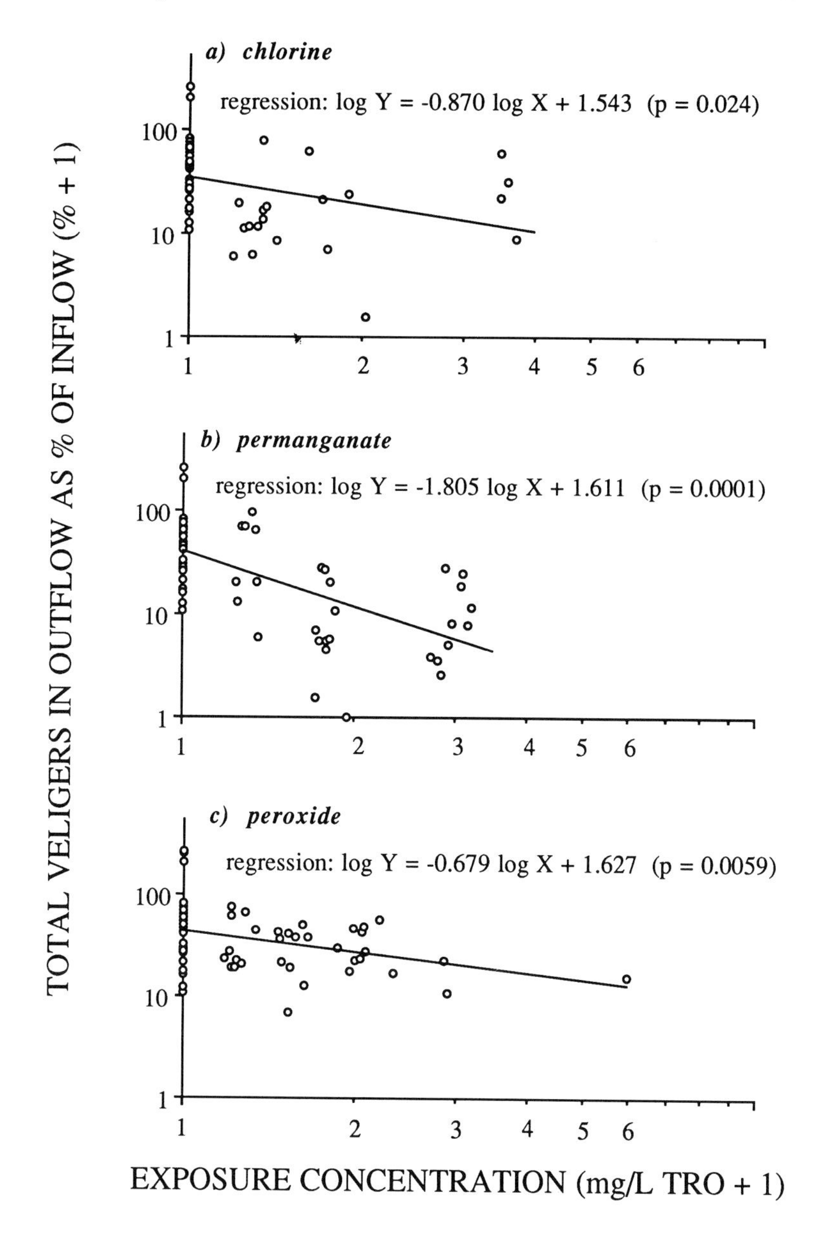

Figure 9. *Dreissena polymorpha* veligers (live + dead) in outflows, as percentage of densities in inflows, of bioboxes in which veligers were exposed to (a) chlorine, (b) permanganate, or (c) peroxide. Percentages are plotted against the average oxidant concentration measured in inflow and outflow, both on a logarithmic scale.

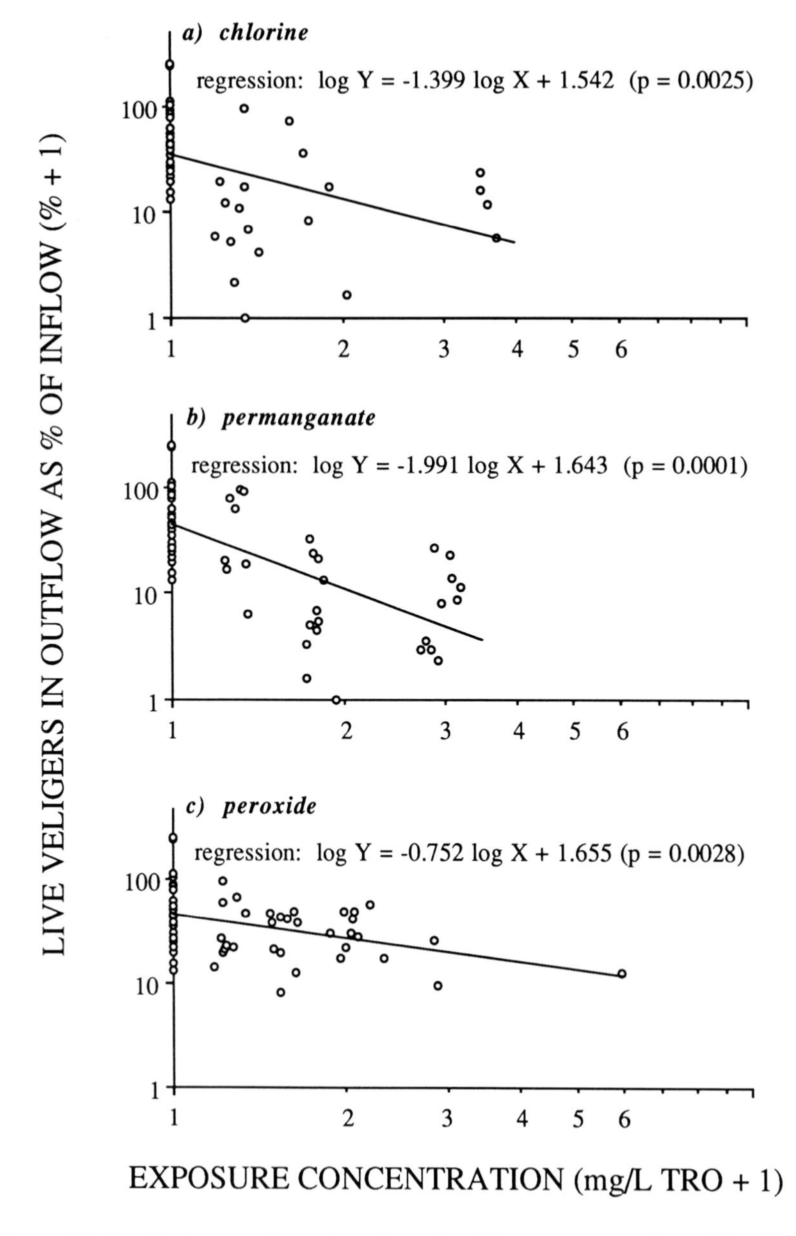

Figure 10. Live *Dreissena polymorpha* veligers in outflows, as a percentage of densities in inflows, of bioboxes in which veligers were exposed to (a) chlorine, (b) permanganate, or (c) peroxide. Percentages are plotted against the average oxidant concentration measured in inflow and outflow, both on a logarithmic scale.

Another effect of oxidants observed in flow-through exposures was veliger inactivation. This reduction of veliger densities in the water column as the water passed through bioboxes was superimposed on the normal density reduction that occurred in all bioboxes including the controls. The reduction in veliger densities was as high as 93% (in the 1.0 and 2.5 mg/L permanganate exposures). We hypothesize that veligers closed their shell upon contact with the oxidants, thereby losing mobility and sinking to the bottom of bioboxes. However, an avoidance response leading to movement out of the water column or away from the water surface could also lead to the same reduction in outflow densities and would be more consistent with the observation that dead veligers were found in outflows. In this study the term ''inactivation'' is used in the absence of information on the processes leading to the reduced outflow densities. In continuous treatments, inactivation is likely to result in mortality; veligers that stay in a treatment container will be exposed to the oxidant for a longer period of time. However, only a few veligers were found in an inspection of material on the bottom of a biobox, although this was done during a period that a lot of sediment was accumulating in bioboxes. A loss of mobility during chemical exposures has also been reported for larvae of the marine mussel, *Mytilus edulis*, exposed to ozone in a static exposure system (Toner and Brooks, 1977). Lower densities of veligers in biobox outflows relative to inflows were not due to settlement of mussels within the bioboxes, as virtually no settlement occurred there for the duration of the flow-through exposures. It does not appear that the oxidant-induced reduction in veliger density is entirely a consequence of veliger mortality, followed by settling out of dead veligers. Dead veligers did show up in the outflows, and the pattern in reduction among the three oxidants (permanganate highest, followed by chlorine, and peroxide) was different from the pattern observed for percentages of dead veligers in outflows. The lowered densities may have been a consequence of low flow rates in the bioboxes and may not be representative of events in water intake pipes.

Flow-through exposures were not successful in determining effects of oxidants on postveliger settlement. Virtually no settlement occurred in these bioboxes; only two mussels were found on settling plates, both in control boxes. Several factors may be responsible for this absence of settlement in the bioboxes during a period in which settlement occurred in the area of the source of the water. It has been shown that veliger densities in intake systems on Lake Erie are lower than densities in the lake itself (Fraleigh et al., unpublished data), possibly because of predation while veligers passed through intake pipes. Recent evidence indicates that adult mussels may prey on veligers (MacIsaac and Sprules, 1990). Combined with the normally high mortality at the postveliger stage (Stanczykowska, 1977), veliger densities in our exposure system may have become too low for any significant settlement to occur. In addition, cursory observations on the size of veligers in samples

TABLE 3. Summary of Effects (with 95% Confidence Limits) of 1 mg/L Total Chlorine, Permanganate, and Peroxide[a] on Veligers in Flow-Through Exposures

	Chlorine	Permanganate	Peroxide
Dead veligers in outflow (%)	30 (25–36)	10 (6–15)	7 (4–11)
Reduction from inflow to outflow			
All veligers (%)	59 (34–74)	76 (65–83)	42 (24–56)
Live veligers (%)	74 (55–85)	79 (69–85)	45 (28–58)

Note: Effects, corrected for control effects, were estimated from regressions of effects on residual oxidant concentrations (Figures 8–10).

[a] Estimated applied oxidant concentrations were 1.04 mg/L total Cl_2, 1.38 mg/L $KMnO_4$ and 0.93 mg/L H_2O_2.

from bioboxes indicated that very few of them were near the size expected for the settling postveliger stage. It appeared that the postveligers that were ready to settle might have done so in the intake pipe before reaching our experimental bioboxes. Moreover, recent observations indicate that zebra mussel settlement in 1990 may have been late and at low rate for western Lake Erie (unpublished data).

Treatment effects of the three oxidants at 1 mg/L total residual in the flow-through exposures were determined to compare the effectiveness of oxidant concentrations (Table 3). This showed that chlorine and permanganate were generally more effective against veligers than peroxide. Results of long-term exposures of adult mussels to peroxide also showed a small effect; concentrations of 5 mg/L H_2O_2 with 1.25 mg/L Fe resulted in 30% mortality after 8 weeks, whereas mortalities in concentrations up to 2 mg/L peroxide did not exceed control mortalities (Klerks and Fraleigh, 1991). This small effect of peroxide contrasts with results obtained for the control of the marine mussel, *Mytilus edulis* (Ikuta et al., 1988). However, that study looked at the effectiveness against new mussel settlement. This still leaves open the possibility that a treatment with peroxide may be effective at inhibiting settlement of the postveliger stage. In comparing the effectiveness of permanganate to that of chlorine, differences in their effects have to be taken into account. Chlorine had a more consistent and stronger effect on veliger survival than did permanganate. This difference was found in both the static and flow-through systems. The low mortality in exposures to permanganate agrees with results obtained for juveniles and adults of the Asiatic clam, *Corbicula fluminea* (Cameron et al., 1989; Chandler and Marking, 1979), while chlorine had a much stronger effect on the larval zebra mussels (a median lethal concentration of less than 0.5 mg/L for a 2-hr exposure) than on larval *Corbicula* (MLC of 6.7 mg/L after 8 hr) (Sickel, 1976 — cited in Doherty and Cherry, 1988). Experiments with adult zebra mussels (Klerks and Fraleigh, 1991) also showed a stronger mortality with chlorine than with permanganate, though the difference in effectiveness of these two oxidants was much less pronounced for adult mussels. The situation was different for the inactivation effects of chlorine and permanganate on veligers that was observed in flow-through exposures, where it appears that permanganate had a stronger inactivation effect than chlorine (Table 3). The net effect of the two oxidants, determined as the number of live veligers coming out of bioboxes, was rather similar.

Our results do not allow an evaluation of a control strategy aimed at inhibiting the settlement process, because no settlement occurred in bioboxes. Strategies aimed at controlling veligers and adults can be contrasted, however. At oxidant doses ranging from 0.5 to 2.5 mg/L, a complete kill was never achieved in the 27-min durations of exposures in bioboxes. A mortality of 100% occurred between 1 and 2 hr in the static exposures to chlorine. Effects on veligers, therefore, do not appear to occur very rapidly or to be achievable at low oxidant levels. Control of veligers may be possible in systems with a residence time of several hours, but does not seem feasible in situations where water passes through an intake system in 30 min or less. In addition, a control program aimed at the veliger stage has to be continued for the entire reproductive season (or at least for the time that veliger settling occurs). A treatment directed toward the attached mussels does not require an extremely rapid kill, since these mussels reside in the system and will be exposed as long as treatment continues. Experiments with settled mussels have shown that a treatment with concentrations as low as 0.25–0.3 mg/L total residual chlorine (Jenner, 1985; Klerks and Fraleigh, 1991) or 0.6 mg/L residual MnO_4^- (Klerks and Fraleigh, 1991) will eventually kill attached mussels. For a complete kill, treatment times of 1 week to several months will be required, depending on water temperature. In situations where some buildup of attached mussels can be tolerated, one or two short treatment periods per year (e.g., a short treatment in the summer when water temperatures are high, followed by a longer treatment late in the reproductive season) may be an effective control measure.

CONCLUSION

With respect to toxicity to veligers, chlorine is probably a better antifoulant than permanganate, which in turn showed more promise than peroxide. Chlorine had a much stronger effect on veliger survival than did permanganate and peroxide at comparable oxidant concentrations. All three oxidants, and especially permanganate, caused veliger inactivation in the flow-through systems. This effect would enhance the effectiveness of permanganate and peroxide relative to chlorine, but this more favorable opinion on permanganate and peroxide awaits demonstration of the generality of the inactivation phenomenon in water intake structures. Since exposure times well over 30 min are needed for a complete kill of veligers, the relative advantage of veliger-directed control over control aimed at the settled stage will depend on the residence time of the water in a water intake system. In long water intakes, veliger-directed control with chlorine, permanganate, or peroxide is expected to be similar in effectiveness to control aimed at the settled stage. In short water intakes, control aimed at the settled stage will probably be more effective.

ACKNOWLEDGMENTS

This research was funded, in part, by grants from the City of Toledo and the American Water Works Association Research Foundation (AWWARF). Other participants in this AWWARF grant to the city of Toledo were Finkbeiner, Pettis and Strout, Ltd. (consulting engineers), the City of Cleveland, and Case Western Reserve University. Assistance from Toledo Edison and the City of Oregon was greatly appreciated. This work would not have been possible without the technical assistance of O. Alford, H. Batra, H. Bennett, W. Branford, P. Cherry, J. Desouza, T. Go, M. Haq, K. Kupcak, C. Paliobeis, S. Patel, S. Kiefer, M. Mahato, C. Sims, R. Sreekumar, and B. Stough.

REFERENCES

American Public Health Association (APHA), American Water Works Association (AWWA) and Water Pollution Control Federation (WPCF). *Standard Methods for the Examination of Water and Wastewater,* 16th ed. (Washington, DC: 1985).

Baldry, M. G. C. "The Bactericidal, Fungicidal and Sporicidal Properties of Hydrogen Peroxide and Peracetic Acid," *J. Appl. Bacteriol.* 54:417–423 (1983).

Cameron, G. N., J. M. Symons, S. R. Spencer, and J. Y. Ma. "Minimizing THM Formation During Control of the Asiatic clam: A Comparison of Biocides," *J. Am. Water Works Assoc.* 81:53–62 (1989).

Chandler, J. H., Jr., and L. L. Marking. "Toxicity of Fishery Chemicals to the Asiatic clam, *Corbicula manilensis*," *Prog. Fish-Cult.* 41:148–151 (1979).

Clarke, K. B. "The Infestation of Waterworks by *Dreissena polymorpha*, a Freshwater Mussel," *J. Inst. Water Eng.* 6:370–379 (1952).

Cleasby, J. L., E. R. Baumann, and C. D. Black. "Effectiveness of Potassium Permanganate for Disinfection," *J. Am. Water Works Assoc.* 56:466–474 (1964).

Doherty, F. G., and D. S. Cherry. "Tolerance of the Asiatic Clam *Corbicula* spp. to Lethal Levels of Toxic Stressors — A Review," *Environ. Pollut.* 51:269–313 (1988).

Greenshields, F., and J. E. Ridley. "Some Researches on the Control of Mussels in Water Pipes," *J. Inst. Water Eng.* 11:300–306 (1957).

Ikuta, S., K. Nishimura, T. Yasunaga, S. Ichikawa, and Y. Wakao. "Biofouling Control Using a Synergistic Hydrogen Peroxide and Ferrous Ion Technique," 49th Annual Meeting, International Water Conference, Pittsburgh, PA (1988).

Jenner, H. A. ''Control of Mussel Fouling in The Netherlands: Experimental and Existing Methods,'' in I. A. Diaz-Tous, M. J. Miller, and Y. G. Mussalli, Eds. *Symposium on Condensor Macrofouling Control Technologies: The State of the Art* (Palo Alto, CA: Electric Power Research Institute, 1983), 18.1–18.13.

Jenner, H. A. ''Chlorine Minimization in Macrofouling Control in The Netherlands,'' in *Water Chlorination. Chemistry, Environmental Impacts and Health Effects, Vol. 5*, R. L. Jolley, R. J. Bull, W. P. Davis, S. Katz, M. H. Roberts, Jr., and V. A. Jacobs, Eds. (Chelsea, MI: Lewis Publishers, 1985), pp. 1425–1433.

Klerks, P. L., and P. C. Fraleigh. ''Controling Adult Zebra Mussels with Oxidants,'' *J. Am. Water Works Assoc.* 83:92–100 (1991).

Kovalak, W. P., G. D. Longton, and R. D. Smithee. ''Infestation of Monroe Power Plant by the Zebra Mussel (*Dreissena polymorpha*),'' *Proc. Am. Power Conf.* 52:998–1000 (1990).

MacIsaac, H. J., and W. G. Sprules. ''Direct Suppression of Lake Erie Zooplankton by Zebra Mussels (*Dreissena polymorpha*),'' presented at International Zebra Mussel Research Conference, Columbus, OH, December 5–7, 1990.

Sokal, R. R., and F. J. Rohlf. *Biometry*. (San Francisco: W. H. Freeman & Company Publishers, 1981)

Stanczykowska, A. ''Ecology of *Dreissena polymorpha* (Pall.) (Bivalvia) in Lakes,'' *Pol. Arch. Hydrobiol.* 24:461–530 (1977).

Toledo, R. T., F. E. Escher, and J. C. Ayres. ''Sporicidal Properties of Hydrogen Peroxide Against Food Spoilage Organisms,'' *Appl. Microbiol.* 26:592–597 (1973).

Toner, R. C., and B. Brooks. ''The Effects of Ozone on the Larvae and Juveniles of the Mussel *Mytilus edulis*,'' in *Biofouling Control Procedures*, L. D. Jensen, Ed. (New York: Marcel Dekker, 1977), pp. 19–22.

Von Bockelmann, I., and B. Von Bockelmann. ''The Sporicidal Action of Hydrogen Peroxide — A Literature Review,'' *Lebensm. Wiss. Technol.* 5:221–225 (1972).

Waite, T. D., and J. R. Fagan. ''Summary of Biofouling Control Alternatives,'' in *Condenser Biofouling Control*, J. F. Garey, R. M. Jorden, A. H. Aitken, D. T. Burton, and R. H. Gray, Eds. (Ann Arbor, MI: Ann Arbor Science, 1980), pp. 441–462.

CHAPTER 37

The Use of Endod (*Phytolacca dodecandra*) to Control the Zebra Mussel (*Dreissena polymorpha*)

Harold H. Lee, Aklilu Lemma, and Harriett J. Bennett

Experiments using a static bioassay system as a basis to develop a control method for *Dreissena polymorpha* illustrate the potential usefulness of the plant molluscicides, Lemmatoxins, from Endod, *Phytolacca dodecandra*. Endod at a dose higher than 15 mg/L is lethal to adult zebra mussels, while a lower dose prevents their adhesion and aggregation. Since Endod plants have been successfully grown as monoculture, demands for large quantities of Endod for use in water intakes should stimulate further agricultivation. Since infestation of zebra mussels is a long-term problem and waterworks vary in design and environment, a conceptual methodolgy for mitigation is suggested using Endod as the primary agent in combination with mechanical and chemical means to remove adult mussels from, and to prevent aggregation in, water intake pipes.

0-87371-696-5/93/$0.00 + $.50

INTRODUCTION

The long-term economic and ecological effects by invasion of the zebra mussel, *Dreissena polymorpha* (Pallas), to the Great Lakes and possibly other waterways of North America are understandably difficult to determine. Control of the population in open waters is virtually impossible because of the vastness of the water body. The highly adaptive ability of the organism and the free-swimming larvae (i.e., veligers) have resulted in wide distribution and infestation in many water systems (Griffiths et al., 1989). Because of these factors, the mussel infestation of water intakes in various waterworks (and to some degree in cooling systems of boats) is a serious, rapidly expanding, and continuous problem. Effects include the reduction of and (at worst) stoppage of water flow. This undersirable economic effect may in time emigrate to inland waterways. It is therefore vital, especially to the Great Lakes community at this time, to find ways to control the population of zebra mussels in intake pipes. Due to differences in geography, fauna, and flora, much of the experience of *Dreissena* in Europe is not readily applicable to North America. Therefore, new methods are needed to prevent or minimize economic loss.

In this chapter, we present brief arguments for the use of natural products as a primary means, in combination with others, for control of zebra mussels in water intakes. A conceptual working model using a multiple approach is presented based on positive results of the experiments to control *Dreissena* with Endod, *Phytolacca dodecandra*, the African soapberry.

Selection of Control Methods

Conditional to the selection of methods for control of zebra mussel populations in restricted localities, such as intake pipes of water treatment plants and cooling systems of power plants and various industry establishments, are several factors that are of concern to the economy in a long-term context. The pros and cons of using mechanical, synthetic chemicals, and natural products for control of zebra mussels are discussed as follows.

Mechanical Control

Mechanical removal of aggregated and adhered adult mussels from the inside surface of intake pipes requires remote control machines that not only are powerful enough to dislodge the animals but also are able to penetrate into intake pipes as long as one or more kilometers. Because length and diameters of pipes and designs of intakes are different, there is no universal machine that can be used for all water intake systems. Assuming removal by mechanical means could be effective in that many animals would be crushed, many living mussels still will remain being able to reaggregate to each other, to readhere to the substratum, and to continue to reproduce if they are not immediately isolated from the system. Even though removal can be done prior to readherence, transportation of living mussels to distant waste dump sites may result in an unintentional wider distribution. Similar concern is also on the unintentional transport of animals into inland waterways.

Synthetic Chemicals Control

Synthetic chemicals in pest control ideally use specific actions that affect target organisms with little or no effects on nontarget organisms. There are several synthetic and inorganic chemicals available that have general molluscicidal properties (such as niclosamide [Bayluscide], copper sulfate, sodium hypochlorite, permanganate, and surface-coating agents), which prevent the attachment of mollusks. Zebra mussels appear to be sensitive to oxidants like sodium hypochlorite at a mean concentration of 1 mg/L chlorine at which animals survived up to 5 days with continuous exposure (Fraleigh and Klerks, 1990). Since some of these chemicals are toxic and not readily degradable, their accumulation will lead to the formation of undesirable compounds, such as chlorinated compounds. This is of special concern for water intakes of drinking water systems and effluents to the environment from cooling systems. Their uses are both restrictive and therefore negatively cost-effective in a long-term context.

Natural Products Control

Natural products in pest control have been proven to be effective and have little or no undesirable side effects on the environment and nontarget animals. Such a product should have the following properties:

1. Molluscicidal actions under a wide range of conditions, such as pH and temperature
2. Easily degraded under a wide range of environmental conditions and in a relatively short period of time with no residual accumulation
3. Cost-effective being economically produced, stored, transported, and applied
4. Low dose requirement
5. Nontoxic to mammals, including humans at dosage lethal to zebra mussels
6. Noncorrosive to existing structures in various waterworks such as metal pipes

Based on these premises and extensive studies on snails and the control of schistosomiasis, the African soapberry, *Phytolacca dodecandra* (commonly known as Endod), appears to be the most promising candidate among several plants, such as *Ambrosia maritima* and *Swatzia madagascariensis*.

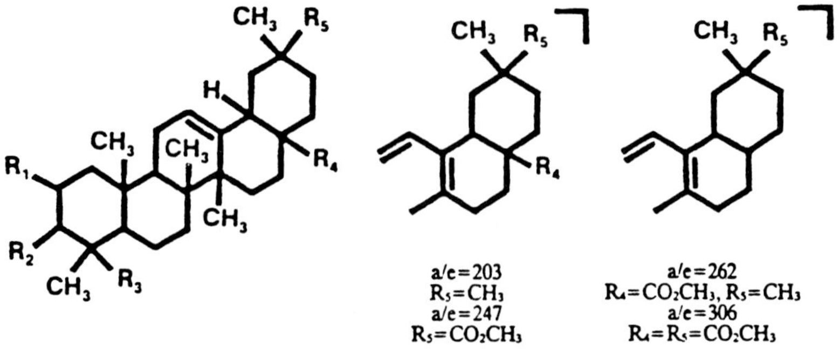

a/e=203
$R_5=CH_3$
a/e=247
$R_5=CO_2CH_3$

a/e=262
$R_4=CO_2CH_3$, $R_5=CH_3$
a/e=306
$R_4=R_5=CO_2CH_3$

	R_1	R_2	R_3	R_4	R_5
A : oleanolic acid	H	OH	CH_3	CO_2H	CH_3
B : 2-hydroxyleanolic acid	OH	OH	CH_3	CO_2H	CH_3
C : hederogenin	H	OH	CH_3OH	CO_2H	CH_3
D : bayogenin	OH	OH	CH_2OH	CO_2H	CH_3
E : phytolaccagenin	OH	OH	CH_2OH	CO_3H	CO_2CH_3
F : 2-deoxyphytolaccagenin	H	OH	CH_2OH	CO_3H	CO_2CH_3
G : serjanic acid	H	OH	CH_3	CO_2H	CO_2CH_3
H : jaligosic acid	OH	OH	CH_2OH	CO_2H	CO_2H

Figure 1. Aglycones of *Phytolacca* species. (From Parkhurst, R. M. *Phytolacca dodecandra,* Endod [Dublin, Ireland: Tycooly International Publishing, Ltd., 1984] p. 139. With permission.)

ENDOD DISCOVERY AND ITS CHEMISTRY

For thousands of years in Ethiopia, the berries of Endod plants have been used as laundry soap (hence the name soapberry) in streams that constitute the major source of drinking water for humans as well as for other animals. These streams also harbor the source of schistosoma that depend on snails as their immediate host prior to infection of some 300 million people in Africa (Lemma, 1965). During an epidemiology survey of schistosomiasis in Northern Ethiopia, Lemma noticed that many snails died downstream from where laundry was done. Subsequent laboratory (Lemma, 1970) and field investigations (Lemma, 1971) revealed that Endod possessed molluscicidal compounds that are saponins found in almost all green plants. These saponins (called Lemmatoxins) with glucose and/or galactose molecules are synthesized in Endod berries (Lemma, 1984; Parkhurst et al., 1973a, 1973b, and 1974).

The chemical structures of only three molluscicidal saponins of Endod have been determined: oleanoglycotoxin A, Lemmatoxin, and Lemmatoxin C. Each of these compounds has a glucose unit attached directly to the 3-hydroxy group of oleanolic acid (see Figure 1). Oleanoglycotoxin A has two other glucose units attached at the 2 and 4 positions of the first glucose ($[Glu]_2^{2,4}Glu^3Ole\text{-}28COOH$). Lemmatoxin has a galactose and a glucose unit attached to the 3 and 4 positions of the first glucose ($[Gal]^3[Glu]^4Glu^3Ole$-28COOH) while Lemmatoxin C has a glucose attached to the 2 position of the first glucose that is further substituted by a rhamnose at its 2 position ($Rha^2Glu^2Glu^2Ole$-27COOH). All these materials are active at the 1.5–3.0 ppm (LD_{90}/24 hr) range and have a free 28-carboxylic acid group.

Kloos and McCullough (1984) reported that more than 1000 plants have been screened for molluscicidal activities and many remained to be tested. The potency of plants and various plant parts are different. Currently, immature berries of *P. dodecandra* have been found to have the most potent molluscicides.

Systemic screening selection of some 600 wild types of Endod plants indicated that berries of a *Phytolacca* species, type 44, in Ethiopia contained as much as 25% by weight of saponins, from which the molluscicides, Lemmatoxins, have been isolated and purified with organic solvents (Lemma et al., 1972). Because of the detergent properties of Endod plants in combination with additives, Endod powder has been used to produce laundry detergents. Type 44 has also been successfully cultivated as monoculture in large-scale farming.

Since the discovery of Endod in 1965, there have been extensive studies on the chemistry, toxicity, and epidemiology of Lemmatoxins, together with cultivation of the Endod plant (Wolde-Yohannes et al., 1987). Related studies indicate the potential of Endod as a mitigation agent for undesirable mollusks, including zebra mussels. However, there has been no systemic study of mechanism of action of Lemmatoxins at cellular or physiological levels.

METHODS

Endod powder of approximately 250 μm particle size of *P. dodecandra* (type 44) was obtained from the Institute of Pathobiology, Addis Ababa University, Ethiopia. Stock solutions of 1000 mg/L were prepared by dissolving the powder in aged and aerated tap water. Insoluble cellular debris was not removed. Working solutions therefore contained insoluble cellular components that are essential for the synthesis of the Lemmatoxins (Parkhurst et al., 1973a and 1973b). Freshly made Endod solutions are not as active unless they have been incubated at 37°C for 1 hr or at room temperature for 16 hr. Longer incubation will lead to reduced molluscicidal potency, presumably due to biodegradation by microbial activity (Lemma, 1970; Lemma and Yau, 1974). Endod stock solutions used in this study were incubated in either manner and stored at 4°C until used. Although no potency tests were done, stock solutions older than 1 month were not used.

To test biodegradability or modification of the molluscicidal activity of Endod, solutions of 25 mg/L were filter-sterilized through 0.45-mm millipore filters and were stored in sterilized bottles for 3 days at room temperature before being used.

Between June and September 1990, adult zebra mussels were collected in Lake Erie near the Toledo and Oregon water intake stations. Animals were scraped off rocks, brought to the laboratory, and then rinsed with water and placed in half-gallon fish bowls which were immersed into an 80-gal recirculating aquarium at 10°C. Mussels were fed every other day with the unicellular alga *Chlamydomonas* spp.

Prior to using mussels in experiments, the fish bowls containing animals were placed overnight at room temperature with continuous aeration. For all bioassays, animals were separated from clusters. Unless stated otherwise, mussels of approximately 10- to 15-mm length were placed in appropriate vessels and allowed to acclimate for at least 24 hr before the addition of Endod. All experiments were done in tap water and in lake water as indicated. Unless otherwise indicated, water containing Endod was replaced with clean water at 24 hr after treatment. Animals were not fed during exposures to Endod, but were fed afterward.

To assay the effect of Endod in a static system, animals were placed in beakers (either 250 or 400 mL, depending on the number of animals used) with aeration. Animals were considered dead when they failed to close their shells upon mechanical stimulation.

Two tests were conducted to determine the biogradability of Endod. In the first test, 10 mussels were placed in four concentrations of Endod solution (2.5, 5.0, 10.0, and 20.0 ppm). After 24 hr, the solutions were replaced with untreated water (no Endod) or with fresh concentrations of Endod, or left as is. Mortality was monitored at 24 and 48 hr. In the second test, two 20-ppm solutions of Endod were prepared with Lake Erie water. The solutions were placed in capped bottles at room temperature. At 24, 48, and 72 hr, 200-mL aliquots from each replicate solution were taken to determine effectiveness on zebra mussels. Twenty mussels were used in each assay and mortality; attachment and clustering behavior were examined at 24, 48, 72, and 96 hr.

RESULTS

Effects of Endod

Recent studies (Lemma et al., 1991) have indicated that powder from dried Endod berries was lethal to zebra mussels in a static bioassay system with a 24-hr LC_{90} and LC_{50} of 20.0 and 8.8 ppm, respectively. Those that were treated at low doses of 5–10 ppm but did not die failed to reaggregate to each other and attach to Pyrex(TM) glass. The fact that more than 50% of the mussels were killed when they were exposed to only 4–8 hr of Endod indicated that there was an effective exposure time during which not all

TABLE 1. Mortality (%) of Zebra Mussels at Different Concentrations (ppm) of Solutions Derived from Various Plant Species

		P. dodecandra			*P. americana*					*A. maritima*	
Hr	Control	5	10	20	5	10	20	40	100	100	200
24	0	0	60	80	0	0	0	0	0	0	0
48	0	20	100	100	0	0	0	0	0	0	0
72	0	30	—	—	0	0	0	0	0	0	0
	+	−	−	−	+	+	+	+	+	+	+

animals were killed, but death did continue to occur later. This finding is important because of the implication that in a field situation, for example, application into water pipes could be short but with a lasting effect. This latent physiological response of zebra mussels to Endod has not been reported for snails or with other chemicals currently used for zebra mussel control studies.

Effects of Other Plant Extracts

To compare the effects of aqueous extracts from the ground powders of *Phytolacca americana* berries and *Ambrosia maritima* leaves on zebra mussels, a static experiment was done with solutions made in the same manner as that of the powder of the berries of *P. dodecandra*. Results in Table 1 show that both *P. americana* and *A. maritima* at doses as high as 100 and 200 ppm, respectively, did not affect zebra mussels. Endod at doses as low as 5 ppm had significant effects.

Degradability of Endod

The first test of the degradability of Endod confirmed previous studies that 100% mortality occurred at 10- to 20-ppm Endod. When Endod solution was replaced with fresh water, with fresh Endod, or with no change after animals were treated for 24 hr, mortality for all three sets were similar (Table 2). This result indicated that: (1) once zebra mussels had been exposed to an effective concentration in sufficient duration, additional Endod was not required and (2) activity of the Endod decreases after 24 hr (Table 2, group III).

Results of the second test (Table 3) demonstrate that there is a loss of molluscicidal activities of Endod, i.e., degradability, by merely storing it at room temperature for 24 hr or longer. However, the degradability differs in different water samples. The difference may be attributed to the chemical and biological composition of water samples. Although no chemical or biological tests were conducted other than pH (both replicates had a pH of 7.2), sample I was more turbid.

TABLE 2. Mortality (%) of *Dreissena polymorpha* at Different Concentrations of Endod

Hr	ppm	I	II	III
24	2.5	0	0	0
	5.0	10	10	0
	10.0	50	70	30
	20.0	100	100	90
48	2.5	0	0	0
	5.0	20	50	50
	10.0	70	100	100
	20.0	—	—	100

Note: Roman numerals refer to different treatment conditions: I = Endod solution replaced with fresh water, II = Endod solution replaces with a fresh solution of Endod at the same concentration, III = no change from original solution.

TABLE 3. Effect of Endod on the Mortality (%) Attachment, and Clustering of *Dreissena polymorpha* Under Different Degrees of Endod Degradability

	Controls			Endod Solution Time after Preparation					
Time Post-treatment	No Endod		Fresh Endod	24 hr		48 hr		72 hr	
(hr)	I	II	II	I	II	I	II	I	II
24	0	0	10	5	0	0	0	0	0
			NA	NA	NA		NA		
48	0	0	45	15	15	0	40	0	10
			NA	WA	NA		NA		
72	0	0	45	30	30	0	40	0	10
			WA	WA	NA		WA		
96	0	0	45	70	30	0	50	0	10
			WA						

Note: A solution of 20 ppm Endod was assayed at 24, 48, and 72 hr after the initial solution was made. Roman numerals refer to each of the two initial solutions. Attachment and clustering behavior was normal unless indicated. Normal behavior was defined as >75% of the individuals attached and >70% clumped. WA = weak attachment (<30% attached), NA = no attachment.

DISCUSSION

Safety Evaluation

Besides the clearly lethal effect of Endod to zebra mussels, as illustrated above and by other studies in our laboratory, there are other factors that indicate Endod would be an ideal candidate as a natural product for zebra mussel control. A consortium of laboratories in North America and Europe

(Lambert et al., 1991) undertook a Tier 1 EPA study with the Organization for Economic Cooperation in Development (OECD) in accordance with Good Laboratory Practice guidelines and indicated that Endod at 10 ppm was not toxic to mammals, although it was lethal to the snails *Biomphalaria glabrta* and *Bulinus truncatus*. These tests included a 28-day oral administration to rats and irritant tests for eyes and skin. An eye irritant test showed that Endod irritates eyes; therefore, eye protection should be used during preparation and application of the Endod powder. Together with previous tests by Lemma and Ames (1975), which indicated that Endod was neither mutagenic nor carcinogenic, Endod has been considered to be safe in large systemic scale field tests in African streams where people perform their daily chores, in addition to being the drinking water sources for many. Presently, usage in water intakes in North America appears to be acceptable with respect to mammalian toxicity. It is more attractive when only 1–2 ppm of Lemmatoxins, the active molluscicide in Endod, is needed to deliver lethal effects to zebra mussels. The rest are pulps of plant materials and oleanolic acids that are readily degraded. Nevertheless, a multitude of test requirements by the EPA, for example, absorption/adsorption tests with sediments or charcoal, must be done before Endod is accepted as a control agent for zebra mussels in water intakes of North America.

Biodegradability

It has been reported that the molluscicidal activity in Endod disappeared in 1 or 2 days in field trials in Africa (Lemma and Yau, 1974). Repeated application was therefore required to kill snails that carried schistosoma larvae. These studies indicate that Endod is biodegradable in North American waters, as indicated in the current study. The loss or decrease of molluscicidal activities due to biodegradation implies that Endod toxicity to nontarget organisms would elicit a minimal risk, an advantage over the use of nondegradable compounds. Unlike other molluscicides which accumulate (i.e., gradually increase in concentration) Lemmatoxins will disappear after they exert lethal effects to zebra mussels and the insoluble material decays. Although no biochemical analyses have yet been done on its degradation, it might be due to the breaking of the glycosidic bonds by enzymatic procedures since glycosidase is a common cellular enzyme. The nontoxicity on mammals (Lambert et al., 1991) discussed below may also be due to enzymatic degradation of Endod in animal systems.

Practicality and Efficacy

The use of Endod is practical for the following reasons. Agrobotanical studies indicate that monocultivation of the most potent variety, type 44, of

P. dodecandra in Ethiopia has been successful with a yield of about 1.5 metric ton/ha of berries, which is translated into about 1 ton of dried berry powder that can be used as such in the field (Wolde-Yohannes et al., 1987). Mass cultivation will lower the costs of production. Under optimum conditions, as in Ethiopia highlands, there can be two harvests annually. With improvement of fertilizers, pesticides, and intercropping techniques under testing, in combination with the existing pruning techniques, the cost of Endod powder will be lowered further. Growing Endod is more practical because it is a perennial bush with no known pests. Since there are no known organisms using dried Endod berries as a foodstuff, there will be no loss in storage and transportation as there is with grains. Although butanol extract of Lemmatoxins is a more potent molluscicide than water extract (Lemma et al., 1972), the cost of using butanol and its disposal later may not be economical or environmentally desirable.

Another argument for the practicality of using Endod is the removal of residual plant materials and dead zebra mussels. Dead mussels and those exposed to sublethal doses of Endod attach loosely to the substratum and each other. Dead mussels can be flushed out of water systems with less mechanical force than living, firmly adhered mussels. The residue of Endod solution is natural plant material which is largely insoluble and can be adsorbed by activated charcoal (Monkieje, 1990; unpublished data).

The practicality of using Endod powder could lead to the development of a cost-effective mitigation method for water intakes. For example, at an effective dose range of 10–20 mg/L, approximately 20 kg is needed for 10^6 gal of water. Given an intake pipe of 2-m diameter and 5-km length, approximately 350 kg of Endod powder would be needed to achieve a theoretical mortality of 50–90% of adult mussels. Although no current production cost is available, Lugt (1981) estimated that the cost of production of 1 metric ton of Endod in Ethiopia was about U.S. $500, while the comparable purchase price for Bayluscide was U.S. $25,000. Assuming four applications per year, the cost would be approximately U.S. $1,726, compared to $100,000 for Bayluscide and $30,000 for sodium hyperchorite. If the cost for Endod is 10 times the cost in 1981, the cost would be only U.S. $17,260. This calculation is theoretical, but does indicate that using Endod may be a cost-effective control method.

Low Dose Requirement

Endod solution made from the berries of type 44 Endod plants contain about 20–25% saponins, and the molluscicidal components in these saponins are no more than 25%. The actual concentration of Lemmatoxins in a 20-ppm Endod solution from the berry powder, which gives a LC_{90} in 24 hr, are close to 1–2 ppm! The rest of the components in the Endod solution are mostly soluble cellular components of the berries. Unlike chloride, continuous presence of Endod is not necessary. This latent effect of Endod on zebra mussels would reduce the usage of Endod. Since most animals would close the shell in the presence of Endod and would die later, either Endod is hypothetically permeable or the animals do not discriminate Endod from other plant materials in the water.

TOWARD CONTROL

Existing mussels clustering in water pipes evidently reduce the flow rate because of the decreased inside diameter of the pipes and the friction created by the aggregates changing the flow dynamics. If mussels are killed by either Endod or other agents, they can still cling (even though loosely) to each other by the entangling byssal threads. Endod is able to loosen and even partially disperse the clusters.

This action will facilitate the removal of dead mussels by mechanical means. However, since not all dead mussels will detach from each other and the substratum because of entanglement and since not all of them are killed, mechanical removal is still a necessary step. Endod also prevents reaggregation. Since attachment and clustering become weaker after treatment with Endod, less mechanical force will be required to remove the mussels.

Transport of living mussels through land or waterways increases the chance of unintentional distribution, as in the case of the dumping of ballast water which is the source of the current problem. We suggest here that even after their removal from the effluent of water intake pipes or similar structures and before transportation to disposal sites, these materials should be again treated with Endod and oxidants to kill survivors. Treatment with Endod is more desirable than with other molluscicides because it is biodegradable and is not toxic to other organisms. Other compounds are just as efficient in killing mussels, but they do not have these desirable properties.

Since pulps from plants are natural products, the Endod powder need not be treated with any chemicals prior to its application. Its decomposition, either biological or chemical, is a natural process. To accelerate the decomposition of Endod and soft tissues of mussels, other chemical agents may also be needed. Chlorine at low dosage and adsorption with activated charcoal can be used in water treatment processes and can improve water quality at the same time. Technologies are available to deal with various waste removal and transport in an environmentally acceptable manner. The mixture of Endod and mussels, both being natural materials, may not pose environmental problems that are present when toxic chemicals are used.

In addition to the problem of control of adult mussels that are already in aggregates, continuous addition of young veligers to existing aggregates is a long-term problem. Prevention of veliger attachment is another point of control. Intervention at this stage will not allow establishment of a new population. Since Endod has been demonstrated to inhibit the attachment of adult mussels, it is logical to assume that Endod will also inhibit the attachment of veligers. Nevertheless, this has to be tested because initial attachment of veligers to a substratum may differ from that of the byssal threads of adult zebra mussels.

CONCLUSION

We have presented arguments and evidence that Endod, because of its multifunctional characteristics, has the potential to be an effective control agent of adult *Dreissena polymorpha*. This compound loosens mussel attachment and kills the animal to facilitate removal by existing technologies. Considering the diverse types of water systems, environmental conditions, and various usages of the water, we suggest that a multiple methodological approach should be worked out because the accumulation of zebra mussels in water intakes is an ongoing process. Workable application schedules of Endod as a primary controlling agent, possibly in combination with other compounds, removal of dead mussels to waste disposal sites, recycling of shells, and treatment of effluent water are necessarily coordinated efforts to minimize the economic and ecological impact on the Great Lakes states, and to prevent possible similar occurrence in inland waters.

ACKNOWLEDGEMENTS

This research was supported in part by the Ohio Sea Grant Program, project R/PS-7-PD.

REFERENCES

Fraleigh, P. C., and P. L. Klerks. "Zebra Mussel Interim Report to Toledo Water Division," Toledo, OH, (1990).

Griffiths, R. W., W. P. Kovalak, and D. W. Schloesser. "The Zebra Mussel, *Dreissena polymorpha* (Pallas, 1777), in North America: Impact on Raw Water Users," in Symposium: *Service Water System Problems Affecting Safety-Related Equipment* (Palo Alto, CA: Nuclear Power Division, Electric Power Research Institute, 1989), 11–27.

Kloos, H., and F. S. McCullough. "Plant Molluscicides," in A. Lemma, D. Heyneman, and S. M. Silangwa, Eds. *Phytolacca dodecandra (Endod)* (Dublin, Ireland: Tycooly International Publishing Ltd., 1984), pp. 227–255.

Lambert, J. D. H., J. H. M. Temmink, J. Marquis, R. M. Parkhurst, C. B. Lugt, A. J. M. Schoonen, K. Holtze, J. E. Warner, G. Dixon, L. Wolde-Yohannes, and D. DeSavigney. "Endod: Safety Evaluation of a Plant Molluscicide," *Regul. Pharm. Toxicol.* 14:189–201 (1991).

Lemma, A. "A preliminary Report on the Molluscicidal Property of Endod (*Phytolacca dodecandra*)," *Ethiop. Med. J.* 3:187–190 (1965).

Lemma, A. "Laboratory and Field Evaluation of the Molluscicidal Properties of *Phytolacca dodecandra*," *Bull. W.H.O.* 42:597–617 (1970).

Lemma, A. "Present Status of Endod as a Molluscicide for the Control of Schistosomiasis," *Ethiop. Med. J.* 9:113–118 (1971).

Lemma, A. "Background and Historical Review," in *Phytolacca dodecandra (Endod)* A. Lemma, D. Heyneman, and S. M. Silangwa, (Dublin, Ireland: Tycooly International Publishing Ltd., 1984), pp. 12–44.

Lemma, A., and B. N. Ames. "Screening for Mutagenic Activity of Some Molluscicides," *Trans. R. Soc. Trop. Med. Hyg.*, 69:167–168 (1975).

Lemma, A., G. Brody, G. W. Newell, R. M. Parkhurst, and W. A. Skinner. "Endod (*Phytolacca dodecandra*), a Natural Product Molluscicide: Increased Potency with Butanol Extraction," *J. Parasitol.* 58:104–107 (1972).

Lemma, A., L. Wolde-Yohannes, P. C. Fraleigh, P. L. Klerks, and H. H. Lee. "Endod is Lethal to Zebra Mussels and Inhibits Their Attachment," *J. Shellfish Res.* 10:361–365 (1991).

Lemma, A., and P. Yau. "Studies on the Molluscicidal Properties of Endod (*Phytolacca dodecandra*). III. Stability and Potency under Different Environmental Conditions," *Ethiop. Med. J.* 13:115–124 (1974).

Lugt, C. B. "Development of Molluscicidal Potency in Short and Long Staminate Racemes of *Phytolacca dodecandra*," *Planta Med.* 38:68–72 (1981).

Monkieje, A. "Laboratory and Simulated Field Evaluation of the Plant Molluscicide, *Phytolacca dodecandra* (Endod-44), as It Relates to Schistosomiasis Control in Cameroon," PhD Thesis, Tulane University School of Public Health, New Orleans, LA (1990).

Monkieje, A., J. H. Wall, A. J. Englande, and A. C. Anderson. "A New Method for Determining Concentrations of Endod-S (*Phytolacca dodecandra*) in Water During Mollusciciding," *J. Environ. Sci. Health* B25(6):777–786 (1990).

Parkhurst, R. M. "The Chemistry of *Phytolacca dodecandra*," in *Phytolacca dodecandra, Endod* A. Lemma, D. Heyneman, and S. M. Silangwa, Eds. (Dublin, Ireland: Tycooly International Publishing, Ltd., 1984), p. 139.

Parkhurst, R. M., D. W. Thomas, W. A. Skinner, and L. W. Cary. "Molluscicidal Saponins of *Phytolacca dodecandra*: Oleanoglycotoxin-A," *Phytochemistry* 12:1437–1442 (1973a).

Parkhurst, R. M., D. W. Thomas, W. A. Skinner, and L. W. Cary. *Molluscicidal Saponins of Phytolacca dodecandra*: Lemmatoxin-C," *Indian J. Chem.* 11:1192–1195 (1973b).

Parkhurst, R. M., D. W. Thomas, W. A. Skinner, and L. W. Cary. "Molluscicidal Saponins of *Phytolacca dodecandra*: Lemmatoxin," *Can. J. Chem.* 52:702–705 (1974).

Wolde-Yohannes, L., T. Demsbe, and J. D. H. Lambert. "Cultivation studies of *Phytolacca dodecandra* and Its Role in Schistosomiasis control," in *Endod II,* L. Makhuba, A. Lemma, and D. Heyneman, Eds. United Nations Council on International and Public Affairs, New York (1987), pp. 81–87.

CHAPTER 38

Effect of Acoustic Energy on the Zebra Mussel (*Dreissena polymorpha*)

John J. Kowalewski, Paul H. Patrick, and A. E. Christie

Laboratory experiments were conducted on the effectiveness of using acoustic energy (3–18 kHz) as a potential control measure for zebra mussels (*Dreissena polymorpha*) in water handling facilities such as power plants. Experiments using solidborne sound at sonic frequencies were effective in preventing attachment of juvenile mussels in a pipe section as compared to a control group. In the 8- to 10-kHz range, nearly 100% control (i.e., unattachment) was achieved with vibration amplitudes up to about 150 g′ acceleration units. Mortality in this frequency and amplitude range was estimated to be 75–95%. In the 10- to 12-kHz range, almost 100% unattachment and mortality occurred at vibration amplitudes exceeding 200 g′ acceleration units. These results indicate that with further development acoustic energy may be a practical mitigation strategy against mussel attachment in water handling facilities.

0-87371-696-5/93/$0.00 + $.50

INTRODUCTION

The zebra mussel, *Dreissena polymorpha*, has caused operational problems with utility generating stations in the Great Lakes of North America (Griffiths et al., 1989). Ontario Hydro is one of the largest utilities in the world, operating numerous power generating stations in the Great Lakes region (Figure 1). In 1989, zebra mussels were found at the Nanticoke power station on Lake Erie. By 1990, mussels were reported at Lennox, Thunder Bay, Lakeview, Sir Adam Beck, and Bruce.

The use of sound energy was considered as a control measure since it potentially could provide a cost-effective means of controlling zebra mussels. However, the use of sound to control biofouling organisms has not been well investigated and, in addition, results have been inconsistent (Jenner, 1983; Kawabe, 1983). Little information is available on other frequencies to either resonate the attachment site or the animal itself (Field et al., 1987). Sound has advantages over other potential methods such as chemicals in that it is less likely to kill nontargeted organisms, has no obvious residue effects, and equipment can be installed relatively easily. The objective of this study is to determine the effectiveness of sound as a control measure for zebra mussels.

METHODS

Preliminary experiments were conducted with mussels (3–12 mm in length) by exposing them to sound projected from three different sources (airborne, waterborne, and solidborne). Airborne experiments involved placing several mussels inside water-filled beaks which were exposed to sinusoidal tones (170 dB at 1 μPa) produced by a high intensity compression driver operating in air. At various tones, mussels reacted quickly with rapid shell closure. Although mussels remained closed for about 20 hr, there appeared to be no apparent mortality relative to controls. Further experiments involved exposing mussels to low frequency, waterborne tones ranging in frequency from 40 to 200 Hz (180 dB at 1 μPa). Although mussels remained closed during and a few minutes after exposure, they eventually opened their shells. There was no evidence of mortality relative to the control group. A third experiment, which involved introducing solidborne sound (sonic vibration) into a short aluminum rod (<20 cm), indicated that individual mussels could be removed. Thus, further studies focused on estimating effects due to primarily solidborne acoustic energy.

Zebra mussels used in experiments were juveniles that ranged in size from approximately 1–5 mm in length. Individuals in this size range are extremely mobile and thus well-suited to the experimental design. Any response to the acoustic stimulus by this size range might also be expected by

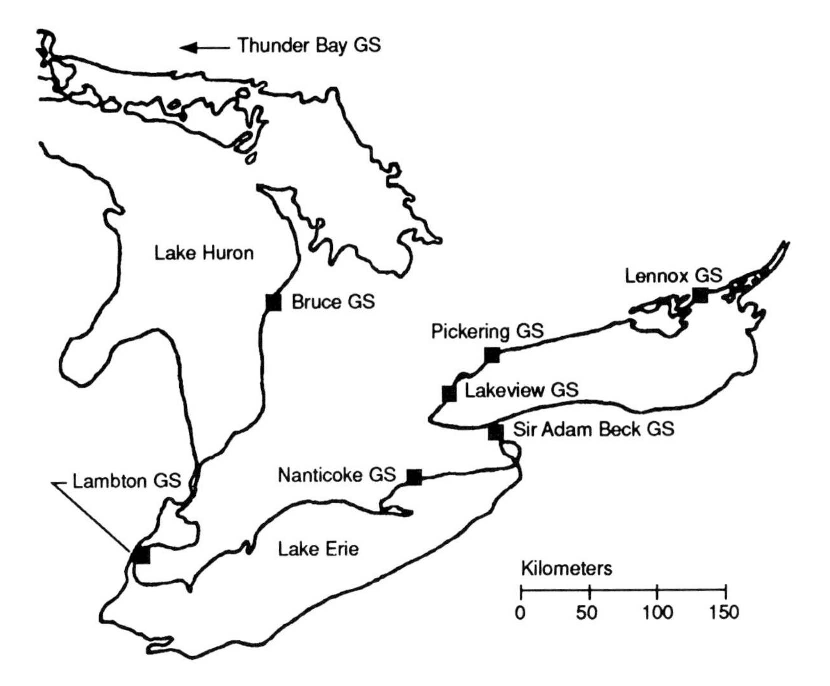

Figure 1. Location of power generating stations of Ontario Hydro in the Great Lakes region.

veligers (perhaps even more so because veligers have no shell and thus would be more vulnerable). Mussels were collected by divers from Lake Erie, near the shore of Nanticoke power station, and transported immediately to the laboratory where they were held in controlled environmental chambers at 4–10°C until the experiments began. Mussels were acclimated to room temperature for several hours prior to use.

Paired tanks, each containing approximately 80 L of dechlorinated water, were used in the experiments (Figure 2). Both tanks were placed inside a anechoic chamber (64 m^3) where all experiments could be conducted under controlled acoustic conditions. Tanks were placed approximately 3 m apart and checked during experiments to determine whether cross-talk occurred between tanks. Aluminum piping (Schedule 40, 55-cm length × 10-cm diameter) was selected as the substrate for evaluating zebra mussel attachment in response to experimental variables. This material provided a structure similar in acoustic properties to that of service and industrial water pipes but without the surface corrosion problems of commonly used carbon steel. The experimental tank contained two pipes whereas the control tank contained one. Solidborne sound was provided by a piezoelectric shaker (5000 N peak force) bolted to the side of the treated pipe (Figure 3). The other pipe in the

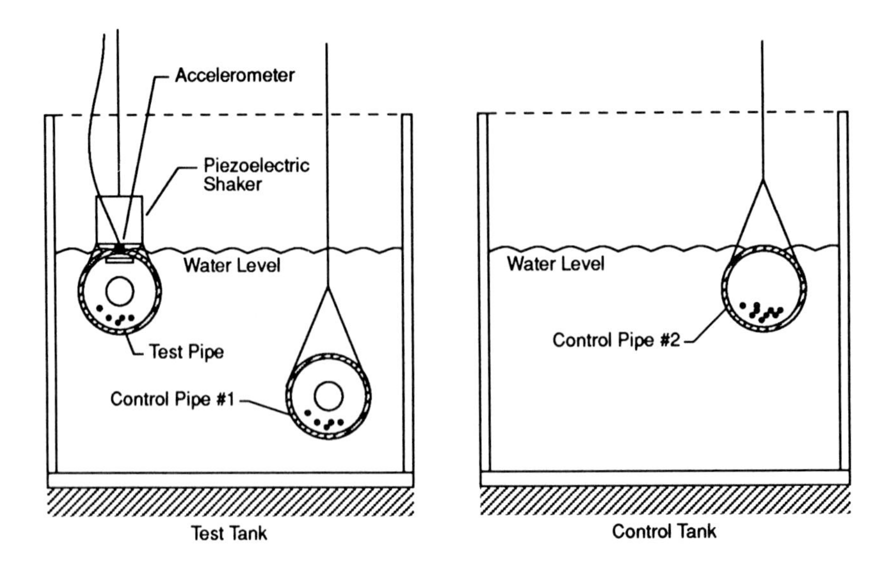

Figure 2. End view of experimental design and equipment used to determine the effect of acoustic energy on *Dreissena polymorpha.*

experimental tank was placed approximately 10 cm from the treated pipe, and acted as another control (to observe animal condition). This arrangement produced waterborne sound in the control pipe by direct coupling with the fluid. Pipe ends were partially sealed with a thin, semitransparent plastic sheet which prevented mussels from migrating or falling out of the pipe, but still allowed the free exchange of water (Figure 3).

A total of 29 experiments were conducted at different frequencies and amplitudes of acoustic energy. For each experiment, three samples of 100 animals were placed in each pipe during the beginning of each experiment. All individuals were visually inspected for general condition and only live, active mussels were used in experiments. Individuals were exposed to the treatment for between 4 and 24 hr.

After each experimental treatment, mussels in each pipe were examined and classified as attached or unattached. This was done by gently decanting both water and unattached mussels from the pipe. Any mussels still holding to the pipe were counted as attached. The distribution of attached animals was determined as near the top, sides, and bottom of the pipe. Mortality was determined for both attached and unattached individuals immediately at the end of an experiment by behavioral probing and by direct observation using a dissecting microscope (magnification × 20–50).

A piezoelectric shaker was designed to generate sinusoidal oscillations sweeping continuously over a frequency range of 1500 Hz every 10 sec. The variance for each 10-sec sweep, however, was relatively constant over the

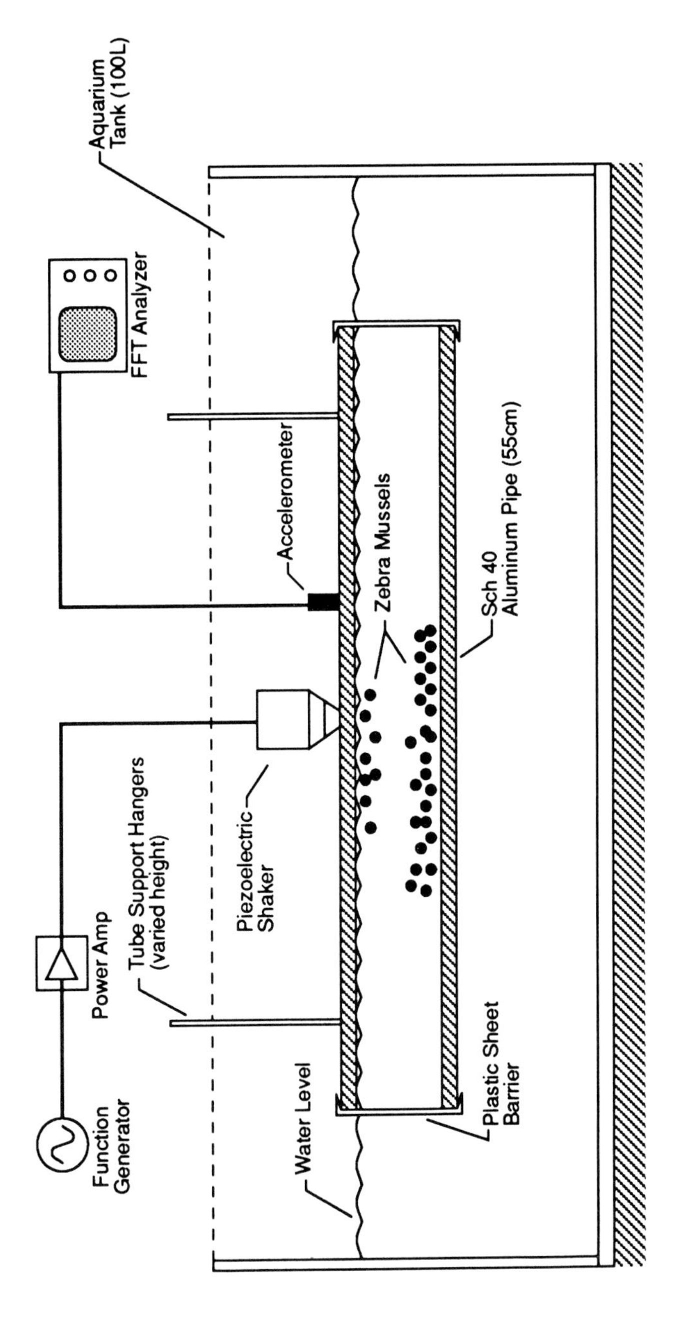

Figure 3. Side view of experimental design and equipment used to determine the effect of acoustic energy on *Dreissena polymorpha.*

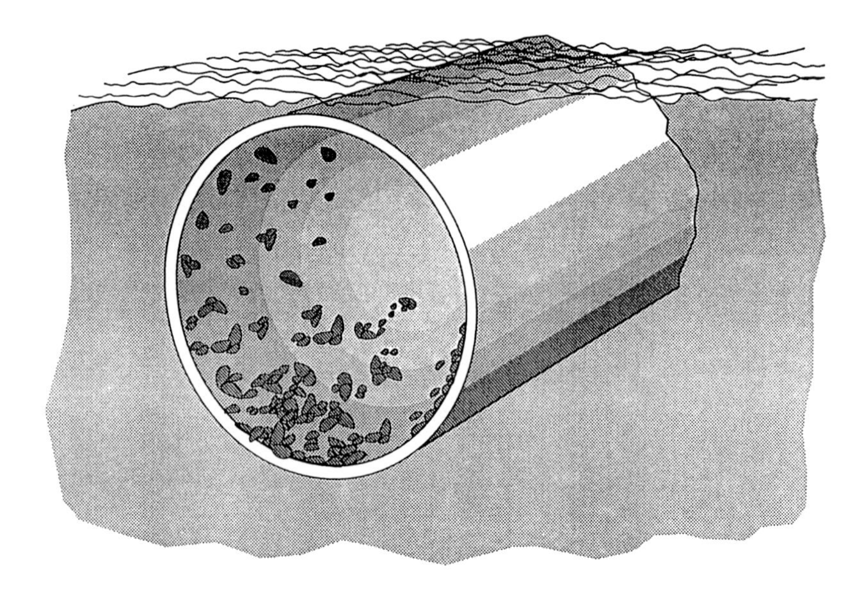

Figure 4. Typical distribution of *Dreissena polymorpha* in a section of pipe during the experiments.

duration (typically 4–24 hr) of each experiment. Drive frequencies used in these experiments started at 3 kHz and ended at 18 kHz.

Experiments were conducted with the shaker drive level set at either 10, 25, or 50% of full power. An accelerometer, mounted at an arbitrary fixed location on the pipe, monitored the vibrational amplitude of the pipe in terms of RMS acceleration units in g′ (the unit "g RMS" refers to the root-mean-square where $g = 9.8 \text{ m/sec}^2$). As expected, pipe response was variable due to the multitude of vibrational modes generated by the structural dynamics of the pipe. The mean acceleration of vibration in these experiments was found to be between 30 and 350 g′. Analysis of variance was used to determine whether significant differences occurred in attachment and mortality between treated and control groups as a function of both frequency and amplitude.

RESULTS AND DISCUSSION

Mussels placed in both control and experimental pipes established themselves quickly (within 1 hr from start of each experiment). A typical distribution of mussels in a control group is shown in Figure 4. Effective prevention of zebra mussel attachment by solidborne sounds was observed at various sonic frequencies in the range of 8–14 kHz when the structural vibration of pipes exceeded a certain minimum value (Figure 5). At these frequencies the mean acceleration of the pipe surface ranged from approximately 6 g′ at 3

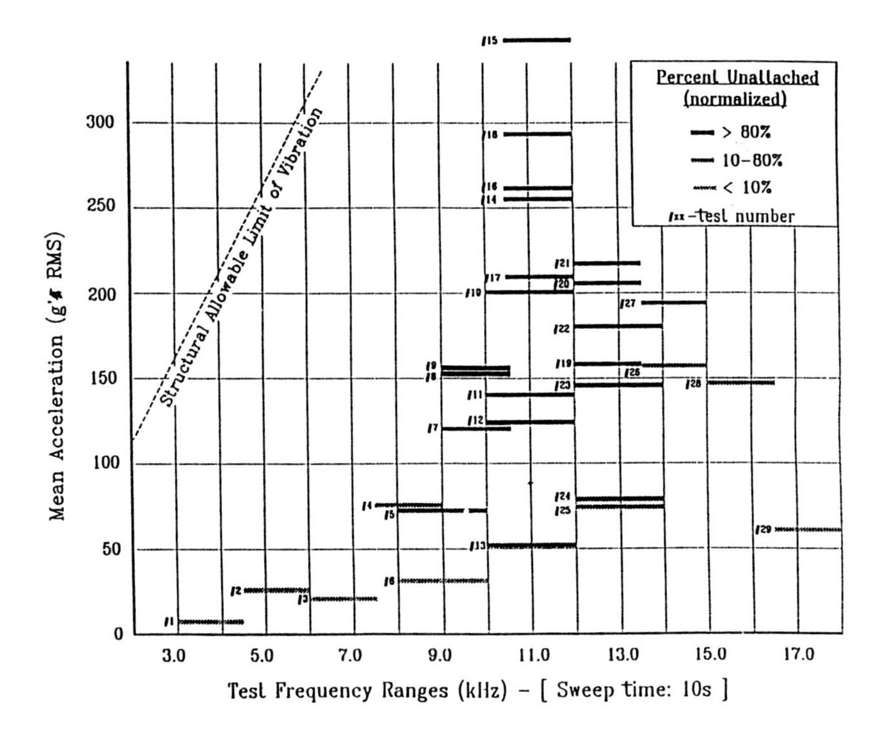

Figure 5. Summary of the results of acoustic experiments for controlling attachment of *Dreissena polymorpha.*

kHz (3.0–4.5 kHz) to approximately 350 g′ at 11 kHz (10.5–12.0 kHz). In the 8- to 10-kHz range, nearly 100% detachment was achieved with vibration amplitudes up to about 150 g′. Mortality in this frequency and amplitude range was estimated to be 75–95%. In the 10- to 12-kHz range, almost 100% detachment occurred at vibration amplitudes exceeding 200 g′ (Figures 5 and 6). With the exception of one test, mortality in this case approached 100%. Finally, in the 12- to 14-kHz range, about 90% detachment was recorded with amplitudes of about 210 g′. Mortality was estimated to be approximately 70%. Higher amplitudes in this frequency range may have yielded higher mortality but were not to be attempted in the present study.

Three-way analysis of variance (control vs experimental, frequency, and amplitude effects) of results reveal that amplitude was significant and that frequency was not significant in affecting mussel detachment (amplitude $F1,19 = 3.2$, $p < 0.05$; frequency $F1,19 = 1.56$, $p > 0.05$). These factors, however, were not as clear for mortality observations. Both amplitude ($F1,19 = 3.07$, $p < 0.05$) and frequency ($F1,19 = 2.67$, $p < 0.05$) were significant variables affecting mortality. High numbers of detached and dead mussels were associated with high amplitude test treatments above 200 g′ (Figure 6). However, it is possible that lower amplitudes (possibly below 100 g′) may also be effective.

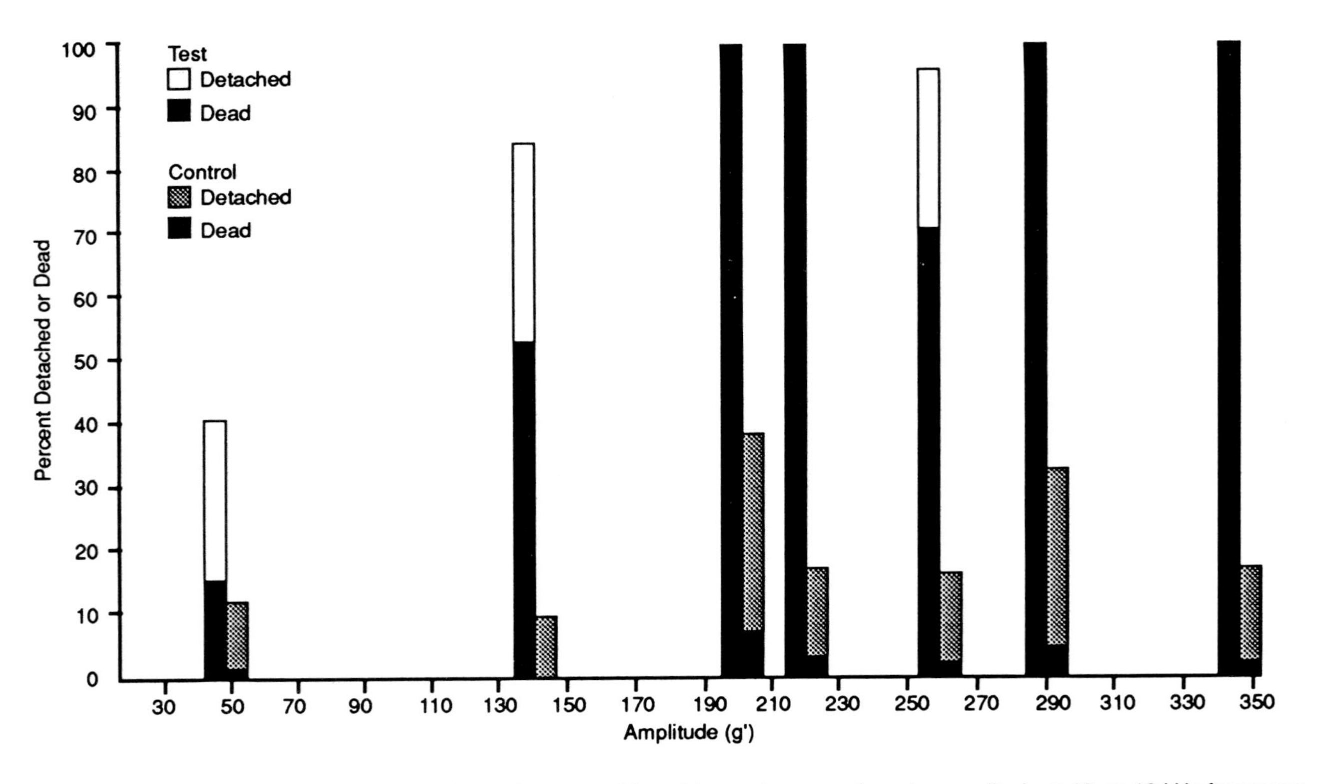

Figure 6. Percent detachment of *Dreissena polymorpha* exposed to solid sound energy of varying amplitude at 10- to 12-kHz frequency.

Although some mortality and an increase in numbers of detached mussels were observed for some frequencies at low amplitude, further delayed effects could also occur. All measurements conducted in this study were done immediately after turning off the sound source.

Observations of the high amplitude tests indicate that some effect occurred in the "control" pipe in the experimental tank (Figure 2). Significant mortality occurred when the acceleration rate exceeded 200 g′. These animals were affected by acoustic energy transmitted by waterborne coupling between the two closely spaced (10 cm) pipes.

In experiments where vibration was moderate (<100 g′), significant "clumping" was observed, with individuals holding by a few byssal threads. At higher vibration levels exceeding 150 g′, some individual mussels were observed to be "skidding" uncontrollably on the inside surface of the pipe. In these tests, there was significant mortality with many mussels being totally disintegrated. These results are unique since they address solidborne sound and were conducted in a controlled environment (Kawabe, 1983; Jenner, 1983). It is one of the few studies which focuses on varying both frequency and amplitude in the audible range.

It must be emphasized that the acoustic energy necessary for preventing mussel attachment in these experiments does not result in vibration levels that would damage equipment structures. Acceptable vibration limits generally increase with frequency and depend on structure composition and shape. At audible sonic frequencies, there is generally less concern with structural damage due to vibration than at lower frequencies, where bending modes and much larger displacements can cause failure. Based on criteria used in nuclear power plant piping systems, the allowable limits for the pipe that was used in these experiments was calculated (ASME/ANSI, 1987); and results indicate that vibration limits for the test pipe were not exceeded. Thus, this stress level is not expected to create a fatigue problem for piping systems found in power plant facilities.

CONCLUSION

Experiments conducted in this investigation demonstrate that effective prevention of zebra mussel attachment is possible in structures by the use of solidborne acoustic energy. The intensity and frequency of acoustic excitation needed was equivalent to about 100 g′ on the surface of the pipe between 8 and 14 kHz. Although the vibration amplitude needed for effectiveness appeared to increase with frequency, these were well within the permissible limits for normally operating equipment such as piping. Significant mortality rates were encountered in higher level exposures. This may be attributed to sonically induced cavitation as evidenced by the formation of small bubbles near the surface of the structure.

ACKNOWLEDGMENTS

The authors are grateful to Dr. F. Spencer, M. Colbert, A. Sakuta, Dr. E. Leitch, B. Sim, and S. McLeod of Ontario Hydro.

REFERENCES

ASME/ANSI Standard. ''Operation and Maintenance of Nuclear Power Plants,'' The American Society of Mechanical Engineers (1987).

Griffiths, R. W., W. P. Kovalak, and D. W. Schloesser. ''The Zebra Mussel, *Dreissena polymorpha* (PALLAS, 1771), in North America: Impact on Raw Water Users,'' in Symposium: *Service Water Systems Affecting Safety-Related Equipment* (Palo Alto; CA: Nuclear Power Division, Electric Power Research Institute, 1989), 11–27.

Field, L. H., A. Evans, and D. L. MacMillan. ''Sound Production and Stridulatory Structure in Hermit Crabs of the Genus *Trizopagurus*,'' *Mar. Biol.* 65:230–235 (1987).

Jenner, H. A. ''Control of Mussel Fouling in The Netherlands: Experimental and Existing Methods,'' in Proceedings of Symposium on Condenser Macrofouling Control Technologies, I. A. Diaz-Tous, M. J. Miller, and Y. G. Mussali, Eds. Hyannis, MA (1983) 18-1 to 18-13.

Kawabe, A. ''Control of Macrofouling in Japan: Existing and Experimental Methods,'' in Proceedings of Symposium on Condenser Macrofouling Control Technologies, I. A. Diaz-Tous, M. J. Miller, and Y. G. Mussali, Eds. Hyannis, MA (1983) 23-1 to 23-42.

CHAPTER 39

Upper Lethal Temperatures of Adult Zebra Mussels (*Dreissena polymorpha*)

Stanley Iwanyzki and Robert W. McCauley*

Upper lethal temperatures of adult zebra mussels, *Dreissena polymorpha*, acclimated to 2.5, 11.0, 15.0, 20.0 and 25.0°C were determined by exposing mussels to constant high temperatures and calculating median survival times. Close correlations between upper lethal and acclimation temperatures were found. Plotted data yielded a family of nearly parallel semilogarithmic lines, each line corresponding to an acclimation level. The ultimate upper incipient lethal temperature appears to be approximately 30°C — a value several degrees higher than the normal maximum temperature where mussels are currently found in the Great Lakes. Results of the present study may be useful in determining times and temperatures for treatment of intake pipes colonized by mussels.

INTRODUCTION

The recent invasion of the zebra mussel (*Dreissena polymorpha*) in the Laurentian Great Lakes has raised concern about its ecological and economic

* To which correspondence should be addressed.

0-87371-696-5/93/$0.00 + $.50

consequences and has stimulated research on mussel life history and control methodology (Hebert et al., 1989; Cooley, 1991). Some of this research is focused on methods of defouling the inner surfaces of pipes through which lake and river water are pumped. Pipes can become fouled with mussels to the extent that the flow of water is seriously reduced (Griffiths et al., 1991). One possible technique to control zebra mussels is to backflush fouled pipes with heated water. This technique may be a preferred treatment in some cases because, after passage through pipes, the heat is quickly dissipated into the environment. Temperatures of heated water during application must be sufficiently high to kill mussels without causing structural damage to pipes by thermal expansion and contraction. In this study, we describe the upper lethal temperature relations of adult zebra mussels to provide data useful in planning thermal treatments, and also to contribute to the understanding of the thermal ecology of this species.

METHODS

Specimens of zebra mussels attached to rocks were collected from Lakes Erie and St. Clair in early winter 1989 and transported to the laboratory. Rocks and attached mussels were placed in 50-L aquariums containing well-aerated, dechlorinated tap water which was replaced daily. Mussels were fed yeast daily, and uneaten food and detritus were removed on a regular basis by siphoning. Mussels 10–20 mm in length were used in the experiments. Groups of 150 mussels were acclimated to 2.5, 11.0, 15.0, 20.0 and 25.0°C. Mussels were acclimated to temperatures exceeding those at which they were collected by increasing temperatures of rearing baths at rates not exceeding 1°C/day. Test animals were detached from substrates in rearing tanks by severing their byssal threads with a razor. Upper lethal temperatures were determined on subsamples of 10 mussels, each placed in a series of test baths consisting of 20-L aquariums containing well-aerated water and maintained at a constant temperature (± 0.1°C) by an immersion electric heater controlled by a thermostat. After introduction to the test bath, mussels were continuously observed during the first 16 min. After this period, observations occurred at specific time intervals, usually in a geometric series depending on the rate of mortality. The criterion of death was that used by Basedow (1969). Mussels undergoing heat death expire with their valves open. To distinguish this state from the normal gaping behavior, mussels were tapped several times at each observation period with a probe to elicit a closing reflex. Lack of a response to this stimulus was taken as a sign of mortality. Times of individual deaths were estimated by interpolation between successive observation periods. Animals recorded as dead were returned to acclimation baths for further observation and, in all but a few instances, no recovery was noted. Experiments

TABLE 1. Mean Resistance Times (min) of *Dreissena polymorpha* Exposed to Elevated Test Temperatures

Test	Acclimation Temperature (°C)				
Temperature	2.5	11.0	15.0	20.0	25.0
30	6820	3777	—	3842	5702
31	196	1010	1066	1989	3157
32	20	182	—	1730	2812
33	13	69	316	765	1048
34	10	27	34	110	253
35	—	—	—	32	41
36	10	6	11	50	39

TABLE 2. Regression of Logarithmic Mean Survival Time on Lethal Test Temperatures of *Dreissena polymorpha*

Acclimation Temperature (°C)	Regression line[a]	N	r
2.5	Y = 15.57 − 0.4231 X	60	−0.81
11.0	Y = 17.46 − 0.4688 X	60	−0.98
15.0	Y = 16.07 − 0.4194 X	40	−0.97
20.0	Y = 14.91 − 0.3728 X	70	−0.95
25.0	Y = 16.12 − 0.4041 X	70	−0.95

[a] Y, in min; X, °C.

at each test temperature were continued until all individuals were dead. This occurred between 20 min and 1 week.

Median survival times for each test at a specific temperature were estimated from geometric means of individual resistance times because plotting these on logarithmic probability paper revealed that the logarithms of individual times to death were normally distributed about the mean (Fry et al., 1946). An estimate of the ultimate upper incipient lethal temperature was made from inspection of the whole data set (Brett, 1952).

Results and Discussion

Mean survival times of mussels at upper lethal temperatures are summarized in Table 1. Regressions between test temperatures and mean survival times were similar for each acclimation temperature (Table 2). This is unusual and may be the result of some upward acclimation in test baths, especially at experiments of relatively long duration. Basedow (1969), for instance, showed that *Dreissena polymorpha*, when acclimated to 10.0°C and transferred directly to 27.5°C, thermally adjusted to the new temperature in several hours. Our first approximation of the ultimate upper incipient lethal temperature of zebra mussels based on the inspection of the whole data set is 30.0°C.

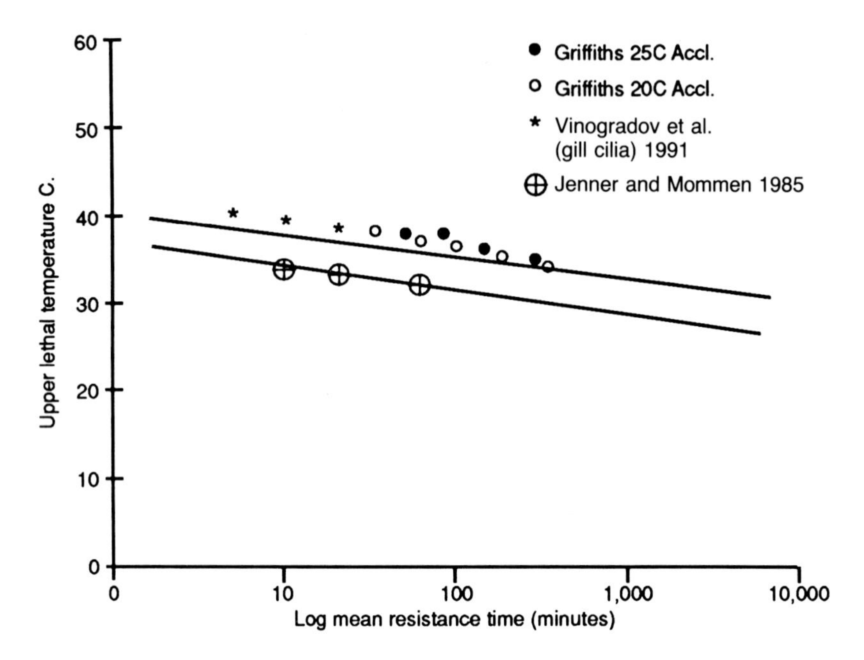

Figure 1. Upper lethal temperatures and log mean survival times of adult *Dreissena polymorpha*. Regression lines represent the relationship for data collected in this study: lower line, mussels acclimated to 2.5°C; upper line, mussels acclimated to 25.0°C.

With the exception of Griffiths (1991), there are relatively few determinations or estimates of the thermal resistance of *Dreissena*. Techniques of testing and the criteria for death vary among different studies, making comparisons difficult. In spite of this lack of standardization, there is good general agreement between data from various sources; all values lie within a relatively narrow range in the upper limit of the biokinetic tolerance of the species. Results of our experiments are compared to results of other studies in Figure 1. At upper lethal temperatures, survival times in this study were shorter then those found by Griffiths (1991). This difference probably was a result of different criteria for death. Griffiths (1991) determined death as the cessation of a response of the somatic muscles to a mild electrical stimulus. Since individual tissues of an animal can survive long after the integrity of the whole organism has been destroyed, this endpoint overestimates the time when death actually occurs. The method of Griffiths (1991) was conservative since his study had the purpose of devising an effective, thermal treatment which would kill mussels in pipes of mussels with some margin of assurance. Similarly, survival times at upper lethal temperatures of gill tissues taken from mussels acclimatized to warm summer temperatures in Rybinsk reservoir (Vinogradov et al., unpublished data) were slightly higher than found in this study. Again, this difference may be explained by the higher thermal resistance displayed by some body cells over that of the intact animal.

Since *Dreissena* survives winter temperatures in the northern part of its range, the lower lethal temperature probably is just above 0°C (Strayer, 1991). When upper and lower lethal temperatures are both taken into account, the range of thermal tolerance is about 30°C. This range is characteristic of a mesothermal species and is consistent with the thermal regimes of the water bodies in which *Dreissena* inhabits. Its range in Europe extends from southern Sweden to the shores of the Mediterranean (Mackie et al., 1989; Strayer, 1991). In Lake St. Clair, mean maximum nearshore temperature in the summer is about 25°C — several degrees below the final upper lethal temperature.

Upper lethal temperatures determined in the present study are based on the abrupt transfer of test animals from acclimation tanks to test baths maintained at a lethal or near lethal constant temperature. Since lethal warm temperatures cannot be produced instantaneously during thermal backflushing treatments, mussels are subjected to increased heat over several hours. Temperatures rise from the acclimation level: first through the zone of thermal tolerance and then, once the lethal temperature threshold is crossed, into the lethal zone of thermal resistance. During this time period, mussels increase their heat tolerance through acclimation. The acclimation process can proceed rapidly, and it is probable that animals will become acclimated to their fullest extent during this period. With this in mind, a relationship between lethal temperatures and the time needed to produce 100% mortality was developed using values from various sources (Figure 2). Since these values, in some instances, are based on median resistance times, temperatures necessary for 100% mortality were sometimes calculated. Accrued lethal effects during the period of time needed to reach lethal temperatures have been ignored to allow for a certain margin of error. The regression relating time to 100% mortality and lethal temperature is: $\log Y = 13.2 - 0.3 X$ where Y = time to 100% mortality and X = temperature (°C). This ''Delphi Approach'', in which experimental data from different investigators are combined, results in information from which a heat treatment can be planned with confidence. To ensure the maximum likelihood of complete mortality, times to 100% determined from this regression should be considered minimum values.

ACKNOWLEDGMENTS

We thank S. Griffiths of Ontario Hydro for sharing data on upper lethal temperatures with us and the Biology Department of Wilfrid Laurier University for financial support.

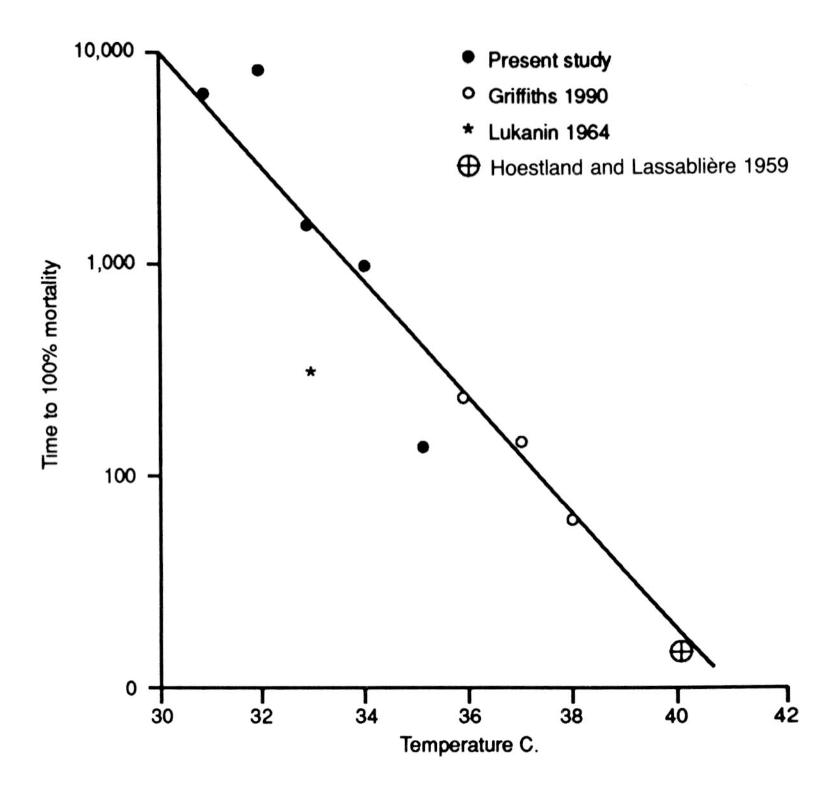

Figure 2. Recommended time-temperature treatments to obtain 100% mortality of *Dreissena polymorpha* based on data from different sources. Values given for the present study were derived from the pattern of mortality of mussels acclimated to 25.0°C.

REFERENCES

Basedow, T. "Uber die Auswirkung von Temperaturschocks auf die Temperaturresistenz poikilothermer Wassertiere. Eine Untersuchung zum Problem der thermischen Schockanpassung bei Tieren," *Int. Rev. ges. Hydrobiol.* 54:765–789 (1969).

Brett, J. R. "Temperature Tolerance of Young Pacific Salmon Genus *Oncorhynchus*," *J. Fish. Res. Board Can.* 9:265–323 (1952).

Cooley, J. M. "Zebra Mussels," *J. Great Lakes Res.* 17:1–2 (1991).

Fry, F. E. J., J. S. Hart, and K. F. Walker. "Lethal Temperature Relations of a Sample of Young Speckled Trout," University of Toronto Studies, Biological Series No. 54, *Publ. Ont. Fish. Res. Lab.* 66:1–35 (1946).

Griffiths, J. S. "Zebra Mussel Thermal Shock — Research," Ontario Hydro Internal Report, Toronto, Ontario (1991).

Hebert, P. D. N., B. W. Muncaster, and G. L. Mackie. ''Ecological and Genetic Studies on *Dreissena polymorpha* (Pallas): A New Mollusc in the Great Lakes,'' *Can. J. Fish. Aquat. Sci.* 46:1587–1591 (1989).

Hoestlandt, H., and J. Lassabliere. ''Destruction thermique de la moule d'eau douce,'' *L'Eau* 46:38–41 (1959).

Jenner, H. A., and J. P. M. Mommen. *Driehoeksmosselen en aangroeiproblemen 'H$_2$O',''* *Tijdschr. Watervoorz. Afvalwaterbeh.* 18:2–6 (1985).

Lukanin, V. S. ''Survival of Adult *Dreissena* in Copper Sulphate Solutions of Different Concentrations and Temperature,'' in Biology and Control of *Dreissena*, B. K. Shtegman, Ed. *Tr. Inst. Biol. Vnut. Vod. Akad. Nauk* (1964), pp. 69–70.

Mackie, G. L., W. N. Gibbons, B. B. Muncaster, and I. M. Gray. ''The Zebra mussel, *Dreissena polymorpha*. A Synthesis of European Experience and a Preview for North America,'' Ontario Ministry of the Environment, Water Resources Branch, Great Lakes Section, Queen's Printer, Toronto, Ontario (1989).

Strayer, D. L. ''Projected Distribution of the Zebra Mussel, *Dreissena polymorpha*, in North America,'' *Can. J. Fish. Aquat. Sci.* 48:1389–1395 (1991).

SECTION V

General

CHAPTER 40

Dispersal Mechanisms of the Zebra Mussel (*Dreissena polymorpha*)

James T. Carlton

The dispersal of the zebra mussel (*Dreissena polymorpha*) is mediated by 3 natural (currents, birds, and other animals) and 20 human-related mechanisms. Human mechanisms include those related to waterways, vessels, navigation, and fishery activities, and a wide variety of miscellaneous vectors (e.g., including intentional movements, aquariums releases, and scientific research). All mechanisms transport juveniles and/or adults, but fewer mechanisms transport eggs and larvae (e.g., currents, perhaps animals, canals, ballast water, other vessel water, fish stocking, bait bucket, firetruck water, aquarium releases, amphibious planes and scientific research). Three lines of evidence lead to the conclusion that ballast water in transoceanic vessels transported zebra mussels from Europe to the Laurentian Great Lakes: (1) the common presence of bivalve larvae in ballast water combined with patterns of vessel traffic into the lakes; (2) the likelihood that ballast water was the mechanism for the arrival in the same decade of other European aquatic organisms (fish and crustaceans); and (3) the inability to identify any other probable mechanism. Rapid upstream dispersal can be achieved by all mechanisms except for current flow, and thus water direction serves as no barrier to zebra mussels. Downstream movement is achieved by all mechanisms. Overland dispersal is achieved by all mechanisms except for currents, fish and

0-87371-696-5/93/$0.00 + $.50

some other animals, canals, ballast water, and fishing vessel wells. Much of the initial movement of the zebra mussel out of the Great Lakes basin region has been by canals and by transport of boats on trailers to distant overland sites. The Great Lakes are now a new major global donor region for zebra mussels as a result of ballast water export.

INTRODUCTION

Freshwater and marine invertebrates possess a remarkable variety of dispersal mechanisms. Natural mechanisms are considered by Ekman (1953), Briggs (1974), Vermeij (1978), van den Hoek (1987), Scheltema (1988), Locke and Corey (1989), Pennak (1990), and others. Human-mediated mechanisms are summarized by Carlton (1985, 1989, 1992) and Zibrowius (1983). Dispersal by natural means is generally assumed to be the primary mode of both inter- and intracontinental dissemination of aquatic organisms. It is, however, often difficult to determine whether the dispersal history of any given species is due to natural or human-mediated processes.

The introduction of the zebra mussel, *Dreissena polymorpha*, into North America in about 1986 started an irreversible process leading to the conclusion that its eventual spread to every habitable region on the continent is inevitable. There are, nevertheless, a number of reasons to understand the dispersal mechanisms available to *Dreissena*. First, an understanding of these mechanisms is indispensable to predicting the direction and rate of spread of this species. Second, containment or control measures to slow or prevent the dispersal of *Dreissena* by human activities must take into account the full range of mechanisms available. Third, knowledge of dispersal mechanisms is fundamental to understanding the potential for gene flow between populations for ecological and evolutionary studies.

There are no previous reviews of the dispersal mechanisms of zebra mussels. Based upon largely anecdotal remarks in the literature and upon unpublished speculation following the discovery of the zebra mussel in North America in 1988, a table of 23 natural and human-mediated mechanisms has been assembled (Table 1). For the sake of space, some mechanisms listed separately in Table 1 are treated together in the text. Included, by way of contrast, are transport modes believed to have led to the dispersal of the Asian clam (*Corbicula fluminea*) the other most important freshwater mollusk to have invaded North America.

NATURAL MECHANISMS

Eggs and larvae of zebra mussels are carried by currents and downstream flow (Griffiths et al., 1991) and may remain in the plankton over 30 days (Morton, 1969; Sprung, 1989), although many workers report the typical

TABLE 1 Potential Dispersal Mechanisms of Different Life Stages of the Zebra Mussel (*Dreissena polymorpha*)

	Life Stage		
	Planktonic	Sedentary	
Dispersal Mechanisms	Eggs and Larvae	Juveniles	Adults
Natural			
1. Currents[a]	X	X	X
2. Birds[b]	X?	X	X
3. Other animals	X?	X	X
Human-Mediated			
Waterways			
4. Canals[a]: vessels, irrigation	X	X	X
Vessels/Navigation			
5. Ballast water [b]	X	X	
6. Vessel exteriors[a]		X	X
7. Vessel interiors	X	X	
8. Fishing vessel wells	X	X	
9. Navigation buoys[c]		X	X
10. Marina/boatyard equipment[c]		X	X
Fisheries			
11. Fishing equipment		X	X
12. Fish cages[c]		X	X
13. Fish stocking water	X	X	
14. Bait and bait-bucket water	X	X	X
Other Industry			
15. Commercial products[a]		X	X
16. Marker buoys and floats[c]		X	X
17. Firetruck water	X	X	X
Other			
18. Intentional movements[a]		X	X
19. Aquarium releases[a]	X	X	X
20. Amphibious planes	X?	X	X
21. Recreational equipment		X	X
22. Litter (garbage)[c]		X	X
23. Scientific research	X	X	X

Note: ? = Uncertain whether this life stage would be associated with the indicated mechanism.

[a] Dispersal by this mechanism has been documented.
[b] Dispersal by this mechanism is suspected to have occurred.
[c] Zebra mussels known to occur on the animal or object.

length of planktonic life to be less than 16 days (Mackie et al., 1989; Griffiths et al., 1991). Mussels can actively detach and undertake "bysso-pelagic transport" (Rossmassler, 1865; Korschelt, 1892, p. 168, footnote; Oldham, 1930; Mackie et al., 1989; Griffiths et al., 1991) or be passively detached (resuspended) by storm activity. Thus, newly settled and small juvenile mussels can resume planktonic dispersal and considerably extend the length of current transport. Rafting (Scheltema, 1977) on floating driftwood (Hesse et al., 1937, p. 60; Mackie et al., 1989), aquatic plants (Lewandowski, 1983), detached floats and buoys, plastic garbage, and other natural detritus or human debris may be equally effective means of long-distance dispersal within and between connected bodies of water. Postlarval, juvenile, and adult *Corbicula* are similarly believed to be carried downstream by being swept along the bottom or by floating on mucous threads, algal mats, or driftwood (Stein, 1962; Sinclair and Isom, 1963; Eng, 1979; Britton and Morton, 1979; Prezant and Chalermwat, 1984; Counts, 1986).

Waterspouts (Mooney, 1990) are another natural mechanism that theoretically could move living bivalve mollusks between water bodies at any life stage. Fair-weather spouts form only over water and typically lift water to a height of about 3 m (Grant, 1944); while tornadic spouts start on land, cross over water, and can move or draw up considerable volumes of water (Ludlum, 1970). Kew (1893) notes cases of powerful storms carrying fish and frogs for some distances; this includes a record of a tornado dropping hundreds of mussels (*Anodonta*), presumably drawn up from a nearby river, on the streets of Paderborn, Germany.

Dispersal of living freshwater mollusks (including bivalves) by birds (Russell Hunter et al., 1964; Rees, 1965; Malone, 1965; Proctor and Malone, 1965; Mackie, 1979; Boag, 1986), amphibians (Rees, 1952), insects (Rees, 1965), and fish (Voskresenskii, 1966) is a well-known phenomenon (Table 1, no. 2 and 3). Maguire (1963) provides an early review. Turtles may also carry other aquatic organisms (Frazier, 1986); aquatic organisms could be transported short distances between ponds and lakes in the wet fur of semiaquatic mammals such as beavers and muskrats (Peck, 1975), or on larger mammals such as moose (Linza, Personal communication). *Corbicula* are eaten by mink and muskrat (van der Schalie, 1973) and raccoons (Taylor and Counts, 1977). Short-distance transport of aquatic bivalves by mammals (presumably by being carried, rather than by survival through the alimentary track) may be possible, although this is thought unlikely by Isom (1986).

Zebra mussels are known to settle on live crayfish (Astacidae) (Rossmassler, 1865; and observations in Lake Erie [Lach, Personal communication]), and crayfish have been implicated in local dispersal events of small crustaceans (Moore and Fausti, 1972). While zebra mussels may be carried wherever the host crayfish moves it seems probable that such movements would be spatially limited. However, the movement of zebra mussels attached to crayfish used as bait (Table 1, no. 14) has the potential to release zebra mussels into waters hundreds of kilometers away from where they were collected.

Of all animal vectors, birds may be the most important dispersal agents. It is for this reason that birds are distinguished as a mechanism separate from "other animals." Mussels could be dispersed by aquatic waterfowl in three ways: attached externally, carried internally, and carried directly.

Mussels could (1) adhere externally to birds in and on the plumage, perhaps being confounded in debris and mud; (2) be carried as eggs and larvae in water on the body; (3) be transported when attached to algae or other debris wrapped around a bird's feet, legs, or neck; and (4) be attached directly to the feet or legs by byssal threads (should opportunity arise for attachment). However, Kew (1893, p. 49) noted that it remained to be determined whether the "free-roving pelagic larvae" of *Dreissena* could withstand atmospheric exposure such as might be experienced during bird flight.

Live mussels could be defecated, regurgitated, and (as noted by Rees [1965] drawing from earlier comments by Darwin and Lyell) released from a bird's crop after the bird had been attacked by a bird of prey and torn apart. A corollary to this would be the capture and subsequent dropping of a mussel-eating fish by piscivorous birds such as cormorants and ospreys. Live passage of mussels of any life stage through a bird's gastrointestinal system is uncertain. Studies with aquatic snails and the clam *Corbicula* would suggest that this is unlikely (Thompson and Sparks, 1977). Live regurgitation of *Dreissena* remains to be investigated. Mackie (1979) reported that sphaeriacean clams may be regurgitated alive by migratory waterfowl, but only if they are regurgitated "after a few hours" (Mackie — in Counts, 1986).

With respect to mollusks being carried directly on birds, Davis (1982) has noted that birds may fly along river courses with small sticks "a considerable distance" and that "presumably some of these sticks will fall back into the water." Such sticks, Davis noted, are frequently covered with the eggs of aquatic snails. The comparative behavior of different bird species relative to their propensity for transporting sticks or other aquatic materials (such as might be encrusted with zebra mussels) may be important in establishing the potential significance of this mechanism.

It is curious that no European naturalist has recorded any life stage of *Dreissena* on birds; although as Kew (1893, pp. 49–50) has noted, this may be in part due to no one ever having particularly looked. Should bird transport be observed for *Dreissena,* the distance that could be achieved is a central issue. Boag (1986) found that aquatic gastropods were likely to survive transport on a mallard flying for distances up to 10 km. Dispersal by birds is, for most aspects noted here, clearly amenable to quantitative observation and experimentation (Mackie, 1979; Boag, 1986).

Transport of *Corbicula* by birds, in the form of small byssate juveniles attached to feet and feathers and within gastrointestinal tracts, has been suggested by numerous authors (Ingram et al., 1964; Clench, 1970; Britton and Morton, 1979; McMahon, 1982). In addition, fish (e.g., the mollusk-feeding drum *Aplodinotus grunniens*) have been suggested as a possible *Corbicula* dispersal mechanism (Parmalee, 1965). Isom (1986) regarded transport of unionids and *Corbicula* by birds and fish highly unlikely. He noted that while both juvenile unionids and *Corbicula* have byssal attachment organs, the historical biogeographic evidence, particularly for native unionids, does not support bird transport. To date, there have been no documented cases of regional and interregional dispersal of *Dreissena* attributable to aquatic animals.

HUMAN-MEDIATED MECHANISMS

Waterways

Canal systems provide waterways for the dispersal of freshwater organisms (Marshall, 1892), usually serving to bypass natural barriers (Table 1, no. 4). Whether by larval dispersal, resuspension of young mussels, or by rafting, *Dreissena* is believed to have achieved most of its dispersal across Europe ''naturally'' via the canal systems that developed in the 18th and 19th centuries (Zhadin, 1952; Kerney and Morton, 1970). *Corbicula* has achieved a large amount of its dispersal via human waterways, often gaining access into agricultural lands and covering the bottoms of irrigation canals for many kilometers (Hanna, 1966; McMahon, 1982; Counts, 1986). Similarly, much of the initial movement of zebra mussels out of the Great Lakes basin region appears to have been by canals.

A subcategory of mussel dispersal through waterways is dispersal in pipes. For example, industrial pipeline construction includes the hydrostatic testing of new gas lines, where water is drawn in large volumes from a water source and transported many kilometers before being released into city drains or into another water body (Bimber, Personal communication).

Vessels and Navigation

Ballast water as a means of biotic dispersal (Table 1, no. 5) has been considered by Carlton (1985) and Williams et al. (1988). Ballast water is taken up by cargo and other vessels for trim, stability, and other purposes. Vessels without cargo may carry large amounts of ballast water, with some vessels carrying tens of thousands of metric tons of water. Vessels with cargo may also carry ballast water, but in lesser quantities.

Three lines of evidence have led to the conclusion that ballast water contained in transoceanic vessels transported zebra mussels to the Laurentian Great Lakes:

1. Ballast water is known to transport large numbers of living organisms across oceans on a regular basis; commonly encountered are the larvae of bivalve mollusks (Carlton et al., 1990). More specifically, ballast water is known to carry living molluscan larvae from Europe to the Great Lakes (see Appendix A). Large amounts of ballast water have been transported to the Great Lakes from all regions of the world for decades, although no historical analysis of this phenomenon is available. The amount and frequency of ballast water re-

leased into the Great Lakes increased with the opening of the St. Lawrence Seaway system in 1959. Up until May 1989, water ballast from Europe was released on a regular basis throughout the Great Lakes with no restrictions.

2. A number of other European freshwater species have preceded the arrival of the zebra mussel in the Great Lakes in the 1980s. All of these species have a life stage that could be transported in water and are therefore believed to owe their North American presence to ballast water transport. Thus, a pattern of ballast water invasions was established prior to the appearance of *Dreissena*. These include the mitten crab *Eriocheir sinensis* and the flounder *Platichthys plesus* (neither established), the spiny water flea *Bythotrephes cederstroemi*, and the ruffe *Gymnocephalus cernuus*.
3. Of the mechanisms listed in Table 1, only ballast water could have been the transport mechanism for *Dreissena* across the ocean. Arrival as a fouling organism is excluded by the inability of the zebra mussel to live in full saltwater for the length of time (i.e., more than 10 days) a vessel would require to cross between Europe and North America (Kew, 1893; Scharff, 1899; Kerney and Morton, 1970). There is no evidence for the other four potential mechanisms (i.e., anchor systems, commercial products, water used to transport freshwater fish or other aquacultural stock, and intentional movement).

Ballast water may have played two roles in the dispersal of zebra mussels. First, it may have led to the movement of mussels between water basins in Europe; second, it may have led to the transoceanic movement of mussels to North America. *Dreissena* eggs and larvae could easily be pumped into ballast tanks from one river system in Europe and then be released into another. Whether new populations in any river system in Europe owe their occurrences to ballast water is not known. It would appear likely, however, given the probability that ballast water was the means of introduction of *Dreissena* to North America that eggs and larvae are similarly taken up and discharged throughout various European watersheds as well. It would appear equally likely that *Dreissena* continues to be transported by ballast water from Europe, water destined to be released in or near freshwater drainages in continents around the world.

Ballast water is presumed to have been the means by which *Dreissena* as eggs and/or larvae (trochophores, veligers, and/or postveligers) were transported from Europe to North America (Hebert et al., 1989). The length of a typical trans-Atlantic ship voyage (10–15 days) does not exceed the mussel's planktonic larval life. However, there is no reason to exclude the transport of juveniles in ballast water by two other means: (1) wood or plant fragments bearing newly settled mussels could be entrained into a ballast tank and later pumped out, and (2) larvae could metamorphose and settle upon floating debris inside a ballast tank. Thus, ballast water transport of considerably longer than 2 weeks could effectively transport zebra mussels. I have frequently observed metamorphosed juveniles, only a few days old, of a wide variety of benthic organisms (e.g., spionid polychaetes, bivalve mollusks, echinoid echinoderms, ascidians) in the ballast water of vessels arriving in Coos Bay, Oregon, having completed a 13–15 day voyage from Japan.

Commercial vessels operating within the Great Lakes (''lakers'') are probably a major dispersal mechanism that has and will continue to transport zebra mussels and other aquatic species (e.g., the spiny water flea *Bythotrephes cederstroemi*) within and between the Great Lakes (Griffiths et al., 1991). Small sailing vessels may also have ballast water (Nouse, 1988). There are no formal studies on the actual role lakers play in moving planktonic organisms (holoplankton or, as with mussel eggs and larvae, meroplankton) across the lakes, and thus no insights are available into the temporal patterns or scale of this phenomenon.

Ballast water obtained from a coal-ore-grain bulk carrier (ballast capacity 11,153 metric ton) in October 1989 provides insight into the potential for rapid interlake movement of planktonic organisms. The bulk carrier had taken on ballast water the day before in Hamilton Harbor, Lake Ontario. Samples were collected from the aftpeak ballast tank using 53- and 100-μm mesh nets. Living organisms in the water included nematodes; the rotifer *Keratella cochlearis*; cladocerans *Bosmina longirostris, Eubosmina coregoni,* and *Daphnia galeata*; copepods *Mesocyclops edax* and *Eurytemora affinis*; blue-green algae *Microcystis aeroginosa* and *Aphanizomenon flos-aquae*; diatoms *Nitzschia acicularis, Fragilaria crotonensis,* and *Stephanodiscus niagarae*; and the green algae *Spirogyra, Stigeoclonium, Pediastrum,* and *Coelastrum*. This ballast water was released into Thunder Bay, Lake Superior within 84 hr of sampling.

Outbound lakers and foreign vessels frequently ''ballast up'' in the Great Lakes. There can be little doubt that such vessels are exporting zebra mussels to other freshwater regions in Canada (such as Quebec City), the United States (such as the heads of Chesapeake and Delaware Bay or the lower Mississippi River), or other freshwater systems in the world. In addition to the continued export of *Dreissena* from Europe, the establishment of zebra mussels in North America provides a second major source region for global dispersal. Given the different routes and trade patterns of shipping to-and-from the Great Lakes (as compared to those of western Europe), this new source may provide novel and unexpected opportunities for the inoculation of zebra mussels to other continents.

An older dispersal mechanism, not shown in Table 1, was the possible movement of zebra mussels by ''hard'' ballast such as rocks, stones, and sand (Carlton, 1985). Kerney and Morton (1970) have noted that such ballast provides an alternative mechanism for the arrival of the zebra mussel in Britain in the 1820s. Such movement may have been rare, since *Dreissena* is primarily a sublittoral animal, and presumably most ballast rock came from the shore.

Zebra mussels are typically found on the bottoms of vessels (Table 1, no. 6), colonizing hulls, hull openings (such as water intakes, exhaust outlets, and ballast intakes), and sea chests (Kew, 1893; Djakonoff, 1925; Alibekova et al., 1985 and 1986). On recreational vessels, zebra mussels settle inside outboard and inboard motor systems (e.g., outdrive units, trim tabs and plates, hydraulic cylinders, trolling plates, prop guards, and transducers); pumping systems (including waste and bilge); anchors and hausepipes; and on rudders, propellers, shafts, and centerboards. Transport trailers may also become fouled. The movement of vessels fouled with zebra mussels around the Great Lakes, or in canals leaving the Great Lakes, may be a primary means of dispersal. Bartsch (1954) suggested that the aquatic snail *Goniobasis* was transported (and populations thus mixed) by coal barges in the Chesapeake-Ohio Canal, while Koch (1989) suggest that the American dreissenid *Mytilopsis leucophaeata* was transported up the Mississippi River by barges. Counts (1986) and other have noted that small bysally attached *Corbicula* could similarly be transported by barges.

Overland movement of small, trailerable boats fouled with mussels has probably led to interlake dispersal, and may be a primary means of long distance overland dispersal (Griffiths et al., 1991). Walz (1989) commented that ''growth of tourism and transfer of tourist boats from lake to lake are responsible for the revival of spreading'' of *Dreissena* in Europe. Since its introduction to North America, zebra mussels have been extensively reported from boat bottoms throughout the Great Lakes, including from boats transported out of the Great Lakes to lakes in Ohio and Tennessee (Paxton and Kraft, 1991; Drake and Kraft, 1990). Easily overlooked in control programs to prevent private boat bottom dispersal of mussels would be the movement of work boats and barges by industrial contractors working on bridges, in reservoirs, etc. (Hudak, Personal communication). Zebra mussels may survive in cool, moist conditions out of water for 4 days (Griffiths et al., 1991) and perhaps 21 days or longer (Gray, 1825; Kew, 1893). This ability would provide strong potential for long-distance interlake transport. However, this potential may be modified by actual transport conditions (desiccation induced by airflow over the shell, for example), and thus stationary survival in a moist mat of shaded mussels may not fully describe the limitations to dispersal when out of water.

Standing water often occurs in ''wet wells,'' scuppers, and bilges of small recreational vessels (Table 1, no. 7); other internal compartments, such as catamaran pontoons, also take up water. As with any water taken up where *Dreissena* larvae or juveniles are present, the potential thus exists for the unintentional transport of the mussels. Sinclair (1964) suggested that a ''rapid means of transportation'' of *Corbicula* is ''bilge water being hauled from one reservoir to another'' in boats.

Overlapping the categories of vessel exteriors (Table 1, no. 7) and vessel interiors (Table 1, no. 8) are anchor systems. Anchor systems include the anchor itself, the anchor chain hause, and the anchor locker. Anchors and chains dropped into and dragged through zebra mussel patches may entrain mussels which would then be released when the anchor system is used in another water body. Zebra mussels carried into chain lockers could conceivably survive for many days, depending on the temperature, relative humidity, and length of time involved (McMahon, Personal communication). Anchors and chains have apparently been brought aboard small pleasure craft with zebra mussels attached (Marsden, Personal communication). Anchor systems are not the likely means of mussel transport from Europe to North America because most transoceanic vessels tie up at docks and do not anchor. The few that do anchor (while waiting for dock space, for example) usually do so in harbors which, in general, have soft bottoms; such bottoms support few mussels. Exceptions would occur, of course, when the anchor or chain is scraped over hard, emergent debris. However, chains on most vessels are washed thoroughly thoroughly either automatically in hausepipes or manually with fire hoses in order to avoid any accumulation of mud or debris in the anchor locker. Mussels that might pass through the washing would likely be crushed by the chain as it piles up. Finally, saltwater may enter anchor lockers while the vessel is at sea and exit through bottom drains into bilges. Freshwater organisms might thus be irregularly exposed to saltwater.

Fishing vessels may obtain water containing zebra mussels for use in live fish or bait wells (Table 1, no. 8). This is distinguished from mechanism no. 7 (i.e., vessel interior) because (1) much larger amounts of water are involved, (2) the water is intentionally taken up (and thus offers a greater potential measure of control), (3) the holds used are designed to sustain living organisms (as opposed to the wet walls or bilges of recreational vessels), and (4) recreational vessels and fishing vessels differ dramatically in their relative ability to be regularly transported overland.

Large navigation buoys (Table 1, no. 9), as opposed to smaller marker bouys and floats, may accumulate large numbers of fouling organisms; these include mollusks (WHOI, 1952; Merrill, 1963; Merrill and Edwards, 1976) and *Dreissena* (Zevina and Starostin, 1961; Kraft, 1991). Movement of such buoys between water basins may be an effective mechanism of zebra mussel dispersal. A wide variety of structures support the recreational boating industry (Table 1, no. 10), such as docks, dock guards, lines, and buoys. Along continuous shorelines, it is not uncommon to find that such equipment and gear may be transported, transferred, and towed between small harbors.

Fisheries

Fishing equipment, cages, stocking water, bait, and bait-bucket water (Table 1, no. 11–14) have potential for short- and long-distance transport of zebra mussels. Sport fishermen may transport fouled tackle, nets, traps (for minnows, crayfish, etc.), and other gear from one lake to another. Similarly, cages used to confine fish may become fouled with mussels and then may be transported between water bodies. Water is typically pumped from lake margins into holding tanks aboard trucks for transport of fish for stocking programs, urban fishing, and hatchery use. Water moved in fish stocking programs has similarly been invoked for the dispersal of *Corbicula* (McMahon, 1982).

The use of Dreissena as fish bait is centuries old. Indeed, the zebra mussel's discovery in England was by James Bryant, Esq., "who was in the habit of using the animal as a bait for perch-fishing" (Kew, 1893). Sport fishermen may thus play a role in moving zebra mussels within Europe and North America (Griffiths et al., 1991), although transport is more likely in bait bucket water (eggs and larvae) than directly as bait. As discussed above (mechanism no. 3), zebra mussels attached to bait crayfish provide an indirect form of this activity. *Corbicula* is believed to have achieved a broad portion of its dispersal in the United States as discarded bait (Fitch, 1953; Ingram et al., 1964; Britton and Morton, 1979).

Other Mechanisms

A wide variety of other mechanisms, private and governmental, may contribute to the dispersal of zebra mussels. These can be grouped into two categories: those mechanisms that could disperse all life stages and those that could disperse only juveniles and adults.

Water taken up and discharged between lakes and rivers by firetrucks (Table 1, no. 17), water released from aquariums (no. 19), water in the pontoons of amphibious planes (no. 20), and water released from research projects (no. 23) could lead to the dispersal of zebra mussel eggs and larvae. These four mechanisms and all of those remaining (no. 15, 16, 18, 21, and 22) could also transport juvenile and adult mussels. Thus, commercial goods (no. 15) that have come into contact with zebra mussel populations and then transported under submerged or moist conditions could play a role in mussel transport (Kew, 1893). Lumber and timber transported from the European mainland in the wet holds of ships in the early 1820s has been the widely held means of probable introduction of *Dreissena* to Britain, as first suggested by Sowerby (1825) and followed by most later authors (Gray, 1825; Bell, 1843, and others; see Kerney and Morton [1970] for review). Indeed, there appears to be direct evidence for this, as described by Gray (1825 — quoted in Kerney and Morton, 1970):

> I am now confirmed in the idea that this is the way in which they were introduced, as a friend has informed me that he has seen them sticking to the logs of Baltic timber before they were unloaded from the ship.

Commercial goods are still a viable mechanism because it is conceivable that similar products (including lumber and timber) could be so fouled and transported. Similarly, Sinclair and Isom (1961) and Ingram et al. (1964) have noted the potential for *Corbicula* to be moved hundreds of kilometers in sand and gravel to be used as cement aggregation material.

Small marker buoys and floats with attached mussels could be moved around and between bodies. Such buoys are used to mark channels (and thus could be included under ''Vessels/Navigation''); fish cages and traps (and thus could be included under ''Fisheries'', no. 11 and 12); private moorings (and thus could be under ''Marina/Boatyard Equipment'', no. 10); drilled wells; underwater pipes; etc. Such arbitrary mechanism designations show the difficulties of arriving at exclusive dispersal categories. Mechanisms no. 9 and 16 are distinguished for practical reasons. Large navigation buoys weighing several tons are likely to be moved by government agencies (such as coast guards) within a navigable waterway and between connected waterways (such as through lock systems) rather than between isolated water bodies.

It remains highly possible that individuals can and will intentionally distribute zebra mussels to new localities (Table 1, no. 18). In the 19th century, it was common practice throughout Europe and North America to intentionally move mollusks to new localities (Kew, 1893). Strickland (1838) noted that *Dreissena* was planted near Bristol, England; and Bell (1843) recorded that it was planted near Scarborough — all perhaps to disseminate this ''beautiful mollusk'' with the aim of improving the diversity of nature. The intentional movement of *Corbicula* for food by Asian immigrants in the 1930s is a favored hypothesis to explain the introduction of this clam to North America (Britton and Morton, 1979; Counts, 1986). Ingram et al. (1964) and Sinclair (1964) have further suggested that *Corbicula* could be dispersed by private individuals as a novelty, being transported by tourists as curios. *Corbicula* can also be shipped long distances as a commercial shellfish product (Burch, 1978). Reasons to intentionally disperse zebra mussels in North America include the desire to enhance local water quality (due to the removal of plankton by the mussel's filtering activities), to provide food for waterfowl populations, and to be used in aquaria.

It is predictable that zebra mussels held in private and school aquariums will be released into local rivers, lakes, and ponds where the species is not yet established (Table 1, no. 19). One such incident has been reported in Ontario (O'Neill, 1991). It is of interest to note that both Johnson (1921) and Sinclair and Ingram (1961) independently suggested that *Dreissena* and *Corbicula*, respectively, could be transported to North America in private aquariums. Johnson noted that there has been ''entirely too much reckless dumping of aquariums into our ponds and streams.'' Similarly, Sinclair and Ingram (1961) thought that the dispersal of *Corbicula* from Asia ''would first incriminate the dumping of aquariums, and fish bowls that contained introduced 'aquaria rarities'.'' Ingram et al. (1964) and Abbott (1975) have suggested that aquarium hobbyists would serve as a transport mechanism on a local basis.

Amphibious planes may lead to the dispersal of zebra mussels by means of pontoon water and fouling (Table 1, no. 20), the latter mode presumably being modified in the same ways as that of transport by birds, although at considerably greater velocities. Nevertheless, the attachment of mussels in cracks and crevices of pontoons might lead to mussel survival. Fouled recreational equipment (Table 1, no. 21) such as scuba gear, windsurfers, and water-skis; and debris (Table 1, no. 22) such as bottles, wood, and cartons may serve to transport mussels, although one would suspect these modes to be relatively infrequent.

Scientific research (Table 1, no. 23) could indirectly and directly lead to the accidental dispersal of zebra mussels. Scientific equipment, ranging from sampling gear to underwater cameras, are left for varying lengths of time submerged in waters where they would be susceptible to zebra mussel entrainment and colonization. Lack of attention to the presence of mussels and the transfer of such equipment between water bodies could serve as an effective transport mechanism. Direct introductions could occur by means of zebra mussels being transported to distant laboratories, and then being accidentally released.

DISCUSSION

Life History, Direction, and Overland Movement

The potential dispersal mechanisms reviewed here are related differentially to the various life history stages of the zebra mussel (Table 1). Eggs and larvae are moved by currents and, somewhat surprisingly, by at least 9 human-mediated mechanisms (no. 4, 5, 7, 8, 13, 14, 17, 19, and 23). Conversely, juvenile and adult mussels can potentially be moved by all mechanisms, with the exception of adults by mechanisms no. 5, 7, 8, and 13 (that is, four of those mechanisms that move only larvae and juveniles).

These mechanisms can further be grouped by their potential for moving with or against current direction (upstream vs downstream movement) and by their potential for bridging barriers (overland vs intercontinental transport) (Table 2). Rapid upstream dispersal can be achieved by all mechanisms except current flow, and thus water direction serves as no barrier to zebra mussels. Downstream movement is achieved by all mechanisms. Only water movement along natural and man-made corridors (no. 1, 4), fish and some other animals (no. 3), ballast water (no. 5), and fishing vessel wells (no. 8) would *not* mediate overland transport. Intercontinental transport is the most difficult to

TABLE 2. Direction of Potential Dispersal Mechanisms of *Dreissena polymorpha*

Dispersal Mechanism	Direction			
	Upstream	Downstream	Overland	Intercontinental
Natural				
1. Currents		X		
2. Birds	X	X	X	
3. Other animals	X	X	X	
Human-Mediated				
4. Canals	X	X		
5. Ballast water	X	X		X
6. Vessel exteriors	X	X	X	
7. Vessel interiors	X	X	X	X
8. Fishing vessel wells	X	X		
9. Navigation buoys	X	X	X	
10. Marina/boatyard equipment	X	X	X	
11. Fishing equipment	X	X	X	
12. Fish cages	X	X	X	
13. Fish stocking water	X	X	X	X
14. Bait, bait-bucket water	X	X	X	
15. Commercial products	X	X	X	X
16. Marker buoys and floats	X	X	X	
17. Firetruck water	X	X	X	
18. Intentional movements	X	X	X	X
19. Aquarium releases	X	X	X	
20. Amphibious planes	X	X	X	
21. Recreational equipment	X	X	X	
22. Litter (garbage)	X	X	X	
23. Scientific research	X	X	X	

achieve, and probably accounts for the lack of colonization by *Dreissena* in North America prior to 1986 and for its failure to colonize other continents.

Risk Assessment and Dispersal Mechanisms

Humans and human modifications to the landscape, such as canals, are the primary cause of widespread movement of zebra mussels in Europe and North America. Risk assessment and risk characterization methodology (National Research Council, 1983; National Academy of Sciences, 1986; Paustenbach, 1989) may provide a useful framework to quantitatively and experimentally assess the relative roles and importance of these many dispersal mechanisms in mediating the dispersal of *Dreissena* into new localities. This methodology includes identifying the hazardous agent (in this case, the mechanism of dispersal), quantitatively assessing its scale and rate of zebra mussel inoculation into a new locality, and then arriving at an overall characterization of the relative risk of each mechanism. Subsequently, regulatory options (risk management) could be developed for each mechanism where some measure of control could be achieved.

Few workers have attempted to quantify interlake and interregional human transport of aquatic organisms. Johnstone et al. (1985) examined interlake dispersal of freshwater macrophytes by recreational boat traffic in New Zealand. Scales and Bryan (1979) and Dove and Malcolm (1980) reported on a boat inspection and quarantine program involving the tracking of the dispersal of milfoil (*Myriophyllum*) by small recreational craft in Canada. To date, there have been no such studies with the zebra mussel. Extensive quantitative and experimental data sets will be required to assess and rank the relative roles of each of the known mechanisms of zebra mussel dispersal. By seeking actual field data for specific mechanisms coupled with experimental studies to test short- and long-term survival of mussels under different dispersal conditions, data would be available to assess the probability of mussel transport to certain localities, to formulate integrated control strategies, and to create predictive models on the rates and directions of spread.

It is assumed by most workers that *Dreissena* will move along downstream corridors as rapidly as environmental conditions permit. However, because of the many human mechanisms of dispersal, such movement will occur faster than would be predicted by a simple comparison of planktonic larval life with current speed. The inadvertent movement of *Dreissena* as a fouling organism on boats and other substrates will lead to rapid ''hopscotching'' over suitable stretches or bodies of water, with ''backfilling'' likely to occur.

Assessing the relative likelihood of these mechanisms to facilitate overland movements to isolated and semi-isolated bodies of water will be difficult, but perhaps lead to a better understanding of control methods. Similarly, just as fewer mechanisms serve to transport zebra mussels overland, even fewer are available to move mussels between continents. This fact should provide strong focus for those countries and agencies seeking to prevent the arrival of *Dreissena polymorpha* upon their shores.

ACKNOWLEDGMENTS

I thank numerous colleagues for suggestions and unbounded speculation. David MacNeill provided much advice and key references. Margaret Dochoda, Joseph Leach, Edward Mills, Gregory Ruiz, and Gary Sprules have provided their usual useful guidance. David Blimber, Peter Blaisdell, Gail Linza, Ellen Marsden, Robert McMahon, and Annette and John Motak contributed useful information at the Second International Zebra Mussel Research Conference in November 1991. David Policansky introduced me to ecological risk assessment. We are indebted to Captain Joseph Craig of the St. Lawrence Seaway Authority for facilitating our boarding of a bulk carrier and arranging for us to secure ballast water samples — the plankton of which was collected by Sprules and Carlton and identified by Muni Munawar and by Leach, Mills, and Sprules.

Much of our knowledge of the role of ballast water, and the information about ballast water provided here comes as a result of studies supported by NOAA Oregon Sea Grant R/EM-19. The late William Wick (former Oregon Sea Grant Director), his staff, and site review teams supported and encouraged our ballast water studies 2 years before zebra mussels appeared in North America, pre-arming the not-yet-born North American zebra mussel community with a strong foundation and understanding of the role of ballast water.

REFERENCES

Abbott, R. T. "Beware the Asiatic Freshwater Clam," *Trop. Fish Hobbyist* 23:15 (1975).

Alibekova, I. I., R. M. Bagirov, and G. M. Pyatakov. "Overgrowing of Ships Sailing the Caspian Sea, USSR," *Izv. Akad. Nauk Az. SSR, Ser. Biol. Nauk.* 4:47–50 (1985) (in Russian; as cited in Mackie et al., 1989).

Alibekova, I. I., R. M. Bagirov, and G. M. Pyatakov. "Overgrowth of Vessels in the Caspian Sea, USSR," *Izv. Akad. Nauk Az. SSR, Ser. Biol. Nauk* 2:64–67 (1986) (in Russian; as cited in Mackie et al., 1989).

Bartsch, P. "Hybridization Among Mollusks," *Ann. Rep. Am. Malacol. Union* 20:4 (1954).

Bell, R. J. "Note on the Rapid Increase of the Polymorphous Muscle [sic] (*Dreissena polymorpha*) in Great Britain," *Zoologist* 1:253–255 (1843).

Bio-Environmental Services, Ltd. "The Presence and Implication of Foreign Organisms in Ship Ballast Waters Discharged into the Great Lakes, Vol. 1 and 2," Prepared by Bio-Environmental Services Ltd. for the Water Pollution Control Directorate, Environmental Protection Service, Environment Canada (1981).

Boag, D. A. "Dispersal in Pond Snails: Potential Role of Waterfowl," *Can. J. Zool.* 64:904–909 (1986).

Briggs, J. C. *Marine Zoogeography* (New York: McGraw-Hill Book Company, 1974).

Britton, J. C., and B. Morton. "Corbicula in North America: the Evidence Reviewed and Evaluated," in Proceedings, First International Corbicula Symposium, Texas Christian University, Fort Worth, TX (1979) 249–287.

Burch, B. "Asian Clam, *Corbicula*, Threatens Hawaii," *Nautilus* 92:54 (1978).

Carlton, J. T. "Transoceanic and Interoceanic Dispersal of Coastal Marine Organisms: The Biology of Ballast Water," *Ocean. Mar. Biol. Annu. Rev.* 23:313–374 (1985).

Carlton, J. T. "Man's Role in Changing the Face of the Ocean: Biological Invasions and Implications for Conservation of Near-shore Environments," *Conserv. Biol.* 3:265–273 (1989).

Carlton, J. T. "Dispersal of Living Organisms into Aquatic Ecosystems: The Mechanisms of Dispersal as Mediated by Aquaculture and Fisheries Activities," in *Dispersal of Living Organisms into Aquatic Ecosystems,* A. Rosenfield and R. Mann, Eds. (College Park, MD: University of Maryland and Maryland Sea Grant College Program, 1992), pp. 13–45.

Carlton, J. T., J. K. Thompson, L. E. Schemel, and F. H. Nichols. "Remarkable Invasion of San Francisco Bay (California, USA) by the Asian Clam *Potamocorbula amurensis*. I. Introduction and Dispersal," *Mar. Ecol. Prog. Ser.* 66:81–94 (1990).

Clench, W. J. "*Corbicula manilensis* (Philippi) in Lower Florida," *Nautilus* 84:36 (1970).

Counts, C. L. "The Zoogeography and History of the Invasion of the United States by *Corbicula fluminea* (Bivalvia:Corbiculidae)," *Bull. Am. Malacol. Union, Spec. Ed.* 2:7–39 (1986).

Davis, G. M. "Historical and Ecological Factors in the Evolution, Adaptive Radiation, and Biogeography of Freshwater Mollusks," *Am. Zool.* 22:375–395 (1982).

Djakonoff, F. F. "Einige Beobachtungen uder den Bewuchs an den Dampfern der unteren Wolga," *Arb. Biol.* 8:135–156 (1925).

Dove, R., and B. Malcolm. The 1979 Aquatic Plant Quarantine Project. Studies on Aquatic Macrophytes 32. Water Investigations Branch, Ministry of Environment, British Columbia (1980).

Drake, J. A., and C. Kraft. "Mussels Hitching Rides to Tennessee?" *Zebra Mussel Update, Wis. Sea Grant Ad. Serv., Green Bay, Wis.* 5:3 (1990).

Ekman, S. *Zoogeography of the Sea* (London, England: Sidgwick & Jackson Ltd. 1953).

Eng, L. L. "Population Dynamics of the Asian Clam, *Corbicula fluminea* (Muller) in the Concrete-Lined Delta-Mendota Canal of Central California," in Proceedings First International Corbicula Symposium, Texas Christian University, Fort Worth, TX (1979) 39–68.

Fitch, J. E. "*Corbicula fluminea* in the Imperial Valley," *Minutes Conchol. Club South. Calif.* 130:9–10 (1953).

Frazier, J. G. "Epizoic Barnacles on Pleurodiran Turtles: Is the Relationship Rare?" *Proc. Biol. Soc. Wash.* 99:472–477 (1986).

Grant, H. D. *Cloud and Weather Atlas* (New York: Coward, McCann, Inc., 1944).

Gray, J. E. "A List and Description of Some Species of Shells Not Taken Notice of by Lamarck," *Ann. Phil. (n.s.)* 9:134–140 (1825) (as cited by Kerney and Morton, 1970).

Griffiths, R. W., D. W. Schloesser, J. H. Leach, and W. P. Kovalak. "Distribution and Dispersal of the Zebra Mussel (*Dreissena polymorpha*) in the Great Lakes region," *Can. J. Fish. Aquat. Sci.* 48:1381–1388 (1991).

Hanna, G D. "Introduced Mollusks of Western North America," *Occas. Pap. Calif. Acad. Sci.* 48:1–108 (1966).

Hebert, P. D. N., B. W. Muncaster, and G. L. Mackie. "Ecological and Genetic Studies on *Dreissena polymorpha* (Pallas): A New Mollusc in the Great Lakes," *Can. J. Fish. Aquat. Sci.* 46:1587–1591 (1989).

Hesse, R., W. C. Allee, and K. P. Schmidt. *Ecological Animal Geography* (New York: John Wiley & Sons, Inc., 1937).

Ingram, W. M., L. Keup, and C. Henderson. "Asiatic Clams at Parker, Arizona," *Nautilus* 77:121–124 (1964).

Isom, B. G. "Historical Review of Asiatic Clam (*Corbicula*) Invasion and Biofouling of Waters and Industries," *Bull. Am. Malacol. Union, Spec. Ed.* 2:1–5 (1986).

Johnson, C. W. "*Crepidula fornicata* in the British Isles," *Nautilus* 35:62–64 (1921).

Johnstone, I. M., B. T. Coffey, and C. Howard-Williams. ''The Role of Recreational Boat Traffic in Interlake Dispersal of Macrophytes: A New Zealand Case Study,'' *J. Environ. Manage.* 20:263–279 (1985).

Kerney, M. P., and B. S. Morton. ''The distribution of *Dreissena polymorpha* (Pallas) in Britain,'' *J. Conchol.* 27:97–100 (1970).

Kew, H. W. *The Dispersal of Shells. An Inquiry into the Means of Dispersal Possessed by Fresh-Water and Land Mollusca* (London, England: Kegan Paul, Trench, Trubner and Company, Ltd., 1893).

Koch, L. M. ''*Mytilopsis leucophaeta* (Conrad, 1831) from the Upper Mississippi River (Bivalvia:Dreissenidae),'' *Malacol. Data Set* 2:153–154 (1989).

Korschelt, E. ''On the Development of *Dreissena polymorpha*, Pallas,'' *Ann. Mag. Nat. Hist.* 9:157–168 (1892).

Kraft, C. ''New Sightings in Wisconsin; Oh Buoy, Missed Mussels Found.'' *Zebra Mussel Update, Wis. Sea Grant Advis. Serv., Green Bay, Wis.* 7:1–2 (1991).

Lewandowski, K. ''Occurrence and Filtration Capacity of Young Plant-Dwelling *Dreissena polymorpha* Pall. in Majcz Wielki Lake, Poland,'' *Pol. Arch. Hydrobiol.* 30:255–262 (1983).

Locke, A., and S. Corey. ''Amphipods, Isopods and Surface Currents: A Case for Passive Dispersal in the Bay of Fundy, Canada,'' *J. Plankton Res.* 11:419–430 (1989).

Ludlum, D. M. *The History of American Weather. Early American Tornadoes 1586–1870* (Boston MA: American Meteorological Society, 1970).

Mackie, G. L. ''Dispersal Mechanisms in Sphaeriidae (Mollusca: Bivalvia),'' *Bull. Am. Malacol. Union* 1979:17–21 (1979).

Mackie, G. L., W. N. Gibbons, B. W. Muncaster, and I. M. Gray. ''The Zebra Mussel, *Dreissena polymorpha*: A Synthesis of European Experiences and a Preview for North America,'' A report to the Ontario Ministry of the Environment, Water Resources Branch, Great Lakes Section, Queen's Printer, Toronto, Ontario (1989).

Maguire, B. ''The Passive Dispersal of Small Aquatic Organisms and Their Colonization of Isolated Bodies of Water,'' *Ecol. Monogr.* 33:161–185 (1963).

Malone, C. R. ''Dispersal of Aquatic Gastropods via the Intestinal Tract of Water Birds,'' *Nautilis* 78:135–139 (1965).

Marshall, W. B. ''Notes on the Colonization of Fresh-Water Shells,'' *Nautilus* 5:133–134 (1892).

McMahon, R. F. ''The Occurrence and Spread of the Introduced Asiatic Freshwater Clam, *Corbicula fluminea* (Muller), in North America: 1924–1982,'' *Nautilus* 96:134–145 (1982).

Merrill, A. S. ''Mollusks from a Buoy off Georgia,'' *Nautilus* 77:68–70 (1963).

Merrill, A. S., and R. L. Edwards. ''Observations on Mollusks from a Navigation Buoy with Special Emphasis on the Sea Scallop *Placopecten magellanicus*,'' *Nautilus* 90:54–61 (1976).

Mooney, M. J. ''Waterspout Watch,'' *Marin. Weather Log* 34:12–16 (1990).

Moore, W. G., and B. F. Fausti. ''Crayfish as Possible Agents of Dissemination of Fairy Shrimp into Temporary Ponds,'' *Ecology* 53:314–316 (1972).

Morton, B. S. ''Studies on the Biology of *Dreissena polymorpha* Pall. III. Population Dynamics,'' *Proc. Malacol. Soc. London* 38:471–482 (1969).

National Academy of Sciences. *Ecological Knowledge and Environmental Problem-Solving* (Washington, DC: National Academy Press, 1986).

National Research Council. *Risk Assessment in the Federal Government: Managing the Process*, Committee on the Institutional Means for Assessment of Risks to Public Health, Commission on Life Sciences (Washington, DC: National Academy Press, 1983).

Nouse, D. ''Variable Displacement' (Water Ballast) for Racing and Cruising Boats?,'' *Pract. Sailor* 14:4–7 (1988).

Oldham, C. ''Locomotive Habit of *Dreissena polymorpha*,'' *J. Conchol.* 19:25–26 (1930).

O'Neill, C. R. ''Aquarium Release Places Zebra Mussels in Reservoir,'' *Dreissena polymorpha Inf. Rev.* 2:5 (1991).

Parmalee, P. W. ''The Asiatic Clam (*Corbicula*) in Illinois,'' *Trans. Ill. Acad. Sci.* 58:39–45 (1965).

Paustenbach, D. J. Ed., *The Risk Assessment of Environmental and Human Health Hazards: A Textbook of Case Studies* (New York: Wiley Interscience, 1989).

Paxton, K., and C. Kraft. ''Mussels Sighted on Boats at Inland Lake.'' *Zebra Mussel Update, Wis. Sea Grant Advis. Serv., Green Bay, Wis.* 6:1–2 (1991).

Peck, S. B. ''Amphipod Dispersal in the Fur of Aquatic Mammals,'' *Can. Field-Nat.* 89:181–182 (1975).

Pennak, R. W. *Fresh-water Invertebrates of the United States*, 3rd ed. (New York: John Wiley & Sons, Inc., 1990).

Prezant, R. S., and K. Chalermwat. ''Flotation of the Bivalve *Corbicula fluminea* as a Means of Dispersal,'' *Science* 225:1491–1493 (1984).

Proctor, V. W., and C. R. Malone. ''Further Evidence of the Passive Dispersal of Small Aquatic Organisms via the Intestinal Tract of Birds,'' *Ecology* 46:728–729 (1965).

Rees, W. J. ''The Role of Amphibia in the Dispersal of Bivalve Molluscs,'' *Br. J. Herpetol.* 1:125–129 (1952).

Rees, W. J. ''The Aerial Dispersal of *Mollusca*,'' *Proc. Malacol. Soc. London* 36:269–282 (1965).

Rossmassler, E. A. ''On *Dreissena* Detaching from the Substrate,'' in ''Eine Eingewanderte Muschel. Der Zoologische Garten,'' E. V. Martens, 6 Jahrg., 50–59, 89–95, (1865). (See also English Abstr., *Ann. Mag. Nat. Hist.* (Ser.3) 18:493–494 [1866], and footnote in Korschelt [1892, p. 168].)

Russell Hunter, W., P. S. Maitland, and P. K. H. Yeoh. ''*Potamopyrgus jenkinsi* in the Loch Lomond area, and an authentic case passive dispersal,'' *Proc. Malacol. Soc. London* 36:27–32 (1964).

Scales, P., and A. Bryan. ''Transport of *Myriophyllum spicatum* Fragments by Boaters and Assessment of the 1978 Boat Quarantine Program,'' Studies on Aquatic Macrophytes 27, Water Investigations Branch, Ministry of Environment, British Columbia (1979).

Scharff, R. F. *The History of the European Fauna* (London, England: Walter Scott, Ltd., 1899) and (New York: Charles Scribner's Sons, 1899).

Scheltema, R. S. ''Dispersal of Marine Invertebrate Organisms: Paleobiogeographic and Biostratigraphic Implications,'' in *Concepts and Methods of Biostratigraphy,* E. G. Kaufman and J. E. Hazel, Eds. (Stroudsburg, PA: Dowden, Hutchinson, & Ross, 1977), pp. 73–108.

Scheltema, R. S. "Initial Evidence for the Transport of Teleplanic Larvae of Benthic Invertebrates Across the East Pacific Barrier," *Biol. Bull.* 174:145–152 (1988).

Sinclair, R. M. "Clam Pests in Tennessee Water Supplies," *J. Am. Water Works Assoc.* 56:592–599 (1964).

Sinclair, R. M., and W. M. Ingram. "A New Record for the Asiatic Clam in the United States, the Tennessee River," *Nautilus* 74:114–118 (1961).

Sinclair, R. M., and B. G. Isom. "A Preliminary Report on the Introduced Asiatic Clam *Corbicula* in Tennessee," Tennessee Stream Pollution Control Board, Tennessee Department of Public Health, Nashville, TN (1961).

Sinclair, R. M., and B. G. Isom. "Further Studies on the Introduced Asiatic Clam *Corbicula* in Tennessee," Tennessee Stream Pollution Control Board, Tennessee Department of Public Health, Nashville, TN (1963).

Sowerby, J. De C. "Extracts from the Minute-Book of the Linnean Society for November 2, 1824," *Trans. Linn. Soc. London* 14:585 (1825) (as cited by Kerney and Morton, 1970; see Kew, 1893, pp. 210–211 and footnotes).

Sprung, M. "Field and Laboratory Observations of *Dreissena polymorpha* Larvae: Abundance, Growth, Mortality and Food Demands," *Arch. Hydrobiol.* 115:537–561 (1989).

Stein, C. B. "An Extension of the Known Range of the Asiatic Clam *Corbicula fluminea* (Muller) in the Ohio and Mississippi Rivers," *Ohio J. Sci.* 62:326–327 (1962).

Strickland, H. E. [on *Dreissena* in England]. *Ann. Mag. Nat. Hist.* 2:362 (1838) [as cited by Bell, 1843].

Taylor, R. W., and C. L. Counts. "The Asiatic Clam, *Corbicula manilensis*, as a Food of the Northern Raccoon, *Procyon lotor*," *Nautilus* 91:34 (1977).

Thompson, C. M., and R. E. Sparks. "Improbability of Dispersal of Adult Asiatic clams, *Corbicula manilensis*, via the Intestinal Tract of Migratory Waterfowl," *Am. Mid. Nat.* 98:219–223 (1977).

van den Hoek, C. "The Possible Significance of Long-Range Dispersal for the Biogeography of Seaweeds," *Helgol. Meeresunters.* 41:261–272 (1987).

van der Schalie, H. "The Mollusks of the Duck River Drainage in Central Tennessee," *Sterkiana* 52:45–55 (1973).

Vermeij, G. J. *Biogeography and Adaptation* (Cambridge, MA: Harvard University Press, 1978).

Voskresenskii, K. A. "Dispersal of Bivalvia by Fishes," *Zool. Zh.* 45:1097–1098 (1966) (in Russian).

Walz, N. "Spreading of *Dreissena polymorpha* (Pallas) to North America," *Heldia* 1:196 (1989).

Williams, R. J., F. B. Griffiths, E. J. Van der Wal, and J. Kelly. "Cargo Vessel Ballast Water as a Vector for the Transport of Non-Indigenous Marine Species," *Estuarine Coastal Shelf Sci.* 26:409–420 (1988).

Woods Hole Oceanographic Institution (WHOI). *Marine Fouling and Its Prevention* (Annapolis, MD: United States Naval Institute, 1952).

Zevina, G. B., and I. V. Starostin. "Qualitative and Quantitative Changes in the Fouling of the Caspian Sea after the Opening of the Volga-Don Canal," in Marine Fouling and Borers I. V. Starostin, Ed. *Tr. Instituta Okean.* 49:97–108 (1961) (translated 1968, Israel Program for Scientific Translations, Jerusalem).

Zhadin, V. I. ''Mollusks of Fresh and Brackish Waters of the USSR,'' *Keys to the Fauna of the USSR*, Zoological Inst. Sci., Leningrad, 46 (1952).
Zibrowius, H. ''Extension de l'aire de repartition favorisee par l'homme chez les invertebres marins,'' *Oceanis* 9:337–353 (1983).

Appendix

The 1980 Great Lakes Ballast Water Sampling Study

Workers following the North American *Dreissena* literature will note occasional reference to a 1980 study (by a private environmental consulting firm under a Canadian government contract) that documented the occurrence of living planktonic organisms in ballast water arriving and being released in the Great Lakes. This work is cited here as Bio-Environmental Services (1981). This is the same report cited as Howarth (1981) by Carlton (1985) and as EPS (1981) by Hebert et al. (1989); the suggested correct citation is the one used here.

Because the Bio-Environmental Services report is unpublished and largely unavailable outside of North America (although many copies have been widely distributed since the arrival of the zebra mussel), the methods used in this study and the results relative to the collections of living molluscan larvae are worth recording. Fifty-five ships were sampled by one of two methods: ''For one-half of the ships sampled, the aft manhole cover of a double bottom tank was removed to collect a grab sample of ballast water and side-wall encrustations. On the remaining vessels, ballast water was obtained from the discharge side of the ship's water pump which pumped water from the bottom of the tank.'' Which ships were sampled by which method is not indicated. ''Up to 100 litres of sample water'' per ship were processed; however, the data analysis sheets later record this as always being 100 L. A ''zooplankton'' fraction was removed on a 200-μm filter, thus inadvertently eliminating all but the largest bivalve larvae. A ''phytoplankton'' fraction was retained on an 80-μm filter, from which two aliquots of 0.85 mL each were removed and examined under a compound microscope (''any zooplankton material found in the samples was also analyzed'').

The single sample taken from each ship and the way in which the samples were subsequently analyzed rendered the probability of finding bivalve mollusk larvae (and indeed many other taxa) lower than might have been generated through other sampling and analytical designs. Of the 55 vessels sampled 43 were from European waters; of these, 2 (1 with ballast from the Mediterranean and 1 with ballast from Dunkirk and the North Sea) were reported to have unidentified bivalve larvae (Hebert et al., 1989 — report that ''10% of the ships surveyed'' contained veliger larvae). Four vessels had molluscan larvae from Europe, and one contained larvae from Baltimore, MD). While Hebert et al. (1989) report that ''veliger larvae were numerous'' in these vessels (and indeed they may have been), it is difficult to make this conclusion based upon one sample per ship. Only 1 of the 55 ships contained fresh water as ballast. This vessel (from Leningrad) contained phytoplankton, the rotifer *Keratella quadrata*, and copepods (it is regrettable that this vessel was not more thoroughly sampled). The samples collected in this 1980 survey no longer exist (Sprules, Personal communication).

CHAPTER 41

Recent Introductions of *Dreissena* and Other Forms to North America — the Caspian Sea/Black Sea Connection

Michael L. Ludyanskiy

Considerable interest in the origins of North American populations of *Dreissena polymorpha* and other recently introduced forms has prompted this examination of evidence which suggests that the source of these populations was likely the Caspian Sea/Black Sea region. Given this evidence, it should be emphasized that other species and varieties of *Dreissena* in this region can potentially be introduced into North America with resulting impacts similar to those for *Dreissena polymorpha*.

In a series of recent papers, Biochino (1989; 1990) investigated the morphological features of *Dreissena polymorpha* from various water bodies in the former European USSR (Figure 1) and classified five distinct populations: Aral-Caspian, Ponto-Caspian, Middle-Russian, North-Eastern, and Baltic.

0-87371-696-5/93/$0.00 + $.50

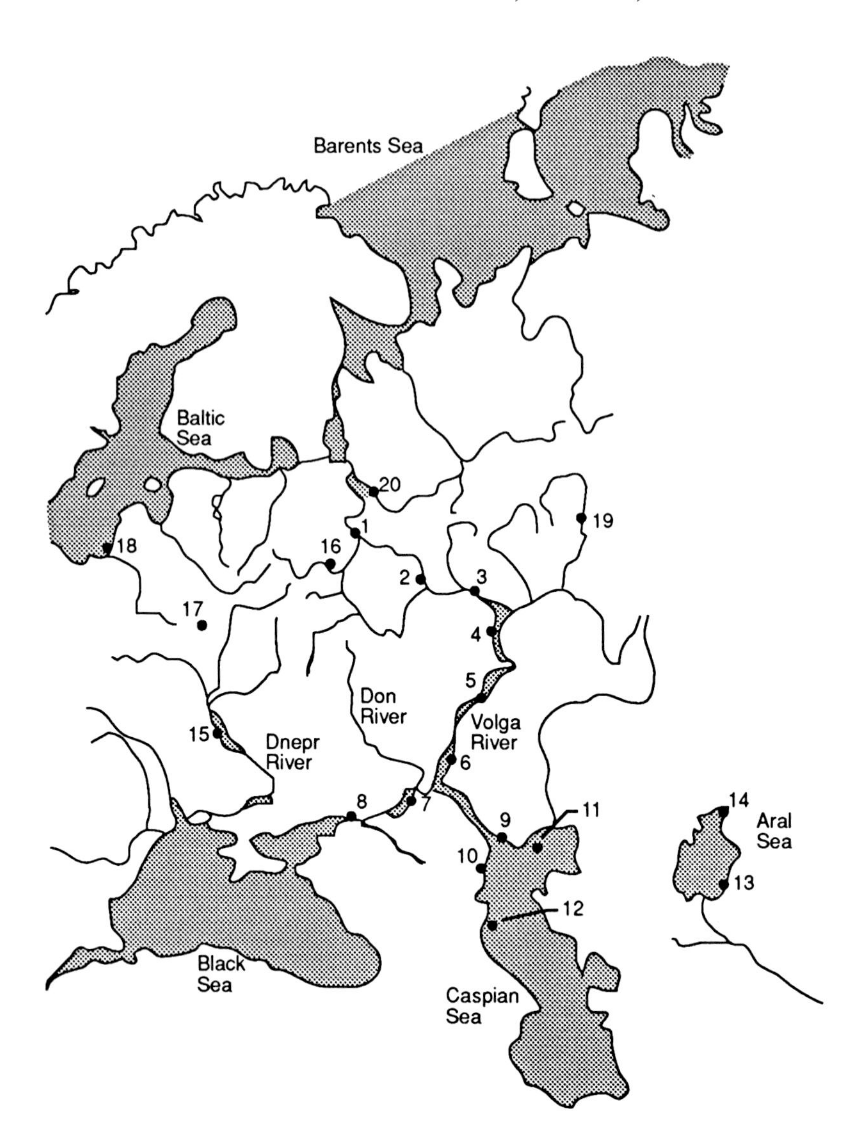

Figure 1. Location of various water bodies in European USSR where specimens of *Dreissena polymorpha* were collected for analysis of relative proportions of phenotypes. 1, Rybinsk reservoir; 2, Gor'kii reservoir; 3, Cheboksary reservoir; 4, Kuibyshev reservoir; 5, Saratov reservoir; 6, Volgograd reservoir; 7, Tsimlyansk reservoir; 8, Mouth of the Don River; 9, Delta of the Volga River; 10, Volgo-Caspian canal; 11, Northeastern Caspian; 12, Western Caspian; 13, Lake Kara-Teren (Aral Sea); 14, Lake Kamyshlybash (Aral Sea); 15, Kremenchug reservoir; 16, Ivan'kov reservoir; 17, Lake Lukomskoye; 18, Kurshskiy Bay; 19, Nizhnekamsk reservoir; 20, Lake Siverskoye. (Modified from Biochino, G. T. *Inst. Biol. Vnut. Vod. Tr., Ryb.* 59:143–158 [1990].

TABLE 1. Relative Proportions of Different Phenotypes of *Dreissena polymorpha* in Populations from Various Regions in the Former European USSR

	Phenotype				
Region and Water Body	**DD**	**OO**	**AA**	**RR**	**CC**
Baltic					
Rybinsk reservoir (1)	—	0.09	—	—	0.91
Ivan'kov reservoir (16)	—	—	0.03	—	0.97
Lake Lukomskoye (17)	—	—	0.01	—	0.99
Kurshskiy Bay (18)	—	—	0.04	—	0.96
North-Eastern					
Nizhnekamsk reservoir (19)	—	0.43	0.02	—	0.55
Lake Siverskoye (20)	—	0.26	0.13	—	0.61
Middle Russian					
Cheboksary reservoir (3)	—	0.15	0.11	—	0.74
Kuibyshev reservoir (4)	—	0.27	0.12	—	0.61
Saratov reservoir (5)	—	0.09	0.06	—	0.85
Volgograd reservoir (6)	—	0.02	0.21	—	0.77
Ponto-Caspian					
Volgo-Caspian canal (10)	0.04	0.05	0.11	—	0.80
Mouth of Don River (8)	0.01	0.05	0.18	—	0.76
Delta of Volga River (9)	—	—	0.24	—	0.76
Tsimlyansk reservoir (7)	—	—	0.12	—	0.88
Kremenchug reservoir	—	—	0.12	—	0.88
Aral-Caspian					
Northeastern Caspian (11)	0.09	0.06	0.38	0.19	0.28
Western Caspian (12)	0.06	0.01	0.63	0.10	0.20
Lake Kara-Teren (13)	0.04	0.07	0.36	0.21	0.32
Lake Kamyshlybash (14)	0.03	0.08	0.55	0.17	0.17
Intermediate					
Gor'kii reservoir (2)	—	—	0.31	—	0.69

Source: Modified from Biochino, 1990.

Note: Number in parentheses indicates location in Figure 1.

These distinctions were based on the relative proportions of five different phenotypes designated as DD, OO, AA, RR, and CC (Table 1). The shell morphology of these phenotypes is presented in Chapter 13, Figure 5 of this volume. As shown in Table 1, the RR phenotype was only found in the Aral-Caspian Sea region. Several scientists familiar with *Dreissena* populations in western Europe (A. bij de Vaate and N. Walz, Personal communication) have never seen this phenotype and are unaware of any references to it in western European literature. Despite this, all five phenotypes including the RR variety have been found in the Great Lakes (Marsden, Personal communication). My own investigations have shown that the RR phenotype can be found in the Erie Canal, New York. (Ludyanskiy, Unpublished data). In a more recent study, Smirnova et al. (Chapter 13, this volume) have shown that the RR phenotype also occurs in small proportions in other regions of the former European USSR besides the Aral-Caspian region, including sites in the Ponto-Caspian Region (area of Black Sea). In their study, similarities between specimens from populations in Lake Erie and specimens from populations in the Ponto-Caspian region were emphasized, with both populations having the DD phenotype. However, the authors concluded that, since the proportion of phenotypes from the various regions of the former European USSR were more similar to each other than to those from populations in Lake Erie, these regions could not have been the source of *Dreissena* in North America. This conclusion, of course, does not contradict the possibility that the Aral-Caspian region was the origin of North American populations. Unfortunately, the Smirnova et al. study (this volume) did not include specimens from the Aral-Caspian Sea Region. Comparisons between relative phenotype proportions in populations from Lake Erie to previously published data on phenotypes from the Aral-Caspian region (Table 1) are difficult. The reason is that the Smirnova et al. (this volume) included a new phenotype, MM-spotted, for which there are no data available for populations in the Aral-Caspian region. This MM phenotype may have been separated from the AA phenotype of earlier works (Biochino, 1989; 1990). At any rate, it is evident that the proportion of phenotypes in Lake Erie populations (Smirnova et al., this volume) are closer to those in populations from the Aral-Caspian region (Table 1) than to populations from any other region in the former European USSR, especially the DD, RR, and CC phenotypes.

Recently, two benthic species of Gobiidae have been reported from North America. The round goby, *Neogobius melanostromus,* and the tube nose goby, *Proterorhinus marmoratus,* have been collected in the St. Clair River and have established reproducing populations (Jude et al., 1992). Both of these species are native to the Caspian Sea/Black Sea region. Interestingly, laboratory experiments have shown that the round goby feeds heavily on *Dreissena* (Ghedotti, 1992). The goby feeds on mussels by extracting the soft tissue from the shell rather than ingesting the whole animal and then crushing the shell; this more efficient method of feeding is typical of an adaptive predator/prey relationship of co-occurring species. It is believed that the only likely explanation for the introduction of these goby species is that they were transported in the ballast water of ships originating from this region.

A second form of *Dreissena,* given the working name of "quagga", has been recently found in Lake Ontrario (Marsden, Personal communication). To date, this form has not been identified, in part because the taxonomy of the Dreissenidae is not well resolved. Being familiar with the published literature on the various species and subspecies of Dreissenidae from the Caspian Sea/Black Sea region (Zhadin, 1952; Logvinenko and Starobogatov, 1968; Starobogatov, 1970), my own examinations of the quagga mussel indicate that it most closely resembles either *D. polymorpha andrusovi,* or *D. rostriformis bugensis.* The former form, until recently, was considered a species — *D. andrusovi* (Brusina) Andr. (Zhadin, 1952; Karpevitch, 1955) — rather than a subspecies. This form inhabits the northern part of the Caspian Sea. In this area of the Caspian there is apparently a gradual gradient in relative abundances of the more brackish *D.p. andrusovi* and the freshwater *D.p. polymorpha* (Spasskiy, 1948). In addition, Logvinenko (1965) showed that communities in the northeastern Caspian, in the area of the Volga River delta, are comprised of many transitional forms ranging from freshwater to brackish, and the two subspecies *D.p. polymorpha* and *D.p. andrusovi* are found at opposite ends of the salinity gradient. The other species, *D. rostriformis bugensis,* is found in the South Bug-Dnieper River basin of the Black Sea and is mostly a freshwater form (Ludyanskiy, 1991). Since genetic evidence from the quagga mussel places it as a separate species in relation to *D.p. polymorpha* (Marsden and May, in review), certainly more quantitative genetic, morphometric, and anatomic data will have to be obtained from *D.p. andrusovi* and *D. rostriformis bugensis* in order to establish their relationship to the quagga mussel.

Considering the evidence that several recently introduced species had origins in the Caspian Sea/Black Sea region, potentially other *Dreissena* varieties may also have been transported to North America. At least two more varieties of *D. polymorpha* and some varieties of *D. rostriformis* are native to this region. The latter species is able to tolerate more brackish conditions than *D. polymorpha* and, if introduced into North America (if not already present), may colonize areas not suited for *D. polymorpha.* Future research should compare genetic and morphological characteristics of *Dreissena* species in North America and in the Caspian Sea/Black Sea region in order to better predict the future range of *Dreissena* infestations in North America.

REFERENCES

Biochino, G. T. "A phenetic method of *Dreissena polymorpha* (Pallas) Investigation," *Nauchn. dokl. Vyssh. Shk.* 10:36–41 (1989).

Biochino, G. T. "Polymorphism and Geographical Variability of *Dreissena polymorpha* (Pallas)," in Mikroevolutsiya presnovodnykh organisnov (microevolution of freshwater organisms) *Inst. Biol. Vnut. Vod. Tr.* (papers), *Ryb.* 59:143–158 (1990).

Ghedotti, M. "Feeding preference of an introduced mollusk predator Negobius melanostromus in the Great Lakes," *Biocurrents* 8:2 (abstract) (1992).

Jude, D. J., R. H. Reider, and G. R. Smith. "Establishment of Gobiidae in the Great Lakes Basin." *Can. J. Fish Aquat. Sci.* 49:416–421 (1992).

Karpevitch, A. F. "Some data on the formation of species in Bivalvea," *Zoologicheskii Zhurnal* 34:40–67 (1955).

Logvinenko, B. M. "The Changes in the Fauna of Caspian Molluscs *Dreissena* After Intrusion of *Mytilaster lineatus* (Gmelin)," *Nauchn. Dokl. Vyssh. Shk., Biol. Nauk.* 2:13–16 (1965).

Logvinenko, B. M., and Ya. I. Starobogatov. "Type Molluski. Mollusca," in *Atlas bespozvonochnykh Kaspiyskogo Morya (Atlas of Invertebrates of Caspian Sea),* Ya. A. Birshtein et al., Eds. (Moscow: Izd-vo Pishchevaya promyshlennost, 1968), pp. 368–385.

Ludyanskiy, M. L. "Are *Dreissena polymorpha* and *Dreissena bugensis* synonymous?" *Dreissena polymorpha luf. Rev.*, NY Sea Grant; 2:2–3 (1991).

Marsden, J. E., and B. May. Genetic identification and implications of another invasive species of Dreissenid mussel in the Great Lakes, Can. J. Fish Aquat. Sci., in press (1992)..

Spasskiy, N. N. "The Variability of *Dreissena polymorpha* in the Northern Caspian Sea and Feeding Significance of Its Varieties for Roach," in *Tr. Volgo-Caspiyskoi Nauchn. Rybokhoz. St. (Proc. Volgo-Caspian Res. Fish. Stn.)* 10:117–128 (1948).

Starobogatov, Ya. I. *Mollusc Fauna and Zoogeographical Partitioning of Continental Water Reservoirs of the World.* Izd. Nauka, Leningrad, p. 372 (1970).

Zhadin, V. I. *Mollusks of Fresh and Brackish Waters of the USSR: keys to the Fauna of the USSR. Izdatelstvo Akademii Nauk USSR,* Moskva-Leningrad, 46, 376 (1952).

CHAPTER 42

Early Detection of the Zebra Mussel (*Dreissena polymorpha*)

Clifford Kraft

During 1990 and 1991, a coordinated plankton and substrate sampling program was conducted to determine when zebra mussels would invade Wisconsin waters of the Great Lakes. In conjunction with this sampling effort, an intensive effort was made to inform municipal and industrial water users and the general public about zebra mussels. At most sampling sites, the presence of zebra mussels was first reported from unplanned observations that were not part of the coordinated sampling program. In most instances, such reports were soon followed by abundance estimates from the sampling program, providing information on rates of colonization. These results suggest that: (1) educated observers are likely to provide the first reports of mussels in a given area, (2) intensive and costly sampling for veligers provide little (2 weeks) preinvasion warning of settled mussels, and (3) a standard sampling program can provide useful information once a local reproducing population of zebra mussels has become established.

INTRODUCTION

In summer 1990, public and private water users along the Great Lakes shorelines of Wisconsin were faced with the realization that zebra mussels would soon colonize Wisconsin waters of Lakes Michigan and Superior. Key questions being asked were (1) when would this invasion materialize, and (2) could a coordinated sampling effort provide more useful information about the invasion of zebra mussels than unplanned observations? These questions are similar to those that will be confronted by water users throughout North America in areas where the zebra mussel is likely to spread (Strayer, 1991; Ramcharan et al., 1992). To better approach these questions, a variety of monitoring protocols for zebra mussels have been developed (Marsden, 1992), but information on the relative effectiveness of different techniques is scarce.

Documented observations from established monitoring programs for early detection of zebra mussels are necessary to provide a base for management decisions in both the public and private sectors. The great public interest in zebra mussels has further contributed to the need to determine the value of a planned monitoring program, since public perceptions of the distribution of zebra mussels have at time exceeded their prodigious dispersal capacity (Kraft et al., 1991).

Results presented here address the effectiveness of zebra mussel monitoring programs. Comparisons are made between reports of mussel distributions obtained from a planned monitoring program and reports obtained from unplanned observations. Such comparisons are instructive for water users and managers deciding how to detect zebra mussels at the leading edge of their invasion.

METHODS

Zebra Mussel Watch

In spring 1990, a sampling program — termed "Zebra Mussel Watch" — was initiated in the Wisconsin waters of the Great Lakes to provide information on the distribution and abundance of zebra mussels. The program was designed to determine the presence of zebra mussels as planktonic veligers in water samples and as postveligers settling on artificial substrates. This monitoring program was implemented concurrently with an outreach effort designed to provide training and education regarding zebra mussels.

During summer 1990, weekly water column and substrate samples were examined at 29 stations in 10 harbors in the Wisconsin waters of the Great Lakes (Figure 1). Water samples from intakes at 11 power plants, 16 municipal

Figure 1. Harbors in Wisconsin waters of Lakes Michigan and Superior where water column and substrate samples were collected in 1990 and 1991.

TABLE 1. Number of Samples Collected as Part of the Zebra Mussel (*Dreissena polymorpha*) Watch Program in Wisconsin Waters of Lakes Michigan and Superior in 1990 and 1991

	1990		1991	
	Lake Michigan	Lake Superior	Lake Michigan	Lake Superior
Plankton samples				
Harbor	295	116	140	72
Intake	305	38	298	47
Substrate samples	230	167	180	109

water utilities, and 5 industrial facilities were included in the monitoring program. During summer 1991, water and substrate samples were examined from 26 stations in 11 harbors; and water samples were examined from intakes at 11 power plants, 20 municipal water utilities, and 5 industrial facilities. A summary of samples examined in 1990 and 1991 is shown in Table 1.

General education efforts were prepared for the public at large. These included presentations (86) to people (7,000) representing a variety of groups, media contacts (81), publication and distribution of wallet-size zebra mussel identification cards (34,000), and publication of a ''Zebra Mussel Update'' newsletter (11 issues, Wisconsin Sea Grant Institute, Madison, Wisconsin). This newsletter became the official record for reporting confirmed zebra mussel sightings in Wisconsin. The accuracy of reported sightings was maintained by implementing two programs: (1) personnel at five university laboratories throughout the state confirmed reported observations and examined water intake and harbor samples, and (2) training workshops in zebra mussel identification were provided in 1991 to utility and agency personnel.

Nearshore Water Column Samples

Sampling efforts were concentrated in 10 harbors and nearshore areas likely to be exposed to zebra mussel vectors (Griffiths et al., 1991). Zooplankton samples were collected by vertical tows with a conical plankton net (0.5-m diameter × 2.5-m length, 64-μm mesh). Plankton tow depths ranged from 2 to 7 m. During summer 1990, plankton samples were preserved in either 95% ethanol or 10% sugar buffered formalin, and examined within 1 week of sampling. During 1991, most of the plankton samples were examined live within 1 day of collection, although a small proportion of samples were preserved in a 10% sugar buffered formalin solution.

An initial scan was performed to detect the presence of veligers by examining three 1-mL subsamples (at magnification × 40) of settled sample material. When veligers were detected by this initial screening, five 1-mL subsamples were enumerated after mixing the contents of the sample.

Newly settled juveniles were collected on artificial substrates placed at readily accessible locations near water column sample sites, using a sampling device constructed from PVC pipe (Figure 2). Two samples were placed at each location and removed for inspection at 4-week intervals; another sampler was left in place for the entire season. The entire inside surface of each sampler was enumerated under a dissecting microscope (magnification × 10–30). A smaller area was enumerated when large numbers of mussels (>3000) were found on a sampler.

Water Intake Samples

Analytical services were offered to public electric and water utilities and industries that draw water from Lakes Michigan and Superior. Procedures for transporting field samples for laboratory analysis were provided, along with assistance in obtaining sampling equipment. Plant personnel collected water samples according to a given protocol and then sent the samples to academic institutions for veliger analysis. Sample enumeration procedures were the same as for harbor plankton samples.

Most water intake samples were collected weekly or biweekly with a conical zooplankton net (64-μm mesh) suspended inside a 225-L steel drum with outlets at the top and bottom. The device was used in either of two ways: (1) water at a known and constant flow rate was poured through the net or (2) the drum was repeatedly filled and emptied. Several water intakes

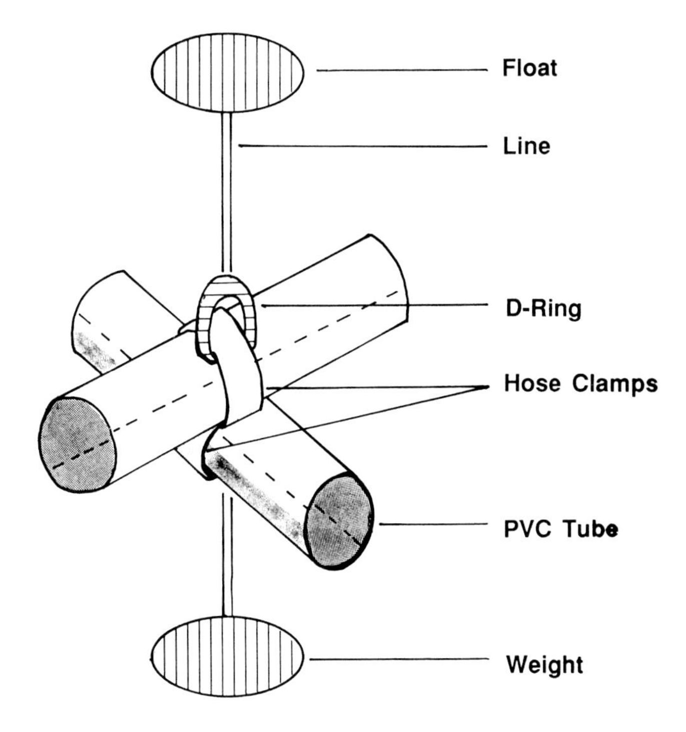

Figure 2. Artificial substrate sampler used to collect settled postveliger zebra mussels (*Dreissena polymorpha*) in Wisconsin waters of Lakes Michigan and Superior.

had wet wells deep enough and with low enough current velocity to permit taking a vertical plankton tow. In most cases, water volume sampled was about 1 m^3.

RESULTS

Regular reports of zebra mussels in Wisconsin waters of Lake Michigan began in 1989 with mussels being found on commercial ships undergoing maintenance at shipyards in Sturgeon Bay, Wisconsin. During summer 1990, small populations of settled zebra mussels (<200 individuals at a given location) were found at numerous sites in southern Lake Michigan waters of Illinois, Indiana, and Michigan. The discovery of a single adult mussel on flotsam in Kenosha harbor, and a late season report of two adult mussels on a Sheboygan water intake provided the first indication that zebra mussels had entered Wisconsin waters of Lake Michigan. However, no zebra mussels were detected in 600 Lake Michigan plankton samples (harbor and water intake) or 230 substrate samples collected in 1990.

TABLE 2. Date of First Sighting of *Dreissena polymorpha* in Plankton and Substrate Samples in Wisconsin Waters of Lake Michigan in 1991

Location	Plankton[a]	Substrate
Kenosha	July 31 (harbor and intake)	August 14
Racine	July 31 (harbor and intake)	August 14
Milwaukee	August 2 (harbor)	August 5
Port Washington	August 21 (harbor and intake)	September 4
Sheboygan	September 13 (harbor and intake)	September 27
Manitowoc	August 30 (harbor)	September 13
Green Bay	August 28 (harbor)	August 14

[a] Locations from which plankton samples containing veligers were collected are shown in parentheses.

Zebra mussels were first reported from Lake Superior waters in April 1990 when mussels were found attached to several navigation buoys that were retrieved the previous fall from Superior harbor. In August 1990, veligers were detected in water samples from Superior harbor, followed 1 week later by the observation of recently settled zebra mussels on substrate samplers. Zebra mussel veligers were detected in 6 of 154 Lake Superior plankton samples and 8 of 167 substrate samples collected in 1990.

In 1991, occurrences of zebra mussels in Lake Michigan samples contrasted dramatically with those in summer 1990. Veligers were detected in 73 of 438 plankton samples from harbors and water intakes, and settled mussels were observed on 57 of 180 substrates. The southern half of Lake Michigan had the greatest concentration of mussels, with veligers detected in water samples collected at every sampling location south of Sheboygan. In contrast, occurrences of zebra mussels in Lake Superior in 1991 were very similar to those in 1990. Veligers were detected in 5 of 119 plankton samples collected from Superior harbor and several water intakes, and settled mussels were found on 12 of 109 substrates.

The results of 1990–1991 efforts provided useful information regarding early detection of zebra mussels. At most locations except Green Bay, zebra mussels were detected in plankton samples prior to detection on substrate samplers (Table 2). A 2-week time lag between the detection of veligers and the detection of mussels on substrate samplers occurred at five of six locations where veligers were detected first. Settled postveligers were found at all locations where veligers were found. Plankton samples from water intakes and harbor tows were equally effective for detecting zebra mussels. Of the three harbors where veligers were not detected in harbor and intake samples

TABLE 3. Dates of the First Report of Zebra Mussels, *Dreissena polymorpha*, in Wisconsin Waters of Lakes Michigan and Superior

Location	First Sighted[a]	First Sampled[b]
Lake Michigan		
Kenosha	June 19, 1990 (angler)	July 31, 1991
Racine	April 14, 1991 (intake personnel)	July 31, 1991
Milwaukee	July 3, 1991 (boater)	August 2, 1991
Port Washington	August 21, 1991 (intake sample)	August 21, 1991
Sheboygan	October 16, 1990 (intake personnel)	September 13, 1991
Manitowoc	February 21, 1991 (intake personnel)	August 30, 1991
Green Bay	June 9, 1991 (bird watcher)	August 14, 1991
Lake Superior	April 12, 1990 (resource agency personnel)	August 21, 1990

[a] Includes all reported observations, with observer noted in parentheses.
[b] Includes observations of mussels in plankton samples and on substrate samplers.

on the same date, two had no harbor water intakes (Milwaukee and Green Bay) and the third (Manitowoc) had very low densities of zebra mussels.

At all locations except Port Washington, the presence of zebra mussels had been reported prior to their detection in the monitoring program (Table 3). Such reports resulted from supplemental observations by personnel at power plants and water utilities; and by anglers, boaters, and bird watchers. In some instances, these reports were made by personnel or individuals who attended the training workshops. In other cases, sightings were referred to a coordinating office through contacts resulting from the general education efforts, such as media exposure or distributed material. This provides strong evidence for the value of workshops for municipal and industrial personnel, and the general public education campaign.

DISCUSSION

Isolated observations provided the first evidence that zebra mussels were present at most locations, but these observations did not provide information on the extent of infestation and timing of reproduction. The presence of mussel reproduction with potential for nuisance fouling was detected by finding zebra mussels in plankton and substrate samples, as can be seen by the rapid increase in mussel abundance following their first detection on artificial substrates (Table 4). During periods when zebra mussel densities increased substantially

TABLE 4. Densities of Zebra Mussels (*Dreissena polymorpha*) on Substrates Placed in the Harbor at Kenosha, Wisconsin, on Various Dates in 1991

Date	Density (no./m^2)
July 31	0
August 14	520
August 28	1,400
September 11	28,000
September 25	14,000

(e.g., Kenosha harbor, July 31 to September 25, 1991), unplanned sightings did not show a commensurate increase, probably because newly settled mussels are smaller than the limit of detection for most casual observers (unsolicited reports of zebra mussel sightings seldom included individuals less than 10 mm in length).

At locations where zebra mussels had not been previously found, a training program and general educational campaign provided the best system for early detection. Targeted users included municipal and industrial personnel from facilities with water intakes. The general education campaign was directed at individuals active in water-related activities, including divers, boaters, commercial fishermen, anglers, and natural resource agency personnel. Following the first confirmed reports of zebra mussels in previously uninfested waters, a coordinated plankton and/or substrate sampling effort provided important information on the timing and extent of reproduction.

Although comparisons between planned monitoring programs and unplanned observations in other states have been less systematic, similar conclusions have been noted by others (for Illinois — Blodgett, Personal communication; for New York — Lange, Personal communication). One notable exception occurred when veligers were first detected in water samples from a power station on the Susquehanna River near Johnson City, New York, prior to any observation of zebra mussels in that watershed. This sequence of detection — veligers found in a water sample prior to any detection of adults — was unusual and might have been due to a lack of water users familiar with zebra mussels in the Susquehanna River system at the time of detection.

A comparison of when zebra mussels were detected in plankton samples and when detected on artificial substrates suggests that the timing and extent of colonization can be reliably determined with either method. Evidence of reproduction determined from substrates usually lagged 2 weeks behind plankton samples. For many water users, implementation of a substrate sampling program would require less expense and effort than mounting an effective plankton sampling program.

Intake samples can be effectively used to determine the presence of veligers, since harbor and intake plankton samples were equally effective for detecting the presence of veligers. This corroborates the observation by Evans and Flath (1984) that zooplankton collected from water intake samples were similar to nearshore Lake Michigan zooplankton collected by plankton tows at the same time. Makarewicz (1991) found differences in zooplankton collected from Lake Ontario water intake and nearshore samples. He attributed these differences to the duration of sample collection.

SUMMARY

The present study demonstrates that: (1) zebra mussels are not likely to be detected by standard sampling protocols at the earliest stage of colonization, (2) educated observers can provide useful clues regarding early infestations by zebra mussels, and (3) a standard monitoring program can provide useful information once a local population of zebra mussels has become established. Therefore, it is not necessary to implement a monitoring program to examine plankton samples and artificial substrates until some indication of the presence of zebra mussels at a given location has been obtained. Education and training of water users, along with development of a central information authority, can provide suitable early warning for the presence of zebra mussels at most locations. A single observer can monitor thousands of square meters of substrate at a given location in a short period of time, covering a much larger surface area than any substrate sampler. Following the discovery of mussels, a planned monitoring program can provide information on the level of infestation and timing of reproduction.

ACKNOWLEDGMENTS

The work described in this chapter resulted from the efforts of many people who need to be recognized. Co-investigators A. H. Miller and S. Wittman provided direction and management to the project, gave many public presentations, worked with the media, coordinated many aspects of the efforts reported here, and have provided helpful comments on a draft of this manuscript. J. Jonas provided invaluable assistance in preparing educational materials, conducting training programs for water users, making presentations to the general public, and in her "spare" time collecting and analyzing water and substrate samples from the Green Bay area. Collection and analysis of samples from other areas of the state were the responsibility of M. Balcer, A. Brooks, S. Dodson, and H. Pearson, who have been helped by many student assistants in confirming reported zebra mussel sightings and collecting and analyzing harbor and substrate samples. Numerous water intake personnel also helped in the collection of intake plankton samples. The Green Bay

Metropolitan Sewerage District provided additional field support in gathering Green Bay harbor samples. This work was funded by the University of Wisconsin Sea Grant Institute under grants from the National Sea Grant College Program, National Oceanic and Atmospheric Administration, U.S. Department of Commerce, and the State of Wisconsin. Federal grants include NA16RG0273 and NA90AA-D-SG469; and projects A/AS-28, A/AS-1, A/AS-2, and M/SGA-1.

REFERENCES

Evans, M. S., and L. E. Flath. ''Intakes as Sampling Locations for Investigating Long-Term Trends in Zooplankton Populations,'' *Can. J. Fish. Aquat. Sci.* 41:1513–1518 (1984).

Griffiths, R. W., D. W. Schloesser, J. H. Leach, and W. P. Kovalak. ''Distribution and Dispersal of the Zebra Mussel (*Dreissena polymorpha*) in the Great Lakes Region,'' *Can. J. Fish. Aquat. Sci.* 48:1381–1388 (1991).

Kraft, C. E., M. Balcer, J. Jonas, A. H. Miller, H. Pearson, and C. W. Ramcharan. ''Watching for Zebra Mussels: Why Can't We Find Them If They're Already Here?'' (Abstr.) *J. Shellfish Res.* 10:257 (1991).

Makarewicz, J. C. ''Feasibility of Shoreside Monitoring of the Great Lakes,'' *J. Great Lakes Res.* 17:344–360 (1991).

Mardsen, J. E. ''Standard Protocols for Monitoring and Sampling Zebra Mussels,'' *Ill. Nat. Hist. Surv. Biol. Notes.* 138. 40pp. (1992).

Ramcharan, C. W., D. K. Padilla, and S. I. Dodson. A Multivariate Model for Predicting Population Fluctuations of *Dreissena polymorpha* in North American Lakes. *Can. J. Aquat. Sci.* 49:150–158 (1992).

Strayer, D. L. ''Projected Distribution of the Zebra Mussel, *Dreissena polymorpha*, in North America,'' *Can. J. Fish. Aquat. Sci.* 48:1389–95 (1991).

CHAPTER 43

Distribution of the Zebra Mussel (*Dreissena polymorpha*) in Estuaries and Brackish Waters

David L. Strayer and Lane C. Smith

The zebra mussel, *Dreissena polymorpha*, is widespread in estuaries and inland brackish waters of Europe. Within these habitats, its distribution and abundance are limited by salinity, availability of hard substrata, ice scour, exposure, and perhaps turbidity. High salinity is a preeminent limiting factor in many brackish waters. The upper tolerance limit ranges from about 0.6 to 12‰ of salinity, depending on the short-term variability and composition of the salinity (e.g., the ratio of monovalent to divalent ions). As an exercise, we use these results to predict the distribution and abundance of *D. polymorpha* in the Hudson River estuary. Our analysis of European literature suggests that *D. polymorpha* will become widespread and abundant in estuaries and brackish waters in North America.

0-87371-696-5/93/$0.00 + $.50

INTRODUCTION

Since the initial introduction of the zebra mussel (*Dreissena polymorpha*) into North America (Hebert et al., 1989), much attention has been focused on predicting the distribution and effects of this species (e.g., Mackie et al., 1989; McMahon and Tsou, 1990; Griffiths et al., 1991; Strayer, 1991). Most of this work focuses on freshwater ecosystems, because *D. polymorpha* is primarily a freshwater animal. However, it has a considerable tolerance to salinity and is likely to invade estuaries and brackish waters as well as fresh waters in North America. The purpose of this chapter is to describe the distribution of *D. polymorpha* in estuaries and brackish waters in Europe, to discuss the factors that may limit its distribution and abundance in such waters, and to speculate on which North American brackish waters and estuaries are most likely to support mussel populations.

REVIEW

Distribution in Europe

Dreissena polymorpha is found in estuaries and brackish waters throughout its European range. It often is a dominant member of the zoobenthic community in these waters, frequently reaching densities and biomasses of 100–10,000/m^2 and 1–100 g/m^2 (wet weight including shells), respectively (Mordukhai-Boltovskoi, 1960; Wiktor, 1963; Zenkevitch, 1963; Andreeva and Andreev, 1990).

In reviewing the distribution of *D. polymorpha* in brackish and estuarine European waters, it is useful to analyze four broad regions separately because the relevant environmental characteristics (and therefore the distribution of this species) of these regions are different. These regions are estuaries of the eastern Atlantic, the Baltic Sea and its estuaries, the Black Sea and its estuaries, and the Caspian and Aral Seas.

D. polymorpha lives in estuaries of the eastern Atlantic and North Sea from Great Britain and Germany to France (Germain, 1931; Wolff, 1969; Morton, 1969). Wolff's (1969) analysis of distributions in Dutch estuaries is the most detailed study of zebra mussels in this region. He found mussels in most tidal estuaries where the mean salinity was less than 0.6‰ (parts per thousand) but not in saltier waters (Figure 1). In The Netherlands, *D. polymorpha* has been reported from areas with salinities of 1–4‰, but these apparently are from nontidal habitats with no daily fluctuations in salinity. Storm surges that carry saline water inland periodically push the distribution of zebra mussels eastward, away from the coast.

D. polymorpha was rarely found in the intertidal zone of Dutch estuaries, probably because it cannot tolerate extended exposure to air. In addition to

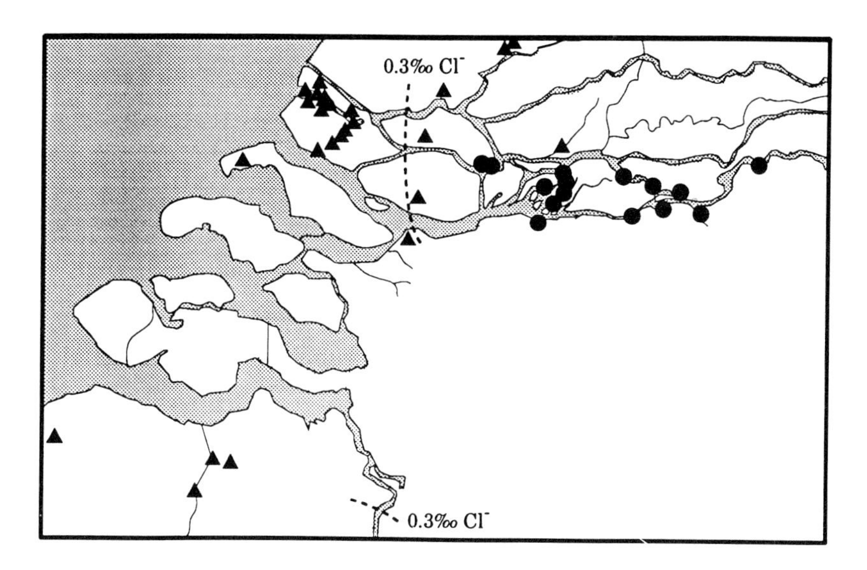

Figure 1. Distribution of *Dreissena polymorpha* in the Delta area of The Netherlands. Circles show populations in tidal waters; triangles show populations in nontidal waters. The isoline of 0.3‰ of chloride corresponds to about 0.54‰ of salinity. (Redrawn from Wolff, W. J. *Basteria* 33:93–103 [1969]. With permission.)

salinity, Wolff (1969) believed that the availability of hard substrata (e.g., bricks, wood, unionid mussels), strong currents, and pollution limit the distribution and abundance of this species in Dutch estuaries.

D. polymorpha is found in many bays and lagoons along the Baltic Sea from the Gulf of Finland southwestward to Poland and Germany, but not in the open Baltic itself (Schlesch, 1930; Ehrmann, 1933; Zenkevitch, 1963; Zhadin, 1965; Theede, 1974). One of these bays, the Szczecin lagoon, supports the densest population (114,000/m^2) reported in Europe (Wiktor, 1963). In most Baltic estuaries, *D. polymorpha* is found only near river mouths at low salinities: the eastern Gulf of Riga (<1‰) and the extreme eastern Gulf of Finland (<2‰), for example (Segerstråle, 1957; Zenkevitch, 1963; Zhadin, 1965). In contrast, Klimowicz (1958) found it throughout the Vistula delta and lagoon, including the most saline part of the lagoon (3.3–4.8‰). Similarly, Reshöft (1961) collected it from the North-Baltic Sea canal at salinities of 3.8 and 6.2‰. Klimowicz noted that *D. polymorpha* (and other freshwater mollusks) was conspicuously stunted in more saline areas (Table 1), suggesting physiological stress.

D. polymorpha is widespread and abundant in the upper reaches of many of the estuaries bordering the Black Sea (Markovsky, 1953, 1954, 1955; Mordukhai-Boltovskoi, 1960, 1964, 1979; Zenkevitch, 1963; Zhadin, 1965). It is found primarily at mean annual salinities of less than 2‰, where it is often extremely abundant (Figure 2). There is some indication that it is found

TABLE 1. Shell Length of *Dreissena polymorpha* from Various Zones of the Vistula Delta and Lagoon in Poland, as a Function of Salinity

Zone	Salinity (‰)	Shell Length (mm)
I	0.05–1	27–32
III	2.9–3.3	15–20
IV	3.3–4.8	9–14

Source: Klimowicz, 1958.

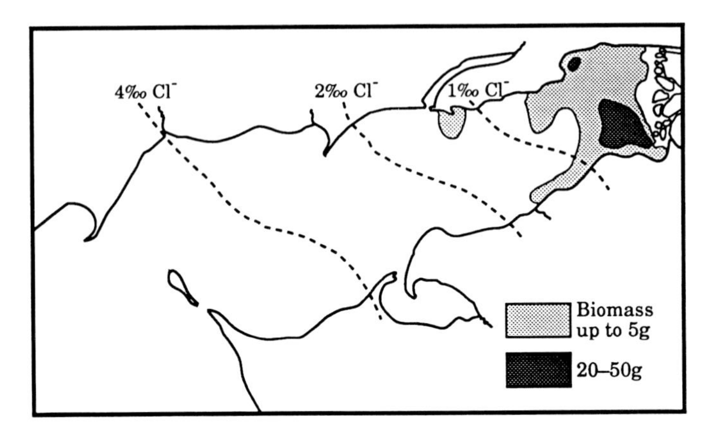

Figure 2. Biomass (wet weight, including shells) of *Dreissena polymorpha* in Taganrog Bay, a bay of the Sea of Azov along the north shore of the Black Sea, in relation to salinity. The isolines show chloride concentrations; 1‰ chloride = ca. 1.8‰ salinity. (Redrawn from Zenkevitch, L. *Biology of the Seas of the U.S.S.R.* [New York: Interscience Publishers, 1965] after Mordukhai-Boltovskoi).

in more saline waters in the eastern estuaries of the Black Sea (e.g., Taganrog Bay) than those to the west (e.g., the Danube) (Markovsky, 1953, 1954, 1955; Zenkevitch, 1963; Mordukhai-Boltovskoi, 1964). Mordukhai-Boltovskoi (1960) believed that *D. polymorpha* was limited in Black Sea estuaries by turbidity and availability of hard substrata, as well as by salinity. He noted that abundances in the Black Sea region were directly related to transparency of the water; however, no quantitative data were presented. He also reported that abundances were higher on stony or silty bottoms compared to other substrata (Figure 3). Relatively high abundances on silty bottoms was surprising, but was likely due to settling on the shells of burrowing bivalves (e.g., Unionidae) that often are abundant on such substrata.

The salinity of Black Sea estuaries varies among years, in response to both natural fluctuation in freshwater discharge and human diversions of waters from tributaries. In response, the distribution of *D. polymorpha* in Taganrog Bay shifted between years in response to changing freshwater inputs

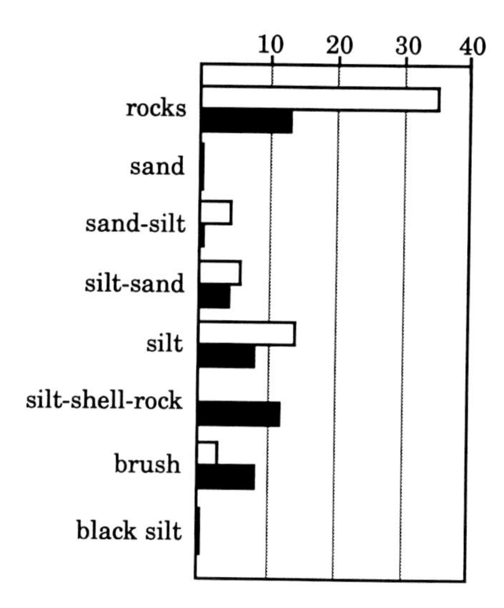

Figure 3. Relative abundance of *Dreissena polymorpha* on various substrata in the delta regions of the Dnieper (white bars) and Don (black bars) rivers in the U.S.S.R. There are no data for the silt-shell-rock substratum in the Dnieper River. (Redrawn from Mordukhai-Boltovskoi, F. D. *Akad. Nauk U.S.S.R., Moscow* [1960] [In Russian]. With permission.)

to the bay (Mordukhai-Boltovskoi, 1960). More recently, diversions of fresh water from the Don River for irrigation have increased the salinity of Taganrog Bay and reduced its range in this location (Nekrasova, 1972; Mordukhai-Boltovskoi, 1979).

The situation in the brackish Caspian and Aral Seas is quite different. Here, *D. polymorpha* occurs at much higher salinities than elsewhere in its range. This species is a dominant member of benthic communities throughout the shallow northern Caspian Sea on coarse shell-gravels, at salinities of 6–9‰ (Zenkevitch, 1963; Zhadin, 1965). It does not occur in the main body of the Caspian Sea (12.8‰ salinity). In the Aral Sea, it was found in areas with still higher salinities. *D. polymorpha* was dominant throughout the Aral Sea prior to large irrigation diversions from the tributaries, when the salinity was 10.2‰ (Zenkevitch, 1963). Due to large losses of fresh water from irrigation projects, the salinity of the Aral Sea has risen dramatically and is now 28‰ (Kotlyakov, 1991). Although the indigenous subspecies of *D. polymorpha* (*D. polymorpha aralensis* and *D. polymorpha obtusecarinata*) tolerated the initial increase in salinity, they began to decline when the salinity reached 12‰, and had "virtually disappeared" by the time the salinity reached 14‰ (Andreeva and Andreev, 1990). Although increasing salinity is the most likely explanation for the disappearance of *D. polymorpha* from the Aral Sea, Mordukhai-Boltovskoi (1972) believed that the introduction of exotic species

TABLE 2. Ionic Composition (Approximate Annual Means, as mg/L) of the Most Saline Waters that Support Populations of *Dreissena polymorpha* in Various Regions

	Eastern Atlantic[a]	Black Sea[b]	Baltic Sea[c]	Caspian Sea[d]	Aral Sea[e]
Ca^{+2}	7	30	100	220	600
Mg^{+2}	20	75	210	460	650
K^{+}	6	25	60	60	100
Na^{+}	170	610	1,800	2,000	2,600
SO_4^{2-}	42	150	480	1,900	3,800
Cl^{-}	300	1,100	3,300	3,300	4,100
Salinity	540	2,000	6,000	8,000	12,000

[a] Wolff (1969).
[b] Mordukhai-Boltovskoi (1960).
[c] Klimowicz (1958); Reshöft (1961).
[d] Zenkevitch (1963); Zhadin (1965).
[e] Andreeva and Andreev (1990).

and the loss of nutrients from the diverted tributaries were also important contributing factors.

The usual explanation for the tolerance of high salinities by *D. polymorpha* (and other organisms) in the Caspian and Aral Seas is that these waters are relatively richer in calcium and sulfate and lower in sodium and chloride than are oceanic waters (e.g., Mordukhai-Boltovskoi, 1964). It may also be important that waters of the Caspian and Aral Seas do not undergo such wide and unpredictable short-term swings in salinity as do most estuaries.

Limiting Factors

The most obvious factor limiting *D. polymorpha* in estuaries and brackish waters is salinity. Most regional studies of the distribution of this species have identified critical thresholds of salinity beyond which it is rarely found (e.g., Mordukhai-Boltovskoi, 1960 and 1964; Reshöft, 1961; Wolff, 1969; Andreeva and Andreev, 1990). Unfortunately, the thresholds identified for various regions are very different, ranging from 0.5 to 12‰ (Table 2).

Salinity tolerance has been studied experimentally in the laboratory by Karpevich (1947) and Reshöft (1961). *D. polymorpha* clearly is sensitive to sudden changes in salinity. An increase in salinity of ca. 5‰ above ambient levels causes marked decreases in respiration (Figure 4; Karpevich, 1947) or morphological damage (Reshöft, 1961). Reshöft (1961) found evidence that *D. polymorpha* may be damaged by sudden decreases in salinity. The species is much more tolerant of gradual changes in salinity. Karpevich (1947) found that even salinities as high as 17‰ did not depress respiration rates (Figure 4), although individuals could not survive indefinitely above 10‰. Karpevich also found a difference in the response to acute (but not chronic) salinity change between animals collected from habitats with different salinities (Fig-

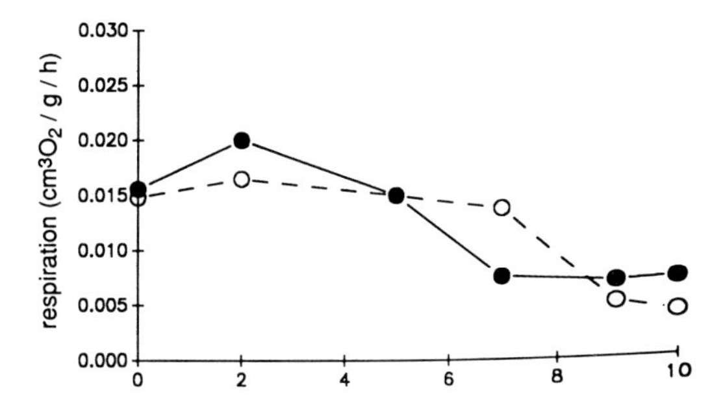

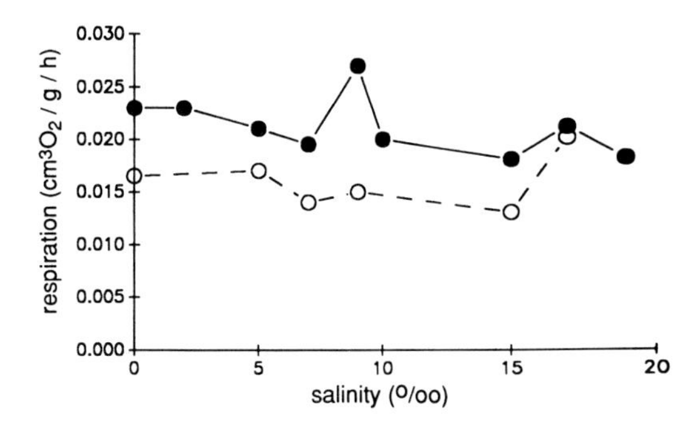

Figure 4. Response of the respiration rate of *Dreissena polymorpha* to abrupt (upper panel) and gradual (lower panel) changes in salinity. Gradual change is an increase of 2‰ every 2 days. The two populations studied are from the fresh waters of the Volga River delta (solid line) and oligohaline waters (<2‰) near the Charyr'ya Spit (dashed line). Note the difference in scale between the two abscissas. (From data of Karpevich, A. F. *Zool. Zh.* 26:331–338 [1947] [In Russian]. With permission.)

ure 4), raising the possibility of genetic or long-term physiological adaptation to salinity stress.

Although it is as difficult to identify an absolute upper salinity tolerance from laboratory studies as it is from distributional data, these laboratory results are broadly consistent with the observed distribution of *D. polymorpha* in European brackish waters. Wolff (1969) pointed out this species apparently has trouble dealing with short-term fluctuations in salinity (in accord with Karpevich and Reshöft's results on acute salinity stress) and is, therefore, unable to penetrate as far seaward in the variable Dutch estuaries as in the more stable estuaries of the Baltic Sea (Table 2). In addition, distributional data suggest that it may tolerate salinity in the form of divalent ions better than monovalent ions (Table 2), a supposition not yet tested in the laboratory.

TABLE 3. Projections of the Population of *Dreissena polymorpha* to Be Expected in the Hudson River Estuary, New York, Based on a Salinity Tolerance of About 2‰

Expected range (km above The Battery)	80–247[a]
Area of range (km^2)	150[b]
Area of hard substratum within range (km^2)	8–40[c]
Expected density on suitable substrata (m^{-2})	10–1000[d]
Expected population on sediments	$0.1–40 \times 10^9$
Population of unionid mussels within range	9 × 109[e]
Zebra mussels per unionid mussel	5–25[f]
Expected population on unionid mussels	$45–225 \times 10^9$
Total population in estuary	$45–265 \times 10^9$

Note: Estimates are approximate, and are intended chiefly for illustrative purposes.

[a] River km 247 is the head of the estuary at Troy.
[b] Morphometry from Gladden et al. (1988).
[c] Suitable sediments are those coarser than sand plus rip-rap and bedrock shores below the low-water mark; data on sediment types in the Hudson estuary from Coch (1986) and Ellsworth (1986).
[d] Mordukhai-Boltovskoi (1960), Stanczykowska (1977).
[e] Simpson et al. (1984, 1986).
[f] Lewandowski (1976), Schloesser and Kovalak (1990).

In fact, Zhadin (1965) suggested that the chloride content of water, rather than its total salinity, may be a critical limiting factor.

Because *D. polymorpha* requires firm substrata on which to attach, it may be severely limited by the availability of suitable substrata in estuaries, which often are largely soft bottomed. As Mordukhai-Boltovskoi (1960) has demonstrated, however, even these soft-bottomed areas may support many zebra mussels that are living on isolated firm substrata (e.g., unionid clams, bricks, wood, etc.) embedded in ''soft'' bottoms. The importance of these isolated firm substrata should not be underestimated (cf. Table 3).

Other physical factors may be important as well. Mordukhai-Boltovskoi (1960) suggested that high turbidity may limit estuarine populations by interfering with feeding activities, but his hypothesis seems to have received little attention. Zebra mussels may be excluded from the shallow waters of estuaries by harsh physical factors: excessive exposure to air and sun in the intertidal zone (Wolff, 1969) and ice scour in the intertidal and upper subtidal zones of northern waters (cf. Theede, 1974).

Finally, there is the possibility that biological interactions with either sessile benthic competitors or predators might limit the distribution of zebra mussels in estuaries, although we have seen no evidence that this occurs. There are a few bivalves with distribution overlapping that of *D. polymorpha* in estuarine waters (e.g., *Mytilopsis, Mytilaster*). In fresh waters, waterfowl sometimes eat enough zebra mussels to affect their abundance, at least locally (e.g., Suter, 1982).

A Speculative Case Study

It is possible to use the information just presented on the salinity tolerance and substratum preferences of the zebra mussel to make rough projections of the distribution and abundance of this species in North American estuaries. Table 3 shows the results of such an exercise for the Hudson River estuary in New York. We expect *D. polymorpha* to range from the head of the estuary at Troy to the middle of the oligohaline zone near West Point (river km 80), although zebra mussels probably will occur near the mouths of tributaries further downriver (Figure 5).

As suggested by Mordukhai-Boltovskoi (1960), a major part of the population in the estuary probably will live on shells of living unionid mussels, rather than on substrata mapped as ''hard'' (i.e., bedrock, cobbles, etc.) In fact, our calculations probably underestimate the number of zebra mussels living on ''soft'' bottoms, because we have not been able to estimate the number of animals living on snags, coal, clinkers, stones, and other hard objects embedded in mud and sand.

Finally, the results shown in Table 3 can be used to make a very rough estimate of the potential impact of *D. polymorpha* on the Hudson River ecosystem. From published filtering rates (Kryger and Riisgard, 1988), the projected zebra mussel population in the Hudson River estuary was calculated to filter 70–400 $\times$ 10^6 m^3/day of water during the summer, a volume equal to 5–30% of the volume of the entire estuary. Therefore, we conclude that zebra mussels might be expected to have a moderate to major impact on the ecosystem.

Prospects for North America

D. polymorpha will spread widely into estuaries and brackish waters in North America. The amount, composition, and variability of salinity will exert a primary control on the distribution of this species on our continent. In estuaries where there is considerable short-term variation in salinity due to tides or fluctuation in freshwater inputs, the seaward limit of zebra mussels will probably be at only 0.4–2‰ of salinity. In chemically more stable estuaries (e.g., nontidal lagoons), it may occur to ca. 6‰, as it does in the nontidal lagoons of the Baltic. Finally, it will probably do well in brackish lakes, such as those of the Dakotas and Canadian Prairie provinces, where it may tolerate as much as 10–14‰ of these sulfate-rich waters (cf. Rawson and Moore, 1944).

We have assumed that there is no genetic adaptation to salinity (an unlikely prospect), so that North American zebra mussels will behave like a composite of studied European populations. It would therefore be very desirable to test

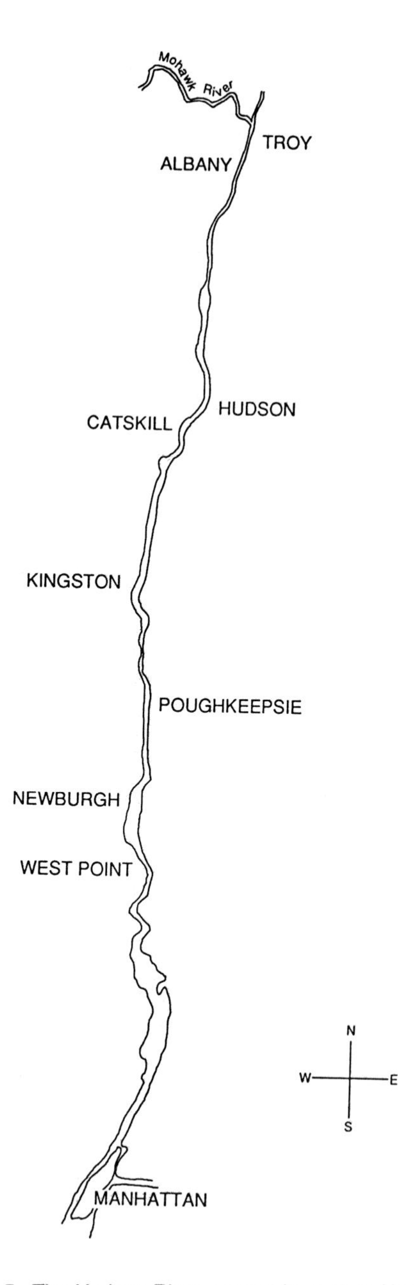

Figure 5. The Hudson River estuary in eastern New York.

the tolerances of North American zebra mussels to both constant and varying salinity of various compositions (i.e., both seawater and sulfate-rich waters).

Within chemically suitable habitats, we expect *D. polymorpha* to colonize hard substrates from the upper subtidal zone to the deepest oxygenated sediments. In the coming decades, this species will become a conspicuous and dominant animal in many North American estuaries and brackish waters.

ACKNOWLEDGMENTS

We thank Ninele Slonin for translating relevant Russian literature; Annette Frank for locating hard-to-find literature; Sharon Okada for help with graphics; and Nina Caraco, Dean Hunter, and the editors for their helpful reviews of the manuscript. Financial support was provided by the Hudson River Foundation. This is a contribution to the program of the Institute of Ecosystem Studies of The New York Botanical Garden.

REFERENCES

Andreeva, S. I., and N. I. Andreev. "Trophic Structure of Benthic Communities in Aral Sea Under Conditions of Changed Regime," *Sov. J. Ecol.* 21:94–98 (1990).

Coch, N. K. "Sediment Characteristics and Facies Distributions in the Hudson System," *Northeast. Geol.* 8:109–129 (1986).

Ehrmann, P. "Mollusken (Weichtiere)," in *Die Tierwelt Mitteleuropas, Band 10,* Liefierung 1, P. Brohmer, P. Ehrmann, and G. Ulmo, Eds. von Quelle and Meyer. Leipzig. (1933).

Ellsworth, J. M. "Sources and Sinks for Fine Grained Sediments in the Lower Hudson River," *Northeast. Geol.* 8:141–155 (1986).

Germain, L. "Mollusques terrestres et fluviatiles, deuxième partie," *Faune de France, Vol. 22* (Paris, France: Paul Lechevalier, 1931).

Gladden, J. B., F. R. Cantelmo, J. M. Croom, and R. Shapot. "Evaluation of the Hudson River Ecosystem in Relation to the Dynamics of Fish Populations," *Am. Fish. Soc. Monogr.* 4:37–52 (1988).

Griffiths, R. W., D. W. Schloesser, J. H. Leach, and W. P. Kovalak. "Distribution and Dispersal of the Zebra Mussel (*Dreissena polymorpha*) in the Great Lakes Region," *Can. J. Fish. Aquat. Sci.* 48:1381–1388 (1991).

Hebert, P. D. N., B. W. Muncaster, and G. L. Mackie. "Ecological and Genetic Studies on *Dreissena polymorpha* (Pallas): A New Mollusc in the Great Lakes," *Can. J. Fish. Aquat. Sci.* 46:1587–1591 (1989).

Karpevich, A. F. ["The adaptabiilty of metabolism in north Caspian mussels (genus *Dreissena*) to variations in salinity regime"]. *Zool. Zh.* 26:331–338 (1947) (In Russian).

Klimowicz, H. [''The Molluscs of the Vistula Lagoon and the Dependence of Their Distribution on the Water Salinity''] *Pol. Arch. Hydrobiol.* 5:93–121 (1958) (In Polish with English summary).

Kotlyakov, V. M. ''The Aral Sea Basin: A Critical Environmental Zone,'' *Environment* 33:4–9, 36–38 (1991).

Kryger, J., and H. U. Riisgard. ''Filtration Rate Capacities in Six Species of European Freshwater Bivalves,'' *Oecologia (Berlin)* 77:34–38 (1988).

Lewandowski, K. ''Unionidae as a Substratum for *Dreissena polymorpha* Pall.,'' *Pol. Arch. Hydrobiol.* 23:409–420 (1976).

Mackie, G. L., W. N. Gibbons, B. W. Muncaster, and I. M. Gray. ''The Zebra Mussel, *Dreissena polymorpha*: A Synthesis of European Experiences and a Preview for North America,'' Report prepared for the Ministry of Environment, Water Resources Branch, Great Lakes Section, Queen's Printer, Toronto, Ontario (1989).

Markovsky, J. M. [''Invertebrate Fauna of the Lower Course of Ukrainian Rivers. I. Water-bodies of the Dniestr Delta and of the Dniester Liman''], *Akad. Nauk Ukr. S.S.R.* (1953) (In Russian; not seen, cited by Mordukhai-Boltovskoi, 1964).

Markovsky, J. M. [''Invertebrate Fauna of the Lower Course of Ukrainian rivers. II. Dniepr-Bug liman''], *Akad. Nauk. Ukr. S.S.R.* (1954) (In Russian; not seen, cited by Mordukhai-Boltovskoi, 1964).

Markovsky, J. M. [''Invertebrate Fauna of the Lower Course of Ukrainian rivers. III. Water-bodies of the Kilia-delta of Danube''], *Akad. Nauk Ukr. S.S.R.* (1955) (In Russian; not seen, cited by Mordukhai-Boltovskoi, 1964).

McMahon, R. F., and J. L. Tsou. ''Impact of European Zebra Mussel Infestation to the Electric Power Industry,'' *Proc. Am. Power Conf.* 52:988–997 (1990).

Mordukhai-Boltovskoi, F. D. [''The Caspian fauna in the Azov-Black Sea basin,''] *Akad. Nauk U.S.S.R. Moscow* (1960) (In Russian).

Mordukhai-Boltovskoi, F. D. ''Caspian Fauna Beyond the Caspian Sea,'' *Int. Rev. Ges. Hydrobiol.* 49:139–176 (1964).

Mordukhai-Boltovskoi, F. D. ''Current Status of the Aral Sea fauna,'' *Hydrobiol. J.* 8:8–13 (1972).

Mordukhai-Boltovskoi, F. D. ''Composition and Distribution of Caspian Fauna in Light of Modern Data,'' *Int. Rev. Ges. Hydrobiol.* 64:1–38 (1979).

Morton, B. S. ''Studies on the Biology of *Dreissena polymorpha* Pall. IV. Habits, Habitats, Distribution and Control,'' *Water Treat. Exam.* 18:233–240 (1969).

Nekrasova, M. Y. [''Zoobenthos of the Azov Sea after the Reconstruction of the Don River,''] *Zool. Zh.* 51:789–797 (1972) (In Russian, not seen, cited by Mordukhai-Boltovskoi, 1979).

Rawson, D. S., and J. E. Moore. ''The Saline Lakes of Saskatchewan,'' *Can. J. Res. Sect. D* 22:141–201 (1944).

Reshöft, K. ''Untersuchungen zur zellulären osmotischen und thermischen Resistenz verschiedener Lamellibranchier der deutschen Küstengewässer,'' *Kiel. Meeresforsch.* 17:65–84 (1961).

Schlesch, H. ''Ueber die Verbreitung von *Dreissensia polymorpha* Pall. im Norden,'' *Folia Zool. Hydrobiol.* 2:20–22 (1930).

Schloesser, D. W., and W. P. Kovalak. ''Infestation of Native Unionids by *Dreissena polymorpha* in a Power Plant Canal in Lake Erie,'' *Bull. N. Am. Benthol. Soc.* 7:107 (1990).

Segerstråle, S. G. ''Baltic Sea,'' in *Treatise on Marine Ecology and Paleoecology, Vol. 1,* J. W. Hedgpeth, Ed. *Mem. Geol. Soc. Am.* 67:751–800 (1957).
Simpson, K. W., R. W. Bode, J. P. Fagnani, and D. M. DeNicola. ''The Freshwater Macrobenthos of the Main Channel Hudson River. Part B: Biology, taxonomy, and distribution of resident macrobenthic species,'' Final report to the Hudson River Foundation, Grant 8/83A/39 (1984).
Simpson, K. W., J. P. Fagnani, R. W. Bode, D. M. DeNicola, and L. E. Abele. ''Organism-Substrate Relationships in the Main Channel of the Lower Hudson River,'' *J. N. Am. Benthol. Soc.* 5:41–57 (1986).
Stanczykowska, A. ''Ecology of *Dreissena polymorpha* (Pall.) (Bivalvia) in Lakes,'' *Pol. Arch. Hydrobiol.* 24:461–530 (1977).
Strayer, D. L. ''Projected Distribution of the Zebra Mussel, *Dreissena polymorpha,* in North America,'' *Can. J. Fish. Aquat. Sci.* 48:1389–1395 (1991).
Suter, W. ''Der Einfluss von Wasservögeln auf Populationen der Wandermuschel (*Dreissena polymorpha* Pall.) am Untersee/Hochrhein (Bodensee),'' *Schweiz. Z. Hydrol.* 44:149–161 (1982).
Theede, H. ''Die Tierwelt I. Ökologie,'' in: *Meereskunde der Ostsee* L. Magaard and G. Rheinheimer, Eds. (Berlin, Germany: Springer-Verlag, 1974), pp. 171–188.
Wiktor, J. ''Research on the Ecology of *Dreissena polymorpha* Pall. in the Szczecin Lagoon (Zalew Szczeciński),'' *Ekol. Pol. (Ser. A)* 11:275–280 (1963).
Wolff, W. J. ''The Mollusca of the Estuarine Region of the Rivers Rhine, Meuse and Scheldt in Relation to the Hydrography of the Area. II. The Dreissenidae,'' *Basteria* 33:93–103 (1969).
Zenkevitch, L. *Biology of the Seas of the U.S.S.R.* (New York: Interscience Publishers, 1963).
Zhadin, V. I. ''Mollusks of Fresh and Brackish Waters of the U.S.S.R. Keys to the Fauna of the U.S.S.R., *Acad. Nauk U.S.S.R., Moscow* No. 46, (1965) [English translation by the Israel Program for Scientific Translations, Jerusalem].

CHAPTER 44

Perspectives on the Ecological Impacts of the Zebra Mussel (*Dreissena polymorpha*) in the Former European USSR and in North America

Valery N. Karnaukhov and Alexei V. Karnaukhov

The mussel *Dreissena polymorpha* formed as a species millions of years ago in the huge saline water basin which includes the present aral, Caspian, Azov, and Black Seas. It was in the mouths of the rivers that flowed into these seas, such as the Volga, Don, Dnepr, and Danube, that there was a stable relationship between *Dreissena* and other organisms in the area. About 200 years ago, evidently in connection with the construction of canals and increased eutrophication of the water basins, *Dreissena* began to increase its range into North European rivers that emptied into the Baltic Sea. Similarily, about 40–50 years ago, this mollusk began moving northward along the Volga, Don, and Dnepr Rivers when large regions were flooded as a result of the construction of a series of reservoirs for hydroelectric stations.

0-87371-696-5/93/$0.00 + $.50

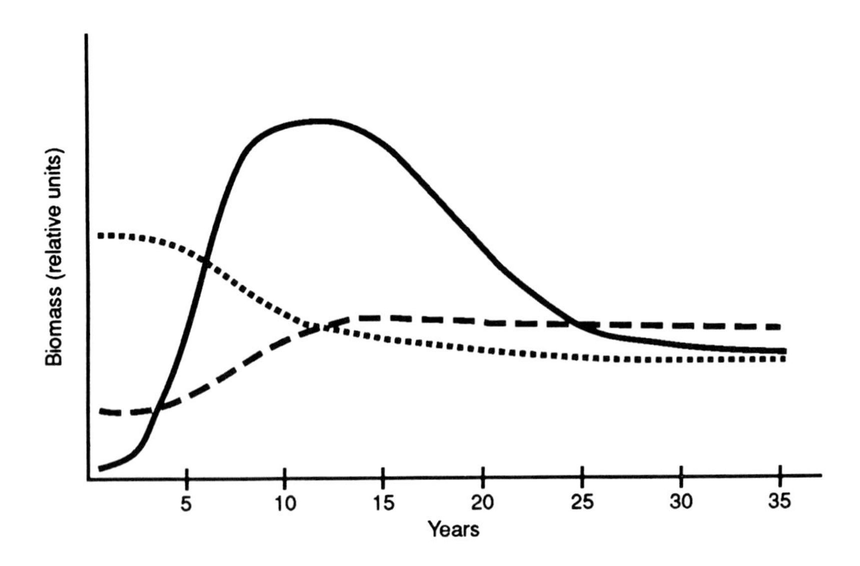

Figure 1. Mathematical model depicting long-term changes in populations of *Dreissena*, plankton, and fish after *Dreissena* first becomes established. The model is based on observations and data collected from various reservoirs in the former European USSR. *Dreissena* = solid line, plankton = dotted line, fish = dashed line.

Based on observations and data from a number of reservoirs in the former European USSR, a mathematical model was developed to typify long-term changes in numbers of *Dreissena* and other ecosystem components after the introduction of this species. After first being introduced into a system, the number of *Dreissena* increases relatively slowly at first, then increases rapidly to a plateau about 10–15 years after becoming established, and finally decreases to a stable lower population level (Figure 1). The decline in numbers is a result of both a decrease in food resources (microplankton), since this species is a very active filter feeder, and also a result of predation by various species of fish. Typically, when the biomass of *Dreissena* increases, the biomass of the fish population also increases (Figure 1). The dominant fish in the rivers and reservoirs of the former European USSR is the river roach, *Rutilus rutilus*. This fish species begins to feed on *Dreissena* when it reaches a length of about 13–15 cm. In addition to the river roach, other abundant roach species also feed heavily on *Dreissena* including: *Rutilus caspicus, Rutilus heckeli, Rutilus frisii,* and *Rutilus frissi cutum*. Nonroach species that also feed on mollusks are the gustera (*Blicka bioerkna*), the Aral barbel (*Barbus brachycephalus*), and sturgeons (Acipensetidae).

The increase in numbers of *Dreissena* in Lake St. Clair and Lake Erie in the first few years after its introduction far exceeds those increases noted in water bodies of the former European USSR. When the mathematical model

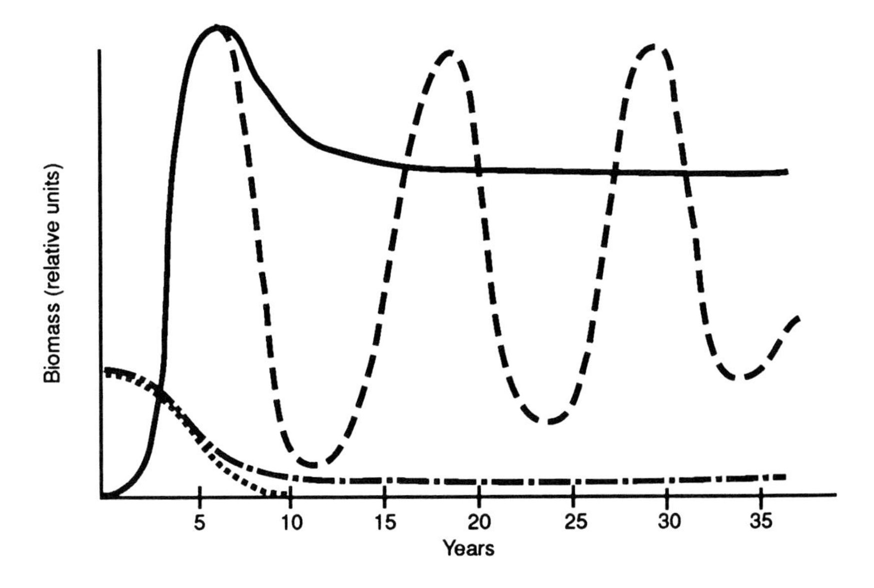

Figure 2. Mathematical model depicting long-term changes in populations of *Dreissena* and fish under two different scenarios in waters of North America, including the Great Lakes. In scenario one, the *Dreissena* population is food limited (solid line); in scenario two, the *Dreissena* population is self-limiting (dashed line). The corresponding response of the fish population is shown for scenario one (dot-dashed line) and scenario two (dotted line), respectively.

developed for predicting ecosystem changes in the former European USSR was applied to changes in the North American Great Lakes, the results indicated that the relatively rapid, explosive increase in *Dreissena* could only have occurred if there were no (or almost no) fish in the Great Lakes able to feed on the mussels. This indeed appears to be the case. In Europe, most fish species that feed on *Dreissena* are of the variety that have several rows of pharyngeal teeth and are able to crush the mussel shells. Few fish species with this characteristic are found in North America. Fish biomass in the water bodies of North America should decline rather than increase after *Dreissena* becomes established. With few predators, large populations of *Dreissena* will develop and proceed to filter the water of plankton, thereby limiting the food resources of larval fish. Over the long term, the *Dreissena* population in North American water bodies will either be limited by the amount of available food, or will be limited by parasites, bacteria, disease, etc. In the former case, a stable population level will become evident about 10–15 years after the species first becomes established; in the latter case, the population will fluctuate broadly over periodic intervals (Figure 2). Either way, the fish population will probably decline and remain at low levels.

CHAPTER 45

Maintenance of the Zebra Mussel (*Dreissena polymorpha*) Under Laboratory Conditions

S. Jerrine Nichols

Zebra mussels (*Dreissena polymorpha*) with a shell length >1 mm are adaptable to laboratory conditions if ammonia levels are low (<1 mg/L) and water temperatures range between 4 and 24°C. The most difficult aspect of maintaining mussels is providing acceptable food that will support growth and survival over a long period of time. Several standard invertebrate foods, such as yeast, bacterial infusions, and live and dried algae were tested for initial acceptance and long-term survival. The mussels accepted and survived best on green algae (either live *Chlamydomonas reinhardtii* or dried *Chlorella* sp.). A ration of 3.2 g of algae (dry weight) per 1000 mussels (shell length >10 mm) resulted in an average growth rate of 2 mm/month.

INTRODUCTION

The recent invasion of the Great Lakes region by zebra mussels (*Dreissena polymorpha*) emphasizes the ability of these mussels to rapidly colonize an area. In less than 3 years, mussels in the western basin of Lake Erie reached an average density of over 20,000/m^2 and a maximum of 700,000/m^2 (Griffiths et al., 1989). This invasion has stimulated research on the ecological impact, physiology, and control of this species. However, the ability to conduct experimental research in the laboratory has been hampered by the lack of efficient and economical maintenance methods. Moreover, research depending on field collections is difficult because zebra mussel growth and reproduction is reduced during winter. The present study examines the feasibility of transporting, maintaining, and feeding zebra mussels under laboratory conditions.

METHODS

Growth

Colonies of *Dreissena polymorpha* were collected by Ponar grab from areas less than 3 m in water depth from May through September 1989 in Lake St. Clair and Lake Erie. Colonies were initially rinsed in the field. The mussels were transported from the field to the laboratory in less than 3 hr. Once the mussels arrived at the laboratory, they were rinsed again and all damaged or dying mussels were discarded.

The mussels were divided into three groups based on laboratory treatment: (1) general laboratory stock, (2) cold-room stock, and (3) experimental stock. The general laboratory stock (Figure 1) were kept under ''normal'' laboratory conditions, that is, a water temperature of 20°C and a photoperiod of 12 L:12 D. These mussels were held in aquariums that varied in size from 4 to 80 L and had outside power filters (no filtration material) or airstones to maintain water flow. Water circulation in aquariums with power filters was 400 L/hr for each 40 L of water (e.g., a 40-L aquarium used one outside filter [400 L/hr], an 80-L aquarium used two filters). The amount of water flow in aquariums using airstones (1 airstone per 8 L of water) was not measured. Colonies were not removed from the substrate to which they were initially attached. Stocking densities ranged from 10 to 50 mussels (>10 mm) per liter. All aquariums were cleaned three times per week, and dead mussels were removed from each aquarium and counted. Water was nonchlorinated well water with oxygen levels above 7 mg/L and with a pH of 7.0–7.5. Ammonia levels were determined by the Nesslerization method (APHA, 1989) and were below 1 mg/L. Calcium levels were determined by using a multielement plasma spectrophotometer (detection limit 0.03 mg/L with a relative

Figure 1. Colonies of *Dreissena polymorpha* held under general laboratory conditions.

standard deviation of 0.08%) and ranged from 115 to 142 mg/L (average 125 mg/L). General laboratory stock were fed a diet of dried algae at a rate of 3.2 g dry weight per 1000 animals per day. This was based on Walz's (1978a) estimate of 75 μg of carbon per mussel per hour as minimum dietary needs.

Mussels designated cold-room stock were held at a constant temperature of 4°C without food to determine long-term survival under conditions requiring minimal care. These mussels were kept in 4-L aquariums in which complete water changes were made every 3 weeks; dead mussels were counted when the water was changed. No aeration was provided, and the aquariums were kept in complete darkness. With the exception of temperature, water quality conditions provided for the cold-room stock were the same as for the general laboratory stock.

Experimental stock consisted of groups of mussels exposed to constant handling, rapid temperature changes, and low calcium. The group exposed to constant handling consisted of 100 mussels isolated in test tubes. Each day for 6 weeks, mussels were detached (byssus threads cut), removed from the test tube, and left out to dry for 1 hr. They were then replaced in water of the same temperature, and the number of dead mussels was noted on a daily basis. Mussel response to rapid changes in water temperature was determined by acclimating 175 mussels to 4°C and 175 mussels to 20°C for 2 weeks. After 2 weeks, 25 mussels from the 4°C group and 25 from the 20°C group were placed in test tubes containing 0, 4, 10, 15, 20, 24, and 30°C water with no acclimation period. The general appearance of the mussels and the number of dead individuals were determined after 24 hr.

Literature sources indicate that calcium levels limit zebra mussel distribution under field conditions and that there is some discrepancy in actual levels required (Stanczykowska, 1977). Two tests were used to investigate calcium needs of a third group of experimental mussels under laboratory conditions. The first test involved isolating 100 mussels in a 4-L aquarium containing distilled water (0 mg/L calcium) and monitoring mortality. The second test examined calcium use under general laboratory conditions. Calcium concentrations were measured over three 72-hr periods in two 80-L aquariums with 1000 mussels (>15 mm) in each. One aquarium contained mussels that had been in captivity for 6 weeks; the other contained mussels that had been in captivity for 6 months. Each aquarium initially contained 125 mg/L calcium. Identical aquariums without zebra mussels were used as controls. Calcium levels in all aquariums were monitored daily for three 72-hr periods. Reported test results are an average of all three 72-hr tests.

Zebra mussel response to maintenance methods and experimental procedures was monitored by counting the number of dead animals and checking live mussels for signs of stress. The degree of stress was characterized by the width of the shell opening, the distance that the siphons extended past the shell margin, the extension of tentacles on the siphons, overall response to external stimuli, and willingness to attach to a substrate. Nonstressed mussels were open with siphons and tentacles extended past the shell margin for at least 16 hr each day. The siphons were rapidly retracted back into the shell in response to external stimuli such as overhead shadows and changes in water movement. Mussels either remained attached to the substrate or reattached within 24 hr after being removed from a substrate. Lightly stressed mussels remained closed or, if shells were open, had little or no siphon extension. Moderately stressed mussels showed some shell gape, retracted tentacles and siphons, and responded only to strong external stimuli (such as a sharp prod to the shell). Lightly and moderately stressed mussels were often found loose on the bottom of aquariums and did not reattach to the substrate within 24 hr. Severely stressed mussels had a wide shell gape, totally retracted tentacles and siphons, and wide siphon openings that exposed the entire inner region. These mussels responded only to a direct prodding of the body tissue. Mussels were considered dead when the shell gape was wide, tentacles and siphons were totally retracted, no response was noted to any stimuli, and body tissues showed signs of disintegration. Tissue disintegration occurred within 12 hr after a mussel was considered dead; and frequently the entire body mass sloughed away from the shell and floated in the water column.

Feeding*

Various types of food were tested for initial acceptance (removal of material from the water over a 24-hr period) and long-term growth and survival

* Mention of trade names or manufacturers does not imply U.S. Government endorsement of commercial products.

(over a 5- to 8-month period). Food types included: a live green alga (*Chlamydomonas reinhardtii*), bacterial infusions from boiled rice, live yeast, dried yeast suspension (sold as marine invertebrate food), hex mix (ground fish food and alfalfa juice powder), ABM MicroMac (powdered food supplement used to rear marine oyster larvae), a dried blue-green alga (*Spirulina* sp.), and a dried green alga (*Chlorella* sp.). All diet tests were conducted at 20°C. Mussels used in the feeding experiments were 15–18-mm long at the beginning of the test and were measured on a monthly basis for the duration of the experiments.

Mussels used in the diet acceptance tests were acclimated to the testing procedures by being placed in the test chamber, fed, left for 12 hr, and then removed to the general laboratory aquariums. This procedure was completed twice before the actual diet test was run. Without this acclimation period, mussels frequently did not feed or continually crawled around the test chamber.

The initial acceptance of the various food types was tested over a 24-hr period. Single mussels or small clusters of 4 to 10 individuals were isolated in 200-mL beakers or in 4-L aquariums with enough water flow to keep food particles and pseudofecal material suspended in the water column. Similar aquariums or beakers without zebra mussels were set up at the same time to serve as controls. Food was added to both mussel and control chambers; food concentrations were measured, checked hourly for 12 hr, and checked again at 24 hr. Concentrations of live algae were measured by counting cells per milliliter. Concentrations of other food sources were determined by measuring turbidity with a nephelometer and reporting results in nephelometric turbidity units (NTU) (APHA, 1989). All diet tests were repeated three times over a 2-week period, and the test results are reported as an average of the three replicates.

Mussels were divided into groups of 100 individuals and each group was placed separately in a 4-L aquarium. Each group was fed a different food type (eight total) while one group was not fed at all. Survival and growth rates were determined over a 5-month period. Shell lengths and the number of dead individuals were measured monthly. Mussels were fed twice daily 5 days a week and once 2 days a week. The amount of food provided was again based on an estimate of a daily maintenance ration of 75 μg carbon per mussel with an assumed 24 hr siphoning rate (Walz, 1978a).

Siphoning rates for fed and starved mussels were determined by videotaping colonies for 24 hr and counting the number of hours that the siphons were extended. Videotaping was done in situ, with dim red lights used at night to allow filming without interfering with photoperiodicity. Filming was repeated for three 24-hr periods, and results presented are an average of all three tests.

Analysis of variance and χ^3-tests were performed on results from short- and long-term diet replicates and siphoning rates. Results were considered significant when $p \leq 0.05$.

RESULTS AND DISCUSSION

General Maintenance

Zebra mussels readily adapt to laboratory conditions; no special equipment is required to make them survive and grow. However, it is also possible for an entire collection of mussels to die in less than 24 hr. Reasons for mussel die-off (>1% of the total population) vary depending upon how long the mussels have been in captivity. If mussels die within their first 72 hr in the laboratory, the causative factors are probably a lack of cleanliness during transportation or rapid temperature fluctuation (stage 1, Figure 2). High mortality after the first 72 hr but within the first 4 months is caused by insufficient removal of mussel waste products, high water temperatures, or unsuitable water quality (stage 2, Figure 2). After 4–5 months, increases in mortality are related to improper diet (stage 3, Figure 2).

Poor water quality or temperature fluctuations are usually the cause of mortality in mussels during their first 72 hr in captivity. Problems with poor water quality begin as soon as the mussels are collected. The first three attempts to bring zebra mussels into the laboratory for this study failed. Mussels were rinsed in the field and transported in coolers filled with lake water. However, the mussels showed signs of severe stress less than 2 hr of being removed from coolers and placed in aquariums filled with lake water. Mortality rates were over 80% each time, with mussels dying within a few hours of reaching the laboratory. It was noted that the mussels had their siphons extended during transportation, and field rinsing usually did not prevent the water in the coolers from becoming very turbid. It was assumed that this exposure to turbid water during transportation was causing the high levels of mortality. High silt loads in the water column are known to cause substantial mortality in field populations (Stanczykowska, 1977). After this, all mussels were transported without water. To prevent excessive jostling and damage during transportation, the mussels were placed into plastic bags and then into coolers. This method of transport decreased the mortality rate to <10% of the total.

After reaching the laboratory, water quality remains a critical factor in reducing mortality in the first 72 hr. The visual appearance of water in the aquariums is a good indicator of problems — it must remain clear. Mussels will show signs of stress if the water is very turbid or has a milky white color. Once the mussels are moderately stressed, mortality rates increase rapidly with each hour. Stress reduction and water clarity improvement are accomplished by using a combination of procedures, such as rinsing the colonies, changing aquarium water 1–2 times daily, increasing water flow, or reducing the number of mussels held in the aquarium.

Rapid fluctuations in water temperature are the second most frequent cause of mussel die-off during the first 72 hr in the laboratory, particularly

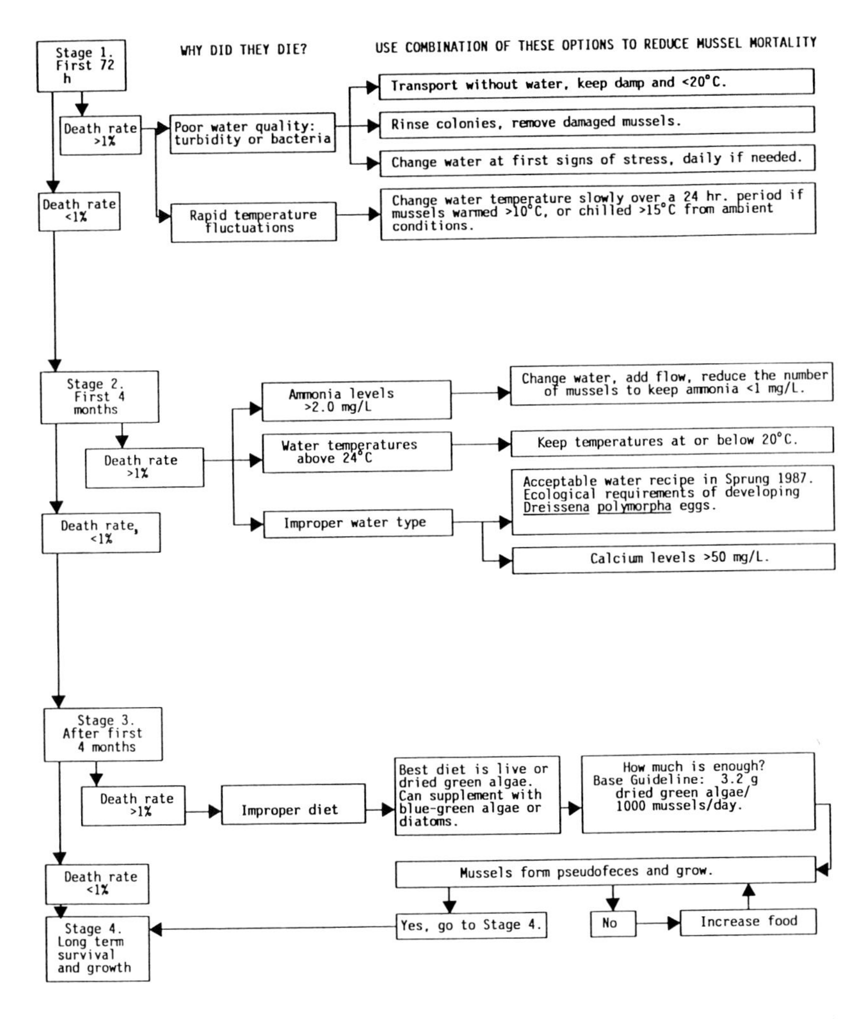

Figure 2. Potential causes of mortality of *Dreissena polymorpha* in the laboratory and options to consider to keep mortality to a minimum. Stage 1, within 72 hr of collection; stage 2, within 4 months of collection; stages 3 and 4, 6 months after collection.

mussels brought in when field temperatures are below 10°C or above 20°C. For example, two of the collections made in early April (water temperatures at 8°C) and one in August (25°C) had a mortality rate of 100%, compared to the normal mortality rates of less than 10%. Tests using experimental stock indicated that although mussels can tolerate a rapid increase of up to 10°C in water temperature with no mortality, increases greater than this increased mortality (Table 1). Mussels are more tolerant of rapid decreases in water temperatures, requiring drops greater than 15°C before mortality occurs. Gradual changes in water temperatures over a period of 24 hr did not seem to

TABLE 1. Mortality (Number Dead) of *Dreissena polymorpha* Exposed to Rapid Temperature Fluctuations Under Laboratory Conditions

Initial Water Temperature (°C)	New Water Temperature (°C)	Mortality (n = 25)
4	0	0
4	10	0
4	15	1
4	20	10
4	24	16
4	30	25
20	0	0
20	10	0
20	15	0
20	24	0
20	30	22

Note: Length of each test was 24 hr.

cause stress in the zebra mussels and enabled them to be brought into the laboratory at any season with less than 10% die-off.

At the end of the first 72 hr in the laboratory, all the mussels should have siphons and tentacles extended with overall mortality rates <1%. Mussels should rapidly retract siphons in response to any external stimuli, including overhead shadows. Any loose mussels should be in the process of reattaching to a substrate or crawling about the aquarium hunting for a suitable site.

Once the 72-hr introductory period is over, any increase in the number of dead mussels can be directly related to four factors — increase in ammonia levels, water temperatures above 24°C, wrong water type, or lack of food (stage 2, Figure 2). Under laboratory conditions, zebra mussels are extremely sensitive to ammonia levels. Ammonia levels of 2 mg/L cause severe stress, and levels of 3 mg/L caused 90–100% mortality. Ammonia levels can rise to lethal concentrations within 24 hr. Changes in ammonia are usually due to either a dead mussel decomposing in the aquarium or a buildup of waste products. Daily testing of ammonia indicated that under conditions provided in this study, all aquariums had to be completely cleaned three times a week to keep ammonia levels below 1 mg/L.

Water temperatures above 24°C are another cause of mussel die-off. Mortality in the general maintenance stock held at 20°C and in the cold-room stock held at 4°C averaged less than 0.04%/week (mussels >10 mm in length) from May 1989 to August 1990. However, water temperatures in the general maintenance aquariums rose to 25°C for 1 week and 27°C for another week. The two aquariums without water flow showed an estimated 95% mortality within 24 hr, and an estimated 50% of the mussels in all other aquariums showed signs of moderate stress. Many of the mussels under 10 mm in length moved upward to the air-water interface. Mortality rates during the two separate weeks went from an average of 3 to 105 mussels per week. As soon as

TABLE 2. Calcium Depletion after 24, 48, and 72 hr in 80-L Aquariums Containing 1000 *Dreissena polymorpha* (>10 mm) Each

	Calcium (mg/L)		
	24 hr	48 hr	72 hr
Group A[a]	3	2	3
Group B[b]	4	3	4

[a] Had been in the laboratory for 6 weeks.
[b] Had been in the laboratory for 6 months.

temperatures dropped to 20°C, mortality rates decreased to an average of three per week. Dissolved oxygen was measured during these two events and remained above 8 mg/L.

The type of holding water plays a major role in the long-term survival of zebra mussels. Tests were conducted to determine the feasibility of using distilled water as a holding medium. A group of 100 individuals ranging in size from 5 to 20 mm showed very high mortality when isolated in a 4-L aquarium filled with distilled water. Mortality was 10% after 5 days, 25% after 10 days, 85% after 15 days, and 100% after 30 days. All mussels with a shell length <10 mm died within 1 week. None of the controls held in well water died in that time period. The lack of calcium in the distilled water (0 mg/L) was assumed to be the cause of mortality. Minimum aqueous concentrations of calcium are described in the literature as 30 mg/L for growth (Stanczykowska, 1977) and 40–50 mg/L for egg development (Sprung, 1987), although time limits for these experiments are not presented. Zebra mussels removed between 2 and 4 mg/L/24 hr/1000 mussels in the 80-L experimental aquariums (Table 2). Calcium depletion averaged 2 mg/L/24 hr/1000 mussels on mussels in the laboratory for 6 weeks that were not growing and 4 mg/L/24 hr/1000 mussels for those in the laboratory for 6 months with an average growth rate of 1.5 mm/month. If the amount of calcium depleted daily is projected on a monthly basis, all calcium would be removed from this aquarium in less than 30 days (assuming a growing population and no water replacement).

The formula for reconstituted fresh water for invertebrates presented in APHA (1989) will not support zebra mussels. The potassium concentration recommended in this recipe kills zebra mussels within 24 hr (Fisher, Ohio State University, Personal communication). However, the standardized fresh water prepared according to Sprung (1987) will support zebra mussels in all stages of their life cycle.

In general, zebra mussels are very tolerant of the continual handling and rapid change in physical surroundings that experimental animals are exposed to. For example, 100 mussels that were isolated in test tubes, removed daily with all byssus threads cut, dried out for 1 hr, and replaced in completely new water showed no signs of stress or mortality over a 6-week period. Mussels were open and siphoning within 15 min of being replaced in the tubes.

Feeding

Long-term survival of mussels for greater than 4 months in the laboratory is related to the type and quantity of food provided (stage 3, Figure 2; Table 3). Over a 1-year period, the only food for which survival remained at 100% was green algae (live or dead). Nonchlorophyllous foods (ABM MicroMac, bacterial infusions from boiled rice, hex mix, live and dead yeasts) proved unsuitable, although growth occurred on these foods for 2–3 months. One problem in determining success of a food is that zebra mussels take many months to starve to death. In the present experiments, mortality in mussels was apparent only after 5 months, and some survived for up to 11 months without food (Table 3). It is also difficult to determine the success of a diet by the external appearance of mussels. Mussels larger than 10 mm kept without food showed no visible signs of stress right up until they died, although Bieleseld (1991) has shown that internal changes occur within 10 days in mussels kept without food. A comparison of siphoning rates between starved and fed animals showed that starved mussels siphoned for an average of 20 out of 24 hr (range 18–23), and fed mussels siphoned 18 hr (range 16–21). There were no significant differences in siphoning rates between the fed group and the starved group or between single mussels and colonies. Mussels under 10 mm did not survive as long without food and did show signs of stress within 2 or 3 months at 20°C; most of the starved mussels that died in September and November were less than 10 mm (Table 3). Increased mortality or a reluctance to attach to a substrate occurred in mussels less than 10 mm kept at 4°C (cold-room stock was not fed) after 5–6 months. However, it appears that, for short-term experiments of less than 3–4 months, zebra mussels >10 mm can be handled with minimal feeding at either temperature without altering their basic behavior.

In diet acceptance tests, zebra mussels usually removed only minor amounts of the nonchlorophyllous food types after 24 hr, and long-term survival on these foods was poor (Table 3). The amount of a dried yeast suspension removed after a 24-hr period is shown in Figure 3; the amount removed was not significantly different from a control suspension without zebra mussels. The same basic pattern was also found for the other four nonchlorophyllous food sources.

Short-term experiments (24 hr) using *Chlamydomonas reinhardtii* as a food source indicated that zebra mussels willingly consume this food (Figure 4) and long-term tests showed 100% survival (Table 3). These results confirm findings by other researchers showing that green algae are a preferred food (Morton, 1971; Stanczykowska, 1977; Walz, 1978a and 1978b; Sprung and Rose, 1988). In the 24-hr tests, the initial concentration of 3.2×10^6 cells per milliliter was reduced in the first 4hr to 1.3×10^6 cells per milliliter. The 24-hr average algal consumption rate by a single mussel was 3.6×10^7

TABLE 3. Survival (Percent) and Mean Growth (mm) in Parentheses of *Dreissena polymorpha* on a Variety of Standard Diets Used in Invertebrate Culture or on No Food

	1989				1990			
	May	July	Sept.	Nov.	Jan.	March	May	July
No food	100 (0.5)	100 (0)	95 (0)	70 (0)	53 (0)	20 (0)	0 (—)	0 (—)
ABM	100 (1.0)	100 (0.6)	80 (0.2)	59 (0)	44 (0)	37 (0)	22 (0)	0 (—)
Bacteria	100 (0.7)	100 (0)	90 (0)	54 (0)	40 (0)	29 (0)	10 (0)	0 (—)
Hex mix	100 (0.5)	100 (0.5)	82 (0.3)	74 (0)	36 (0)	23 (0)	20 (0)	5 (0)
Live yeast	100 (0)	100 (0)	23 (0)	12 (0)	3 (0)	0 (—)	0 (—)	0 (—)
Dried yeast	100 (0.2)	95 (0.2)	71 (0.2)	62 (0)	62 (0)	43 (0)	12 (0)	0 (—)
Chlamydomonas	100 (1.0)	100 (1.0)	100 (1.5)	100 (1.6)	100 (1.8)	100 (1.8)	100 (1.8)	100 (1.8)
Spirulina	—	—	—	100 (1.0)	100 (0.7)	90 (0.7)	89 (0.4)	68 (0.3)
Chlorella	—	—	—	100 (0.4)	100 (1.1)	100 (1.3)	100 (1.5)	100 (2.0)

Note: Water temperature was a constant 20°C.

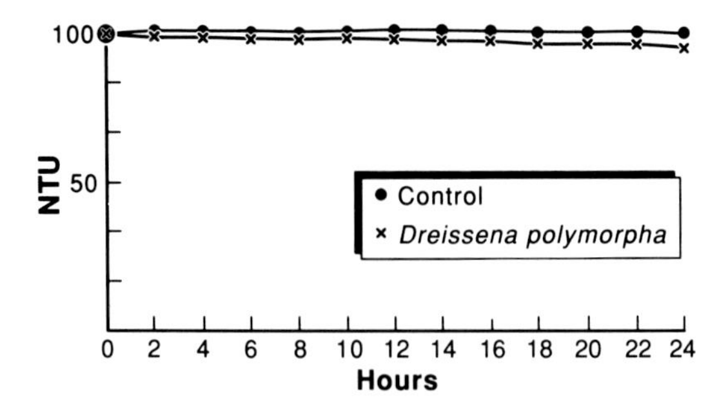

Figure 3. Consumption of dried yeast suspension by *Dreissena polymorpha* during a 24-hr feeding experiment. NTU are hephelometric turbidity units.

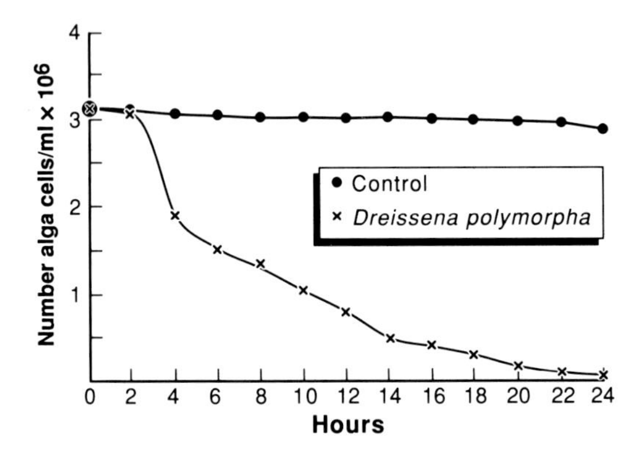

Figure 4. Consumption of a live green algae. *Chlamydomonas reinhardtii*, by *Dreissena polymorpha* during a 24-hr feeding experiment.

cells/hr (range 3.4–3.9 $\times$ 10^7 cells per hour). The average number of algal cells consumed per hour per mussel was the same in all three test replicates, regardless of number of mussels being tested. A rate of 3.6 $\times$ 10^7 cells per hour per mussel as determined from the 24-hr study was initially planned as a base feeding rate to check for growth and mortality. However, it proved impossible to provide this amount of live algae for the 7000 animals under cultivation.

Since the number of live algae required could not be provided and the nonchlorophyllous diets were not successful, two types of dried algae (*Spirulina* sp. and *Chlorella* sp.) were tested as food sources. Although *Spirulina* sp. is a blue-green alga with a chainlike form, visual examination of a blended suspension of this alga shows that about 50% of the particles are less than 50 μg in size and therefore could be eaten by the zebra mussels. However, in the short-term experiments, zebra mussels consumed less than 25% of the

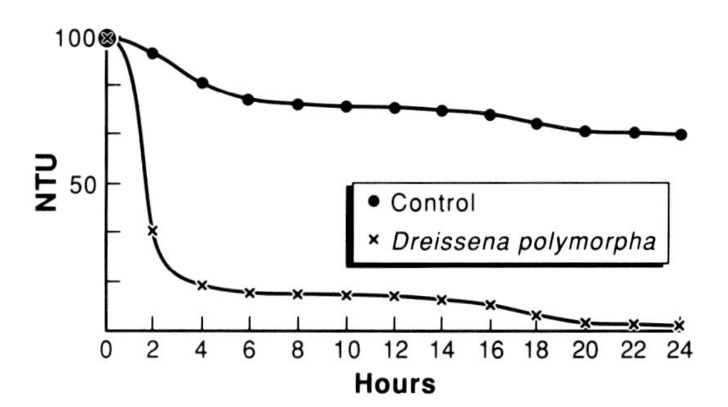

Figure 5. Consumption of a dried green algae. *Chlorella* sp., by *Dreissena polymorpha* during a 24-hr feeding experiment. NTU are nephelometric turbidity units.

Spirulina sp. suspension; and this consumption rate did not change regardless of the number of times they were exposed to the test food or the density of the food. This reluctance to feed in blue-green algae has been noted in natural waters (Stanczykowska, 1977). Another problem with using this blue-green alga even as a food supplement in the laboratory is that *Spirulina* sp. suspensions froth when exposed to water flow, which removes much of the algal particles from the water column. Mussels fed dried *Spirulina* sp. had greater long-term survival than those fed nonchlorophyllous feeds, but did not survive as well as those fed green algae (Table 3). Of the 100 mussels fed *Spirulina* sp. in November 1989, 68% were still alive in July 1990 (compared to 100% survival on live algae). However, growth rates were less than 0.3 mm/month, compared to nearly 2 mm/month on live or dried green algae.

The second type of dried alga tested, *Chlorella* sp., proved to be a readily accepted, easily handled food source. Both 24-hr and long-term diet experiments indicated that initial consumption and long-term survival were almost identical to those of *Chlamydomonas reinhardtii* (Figures 4 and 5; Table 3). Most of the *Chlorella* was removed from suspension by the mussels within the first 6 hr of the 24-hr experments (Figure 5). These results agree with feeding assimilation studies by others, which indicate that this dried alga is assimilated at rates equal to and even greater than live green algae (Garton, Ohio State University, Personal communication).

Large amounts of dried *Chlorella* sp. cannot be rapidly added to aquarium water without settling out of the water column and becoming unavailable to the mussels. However, it is possible to set up a drip-feeding mechanism for delivery of small portions of food over a period of several hours. Suspensions of dried *Chlorella* freeze with little alteration in cellular appearance when compared to nonfrozen samples (visual observation). The amount of algae that is to be fed during 24-hr period is split in half; each half is mixed with 500 mL of water, and is frozen in a styrofoam cup. One half of the day's ration (1 cup) is fed in the morning and the other half, in the afternoon. The frozen food suspension is removed from the cup and suspended in a funnel over the aquarium. This frozen food takes almost 8 hr to melt at room temperature (20°C). Despite the drying of about 10% of the algal material in the delivery process, it is a very convenient method that provides a continuous source of food for mussels with minimal required maintenance.

The amount of dried *Chlorella* sp. recommended for feeding in this study is presented as a guideline. Different laboratory conditions and overall physical status of the mussels will alter the amount of food required. If the mussels are given enough food, pseudofeces form, growth occurs, and mussels survive longer than 6 months. Although some authors (Morton, 1971; Sprung and Rose, 1988) indicate that pseudofecal formation is minimal on single-cell green algae, Walz (1978a) indicates that regardless of diet a mussel ingests about 44% of the food and extrudes the rest as pseudofeces. Under conditions used in this study, growth was minimal if colonies did not form pseudofeces (visual observation), and mortality (particularly of mussels less than 10 mm) increased after 5–6 months. Conversely, if the food ration was increased until pseudofecal formation occurred, growth rates increased and long-term survival was good. A ration of 3.2 g of dried *Chlorella* sp. per 1000 mussels daily produced pseudofeces, growth, and long-term survival.

CONCLUSIONS

After the initial 72-hr introductory period, zebra mussels readily adapt to laboratory conditions. They tolerate continual handling, frequent water changes, and minimal food. Long-term survival of the mussels is determined by the type and quantity of food provided. Although it would probably be better to mimic natural conditions by feeding mixed cultures of live algae, zebra mussels will survive and grow on monotypic cultures of live or dried green algae. Even on a diet of dried *Chlorella* sp. alone, colonies of mussels produced pseudofeces, grew, and survived for over 1 year.

ACKNOWLEDGMENTS

Contribution 811 of the National Fisheries Research Center-Great Lakes, Ann Arbor, MI.

REFERENCES

American Public Health Association (APHA). *Standard Methods for the Examination of Wastewater Water,* 17th ed. (Washington, DC: American Public Health Association, 1989).

Bielefeld, U. "Histological Observation of Gonads and Digestive Gland in Starving *Dreissena polymorpha* (Bivalvia)," *Malacologia* 33:31–42 (1991).

Griffiths, R., W. P. Kovalak, and D. W. Schloesser. "The Zebra Mussel, *Dreissena polymorpha* (Pallas, 1771), in North America: Impact on Raw Water Users," in Symposium: Service Water System Problems Affecting Safety-Related Equipment (Palo Alto, CA: Nuclear Power Division, Electric Power Research Insittute, 1989), 11–26.

Morton, B. "Studies on the Biology of *Dreissena polymorpha* Pall. V. Some Aspects of Filter-Feeding and the Effect of Micro-organisms upon the Rate of Filtration," *Proc. Malacol. Soc. London* 39:289–301 (1971).

Sprung, M. "Ecological Requirements of Developing *Dreissena polymorpha* Eggs," *Arch. Hydrobiol. Suppl.* 79:69–86 (1987).

Sprung, M., and V. Rose. "Influence of Food Size and Food Quality on the Feeding of the Mussel *Dreissena polymorpha*," *Oceologia (Berlin)* 77:526–532 (1988).

Stanczykowska, A. "Ecology of *Dreissena polymorpha* (Pall.) (Bivalvia) in Lakes," *Pol. Arch. Hydrobiol.* 24:461–530 (1977).

Walz, N. "The Energy Balance of the Freshwater Mussel *Dreissena polymorpha* Pallas in Laboratory Experiments and in Lake Constance. I. Pattern of Activity, Feeding, and Assimilation Efficiency," *Arch. Hydrobiol., Suppl.* 55:83–105 (1978a).

Walz, N. "The Energy Balance of the Freshwater Mussel *Dreissena polymorpha* Pallas in Laboratory Experiments and in Lake Constance. IV. Growth in Lake Constance," *Arch. Hydrobiol., Suppl.* 55:142–156 (1978b).

CHAPTER 46

Trematode Parasites of the Zebra Mussel (*Dreissena polymorpha*)

Cees Davids and Michiel H. S. Kraak

The trematode parasites of the zebra mussel (*Dreissena polymorpha*) are discussed. Two genera are of importance, *Phyllodistomum* and *Bucephalus*. The highest incidence of infestation found in *Dreissena* by *Phyllodistomum* was 10%. At most sample sites, however, it was only about 1%. Groups of sporocysts, containing metacercariae, are released from the mussels, float in the water, and are eaten by fish. Metacercariae fed to carp were found in the ureters as adult flukes. Seven weeks after infestation trematodes were sexually mature. Infestation by *Phyllodistomum folium* has a devastating effect on *Dreissena*. Infested mussels have lower dry weights and contain higher concentrations of heavy metals (Zn, Cu, Cd, and Pb) than noninfested mussels. The possible effects of trematode infestations on fish populations in dense *Dreissena* populations are discussed.

0-87371-696-5/93/$0.00 + $.50

INTRODUCTION

Mollusks, usually gastropods, are used by digenean trematodes as the first intermediate host. Cercariae, advanced larva stages, generally emerge from the first intermediate host and take up a short, free-swimming existence. Some encyst and become metacercariae on vegetation likely to be eaten by herbivorous final hosts. Others penetrate a second intermediate host, which may be either an invertebrate or a vertebrate. When the second intermediate host has been eaten by the carnivorous final host, the metacercariae develop into adults and sexual maturity is reached. Contrary to many other trematodes, cercariae belonging to the genus *Phyllodistomum* do not leave the first intermediate host but encyst within the sporocysts and transform into metacercariae.

A few species in the families Aspidogastridae, Gorgoderidae, and Bucephalidae use bivalves like the zebra mussel, *Dreissena polymorpha*, as the first intermediate host (Dawes, 1956). Most of the species are well-known parasites of fish. The effects on bivalves are much less studied. The aim of this study is to examine the effects of these trematodes on the bivalves which serve as intermediate hosts. Experimental data are presented for a species which belongs to the family Gorgoderidae. In addition, we reviewed the literature data on trematode parasites of the zebra mussel. Most of these studies originate from the first half of this century and consequently these data are relatively obscure and difficult to obtain.

METHODS

In 1983, initial observations of trematode sporocysts in the gills of zebra mussels were made on individuals collected in a canal near the village of St. Maartensvlotbrug, 60 km north of Amsterdam, The Netherlands. After a literature study and comparing the organisms with given descriptions, we came to the conclusion that the species must belong to the genus *Phyllodistomum*. This species lives as adults in the ureters of freshwater fish. In 1988, the incidence of infestation was determined by collecting a number of zebra mussels from near St. Maartensvlotbrug and also in Lake Maarsseveen, which is located near the city of Utrecht, some 35 km southeast of Amsterdam. Since species-level identification of trematodes is only possible in the adult stage, young carp (*Cyprinus carpio* L.) were infested by feeding them mussels with sporocysts and also feeding them pieces of mussel gills containing metacercariae. Every 2 weeks a number of carp were sacrificed to study the parasitic infection. Carp used in these experiments were 10 cm long and bred from eggs; they were, therefore, free from trematode parasites prior to the experiments.

Figure 1. Gill of *Dreissena polymorpha* infested with sporocysts of *Phyllodistomum folium.*

RESULTS

Zebra mussels collected in 1983 near St. Maartensvlotbrug had an incidence of infestation of about 10% (n = 150). Mussels collected in Lake Maarsseveen in 1988 had an incidence of infestation of 1.4% (n = 1400). Infestations of *Phyllodistomum* in zebra mussels were found from other water bodies as well, with the exception of Lake Markermeer, which is a part of the former Zuiderzee. Seasonal patterns of infestation were not observed in any of the water bodies.

In infested mussels, small yellowish stripes could be observed in gills with the unaided eye (Figure 1). After closer examination, sporocysts could be discerned in different stages of development. Sporocysts containing metacercariae had a mean length of 3.6 mm (range 2.0–6.1 mm) and a mean width of 0.7 mm (range 0.59–0.92) (n = 24). Most observed sporocysts were in the final stage, each containing eight metacercariae.

A number of mussels were kept in tanks; during the following 2 weeks, cloudy and whitish lumps of tissue were found floating in the tanks. These appeared to be groups of sporocysts containing metacercariae that were released from the gills. Metacercariae freed from these lumps of tissue and from the gills unrolled themselves and started creeping as juvenile trematodes. Mean length and width of these juveniles were 1.1 mm (range 0.9–1.9 m) and 0.2 mm (range 0.2–0.3 mm), respectively (n = 33).

Two weeks after infestation, several juvenile trematodes were found in the ureters of the carp. The characteristic shape of adult flukes could already

TABLE 1. Mean Concentration (μg/g dry weight) of Zinc, Copper, Cadmium, and Lead in Gills and Total Soft Tissues of Zebra Mussels Infested with *Phyllodistomum folium* and in Noninfested Mussels (n = 31)

	Infested		Noninfested	
Metal	Gill	Soft Tissue	Gill	Soft Tissue
Zinc	122.0	105.0	102.0	95.0
Copper	12.8	14.9	9.2	12.4
Cadmium	2.6	2.8	1.3	1.9
Lead	5.9	13.0	4.1	7.9

be observed. After 7 weeks the flukes started to produce eggs. By the characteristic shape of the testes and ovary, the trematode could be identified with the information in Dawes (1956) as *Phyllodistomum macrocotyle* (Luhe). According to the keys of Pavlovskii (1964), however, the trematodes were identified as *P. folium* (Olfers) (*P. macrocotyle* is not included by Pavlovskii). Luhe (1909) mentioned that the number of metacercariae per sporocyst was 8 in *P. folium* and 12–14 in *P. macrocotyle*. Based on this information, the trematodes had to be *P. folium*. Taking into account the knowledge about infestations by miracidia, it is highly probable that all these daughter sporocysts are the descendants of one miracidium.

Devastation of the mussel gill is clear from Figure 1. The function of the gills must be hampered severely considering the loss of weight. Total dry weight of infested mussels decreased dramatically compared to noninfested ones. In mussels with a length of 16–20 mm, the dry weight of soft tissues decreased from 15.0 (S.D. 4.4, n = 31) to 10.2 mg (S.D. 3.7, n = 19) per mussel. In addition, under macroscopic examination, the gonad tissue of infested mussels was very reduced compared to noninfested mussels, indicating the fecundity of infested mussels may also be drastically affected. The decrease in gonads was likely related to the loss of gill function and the subsequent decrease in filtering potential and food intake.

Infested and noninfested mussels from Lake Maarsseveen were analyzed for the heavy metals Cu, Zn, Cd, and Pb. As Table 1 shows, concentrations of these metals in total soft tissues and specifically in the gills were higher in infested than in noninfested mussels (Kraak and Davids, 1991). In the infested state, mussels appear to be weakened by a biotic stress factor and contain higher concentrations of metals than healthy individuals do.

DISCUSSION

Kluczycka (1939), Wisniewski (1958), Kinkelin et al. (1968), and Baturo (1977) have surveyed parasites of zebra mussels in various water bodies. There is some evidence (Wisniewski, 1958; Stanczykowska, 1977) that mussels are also used as a second intermediate host by encysting cercariae (e.g., cercariae of *Echinostomum* sp.). In the second intermediate host, metacercariae can be seen as resting stages, linked between the first intermediate host and the final host. It is assumed that the detrimental effects of metacercariae on their second intermediate hosts are, in general, rather insignificant. Therefore, this review is limited to parasites using zebra mussels as the first intermediate host and, subsequently, having a detrimental effect on them.

Aspidogastridae

Aspidogaster conchicola (Family Aspidogastridae, Subclass Aspidobothrea), a small to medium-sized trematode (ca. 2.5 mm in length) with the ventral surface of the body forming a large alveolated sucker, is a common parasite of species in the family Unionidae. The entire life cycle occurs in the mussel host (Bakker and Davids, 1973). The fluke can be found in the pericard and renal cavities feeding on hemolymph. Only Kulczycka (1939) reported an infestation of *A. conchicola* in *Dreissena*. The significance of zebra mussels as a host for *A. conchicola* is unknown.

Gorgoderidae

The species *Phyllodistomum folium* belongs to the family Gorgoderidae; our findings clearly show the deleterious effects of this parasite on the zebra mussel, that is, total dry weight decreased dramatically probably as a result of damage to the gills. In agreement with these results, Brown and Pascoe (1989) observed that *Gammarus pulex* infested with the acanthocephalan *Pomphorhynchus laevis* consumed only 17–21% of the food eaten by non-infested animals. Pennycuick (1971) demonstrated that the condition factor (weight/length) was lower in three-spined sticklebacks (*Gasterosteus aculeatus*) infested with three different parasite species than in healthy animals.

A number of parasitic species of the Gorgoderidae have been described (Dawes, 1956; Yamaguti, 1958). It is known that *P. macrocotyle* (Luhe, 1909; Odhner, 1911) and *P. folium* (Olfers, 1817) use *Dreissena*, as well as species belonging to the family Unionidae, as the first intermediate host. The intermediate hosts of other *Phyllodistomum* spp. are relatively unknown. Thomas (1956) found sporocysts of *P. simile* Nybelin between the gill lamellae of the fingernail clam *Sphaerium corneum* L. and the adult fluke in rainbow trout *Salmo trutta* L.

P. folium is a small-sized trematode, whose body is divided into a narrowed anterior motile part and a widened posterior part. It usually parasitizes the urinary bladder and ureters of fish and other vertebrates. Ssinitzin (1901) gave a first description of the development of *Distomum* (= *Phyllodistomum*) in *Dreissena*. It is assumed that miracidia enter the mantle cavity via the inhalent siphon. Next, the miracidium attaches to the gills; and after shedding the ciliated covering, it transforms directly into a saclike body, the mother sporocyst. The growing sporocyst reaches a length of 0.5–1.4 mm and daughter sporocysts develop from 12 to 14 germinal cells. These sporocysts penetrate the sporocyst wall and spread themselves out in the gill tissue. These daughter sporocysts can give rise to a second, third, or even a fourth generation of daughter sporocysts until all space in the gills is filled with daughter sporocysts. The total number of sporocysts per mussel varies between 200 and 300. Cercariae are formed, but they do not move and encyst as metacercariae. Finally, the sporocysts will crowd out at the free edge of the gills and leave the mussel via the exhalant siphon. Floating groups of metacercariae will then be taken by fish as food. According to Ssinitzin (1901), sporocysts with metacercariae have a length of up to 8 mm. This is larger than in our observations. He also found 12–14 metacercariae per sporocyst while we found only 8. This indicates that Ssinitzin was dealing with another species. Incidence of infestation in mussels was 10%. Ssinitzin (1901) infested carp and bream by offering sporocysts as food. One hour after infestation, he sacrificed two fishes and found juvenile trematodes in the midgut. After 24 hours, trematodes were found in the ureters. Obviously, trematodes creep through the gut and reach the opening of the ureters. Two weeks later, the trematodes had developed gonads but did not have egg production. From our observations, egg production starts within 7 weeks at a temperature of 13°C. We found one literature source (Beilfuss, 1954) concerning the time required between miracidial entrance and cercarial production; this source indicated that the process takes about 2–3 months in *P. caudatum*.

Adult *Phyllodistomum* spp. are known from the urinary bladder and ureters of many European fishes including members of the families Salmonidae, Cyprinidae, Esocidae, Percidae, and Cottidae. Several species of this trematode genus are also known from fish species in North America, including fishes of the Great Lakes (Fischtal, 1942; Hoffman, 1967). There are no literature data on the effects of *Phyllodistomum* spp. on their final hosts.

Miracidia of *Phyllodistomum folium* can apparently penetrate *Dreissena* as well as species belonging to the family Unionidae. Consequently, it is possible that American species of *Phyllodistomum*, having unionids as a first intermediate host, can also use *Dreissena* as a host. Dense populations of the first intermediate host can increase the incidence and intensity of infestation of the final host.

Bucephalidae

Two species are of special interest: *Bucephalus polymorphus* (Baer, 1827) and *Rhipidicotyle illense* (Vejnar, 1956; Figure 2). Sporocysts of *Bucephalus* can be found in the gonad and sometimes also in the digestive gland and gills

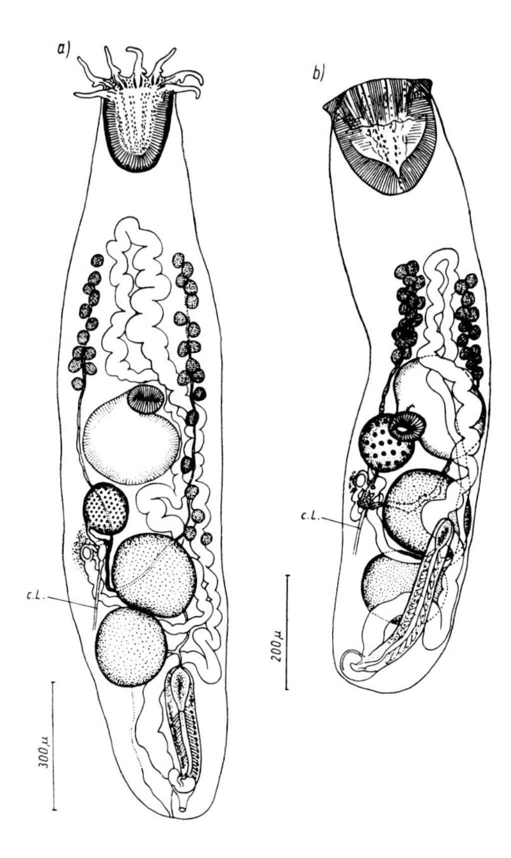

Figure 2. Trematode parasites of the zebra mussel, (a) *Bucephalus polymorphus*; (b) *Rhipidicotyle illense*; ventral view c.L. Laurer's canal. (From Kozicka, J. *Acta Parasitol. Pol.* 7:1–68 [1959]. With permission.)

of *Dreissena polymorpha*. Gonads of infested individuals are infested to the point where sterilization is a possibility. The color of the sporocyst is whitish, bearing numerous branches and protrusions and contains cercariae at different stages of development (Baturo, 1979). Furcocercariae develop from the sporocyst and swim around in search for the second intermediate host, a fish. They then encyst in the fins, gills, or mouth and transform into metacercariae. They are still alive 20 months after infestation. Metacercariae are rather harmless if few in number. In a mass, however, they can be dangerous to the second intermediate host (Wallet and Lambert, 1986). Wallet et al. (1985) studied the shedding pattern of the cercariae. The second intermediate host must be swallowed by the definitive host. The parasite reaches sexual maturity in the gut of the latter.

Cercariae of *Bucephalus* are characterized by two long furcae arising from a large bulbous structure. Baturo (1977) gave a good description of the cer-

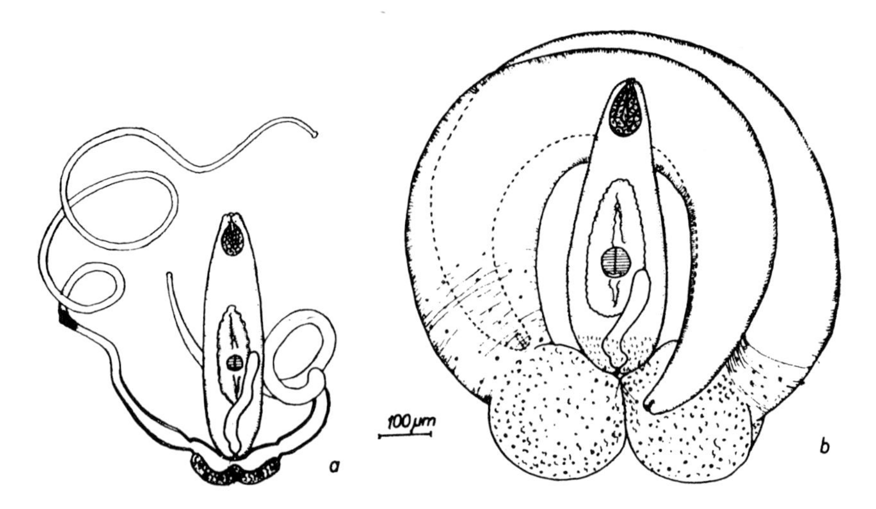

Figure 3. (a) Cercariae of *Bucephalus polymorphus*; (b) cercariae of *Rhipidicotyle illense*. (From Baturo, B. *Acta Parasitol. Pol.* 24:203–220 (1977). With permission.)

cariae and the life cycles of *Bucephalus polymorphus* and *Rhipidocotyle illense* (Figure 3). Based on experimental results, Baturo found that cercaria originally described as *Bucephalus polymorphus*, and until then regarded as a cercaria of the very common fish trematode under the same name, is in actuality a larval stage of *Rhipidocotyle illense*. Baturo (1979) discussed the highly complicated problem of taxonomy and synonymy of both species. The trematodes were found to show a narrow specificity to their first intermediate hosts. She demonstrated that *Bucephalus* sporocysts can be found in *Dreissena* while those of *Rhipidocotyle* can be found in unionids.

Wallet and Lambert (1986) gave epidemiological evidence of the larval infection of fish by *Bucephalus polymorphus*, in southeast France. Kinkelin et al. (1968) pointed out that an explosion of cercariae attacking fish is highly pathological, causing necrotic spots and lesons and resulting in weakening or death of the fish. They reported that in the Seine basin up to 95% of the fish were infested; maximum number of metacercarial cysts found in a single fish was over 10,000. This high rate of fish infestation was attributed to the high number of zebra mussels in this region; the incidence of mussel infestation was between 15 and 40%. Other literature sources indicate that the infestation intensity in fish is usually lower than 100 cysts per individual. According to Baturo (1979), the incidence of infection of *Dreissena* with *Bucephalus* sporocysts was 0.6–3.7% in Polish lakes. The incidence of infestation in fish (e.g., *Rutilus, Scardinius, Blicca,* and *Alburnus*) was 22 to 87%.

Metacercariae can be found underneath the epithelium of the gills, mouth and fins of fish — the second intermediate host. Infestation in various families are known. Final hosts are fish predators. These fish species are of the genera: *Lucioperca, Perca, Anguilla, Esox,* and *Acerina* (Wisniewski, 1958; Luhe, 1909). The pike perch (*Lucioperca*) seems to be very susceptible (Kinkelin et al., 1968).

Kozicka (1959) found *Rhipidicotyle illense* to be the dominant fish parasite at Lake Druzno, while *Bucephalus polymorphus* was rarely found. Several hundred (i.e., 300–600) of *Rhipidicotyle illense* were found in infested *Esox* and up to a hundred in *Perca*.

In the Great Lakes region of North America, adult flukes of *Bucephalus elegans* occur commonly in the cecal pouches of rock bass (*Ambloplites rupestris*) (Woodhead, 1930). First intermediate host are the unionids *Eurinia* and *Elliptio*. If certain bucephalids can use not only native unionids as a intermediate host but also *Dreissena*, the establishment of zebra mussel populations in the Great Lakes can bring about an explosive development of parasitic infestation in fish (Wallet et al., 1985). If there is a narrow specificity for their first intermediate host, as for *Bucephalus polymorphus* in Europe, the zebra mussel will not be infested by other species of *Bucephalus* and related genera in the Great Lakes.

ACKNOWLEDGMENTS

The authors gratefully acknowledge the editor of the *Acta Parasitologica Polonica* for the permission to reproduce figures of Kozicka and Baturo. Special thanks are due to Chris de Groot for his practical assistance.

REFERENCES

Bakker, K. E., and C. Davids. "Notes on the Life History of *Aspidogasterconchicola* Baer, 1826 (Trematoda:Aspidogastridae)," *J. Helminthol.* 47:269–276 (1973).

Baturo, B. "*Bucephalus polymorphus* Baer, 1827 and *Rhipidicotyle illense* (Ziegler, 1883) (Trematoda:Bucephalidae): Morphology and Biology of Developmental Stages," *Acta Parasitol. Pol.* 24:203–220 (1977).

Baturo, B. "*Bucephalus* Baer, 1827, and *B. polymorphus* Baer, 1827 (Trematoda); Proposed Use of the Plenary Powers to Conserve These Names in Accordance with General Use, Z.N.(S.) 2251," *Bull. Zool. Nomenclature* 36:30–36 (1979).

Beilfuss, E. R. "The Life Histories of *Phyllodistomum lorenzi* Loewen, 1935, and *P. caudatum* Steelman, 1938 (Trematoda:Gorgoderinae)." *J. Parasitol.* 40 Sect. 2 Suppl. 44 (1954).

Brown, A. F., and D. Pascoe. "Parasitism and Host Sensitivity to Cadmium: An Acanthocephalan Infection of the Freshwater Amphipod *Gammarus pulex*," *J. Appl. Ecol.* 26:473–487 (1989).

Dawes, B. "*The Trematodes*" (London, England: Cambridge University Press, 1956).

Fischtal, J. H. "Three New Species of *Phyllodistomum* (Trematoda:Gorgoderidae) from Michigan Fishes," *J. Parasitol.* 28:268–275 (1942).

Hoffman, G. L. *Parasites of North American Freshwater Fishes* (Berkeley, CA: California Press, 1967).

Kinkelin, P., G. Tuffery, G. Leynaud, and J. Arrinnon. "Etude epizootiologique de la bucephalose larvaire a *Bucephalus polymorphus*, (Baer 1827) dans le peuplement piscicole du bassin de la Seine," *Rech. Vet.* 1:77–98 (1968).

Kozicka, J. "Parasites of Fishes of Druzno Lake," *Acta Parasitol. Pol.* 7:1–68 (1959).

Kraak, M. H. S., and C. Davids. "The Effect of the Parasite *Phyllodistomum folium* (Trematoda) on Heavy Metal Concentrations in the Freshwater Mussel *Dreissena polymorpha*," *Neth. J. Zool.* 41:69–76 (1991).

Kulczycka, A. "Contributions a l'etude des formes larvaires des Trematodes chez les Lamellibranches aux environs de Warszawa," *C.R. Seances Soc. Sci. Lett. Varsovie Cl.* 4(32):80–82 (1939) (In French).

Luhe, M. "Parasitische Platwurmer I: Trematoda," in *Die Susswasserfauna Deutschlands*, Brauer, A., Ed. Heft 17 (Verlag von Gustav, Jena., 1991).

Pavlovskii, E. N. "Key to Parasites of Freshwater Fish of the U.S.S.R.," in Keys to the Fauna of the U.S.S.R. Academy of Sciences of the U.S.S.R. Israel Programs for Scientific Translations, No. 80, Jerusalem (1964).

Pennycuick, L. "Quantitative Effects of Three Species of Parasites on a Population of Three-Spined Sticklebacks, *Gasterosteus aculeatus*," *J. Zool.* 165:143–162 (1971).

Ssinitzin, D. Th. "Einige beobachtungen uber die entwicklungsgeschichte von *Distomum folium* Olf," *Zool. Anz.* 24:689–694 (1901).

Stanczykowska, A. "Ecology of *Dreissena polymorpha* (Pall). (Bivalvia) in Lakes," *Pol. Arch. Hydrobiol.* 24:461–530 (1977).

Thomas, J. D. "Life History of *Phyllodistomum simile* Nybelin, *Nature (London)* 178:1004 (1956).

Wallet, M., and A. Lambert. "Enquete sur la repartition et l'evolution du parasitisme a *Bucephalus polymorphus* Baer, 1827 chez le mollusque *Dreissena polymorpha* dans le sud-est de la France," *Bull. Fr. Piscic.* 300:19–24 (1986).

Wallet, M., A. Theron, and A. Lambert. "Rythme d'emission des cercaires de *Bucephalus polymorphus* Baer, 127 (Trematoda, Bucephalidae) en relation avec l'activite de *Dreissena polymorpha* (Lamellibranche, Dreissenidae) premier hote intermediaire," *Ann. Parasitol. Hum. Comp.* 60:675–684 (1985).

Wisniewski, W. L. "Characterization of the Parasitofauna of an Eutrophic Lake," *Acta Parasitol. Pol.* 6:1–61 (1958).

Woodhead, A. E. "Life History Studies on the Trematode Family Bucephalidae. No. II," *Trans. Am. Microsc. Soc.* 49:1–17 (1930).

Yamaguti, S. *Systema Heminthum, Vol. 1, Digenetic Trematodes* (New York: Interscience, 1958).

CHAPTER 47

A Photographic Guide to the Identification of Larval Stages of the Zebra Mussel (*Dreissena polymorpha*)

Gordon J. Hopkins and Joseph H. Leach

The recent invasion of the zebra mussel, *Dreissena polymorpha*, into North America has created concern about the spread of this organism and its impact on water users and aquatic invertebrate, plant and fish communities. Early identification of the organism is important to managers who are developing prevention and control strategies for hydrotechnical installations and to researchers planning ecological impact studies. Because the early life history stages of *D. polymorpha* are planktonic, they can be easily collected with a fine-mesh net and identified at relatively low microscopic magnification. This chapter represents a photographic guide to the morphology and sizes of various larval stages of zebra mussels found in the plankton of western Lake Erie.

0-87371-696-5/93/$0.00 + $.50

INTRODUCTION

Early identification of zebra mussels (*Dreissena polymorpha*) is an important first step in planning ecological studies and prevention and control programs for the organism in water pumping installations. This chapter provides an outline of the life history stages and a photographic aid to the identification of planktonic larval stages of the zebra mussel.

METHODS

Larvae (veligers) were collected in August 1989 from the western basin of Lake Erie with a 50-cm diameter conical net having a mesh size of 76 μm. Observations were made on: (1) fresh samples, (2) samples preserved with 4% sugared formalin (Haney and Hall, 1973), and (3) samples preserved with Lugol's iodine solution. Lengths of larval stages were measured at magnification × 40 with a digitizing pad (Roff and Hopcroft, 1986). Specimens were isolated for observation and photography by placing several drops of sample containing larvae on a depression slide.

Photographs were taken using Kodak color slide film (ASA 400) with a Nikon Microflex AFX photographic attachment (Nikon Nippon Kogaku K.K.) fitted to a Leitz Dialux compound microscope (Ernst Leitz GMBH Wetzler). The photographs were processed as black and white glossy prints.

LIFE HISTORY STAGES

The life cycle of the zebra mussel has been studied extensively in Europe (e.g., Stanczykowska, 1977) with the larval phase addressed specifically by Kirpichenko (1964), Lewandowski (1982), and Sprung (1987, 1989, 1990). A summary of Sprung's interpretation of the early life history stages of the zebra mussel is shown in Table 1.

Figure 1 is reproduced from Mackie et al. (1989) and presents a useful summary and comparison of the life cycle of *Dreissena polymorpha* from egg, through the veliger, postveliger, and adult stages. In the early stage, the veliger looks like a rotifer and swims upward with the aid of a ciliated velum. The early shell is D-shaped and is easily identified. An intestine is formed at this stage. In the postveliger stage, the umbonal hinge and the foot develop as the shell enlarges. In the settling stage, metamorphosis leads to a disappearance of the velum, development of labial palps, and secretion of byssus threads. Figure 1 provides an overall perspective to the ensuing photomicrographs.

TABLE 1. Summary of Early Life History Stages of *Dreissena polymorpha*

Period	Stage	Length (μm)	Approximate Time to Complete Period	Features
Lecithotrophic	Fertilized egg to trochophore	70	6–20 hr	Cilia and velum formed; begins to swim
Planktotrophic	Velichoncha (shelled veliger)	90–100	2–4 d	D-shaped shell, intestine formed, begins to filter feed
Settling	Postveliger	220–300	8–35 d	Foot developed

Source: Sprung, 1989; Walz, 1973 and 1975.

Measurements of larvae from western Lake Erie, preserved in sugared formalin, are given in Table 2. The trochophore and D-shaped veligers are distinct, and have the narrowest ranges of size. All larvae without straight-hinged shells and with an umbo were included in the postveliger stage. The size range in this group is wide (150–263 μm), and the largest members are in the settling phase. In 1990, zebra mussel larvae settled on offshore settling chambers in western Lake Erie were about 250 μm in length (Leach, Unpublished data); this length agrees with those reported from Europe (Table 1).

PHOTOMICROGRAPHY

Microscopic examination of freshwater plankton samples for zebra mussel larvae will invariably include many other organisms with similar size ranges as veligers. Ciliated protozoa, rotifers, and other microcrustaceans could easily be confused with resting veligers (Figures 3, 8, and 10 include some of these other planktonic forms). If there is any doubt or confusion with ciliated protozoa or constricted rotifers, confirmation may be made by moving up to a higher magnification (× 120–150) or possibly by probing the veliger to reposition it into a different view. It would be advisable for the novice to obtain preserved veliger specimens and have live veliger samples confirmed by someone with experience. Veligers may be readily identified at magnification × 50–70.

Figures 2 and 3 give lateral views of a live trochophore veliger. At this stage the veliger is very active and motile, and the velum may or may not be extended. The cilia are in constant motion so that only a few will be in focus (Figure 2 — 1 o'clock position). The velum (Figure 2) is a densely flattened extension crowned with cilia. In Figure 3, the size of an early trochophore may be compared to the diatom *Asterionella formosa* (A — upper left corner of figure) and a shelled ciliated protozoan (3 o'clock). Figure 4 gives a lateral view of a single live planktotropic veliger, approximately

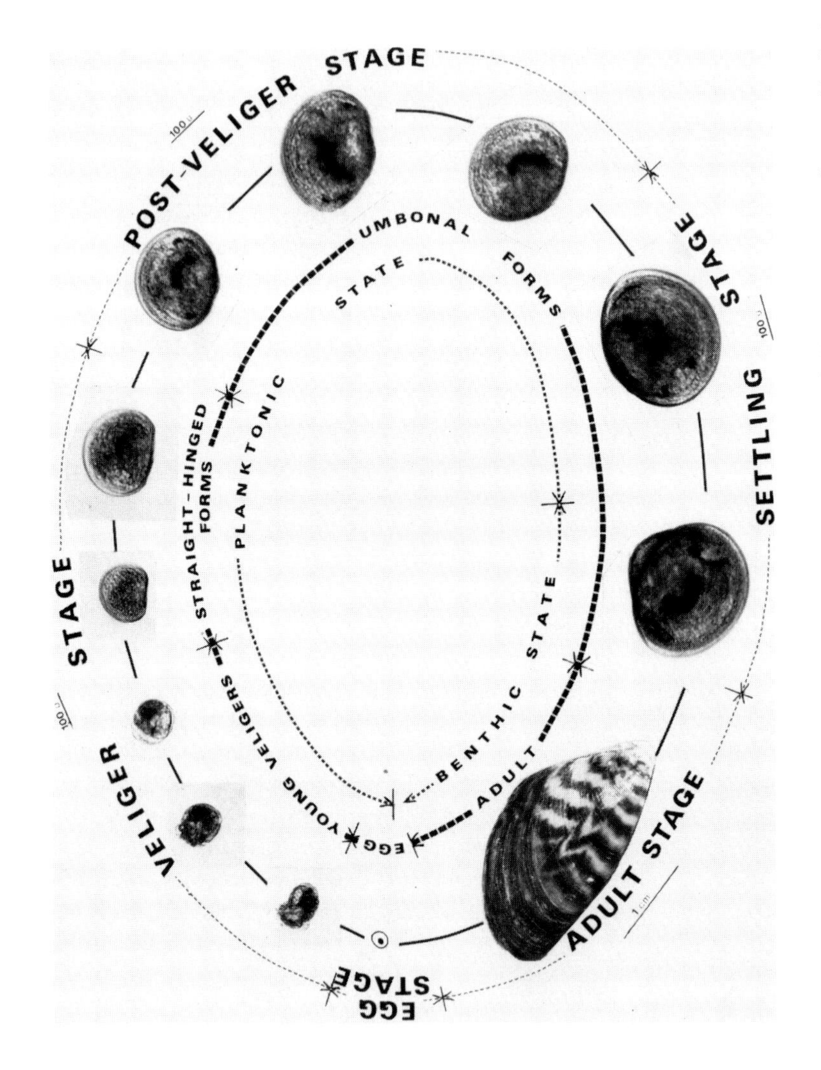

Figure 1. Life cycle stages of *Dreissena polymorpha*. Time periods for approximate life history stages are discussed in the text. (Photograph courtesy of Dr. Gerry Mackie, University of Guelph.)

TABLE 2. Lengths (μm) of Larval Stages of *Dreissena polymorpha* from Western Lake Erie 1990

Stage	Mean Length	Standard Deviation	n	Maximum Length	Minimum Length
Trochophore	81.7	12.4	50	108	59
D-shaped veliger	115.2	15.1	50	148	84
Postveliger	196.9	26.1	50	263	150

150 μm in size. The upper and lower edge of the hinged shell may be observed with the velum projecting out over two thirds of the open edge showing numerous cilia in focus.

Figure 5 gives an anterior view of a live postveliger (approximately 180 μm in size) with the hinged edge of the shell facing the camera and slightly open. The shell (S) is slightly out of focus but the velum (V) and cilia (C) are fully extended. Figure 6 is a lateral view (L) of a live planktotrophic (D-shaped shell) veliger wide open with the muscular velum fully extended. In this extended position, the velum is quite transparent and flows out in an amoeboid fashion.

Preserved specimens in various stages of development, as shown in Figure 7, may be separated into veliger and postveliger stages based on size, structural outline, and density of internal protoplasm. In this photo (taken at magnification × 70) the V arrows point to the small D-shelled veligers which are less than 100 μm and are straight hinged. The larger postveligers (Pv) are developing the umbonal hinge. There is one postveliger to the left of the scale bar in the anterior position. It may also be noted in this photograph that the planktonic alga *Pediastrum* (Pe) (five specimens) and the many filaments of *Melosira* (M) have much less pigmentation than the veligers.

Figures 8 shows a planktotrophic straight-hinged form preserved in Lugol's iodine solution. The killing action of this preservative causes the veliger to open up, the velum to disintegrate, and the cilia to detach. In this figure, there are several other planktonic forms which serve as guides to size comparisons (A — *Asterionella formosa*; M — *Melosira*; Pe — *Pediastrum*; Po — *Polyarthra*).

Figure 9 shows preserved specimens (Lugol's solution) at magnification × 200. Note the distinct D-shell form of the veliger (V) and the large umbonal hinge that can be readily observed on the postveliger (Pv). Note also the discharged fatty globules and protoplasmic material from the ruptured velum in the D-shell veliger.

Figure 10 shows a dense collection of veligers, postveligers, phytoplankton, rotifers, and microcrustaceans in a sugar-formalin preserved sample. Note the two large *Cyclops* (Cy). The filaments in this figure are *Melosira* sp.

Figure 11 shows a well-developed postveliger at the settling stage with a protruding foot (F); Figure 12 shows two young adult zebra mussels, the smaller of which has the foot (F) protruding with a postveliger (Pv) attached to it.

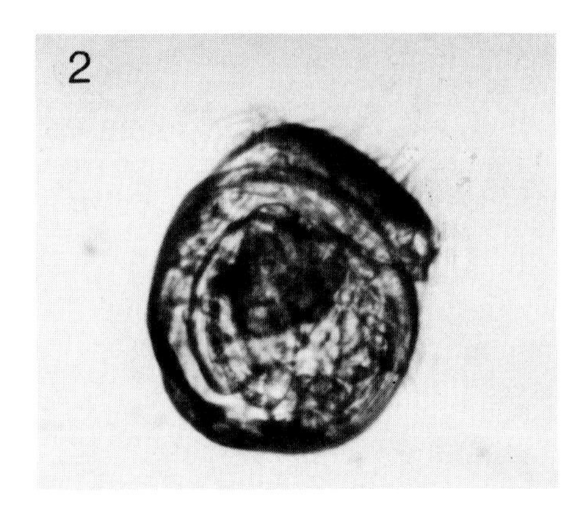

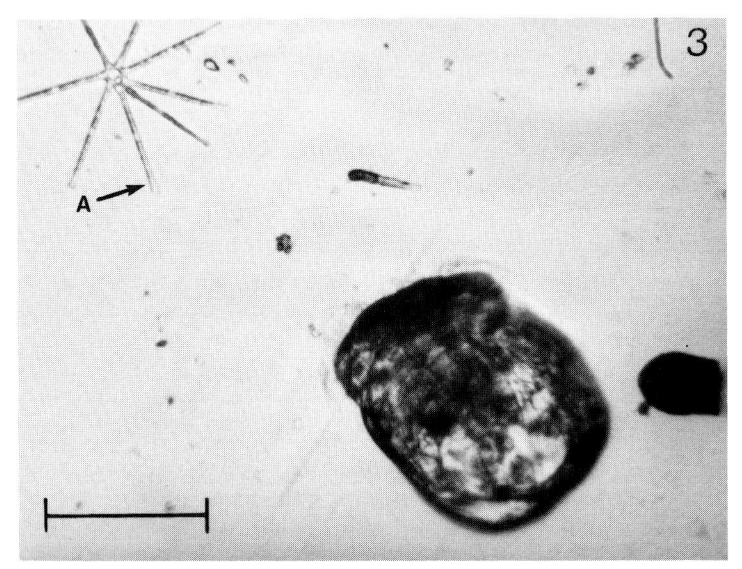

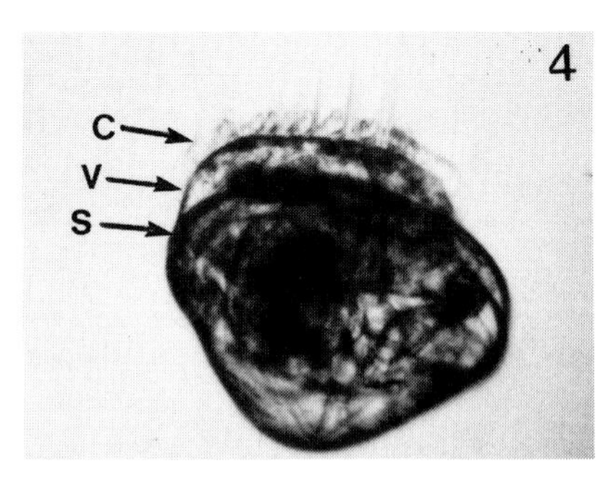

Figures 2, 3, 4. Live *Dreissena polymorpha* veligers and postveligers. All lateral views. A, Diatom *Asterionella formosa; C, cilia; V, velum; S, shell. (Bar = 100 μm.)*

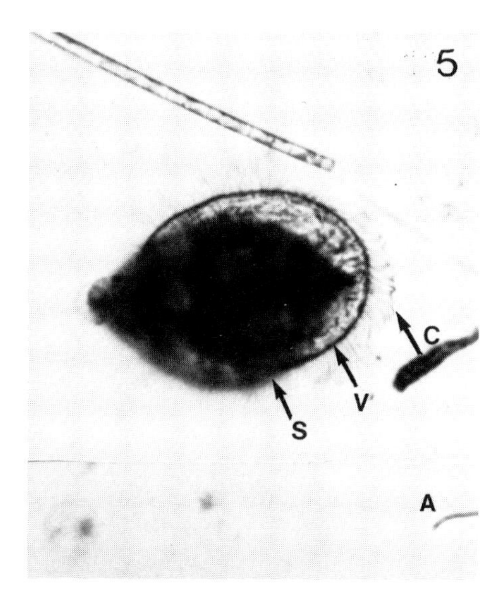

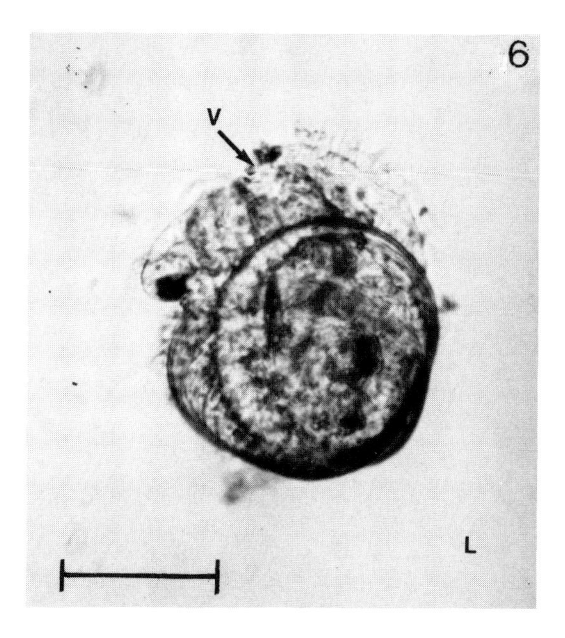

Figures 5, 6. Live *Dreissena polymorpha* veligers and postveligers. Figure 5 anterior (A) view of postveliger with cilia extended. Figure 6 lateral (L) view of postveliger with velum fully extended; C, cilia; S, shell; V, velum. (Bar = 100 μm.)

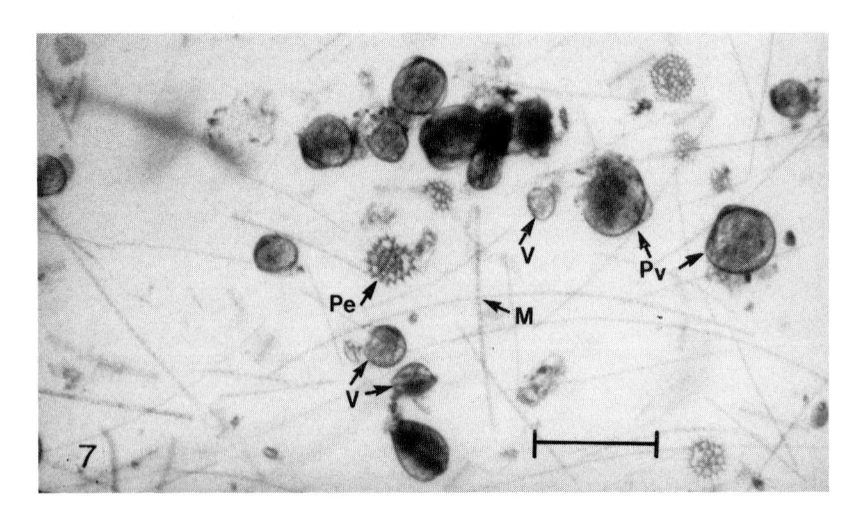

Figure 7. Preserved plankton sample containing veligers (V), postveligers (Pv), *Pediastrum* (Pe), and *Melosira* (M). (Bar = 200 μm.)

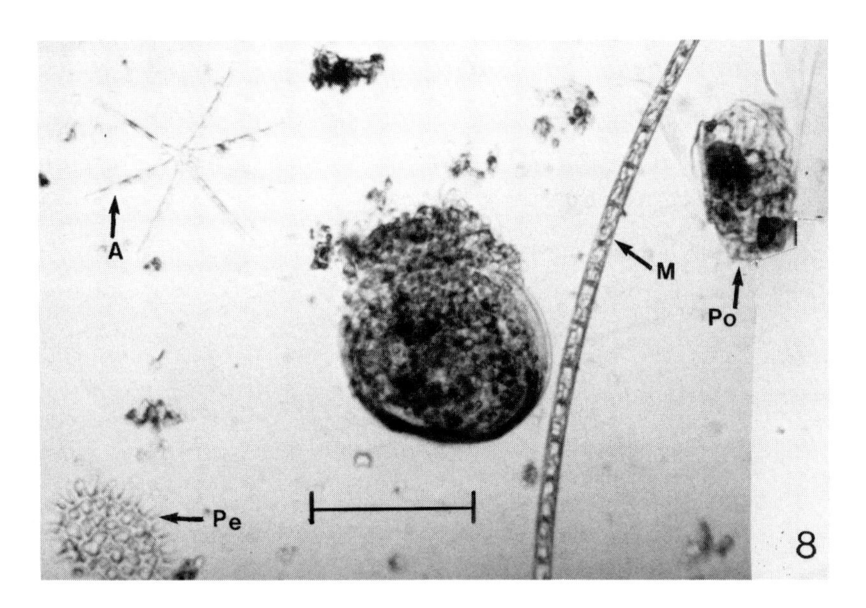

Figure 8. Veligers "preserved" with Lugol's solution. Note ruptured velum and dislodged cilia. A, *Asterionella formosa*; M, *Melosira*; Pe, *Pediastrum*; Po, *Polyarthra* (a rotifer). (Bar = 100 μm.)

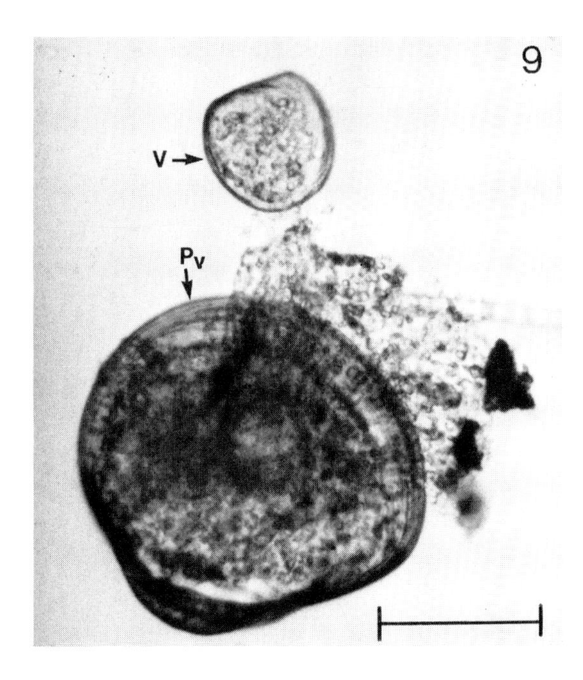

Figure 9. *Dreissena polymorpha* veligers and postveligers, "preserved" with Lugol's solution. Note ruptured velum with spherical fatty globules extruding from shell. V, veliger; Pv, postveliger. (Bar = 100 μm.)

It is hoped that the explanation here will assist others in recognizing larvae of the zebra mussel in areas of North America where few people are familiar with this exotic organism.

ACKNOWLEDGMENTS

We express our appreciation to Ken Nicholls for reviewing the manuscript and to Nancy Robson and Grant Gaspari for typing it. Special thanks to Dr. G. Mackie for providing us with the excellent life cycle photograph. The photomicrographs included here are from an earlier technical report printed by the Ontario Ministry of the Environment (Hopkins, 1990). Contribution No. 91-15 of the Ontario Ministry of Natural Resources, Research Section, Fisheries Branch, Box 5000, Maple, Ontario, Canada L6A 1S9.

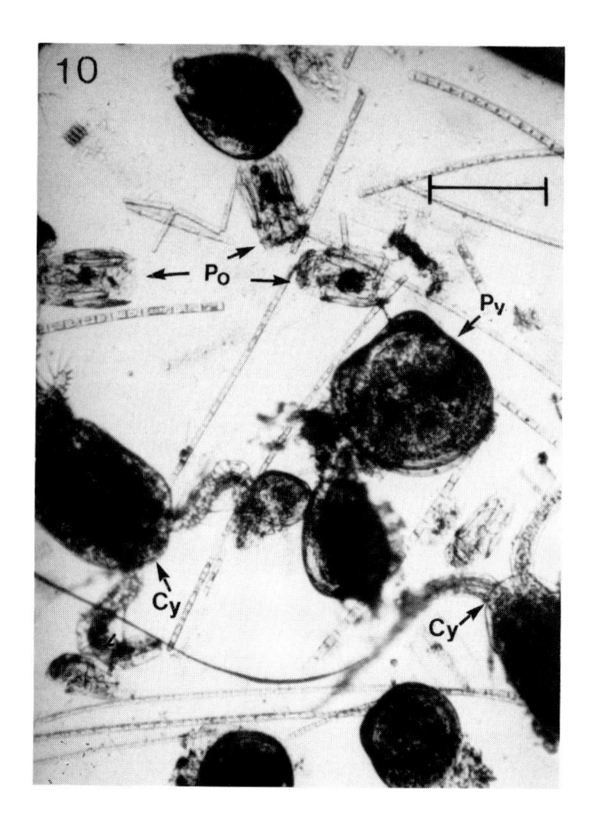

Figure 10. Plankton sample "preserved" with sugar-formalin containing veligers, postveligers, phytoplankton, rotifers, and microcrustaceans. Pv, postveliger; Po, *Polyarthra*; Cy, *Cyclops*. (Bar = 200 μm.)

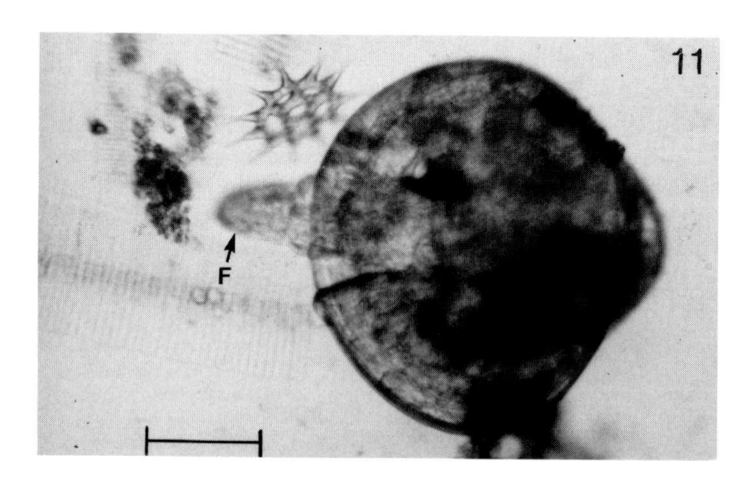

Figure 11. Postveliger (live specimen). Showing the appearance of a shell with an extended foot (F). (Bar = 200 μm.)

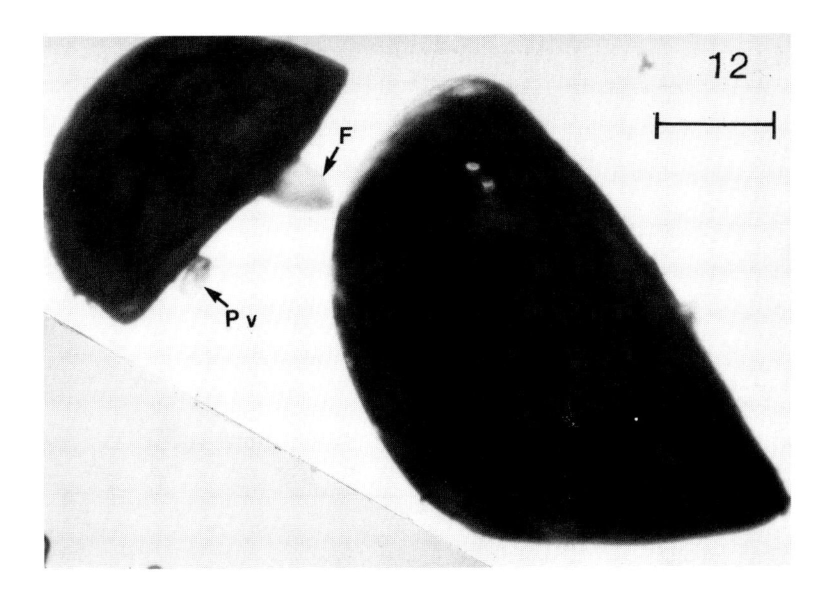

Figure 12. Young-of-the-year *Dreissena polymorpha*. Note postveliger attached to smaller mussel. F, foot; Pv, postveliger. (Bar = 200 μm.)

REFERENCES

Haney, J. F., and D. J. Hall. "Sugar-coated Daphnia: A Preservation Technique for Cladocera," *Limnol. Oceanogr.* 18:331–333 (1973).

Hopkins, G. J. "The Zebra Mussel, *Dreissena polymorpha*: A Photographic Guide to the Identification of Microscopic Veligers," Ontario Ministry of the Environment, Water Resources Branch, Queen's Printer, Toronto (1990).

Kirpichenko, M. Y. "Phenology, Abundance, and Growth of *Dreissena* Larvae in the Kujbyshev Reservoir," in *Biologija Drejsseny i borba z nej,* Sbornik, Ed. (Moskva: 1964), pp. 19–30 (In Russian).

Lewandowski, K. "The Role of Early Developmental Stages in the Dynamics of *Dreissena polymorpha* (Bivalvia) Populations in Lakes. II. Settling of Larvae and the Dynamics of Numbers of Settled Individuals," *Ekol. Pol.* 30:223–286 (1982).

Mackie, G. L., W. N. Gibbons, B. W. Muncaster, and I. M. Gray. "The Zebra Mussel, *Dreissena polymorpha*: A Synthesis of European Experiences and a Preview for North America," Report prepared for Ontario Ministry of the Environment, Water Resources Branch, Great Lakes Section, Queen's Printer, Toronto (1989).

Roff, J. C., and R. R. Hopcroft. "High Precision Microcomputer Based Measuring System for Ecological Research," *Can. J. Fish. Aquat. Sci.* 43:2044–2048 (1986).

Sprung, M. ''Ecological Requirements of Developing *Dreissena polymorpha* Eggs,'' *Arch. Hydrobiol. Suppl.* 79:69–86 (1987).

Sprung, M. ''Field and Laboratory Observations of *Dreissena polymorpha* larvae: Abundance, Growth, Mortality and Food Demands,'' *Arch. Hydrobiol.* 115:537–561 (1989).

Sprung, M. ''Costs of Reproduction: A Study on Metabolic Requirements of the Gonads and Fecundity of the Bivalve *Dreissena polymorpha*,'' *Malacologia* 32:267–274 (1990).

Stanczykowska, A. ''Ecology of *Dreissena polymorpha* (Pall.) (Bivalvia) in Lakes,'' *Pol. Arch. Hydrobiol.* 24:461–530 (1977).

Walz, N. ''Untersuchungen zur biologie von *Dreissena polymorpha* Pallas im Bodensee,'' *Arch. Hydrobiol., Suppl.* 42:452–482 (1973).

Walz, N. ''Die besiedlung von kunstlichen substraten durch larven von *Dreissena polymorpha*,'' *Arch. Hydrobiol., Suppl.* 47:423–431 (1975).

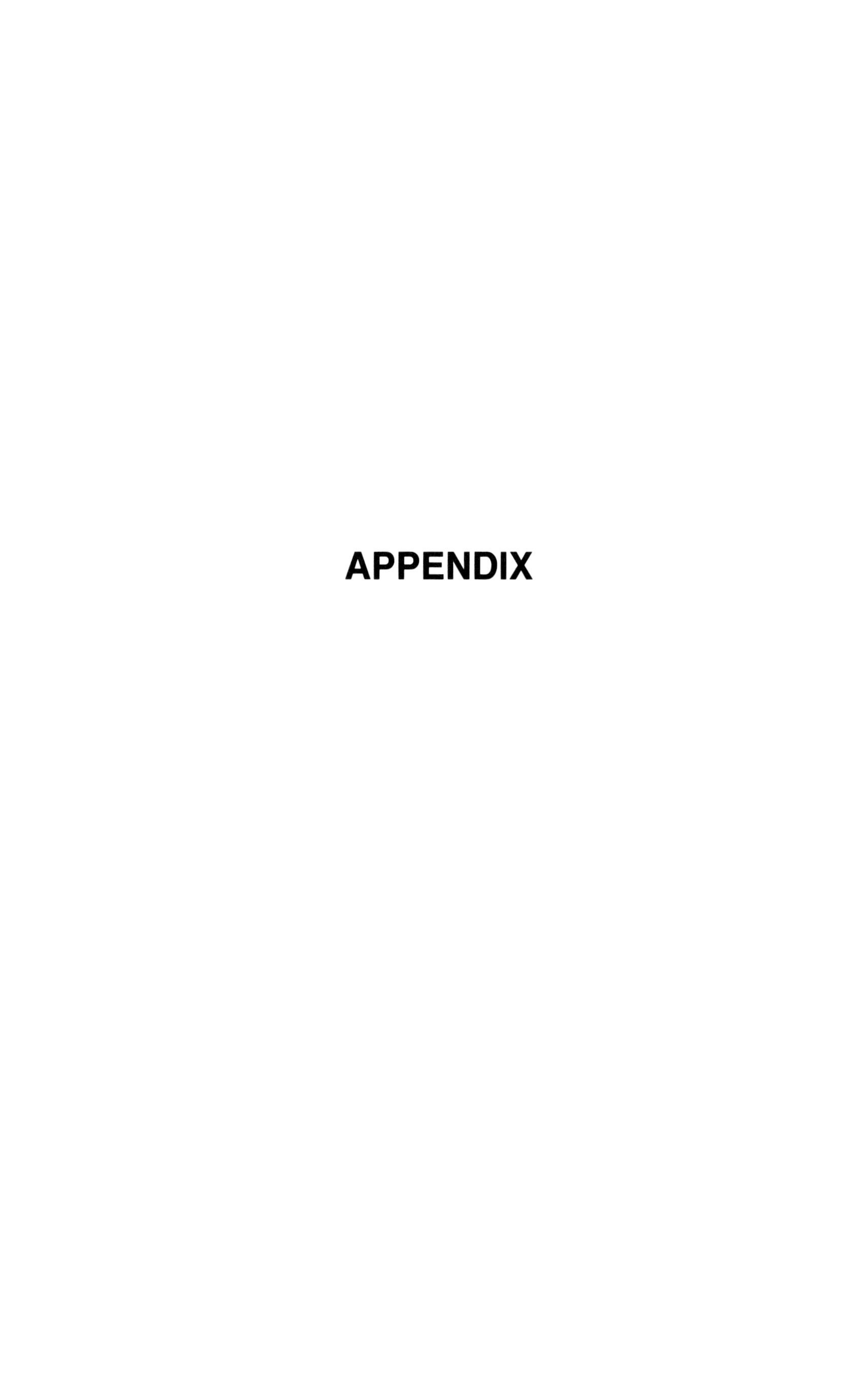

APPENDIX

Contributors

Josef D. Ackerman
Department of Mechanical Engineering
University of Toronto
5 King's College Road
Toronto, Ontario, Canada M5S 1A4

D. Grant Allen
Department of Chemical Engineering and Applied Chemistry
University of Toronto
200 College Street
Toronto, Ontario, Canada M5S 1A4

Louise K. Barton
Cleveland Electric Illuminating Company
Perry Nuclear Power Plant
10 Center Road
Perry, Ohio 44081

H. Bennett
Department of Biology
University of Toledo
Toledo, Ohio 43606

Abraham bij de Vaate
Institute for Inland Water Management and Waste Water Treatment
P.O. Box 17, NL-8200 AA
Lelystad, The Netherlands

G. I. Biochino
Institute of Inland Waters
Nekovskiy Raion
Yaroslavskaya Oblast
Borok, Russia

Marc G. Boileau
Department of Zoology
University of Guelph
Guelph, Ontario, Canada N1G 2W1

Jost Borcherding
Department of Zoology (Physiological Ecology)
University of Köln
Weyertal 119, D-5000 Köln 41
FRG-Germany

C. T. Bowmer
TNO Environmental and Energy Research
Ambachtsweg 8a, P.O. Box 57
1780 AB Den Helder
The Netherlands

Thomas J. Brady
Department of Biological Sciences
University of Buffalo
119 Hockstetter Hall
Buffalo, New York 14260

A. A. Bruznitsky
Institute of Inland Waters
Nekovskiy Raion
Yaroslavskaya Oblast
Borok, Russia

Kathleen A. Bruner
Department of Zoology
Ohio State University
1735 Neil Avenue
Columbus, Ohio 43210

Michael T. Bur
U.S. Fish and Wildlife Service
National Fisheries Research Center-Great Lakes
1451 Green Road
Ann Arbor, Michigan 48105

James T. Carlton
Maritime Studies Program
Williams College-Mystic Seaport
Mystic, Connecticut 06355-0990

A. E. Christie
Ontario Hydro
700 University Avenue
Toronto, Ontario, Canada M5G 1X6

Renata Claudi
Ontario Hydro
700 University Avenue A7A4
Toronto, Ontario, Canada M5G 1X6

Cees Davids
Department of Aquatic Ecology
University of Amsterdam
Kruislaan 320
1098 SM Amsterdam
The Netherlands

W. Chr. de Kock
TNO Environmental and Energy Research
Ambachtsweg 8a
P.O. Box 57
1780 AB Den Helder
The Netherlands

Ronald Dermott
Fisheries and Oceans Canada
P.O. Box 5050, CCIW
Burlington, Ontario, Canada L7R 4A6

Jaap Dorgelo
Department of Aquatic Ecology
University of Amsterdam
Kruislaan 320
1098 SM Amsterdam
The Netherlands

Larry E. Eckroat
Pennsylvania State University at Erie
The Behrend College
Station Road
Erie, Pennsylvania 16563

C. Ross Ethier
Department of Mechanical Engineering
University of Toronto
5 King's College Road
Toronto, Ontario, Canada M5S 1A4

David W. Evans
Ontario Hydro
700 University Avenue A7A4
Toronto, Ontario, Canada M5G 1X6

Elise Fear
Fisheries and Oceans Canada
P.O. Box 5050, CCIW
Burlington, Ontario, Canada L7R 4A6

Thomas A. Ferro
Department of Biology
SUNY College at Buffalo
1300 Elmwood Avenue
Buffalo, New York 14222

Susan W. Fisher
Department of Entomology
Ohio State University
1735 Neil Avenue
Columbus, Ohio 43210

Peter C. Fraleigh
Department of Biology
University of Toledo
Toledo, Ohio 43606

John R. P. French III
U.S. Fish and Wildlife Service
National Fisheries Research Center-
Great Lakes
145 Green Road
Ann Arbor, Michigan 48105

Wayne S. Gardner
Great Lakes Environmental
Research Laboratory, NOAA
2205 Commonwealth Boulevard
Ann Arbor, Michigan 48105

David W. Garton
Department of Zoology
Ohio State University
1735 Neil Avenue
Columbus, Ohio 43210

Wendy M. Gordon
Great Lakes Environmental
Research Laboratory, NOAA
2205 Commonwealth Boulevard
Ann Arbor, Michigan 48105

Duane C. Gossiaux
Great Lakes Environmental
Research Laboratory, NOAA
2205 Commonwealth Boulevard
Ann Arbor, Michigan 48105

Ronald W. Griffiths
Ministry of the Environment
Water Resources Assessment Unit
985 Adelaide Street South
London, Ontario, Canada N6E 1V3

Gerald Gubanich
Cleveland Division of Water
Crown Water Treatment Plant
955 Clague Road
Westlake, Ohio 44145

Wendell R. Haag
Department of Zoology
Ohio State University
1735 Neil Avenue
Columbus, Ohio 43210

G. Douglas Haffner
Great Lakes Institute
Department of Biological Sciences
University of Windsor
Windsor, Ontario, Canada N9B 3P4

Paul D. N. Hebert
Department of Zoology
University of Guelph
Guelph, Ontario, Canada N1G 2W1

Gordon J. Hopkins
Ministry of the Environment
Water Resources Branch
125 Resources Road
Rexdale, Ontario, Canada M9M 5L1

Stanley Iwanyzki
Department of Biology
Wilfrid Laurier University
Waterloo, Ontario, Canada N2L 3C5

Brigitte Jantz
Department of Zoology
(Physiological Ecology)
University of Köln
Weyertal 119
D-5000 Köln 41, FRG-Germany

Henk A. Jenner
KEMA
Utrechteseweg 310, 6812 AR
Arnhem
P.O. Box 9035
6800 ET Arnhem, The Netherlands

James N. Jensen
Department of Civil Engineering
University of Buffalo
212 Ketter Hall
Buffalo, New York 14260

R. Allan Kamman
Department of Biology
SUNY College at Buffalo
1300 Elmwood Avenue
Buffalo, New York 14222

Alexei V. Karnaukhov
Science-Engineering Center
Leninsky District
4 Smolensky Boulevard
Moscow, 119039, Russia

Valery N. Karnaukhov
Institute of Biophysics
Academy of Sciences USSR
Puschino Moscow Reg. 142292,
Russia

Bruce W. Kilgour
Mackie and Associates Water
Systems Analysts
381 Elmira Road
Guelph, Ontario, Canada NIK 1H3

Paul L. Klerks
Department of Biology
University of
Southwestern Louisiana
P.O. Box 42451
Lafayette, Louisiana 70504

William P. Kovalak
Detroit Edison Company
2000 Second Avenue
Detroit, Michigan 48826-1279

John J. Kowalewski
Ontario Hydro
800 Kipling Avenue
Toronto, Ontario, Canada M8Z 5S4

Michiel H. S. Kraak
Department of Aquatic Ecology
University of Amsterdam
Kruislaan 320
1098 SM Amsterdam
The Netherlands

Clifford Kraft
Sea Grant Advisory Services
ES-105
University of Wisconsin-Green Bay
Green Bay, Wisconsin 54311

Peter F. Landrum
Great Lakes Environmental
Research Laboratory, NOAA
2205 Commonwealth Boulevard
Ann Arbor, Michigan 48105

Daphna Lavy
Department of Aquatic Ecology
University of Amsterdam
Kruislaan 320
1098 SM Amsterdam
The Netherlands

Joseph H. Leach
Ministry of Natural Resources
Lake Erie Fisheries Station
RR#2
Wheatley, Ontario,
Canada N0P 2P0

Harold Lee
Department of Biology
University of Toledo
Toledo, Ohio 43606

Aklilu Lemma
Department of Biology
University of Toledo
Toledo, Ohio 43606

Wilfred L. LePage
Monroe Water Works
City of Monroe
915 E. Front Street
Monroe, Michigan 48161

Krzysztof Lewandowski
Institute of Ecology
Polish Academy of Sciences
Dziekanow Lesny
Poland 05-092

Donald Lewis
Aquatic Sciences Inc.
P.O. Box 2205, STN B
St. Catherines, Ontario,
Canada L2M 6P6

David P. Long
Center for Biological Macrofouling Research
P.O. Box 19498
The University of Texas at Arlington
Arlington, Texas 76019

Gary D. Longton
Detroit Edison Company
2000 Second Avenue
Detroit Michigan 48826-1279

Michael L. Ludyanskiy
Marine Biocontrol Corporation
P.O. Box 636
Sandwich, Massachusetts 02563

Gerald L. Mackie
Department of Zoology
University of Guelph
Guelph, Ontario N1G 2W1

Edwin C. Masteller
Pennsylvania State University at Erie
The Behrend College
Station Road
Erie, Pennsylvania 16563

Gerald Matisoff
Department of Geological Sciences
Case Western Reserve University
112 A.W. Smith Building
Cleveland, Ohio 44106

Robert W. McCauley
Department of Biology
Wilfrid Laurier University
Waterloo, Ontario, Canada N2L 3C5

Robert F. McMahon
Center for Biological Macrofouling Research
P.O. Box 19498
The University of Texas at Arlington
Arlington, Texas 76019

Joanne Mitchell
Fisheries and Oceans Canada
P.O. Box 5050, CCIW
Burlington, Ontario,
Canada L7R 4A6

Joke P. M. Janssen-Mommen
KEMA
Utrechteseweg 310, 6812 AR Arnhem
P.O. Box 9035
6800 ET Arnhem, The Netherlands

Brian Morton
Department of Zoology
The University of Hong Kong
Hong Kong

Ian Murray
Fisheries and Oceans Canada
P.O. Box 5050, CCIW
Burlington, Ontario,
Canada L7R 4A6

Dietrich Neumann
Department of Zoology
(Physiological Ecology)
University of Köln
Weyertal 119
D-5000 Köln 41, FRG-Germany

S. Jerrine Nichols
U.S. Fish and Wildlife Service
National Fisheries Research Center-
Great Lakes
1451 Green Road
Ann Arbor, Michigan 48105

Ruurd Noordhuis
Institute for Inland Water
Management and Waste Water
Treatment
P.O. Box 17, NL-8200 AA
Lelystad, The Netherlands

Paul H. Patrick
Ontario Hydro
800 Kipling Avenue
Toronto, Ontario M8Z 5S4

Wilma H. M. Peeters
Department of Aquatic Ecology
University of Amsterdam
Kruislaan 320
1098 SM Amsterdam
The Netherlands

Michael A. Quigley
Great Lakes Environmental
Research Laboratory
2205 Commonwealth Boulevard
Ann Arbor, Michigan 48105

Jeffery L. Ram
Department of Physiology
Wayne State University
Detroit, Michigan 48201

Harro H. Reeders
Institute for Inland Water
Management and Waste Water
Treatment
P.O. Box 17, NL-8200 AA
Lelystad, The Netherlands

Howard P. Riessen
Department of Biology
SUNY College at Buffalo
1300 Elmwood Avenue
Buffalo, New York 14222

Hans Schoon
Department of Aquatic Ecology
University of Amsterdam
Kruislaan 320
1098 SM Amsterdam
The Netherlands

Jennifer C. Shaffer
Pennsylvania State University
at Erie
The Behrend College
Station Road
Erie, Pennsylvania 16563

Bradley N. Shipman
Center for Biological Macrofouling
Research
P.O. Box 19498
The University of Texas at
Arlington
Arlington, Texas 76019

Nataliya F. Smirnova
Institute of Inland Waters
Nekovskiy Raion
Yaroslavskaya Oblast
Borok, Russia

Lane C. Smith
Institute of Ecosystem Studies
The New York Botanical Garden
Box AB
Millbrook, New York 12545

Richard D. Smithee
Detroit Edison Company
2000 Second Avenue
Detroit, Michigan 48826-1279

Henk Smit
Ministry of Transport and Public Works
Tidal Waters Division
P.O. Box 20907, NL-2500 EX
Den Haag, The Netherlands

V. A. Sokolov
Institute of Inland Waters
Nekovskiy Raion
Yaroslavskaya Oblast
Borok, Russia

Jan K. Spelt
Department of Mechanical Engineering
University of Toronto
5 King's College Road
Toronto, Ontario, Canada M5S 1A4

Martin Sprung
UCTRA, Universidade do Algarve
Campus de Gambelas
Apartado 322
P-8004 Faro Codex, Portugal

Anna Stańczykowska
Department of Ecology and Environmental Protection
Institute of Biology
Agricultural-Pedagogical University
Siedlce, Poland 08-110

Louise M. Steele
Pennsylvania State University at Erie
The Behrend College
Station Road
Erie, Pennsylvania 16563

Robert C. Stevenson
Division of Water
City of Toledo
P.O. Box 786
Toledo, Ohio 43695

David L. Strayer
Institute of Ecosystem Studies
The New York Botanical Garden
Box AB
Millbrook, New York 12545

Merel Toussaint
Department of Aquatic Ecology
University of Amsterdam
Kruislaan 320
1098 SM Amsterdam
The Netherlands

John E. Van Benschoten
Department of Civil Engineering
University of Buffalo
212 Ketter Hall
Buffalo, New York 14260

Egbert H. van Nes
Institute for Inland Water Management and Waste Water Treatment
P.O. Box 17, NL-8200 AA
Lelystad, The Netherlands

Germane A. Vinogradov
Institute of Inland Waters
Nekovskiy Raion
Yaroslavskaya Oblast
Borok, Russia

Tamara L. Yankovich
Great Lakes Institute
Department of Biological Sciences
University of Windsor
Windsor, Ontario, Canada N9B 3P4

Reviewers

Steven Ahlstedt, Tennessee Valley Authority, Norris, Tennessee
Norman Andresen, University of Michigan, Ann Arbor, Michigan
Robert E. Baier, University of Buffalo, Buffalo, New York
Louise K. Barton, Cleveland Electric Illuminating, Perry, Ohio
Robert H. Boyle, Sports Illustrated, Cold Spring, New York
Joe Britton, Texas Christian University, Fort Worth, Texas
Chris Brousseau, Ontario Ministry of Natural Resources, Maple, Ontario
G. Allen Burton, Wright State University, Dayton, Ohio
Earl Chilton, Texas Parks and Wildlife, Ingram, Texas
Renata Claudi, Ontario Hydro, Toronto, Ontario
David Culver, Ohio State University, Columbus, Ohio
Mark Dahlager, Science Museum of Minnesota, St. Paul, Minnesota
George M. Davis, Academy of Natural Sciences of Philadelphia, Philadelphia, Pennsylvania
John A. DeKam, Bay Metro Water Treatment Plant, Bay City, Michigan
Marge Dochoda, Great Lakes Fishery Commission, Ann Arbor, Michigan
Steve Donner, Consumers Power Company, Jackson, Michigan
Elwin Evans, Michigan Department of Natural Resources, Lansing, Michigan
Marlene Evans, National Hydrology Research Institute, Saskatoon, Saskatchewan
Billy D. Fellers, Electric Power Research Institute, Palo Alto, California
Ronald L. Fobes, Consumers Power Company, Jackson, Michigan
Barbara G. Fox, Chicago Department of Water, Chicago, Illinois
Thomas Freitag, U.S. Army Corps of Engineers, Detroit, Michigan
John E. Gannon, U.S. Fish and Wildlife Service, Ann Arbor, Michigan
Wayne S. Gardner, Great Lakes Environmental Research Laboratory, NOAA, Ann Arbor, Michigan
David W. Garton, Ohio State University, Columbus, Ohio
William G. Gordon, New Jersey Sea Grant, Fort Hancock, New Jersey
Ronald W. Griffiths, Ontario Ministry of the Environment, London, Ontario
Matthew Hart, Atlantic Monthly, Toronto, Ontario
Walter Hoagman, Michigan Sea Grant, East Tawas, Michigan
Peter Kauss, Ontario Ministry of the Environment, Toronto, Ontario
Donald Klemm, U.S. Environmental Protection Agency, Cincinnati, Ohio
Clifford Kraft, University of Wisconsin Sea Grant Program, Green Bay, Wisconsin
Gary Kraidman, Margaronics, Inc., East Brunswick, New Jersey
John Lambert, Carleton University, Ottawa, Ontario
Peter F. Landrum, Great Lakes Environmental Research Laboratory, NOAA, Ann Arbor, Michigan
Mark Luttenton, Ferris State University, Big Rapids, Michigan
Michael Lydy, Great Lakes Environmental Research Laboratory, NOAA, Ann Arbor, Michigan

Gerald L. Mackie, University of Guelph, Guelph, Ontario
Eric Mallen, Indiana and Michigan Electric Company, Bridgman, Michigan
Leif L. Marking, U.S. Fish and Wildlife Service, LaCrosse, Wisconsin
J. Ellen Marsden, Illinois Natural History Survey, Zion, Illinois
William T. Mason, U.S. Fish and Wildlife Service, Gainesville, Florida
Edwin C. Masteller, Penn State University - Behrend, Erie, Pennsylvania
Jack Mattice, Electric Power Research Institute, Palo Alto, California
Robert McCauley, Wilfrid Laurier University, Waterloo, Ontario
Robert F. McMahon, University of Texas, Arlington, Texas
Dennis M. McMullen, Technology Applications Inc., Cincinnati, Ohio
Bernie Muncaster, Ecological Services for Planning, LTD, Guelph, Ontario
Kenneth M. Muth, U.S. Fish and Wildlife Service, Sandusky, Ohio
Patrick M. Muzzall, Michigan State University, East Lansing, Michigan
Charles R. O'Neill Jr., New York Sea Grant, Brockport, New York
James Oris, Miami University, Oxford, Ohio
Randall W. Owens, U.S. Fish and Wildlife Service, Oswego, New York
Barry S. Payne, U.S. Army Corps of Engineers, Vicksburg, Mississippi
Michael A. Quigley, Great Lakes Environmental Research Laboratory, NOAA, Ann Arbor, Michigan
Jeffery Reutter, Ohio Sea Grant, Columbus, Ohio
John Schwartz, Michigan Sea Grant, East Lansing, Michigan
W. Gary Sprules, University of Toronto, Mississauga, Ontario
Kenton M. Stewart, State University of New York, Buffalo, New York
Glenn E. Stout, Illinois-Indiana Sea Grant, Urbana, Illinois
David L. Strayer, Institute of Ecosystem Studies, Millbrook, New York
Donald Tillitt, U.S. Fish and Wildlife Service, Columbia, Missouri
John L. Tsou, Electric Power Research Institute, Palo Alto, California
Ray L. Tuttle, New York State Electric and Gas, Binghamton, New York
Henry A. Vanderploeg, Great Lakes Environmental Research Laboratory, NOAA, Ann Arbor, Michigan
G. Thomas Waters, Ohio State University, Columbus, Ohio
Carl M. Way, U.S. Army Corps of Engineers, Vicksburg, Mississippi
Lawrence J. Weider, Max-Planck-Institut fur Limnologie, Plön, Germany
Michael Wiley, University of Michigan, Ann Arbor, Michigan
James D. Williams, U.S. Fish and Wildlife Service, Gainesville, Florida

INDEX

Index